Organic Synthesis Highlights III

Edited by Johann Mulzer and Herbert Waldmann

Further Titles of Interest

Organic Synthesis Highlights I
J. Mulzer, H.-J. Altenbach, M. Braun, K. Krohn, H.-U. Reissig
ISBN: 3-527-27955-5

Organic Synthesis Highlights II
H. Waldmann (Ed.)
ISBN: 3-527-29200-4 (Hardcover)
ISBN: 3-527-29378-7 (Softcover)

Classics in Total Synthesis
K. C. Nicolaou, E. Sorensen
ISBN: 3-527-29231-4 (Hardcover)
ISBN: 3-527-29284-5 (Softcover)

Enzyme Catalysis in Organic Synthesis
K. Drauz, H. Waldmann (Eds.)
ISBN: 3-527-28479-6 Two Volumes

Organic Synthesis Highlights III

Edited by
Johann Mulzer and Herbert Waldmann

 **WILEY-VCH**

Weinheim · New York · Chichester · Brisbane · Singapore · Toronto

Prof. Dr. Johann Mulzer
Institut für Organische Chemie
der Universität
Währinger Straße 38
A-1090 Wien (Vienna)
Austria

Professor Dr. Herbert Waldmann
Institut für Organische Chemie
der Universität
Richard-Willstätter-Allee 2
D-76128 Karlsruhe
Germany

The cover illustration shows the polycyclic ring system of morphine, surrounded by the primary rings. Strategic bonds determined in a retrosynthetic analysis are highlighted.

Library of Congress Card No. applied for
A catalogue record for this book is available from the British Library.

Die Deutsche Bibliothek – CIP-Einheitsaufnahme

Organic synthesis highlights. – Weinheim ; New York ; Chichester ; Brisbane ; Singapore ; Toronto : Wiley-VCH
    3. / Ed. by Johann Mulzer and Herbert Waldmann 1998
    ISBN 3-527-29500-3

© WILEY-VCH Verlag GmbH, D-69469 Weinheim (Federal Republic of Germany), 1998

Printed on acid-free and chlorine-free paper

Composition: Hagedorn Kommunikation, D-68519 Viernheim
Printing: Strauss Offsetdruck GmbH, D-69509 Mörlenbach
Bookbinding: W. Osswald, D-67433 Neustadt/Wstr.
Cover design: Graphik & Text Studio Zettlmeier-Kammerer, D-93164 Laaber-Waldetzenberg
Printed in the Federal Republic of Germany

Preface

Synthesis maintains its central position in Organic Chemistry and as documented by the rapid preparation of very complex molecules such as taxol or brevetoxin has reached an almost unprecedented power and scope. This progress of organic synthesis over the past eight years could be followed in the two volumes of Organic Synthesis Highlights I and II (OSH I and II), which appeared in 1990 and 1995 respectively. In view of the positive response which was met by these two textbooks the third title of this series is now presented. Contrary to the philosophy of OSH I and II, which were both based on the review section "Synthese im Blickpunkt" of the "Nachrichten aus Chemie Technik und Laboratorium", the members' journal of the GDCh, Organic Synthesis Highlights III is a selection of the "Highlights" adapted from Angewandte Chemie, volumes 103–107 (1993–97). Compared to the articles in OSH I and II these highlights are considerably short-

er and instead of trying to review an entire area, they focus on one subject only. This means that the number of articles and hence different topics (and highlights) has been significantly increased. No less than 56 highlights are presented, all updated by their individual authors to 1997. The basic topics have been maintained and as in OSH I/II particular emphasis is laid on stereoselective reactions and reagents, organometallics and the general synthesis of natural and non-natural products. Following current trends articles on the self assembly of biopolymers and on some aspects of combinatorial synthesis are also included. Based on this general editorial philosophy we hope that the new volume will find a similar acceptance in the scientific community as did its two congeners OSH I and II.

Karlsruhe and Vienna, May 1998
Herbert Waldmann and
Johann Mulzer

Contents

Part I. New Methods and Reagents for Organic Synthesis

A. Asymmetric Synthesis

B. Organometallic Reagents

C. Biological and Biomimetic Methods

D. General Methods and Reagents

Part II. Applications in Total Synthesis
Synthesis of Natural and Non-natural Products

List of Contributors

Prof. Dr. M. Beller
Anorganisch-Chemisches Institut
Technische Universität
Lichtenbergstraße 4
D-85747 Garching
Germany

Dr. D. J. Berrisford
Department of Chemistry
University of Manchester
Institute of Science and Technology
PO Box 88
Manchester M60 1QD
United Kingdom

Prof. Dr. C. Bolm
Institut für Organische Chemie
der RWTH Aachen
Professor-Pirlet-Straße 1
D-52074 Aachen
Germany

Prof. Dr. H. Brunner
Institut für Anorganische Chemie
Universität Regensburg
D-93040 Regensburg
Germany

Prof. K. Burgess
Department of Chemistry
Texas A & M University
College Station, TX 77843-3255
USA

Prof. Dr. H. Butenschön
Institut für Organische Chemie
Universität Hannover
Schneiderberg 1B
D-30167 Hannover
Germany

Prof. Dr. Boy Cornils
Hoechst AG
Forschungsleitung
D-65926 Frankfurt
Germany

Prof. Dr. G. Dyker
Institut für Organische Chemie
der Universität-Gesamthochschule Duisburg
FB 6, Institut für Synthesechemie
Lotharstraße 1
D-47048 Duisburg
Germany

Dr. M. Famulok
Inst. für Biochemie der Universität München
Feodor-Lynen-Straße 25
D-81377 München
Germany

Dr. A. Ganesan
Institute of Molecular and Cell Biology
The National University of Singapore
30 Medical Drive
Singapore 117609
Singapore

Priv.-Doz. Dr. A. Giannis
Institut für Organische Chemie
der Universität
Richard-Willstätter-Allee 2
D-76128 Karlsruhe
Germany

Prof. Dr. R. N. Grimes
Department of Chemistry
University of Virginia
Charlottesville VA 22901
USA

Prof. Dr. A. Hirsch
Institut für Organische Chemie
der Universität Erlangen-Nürnberg
Henkestraße 42
D-91054 Erlangen
Germany

Prof. Dr. S. Hoffmann
Institut für Biochemie
FB Biochemie/Biotechnologie
der Martin-Luther-Universität
Kurt-Mothes-Straße 3
D-06120 Halle-Wittenberg
Germany

Prof. Dr. H. Hopf
Institut für Organische Chemie der
Technischen Universität
Hagenring 30
D-38106 Braunschweig
Germany

Prof. Dr. H.-D. Jakubke
Institut für Biochemie
Fakultät für Biowissenschaften,
Pharmazie und Psychologie der Universität
Talstraße 33
D-04103 Leipzig
Germany

Prof. Dr. R. Janoschek
Institut für Theoretische Chemie
Karl-Franzens-Universität Graz
Strassoldogasse 10
A-8010 Graz
Austria

Prof. Dr. G. Kaupp
Fachbereich 9
Organische Chemie I
der Universität
Postfach 2503
D-26111 Oldenburg
Germany

Prof. Dr. G. von Kiedrowski
Lehrstuhl für Organische Chemie I
Ruhr-Universität Bochum
Universitätsstraße 150
D-44801 Bochum
Germany

Prof. Dr. A. J. Kirby
Chemical Laboratory
University of Cambridge
Lensfield Road
Cambridge CB2 1EW
United Kingdom

Prof. Dr. Paul Knochel
FB Chemie
der Philipps-Universität Marburg
Hans-Meerwein-Straße
D-35032 Marburg
Germany

Prof. Dr. U. Koert
Institut für Chemie
der Humboldt-Universität zu Berlin
Hessische Straße 1–2
D-10115 Berlin
Germany

Priv.-Doz. Dr. B. König
Institut für Organische Chemie
der Technischen Universität
Postfach 3329
D-38023 Braunschweig
Germany

Dr. H. B. Kraatz
Department of Chemistry
University of Saskatchewan
110 Science Place
Saskatoon SK S7N 5C9
Canada

Prof. Dr. Karsten Krohn
Fachbereich 13, Chemie und Chemietechnik
der Universität-Gesamthochschule
Warburgerstraße 100
D-33098 Paderborn
Germany

Prof. Dr. H. Laatsch
Institut für Organische Chemie
der Universität
Tammannstraße 2
D-37077 Göttingen
Germany

Prof. Dr. S. Laschat
Institut für Organische Chemie
Technische Universität Braunschweig
Hagenring 30
D-38106 Braunschweig
Germany

Prof. R. M. J. Liskamp
Department of Medicinal Chemistry
University of Utrecht
P.O. Box 80082
NL-3508 TB Utrecht
The Netherlands

Dr. B. B. Lohray
Vice President, Chem. R & D
Dr. Reddy's Research Foundation
Bollaram Road
Miyapur
Hyderabad-500 138
India

Prof. Dr. J. Mulzer
Institut für Organische Chemie
der Universität
Währringerstraße 38
A-1090 Wien
Austria

Prof. Dr. Ulrich Nagel
Institut für Anorganische Chemie
der Universität
Auf der Morgenstelle 18
D-72076 Tübingen
Germany

Prof. Dr. G. R. Newkome
Center of Molecular Design and Recognition
Department of Chemistry, CHE 305
University of South Florida
4202 E. Fowler Ave., ADM 200
Tampa, FL 33620-5950
USA

Prof. Dr. J. Okuda
Institut für Anorganische
und Analytische Chemie
Johannes-Gutenberg-Universität Mainz
J.-J.-Becher-Weg 24
D-55099 Mainz
Germany

Dr. I. Paterson
Chemical Laboratory
University of Cambridge
Lensfield Road
Cambridge CB2 1EW
United Kingdom

Prof. Dr. H. Paulsen
Institut für Organische Chemie
der Universität
Martin-Luther-King-Platz 6
D-20146 Hamburg
Germany

Prof. Dr. O. Reiser
Institut für Organische Chemie
der Universität Regensburg
Universitätsstraße 31
D-93053 Regensburg
Germany

Prof. Dr. H.-U. Reißig
Institut für Organische Chemie
der Technischen Universität Dresden
Mommsenstraße 13
D-01062 Dresden
Germany

Dr. K. Rück-Braun
Institut für Organische Chemie
der Universität
J.-Joachim-Becher-Weg 18–20
D-55099 Mainz
Germany

Prof. J. F. Stoddart
Department of Chemistry and Biochemistry
University of California at Los Angeles
405 Hilgard Avenue
Los Angeles, CA 90095-1569
USA

Dr. Jörg Sundermeyer
Fachbereich Chemie
der Philipps-Universität Marburg
Hans-Meerwein-Straße
D-35032 Marburg
Germany

Prof. Dr. G. Süss-Fink
Institut de Chimie
Université de Neuchâtel
Avenue de Bellevaux 51
CH-2000 Neuchâtel
Switzerland

Prof. Dr. A. Togni
Laboratory of Inorganic Chemistry
Swiss Federal Institute of Technology
ETH-Zentrum
CH-8092 Zürich
Switzerland

Prof. Dr. F. Vögtle
Kekulé-Institut
für Organische Chemie und Biochemie
der Universität
Gerhard-Domagk-Straße 1
D-53121 Bonn
Germany

Prof. Dr. Herbert Waldmann
Institut für Organische Chemie
der Universität Karlsruhe
Richard-Willstätter-Allee 2
D-76128 Karlsruhe
Germany

Dr. L. Wessjohann
Institut für Organische Chemie
Universität München
Karlstraße 23
D-80333 München
Germany

Prof. Dr. Th. Wirth
Institut für Organische Chemie
der Universität Basel
St. Johanns-Ring 19
CH-4056 Basel
Switzerland

Prof. Dr. T. Ziegler
Institut für Organische Chemie
der Universität Köln
Greinstraße 4
D-50939 Köln
Germany

Part I. New Methods and Reagents for Organic Synthesis

A. Asymmetric Synthesis

Catalytic Asymmetric Synthesis Using New Enolato-, Amido-, and Organolithium Chemistry

David J. Berrisford, Carlos Horkan, and Matthew L. Isherwood

Organolithium reagents are a cornerstone in the development of modern organic chemistry. Despite the enormous interest in catalytic asymmetric synthesis, [1, 2] asymmetric catalytic reactions of organolithium reagents are rare. The complex solution equilibria, [3] so characteristic of lithioanions, often frustrate attempts to render these reactions both asymmetric and catalytic with respect to the chiral ligands. One significant problem is that high background reaction rates are often associated with enolate and alkyllithium chemistry. An asymmetric variant must minimize the extent of reaction occuring by an inevitably racemic pathway. Given these difficulties, a number of contributions have made valuable advances in this area and are worthy of highlighting.

Koga et al. have discovered [4] a remarkable catalytic process for the asymmetric benzylation of lithium enolates derived from silyl enol ethers **1** and **3** (Scheme 1). Previously, stereocontrolled lithium enolate alkylation has been restricted to either diastereoselective processes, [2a,c] requiring the covalent attachment of a chiral auxiliary, or to enantioselective reactions requiring stoichiometric amounts of chiral ligands. [2b] Indeed, this new catalytic reaction is a development of an

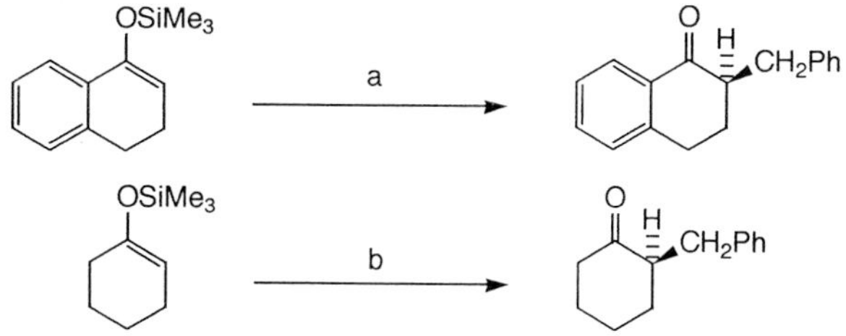

Scheme 1. Asymmetric benzylation of lithium enolates. a) 1. MeLi, LiBr, toluene; 2. 5 mol% **5**, 2 equivs. **6**, 10 equivs. PhCH₂Br, –45 C, 18 h (76 %, 96 % ee). b) 1. MeLi, LiBr, Et₂O; 2. 10 mol% **5**, 2 equivs. **6**, 10 equivs. PhCH₂Br, –45 C, 18 h (52 %, 90 % ee).

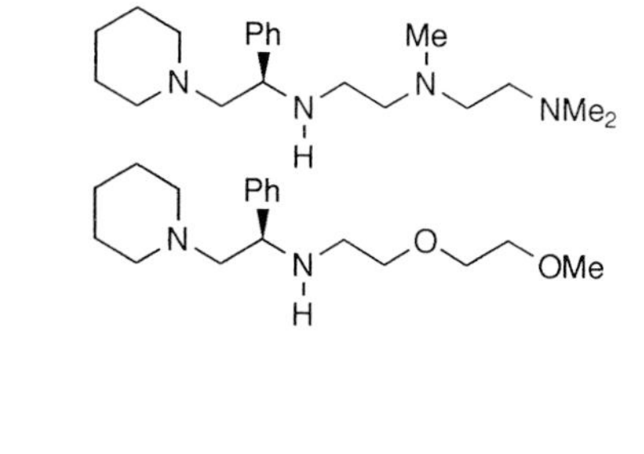

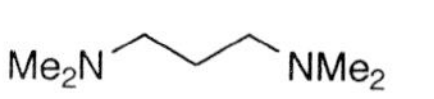

earlier stoichiometric benzylation of **1** which achieved a 92 % enantiomeric excess (ee) of **2** using ligand **7**. [5]

The new process differs from the former stoichiometric procedure in a number of respects. Firstly, the structure of the original chiral ligand **7** has been modified. Furthermore, the chiral ligand must be used in conjunction with an excess of an additional achiral ligand e.g. **6** (Scheme 1). Both polydentate chiral ligands (**5/7**) are prepared [6] by multistep, but straightforward routes from phenylglycine. The new ligand **5** delivers increased enantioselectivity under stoichiometric conditions (up to 97 % ee for **2**) and maintains this performance under defined catalytic conditions. A 96 % ee (76 % yield) of **2** can be achieved using 5 mol% of **5** in toluene. Lower ligand loading (< 5 mol%) has a deleterious effect on both conversion and enantioselectivity.

This study constitutes a further example of how efficient catalytic reactions can be developed despite the presence of complex solution equilibria. Originally, this chemistry was developed in a non-donor solvent rather than using an excess of donor as solvent. By so doing, the critical effects of additives, both chiral and achiral, are more readily appreciated. Indeed, incorporation of lithium bromide and an achiral, bidentate ligand in addition to **5** are essential for selectivity. The effects of these additives are particularly interesting. Tertiary diamines, e.g. **6** and its analogues, [7] substantially accelerate the rate. Without the addition of ligand **6**, conversion drops precipitously under catalytic conditions. It is proposed that the achiral additive sequesters [8] the excess of LiBr that is generated as the alkylation reaction proceeds. It is found that an excess of LiBr slows the reaction prohibitively (< 1 % conversion). However, modest amounts of LiBr benefit the enantioselectivity. Under stoichiometric conditions without additional LiBr, the enantiomeric excess increases with time and conversion. The results implicate a kinetically dominant mixed aggregate

comprised of lithium halide, lithium enolate, and chiral ligand. The achiral ligand modulates this aggregate composition. Related halide effects [9, 10] are reported in other enolate reactions. For example, the addition of lithium chloride has a beneficial effect on the asymmetric deprotonation [11] of ketones by chiral lithium amides. In the latter case, a lithium amide-lithium chloride aggregate is likely to be the active reagent. [12]

This catalytic alkylation reaction is limited as yet to two enolates and a single example of a reactive alkyl halide. The full scope of this process remains to be determined, particularly with respect to the range of compatible substrates and electrophiles. Nevertheless, Koga et al. [4] have discovered a significant example of catalytic lithium enolate chemistry and one that holds considerable promise for future development.

Two further contributions illustrate how chiral lithium amides can be used as catalysts in asymmetric deprotonation reactions (Schemes 2 and 3). The first example of catalytic chiral lithium amide chemistry was reported [13] by Asami (Scheme 2). In this process an achiral base, in this case LDA, provides a stoichiometric reservoir of amidolithium reagent. However, deprotonation of the epoxide is affected primarily by the chiral lithium amide **11** rather than the relative excess of LDA. Turnover is possible since the resulting chiral secondary amine **10** can be deprotonated by the remaining reservoir of LDA thus regenerating the chiral base **11**. For example, the deprotonation of cyclohexene oxide **8** in the presence of DBU as an additive gives the allylic alcohol **9** in 74 % ee (82 % yield) using 50 mol% of chiral base **11**.

Using this concept, the Koga group have developed [14] catalytic asymmetric deprotonation of 4-alkylcyclohexanones (Scheme 3). For example, deprotonation of **12** gives silyl enol ether **13** in good enantioselectivity. The reaction is accomplished by combining 30 mol% of chiral lithium amide **15** along

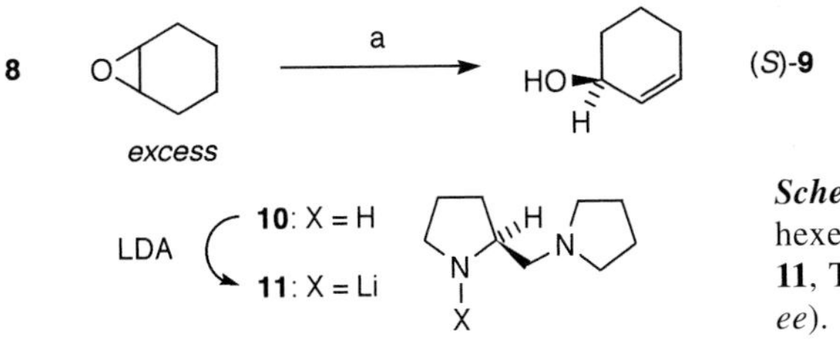

Scheme 2. Asymmetric deprotonation of cyclohexene oxide **8**. a) 1.5 equivs. LDA, 0.5 equivs. **11**, THF, room temperature, 12 h (82 % yield, 74 % *ee*).

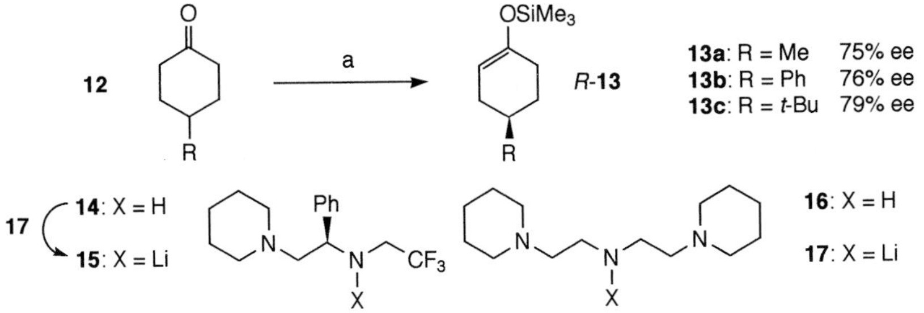

13a: R = Me   75% ee
13b: R = Ph   76% ee
13c: R = t-Bu   79% ee

Scheme 3. Asymmetric deprotonation of 4-alkylcyclohexanones. a) 1. 2.4 equivs. **17**, 30 mol% **15**, 2.4 equivs. HMPA, 1.5 equivs. DABCO, hexane/THF, −78 C, 1.5 h; 2. TMSCl (83–70 %, 75–79 % *ee*).

with an excess of achiral base **17** and trapping the resulting lithium enolate with trimethylsilyl chloride.

Again, the achiral base **17** provides a reservoir of amidolithium reagent to allow catalyst turnover by deprotonation of **14** formed *in situ* (Scheme 3). Clearly, the kinetics of the reaction are such that deprotonation at the ketone α-carbon by the achiral lithium amide **17** is much slower than deprotonation at the 2° nitrogen of the chiral amine **14**. Although the catalytic efficiency is modest, it is remarkable that catalysis of this type can be achieved.

Another recent contribution [15] by Denmark et al. demonstrates an effective method for promoting the nucleophilic addition reactions [2] of organolithium reagents to imines. [16] The chemistry is a development of earlier methodology. [17] The organolithium additions to *N*-aryl imines **18**, are promoted by C_2-symmetric bisoxazoline (BOX) ligands, [18] e.g. **20**, or (−)-sparteine [19] **21** (Scheme 4) with high asymmetric induction.

Notably, BOX ligand **20**, which is prepared [18] from *tert*-leucine, delivers excellent selectivities for additions to aliphatic imines. Substrate enolization, which is often a problem for aliphatic imines, does not compete to a significant extent. Both methyl- and vinyllithium additions can be promoted by **20** with significant enantioselectivities (**19a**, R = Me: 91 % *ee*, Table 1). Using sparteine **21** improves the selectivity of the *n*-butyl- and phenyllithium additions (**19c**, R = *n*-Bu: 91 % *ee*, Table 1). In both cases the structures of the active organometallic complexes involved remains undefined. In certain cases, it is possible to decrease the ligand loading to a practical 20 mol% and still retain a near maximum ee; below this figure the ee drops significantly. The *N*-aryl amine products can be deprotected by using published methodology [17] to afford the corresponding primary amines. Overall, this chemistry offers alternative syntheses of important amino compounds and is complementary to approaches based

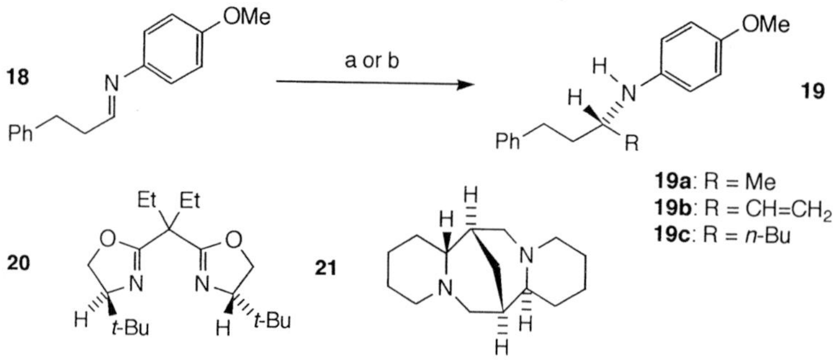

Scheme 4. Asymmetric nucleophilic additions to an aliphatic imine. a) 1. 2 equivs. RLi, toluene, 20 mol% **20**, 1h; b) 2 equivs. RLi, Et$_2$O, 20 mol% **21**, 1 h (for yields and conditions see Table 1).

Table 1. Reaction conditions, yields, and products for the nucleophilic addition of RLi to **18** (Scheme 4).

Product	$T (^\circ C)$	Ligand (equiv.)	Yield (%)	ee (%)
19a	−63	**20** (1.0)	96	91
19a	−63	**20** (0.2)	81	82
19b	−78	**20** (1.0)	95	89
19b	−78	**20** (0.2)	82	82
19c	−94	**21** (1.0)	90	91
19c	−78	**21** (0.2)	91	79

upon catalytic asymmetric hydrogenation [20] of imines and hydrazones.

These studies are significant in two respects. Efficient or improving asymmetric methodology is of primary importance, especially in such a fundamental area. Additionally, this work provides new insights [3] into the complex solution behaviour of these commonplace but valuable main group reagents.

References

[1] a) R. Noyori, *Asymmetric Catalysis in Organic Synthesis*, J. Wiley, New York, **1994**; b) *Catalytic Asymmetric Synthesis* (Ed.: I Ojima), VCH, New York/Weinheim, **1993**.

[2] a) D. Seebach, *Angew Chem.* **1990**, *102*, 1363; *Angew. Chem. Int. Ed. Engl.* **1990**, *29*, 1320; b) K. Tomioka, *Synthesis* **1990**, 541; c) P. G. Willard in *Comprehensive Organic Synthesis*, Vol. 1 (Eds.: B. M. Trost, I. Fleming), Pergamon, Oxford, **1991**, p. 1; d) D.M. Huryn in *Comprehensive Organic Synthesis*, Vol. 1 (Eds.: B. M. Trost, I. Fleming), Pergamon, Oxford, **1991**, p. 49.

[3] D. Seebach, *Angew Chem.* **1988**, *100*, 1685; *Angew. Chem. Int. Ed. Engl.* **1988**, *27*, 1624.

[4] M. Imai, A. Hagihara, H. Kawasaki, K. Manabe, K. Koga, *J. Am. Chem. Soc.* **1994**, *116*, 8829.

[5] a) M. Murakata, M. Nakajima, K. Koga, *J. Chem. Soc. Chem. Commun.* **1990**, 1657; b) Y. Hasegawa, H. Kawasaki, K. Koga, *Tetrahedron Lett.* **1993**, *34*, 1963. For related chemistry using stoichiometric quantities of chiral ligands see: c) H. Fujieda, M. Kanai, T. Kambara, A. Iida, K. Tomioka, *J. Am. Chem. Soc.* **1997**, *119*, 2060; d) M. Uragami, K. Tomioka, K. Koga, *Tetrahedron Asym.* **1995**, *6*, 701; e) K. Yasuda, M. Shindo, K. Koga, *Tetrahedron Lett.* **1996**, *37*, 6343; f) T. Takahashi, M. Muraoka, M. Capo, K. Koga, *Chem. Pharm. Bull.* **1995**, *43*, 1821.

[6] R. Shirai, K. Aoki, D. Sato, H.-D. Kim, M. Murakata, T. Yasukata, K. Koga, *Chem. Pharm. Bull.* **1994**, *42*, 690.

[7] In contrast, addition of ether ligands, e.g. DME, does not accelerate the reaction and has little effect on the ee. Hence, DME can be substituted for toluene as the reaction solvent. Recently, chiral ureas have been used as ligands: K. Ishii, S. Aoki, K. Koga, *Tetrahedron Lett.* **1997**, *38*, 563.

[8] a) Powerful donors such as HMPA are known to suppress mixed aggregation between MeLi and LiCl; H. J. Reich, J. P. Borst, R. R. Dykstra, D. P. Green, *J. Am. Chem. Soc.* **1993**, *115*, 8728. b) For structural characterisation of a 2 amine-enolate complex see: K. W. Henderson, P. G. Williard, P. R. Bernstein, *Angew. Chem. Int. Ed. Engl.* **1995**, *34*, 1117; *Angew. Chem.* **1995**, *107*, 1218.

[9] For reviews see: a) A. Loupy, B. Tchoubar, *Salt Effects in Organic and Organometallic Chemistry*, VCH, Weinheim, **1991**; b) E. Juaristi, A. K. Beck, J. Hansen, T. Matt, T. Mukhopadhyay, M. Simon, D. Seebach, *Synthesis* **1993**, 1271.

[10] a) F. E. Romesberg, D. B. Collum, *J. Am. Chem. Soc* **1994**, *116*, 9187; b) *idem.*, *J. Am. Chem. Soc* **1994**, *116*, 9198; c) J. Corset, F. Froment, M.-F. Lautie, N. Ratovelomanana, J. Seyden-Penne, T. Strzalko, M-C. Roux-Schmidt, *J. Am. Chem. Soc.* **1993**, *115*, 1684; d) K. W. Henderson, A. E. Dorigo, Q-Y. Liu, P. G. Williard, P. von Ragué Schleyer, P. R. Bernstein, *J. Am. Chem. Soc.* **1996**, *118*, 1339.

[11] a) B. J. Bunn, N. S. Simpkins, *J. Org. Chem.* **1993**, *58*, 1847; b) B. J. Bunn, N. S. Simpkins, Z. Spavold, M. J. Crimmin, *J. Chem. Soc. Perkin Trans. 1* **1993**, 3113; c) M. Toriyama, K. Sugasawa, M. Shindo, N. Tokutake, K. Koga, *Tetrahedron Lett.* **1997**, *38*, 567.

[12] F. S. Mair, W. Clegg, P. A. O'Neil, *J. Am. Chem. Soc.* **1993**, *115*, 3388.

[13] M. Asami, T. Ishizaki, S. Inoue, *Tetrahedron Asym.* **1994**, *5*, 793.

[14] T. Yamashita, D. Sato, T. Kiyoto, A. Kumar, K. Koga, *Tetrahedron Lett.* **1996**, *37*, 8195.

[15] S. E. Denmark, N. Nakajima, O. J-C. Nicaise, *J. Am. Chem. Soc.* **1994**, *116*, 8797.

[16] For a recent review see: S. E. Denmark, O. J.-C. Nicaise *J. Chem. Soc. Chem. Commun.* **1996**, 999.

[17] a) K. Tomioka, M. Shindo, K. Koga, *J. Am. Chem. Soc.* **1989**, *111*, 8266; b) I. Inoue, M. Shindo, K. Koga, K. Tomioka, *Tetrahedron* **1994**, *50*, 4429, and references therein.

[18] S. E. Denmark, N. Nakajima, O. J.-C. Nicaise, A-M. Faucher, J. P. Edwards, *J. Org. Chem.*, **1995**, *60*, 4884, and references therein.

[19] For leading references see: a) D. Hoppe, H. Ahrens, W. Guarnieri, H. Helmke, S. Kolazewski, *Pure Appl. Chem.* **1996**, *68*, 613; b) P. Beak, A. Basu, D. J. Gallagher, Y. S. Park, S. Thayumanavan, Acc. *Chem. Res.* **1996**, *29*, 552; c) M. Schlosser, D. Limat, *J. Am. Chem. Soc.* **1995**, *117*, 12342; d) D. Hodgson, G. Lee, *J. Chem. Soc. Chem. Commun.*, **1996**, 1015.

[20] For leading references see: a) M. J. Burk, J. P. Martinez, J. E. Feaster, N. Cosford, *Tetrahedron* **1994**, *50*, 4399; b) X. Verdaguer, U. E. W. Lange, M. T. Reding, S. L. Buchwald, *J. Am. Chem. Soc.* **1996**, *118*, 6784; c) N. Uematsu, A. Fujii, S. Hashiguchi, T. Ikariya, R. Noyori, *J. Am. Chem. Soc.* **1996**, *118*, 4916; d) R. Noyori, S. Hashiguchi, *Acc. Chem. Res.* **1997**, *30*, 97.

Steric and Stereoelectronic Effects in the Palladium Catalyzed Allylation

Oliver Reiser

The development of enantioselective-catalyzed reactions has led to great success over the past few years. Numerous chiral metal catalysts have been tested, which have achieved a wide range of selectivities, never thought possible, for a great number of reactions by using tailored ligands. [1] A problem in the catalysis repertoire was the palladium-catalyzed, enantioselective allylic substitution. [2] Elegant mechanistic investigations have contributed to the development of this reaction as a valued method of synthesis. The understanding of factors such as the initial conformation of the allylic component, [3] the nature of its leaving group, [4] the conformation of the intermediate allylic palladium complexes, [5] the hardness of the nucleophile, [6] and the electronic and steric properties of the ligands bound to the palladium center, [7] today allow such substitution reactions to be carried out with high regio- and diastereoselectivities.

Most of the asymmetric allylic substitution processes start from racemic allylic components *rac*-1-R (where for 1-R, R designates the general group of the compound, e.g. 1-Me is 1 with R = Me), which in the absence of chiral ligands form *meso* complexes of the type 2 with palladium(0). Since a nucleophile can attack at either of the two ends of the allylic component, the enantiomers 3-R and *ent*-3-R are formed. The degree of the enantioselectivity of a reaction depends on how well a chiral ligand in 2 can direct the attack of the nucleophile (Nu) to one of the two allylic termini.

The problem with allylic substitutions, such as reaction (a), is that a soft nucleophile attacks the complexes 2 from the side opposite to the ligands L; as a result the distance between the reaction centers and the chiral inductor is large.

Thus, it is not surprising that ligands which are normally highly selective, such as (+)-

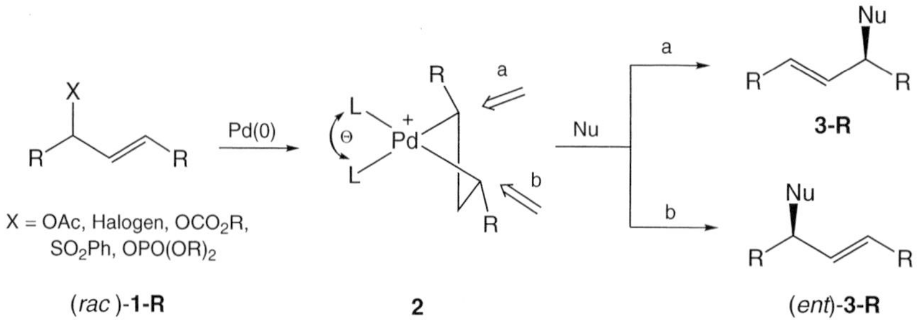

X = OAc, Halogen, OCO₂R,
   SO₂Ph, OPO(OR)₂

(*rac*)-1-R   2   (*ent*)-3-R

Table 1. Enantioselective allylic alkylation of *rac*-**4**-R according to Eq. (a)

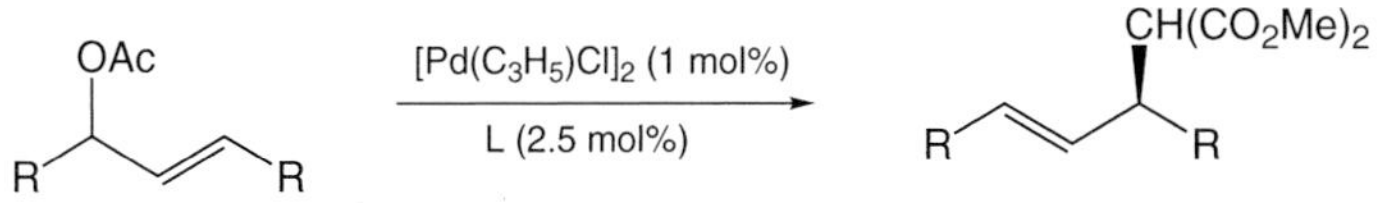

4-R	5-R	L	Yield (%)	% ee	Ref.
4-Me	5-Me	(+)-DIOP [d]	66–88 [a]	22	[9]
4-Me	5-Me	8	98 [b]	92	[14]
4-Me	5-Me	17-*i*Pr	not reported	56	[27]
4-Me	5-Me	17-*t*Bu	98 [c]	71	[21a]
4-Et	5-Et	17-*i*Pr	not reported	74	[27]
4-*n*Pr	5-*n*Pr	17-*t*Bu	96 [c]	69	[21a]
4-*i*Pr	5-*i*Pr	17-*i*Pr	not reported	94	[27]
4-*i*Pr	5-*i*Pr	17-*t*Bu	88 [a]	96	[21a]
4-Ph	5-Ph	6 [e]	85 [a]	96	[11]
4-Ph	ent-5-Ph	13	99 [c]	95	[16b]
4-Ph	5-Ph	14	97 [c]	97	[16b]
4-Ph	5-Ph	16-Se	50–84 [a]	95	[21b]
4-Ph	5-Ph	16-S	89 [b]	90	[19a]
4-Ph	5-Ph	17-Ph	99 [b]	99	[21a]
4-Ph	5-Ph	17-*i*Pr	74 [b]	98.5	[21b]

[a] NaCH(CO$_2$Me)$_2$; [b] CH$_2$(CO$_2$Me)$_2$, Cs$_2$CO$_3$; [c] CH$_2$(CO$_2$Me)$_2$, BSA, KOAc; [d] 100 mol% PdCl$_2$, 200 mol% ligand; [e] 0.5 mol% [Pd(C$_3$H$_5$)Cl]$_2$, 1 mol% ligand.

DIOP, [8] achieved an enantiomeric excess of only 22 % *ee* (Table 1) for the preparation of 5-Me. [9] Even ligands such as BINAP [8], or Chiraphos [8] have so far only shown a high selectivity for few substrates. [10] As a result, ligands were developed which can "reach over" to the *exo* side of the allylic complexes **2**, for example the ferrocene **6** designed by Hayashi. [11] In the allylic complex **7** a flexible hydroxyl group of the phosphane ligand directs the nucleophile. With this ligand 5-Ph was synthesized in up to 96 % *ee*.

A very exciting concept has been developed with the synthesis of ligands which are derived from 2-(diphenylphosphino)benzoic acid [12] or 2-(diphenylphosphino)aniline. [13]

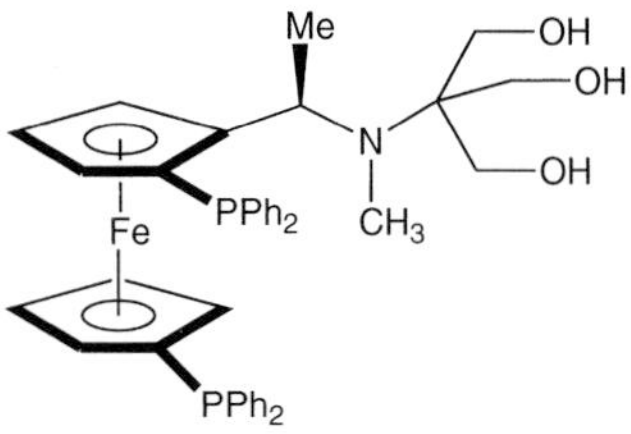

6

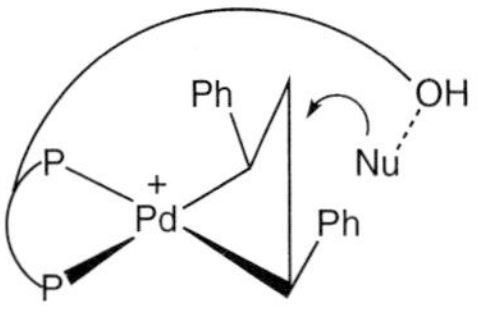

7

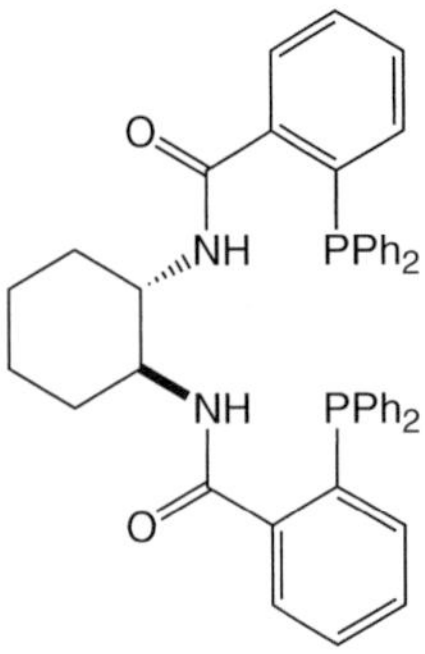

8

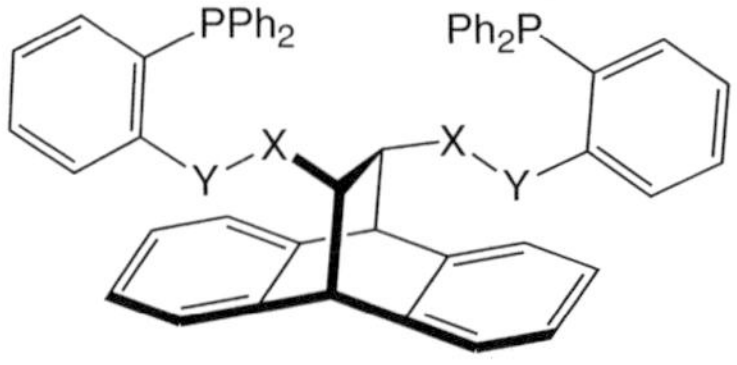

9   a: X = CO, Y = NH
     b: X = NH, Y = CO

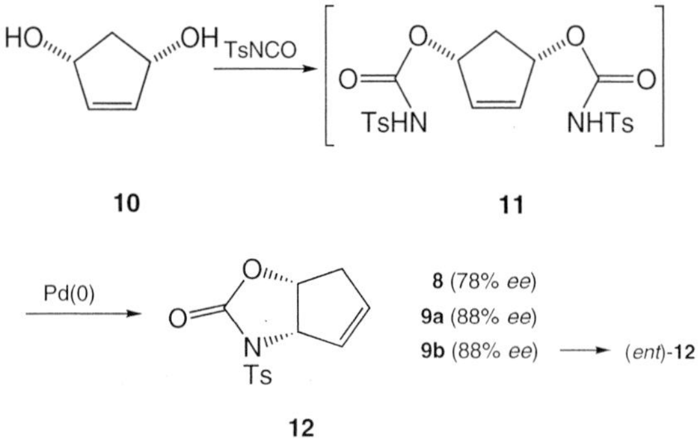

10

11

12

8 (78% *ee*)
9a (88% *ee*)
9b (88% *ee*) ⟶ (*ent*)-12

By the coordination of **8** or **9** to palladium a 13-membered ring is formed each time. This chelate ring extension causes an increase of the bite angle θ in **2**, thus enhancing the depth of the chiral pocket in which the the allylic component is residing.

The *meso*-ditosylcarbamate **11**, which is generated *in situ* from the diol **10**, cyclizes to give the oxazolidinone **12** with up to 88 % *ee*; the highest enantioselectivity for this system obtained so far. Strikingly, with **9b** the opposite sense of induction as with **9a** is observed. [12, 13] With **8** and **9** also excellent results have been obtained for acyclic substrates, however, with the standard system **4**-Ph low yields and only up to 52 % *ee* are achieved. [14] As a possible explanation it has been suggested that the chiral pocket of these ligand is not large enough to accomodate **4**-Ph. In agreement with this proposal, smaller substrates such as **4**-Me give selectivities as high as 92 % *ee*. This study also revealed the importance of the right choice of base being used to deprotonate the malonate and solvent, optimal results are achieved with the combination of caesium carbonate and THF.

C_2 symmetric, chiral semicorrines [15] and bisoxazolines [16] have given a great boost to catalysis research over recent years. They have also proven to be efficient ligands in the enantioselective synthesis of **5**-Ph and *ent*-**5**-Ph, [16b] although these ligands neither "attack" the *exo* side of the complex **2** nor can they increase the bite angle θ in the macrocyclic chelate. A selectivity of 95 % *ee* is achieved for the substitution at **4**-Ph with ligand **13** and with **14** even one of 97 % *ee* (see Table 1) in an almost quantitative reaction. For this substitution, instead of sodium dialkylmalonate as nucleophile, the use of a

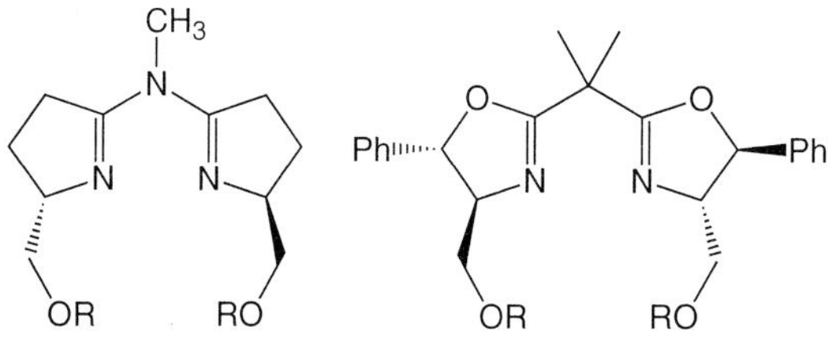

13   R = SiMe$_2$^{t}Bu   14

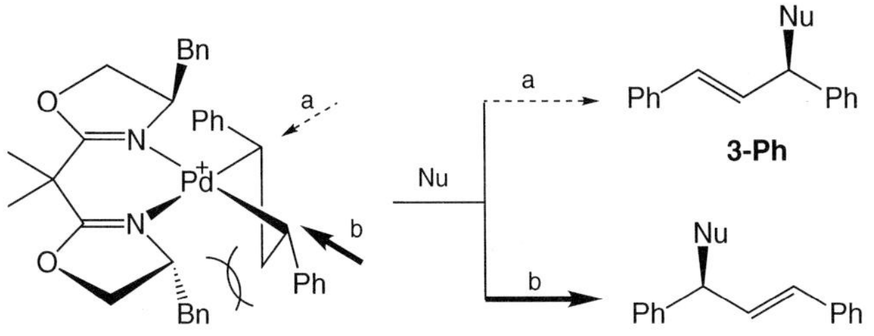

15   (*ent*)-3-Ph

mixture from dimethyl malonate and *N,O*-(bistrimethylsilyl) acetamide (BSA) proved successful. [17]

These excellent results, supported by an X-ray structure analysis of complex **15**, could be explained convincingly: [18] one benzyl group in the ligand and one phenyl substituent at the allylic terminus repel each other considerably. This steric hindrance is, for example, illustrated by a clear twisting of the oxazoline ring in question and from the lengthening of the corresponding Pd-C bonds. The nucleophile attacks at the allylic terminus which is sterically strained, since in this way, in combination with the rupture of the Pd-C bond, the steric strain can be reduced.

If so far the impression was given that the basic condition for an efficient ligand is C$_2$ symmetry, the discovery of new ligands **16** and **17** with S,N [19]-, Se,N [20]- and P,N [21]-coordination at the metal has changed this picture. Especially the ligands **17-R** having a triphenylphosphane and a oxazoline unit showed record selectivities (Table 1) for the substitution of **4**-Ph. Initially, malonates were tested as nucleophiles, but later reports

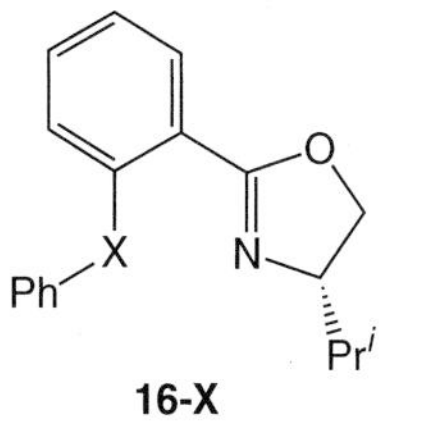

16-X   17-R

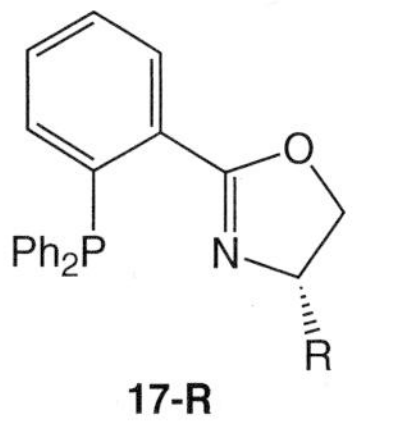

appeared using nitrogen nucleophiles [22] or nitromethane [23] with spectacular results.

Early on it was speculated that a major control element for the effectiveness of these ligands must be centered around stereoelectronic considerations. In an NMR study [24] it was shown that in an allylic palladium complex the carbon atoms of the allylic termini are shifted with increasing acceptor strength of the ligands (e. g., change from nitrogen to phosphorus coordination) to higher field in the ^{13}C NMR spectrum. This effect is particularly marked for the carbon atom in the position *trans* to the ligand. With the necessary caution required when dealing with the correlation of ^{13}C NMR shifts with charge distributions, this is an indication that the electron density of an allylic carbon atom *trans* to the

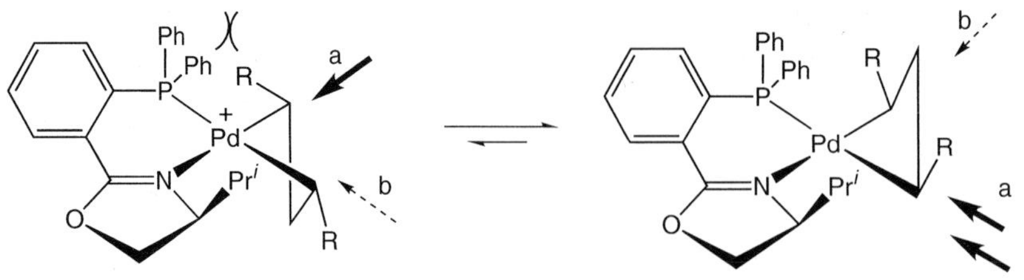

18-R   **19-R**

phosphane ligand, in comparison to the nitro-gen ligand, should be reduced significantly, and consequently being the one which is attacked by the nucleophile. In order to account for these stereoelectronic considerati-ons and the observed stereoselectivity, as the reactive intermediate the palladium complex **19-R** was proposed. [21b, 25]

However, the original suggestion [25] that this diastereomer should be disfavored on ste-ric reasons compared to **18-R** turned out to be incorrect: in most elegant studies [20] using X-ray analysis and nmr spectroscopy Helm-chen et al. could demonstrate that of the two complexes **19-R** is favored in solution with a ratio of 1 : 8. They identified as the major ste-ric interaction to be avoided the close align-ment of the phenyl group in the allyl system with the pseudoequatorial phenyl group of the diphenylphosphino group. The X-ray structure of **19-R** revealed that the longer pal-ladium carbon bond is located at the allylic center *trans* to the phosphorous ligand, thus making this carbon more susceptible for

nucleophilic attack. The interconversion of **18-R** and **19-R** is at least 50 times faster than the reaction with the nucleophile, neverthe-less, in light of the experimental evidence and in agreement with a postulate by Bosnich [26] that the more abundant isomer should be the the more reactive one, it seems plausible that **19-R** ist the decisive intermediate and that stereoelectronic factors – a concept which should find more use in catalysis [27] – direct the nucleophile. The degree of enan-tioselectivity therefore seems to depend on the relative reaction rates at the allylic termini in the complexes **18-R** and **19-R**. NMR stud-ies of the palladium complexes with other allyl substrates revealed that the ratio of **18-R** increases as the substituents on the allylic termini become smaller. [28] Consequently, these substrates give in parallel lower enantio-selectivities.

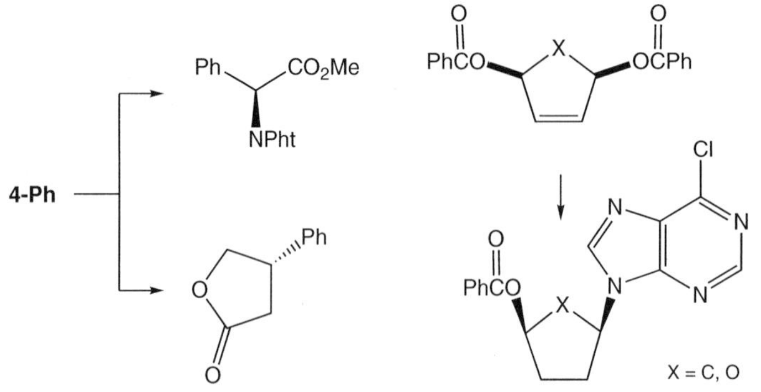

With the chiral ligands such as **8** and **9** and the phoshinooxazolines **17** two powerful and in part complimentary sets of ligands are rea-dily available which allow the effective asymmetric allylic substitution of a broad range of substrates. Moreover, several useful applications of the substitution products such as the synthesis of *a*-amino acids, [21] succinic acids, [2b] butyrolactones, [2b, 28] and nucleosides [29] have been reported, demonstrating the practical usefulness of the methodology developed here.

Acknowledgement: This work was supported by the Winnacker foundation and the Fonds der Chemischen Industrie.

References

[1] a) R. Noyori, M. Kitamura in Modern Synthetic Methods (Ed.: R. Scheffold), Springer, Berlin, **1989**, pp. 115–198; b) H. Brunner, Top. Stereochem. **1988**, *18*, 129.

[2] Recent reviews: a) B. M. Trost, D. L. Vanvranken, *Chem. Rev.* **1996**, *96*, 395–422. b) J. M. J. Williams, *Synlett* **1996**, 705. c) C. G. Frost, J. Howarth, J. M. J. Williams, *Tetrahedron*: Asymmetry **1992**, *3*, 1089; d) S. A. Godleski in Comprehensive Organic Synthesis, Vol. 4 (Ed.: B. M. Trost), Pergamon, Oxford, **1991**, p. 585; e) B. M. Trost, *Angew. Chem.* **1989**, *101*, 1199; *Angew. Chem.* Int. Ed. Engl. **1989**, *28*, 1173; f) G. Consiglio, R. Waymouth, *Chem. Rev.* **1989**, *89*, 257.

[3] T. Hayashi, A. Yamamoto, T. Hagihara, *J. Org. Chem.* **1986**, *51*, 723.

[4] a) F. K. Sheffy, J. K. Stille, *J. Am. Chem. Soc.* **1983**, *105*, 7173; b) B. M. Trost, N. R. Schmuff, M. J. Miller, ibid. **1980**, *102*, 5979; c) J. Tsuji, I. Minami, *Acc. Chem. Res.* **1987**, 20, 140; d) N. Greenspoon, E. Keinan, *Tetrahedron Lett.* **1982**, *23*, 241.

[5] B. Akermark, S. Hansson, A. Vitagliano, *J. Am. Chem. Soc.* **1990**, *112*, 4587.

[6] a) J.-C. Fiaud, J.-Y. Legros, *J. Org. Chem.* **1987**, *52*, 1907; b) N. Greenspoon, E. Keinan, ibid. **1988**, 53, 3723; c) B. M. Trost, J. W. Herndon, *J. Am. Chem. Soc.* **1984**, *106*, 6835; d) B. M. Trost, T. R. Verhoeven, *J. Org. Chem.* **1976**, *41*, 3215.

[7] a) B. Akermark, S. Hansson, B. Krakenberger, A. Vitagliano, K. Zetterberg, *Organometallics* **1984**, *3*, 679; b) B. Akermark, S. Hansson, A. Vitagliano, *J. Organomet. Chem.* **1987**, *335*, 133.

[8] Abbreviations of the ligands: DIOP = 2,3-O-isopropylidene-2,3-dihydroxy-1,4-bis(diphenylphosphino)butane; Chiraphos: 2,3-bis(diphenylphosphino)butane; BINAP 2,2′-bis(diphenylphosphino)-1,1′-binaphthyl.

[9] B. M. Trost, T. J. Dietsche, *J. Am. Chem. Soc.* **1973**, *95*, 8200.

[10] M. Yamaguchi, T. Shima, T. Yamagishi, M. Hida, *Tetrahedron Lett.* **1990**, *35*, 5049–5052.

[11] T. Hayashi, *Pure Appl. Chem.* **1988**, *60*, 7.

[12] a) B. M. Trost, D. L. van Vranken, *Angew. Chem.* **1992**, *104*, 194; *Angew. Chem. Int. Ed. Engl.* **1992**, *31*, 228; b) B. M. Trost, D. L. van Vranken, C. Bingel, *J. Am. Chem. Soc.* **1992**, *114*, 9327–9343.

[13] B. M. Trost, B. Breit, S. Peukert, J. Zambrano, J. W. Ziller, *Angew. Chem.* **1995**, 107, 2577; *Angew. Chem. Int. Ed.* **1995**, *34*, 2386.

[14] B. M. Trost, A. C. Krueger, R. C. Bunt, J. Zambrano, *J. Am. Chem. Soc.* **1996**, *118*, 6520–6521.

[15] a) A. Pfalz in Modern Synthetic Methods (Ed.: R. Scheffold), Springer, Berlin, **1989**, pp. 199–248; b) Chimica **1990**, *44*, 202.

[16] a) D. Müller, G. Umbricht, B. Weber, A. Pfaltz, *Helv. Chim. Acta* **1991**, *74*, 232; b) U. Leutenegger, G. Umbricht, C. Farni, P. von Matt, A. Pfaltz, *Tetrahedron* **1992**, *48*, 2143; c) C. Bolm, K. Weikhardt, M. Zehnder, T. Ranff, *Chem. Ber.* **1991**, *124*, 1173; d) C. Bolm, G. Schlingloff, K. Weickhardt, *Angew. Chem.* **1994**, *106*, 1944; e) D. A. Evans, M. M. Faul, M. T. Bilodeau, *J. Am. Chem. Soc.* **1994**, *116*, 2742; f) S. E. Denmark, N. Nakajima, O. J.-C. Nicaise, *J. Am. Chem. Soc.* **1994**, *116*, 8797; g) T. Fujisawa, T. Ichiyanagi, M. Shimizu, *Tetrahedron Lett.* **1995**, *36*, 5031; h) A. S. Gokhale, A. B. E. Minidis, A. Pfaltz, *Tetrahedron Lett.* **1995**, *36*, 183. i) *Recent Review:* O. Reiser, *Nachr. Chem. Tech. Lab.* **1996**, *44*, 744.

[17] B. M. Trost, S. J. Brickner, *J. Am. Chem. Soc.* **1983**, *105*, 568.

[18] a) P. von Matt, G. C. Lloydjones, A. B. E. Minidis, A. Pfaltz, L. Macko, M. Neuburger, M. Zehnder, H. Ruegger, P. S. Pregosin, *Helv. Chim. Acta* **1995**, *78*, 265–284; b) A. Pfaltz, *Acc. Chem. Res.* **1993**, *26*, 339.

[19] a) J. V. Allen, S. J. Coote, G. J. Dawson, C. G. Frost, C. J. Martin, J. M. J. Williams, *J. Chem. Soc. Perkin Trans. 1* **1994**, 2065; b) J. V. Allen, G. J. Dawson, C. G. Frost, J. M. J. Williams, S. J. Coote, *Tetrahedron* **1994**, *50*, 799; c) C. G. Frost, J. M. J. Williams, *Tetrahedron Lett.* **1993**, *34*, 2015.

[20] a) J. Sprinz, M. Kiefer, G. Helmchen, G. Huttner, O. Walter, L. Zsolnai, M. Reggelin, *Tetrahedron Lett.* **1994**, *35*, 1523–1526; b) H. Steinhagen, M. Reggelin, G. Helmchen, *Angew. Chem.* **1997**, *109*, 2199; *Angew. Chem. Int. Ed. Engl.* **1997**, *36*, 2108.

[21] a) P. von Matt, A. Pfaltz, *Angew. Chem.* **1993**, *105*, 614; *Angew. Chem. Int. Ed. Engl.* **1993**, *32*, 566; b) J. Sprinz, G. Helmchen, *Tetrahedron Lett.* **1993**, *34*, 1769–1772; c) G. J. Dawson, C. G. Frost, J. M. J. Williams, *Tetrahedron Lett.* **1993**, *34*, 3149–3150; d) J. M. Brown, D. I. Hulmes, P. J. Guiry, *Tetrahedron* **1994**, *50*, 4493–4506.

[22] a) P. v. Matt, O. Loiseleur, G. Koch, A. Pfaltz, C. Lefeber, T. Feucht, G. Helmchen, *Tetrahedron Asymmetry* **1994**, *5*, 573; b) R. Jumnah, A. C. Williams, J. M. J. Williams, *Synlett* **1995**, 821.

[23] H. Rieck, G. Helmchen, *Angew. Chem.* **1995**, *107*, 2881; *Angew. Chem. Int. Ed. Engl.* **1995**, *34*, 2881.

[24] B. Akermark, B. Krakenberger, S. Hansson, A. Vitagliano, *Organometallics* **1987**, *6*, 620.

[25] O. Reiser, *Angew. Chem.* **1993**, *105*, 576–578; *Angew. Chem. Int. Engl. Ed.*, **1993**, *32*, 547–549.

[26] B. Bosnich, P. B. Mackenzie, *Pure Appl. Chem.* **1982**, *54*, 189.

[27] T.V. Rajan Babu, A. L. Casalnuovo, *J. Am. Chem. Soc.* **1996**, *118*, 6325.

[28] G. Helmchen, S. Kudis, P. Sennhenn, H. Steinhagen, *Pure Appl. Chem.* **1997**, 513.

[29] a) B. M. Trost, R. Madsen, S. G. Guile, A. E. H. Elia, *Angew. Chem.* **1996**, *108*, 1666; *Angew. Chem. Int. Ed. Engl.* **1996**, *35*, 1569; b) B. M. Trost, Z. P. Shi, *J. Am. Chem. Soc.* **1996**, *118*, 3037.

Asymmetric Alkylation of Amide Enolates with Pseudoephedrine as Chiral Auxiliary – Unexpected Influence of Additives?

Karola Rück-Braun

Asymmetric alkylation of carboxylic acid derivatives has been studied intensively for about 20 years. [1] Numerous auxiliaries, tailor-made structures with high steric demands for effective *Re/Si* face differentiation, have been synthesized and their efficiency tested. [1, 2] In recent years besides the preparative aspects of enolates, physico-chemical investigations *into* their structure-reactivity relationships have gained interest. [3] Crystal structure analyses, osmometric measurements, and NMR studies in solution are helpful in the investigation of the factors that may control enolate reactions. [3–5]

Recently, Myers et al. described a method for alkylation of amide enolates having D-(+)-pseudoephedrine, an industrial product manufactured worldwide in ton quantities, as a chiral auxiliary. [6] Upon treatment with LDA (2 equiv), N-acyl-pseudoephedrine derivatives **1** react with alkyl bromides or iodides efficiently and with high stereoselectivity in the presence of LiCl (6 equiv) (Scheme 1).

Even at 0 °C diastereoselectivities of > 94 % *de* were obtained.

A wide repertoire of cleavage reactions demonstrates the synthetic potential of the pseudoephedrine amides, providing access to chiral α-branched carboxylic acids, aldehydes, ketones or primary alcohols with recovery of the auxiliary (Scheme 2). Moreover, efficient alkylation reactions utilizing epoxides and epoxide-derived electrophiles open up a route to chiral γ-lactones and γ-hydroxy ketones. [7]

Interestingly, the authors note that the reaction time of the alkylation reaction can be shortened and the conversion increased by addition of LiCl. The concentration of LiCl did not seem to affect the diastereoselectivity of the reaction. Under the same reaction conditions the chiral auxiliary L-(–)-ephedrine [8] led to markedly lower diastereoselectivities. [6, 8] Even in the presence of LiCl the alkylation product from N-propionyl-ephedrine **3** and n-butyl iodide was formed

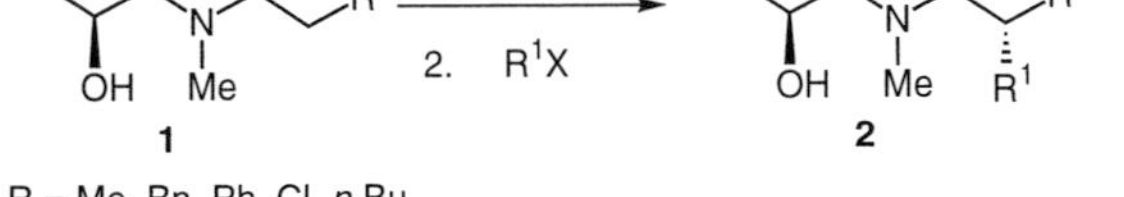

1. 2eq LDA
6eq LiCl

2. R¹X

R = Me, Bn, Ph, Cl, *n*-Bu
R¹X = MeI, EtI, *n*-Bu, BnBr

80 - 99%, 94 - >99% *de*

Scheme 1.
LDA = lithium diisopropylamide.

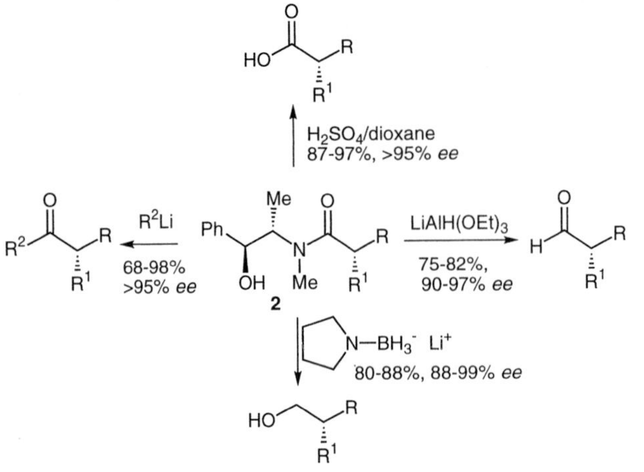

Scheme 2.

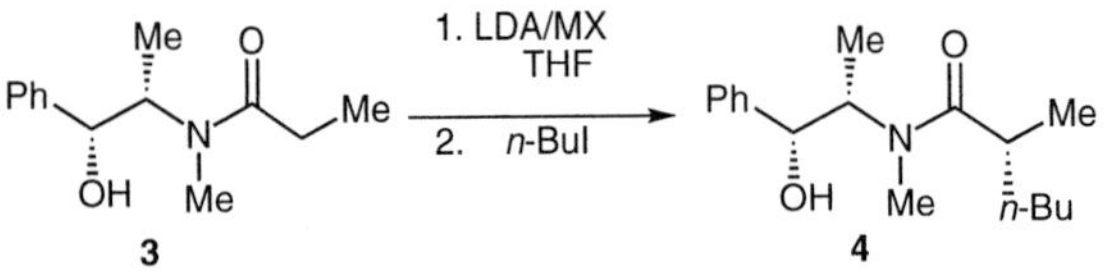

MX	de
LiCl[6]	70%
MgCl$_2$[8]	90%

Scheme 3.

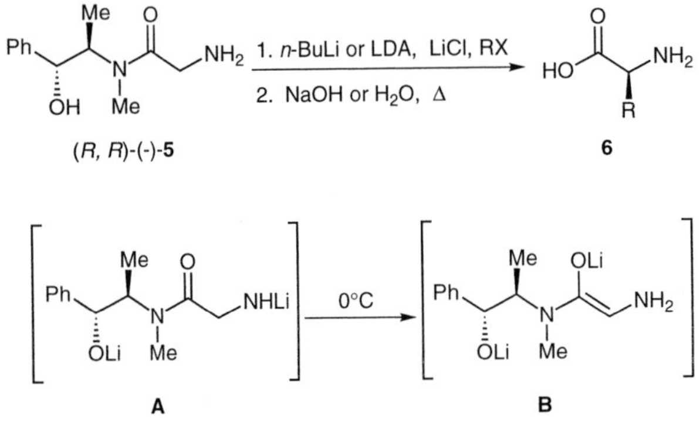

Scheme 4.

with a diastereoselectivity of only 70% de (Scheme 3).

Recently, (+)- and (−)-pseudoephedrine-derived glycinamides [9] proved to be suitable building blocks for the synthesis of D- or L-configurated α-amino acid derivatives. [10] Thus, glycinamides 5, [10b] easily accessible by n-butyl-lithium-promoted condensation of glycine methyl ester with pseudoephedrine in the presence of LiCl, undergo highly diastereoselective alkylation reactions with a wide range of alkyl halides, without protection being necessary for the hydroxyl and amino functionalities present in the molecules (Scheme 4).

Apparently, deprotonation performed at −78 °C is leading to the *O*-, *N*-dianion **A** (Scheme 4). Equilibration of the latter to the thermodynamically more stable *Z*-enolate **B** (Scheme 4) upon subsequent warming to room temperature seems to be reasonable, due to the *C*-alkylated products obtained at 0 °C or even at room temperature. In this temperature range *N*-alkylation, observed at −78 °C, is effectively suppressed. [10d] For lithium enolates derived from the glycinamides **5** an influence of lithium halides on rate enhancement and diastereoselectivity is found. [10] Thus, in the absence of LiCl a significant decrease in diastereoselectivity is observed in the alkylation of **5** with ethyl iodide (82 % *de* without LiCl in comparison to 97 % *de* upon addition of LiCl (6 equiv)). Lithium bromide (6 equiv) was found to accelerate the rate of enolate alkylation, too, but diastereoselectivity was found to be lower (91–93 % *de*).

The practical alkylation procedures published, include efficient hydrolysis protocols providing the advantage to prepare either *N*-*tert*-butyloxycarbonyl (*N*-Boc) or *N*-(9-fluorenyl-methyloxy)-carbonyl (*N*-Fmoc) protected proteinogenic and non-proteinogenic D- or L-configurated *a*-amino acids directly from the alkaline aqueous hydrolysis solution. [10d] Therefore, and due to the low cost and the availability of the chiral auxiliary employed, the method developed by Myers et al. appears to be competitive even for appli-cations in industry. However, the reasons for the high diastereoselectivity observed are neither obvious nor predictable. Considering the investigations of Larcheveque et al., [8] who reported on the alkylation of amide enolates of ephedrine in 1978, one would probably not have been predicted any great success for experiments with pseudoephedrine. Larcheveque and coworkers reported on the deprotonation of *N*-acyl-ephedrines with LDA in the presence of hexamethylphosphoric acid triamide (HMPA) and subsequent reactions with alkyl halides providing products with diastereoselectivities of < 80 : 20. Only when MgCl$_2$ was added, diastereomeric ratios of > 95 : 5 were achieved (Scheme 3). How can the effects brought about by the addition of metal salts be explained?

Since the start of investigations into asymmetric reactions with enolates it has been known that the reactivity and selectivity observed in enolate chemistry is influenced not only by the base employed, but also by the use of cosolvents such as HMPA, and the addition of metal salts or Lewis acids. [2–4, 11] Lithium enolates, in particular, tend to form aggregates by self-assembly. [3, 4] Decisive contributions to the explanation of this phenomenon and its consequences have been made by Seebach et al. by crystal structure analyses of crystalline lithium enolates [12] up to suggestions regarding the structure of the complexes in solution (Scheme 5). [3, 4, 13]

Scheme 5. RCH$_2$Y = alkyl halide.

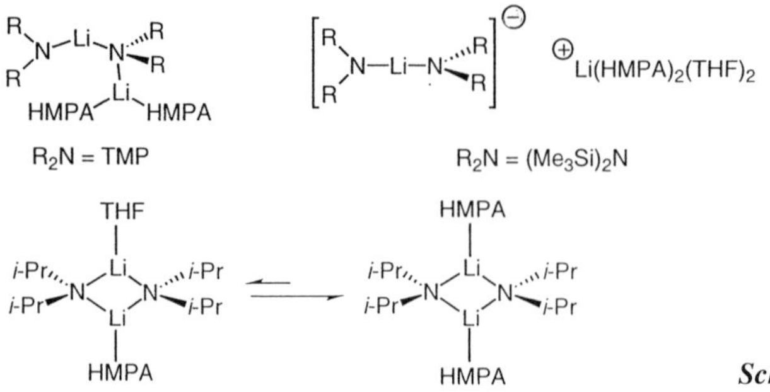

$R_2N = TMP$

$R_2N = (Me_3Si)_2N$

THF

HMPA

HMPA

Scheme 6. R'$_2$N = TMP, R$_2$N = HMDS.

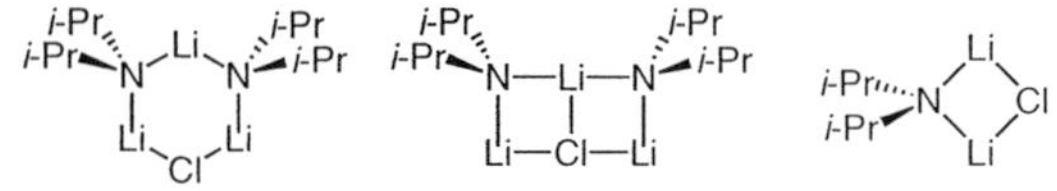

Scheme 7.

The method of choice for structure determination of aggregates in solution is NMR spectroscopy. New investigations are based for instance on ^{6}Li, ^{15}N and ^{31}P measurements with isotopically labeled samples of lithium bases such as lithium hexamethyldisilazane (LiHMDS), LDA, and lithium tetramethylpiperidide (LiTMP). [14] Thus ^{6}Li–^{15}N coupling, for example, permits conclusions to be drawn about the degree of aggregation of LiNR$_2$ derivatives. Even now, despite its carcinogenicity HMPA is often added to reaction mixtures to increase the reactivity of the lithium compounds. Recently, mixed solvates/aggregates between HMPA and lithium bases (LiHMDS, LiTMS, LDA) were proved (Scheme 6). [15–18] The cosolvent does not break up the aggregates [15] but instead promotes formation of open dimers [16] and "triple ions". [17]

In addition, recent spectroscopic studies of Column et al., dealing with the etheral solvation of LiHMDS, revealed no support for the often-cited correlation of reduced aggregation state with increasing strength of the lithium-solvent interaction. [19] Moreover, exact spectroscopic data could be obtained for the structures of complexes of lithium bases and salts and their dependence on the concentra-

tion of the added salt (Scheme 7). [14, 15, 20] A mixed aggregate of LDA and LiCl recently was synthesized and the crystal structure analyzed by Mair et al. [21] Meanwhile, dimeric LDA upon treatment with LiCl is known to form the more reactive mixed aggregate (iPr)$_2$NLi·LiCl. [19, 22]

In the light of increasing knowledge about the nature and reactivity of lithium amide/halide aggregates, information and understanding of lithium enolate/halide aggregates still seem to be rather poor. However, P.v.R. Schleyer's group recently characterized two aggregates consisting of lithium halides and lithium enolates, identified as heterodimers, by X-ray crystallography. [23] Besides the question of the constitution and the reactivity of the aggregates being formed, another problem in analyzing the reactions of enolates with electrophiles is the continually changing concentration of base, enolate, and metal salts during the course of the reaction. Stereochemical investigations and trapping experiments on the enolization of 3-pentanone **7** with LiTMP in THF impressively demonstrate the consequences of the structural variety of the aggregates, thus differing in reactivity (Scheme 8). [15, 20] Whereas at 5 % conversion the E/Z selectivity was 30 : 1, at greater

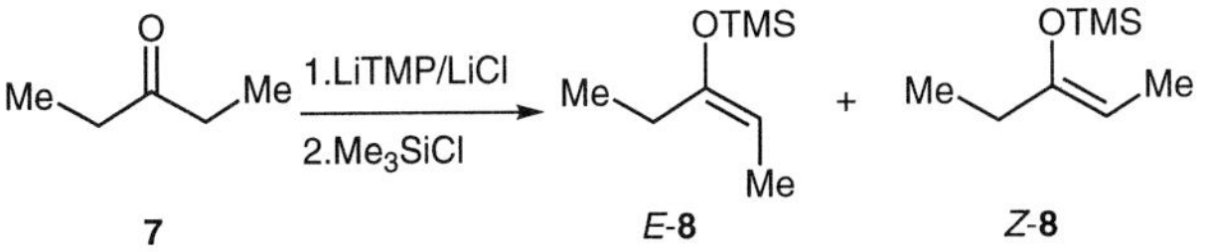

LiCl	E : Z
0.3-0.4 eq	50-60 : 1
1-2 eq	10-20 : 1

Scheme 8. TMS–Cl = Me$_3$SiCl.

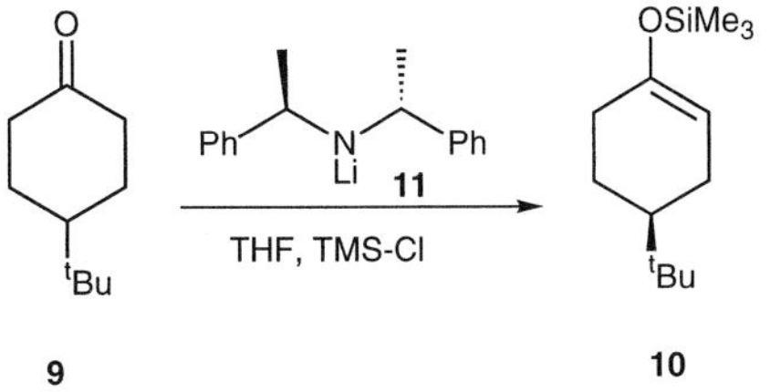

reaction conditions	ee of **10**
ISQ	69
EQ	23
EQ+LiCl	83

Scheme 9.

than 80 % conversion an *E/Z* selectivity of only > 10 : 1 was observed. The addition of 0.3–0.4 equivalents of LiCl led to an *E/Z* ratio of 50 : 1; however, further increase in the amount of LiCl to > 1.0 equivalents produced results identical to those obtained under salt-free conditions.

Remarkable improvements in chiral base-mediated reactions of prochiral ketones under external quench (EQ) conditions with TMS–Cl, furnishing enantiomerically pure enol silanes, were found upon deprotonation in the presence of LiCl. [22, 24] Simpkins et al. studied for instance the conversion of 4-*tert*-butylcyclohexanone **9** into enol silane **10** by employing the chiral amide base **11** (Scheme 9). [24] Applying the TMS–Cl *in situ* quench (TMS–Cl–ISQ) protocol a higher level of enantiomeric excess was observed compared to external quench conditions (EQ). However, under external quench conditions in the presence of LiCl (EQ+LiCl procedure) significantly higher levels of asymme-

tric induction are observed. Simpkins et al. concluded from the results obtained LiCl to be the reason of the "ISQ effect". [24] Thus, the internally available TMS–Cl was proposed to be the source for an increasing LiCl concentration during conversion to the enol silane. Indeed, spectroscopic experiments by Lipshutz and coworkers support TMS–Cl to be the supplier of LiCl as the key component of *in situ* quenching conditions in ketone enolization. [25] Furthermore, the "LiCl effect" observed upon inclusion of LiCl under external quench conditions was concluded to be caused by conversion to the more selective mixed aggregates already mentioned. [19, 22, 23]

Exploring "nonstoichiometric effects" [3a, 5] with organolithium compounds since the early 1980s, these days Seebach et al. are investigating the diastereoselective alkylation of polylithiated open-chain and cyclic peptides in the presence of excess lithium salts and bases (Scheme 10). [26, 27] Thus, highly

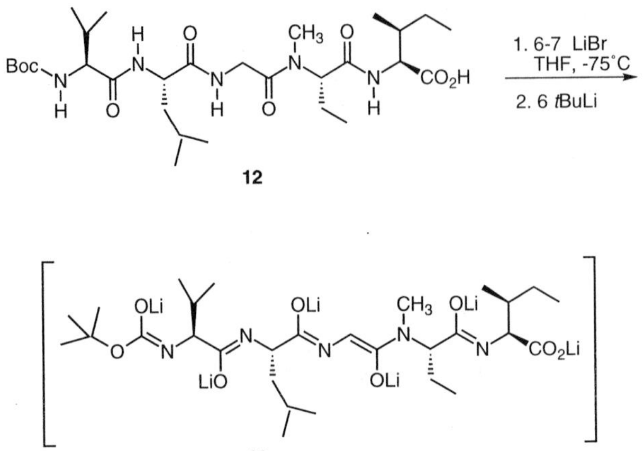

12

13

Scheme 10.

functionalized glycine and sarcosine lithium enolates and dilithium azadienolates of peptide lithium salt complexes were alkylated in yields of up to 90 % with diastereoselectivities ranging from 1 : 1 to 9 : 1.

The work of Myers et al. [6] illustrates the synthetic potential of the use of metal salts (instead of HMPA!) in alkylation reactions of enolates, employing easily accessible amide enolates of the chiral auxiliary pseudoephedrine. It is not surprising that the mechanism of chiral induction is not yet fully understood; further investigations are necessary. Nonetheless, unanswered questions in enolate chemistry remain even for tailor-made, well-established auxiliaries, whose asymmetric induction can be explained convincingly by working models on monomer enolate structures, considering chelation control and steric factors.

Increasing numbers of striking examples for unusual effects of metal salts other than lithium on various metal enolates were reported in recent years. [28, 29] Thus, we await interesting future results of systematic investigations on the influence of metal salts and, thereby, new applications in supposedly well-known reactions.

References

[1] a) D. Caine in *Comprehensive Organic Synthesis,* Vol. 3 (Eds.: B. M. Trost, I. Fleming, G. Pattenden), Pergamon Press, New York, **1991**, p. 1; b) D. A. Evans, J. M. Takacs, *Tetrahedron Lett.* **1980**, *21*, 4233; c) W. Oppolzer, R. Moretti, S. Thomi, *ibid.* **1989**, *30*, 5603; d) K.-S. Jeong, K. Parris, P. Ballester, J. Rebek Jr., *Angew. Chem.* **1990**, *102*, 550; *Angew. Chem. Int. Ed. Engl.* **1990**, *29*, 555.

[2] a) J. D. Morrison, *Asymmetric Synthesis*, Academic Press, New York, **1984**; b) M. Nogradi, *Stereoselective Synthesis*, 2nd ed., VCH, Weinheim, **1995**.

[3] Reviews: a) D. Seebach, *Proc. Robert A. Welch Found. Conf. Chem. Res. 27: Stereospecificity in Chemistry and Biochemistry* (7.–9. Nov. 1983), Houston, TX, USA, **1984**, 93–145; b) D. Seebach, *Angew. Chem.* **1988**, *100*, 1685; *Angew. Chem. Int. Ed. Engl.* **1988**, *27*, 1624; c) G. Boche, *ibid.* **1989**, *101*, 286 and **1989**, *28*, 277.

[4] a) L. M. Jackman, T. S. Dunne, *J. Am. Chem. Soc.* **1985**, *107*, 2805, and references therein; b) P. G. Willard, M. J. Hintze, *J. Am. Chem. Soc.* **1987**, *109*, 5539; c) J. Corset, F. Froment, M.-F. Lautié, N. Ratovelomanana, J. Seyden-

Penne, T. Strzalko, M.-C. Roux-Schmitt, *ibid.* **1993**, *115*, 1684.

[5] For excellent reviews about "some effects of lithium salts", see: a) D. Seebach, A. K. Beck, A. Studer, *Mod. Synth. Methods* **1995**, *7*, 1; b) D. Seebach, A. R. Sting, M. Hoffmann, *Angew. Chem.* **1996**, *108*, 2880; *Angew. Chem. Int. Ed. Engl.* **1996**, *35*, 2708 and references cited therein.

[6] A. G. Myers, B. H. Yang, H. Chen, J. L. Gleason, *J. Am. Chem. Soc.* **1994**, *116*, 9361.

[7] A. G. Myers, L. McKinstry, *J. Org. Chem.* **1996**, *61*, 2428. Interestingly, attack from opposite π-faces of the pseudoephedrine amide enolates is found for epoxides and alkyl halides. In this context, the lithium alkoxide function of the chiral auxiliary, seems to be the crucial moiety, directing the addition of epoxides and operating as a screen in the case of alkyl halides.

[8] a) M. Larcheveque, E. Ignatova, T. Cuvigny, *Tetrahedron Lett.* **1978**, *41*, 3961; b) M. Larcheveque, E. Ignatova, T. Cuvigny, *J. Organomet. Chem.* **1979**, *177*, 5.

[9] a) For the asymmetric alkylation of chiral glycine derivatives, see: R. M. Williams, *Organic Chemistry Series, Volume 7: Synthesis of Optically Active α-Amino Acids* (Eds.: J. E. Baldwin, P. D. Magnus), Pergamon Press, Oxford, **1989**; b) For the role of added salts in enolate alkylation of chiral imines of glycinates, see: A. Solladié-Carvallo, M.-C. Simon-Wermeister, J. Schwarz, *Organometallics* **1993**, *12*, 3743, and references cited.

[10] a) A. G. Myers, J. L. Gleason, T. Yoon, *J. Am. Chem. Soc.* **1995**, *117*, 8488; b) A. G. Myers, T. Yoon, J. L. Gleason, *Tetrahedron Lett.* **1995**, *26*, 4555; c) A. G. Myers, T. Yoon, *Tetrahedron Lett.* **1995**, *26*, 9429; d) A. G. Myers, J. L. Gleason, T. Yoon, D. W. Kung, *J. Am. Chem. Soc.* **1997**, *119*, 656.

[11] A. Loupy, B. Tchoubar, *Salt Effects in Organic and Organometallic Chemistry*, VCH, Weinheim, **1992**.

[12] a) R. Amstutz, W. B. Schweizer, D. Seebach, J. D. Dunitz, *Helv. Chim. Acta* **1981**, *64*, 2617; b) W. Bauer, T. Laube, D. Seebach, *Chem. Ber.* **1985**, *118*, 764.

[13] For the effect of mixed aggregates derived from lithium ester enolates and chiral lithium amides on enantioselective transformations, see: a) E. Juaristi, A. K. Beck, J. Hansen, T. Matt, T. Mukhopadhyay, M. Simson, D. Seebach, *Synthesis* **1993**, 1271; b) M. Uragami, K. Tomioka, K. Koga, *Tetrahedron Asymmetry* **1995**, *6*, 701; c) K. Yasuda, M. Shindo, K. Koga, *Tetrahedron Lett.* **1996**, *37*, 6343, and references therein.

[14] D. B. Collum, *Acc. Chem. Res.* **1993**, *26*, 227.

[15] a) A. S. Galiano-Roth, Y.-J. Kim. J. H. Gilchrist, A. T. Harrison, D. J. Fuller, D. B. Collum, *J. Am. Chem. Soc.* **1991**, *113*, 5053; b) Y.-J. Kim, M. P. Bernstein, A. S. Galiano-Roth, F. E. Romesberg, P. W. Williard, D. J. Fuller, A. T. Harrison, D. B. Collum, *J. Org. Chem.* **1991**, *56*, 4435.

[16] P. L. Hall, J. H. Gilchrist, A. T. Harrison, D. J. Fuller, D. B. Collum, *J. Am. Chem. Soc.* **1991**, *113*, 9575.

[17] F. E. Romesberg, M. P. Bernstein, J. H. Gilchrist, A. T. Harrison, D. J. Fuller, D. B. Collum, *J. Am. Chem. Soc.* **1993**, *115*, 3475.

[18] For amine and unsaturated hydrocarbon solvates of LiHMDS, see: a) B. L. Lucht, D. B. Collum, *J. Am. Chem. Soc.* **1996**, *118*, 2217; b) B. L. Lucht, D. B. Collum, *J. Am. Chem. Soc.* **1996**, *118*, 3529; c) B. L. Lucht, M. P. Bernstein, J. F. Remenar, D. B. Collum, *J. Am. Chem. Soc.* **1996**, *118*, 10707.

[19] B. L. Lucht, D. B. Collum, *J. Am. Chem. Soc.* **1995**, *117*, 9863.

[20] P. L. Hall, J. H. Gilchrist, D. B. Collum, *J. Am. Chem. Soc.* **1991**, *113*, 9571.

[21] F. S. Mair, W. Clegg, P. A. O'Neil, *J. Am. Chem. Soc.* **1993**, *115*, 3388.

[22] For solution structures of chiral lithium amides in the presence of lithium halides, see: K. Sugasawa, M. Shindo, H. Noguchi, K. Koga, *Tetrahedron Lett.* **1996**, *37*, 7377.

[23] K. W. Henderson, A. E. Dorigo, Q.-Y. Liu, P. G. Williard, R. v. R. Schleyer, P. R. Bernstein, *J. Am. Chem. Soc.* **1996**, *118*, 1339.

[24] a) B. J. Bunn, N. S. Simpkins, *J. Org. Chem.* **1993**, *58*, 533; b) B. J. Bunn, N. S. Simpkins, Z. Spavold, M. J. Crimmin, *J. Chem. Soc. Perkin Trans. 1* **1993**, 3113; c) P. Coggins, S. Gaur, N. S. Simpkins, *Tetrahedron Lett.* **1995**, *36*, 1545, and references cited.

[25] B. H. Lipshutz, M. R. Wood, C. W. Lindsley, *Tetrahedron Lett.* **1995**, *36*, 4385.

[26] a) D. Seebach, H. Bossler, H. Gründler, S.-I. Shoda, *Helv. Chim. Acta* **1991**, *74*, 197; b) S. A. Miller, S. L. Griffiths, D. Seebach, *ibid.* **1993**, *76*, 563; c) H. G. Bossler, D. Seebach, *ibid.* **1994**, *77*, 1124.

[27] D. Seebach, O. Bezencon, B. Jaun, T. Pietzonka, J. L. Matthews, F. N. M. Kühnle, W. B. Schweizer, *Helv. Chim. Acta* **1996**, *79*, 588.

[28] K. Rück, *Angew. Chem.* **1995**, *107*, 475; *Angew. Chem. Int. Ed. Engl.* **1995**, *34*, 433, references 17–19 cited therein.

[29] D. C. Harrowven, H. S. Poon, *Tetrahedron Lett.* **1996**, *37*, 4281, and references cited.

Catalytic Asymmetric Carbonyl-Ene Reactions

David J. Berrisford and Carsten Bolm

As one of the fundamental bond constructions, the carbonyl-ene reaction – between an aldehyde and an alkene bearing an allylic hydrogen – attracts considerable attention [1] from the synthetic community. Given the versatile chemistry of the product homoallylic alcohols, both the intra- and intermolecular versions of asymmetric carbonyl-ene reactions are valuable processes. [2] Within the catalytic field, [3] the continuing development of chiral Lewis acids further advances the utility and scope of carbonyl-ene chemistry. We wish to highlight a number of these developments.

Carreira et al. [4] have discovered a titanium-catalyzed asymmetric carbonyl-ene reaction of aldehydes with the cheap commodity chemical 2-methoxypropene (Scheme 1).

Using a catalyst prepared *in situ* from tridentate ligand (*R*)-1 and Ti(O*i*Pr)$_4$ in a 2 : 1 ratio, the yields and enantioselectivities of the new process are generally high (Table 1). The most encouraging results, up to 98 % *ee*, are observed with *a,β*-ynals. Thus, this chemistry provides alternative catalytic syntheses of propargylic alcohols. Unusually for an asymmetric addition process, benzaldehyde gives only modest selectivity – 66 % *ee*. The only *a*-branched aldehyde reported to undergo addition, cyclohexanecarboxaldehyde, affords a product with 75 % *ee*.

The vinyl ether products obtained from asymmetric ene reactions are valuable precursors for a number of enantiomerically-enriched compounds (Scheme 2). Acid hydrolysis affords the corresponding methyl ketones thus

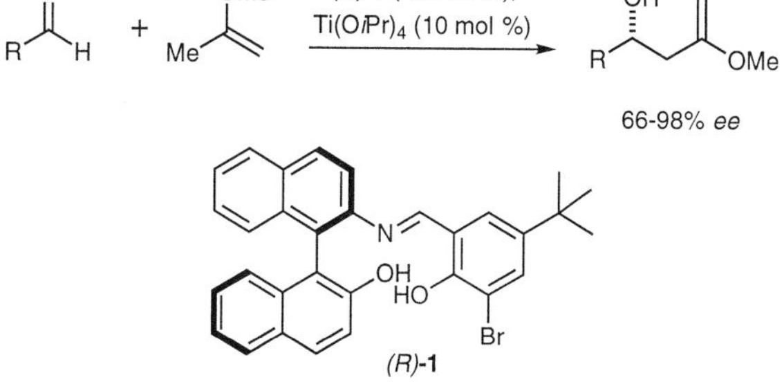

Scheme 1. Asymmetric Ti-catalyzed carbonyl-ene reactions of 2-methoxypropene.

Table 1. Results of the Ti-catalyzed asymmetric carbonyl-ene reaction. [5]

Aldehyde	% Yield[a]	% ee[a]
Ph(CH$_2$)$_3$–C≡C–CHO	99	98
TBSOCH$_2$–C≡C–CHO	85	93
PH–C≡C–CHO	99	91
Ph(CH$_2$)$_2$CHO	98	90
PhCHO	83	66
C$_6$H$_{11}$CHO	79	75

[a] Isolated and analyzed as the corresponding β-hydroxy ketones obtained by treatment of the reaction mixture with Et$_2$O/2N HCl.

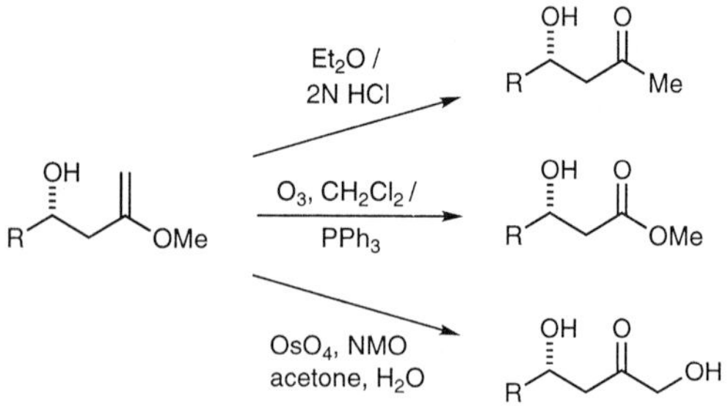

Scheme 2.
Synthetic utility of the carbonyl-ene products.

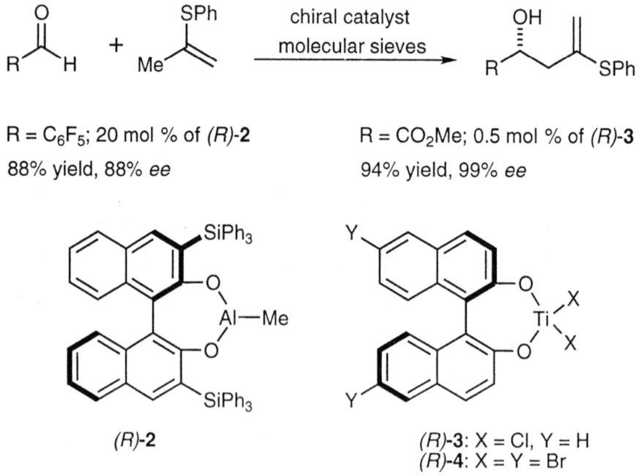

R = C$_6$F$_5$; 20 mol % of (R)-2
88% yield, 88% ee

R = CO$_2$Me; 0.5 mol % of (R)-3
94% yield, 99% ee

(R)-2

(R)-3: X = Cl, Y = H
(R)-4: X = Y = Br

Scheme 3. Asymmetric carbonyl-ene reactions of vinyl sulfides with electron deficient aldehydes.

providing an alternative method to affect asymmetric methyl ketone aldol additions. [5, 6] Oxidative cleavage of the enol ethers with ozone affords the corresponding β-hydroxy esters, and osmium-catalyzed dihydroxylation with N-morpholine-N-oxide (NMO) gives ketodiols.

Carreira's recent work is an extension of earlier studies [5] in which a titanium(IV) complex, prepared in situ from tridentate ligand (R)-1 and Ti(OiPr)$_4$, was found to catalyze Mukaiyama aldol reactions with high enantioselectivities. The chiral ligand used in both the ene and aldol chemistry is prepared from 3-bromo-5-tert-butylsalicylaldehyde and 2-amino-2'-hydroxy-1,1'-binaphthol. This

enantiomerically pure amino alcohol is obtained from a convenient oxidative coupling procedure [7] of Smrcina and Kocovsky et al. We expect that this valuable chiral biaryl ligand will prove popular amongst synthetic chemists.

Usually, intermolecular ene reactions of simple aldehydes with 1,1-disubstituted alkenes bearing no additional activating substituents require stoichiometric quantities of powerful Lewis acids. [1] Therefore, catalytic asymmetric variants using milder Lewis acids have previously been restricted to especially reactive aldehydes in intermolecular processes or to intramolecular reactions. For example, Yamamoto et al. described the use of alumi-

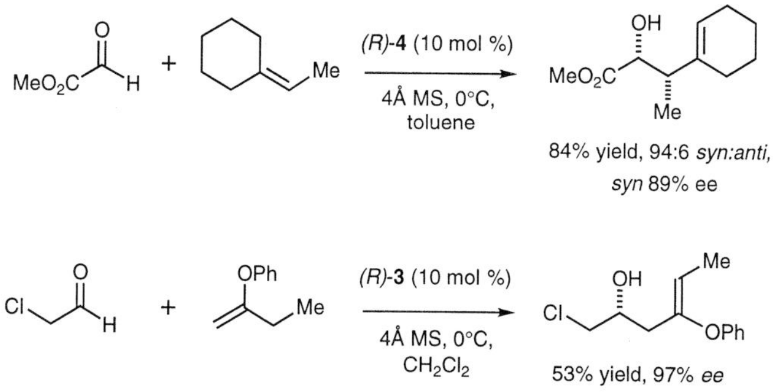

84% yield, 94:6 *syn:anti,*
syn 89% ee

53% yield, 97% *ee*

Scheme 4.
Asymmetric [TiX$_2$BINOL]-cata-
lyzed carbonyl-ene reactions.

67% yield, 95:5 *Z:E,*
Z: >99% *ee*

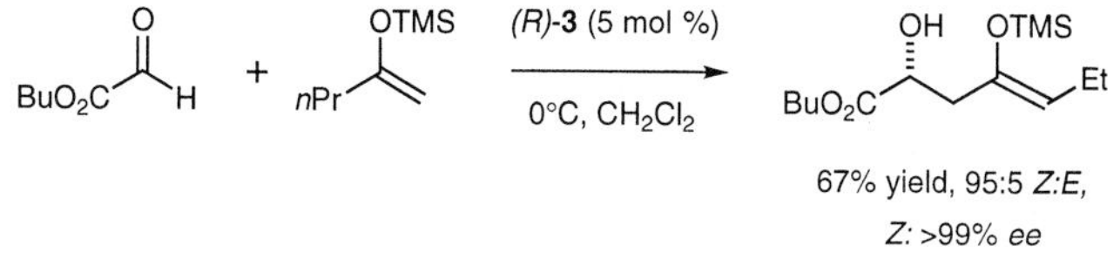

Scheme 5. An asymmetric catalytic glyoxylate ene reaction of a
silyl enol ether.

nium-based chiral Lewis acid (*R*)-**2** in catalyzed asymmetric carbonyl-ene reactions. [8]
However, the choice of enophile is limited to
highly electron deficient aldehydes such as
pentafluorobenzaldehyde and chloral. Both
simple 2,2-disubstituted alkenes and vinyl
sulfides undergo enantioselective reactions
with a maximum *ee* of 88 % even with catalytic quantities (20 mol%) of the Lewis acid
(Scheme 3).

The scope and utility of asymmetric carbonyl-ene reactions has also been advanced by
Mikami et al. [1b, c, e] Readily available
[TiX$_2$BINOL][9] Lewis acids (X = Cl, Br),
e.g. (*R*)-**3**, catalyze ene reactions of glyoxylates with outstanding enantioselectivities.
Recently, it has been demonstrated that
equally enantioselective ene reactions can be
accomplished using vinylogous glyoxylates
[10] and fluoral [11] as the enophiles. Many
of the earlier developments in asymmetric
ene chemistry have been thoroughly reviewed. [1] However, there are a number of
important recent advances worthy of highlighting. [12–18] Even extremely low catalyst
loadings (0.5 mol%) are effective in promo-

ting glyoxylate ene reactions with vinyl sulfides and selenides (Scheme 3). [12] Changing the Lewis acid to (*R*)-**4**, enables certain
trisubstituted alkenes to be used with excellent enantio- and diastereocontrol (Scheme
4). [13] Vinyl ethers, which are more reactive
than their thioether counterparts, undergo ene
reactions with high enantioselectivities using
(*R*)-**3** as catalyst. [14] Thus, addition of 2-
phenoxybutene to chloroacetaldehyde in the
presence of 10 mol% of (*R*)-**3** gives the corresponding ene product in 53 % yield with 97 %
ee (Scheme 4). [14]

The impressive levels of stereocontrol
obtained with this methodology have led to a
number of concise syntheses [1] of important
bioactive targets. Recent studies by the
Mikami group include the development of a
new synthetic route to insect pheromones
using isoprene as the ene component, [15]
the preparation of two fragments of the immu-
nosuppressant rapamycin, [16] and the
synthesis of prostacyclin analogues including
isocarbacyclin. [17]

An ene mechanism is also implicated in the
asymmetric "aldol" additions of ketene silyl

acetals of thioesters to aldehydes [18] and of silyl enol ethers to glyoxylate esters (Scheme 5). [19] The latter process can form part of a novel tandem addition reaction.[19c] The lack of accompanying silyl transfer is in contrast to other asymmetric Mukaiyama-type aldol reactions. [20]

The new methodologies developed by Carreira, Mikami and others widen the scope of asymmetric catalytic ene reactions. Studies on new catalysts [21] often provide the necessary mechanistic insights from which further synthetically-useful developments follow. An excellent example is provided by the recent report [22] of asymmetric catalysis of the glyoxylate ene reaction by a titanium catalyst formed from *racemic* [Ti(O*i*Pr)$_2$BINOL] [9] and a catalytic quantity of an enantiopure activator. We are confident that even more practical chiral Lewis acid catalysts, displaying wider substrate tolerance, and requiring lower catalyst loadings, will emerge in the future.

References

[1] a) B. B Snider in *Comprehensive Organic Synthesis*, Vol. 2 (Eds.: B. M. Trost, I. Fleming), Pergamon, Oxford, **1991**, p. 527; *Vol. 5*, p. 1; b) K. Mikami, M. Shimizu, *Chem. Rev.* **1992**, *92*, 1021; c) K. Mikami, M. Terada, S. Narisawa, T. Nakai, *Synlett* **1992**, 255; d) R. M. Borzilleri, S.M. Weinreb, *Synthesis* **1995**, 347; e) K. Mikami, *Pure Appl. Chem.* **1996**, *68*, 639.

[2] a) S. Sakane, K. Maruoka, H. Yamamoto, *Tetrahedron Lett.* **1985**, *26*, 5535; b) *idem.*, *Tetrahedron* **1986**, *40*, 2203; c) K. Narasaka, Y. Hayashi, S. Shimada, *Chem. Lett.* **1988**, 1609.

[3] a) K. Maruoka, H. Yamamoto in *Catalytic Asymmetric Synthesis* (Ed.: I Ojima), VCH, New York/Weinheim, **1993**, p. 413; b) R. Noyori, *Asymmetric Catalysis in Organic Synthesis*, J. Wiley, New York, **1994**; c) K. Narasaki, *Synthesis* **1991**, 1.

[4] E. Carreira, W. Lee, R. A. Singer, *J. Am. Chem. Soc.* **1995**, *117*, 3649.

[5] a) E. Carreira, R. A. Singer, W. Lee, *J. Am. Chem. Soc.* **1994**, *116*, 8837. For an application in total synthesis see: b) S. D. Rychnovsky, U. R. Khire, G. Yang, *ibid.* **1997**, *119*, 2058. For extensions to this work see: c) R. A. Singer, E. M. Carreira, *ibid.* **1995**, *117*, 12360.

[6] Catalyzed Mukaiyama-type aldol reactions of silyl enol ethers or silyl ketene acetals with aldehydes lead to the same products. For recent advances see: a) G. E. Keck, D. Krishnamurthy, *J. Am. Chem. Soc.* **1995**, *117*, 2363; b) M. Sato, S. Sunami, Y. Sugita, C. Kaneko, *Heterocycles* **1995**, *41*, 1435, and references therein.

[7] a) M. Smrcina, M. Lorenc, V. Hanus, P. Sedmera, P. Kocovsky, *J. Org. Chem.* **1992**, *57*, 1917; b) M. Smrcina, J. Polakova, S. Vyskocil, P. Kocovsky, *ibid.* **1993**, *58*, 4534; c) M. Smrcina, S. Vyskocil, J. Polivkova , J. Polakova, P. Kocovsky, *Collect. Czech. Chem. Commun.* **1996**, *61*, 1520.

[8] K. Maruoka, Y. Hoshino, T. Shirasaka, H. Yamamoto, *Tetrahedron Lett.* **1988**, *29*, 3967.

[9] The formulae used in this article merely imply the stoichiometric composition rather than the solution structure. For recent structural investigations of chiral Ti complexes see: a) T. J. Boyle, N. W. Eilerts, J. A. Heppert, F. Takusagawa, *Organometallics* **1994**, *13*, 2218; b) E. J. Corey, M. A. Letavic, M. C. Noe, S. Sarshar, *Tetrahedron Lett.* **1994**, *35*, 7553; c) K. V. Gothelf, R. G. Hazell, K. A. Jørgensen, *J. Am. Chem. Soc.* **1995**, *117*, 4435.

[10] K. Mikami, T. Yajima, T. Takasaki, S. Matsukawa, M. Terada, T. Uchimaru, M. Maruta, *Tetrahedron* **1996**, *52*, 85.

[11] K. Mikami, A. Yoshida, Y. Matsumoto, *Tetrahedron Lett.* **1996**, *37*, 8515.

[12] M. Terada, S. Matsukawa, K. Mikami, *J. Chem. Soc. Chem. Commun.* **1993**, 327; b) K. Mikami, T. Yajima, N. Siree, M. Terada, Y. Suzuki, I. Kobayashi *Synlett* **1996**, 837.

[13] a) M. Terada, Y. Motoyama, K. Mikami, *Tetrahedron Lett.* **1994**, *35*, 6693; b) K. Mikami, Y. Motoyama, M. Terada, *Inorg. Chim. Acta*, **1994**, *222*, 71.

[14] We thank Professor Mikami, Tokyo Institute of Technology, for helpful discussions and disclosure of unpublished material. a) K. Mikami, E. Sawa, M. Terada, unpublished data; b) E. Sawa, Master's Thesis, Tokyo Institute of Technology, **1992**.

[15] M. Terada, K. Mikami, *J. Chem. Soc. Chem. Commun.* **1995**, 2391.

[16] a) K. Mikami, A. Yoshida, *Tetrahedron Lett.* **1994**, *35*, 7793. See also: b) K. Mikami, S. Narisawa, M. Shimizu, M. Terada, *J. Am. Chem. Soc.* **1992**, *114*, 6566; 9242.

[17] For a formaldehyde-ene reaction see: K. Mikami, A. Yoshida, *Synlett* **1995**, 29; and reference [11].

[18] K. Mikami, S. Matsukawa, *J. Am. Chem. Soc.* **1994**, *116*, 4077.

[19] a) K. Mikami, S. Matsukawa, *J. Am. Chem. Soc.* **1993**, *115*, 7039. See also; b) *idem.*, *Tetrahedron Lett.* **1994**, *35*, 3133; c) K. Mikami, S. Matsukawa, M. Nagashima, H. Funabashi, H. Morishima, *ibid.* **1997**, *38*, 579.

[20] T. K. Hollis, B. Bosnich, *J. Am. Chem. Soc.* **1995**, *117*, 4570.

[21] For new catalysts for asymmetric ene reactions see: a) M. Terada, K. Mikami, *J. Chem. Soc. Chem. Commun.* **1994**, 833; b) D. Kitamoto, H. Imma, T. Nakai, *Tetrahedron Lett.* **1995**, *36*, 1861; c) G. Desimoni, G. Faita, P. Righetti, N. Sardone, *Tetrahedron* **1996**, *52*, 12019.

[22] a) K. Mikami, S. Matsukawa, *Nature*, **1997**, *385*, 613. For a related concept see: b) J. W. Faller, D. W. Sams, X. Liu, *J. Am. Chem. Soc.* **1996**, *118*, 1217.

Chiral 2-Amino-1,3-butadienes: New Reagents for Asymmetric Cycloadditions

Karsten Krohn

Since its discovery over sixty years ago [1] the Diels-Alder reaction has lost none of its attraction. [2, 3] It enables, in a one-step inter- or intramolecular reaction, the rapid preparation of cyclic compounds having a six-membered ring. During the course of the [4 + 2] cycloaddition four new stereocenters can be introduced directly, and their stereo-control is a topic of major interest in modern synthetic chemistry. [4–6] In addition, in intermolecular reactions, the relative positions of the reaction partners (regiochemistry) must be taken into account. If a concerted reaction is assumed, both a *cis* addition (suprafacial mode) and a preferred *endo* orientation (Alder rules) can be expected. But how can the *absolute* configuration of the desired product be controlled? There are three basic possibilities: the use of a chirally modified diene, a chirally modified dienophile, or a chiral catalyst. Although the first successes resulted from the attractive, but difficult, catalytic route, [4b, 7] the majority of the investigators are concerned with the stoichiometric

approach using chiral dienophiles (mostly derivatives of acrylic esters). [4, 5, 8–10] Chiral dienes have been used less frequently, although Trost et al. in 1980 had achieved an attention-riveting result with the (1*S*)-butadiene derivative **2** [11] (Scheme 1). This diene reacted regioselectively with juglone (**1**) to yield a single adduct **3** with 97 % *ee*! The well-known π-stacking model was developed from this example, although it does not appear to be applicable to all reactions. [12]

Unfortunately the result obtained by Trost et al. remained unique for more than a decade, at least in terms of enantioselectivity. The slow development in the area of chiral dienes may in part be ascribed to the difficulty of preparing these compounds. [4b] Recently, in quick succession and independently of one another, the research groups of Enders [13] and Barluenga [14] reported on the cycloaddition of chiral 2-aminobutadienes. Inititially, the prospects of 2-aminobutadienes in this reaction were not promising at all, since the results of MINDO3 calculations had indicated

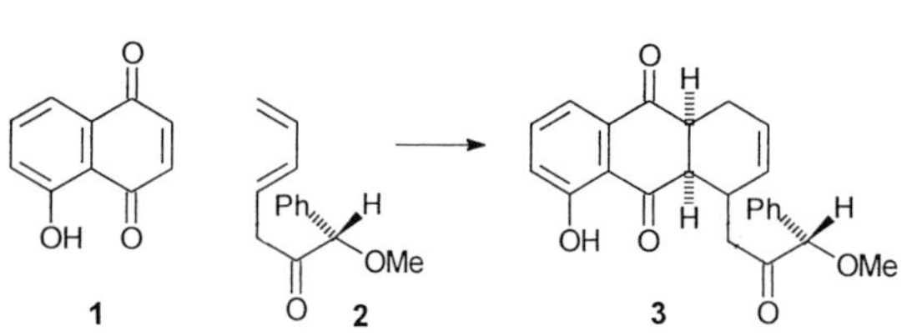

Scheme 1.

that they would not undergo (concerted) Diels-Alder reactions. [15] Nevertheless, Valentin et al. in 1977 [16] and later Gompper et al. in 1979 [17] obtained evidence for the cycloaddition of electron-rich methoxy- and amino-substituted dienes.

Since then, the research groups of Barluenga in Spain and Pitacco and Valentin [18] in Italy have been able to carry out other cycloadditions with activated olefins [19], aldehydes [19, 20], heterocumulenes [21] (e. g. thiophosgenes [22]), aldimines [23], nitrostyrenes [16] or aliphatic nitroolefins. [18] The general reactivity depending on the substituents and conformation of the 2-amino-butadienes (e. g. "enamine" reaction versus cycloaddition) was also intensively investigated. [24] The enamine character of 2-amino-butadienes (or dienamines) is clearly demonstrated by non-Lewis acid catalyzed reactions, and open-chain adducts are frequently isolated alongside the cyclic adducts. [21] Nevertheless, in many cases the stereochemistry of the products indicates a concerted reaction. According to Enders et al. the stereochemistry of the final products can be explained equally plausibly by means of a concerted cycloaddition or a sequential series of ionic addition steps, and the authors leave the question of the actual mechanism open. [13] All these investigations also showed a distinct difference from reactions with the highly successful siloxydienes related to the Danishefsky diene. [25, 26] The hydrolysis of the final cycloadducts (enolethers or enamines) affords the preparatively valuable ketones in both cases (vide infra) but the trivalent aminosubstituent allows the attachment of chiral axiliaries at C-2 much more easily

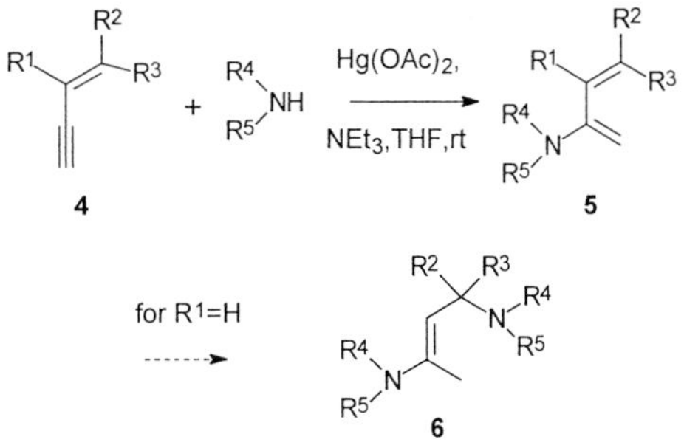

Scheme 2.

than does the silyl ether. This particular aspect is the subject of this overview (for a general review see [27]).

The synthesis of 2-aminobutadienes from ketones is somewhat limited due to difficult control of *E/Z*-stereochemistry. [27] Therefore, the catalytic aminomercuration of 3-alkenyl-1-ynes **4** was a significant improvement of the synthesis of 2-amino-1,3-butadienes **5** (yields 49–75 %). [28] The catalytic aminomercuration of 4-ethoxy-3-alkenyl-1-ynes even leads to electronrich and highly reactive 1,3-diamino-1,3-butadiens which undergo a great variety of cycloadditions. [29] Scheme 2 shows the reaction for a general example (for R^1 = H two secondary amines can add to the enine **4** to yield **6**). The method can easily be extended to the preparation of chiral 2-amino-1,3-butadienes by addition of chiral amines [e. g. (*S*)-2-methoxymethylpyrrolidine (SMP) [27]].

A novel approach to 2-amino-1,3-butadienes **10** was recently presented by Enders, Hekker and Meyer. [30] The synthesis was accomplished by a two step one pot procedure by coupling the alkenyl lithium compounds **8**

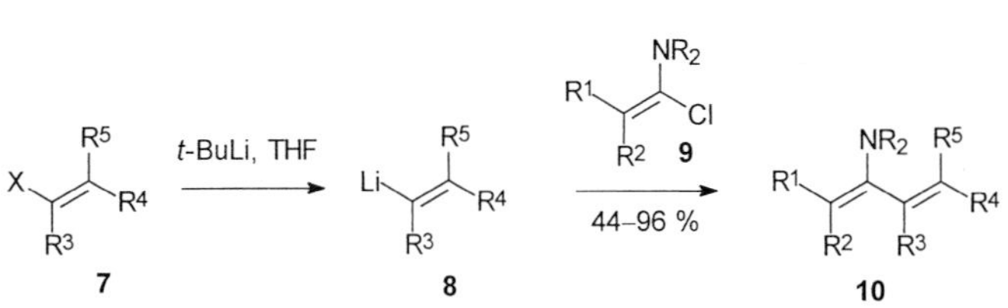

Scheme 3.

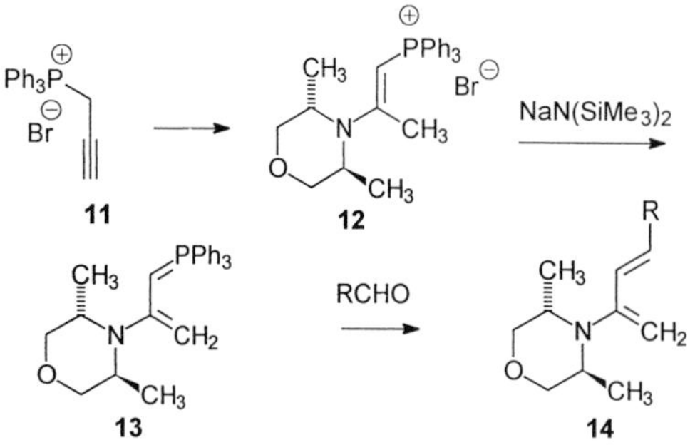

Scheme 4.

(generated by treatment of alkenyl halides **7** with 2 equiv. of *t*-BuLi) with the α-chloro enamines **9**, available from the corresponding amides (Scheme 3). Again, both achiral (amine = diethylamino or piperidino) or chiral (amine = SMP or SDP) dienes were available by this methodology.

Barluenga et al. prepared the desired aminodienes by Wittig olefination of aldehydes with phosphoranes generated *in situ* from β-enamino phosphonium salts. [31] The procedure was extended by Enders et al. to chiral products by introducing (*S*,*S*)-dimethylmorpholine as a novel C_2-symmetric chiral auxiliary as shown in Scheme 4 (reactions Scheme 4, **11** to **14**). [32]

A variation of the Wittig procedure with reversed building blocks served to prepare the chiral 2-aminobutadiene **17** in good yield from diacetyl **15** and (*S*)-2-(methoxymethyl)-

pyrrolidine **16** (SMP, review [32a]), obtained from proline, by means of enamine formation. The cycloaddition of **17** with five different nitrostyrenes **18** (R = H, 4-F, 4-OMe, 4-Me, 3,4-OCH$_2$O) proceeded at room temperature without catalysis and yielded labile cycloadducts **19** (enamine substructure), which decomposed on silica gel to form the corresponding ketones **20**. [13] Careful analysis of the spectra of the products shows that the saponification of the intermediate enamines **19** ("ketonization") does not proceed in a completely uniform manner; epimers are formed at the position α to the carbonyl group. For the reaction in diethyl ether, enantioselectivities from 96 to > 99 % *ee* were obtained for the main products – record values for cycloadditions not catalyzed by Lewis acids. In addition, the reaction is completely regioselective, which, in view of the strongly polarized reaction components, 2-aminobutadiene and nitrostyrene, is compatible with both the frontier orbital model for a concerted reaction and the stepwise ionic course (1,2- and 1,4-addition). The two reactions modes can be described by transition states **A** and **B** (Scheme 6). [13]

Barluenga et al. have extended the scope of the method by reaction with heteroaromatic and aliphatic nitroalkenes and the use of 2-aminobutadienes with a protected hydroxymethyl substituent at C-4 synthesized by the

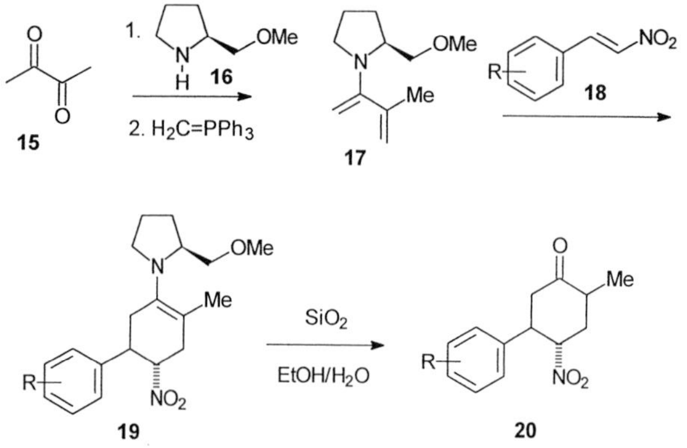

Scheme 5.

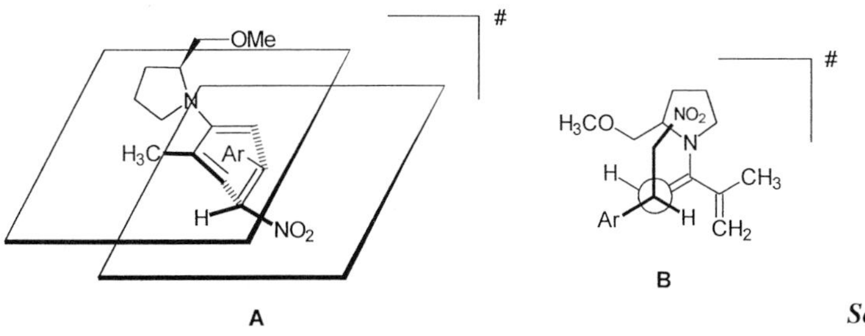

Scheme 6.

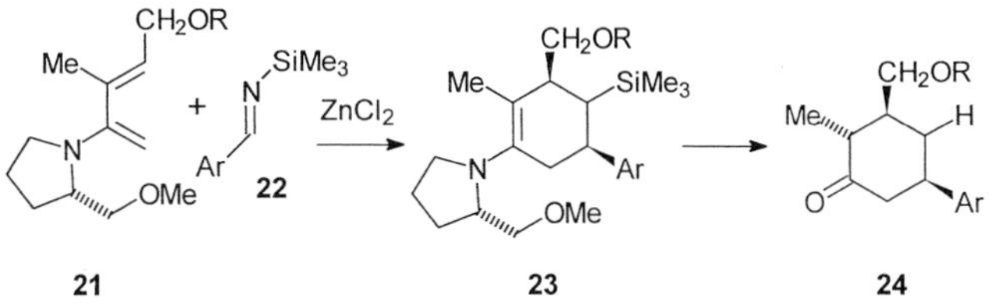

Scheme 7.

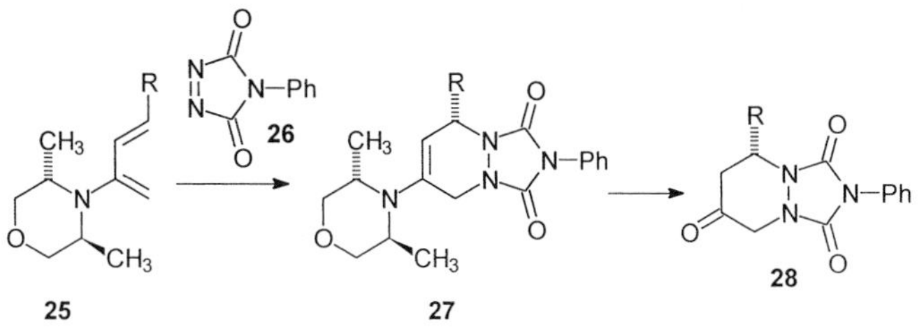

Scheme 8.

addition of SMP to commercially available enynes. [14, 33] Moreover, open chain compounds can also be obtained with high enantioselectivity and an intramolecular cyclization to chiral substituted furanes takes place after acidic hydrolysis of these products.

The Spanish authors also demonstrated two highly interesting reactions which considerably extended the area of application of the new chiral 2-aminodienes such as **21**. One of these is the *hetero*-Diels-Alder reaction with *N*-silylimines (e. g. **22**), whose products **23** can be hydrolyzed to provide substituted piperidones **24** with a high degree of enantioselectivity (85 and 95 % *ee*, Scheme 7). [14, 34] Such piperidones are required for the synthesis of alkaloids and pharmaceuticals.

Enders et al. [32] treated azadienophiles such as phenyltriazoledione **26** with (5,5)-2-(3,5-dimethylmorpholino)butadiene **25**

(R = *n*Pr, *i*Pr, cyclohexyl, cyclopentyl, *t*Bu, Ph−C≡C) to form heterocyclic enamines **27** which, after acid-catalyzed removal of the chiral C_2-symmetrical amine, yielded the hexahydropyridazines **28** (six examples) with an *ee* of 90−91 %. The successful use of another chiral amine, [(S,S)-3,5-dimethylmorpholine], gives further indication of the efficiency of asymmetric induction by means of coupling at the C-2 position of the diene.

Another example from the Spanish authors [14, 35] is especially interesting, as the cycloaddition leads to the formation of a compound with a seven-membered ring (Scheme 9). The literature shows that in recent years a tremendous amount of effort has been expended in trying to prepare carbocyclic five-membered rings by a route as simple and elegant as the Diels-Alder reaction for the synthesis of six-membered rings. [4c] Is a counterpart for seven-membered rings now in sight? The aminodienes were allowed to react with α,β-unsaturated Fischer carbene complexes. [36] The reaction consists of two well-known concerted reactions which occur sequentially: cyclopropanation and a Cope rearrangement of the intermediate divinylcyclopropane (e. g. 30). The principle was first demonstrated by Wulff et al. [37] with the reaction of the Danishefsky diene (1-methoxy-3-trimethylsiloxy-1,3-butadiene) and unsaturated carbene(carbonyl)chromium complexes. The Cope-type rearrangement occurs so rapidly with cis-oriented vinyl groups on the cyclopropane that the intermediate usually cannot be isolated. Barluenga et al. were recently able to show that the analogous reaction with 2-amino-1,3-butadienes in place of the siloxybutadiene leads to the formation of seven-membered rings. [35] The decisive advantage lies, however, in the additional possibility of using a chirally modified diene component. [38] The reaction of 21 with the carbene complex 29 gave the cycloheptadiene 31 and the hydrolysis product 32 in satisfactory yields (52–82 % for the multistep sequence) with total regio- and diastereoselectivity and excellent enantiomeric excesses of 81 and 86 % ee.

More recently, the reaction of chiral 2-amino-1,3-butadienes with cyclic BF$_3$ adducts of vinylcarbene complexes [39], the metathe-

sis reaction [40] and the preparation of enantiomerically pure spiro compounds [41] was investigated (overview [42]). Interestingly, a clean [4+2] cycloaddition to 34 occured if the corresponding tungsten Fischer carbene complexes such as 33 were reacted with the aminodiene 21 (R = SiMe$_3$) instead of their chromium counterparts 29. [43] The endo-selectivity is particularly high with complex 33 (endo : exo = 15 : 1) with 81 % ee for the major endo-product. The tungsten complex 34 can be hydrolysed and oxidized with ceric ammonium nitrate (CAN) to the lactone 35.

What are the advantages of the new chiral 2-aminobutadienes? Firstly, as the well-established SMP is a commercially available chiral reagent [32a], the chances are good that it will find broad application. Secondly, there are established methods for the coupling of SMP with other starting materials to form 2-aminobutadienes. [30–32] Finally, as is shown by the examples, especially the nitroolefins, [13, 33] good control of the regiochemistry is possible because of the strong electron donor in the 2-position. The increased electron density of the diene increases the reactivity (for cycloadditions with "normal" electron requirements), and the reaction can be extended to compounds with, for example C=N bonds. In addition, the reaction with α,β-unsaturated Fischer carbenes opens the door to the

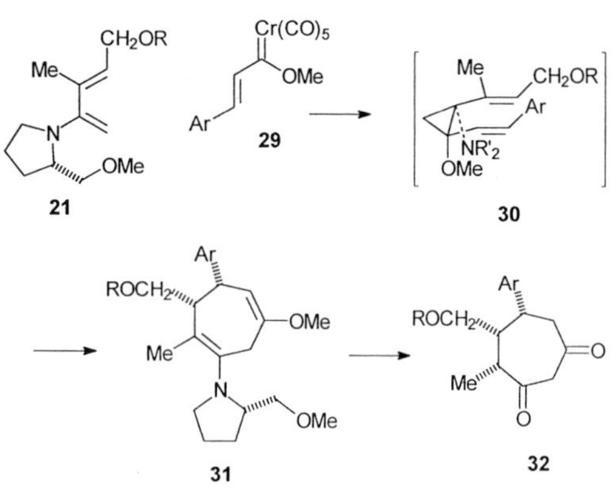

Scheme 9.

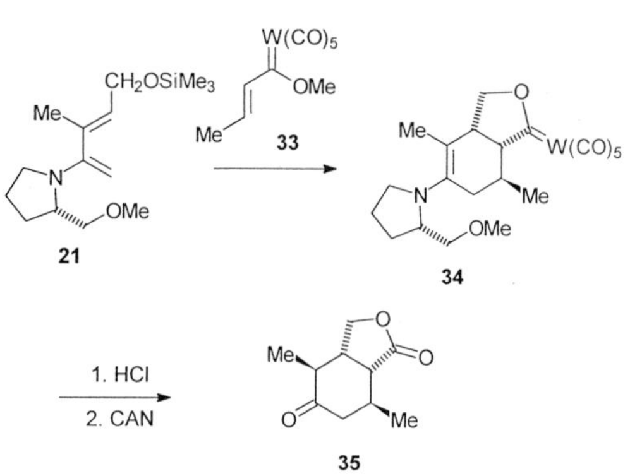

Scheme 10.

single-pot, stereoselective synthesis of carbocyclic seven-membered rings. Especially important is the fact that the chiral component can be removed under very mild conditions and can, in principle, be recovered. Decisive, however, is the excellent enantioselectivity of the reaction which can perhaps be further improved by the choice of the reaction conditions.

References

[1] O. Diels, K. Alder, *Liebigs Ann. Chem.* **1928**, *460*, 98.

[2] F. Fringuelli, A. Taticchi, *Dienes in the Diels-Alder Reaction*, Wiley, New York, **1990.**

[3] W. Carruthers, *Cycloaddition Reactions in Organic Synthesis* in *Tetrahedron Organic Chemistry Series*, Vol. 8, Pergamon Press, Oxford, **1990.**

[4] a) J. Mulzer. H.-J. Altenbach, M. Braun. K. Krohn, H.-U. Reissig, *Organic Synthesis Highlights*. VCH, Weinheim, **1991**; b) K. Krohn in [4a], pp. 5465; c) K. Krohn in [4a], pp. 96–103.

[5] M. J. Taschner, *Asymmetric Diels-Alder Reactions* (Ed.: T. Hudlicky), JAI Press, London, **1989.**

[6] H. B. Kagan, O. Riant, *Chem. Rev.* **1992**, *92*, 1007–1019.

[7] E. J. Corey, T.-P. Loh, T. D. Roper, M. D. Azimioara, M. C. Noe, *J. Am. Chem. Soc.* **1992**, *114*, 8290–8292.

[8] W. Oppolzer, *Tetrahedron* **1987**, *43*, 1969.

[9] W. Oppolzer, *Angew. Chem.* **1984**, *96*, 840–854; *Angew. Chem. Int. Ed. Engl.*, **1984**, *23*, 876–890.

[10] G. Helmchen, A. Goeke, S. Kreisz, A. Krotz, G. H. Lauer, G. Linz, *Cyclopentanoid Natural Products via Asymmetric Diels-Alder Reactions* in *Studies in Natural Products Chemistry* (Ed.: Atta-ur-Rahman), Vol. 8, Elsevier, Amsterdam, **1991**, pp. 139–158.

[11] B. M. Trost, D. O'Krongly, J. L. Belletire, *J. Am. Chem. Soc.* **1980**, *102*, 7595–7596.

[12] C. Siegel, E. R. Thornton, *J. Am. Chem. Soc.* **1989**, *111*, 5722–5728.

[13] D. Enders, O. Meyer, G. Raabe, *Synthesis* **1992**, 1242–1244.

[14] J. Barluenga, F. Aznar, C. Valdés, A. Martín, S. García-Granda, E. Martín, *J. Am. Chem. Soc.* **1993**, *115*, 4403–4404.

[15] L. N. Koikov, P. B. Terent' ev, I. P. Gloriozov, Yu. G. Bundel', *J. Org. Chem. USSR, Engl. Transl.* **1984**, *20*, 832.

[16] G. Pitacco, A. Risalti, M. L. Trevisan, E. Valentin, *Tetrahedron* **1977**, *33*, 3145–3148.

[17] R. Gompper, R. Sobotta, *Tetrahedron Lett.* **1979**, 921–924.

[18] M. Mezzetti, P. Nitti, G. Pitacco, E. Valentin, *Tetrahedron* **1985**, *41*, 1415–1422.

[19] J. Barluenga, F. Aznar, M.-P. Cabal, F. H. Cano, M. de la Conceptión, *J. Chem. Soc., Chem. Commun.* **1988**, 1247–1249.

[20] J. Barluenga, F. Aznar, M.-P. Cabal, C. Valdés, *Tetrahedron Lett.* **1989**, *30*, 1413–1416.

[21] J. Barluenga, F. Aznar, C. Valdés, F. López Ortiz, *Tetrahedron Lett.* **1990**, *31*, 5237–5240.

[22] J. Barluenga, C. Valdés, *Synlett* **1991**, 487–488.

[23] J. Barluenga, F. Aznar, C. Valdés, M.-P. Cabal, *J. Org. Chem.* **1993**, *58*, 3391–3396.

[24] J. Barluenga, F. Aznar, M.-P. Cabal, C. Valdés, *J. Chem. Soc., Perkin Trans. 1* **1990**, 633–638.

[25] S. J. Danishefsky, M. P. DeNinno, *Angew. Chem.* **1987**, *99*, 15–23; *Angew. Chem. Int. Ed. Engl.*, **1987**, *26*, 15–23.

[26] M. Petrzilka, J. I. Grayson, *Synthesis* **1981**, 753.

[27] O. Meyer, D. Enders, *Liebigs Ann.* **1996**, 1023–1035.

[28] J. Barluenga, F. Aznar, C. Valdés, M.-P. Cabal, *J. Org. Chem.* **1991**, *56*, 6168–6171.

[29] J. Barluenga, F. Aznar, M. Fernandez, *Tetrahedron Lett.* **1995**, *36*, 6551–6554.

[30] D. Enders, P. Hecker, O. Meyer, *Tetrahedron* **1996**, *52*, 2909–2924.

[31] J. Barlunenga, I. Merino, F. Palacios, *Tetrahedron Lett.* **1990**, *31*, 6713–6716.

[32] D. Enders, O. Meyer, G. Raabe, J. Runsink, *Synthesis* **1994**, 66–72.

[32a] D. Enders, M. Klatt, *Synthesis* **1996**, 1403–1418.

[33] J. Barluenga, F. Aznar, C. Ribas, C. Valdés, *J. Org. Chem.* **1997**, 6746–6753.

[34] J. Barluenga, F. Aznar, C., Valdés, C. Ribas, M. Fernádez, M.-P., Trujillo, J. Cabal, *Chem. Eur.* **1996**, *2*, 805–811.

[35] J. Barluenga, I. Merino, A. Martín, S. García-Grande, M. A. Salvadó, P. Pertierra, *J. Chem. Soc., Chem. Commun.* **1993**, 319–321.

[36] H. K. Dötz, *Organometallics in Organic Synthesis: Aspects of a Modern Interdisciplinary Field* (Eds.: A. de Meijere, H. tom Dieck), Springer, Berlin, **1988**.

[37] W. D. Wulff, D. C. Yang, C. K. Murray, *J. Am. Chem. Soc.* **1988**, *110*, 2653–2655.

[38] M. Chérest, H. Felkin, N. Prudent, *Tetrahedron Lett.* **1968**, 2199–2204.

[39] J. Barluenga, R.-M. Cantelli, J. Flórez, S. Garcia-Granda, A. Gutiérrez-Rodriguez, *J. Am. Chem. Soc.* **1994**, *116*, 6949–6950.

[40] J. Barluenga, F. Aznar, A. Martin, *Organometallics* **1995**, *14*, 1429–1433.

[41] J. Barluenga, F. Aznar, S. Garcia-Granda, S. Barluenga, C. Alvarez-Rúa, *J. Chem. Soc., Chem. Commun.* **1998**, submitted.

[42] J. Barluenga, F. Aznar, M. Fernandez, *Chem. Eur.* **1997**, 1629–1637.

[43] J. Barluenga, F. Aznar, A. Martin, S. Barluenga, S. Garcia-Granda, A. A. Paneque-Quevedo, *J. Chem. Soc., Chem. Commun.* **1994**, 843–844.

Recent Developments in the Enantioselective Syntheses of Cyclopropanes

Hans-Ulrich Reissig

The hunt for strained molecules is maintained by the competitive ambition to find the most unusual structures and by the enormous synthetic potential of small-ring compounds. Following the general trend of recent years asymmetric syntheses are at the cutting edge of this research. [1] Despite all advances the synthesis of enantiomerically and diastereomerically pure cyclopropane derivatives remains a considerable challenge, especially when particular functional groups are required. The most recent years have provided little that is fundamentally new to add to the very elegant procedures based on asymmetric catalysts. [2] In the main, reaction conditions and ligands were optimized, and the scope and limitations were investigated. Most exciting were examples from Doyle's group who reported on the enantioselective synthesis of macrocyclic lactones by means of intramolecular carbene additions. [3] Whereas the expected intramolecular cyclopropanation to a bicyclic γ-lac-

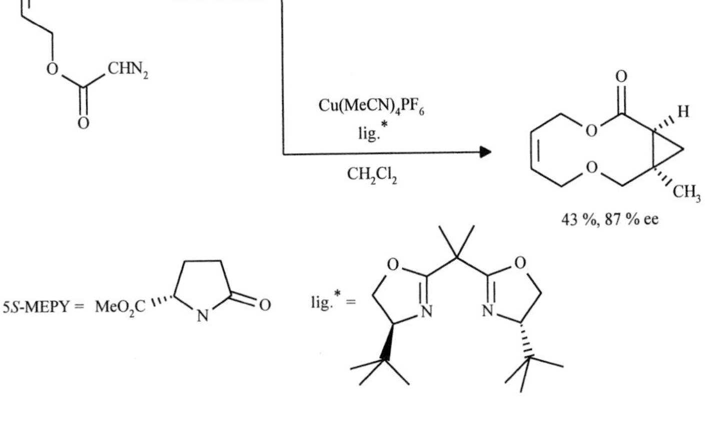

Scheme 1.

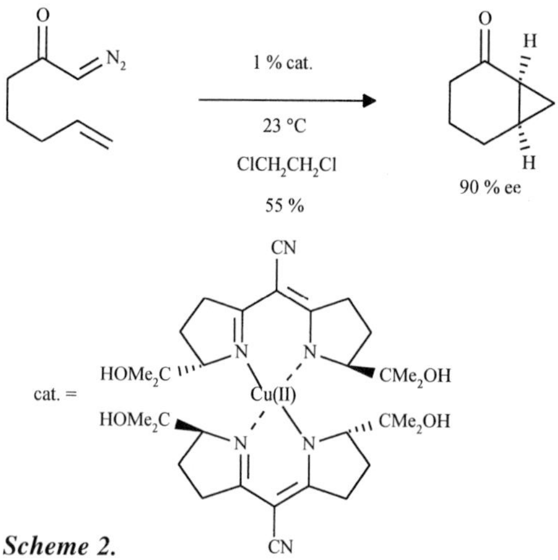

Scheme 2.

tone was observed with a chiral rhodium(II)-MEPY catalyst with 96 % *ee* (Scheme 1), with the Evans copper complex containing a bis(oxazoline) ligand the macrocyclic lactone was strongly favoured (87 % *ee*). Similar, or even higher, regio- and enantioselectivities with astonishingly high yields were also found in other examples. Thus there is now a new strategy, which will certainly find practical uses, for the enantioselective synthesis of macrocycles that are otherwise accessible only with difficulty.

Pfaltz's semicorrin copper complexes – a real breakthrough more then ten years ago – can also be employed for intramolecular [2+1]-cycloadditions (Scheme 2) and achieve between 14 and 95 % *ee* depending on the substrate. [4] The related, otherwise very successful, bis(oxazoline) complexes [5] (see above) do not appear to be as effective here though only one example has been investigated.

It can, however, be said that the ultimate catalyst system for inter- and intramolecular cyclopropanations, which reliably delivers both high enantiomeric excess and good *cis/trans* selectivity, and which is also suitable for highly substituted and/or functionalized olefins is still a dream. [6] Advances, but no breakthroughs, can be seen in the several

variants of the Simmons-Smith cyclopropanation, in particular of allylic alcohols. [7] Although enantioselective catalysis is now developing in this area – with respectable individual successes [8] the auxiliary-controlled procedures are still superior. [9] In general, cyclopropane syntheses controlled by reagent-bound auxiliary groups offer a better guarantee of high enantiomeric purity than those using the more elegant asymmetric catalysts.

A very impressive new example suitable for the synthesis of highly functionalized cyclopropane derivatives, was discovered by the Hanessian group. [10] Whereas in most auxiliary-controlled formal [2+1]cycloadditions [1] the auxiliary group is attached to the olefinic residue, here the chiral information is carried by a chloroallylphosphonic acid amide, such as **1**, which acts as a vinylcarbene equivalent. The reaction of the carbanion derived from **1** with *α,β*-unsaturated carbonyl compounds such as **2** yields the diastereomerically pure bicyclo[3.1.0]hexanone derivative **3** as the *endo*-isomer (90 % yield). By contrast, the reaction between the *cis*-chloroallyl derivative **4** and **2** yields the epimeric bicyclic product **5**. Mechanistically this cyclopropanation is easily understood as resulting from a Michael addition and an intramolecular S_N2 alkylation in which the intermediate **6** could account for the observed diastereoselectivity.

The method seems to be widely applicable as the examples with other cyclic enones, unsaturated lactones, lactams, and *tert*-butyl esters show (Scheme 3). The diastereoselectivities are at least 92 : 8 and usually much better.

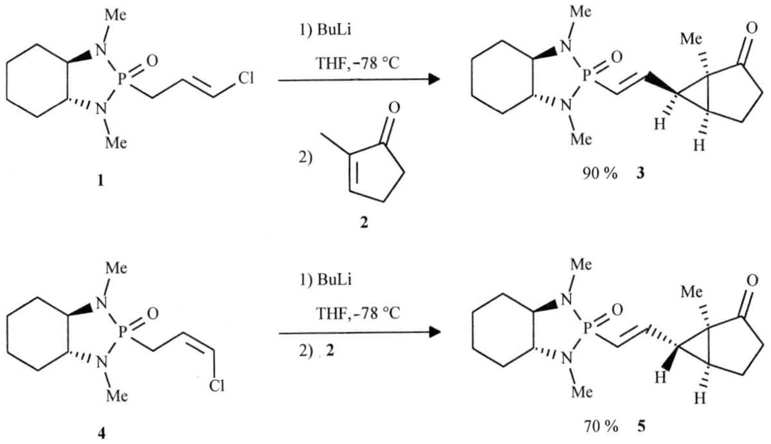

Formel 1.

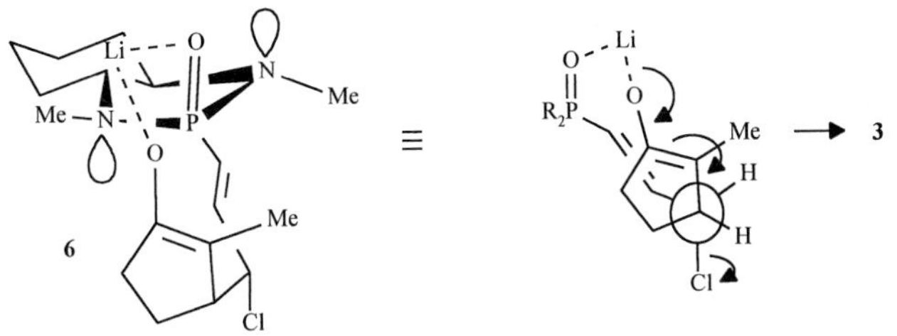

Formel 2.

After reduction of the carbonyl group with sodium borohydride and protective silylation of the resulting hydroxy group, the auxiliary is removed by ozonolysis of the alkenylphosphonamide group. By this method, or other obvious sequences, the primary products are converted into a series of enantiomerically pure, interestingly functionalized cyclopropane derivatives (Scheme 4), which should find applications as synthetic building blocks.

Nevertheless, ozonolysis of the alkenyl substituents destroys some of the preparative potential of the primary products; as vinylcyclopropanes they would be interesting for many other reactions, for example, for 1,3- and 3,3-sigmatropic rearrangements. [11] It can be inferred from the footnotes of the report [10] that a Cope rearrangement of the primary product **8**, derived from the hexadienoic acid ester **7** and **1**, does actually take place at room temperature with the formation of two isomeric cycloheptadiene derivatives.

No details were given so far about the structure of these compounds. One of the most interesting applications of the method could be the synthesis of enantiomerically pure, highly functionalized cycloheptane derivatives, since divinylcyclopropanes should be obtainable by the appropriate rearrangement of other primary products.

On the otherhand, the carbanion derived from the simplier α-chloromethylphosphonamide has been exploited similarly to **1** for the preparation of enantiomerically pure cyclopropane phosphonic acids. [12]

The results described above represent only the classical developments of organic synthesis extended and perfected; however, a publication on antibody-induced cyclopropanation could be the inauguration of what is, in principle, a new path to asymmetric cyclopropane formation (although the example is one with little preparative interest). [13]

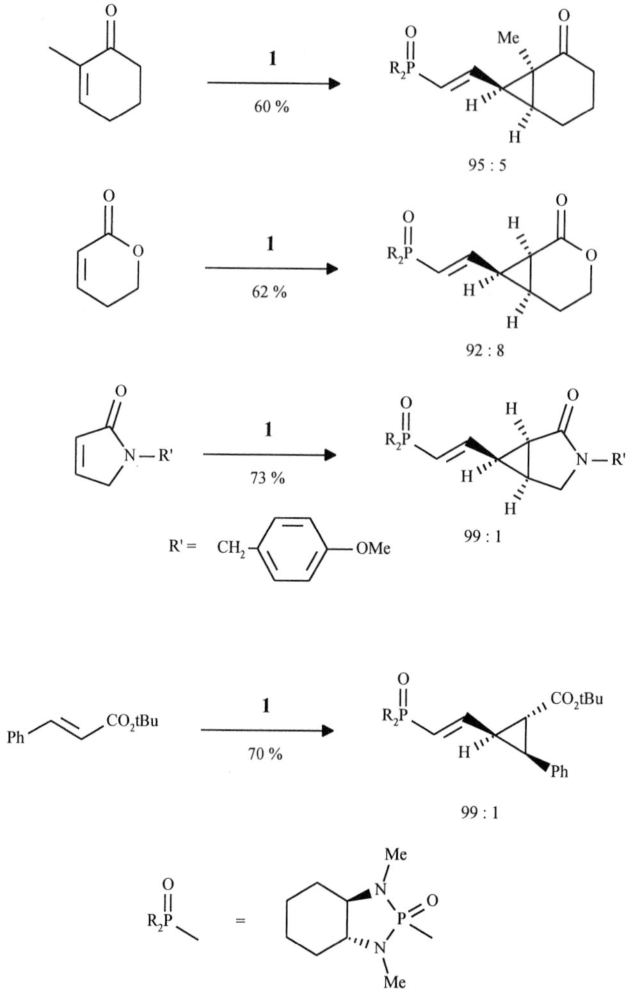

95 : 5

92 : 8

R' = CH₂—⟨aryl⟩—OMe

99 : 1

99 : 1

Scheme 3.

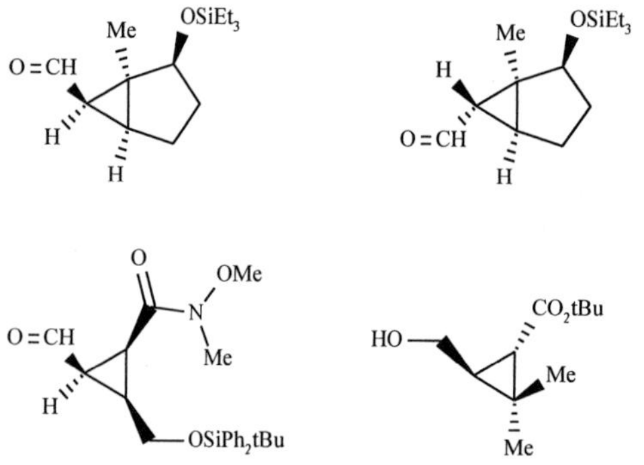

Scheme 4.

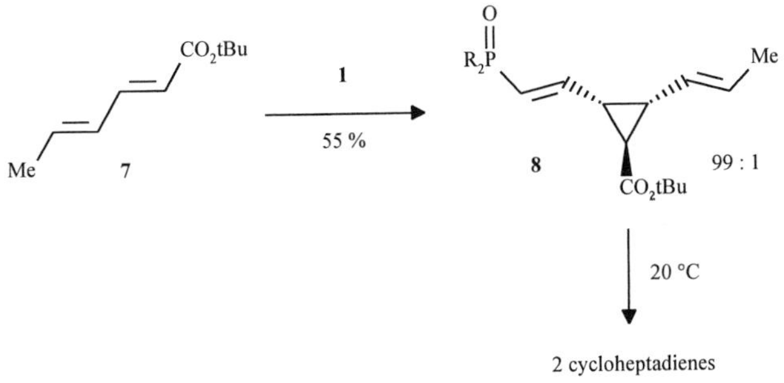

Formel 3.

References

[1] Comprehensive survey of diastereoselective and enantioselective [2+1]cycloadditions covering the literature up to and including 1994: H.-U. Reissig in *Stereoselective Synthesis of Organic Compounds / Methods of Organic Chemistry (Houben-Weyl)*, 4th ed., Vol. E21c (Eds.: G. Helmchen, R. W. Hoffmann, J. Mulzer, E. Schaumann), Thieme, Stuttgart, **1995**, pp. 3179–3270.

[2] M. P. Doyle in *Catalytic Asymmetric Synthesis* (Ed.: I. Ojima), VCH, Weinheim, **1993**, pp. 63–99; M. P. Doyle in *Comprehensive Organometallic Chemistry II*, Vol. 12 (Ed.: L. S. Hegedus), Pergamon Press, New York, **1995**, Ch. 5; V. K. Singh, A. Datta Gupta, G. Sekar *Synthesis* **1997**, 137–149.

[3] M. P. Doyle, C. S. Peterson, D. L. Parker, Jr., *Angew. Chem.* **1996**, *108*, 1439–1440; *Angew. Chem. Int. Ed. Engl.* **1996**, *35*, 1334–1336; see also: M. P. Doyle, M. N. Protopopova, C. D. Poulter, D. H. Rogers, *J. Am. Chem. Soc.* **1995**, *117*, 7281–7282; M. P. Doyle, C. S. Peterson, Q.-L. Thou, H. Nishiyama *Chem. Commun.* **1997**, 211– 212.

[4] C. Piqué, B. Fähndrich, A. Pfaltz, *Synlett* **1995**, 491–492.

[5] Short review: C. Bolm, *Angew. Chem.* **1991**, *103*, 556–558; *Angew. Chem. Int. Ed. Engl.* **1991**, *30*, 542. For the related bis(oxazolinyl)-pyridine ligands see: S.-B. Park, N. Sakata, H. Nishiyama *Chem. Eur. J.* **1996**, *2*, 303–306.

[6] For a systematic study of several catalyst types employed for the cyclopropanation of silyl enol ethers see: R. Schumacher, F. Dammast, H.-U. Reissig *Chem. Eur. J.* **1997**, *3*, 614–619.

[7] Short review: U. Koert, *Nachr. Chem. Tech. Lab.* **1995**, *43*, 435–442.

[8] For a comprehensive introduction to this field see: S. E. Denmark, S. P. O'Connor, *J. Org. Chem.* **1997**, *62*, 584–594. For an interesting application to the quinquecyclo-propane fragment of the inhibitor U-106305 see: W. S. McDonald, C. A. Verbicky, and C. K. Zercher, *J. Org. Chem.* **1997**, *62*, 1215–1222.

[9] Review: A. B. Charette, J.-F. Marcoux, *Synlett*, **1995**, 1197–1207.

[10] S. Hanessian, D. Andreotti, A. Gomtsyan, *J. Am. Chem. Soc.* **1995**, *117*, 10393– 10394.

[11] J. Salaün in *The Chemistry of the Cyclopropyl Group* (Ed.: S. Patai, Z. Rappoport), Wiley, Chichester, **1987**, pp. 809–878.

[12] S. Hanessian, L.-D. Cantin, S. Roy, D. Andreotti, A. Gomtsyan *Tetrahedron Letters*, **1997**, 1103–1106.

[13] T. Li. K. D. Janda, R. A. Lerner, *Nature* **1996**, *379*, 326–327.

Enantioselective Rhodium(II) Catalysts

Henri Brunner

In the complexes [M$_2$(OAc)$_4$] (OAc = CH$_3$COO) two metal(II) cations are bridged by four acetate anions (Fig. 1). In this highly symmetrical framework four oxygen atoms occupy the corners of two ecliptically arranged squares. [1] The valencies of the two octahedrally coordinated metal atoms perpendicular to these squares are oriented to the inside and to the outside of the M$_2$(OAc)$_4$ unit. The metal-metal interaction within the M$_2$(OAc)$_4$ unit depends critically on the electron configuration of the metal atoms. [1] It ranges from a quadruple bond in [Cr$_2$(OAc)$_4$] (Cr^{2+} is a d^4 system) to a weak interaction in [Cu$_2$(OAc)$_4$] (Cu^{2+} is a d^9 system). Further ligands can be bound, but also catalyses can be carried out at the valencies of the M$_2$(OAc)$_4$ unit which point outwards.

[Rh$_2$(OAc)$_4$] is a green, air-stable compound, which is soluble in organic solvents with the retention of its dinuclear structure. It is prepared from RhCl$_3$ and HOAc/NaOAc in boiling ethanol. [2] Other carboxylate ions can also be incorporated in place of acetate ions. Anions of fluorinated carboxylic acids afford particularly stable compounds such as [Rh$_2$(OOCCF$_3$)$_4$]. [Rh$_2$(OAc)$_4$] and its derivatives have for a long time proven valuable as catalysts for carben(oid) reactions, for example, the formation of cyclopropanes from olefins and diazo compounds or the formation of five-membered rings in intramolecular C–H insertions. [3–6] They are also considered to show anti-tumor activity. [1] Recently these dimeric rhodium(II) compounds have created a furore in enantioselective catalysis.

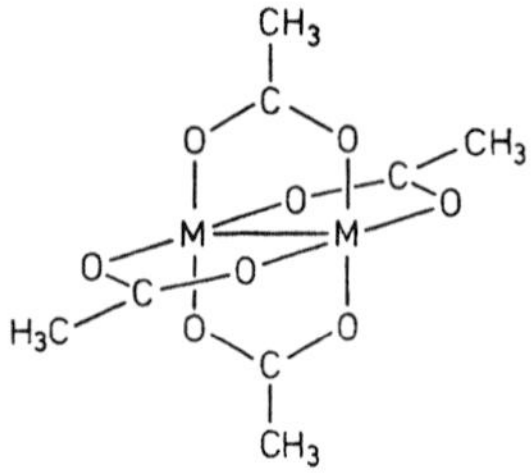

The dimeric rhodium(II) compounds become enantioselective catalysts if they contain optically active ligands. For this the anions of optically active carboxylic acids seem to be most appropriate. The complex in which the Rh$_2$ unit is clamped by four mandelate anions was synthesized and structurally characterized some time ago. [7] As a catalyst, however, this complex results in only small enantiomeric excesses. [8] The reason for this is probably that the asymmetric centers lie in a plane between the two Rh atoms and are thus too far away from the coordination sites directed to the outside, at which the catalysis occurs. Chiral substituents at the nitrogen atoms of carboxamide anions would be considerably closer to these reaction centers, and

Fig. 1.

indeed on this basis a breakthrough has been achieved recently, which in particular, is linked to the [Rh$_2$(5S-mepy)$_4$] complex (Fig. 2). The name of the optically active ligand has, as usual, been abbreviated, and in this case mepy* stands for the methyl ester of the pyrrolidone-5-carboxylic acid, which is deprotonated at the nitrogen atom. In the following the complex [Rh$_2$(5S-mepy)$_4$] and its "capacity" for enantioselective catalysis are introduced before finally discussing why complexes of the type [Rh$_2$(5S-mepy)$_4$] are so successful.

The introduction of carboxamides instead of carboxylates as bridging ligands is achieved by an exchange reaction between [Rh$_2$(OAc)$_4$] and the appropriate amides. In this way [Rh$_2$(5S-mepy)$_4$] is prepared from methyl (−)-(S)-pyrrolidone-5-carboxylate. Similarly, [Rh$_2$(5R-mepy)$_4$] is accessible from methyl (+)-(R)-pyrrolidone-5-carboxylate so that the catalysts are available in both configurations (5S and 5R). In [Rh$_2$(5S-mepy)$_4$] each rhodium atom is in a square-planar environment in which two O atoms and two N atoms are arranged *cis* to each other (Fig. 2).

[Rh$_2$(5S-mepy)$_4$] was first used for cyclopropanation. Enantioselective cyclopropanation is industrially important since synthetic pyrethroids, which are used as insecticides, contain substituted three-membered rings, whose configuration is crucial for their biological effect.[9] Enantioselective cyclopropanation has tradition. It was this reaction type which in 1966 opened up the field of enantioselective homogeneous catalysis with transition metal complexes. The copper(II) complex of the Schiff base from salicylaldehyde and optically active 1-phenylethylamine at that time reached 6 % ee.[10] With optimized opti-

Fig. 2.

cally active amines in the salicylaldimine ligands the industrial group headed by Aratani set early records in the 1970s.[11] New aspects were forthcoming with the introduction of the cobalt(II) semicorrin complexes by Pfaltz et al. 1986, [12, 13] and more recently with the bisoxazoline ligands from Masamune et al. [14] and Evans et al. [15]

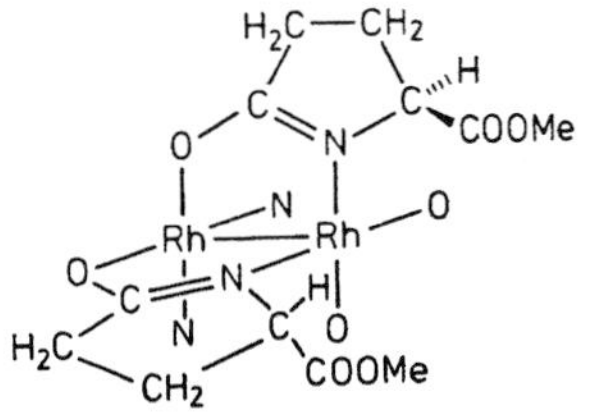

The reaction of styrene with (1S,3S,4R)-menthyl diazoacetate [Eq. (a)] leads to the incorporation of two new stereocenters into the cyclopropane ring. The *cis* and *trans* isomers are formed each consisting of an enantiomeric pair. With the [Rh$_2$(5S-mepy)$_4$] catalyst 86 % ee is achieved in the *cis* series and 48 % ee in the *trans* series.[16]

In the catalysis shown in Equation (a) a double stereoselection is involved. The formation of the new asymmetric centers in the cyclopropane ring is influenced by the menthyl group contained in the substrate (1S,3S,4R)-menthyl diazoacetate and by the mepy ligand contained in the catalyst. The two influences are referred to as substrate and catalyst control, respectively. With regard to the efficiency it has to be noted that whe-

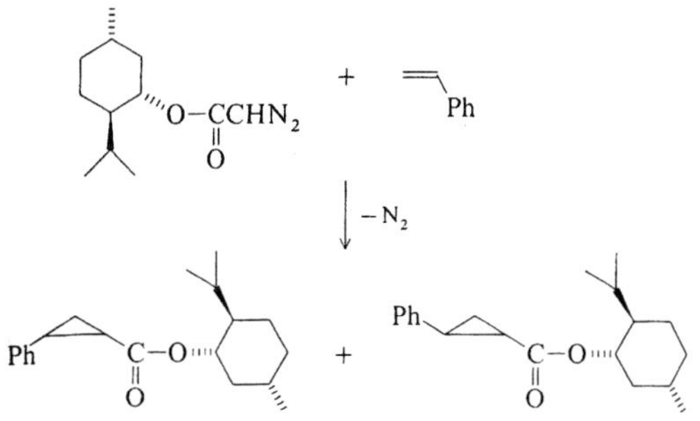

Eq. (a).

---

* The acronym mepy has usually been written in capital letters. However, according to the IUPAC rules for nomenclature abbrevations for ligands should be written with small letters.

reas substrate control requires a stoichiometric amount of the reagent, catalyst control requires a substoichiometric amount of the catalyst. For example, in the case of [Rh$_2$(5S-mepy)$_4$] a catalyst concentration of 1 mol% is sufficient with respect to the substrates.

The reaction in Equation (a) is controlled mainly by the catalyst; substrate control by the menthyl group plays only a minor role. This is evident from the fact that with the achiral [Rh$_2$(OAc)$_4$] catalyst only 9% *ee* can be achieved for the *cis* product and 13% *ee* for the *trans* product. Catalysis with [Rh$_2$(5S-mepy)$_4$] was extended from styrene to other prochiral olefins for which similar results were obtained. [16]

[Rh$_2$(5S-mepy)$_4$] is particularly useful for intramolecular cyclopropanation; in several examples up to 94% *ee* was obtained. [17] Highly substituted cyclopropanes are accessible by this reaction; an example is given in Equation (b). With the two enantiomeric catalysts both enantiomeric products can be obtained from the same allyl diazoacetate. For this intramolecular variant the (Z) configuration of the olefin proves superior to the (E) configuration.

Eq. (b).

As mentioned before, enantioselective cyclopropanation has been known for a long time and there are other efficient catalysts for this reaction apart from dimeric rhodium(II) complexes. This is different to the cyclopropanation of alkynes with diazo compounds. Copper catalysts do not only result in lower enantiomeric excesses but also poor yields, since the high temperatures required for the reaction favor side and consecutive reactions such as the ring opening of the primary products. The improvements brought about by the use of [Rh$_2$(5S-mepy)$_4$] (with regard to yield

and enantioselectivity) are shown for the cyclopropenation of propargyl methyl ether with a series of diazo esters in Equation (c). For the ethyl ester (R = Et) the enantiomeric excess is 69%, for the *tert*-butyl ester (R = tBu) 78%, and for the (+)-menthyl ester (R = 1S,3S,4R-menthyl) 98%. The yields of the reactions range between 43 and 73%. [18]

$$\text{MeOCH}_2-\text{C}{\equiv}\text{C}-\text{H} + \text{N}_2\text{CHCOOR} \xrightarrow[-\text{N}_2]{}$$

Eq. (c).

Metal catalyzed enantioselective C–H insertions of carbenes have so far not been studies in great detail. Copper catalysts are of no use for this type of reaction, rhodium(II) catalysts, however, allow intramolecular C–H insertions, for example, in the alkyl group of diazoacetates with longer chains. The formation of five-membered rings such as γ-lactones is favored. [Rh$_2$(5S-mepy)$_4$] affords 3-methyl-γ-butyrolactone (see Eq. (d)) in 91% *ee* and is thus unprecedented for such reactions. [19]

Eq. (d).

In order to explain the success of [Rh$_2$(5S-mepy)$_4$] a general and a specific argument shall be put forward. Generally for enantioselective catalysts, a reduction of the conformative variety has a favorable effect on the enantioselectivity. In [Rh$_2$(5S-mepy)$_4$] at each of the two planes of the Rh$_2$ unit which point outwards, a chiral framework is built whose rigidity is mainly the result of the incorporation of the amide part into the five-membered ring. In the case of [Rh$_2$(5S-mepy)$_4$] special electronic and steric effects are also involved. Model considerations as well as reactivity and selectivity studies show that during the cleavage of nitrogen from the diazo compound a

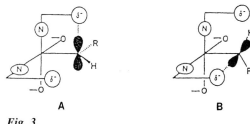

Fig. 3.

carbene-rhodium complex is formed whose empty p orbital at the carbene carbon atom can be stabilized by interaction with a polar substituent. [6] Therefore, as shown in Figure 3, the two *cis* ester groups in [Rh$_2$(5*S*-mepy)$_4$] fix the carbene ligand in the positions **A** and **B**, in which the large substituent R of the carbene adopts the less hindered position.

The attack of the nucleophile at the conformations **A** and **B** occurs from the rear. The observed stereochemistry of the products can be understood when the preferred orientation of the reaction partners with their C=C, C≡C, and C–H bonds in the inter- or intramolecular reactions are taken into consideration.

The article above was written in spring 1992. Since then the field of enantioselective catalysis with rhodium(II) complexes containing mepy and mepy-like ligands has been extended appreciably, in particular by the group of M. P. Doyle. Thus, the recent references in this field can be found by searching for the name of the main author M. P. Doyle. The catalysts [Rh$_2$(5*S*-mepy)$_4$] and [Rh$_2$(5*R*-mepy)$_4$] have been commercialized (REGIS Chemical Company, Morton Grove, IL 60053, USA).

References

[1] F. A. Cotton, R. A. Walton, *Multiple bonds Beteen Metal Atoms*, Wiley, New York, **1982**.

[2] G. A. Rempel, P. P. Legzdins, H. Smith, G. Wilkinson, *Inorg. Synth.* **1972**, *13*, 90.

[3] M. P. Doyle, *Chem. Rev.* **1986**, *86*, 919.

[4] M. P. Doyle, *Acc. Chem. Res.* **1986**, *19*, 348.

[5] G. Maas, *Top. Curr. Chem.* **1987**, *137*, 348.

[6] M. P. Doyle, *Recl. Tray. Chim. Pays-Bas* **1991**, *110*, 305.

[7] P. A. Agaskar, F. A. Cotton, L. R. Falvello, S. Hahn, *J. Am. Chem. Soc.* **1986**, *108*, 1214.

[8] H. Brunner, H. Kluschanzoff, K. Wutz, *Bull. Soc. Chim. Belg.* **1986**, *98*, 63.

[9] D. Arlt, M. Jautelat, R. Lantzsch, *Angew. Chem.* **1981**, *93*, 719; *Angew. Chem. Int. Ed. Engl.* **1981**, *20*, 703.

[10] H. Nozaki, S. Moriuti, H. Takaya, R. Noyori, *Tetrahedron Lett.* **1966**, 5239.

[11] T. Aratani, *Pure Appl. Chem.* **1985**, *57*, 1839.

[12] H. Fritschi, U. Leutenegger, A. Pfaltz, *Angew. Chem.* **1986**, *98*, 1028; *Angew. Chem. Int. Ed. Engl.* **1986**, *25*, 1005.

[13] Short review: C. Bolm, *Angew. Chem.* **1991**, *103* 556; *Angew. Chem. Int. Ed. Engl.* **1991**, *30*, 542.

[14] R. E. Lowenthal, A. Abiko, S. Masamune, *Tetrahedron Lett.* **1990**, *31*, 6005.

[15] D. A. Evans, K. A. Woerpel, M. M. Hinman, *J. Am. Chem. Soc.* **1991**, *113*, 726.

[16] M. P. Doyle, B. D. Brandes, A. P. Kazala, R. J. Pieters, M. B. Jarstfer, L. M. Watkins, C. T. Eagle, *Tetrahedron Lett.* **1990**, *31*, 6613.

[17] M. P. Doyle, R. J. Pieters, S. F. Martin, R. E. Austin, C. J. Oalmann, P. Müller, *J. Am. Chem. Soc.* **1991**, *113*, 1423.

[18] M. N. Protopopova, M. P. Doyle, P. Müller, D. Ene, *J. Am. Chem. Soc.* **1992**, *114*, 2755.

[19] M. P. Doyle, A. v. Oeveren, L. J. Westrum, M. N. Protopopova, T. W. Clayton, Jr., *J. Am. Chem. Soc.* **1991**, *113*, 8982.

Oxazaborolidines and Dioxaborolidines in Enantioselective Catalysis

B. B. Lohray, and Vidya Bhushan

During the last decade, use of oxazaborolidines and dioxaborolidines in enantioselective catalysis has gained importance. [1, 2] One of the earliest examples of oxazaborolidines as an enantioselective catalyst in the reduction of ketones/ketoxime ethers to secondary alcohols/amines was reported by Itsuno et al. [3] in which (S)-valinol was used as a chiral ligand. Since then, a number of other oxazaborolidines and dioxaborolidines have been investigated as enantioselective catalysts in a number of organic transformations *viz* a) reduction of ketones to alcohols, b) addition of dialkyl zinc to aldehydes, c) asymmetric allylation of aldehydes, d) Diels-Alder cycloaddition reactions, e) Mukaiyama Michael type of aldol condensations, f) cyclopropanation reaction of olefins.

Reduction of Ketones

Corey et al. [4] have investigated enantioselective reduction of ketones with THF·BH$_3$ and (S)-diphenyl prolinol-borane adduct as catalyst. They further introduced modification of oxazaborolidines in which R = CH$_3$, *n*-butyl as catalyst **1** and used along with other boranes as reducing agents (Scheme 1). The catalysts **1** are known as CBS catalysts and

are easily accessible in both (R) and (S) forms. Use of CBS reagents in the reduction of ketones afforded alcohols in excellent yields and enantioselectivities (80–99 % *ee*) under very mild reaction conditions (0–25 °C). The reaction requires both the reagents; neither catalyst (S)-**1** nor THF·BH$_3$ alone is able to reduce the ketone in considerable yield. The active reducing species has been postulated to be an intermediate complex **2** which is formed from (S)-**1** and BH$_3$.

Mathre et al. [5a] have modified the preparation of (S)-**1** and used it in the synthesis of MK0471, a carboanhydrase inhibitor. [5b]

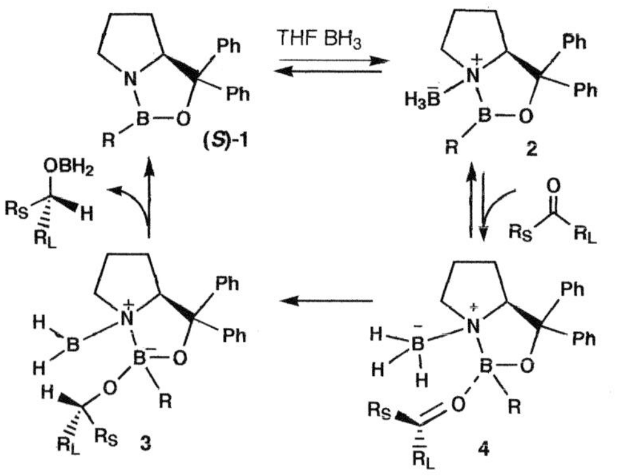

Scheme 1. Possible mechanism of the oxazaborolidine-catalyzed enantioselective reduction of ketones with THF·BH$_3$.

Later, several oxazaborolidines derived from various amino alcohols have been used for the reduction of ketones [6] and ketoimines. [7] In certain cases, diketones were reduced to optically active diols, [8] α,β-unsaturated ketones to allylic alcohols, [9] etc. in varying degree of enantioselectivities. Of particular interest is the reduction of alkyl trichloromethyl ketones [10] which lead to the synthesis of α-amino acids using (S)-**1** and synthesis of substituted biaryls by the reduction of substituted 2-pyrones with THF·BH$_3$ in the presence of oxazaborolidines. [11] More recently, aliphatic dialkyl ketones have been reduced to secondary alcohols in moderate to high enantioselectivity. [12] Quite recently, Corey *et al.* have reduced achiral α,β-ynones using modified oxazaborolidines (Scheme 2) providing highly optically pure propargyl alcohols in excellent yields. [13] Earlier, Corey, and Cimprich had prepared these propargyl alcohols by enantioselective alkynylation process using oxazaborolidines as chiral promoters. [14]

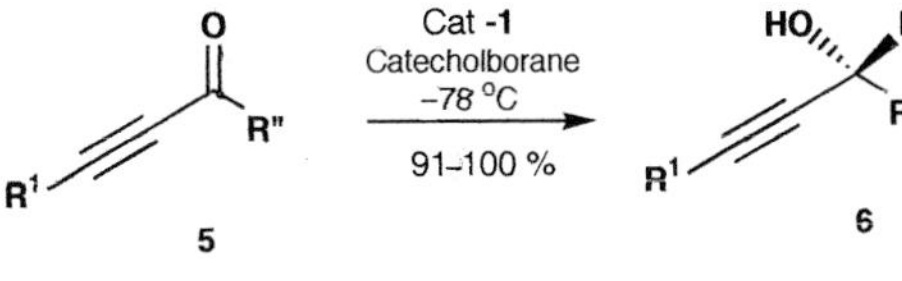

Addition of Dialkyl Zinc to Aldehydes

In 1989, Brown et al. [15] suggested that since B–O and B–N bonds are shorter than metal–oxygen and metal–nitrogen bonds, there is a greater chance that boron complex will be a more effective catalyst. In order to substantiate this hypothesis, they carried out enantioselective addition of diethyl zinc to several aldehydes using (4S,5R)-3,4-dimethyl-5-phenyl-1,3,2-oxazaborolidine as catalyst and achieved very high yield and enantiomeric purity (upto 96 % *ee*) of secondary alcohols.

Asymmetric Allylation of Aldehydes

Yamamoto et al. [16] envisaged that acyloxyboranes might behave as mixed anhydrides because of the electronegative trivalent boron atom and could serve as effective asymmetric catalysts in selected reactions. In the presence of 20 mole% of chiral acyloxy borane (CAB) complex **7** prepared from (2R,3R)-2-O-(2,6-diisopropoxybenzoyl)tartaric acid and BH$_3$·THF, various allyltrimethylsilanes react with achiral aldehydes to afford the corresponding homoallylic alcohols in good yield and high enantio- and diastereoselectivity (Scheme 3).

The reaction proceeds at –78 °C to furnish predominently *erythro* homoallylic alcohols (*erythro* : *threo* = 80 : 20 to 97 : 3) regardless of the configuration of allylsilanes. These observations were explained based on the *re*

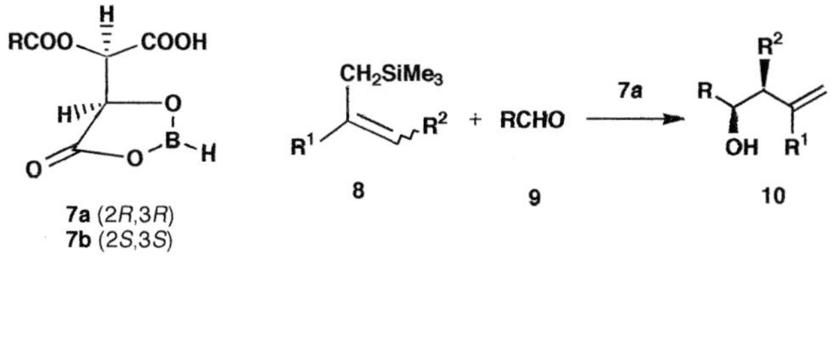

Scheme 2. Enantioselective reduction of achiral α,β-ynones by oxazaborolidines.

Scheme 3. Asymmetric allylation in the presence of 20 mol% **7a**. R = 2,6-dimethoxyphenyl.

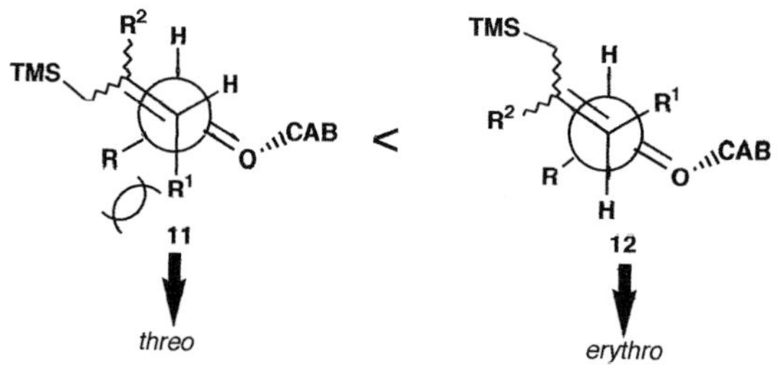

Figure 1. Extended transition state model.

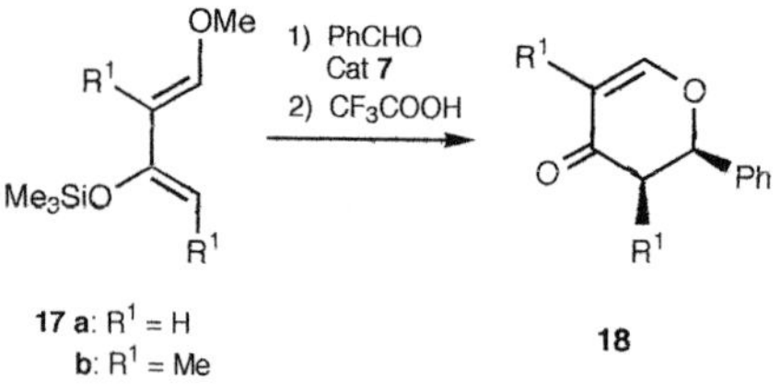

Scheme 5. Asymmetric hetero Diels-Alder reaction catalyzed by dioxaborolidines.

face attack of the nucleophiles on the carbonyl carbon of the aldehyde using an extended transition state model as shown in Figure 1.

Diels-Alder Cycloaddition Reaction

The CAB catalyst used in asymmetric allylation reaction has also been found to be equally effective in asymmetric Diels-Alder reaction of a variety of diones with several dienophiles (a,β-unsaturated acids and aldehydes) under very mild reaction conditions (−78 °C) (Scheme 4). [17] Good to excellent diastereo- (99 : 1 to 88 : 12) and enantioselectivities (88 to 97 % ee) have been observed in most of the cases.

A stable chiral acyloxy borane (CAB) complex **7** is also an effective catalyst for hetero Diels-Alder reaction to produce dihydropyrans in high optical purity. [18] (Scheme 5).

Yamamoto et al. [19] have further extended these studies to Diels-Alder reactions cataly-

sed by *N*-arylsulfonyl-1,3,2-oxazaborolidines **20**. The stereoselectivity with these catalysts is relatively inferior to that observed with **7**. Similar observations have been reported by Helmchen et al. [20] for the Diels-Alder reaction of cyclopentadiene and methacrolein (*exo : endo* = 99 : 1, 64 % ee for the *exo* isomer) or crotonaldehyde (*exo : endo* = 3 : 97; 72 % ee for the *endo* isomer). Use of bulkier aryl substituents such as 2,4,6-triisopropyl and 2,4,6-tri-*t*-butylphenyl had little influence on the diastereoselectivity of the cycloaddition.

In contrast, Corey and Loh [21] have reported cycloaddition reaction of cyclopentadiene and 2-bromoacrolein catalyzed by oxazaborolidine derived from *N*-tosyl-(*S*)-tryptophan in dichoromethane at −78 °C to give a highly diastereo- (*exo : endo* = 97 : 3) and enantioselective (96 % ee) reaction leading to the formation of (*R*)-bromoaldehyde **22** (Scheme 6).

It is interesting to know that the stereoselectivity of the adduct using Corey's catalyst is opposite to that normally observed for oxazaborolidines generated from *N*-tosyl derivative of (*S*)-valine or hexahydrophenyl alanine.

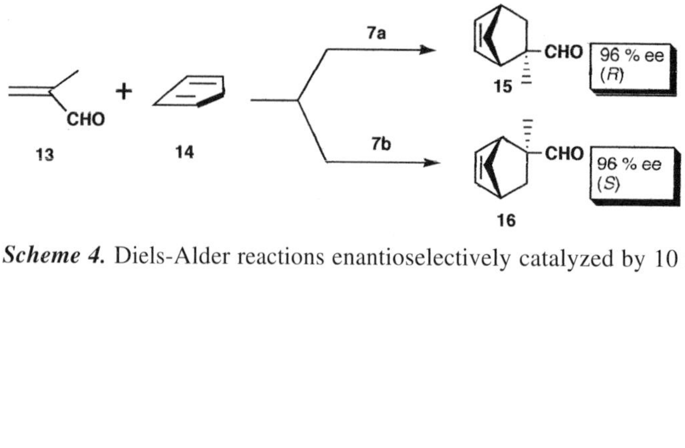

Scheme 4. Diels-Alder reactions enantioselectively catalyzed by 10 mol% **7**.

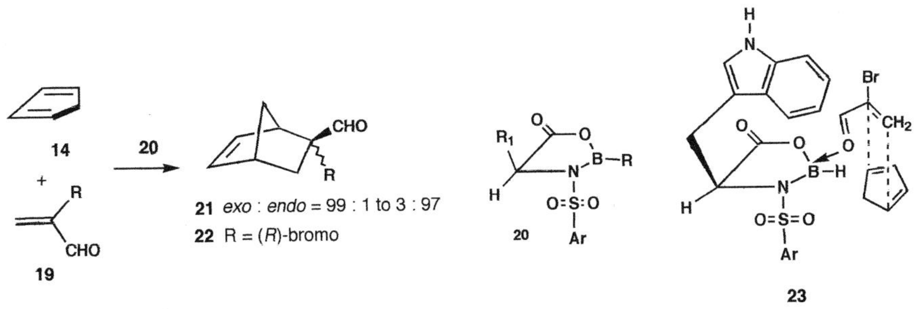

Scheme 6. Asymmetric Diels-Alder reaction catalysed by oxazaborolidine **20**.

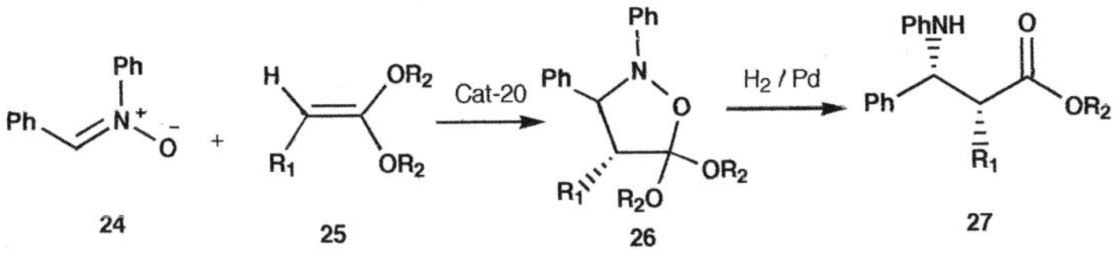

Scheme 7. Enantioselective 1,3-dipolar cycloaddition of nitrones with ketene acetals catalyzed by oxazaborolidines.

This methodology has been used for a simple enantioselective synthesis of (1*S*,4*R*)-Bicyclo[2.2.1]hept-2-ene-2-methanol. [22] These results suggest a transition state **23** in which the dienophile assumes an orientation parallel to the indole ring because of the π-π donor acceptor interaction leading to an unprecedented (200 : 1) enantioselectivity. [21, 23]

Corey et al. [24] have used this oxazaborolidine as an effective catalyst for an efficient synthesis of cassiol and gibberellic acid. Similar high diastereo- (*exo* : *endo* = 99 : 1) and enantioselectivity (96 : 4) was observed in the cycloaddition reaction of furan with 2-bromoacrolein using oxazaborolidine as catalyst. [25]

Chiral oxazaborolidines **20** derived from various amino alcohols have been used as catalysts in asymmetric 1,3-dipolar cycloaddition reaction of nitrones with ketene acetals to give substituted isoxazoles in high yield and stereoselectivity but in moderate enantioselectivity (upto 62 % *ee*). This method has also been used for the synthesis of β-aminoesters

by hydrogenolysis of isoxazoles (Scheme 7). [26]

Mukaiyama-Michael type aldol condensation

The chiral acyloxyborane **7** (CAB) has also been found to be an excellent catalyst for asymmetric Mukaiyama-Michael type aldol reaction between silyl enol ethers and aldehydes (Scheme 8). Yamamoto et al. [27] have used 20 mol % of CAB in propionitrile at −78 °C as a highly efficient catalyst for the condensation of several *E* and *Z* silyl enol ethers and ketene acetals with a variety of aldehydes (yields 49–97 %, 80–97 % *ee*).

Interestingly, regardless of the configuration of the enol ethers, the *erythro* isomer always predominated (*erythro : threo* = 80/20 to > 95/< 5). Aromatic and α,β-unsaturated aldehydes always provided higher diastereo- (*erythro : threo* > 94 : 6) and enantio-

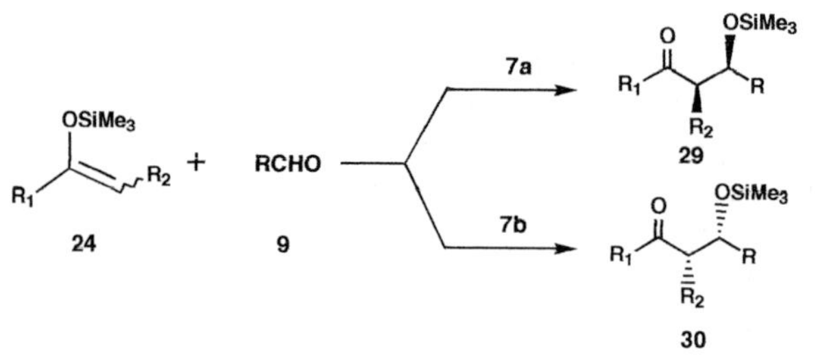

Scheme 8. Enantioselective Mukaiyama-Michael aldol reaction catalyzed by dioxaborolidine.

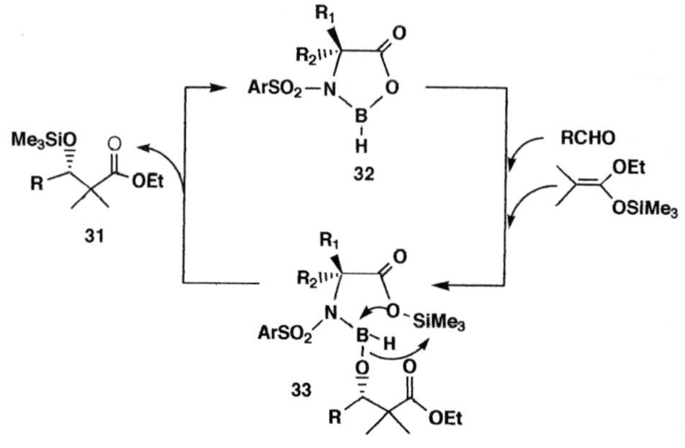

Scheme 9. Enantioselective aldol reaction catalyzed by oxazaborolidine.

selectivities (92–97 % *ee*) than saturated aldehydes. Polar solvents improved the selectivity by decreasing the association of the catalyst with the formation of oligomers as observed by Helmchen. [20] Corey *et al.* have used oxazaborolidines as effective catalysts for the enantioselective Mukiayama aldol and aldol dihydropyrone annulation reaction of trimethylsiloxy olefin and diones. [28]

Kiyooka et al. [29] and Masamune [30] and coworkers have used oxazaborolidines **32** as catalysts for aldol reactions. The latter have suggested that the initial aldol adduct **33** must undergo ring closure (as indicated by arrow in Scheme 9) to release the final product **31** and to regenerate the catalyst. In many cases, slow addition of the aldehyde to the reaction mixture proved beneficial (which permits enough time for **33** to undergo ring closure) for improving the enantioselectivity of the reaction.

Thus, a,a-disubstituted *N*-arylsulfonylglycines were used for the preparation of the oxazaborolidine **32** which resulted in catalytic asymmetric aldol processes providing β-hydroxy esters of > 97 % *ee* from α-unbranched aldehydes (R–CH$_2$CHO) and 84–96 % *ee* with α-branched aldehydes (R$_2$CHCHO). The reaction proceeds smoothly in propionitrile at –78 °C if the aldehyde is added slowly over 3.5 h affording high yields (68–89 %) of the adduct.

Cyclopropanation Reaction

More recently, dioxaborolane derived from (*RR*)-(+)-*N,N,N,'N'*-tetramethyltartaric acid diamide has been used as an efficient chiral controller in Simmons Smith cyclopropanation reaction of allylic alcohols to produce substituted cyclopropyl methanols in high

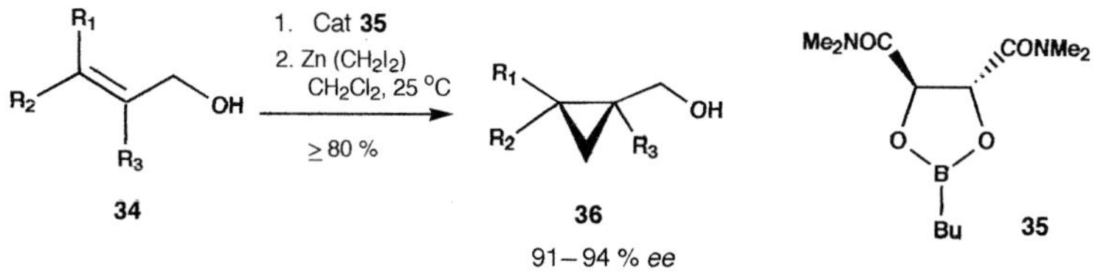

Scheme 10. Enantioselective cyclopropanation of allylic alcohols catalyzed by dioxaborolidines.

enantioselectivities (Scheme 10). [31] This method has been extended for the synthesis of biscyclopropanes [32] and cyclopropanation of polyenes. [33]

Using this cyclopropanation strategy, Yamada et al. have stereoselectively synthesized a side chain segment of an antitumor marine steroid, aragusterol. [34]

Dioxaborolane derived from dimethyl tartrate has found application in enantioselective epoxidation of unfunctionalized alkenes using TBHP as co-oxidant. [35]

The increasing applications of oxazaborolidines and dioxaborolidine derivatives suggest that these class of catalysts will find wide use in synthetic organic and medicinal chemistry to bring about a variety of stereoselective transformations.

References

[1] Lohray BB, Bhushan, V (**1992**) Angew Chem Int Ed Engl *31* : 729.

[2] (a) Singh VK (**1992**) Synthesis 605. (b) Wallbaum S, Martens J (**1992**) Tetrahedron Asymmetry *3* : 1475. (c) Deloux L, Srebnik M (**1993**) Chem Rev *93* : 763.

[3] Itsuno S, Sakurai Y, Ito K, Hirao A, Nakahama S (**1987**) Bull Chem Soc Jpn *60* : 395.

[4] (a) Corey EJ, Bakshi RK, Shibata S (**1987**) J Am Chem Soc *109* : 5551. (b) Corey EJ (**1990**) Pure Appl Chem *62* : 1209. (c) Corey EJ, Bakshi RK (**1990**) Tetrahedron Lett *31* : 611.

[5] (a) Mathre, DJ, Jones TK, Xavier LC, Blacklock TJ, Reamer RA, Mohan JJ, Jones ETT, Hoogsteen K, Baum MW, Grabowki EJJ (**1991**) J Org Chem *56* : 751. (b) Mathre, DJ, Jones TK, Xavier LC, Blacklock TJ, Reamer RA, Mohan JJ, Jones ETT, Hoogsteen K, Baum MW, Grabowki EJJ (**1991**) J Org Chem *56* : 763.

[6] (a) Martens J, Dauelsberg C, Behnen W, Wallbaum S (**1992**) Tetrahedron Asymmetry *3* : 347. (b) Cho BT, Chun YS (**1992**) Tetrahedron Asymmetry *3* : 1539. (c) Quallich GJ, Woodall TM (**1993**) Tetrahedron Lett *34* : 4145. (d) Berenguer R, Garcia J, Gonzalez M, Vilarrasa J (**1993**) Tetrahedron Asymmetry *4* : 13. (e) Kiyooka S, Kaneko Y, Harada Y, Matsuo T (**1995**) Tetrahedron Lett *36* : 2821.

[7] (a) Kawate T, Nakagawa, M, Kakikawa T, Hino T (**1992**) Tetrahedron Asymmetry *3* : 227. (b) Nakagawa M, Kawate T, Kakikawa T, Yamada H, Matsui T, Hino T (**1993**) Tetrahedron *49* : 1739. (c) Hong Y, Gao Y, Nie X, Zepp CM (**1994**) Tetrahedron Lett *35* : 5551.

[8] (a) Quallich GJ, Keavey KN, Woodall TM (**1995**) Tetrahedron Lett *36* : 4729. (b) Prasad KRK, Joshi NN (**1996**) J Org Chem *61* : 3888.

[9] Bach J, Berenguer R, Farras J, Garcia J, Meseguer J, Vilarrasa J (**1995**) Tetrahedron Asymmetry *6* : 2683.

[10] Corey EJ, Link JO (**1992**) J Am Chem Soc *114* : 1906.

[11] (a) Bringmann G, Hartung T (**1992**) Angew Chem Int Ed Engl *31* : 761 (b) Bringmann G, Hartung T (**1993**) Tetrahedron *49* : 7891.

[12] Berenguer R, Garcia J, Vilarrasa J (**1994**) Tetrahedron Asymmetry *5* : 165.

[13] Helal CJ, Magriotis PA, Corey EJ (**1996**) J Am Chem Soc *118* : 10938.

[14] Corey EJ, Cimprich KA (**1994**) J Am Chem Soc *116* : 3151.

[15] Joshi NN, Srebnik M, Brown HC (**1989**) Tetrahedron Lett *30* : 5551.

[16] (a) Furuta K, Mori H, Yamamoto H (**1991**) Synlett 561. (b) Ishihara K, Mouri M, Gao Q, Maruyama T, Furuta K, Yamamoto H (**1993**) J Am Chem Soc *115* : 11490.

[17] (a) Furuta K, Miwa Y, Iwanaga K, Yamamoto H (**1988**) J Am Chem Soc *110* : 6254 (b) Furuta K, Shimizu S, Miwa Y, Yamamoto H (**1989**) J Org Chem *54* : 1481. (c) Ishihara K, Gao Q, Yamamoto H (**1993**) J Am Chem Soc *115* : 10412.

[18] Gao Q, Maruyama T, Miwa T, Yamamoto H (**1992**) J Org Chem *57* : 1951.

[19] Takasu M, Yamamoto H (**1990**) Synlett 194.

[20] (a) Sartor D, Saffrich J, Helmchen G (**1990**) Synlett 197. (b) Sartor D, Saffrich J, Helmchen G, Richards CJ, Lambert H (**1991**) Tetrahedron Asymmetry 2 : 639.

[21] Corey EJ, Loh T -P, (**1991**) J Am Chem Soc *113* : 8966.

[22] Corey EJ, Cywin CL (**1992**) J Org Chem *57* : 7372.

[23] Corey EJ, Loh T -P, Roper TD, Azimioara MD, Noe MC (**1992**) J Am Chem Soc *114* : 8290.

[24] Corey EJ, Guzman-Perez A, Loh T -P (**1994**) J Am Chem Soc *116* : 3611.

[25] Corey EJ, Loh T -P (**1993**) Tetrahedron Lett *34* : 3979.

[26] Seerden J-P G, Scholte op Reimer AWA, Scheeren HW (**1994**) Tetrahedron Lett *35* : 4419.

[27] (a) Furuta K, Muruyama T, Yamamoto H (**1991**) J Am Chem Soc *113* : 1041 (b) Furuta K, Muruyama T, Yamamoto H (**1991**) Synlett 439.

[28] Corey EJ, Cywin CL, Roper TD (**1992**) Tetrahedron Lett *33* : 6907.

[29] (a) Kiyooka S, Kaneko Y, Komura M, Matsuo H, Nakano M (**1991**) J Org Chem *56* : 2276. (b) Kiyooka S, Kaneko Y, Kume K (**1992**) Tetrahedron Lett *33* : 4927. (c) Kaneko Y, Matsuo T, Kiyooka S (**1994**) Tetrahedron Lett *35* : 4107.

[30] Parmee ER, Tempkin O, Masamune S (**1991**) J Am Chem Soc *113* : 9365.

[31] (a) Charette AB, Juteau H (**1994**) J Am Chem Soc *116* : 2651 (b) Charette AB, Prescott S, Brochu C (**1995**) J Org Chem *60* : 1081.

[32] Theberge CR, Zercher CK (**1994**) Tetrahedron Lett *35* : 9181.

[33] Charette AB, Juteau H, Deschenes D (**1996**) Tetrahedron Lett *37* : 7925.

[34] Mitome H, Miyaoka H, Nakano M, Yamada Y (**1995**) Tetrahedron Lett *36* : 8231.

[35] Manoury E, Mouloud HAH, Balavoine GGA (**1993**) Tetrahedron Asymmetry 4: 2339.

Enantioselective Catalytic Hydrogenation

Judith Albrecht and Ulrich Nagel

Enantioselective catalysis is one of the most important tools in asymmetric synthesis. [1, 2] With its assistence biologically active substances can be prepared in enantiomerically pure form – this purity can be a crucial factor with pharmaceutical products. In the field of crop protection the use of enantiomerically pure compounds provides irrefutable advantages for both economic and ecological reasons. [3, 4]

Enantioselective transition metal catalyzed hydrogenation has an important place among the methods of asymmetric synthesis. A large range of substrates can be enantioselectively hydrogenated in this way, which is extremely important for the preparation of natural and also nonnatural amino acids, because it enables the directed synthesis of all possible amino acid derivatives from the many prochiral enamides and ketones. Recently great progress has been made in this field. Very high *ee* values are achieved, and even sterically demanding substrates like *β,β*-disubstituted enamides are able to be hydrogenated in good optical yield. [5, 6]

Since the beginning of the 90's Burk et al. have been exploring the development of novel electron-rich phosphane ligands that give powerful catalysts for enantioselective hydrogenation on complexation with rhodium. [7] The ligands they use each contain two phospholanes *trans* substituted in 2,5-position, whose phosphorus atoms are linked together through different groups as backbone. The carbon atoms adjacent to the phosphorus atoms are chiral and in these catalysts are situated in the immediate vicinity of the rhodium atoms (Scheme 1).

Initially the phospholanes were synthesized by derivatization of homochiral 1,4-diols with mesyl chloride. The dimesylates were then transformed into the 3,5-disubstituted phenylphospholanes with dilithiumphenylphosphide. The phenyl group was cleaved with pure lithium metal, and the resulting lithium phosphide could then be converted into the bridged system with 1,2-dichloroethane, ethyleneglycoldi-*p*-tosylate, or 1,3-dichloropropane. The yields with this synthetic route were moderate and depended strongly on the purity of the lithium metal. [8]

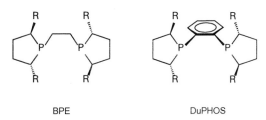

Scheme 1. The bisphospholane ligands BPE and DuPHOS; R = Me, Et, *n*Pr, *i*Pr (to give Me-, Et-, *n*Pr-, and *i*Pr-BPE or -DuPHOS).

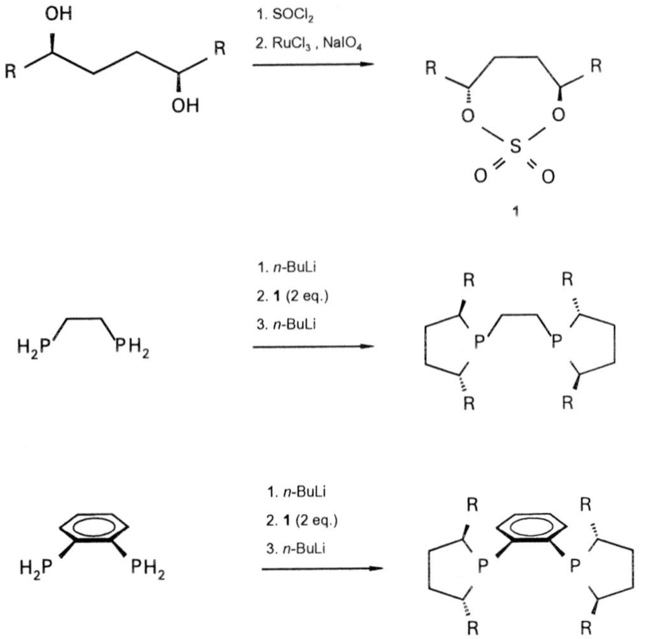

Scheme 2. Synthesis of BPE and DuPHOS ligands.

A later optimized synthesis also starts from the 1,4-diols. They are then converted into cyclic sulfates with thionyl chloride on mediation of ruthenium chloride and sodium periodate. The sulfates are transformed with dilithiumbis(phosphido)ethane into bis(phospholano)ethane (BPE) ligands or with dilithium-1,2-bis(phosphido)benzene into bis(phospholano)benzene (DuPHOS) ligands. Ring closure to form phospholane is achieved by addition of *n*BuLi [9] (Scheme 2).

The differently substituted BPE and DuPHOS rhodium complexes were applied in the hydrogenation of *N*-acetylenamides. Of the BPE ligands, the ethyl-substituted derivative gave the highest *ee* values (93 % for methyl α-acetylaminocinnamate and 98 % for methyl α-acetylaminoacrylate). Higher enantioselectivities were achieved with the DuPHOS ligands. Under optimized conditions some systems afforded *ee* values of over 99 %. In particular the sterically demanding *n*Pr-DuPHOS gave *ee* values of 99.8 % (enantiomer ratio 1000 : 1).

In 1996 Burk et al. also developed a convenient method to synthesize chiral α-1-arylalkylamines through enantioselective hydrogenation of various enamides using their Me–DuPHOS and Me–BPE ligands, respectively. They consistently achieved from 94 to 97 % *ee*. Furthermore their ligands tolerated β-substituents both in (*E*)- and (*Z*)-position. [10]

In all enantioselective hydrogenations the ability of the substrate to form a chelate ring with the catalyst is extremely important. For this reason the enantioselective reductive amination of ketones is always particularly difficult, because these compounds usually do not have a structure suitable for the required chelation. Burk et al. circumvent this problem by reversible derivatization. The ketones are converted into *N*-acetylhydrazones, whose structures resemble those of enamides. [11] The C–N double bond can then be hydrogenated by *n*Pr–DuPHOS–rhodium with *ee* values almost as high as those for C–C double bonds of enamides. The *N*-acetylhydrazines obtained thus can either be transformed into the free hydrazines by acid hydrolysis or into amines by treatment with samarium diiodide. In this way a large number of ketones can be reductively aminated. [12]

To hydrogenate prochiral ketones to the corresponding chiral alcohols, Noyori et al. have developed ternary catalyst systems from [RuCl$_2$(binap)(dmf)$_n$], a chiral diamine, and KOH. [13, 14] By this route methyl(1-naphthyl)ketone can be hydrogenated to 1-(1-naphthyl)ethanol with [RuCl$_2$(binap)(dmf)$_n$], 1,2-diphenylethyldiamine, and KOH in the ratio 1 : 1 : 2 in isopropanol (substrate : catalyst = 500 : 1, 4 bar H$_2$, 28 °C, 6 h) in greater than 99 % yield with 97 % ee.

Other sources of hydrogen can be used instead of elemental hydrogen gas. In transfer hydrogenations a secondary alcohol has been employed as hydrogen donor. Noyori et al. have also made advances in this field. They use a catalyst system from [RuCl$_2$(η^6-mesitylene)]$_2$, *N*-(*p*-toluenesulfonyl)-1,2-diphenylethylenediamine, and KOH in isopropanol. At room temperature acetophenone can be reduced in 15 h with this complex prepared *in situ* (substrate : catalyst = 200 : 1) to 1-phenylethanol in 95 % yield with an optical purity of 97 %. [15, 16] The

same system even works better under otherwise comparable conditions when a formic acid-triethylamine mixture is used as hydrogen source (99 % yield and 98 % ee). [17]

Catalyst systems with a *β*-aminoalcohol as auxiliary are faster, as shown by the reduction of acetophenone to 1-phenylethanol under otherwise identical conditions with [RuCl$_2$(η^6-hexamethylbenzene)]$_2$, 2-methylamino-1,2-diphenylethanol, and KOH within as little as 1 h in 94 % yield and 92 % ee [18] (Scheme 3).

With transfer hydrogenation Noyori et al. were not only able to hydrogenate C–O double bonds but also the C–N double bonds of prochiral imines. As hydrogen donor they used a formic acid-triethylammonium mixture. With the chiral Ru(II)-catalyst systems shown in Scheme 4 they achived ee's up to 96 % in hydrogenating various prochiral imines. [19]

The *N*-acetylhydrazones used by Burk et al. are ketone derivatives, and their structures therefore correspond to those of the *β,β*-disubstituted *N*-acetylaminoacrylic acids. This and the finding that both the (*E*)- and the (*Z*)-enamides are hydrogenated with almost the same enantioselectivity by the BPE and DuPHOS

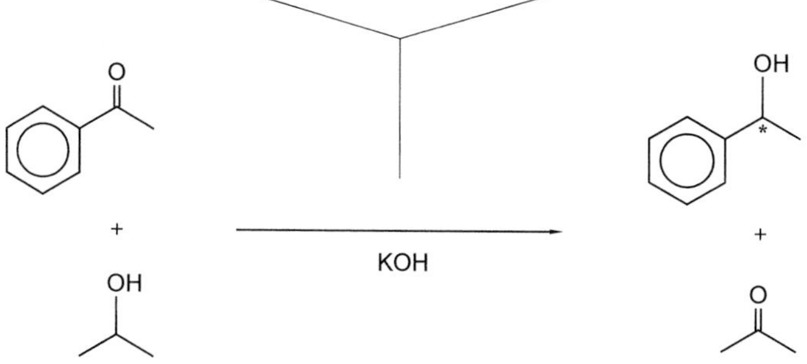

Scheme 3. Transfer hydrogenation of acetophenone.

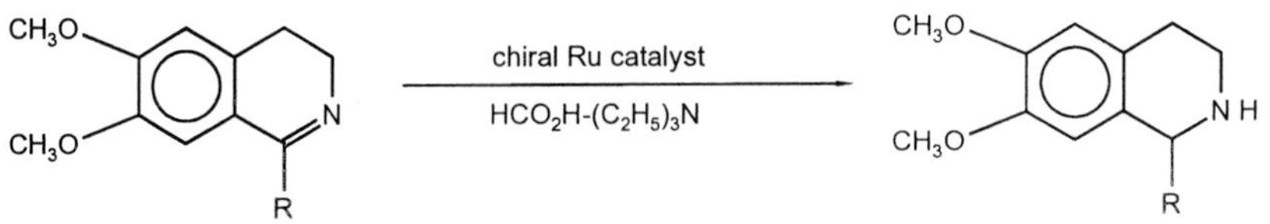

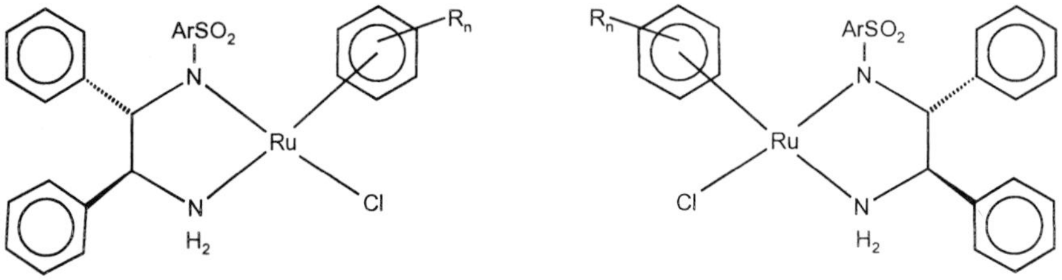

Scheme 4. Enantioselective transfer hydrogenation of imines.

complexes prompted them to investigate the ability of the phospholane ligands to enantio-selectively hydrogenate β,β-disubstituted *N*-acetylaminoacrylic acids. [5]

Attempts to use the ligands Et-, *n*Pr-, and *i*Pr–DuPHOS, which were most successful for the *N*-acetylenamides, gave unsatisfactory results: the *ee* values were 74, 45, and 14 %, respectively. In supercritical CO_2 as solvent [20] they improved to 90 %. Noyori et al. also used supercritical CO_2 as solvent for hydrogenations of α,β-unsaturated carbolxylic acids. With a Ru(II)-BINAP complex as catalyst they thereby reached at best 89 % *ee* in hydrogenating tiglic acid at 20 °C, 180 atm CO_2 and 5 atm H_2. To enhance the solubility of aromatic compounds in supercritical CO_2 and the enantioselectivity of the hydrogenation reaction a fluorinated alcohol was added (here: $CF_3(CF_2)_6CH_2OH$). [21] The use of supercritical CO_2, however, puts a greater demand on the hydrogenation apparatus than conventional conditions, since it must be able to withstand high pressures. The temperature must also be precisely controlled, because in supercricital systems merely a small temperature difference can lead to enormous effects.

The *trans* chelate ligands developed by Ito et al. are also used for enantioselective hydrogenation of β,β-disubstituted *N*-acetylaminoacrylic acids. [6] These ligands fall into the class of TRAP ligands (TRAP = 2,2''-bis[1-(dialkylphosphino)ethyl]–1,1''-biferrocene), with which good *ee* values are achieved. Methyl α-*N*-acetylamino-β,β-dimethylacrylate can be hydrogenated with Bu–TRAP at a catalyst : substrate ratio of 1 : 1000 in 24 h (15 °C, 1 bar H_2) to form the corresponding propane carboxylic acid in 88 % *ee* (Scheme 5).

Further progress in the hydrogenation β,β-disubstituted *N*-acetylenamides was made with the sterically less hindered Me-Du-PHOS–rhodium catalyst. In this system an *ee* value of 92 % was obtained. [5]

The best enantioselectivities till now have been reached with the Me–BPE ligand. With this ligand a large variety of β,β-disubstituted *N*-acetylenamides were hydrogenated with

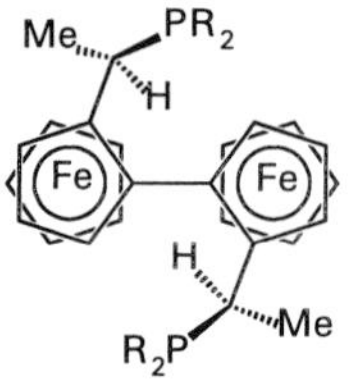

Scheme 5.
Ito's TRAP ligand.

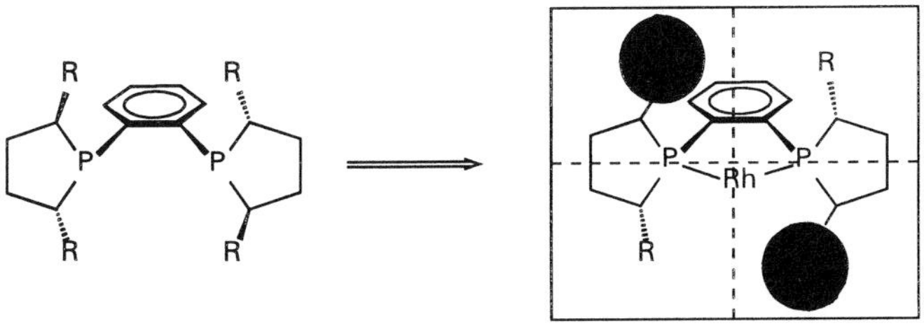

Scheme 6. Blocked quadrants.

very high *ee* values of over 99 % in some cases. For a catalyst : substrate ratio of 1 : 500 under mild conditions (25 °C, 6 bar H$_2$) *β*-methyl- and *β*-ethylphenylalanine derivatives could be prepared from the corresponding *β,β*-disubstituted *N*-acetylaminoacrylates in 12–24 hours with optical yields of 99.4 and 99 %, respectively.

The Pr–DuPHOS ligand is one of the best ligands for the hydrogenation of sterically less congested substrates, whereas for the sterically more demanding *β,β*-disubstituted compounds the ligand Me-BPE is the better choice. The astoundingly different behavior of derivatives of these catalysts is hard to predict, yet because of the opportunities for variation in the synthesis of the bisphospholanes, the broadest possible tailoring of the ligands to a problem at hand is assured.

An old but graphic picture to explain the enantioselectivities with different ligands is that of the blocked quadrants [22] (Scheme 6). The rigid benzene ring in the backbone of the DuPHOS ligands blocks two diagonally oriented quadrants at the metal atom through the substituents on the phospholane ring. The bulkier the substituent and the more rigidly it is held in its position, the better the quadrants are blocked. For most of the sterically less hindered substrates like *β*-monosubstituted acetylaminoacrylic acids the optimum is reached with the *n*Pr-DuPHOS ligand. The ligand in the catalyst may not be so bulky that the binding of the substrate is overly restricted. Thus the *ee* values for hydrogenations with the bulkier *i*Pr–DuPHOS–rhodium catalyst decrease. [9]

In the case of the tetrasubstituted acrylic acids the rigid DuPHOS ligands block both diagonally oriented quadrants so severly that the larger space requirements of these substrates cannot be accommodated. Here the more flexible Me–BPE ligand with its smaller methyl groups performs better.

References

[1] R. Noyori, *Asymmetric Catalysis in Organic Synthesis*, Wiley & Sons, New York, **1994**.

[2] J. Ojima (Ed.), *Catalytic Asymmetric Synthesis*, VCH, Weinheim, **1993**.

[3] W. A. Nugent, T. V. RajanBabu, M. J. Burk, *Science* (Washington D. C.) **1993**, *259*, 479–83.

[4] G. M. Ramos Tombo, D. Belluš, *Angew. Chem.* **1991**, *103*, 1219–41.

[5] M. J. Burk, M. F. Cross, J. P. Martinez, *J. Am. Chem. Soc.* **1995**, *117*, 9375–6.

[6] M. Sawamura, R. Kuwano, Y. Ito, *J. Am. Chem. Soc.* **1995**, *117*, 9602–3.

[7] M. J. Burk, J. E. Feaster, R. L. Harlow, *Organometallics* **1990**, *9*, 2653–5.

[8] M. J. Burk, J. E. Feaster, R. L. Harlow, *Tetrahedron: Asymmetry* **1991**, *2*, 569–92.

[9] M. J. Burk, J. E. Feaster, W. A. Nugent, R. L. Harlow, *J. Am. Chem. Soc.* **1993**, *115*, 10125–38.

[10] M. J. Burk, Y. M. Wang, J. R. Lee, *J. Am. Chem. Soc.* **1996**, *118*, 5142–3.

[11] M. J. Burk, J. E. Feaster, *J. Am. Chem. Soc.* **1992**, *114*, 6266–7.

[12] M. J. Burk, J. P. Martinez, J. E. Feaster, N. Cosford, *Tetrahedron* **1994**, *50*, 4399–428.

[13] T. Ohkuma, H. Ooka, S. Hashiguchi, T. Ikariya, R. Noyori, *J. Am. Chem. Soc.* **1995**, *117*, 2675–6.

[14] T. Ohkuma, H. Ooka, S. Hashiguchi, T. Ikariya, R. Noyori, *J. Am. Chem. Soc.* **1995**, *117*, 10417–8.

[15] S. Hashiguchi, A. Fujii, J. Takehara, T. Ikariya, R. Noyori, *J. Am. Chem. Soc.* **1995**, *117*, 7562–3.

[16] K.–J. Haack, S. Hashiguchi, A. Fujii, T. Ikariya, R. Noyori, *Angew. Chem.* **1997**, *109*, 297–300.

[17] A. Fujii, S. Hashiguchi, N. Uemastu, T. Ikariya, R. Noyori, *J. Am. Chem. Soc.* **1996**, *118*, 2521–2.

[18] J. Takehara, S. Hashiguchi, A. Fujii, S. Inoue, T. Ikariya, R. Noyori, *Chem. Comm.* **1996**, 233–4.

[19] N. Uematsu, A. Fujii, S. Hashiguchi, T. Ikariya, R. Noyori, *J. Am. Chem. Soc.* **1996**, *118*, 4916–7.

[20] M. J. Burk, Shaoguang Feng, M. F. Cross, W. Tumas, *J. Am. Chem. Soc.* **1995**, *117*, 8277–8.

[21] J. Xiao, S. C. A. Nefkens, P. G. Jessop, T. Ikariya, R. Noyori, *Tetrahedron Letters* **1996**, *37* 2813–6.

[22] W. S. Knowles, *Acc. Chem. Res.* **1983**, *16*, 106.

The Sharpless Asymmetric Aminohydroxylation of Olefins

Oliver Reiser

"The time has come for me to discover something new, but there is so much chemistry out there!" With these words Barry Sharpless closed the Merck-Schuchhardt lecture 1995 in Göttingen, Germany [1]. After about 10 years of continuous optimization, the asymmetric dihydroxylation (AD) of olefins had developed into one of the most versatile catalytic asymmetric reaction to date [2].

An "obvious" extension of the AD-process would be the asymmetric transfer of heteroatoms other than oxygen to a carbon carbon double bond. Indeed, the osmium catalyzed [3] or palladium mediated [4] aminohydroxylation of alkenes has been known for 20 years. The resulting β-amino alcohols are an important structural element in biologically active compounds as well as the starting point in the design of many chiral ligands. However, to develop this reaction into a *catalytic, asymmetric* process several problems had to be overcome.

According to the original protocol [3] alkenes can be converted to *racemic N*-tosyl protected β-amino alcohols in the presence of *catalytic* amounts of osmium tetroxide using N-chloramine-T as the nitrenoid source and water as the hydroxyl source. However, unlike the AD-process, the aminohydroxylation of unsymmetrical alkenes can lead to two regioisomeric products which was a drawback in the early stages of its development. Moreover, the direct reaction of osmium tetroxide with the alkene could not be completely surpressed, and diols were sometimes observed as side products.

Formel 1.

Although the first, albeit stoichiometric, example of this asymmetric aminohydroxylation reaction had already been observed in 1980 [5], the discovery of the titanium catalyzed asymmetric epoxidation (AE) [6] at about the same time probably also "interfered" with an earlier development of today's *catalytic* asymmetric aminohydroxylation (AA) process.

Since the discovery of the catalytic AD in 1987, there have been numerous attempts in the Sharpless group to render the old catalytic aminohydroxylation process asymmetric [7]. Until recently, the obvious approach of adding the AD's chiral ligands, but otherwise staying close to the original protocol [3] led to extremely slow catalyst turnover. The initial breakthrough [8] was not due to a sudden conceptual insight but rather to the luck of making a number of small changes, all of which had been tried before but never in concert. The combination of chloramine-T as the nitrogen source, potassium osmate (4 mol%) and the recently discovered phthalazine ligands (Alk*)$_2$PHAL (5 mol%) as the catalyst, *tert*-butylhydroperoide as the stochiometric oxidans, and solvent mixtures *t*-BuOH/H$_2$O (1 : 1) or CH$_3$CN/H$_2$O (1 : 1) proved to be the right recipe. Suddenly the turnover was decent, diol side products were greatly reduced, and good regioselectivity was seen for the first time [7]. Most exciting of all, the products were obtained with moderate to high enantioselectivities, with complimentary discrimination of the enantiotopic faces of the alkene by the two ligands [9], (DHQ)$_2$PHAL and (DHQD)$_2$PHAL. The same sense of asymmetric induction as in the AD is observed, indicating that the chirality transfer occurs by a similar pathway. Moreover, due to the highly crystalline nature of the products they can often be raised to enantiopurity by simple recrystallization.

PHAL

Alk* =

Formel 2.   DHQD   DHQ

Despite these encouraging results, there was much room for improvement of this initial procedure. The obtained selectivites were not satisfactory in many cases, however, perhaps the most substantial disadvantage was the transferal of the nitrogen being tosyl pro-

tected, which is generally difficult to remove. Consequently, the attention of the Sharpless group was directed towards different reagents which could act as the nitrogen source. First, it was discovered that smaller substituents on nitrogen give rise to dramatically improved selectivities [10]. Thus, *N*-chloramine-M, transfering a NSO$_2$Me group proved to be superior for all cases compared to *N*-chloramine-T. Nevertheless, the problem of deprotection was still a draw back for synthetic applications, despite some promising developments of easier removable protecting groups of the sulfonamide type [11]. In parallel it was found that *N*-halo-carbamates are also efficient nitrogen sources [12]. The highest selectivities are observed with *N*-chloro-ethylcarbamate, followed by *N*-chloro-benzylcarbamate, offering the great advantage of giving rise to Z-protected amino alcohols, which can be readily deprotected if desired by standard procedures.

What types of alkenes undergo this reaction? In contrast to the AD reaction also strongly electron poor alkenes are suitable substrates probably due to the greater polarization of the Os=NR group compared to that of the Os=O group. Thus, acrylates in general give good results in the AA, but even dimethyl fumarate (**1**) reacts rapidly – **1** is a very poor AD substrate – to give the 2-hydroxy aspartic acid derivative **2**. Cinnamates **3** can be converted to α-hydroxy-β-amino acids **4**, giving e. g. a most effective entry to the side chain amino acid of taxol [13]. *Para*-substitution of the aryl group in **3** by electron-donating or withdrawing groups is well tolerated, however, *meta*- and in particular *ortho*-substitution results in some loss of enantio- and regioselectivity. One of the most remarkable effects of the chiral ligand is on the regioselectivity. In the absence of the ligand the cinnamates **3** give roughly equal amounts of the regioiosmers (i. e. **4** plus its 2-NHTs, 3-OH isomer) [3]. While most of the reactions to date were carried out with 4 mol%

Substrate	Product	Selectivity [% *ee*] (yield [%])				
		R = Ts	Ms	NCO$_2$Et	NCO$_2$Bn	NCO$_2$t-Bu
1 MeO$_2$C–CO$_2$Me	**2** MeO$_2$C / NHR / OH CO$_2$Me	 77 (65)	95 (76)	–	84 (55)	–
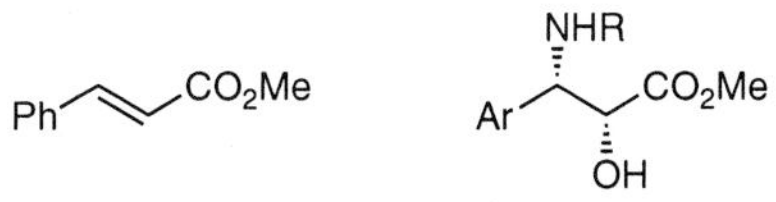 **3** Ph–CO$_2$Me	**4** Ar / NHR / OH CO$_2$Me	82 (60)	95 (65)	98 (70)	94 (65)	78 (71)
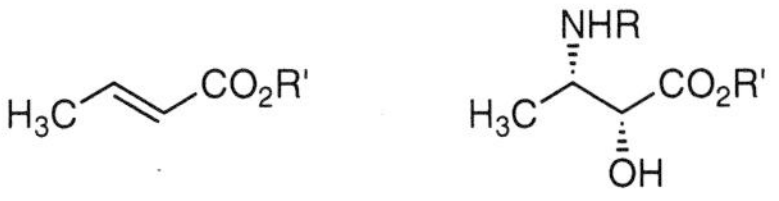 **5** H$_3$C–CO$_2$R'	**6** H$_3$C / NHR / OH CO$_2$R'	74 (52) R' = Et	80 (63) R' = t-Bu	–	–	–
7 Ph–Ph	**8** Ph / NHR / OH Ph	64 (78)	75 (71)	–	91 (92)	–
9 Ph / Ph (cis)	**10** Ph / NHR / OH Ph	50 (57)	–	–	–	–
11 (vinyl naphthalene)	**12** (naphthalene NHR / OH)	–	– (Regioselectivity)	–	99 (70) > 10:1	98 (73) 6.6:1
 **13** (cyclohexene)	**14** (RHN / OH cyclohexane)	45 (64)	66 (49)	–	63 (51)	–

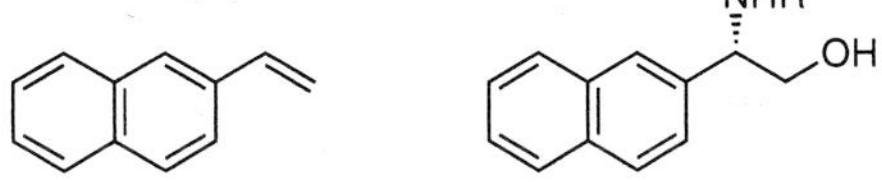

Scheme 1. AA: RNClNa (3–3.5 eq.), *t*-BuOH/H$_2$O 1 : 1, *n*-PrOH/H$_2$O 1 : 1 or CH$_3$CN/H$_2$O 1 : 1, 0C or rt, K$_2$OsO$_2$(OH)$_4$ (4.0 mol%), (DHQ)$_2$-PHAL (5 mol%), (DHQD)$_2$-PHAL leads to the enantiomeric products.

Os/5 mol% ligand, in large scale runs the catalyst amount has been successfully reduced by half, giving similar good results.

Less electron poor alkenes like stilbenes (**7** and **9**) are also viable substrates. Direct comparison of **8** and **10** also indicates, that

(E)-alkenes are – in analogy to the AD process – better substrates than (Z)-alkenes. Nevertheless, it is already clear at this stage that (Z)-alkenes, and in particular symmetrical (Z)-alkenes, are useful substrates in the AA. A syn-dihydroxylation of the latter class of olefins leads to meso-diols and is therefore not suitable in relation of the concept of asymmetric dihydroxylation. Cyclic (Z)-alkenes such as cyclohexene (13) also have given promising results, and heteroatoms can be present as long as they are not directly attached to the double bond [7].

1,3-, 1,4-, and 1,5-dienes as well as certain terminal alkenes also have given encouraging results in the AA, but additional regioselectivity and reactivity problems arise with these substrates. With the discovery of the new nitrogen sources, styrenes such as 11 are converted with excellent enantioselectivity and good regioselectivity to the corresponding amino alcohols. Alkenes which seem to be problematic so far are allylic halides, alcohols and amines, no matter if protected or not. Several allylic acetals have also given poor results [7].

The products obtained by the AA process can be easily transformed into aziridines or into precursors for α,β-diamino acids [7].

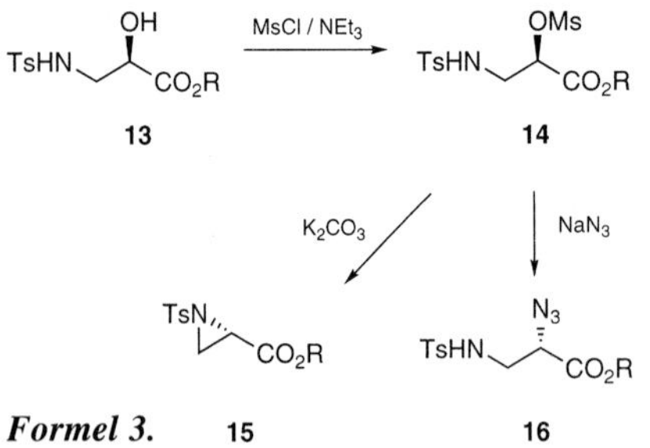

theless, the reaction conditions have to be carefully controlled since little changes regarding the nitrogen source, the solvent and the temperature can have a great effect on selectivity and yield of the products. Thanks to the openness with which Sharpless et al. is presenting this discovery to the synthetic community in that not only short communications but simultaneously full papers with detailed experimental procedures have been published, many exciting applications of this reaction should result in the near future.

Moreover, there are already two other oxidative amination processes on the horizon, namely allylic amination of olefins and 1,2-diamination of 1,3-dienes [14]. Could these useful, but stochiometric transformations be rendered catalytic and perhaps also asymmetric some day?

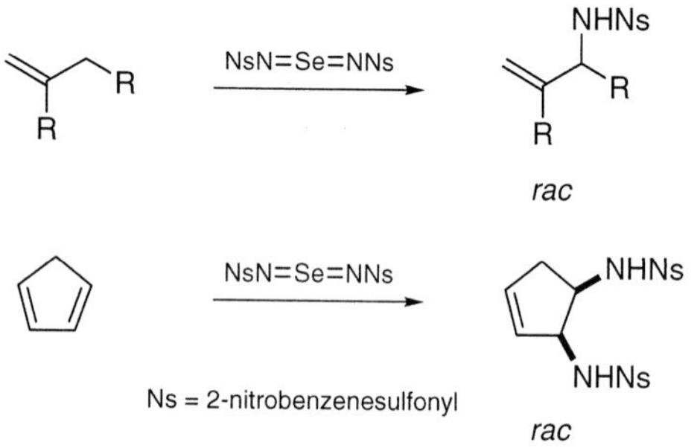

Formel 4.

Acknowledgement: This work was supported by the Winnacker foundation and the Fonds der Chemischen Industrie.

Formel 3.   15   16

In only very little time the new catalytic process of an asymmetric aminohydroxylation of alkenes has been transformed into a practical method with great synthetic potential. The title reaction is easily being carried out, never-

References

[1] University of Göttingen, April 28, 1995. The day of the AA discovery has been June 14, 1995.

[2] H. C. Kolb, M. S. VanNiewenhze, K. B. Sharpless, *Chem Rev.* **1994**, *94*, 2483.

[3] a) K. B. Sharpless, A. O. Chong, K. Oshima, *J. Org. Chem.* **1976**, *41*, 177; b) E. Herranz, K. B. Sharpless, *J. Org. Chem.* **1978**, *43*, 2544.

[4] J. E. Bäckvall, *Tetrahedron Lett.* **1975**, *26*, 2225.

[5] S. Hentges, K. B. Sharpless, *J. Am. Chem. Soc.* **1980**, *102*, 4263.

[6] T. Katsuki, B. Sharpless, *J. Am. Chem. Soc.* **1980**, *102*, 5447.

[7] K. B. Sharpless, personal communication.

[8] G. Li, H.-T. Chang, K. B. Sharpless, *Angew. Chem.* **1996**, *108*, 449; *Angew. Chem. Int. Ed. Engl.* **1996**, *35*, 451.

[9] H. Becker, K. B. Sharpless, *Angew. Chem.* **1996**, *108*, 447; *Angew. Chem. Int. Ed. Engl.* **1996**, *35*, 448.

[10] J. Rudolph, P. C. Sennhenn, C. P. Vlaar, K. B. Sharpless, *Angew. Chem.* **1996**, *108*, 2991; *Angew. Chem. Int. Ed. Engl.* **1996**, *35*, 2813.

[11] T. Fukuyama, C.-K. Jow, M. Cheung, *Tetrahedron Lett.* **1995**, *36*, 6373.

[12] G. Li, H. H. Angert, K. B. Sharpless, *Angew. Chem.* **1996**, *108*, 2995; *Angew. Chem. Int. Ed. Engl.* **1996**, *35*, 2837.

[13] G. Li, K. B. Sharpless, *Acta Chem. Scand.* **1996**, *50*, 649.

[14] M. Bruncko, T.-A. V. Khuong, K. B. Sharpless, *Angew. Chem.* **1996**, *108*, 453; *Angew. Chem Int. Ed. Engl.* **1996**, *35*, 454.

Epoxides in Asymmetric Synthesis: Enantioselective Opening by Nucleophiles Promoted by Chiral Transition Metal Complexes

Ian Paterson and David J. Berrisford

The turn of the millenium will see the 20th anniversary of the seminal discovery of the asymmetric epoxidation [1, 2] of allylic alcohols catalysed by titanium(IV) isopropoxide and tartrate esters. The utility of this transformation largely results from the regio- and stereocontrol possible in subsequent nucleophilic ring opening reactions of the derived epoxy alcohols. Thus, a sequence of asymmetric epoxidation, epoxide opening and further functionalisation leads to a diverse array of molecules in enantiomerically pure form. In comparision, asymmetric epoxidation of unfunctionalised alkenes [3] has yet to match the enantioselectivities which the Ti-tartrate system can deliver with allylic alcohols. The recent discovery of other asymmetric epoxidation reactions [4] suggests that a number of practical options may eventually become available.

A less common strategy for asymmetric synthesis, but one with considerable merit, is the enantioselective opening of *meso* epoxides (Fig. 1a) by achiral nucleophiles in the presence of a chiral catalyst. [5] Similarly, the kinetic resolution of *racemic* epoxides (Fig. 1b), in the best cases, can deliver high enantiomeric excesses in the unreacted epoxide and ring-opened product.

Both processes involve initial coordination to a Lewis acidic metal centre, so activating the epoxide to attack by an external nucleophile (Fig. 2, pathway a). The catalyst functions by complexing the epoxide oxygen atom and the ligand environment should allow discrimination of the formally enantiotopic carbon-oxygen bonds by an appropriate achiral nucleophile.

Enantioselective opening of *meso* epoxides by heteroatom nucleophiles [6–11] has been

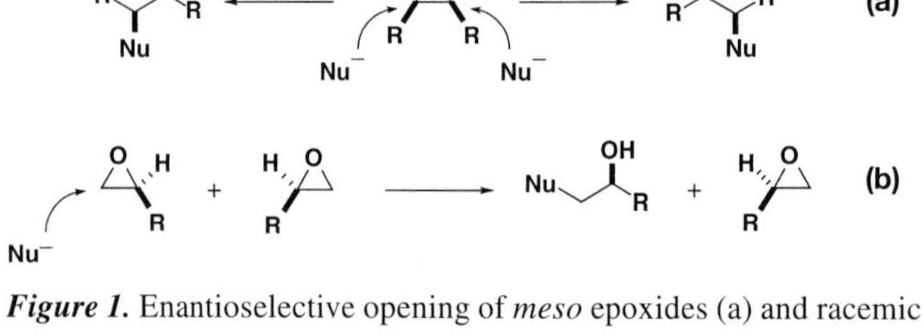

Figure 1. Enantioselective opening of *meso* epoxides (a) and racemic epoxides (b) by achiral nucleophiles.

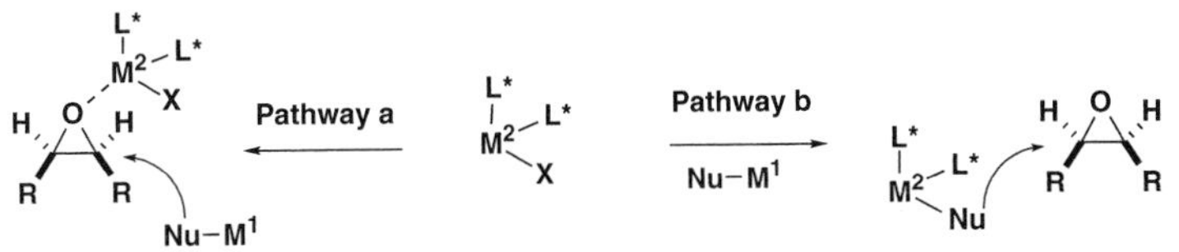

Figure 2. Enantioselective opening of *meso* epoxides. Pathway a: activation of the epoxide by the chiral Lewis acid ($M^2XL^*_2$). Pathway b: activation of the stoichiometric nucleophile (NuM^1) by metathesis with the chiral complex.

investigated by several research groups (Scheme 1). Brown has shown that certain chiral boron Lewis acids display [6] useful levels of asymmetric induction in their stoichiometric reactions with epoxides to give *β*-halohydrins. Heterogeneous zinc and copper tartrates catalyse [7] the opening of simple epoxides with *n*-butyl thiol, trimethylsilyl azide, and aniline with moderate-to-good enantioselectivity. More efficient homogenous catalysts can be derived from titanium alkoxides with chiral ligands. [8, 9] For example, the combination of $TiCl_2(OiPr)_2$ with di-*tert*-butyl tartrate catalyses [10] the opening of cyclohexene oxide with trimethylsilyl azide giving *β*-azido cyclohexanol in 62 % *ee*.

In 1992, Nugent [12] reported a significant advance in the Lewis acid-promoted opening of *meso* epoxides (Scheme 2). A novel chiral zirconium complex catalyses epoxide opening with hindered silyl azides in excellent *ee*. The catalyst is derived from zirconium *tert*-butoxide and the tetradentate C_3-symmetric ligand **1**, available by combination of (*S*)-propene oxide with (*S*)-1-aminopropan-2-ol. Al-

though the ligand **1** possesses C_3-symmetry, the active catalyst is likely to display a less symmetrical structure and the complex will certainly be aggregated in solution. The enantioselectivity of the reaction is enhanced by use of bulkier nucleophilic azides and the trifluoroacetate additive is essential for high selectivity. The reaction is simple to carry out, proceeds at either 0 °C or room temperature and affords excellent yields and enantioselectivities.

In 1995, Jacobsen [13] reported a further advancement in the nucleophilic opening of *meso* epoxides. This discovery, which makes use of Cr-Salen catalysts (Scheme 3), is significant for a number of reasons. Firstly, the catalyst delivers high yields and enantioselectivities using only a 2 mol% loading at 0 °C with trimethylsilyl azide as the nucleophile. Secondly, kinetic investigations [14] reveal that the Cr-Salen complex **4** has an unexpected dual role in the process. Cr-Salen complexes, e. g. **4** and **5**, can act as Lewis acids (cf. Fig. 2, pathway a). In addition, the azide is transferred *in situ* from Si to Cr giving com-

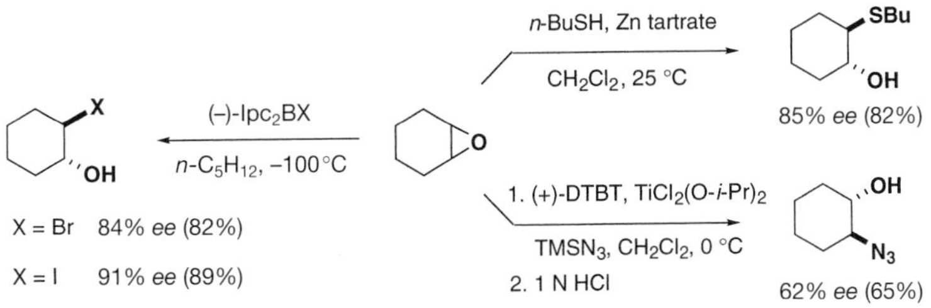

Scheme 1. Enantioselective opening of *meso* epoxides by heteroatom nucleophiles (Ipc = isopinocampheyl; DTBT = di-*tert*-butyl tartrate).

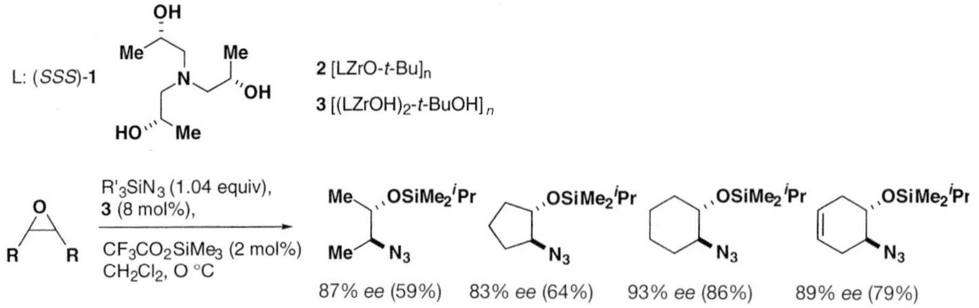

Scheme 2. Enantioselective opening of *meso* epoxides by trialkylsilyl azide catalysed by Zr complex **3**.

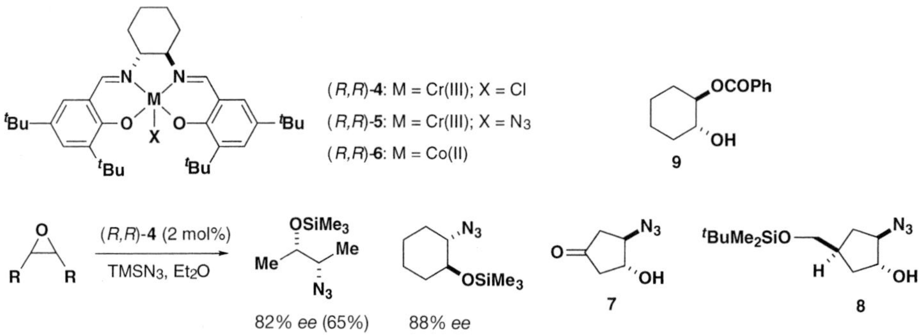

Scheme 3. Enantioselective opening of *meso* epoxides catalysed by chiral Salen complexes.

plex **5**. Thus, a second equivalent of the metal complex acts as a chiral nucleophile (cf. Fig. 2, pathway b). The proposed mechanism [14] combines the two possible roles for the metal and requires two Cr-Salen complexes in the turnover limiting step. Azide complex **5**, rather than the precatalyst **4**, can be used directly in the syntheses [13] of **7** and **8**, useful intermediates for prostaglandins and carbo-cyclic nucleosides. On a practical note, the reactions can be run without solvent and the catalyst can be recycled a number of times. Catalyst recycling involves the potentially hazardous distillation of the neat liquid azides, a procedure which may be incompatible with large scale applications.

Further experimentation [15] using a variety of metal-Salen complexes has enabled the ring opening of *meso* epoxides by benzoic acid to give β-hydroxy benzoates. In this case,

the Co-Salen complex **6** affords the highest enantioselectivities; for example, benzoate **9** may be prepared in 98 % *ee* after recrystallisation. These results give encouragement to chemists to search for other new catalysts which despite apparently similar constitution may possess striking new reactivity profiles.

Catalytic kinetic resolution of racemic epoxides [16, 17] has proven to be an elusive goal. The successful implementation of this strategy by Jacobsen [16] is a highly significant achievement (Scheme 4). Complex **5** catalyses the opening of simple terminal epoxides with trimethylsilyl azide with outstanding enantioselectivities. The unreacted epoxides are evaporated, leading to the β-azido silyl ethers **10** in ≥ 95 % *ee* and near quantitative yields. The catalyst displays near perfect enantioselection; kinetic resolution [18] often requires a much greater sacrifice of yield to

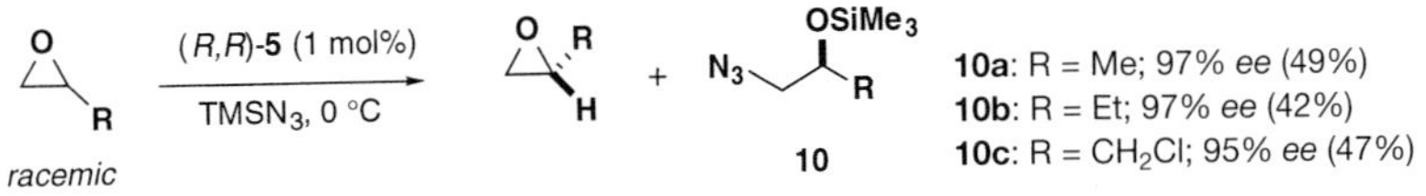

Scheme 4. Kinetic resolution of terminal epoxides catalysed by chiral Cr-Salen complex **5**.

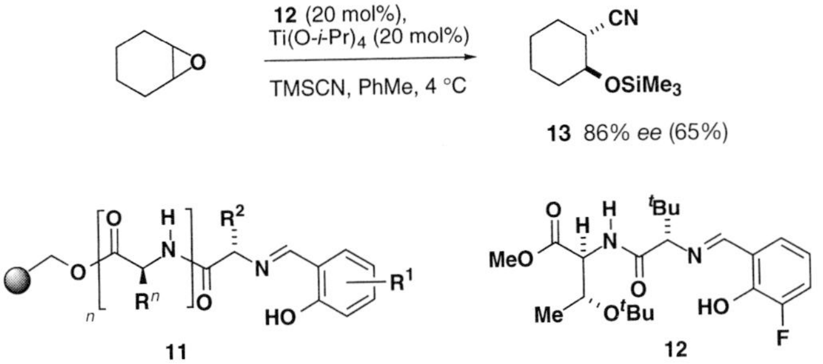

Scheme 5. Combinatorial screening of ligands for Ti-catalysed enantioselective opening of *meso* epoxides by trimethylsilyl cyanide.

attain high *ee* and at > 50 % conversion the starting material is returned with the higher enantiopurity. Given the economy of the process and the ready availability of the epoxide substrates, this discovery is likely to emerge as an established method of asymmetric synthesis.

In 1996, Hoveyda [19] reported an important contribution to catalytic asymmetric epoxide opening through the use of combinatorial library screening. All the foregoing chemistry in this article concerns heteroatom nucleophiles, where the use of carbon nucleophiles is noticeably absent. Previous work [20] had shown that combinations of early transition metals and Schiff base ligands could act as Lewis acids in the opening of epoxides with trimethylsilyl cyanide. Using a library of potential ligands **11**, prepared by solid phase peptide synthesis, a large number of possible catalysts were screened for reactivity and enantioselectivity. This led to the discovery of a process capable of delivering good enantioselectivity for certain *meso* epoxides (Scheme 5). For example, ligand **12** combined with Ti(O*i*Pr)$_4$ gives β-cyanohydrin

13 with 86 % *ee*. Moreover, this study addresses some pertinent issues with respect to screening for asymmetric catalysis by library methods. Clearly, library screening offers considerable promise in both the discovery and optimisation of new catalysts.

Taken together, these results for epoxide opening promoted by chiral metal complexes [21] illustrate how much there is find out about asymmetric catalysis by transition metals. These valuable new processes are further proof, if any is required, of the rewards of searching for new reactivity amongst the transition metals.

References

[1] R. A. Johnson, K. B. Sharpless in *Catalytic Asymmetric Synthesis* (Ed.: I Ojima), VCH, New York/Weinheim, **1993**, p. 103–158.
[2] R. Noyori, *Asymmetric Catalysis in Organic Synthesis*, Wiley, New York, **1994**.
[3] For reviews see: T. Katsuki, *Coord. Chem. Rev.* **1995**, *140*, 189–214; E. N. Jacobsen in *Catalytic Asymmetric Synthesis* (Ed.:

I Ojima), VCH, New York/Weinheim, **1993**, p. 159–202.

[4] V. K. Aggarwal, J. G. Ford, A. Thompson, R. V. H. Jones, M. C. H. Standen, *J. Am. Chem. Soc.* **1996**, *118*, 7004–7005; D. Yang, X-C. Wang, M-K. Wong, Y-C. Yip, M-W. Tang, *ibid.* **1996**, *118*, 11311–11312; Y. Tu, Z-X. Wang, Y. Shi, *ibid.* **1996**, *118*, 9806–9807; C. Bousquet, D. G. Gilheany, *Tetrahedron Lett.* **1995**, *36*, 7739–7742.

[5] I. Paterson, D. J. Berrisford, *Angew. Chem.* **1992**, *104*, 1204–1205; *Angew. Chem. Int. Ed. Engl.* **1992**, *31*, 1179–1180.

[6] M. Srebnik, N. N. Joshi, H. C. Brown, *Isr. J. Chem.* **1989**, *29*, 229–237; N. N. Joshi, M. Srebnik, H. C. Brown, *J. Am. Chem. Soc.* **1988**, *110*, 6246–6248.

[7] H. Yamashita, T. Mukaiyama, *Chem. Lett.* **1985**, 1643–1646; H. Yamashita, *ibid.*, **1987**, 525–528.

[8] M. Emziane, K. I. Sutowardoyo, D. Sinou, *J. Organomet. Chem.* **1988**, *346*, C7–C10.

[9] C. Blandy, R. Choukroun, D. Gervais, *Tetrahedron Lett.* **1983**, *24*, 4189–4192.

[10] M. Hayashi, K. Kohmura, N. Oguni, *Synlett* **1991**, 774–776.

[11] H. Adolfsson, C. Moberg, *Tetrahedron Asym.* **1995**, *6*, 2023–2031.

[12] a) W. A. Nugent, *J. Am. Chem. Soc.* **1992**, *114*, 2768–2769; b) W. A. Nugent, R. L. Harlow, *ibid.* **1994**, *116*, 6142–6148.

[13] L. E. Martinez, J. L. Leighton, D. H. Carsten, E. N. Jacobsen, *J. Am. Chem. Soc.* **1995**, *117*, 5897–5898; J. L. Leighton, E. N. Jacobsen, *J. Org. Chem.* **1996**, *61*, 389–390; L. E. Martinez, W. A. Nugent, E. N. Jacobsen, *J. Org. Chem.* **1996**, *61*, 7963–7966.

[14] K. B. Hansen, J. L. Leighton, E. N. Jacobsen, *J. Am. Chem. Soc.* **1996**, *118*, 10924–10925.

[15] E. N. Jacobsen, F. Kakiuchi, R. G. Konsler, J. F. Larrow, M. Tokunaga, *Tetrahedron Lett.* **1997**, *38*, 773–776.

[16] J. F. Larrow, S. E. Schaus, E. N. Jacobsen, *J. Am. Chem. Soc.* **1996**, *118*, 7420–7421; S. E. Schaus, E. N. Jacobsen, *Tetrahedron Lett.* **1996**, *37*, 7937–7940.

[17] M. Brunner, L. Mussmann, D. Vogt, *Synlett* **1993**, 893–894.

[18] E. L. Eliel, S. H. Wilen, L. M. Mander, *Stereochemistry of Organic Compounds*, Wiley-Interscience, New York, **1994**, p. 395–415.

[19] B. M. Cole, K. D. Shimizu, C. A. Krueger, J. P. A. Harrity, M. L. Snapper, A. H. Hoveyda, *Angew. Chem.* **1996**, *108*, 1776–1779; *Angew. Chem. Int. Ed. Engl.* **1996**, *35*, 1668–1671; *Angew. Chem. Int. Ed. Engl.* **1996**, *35*, 1995.

[20] M. Hayashi, M. Tamura, N. Oguni, *Synlett* **1992**, 663–664. For a review of ligand accelerated catalysis see: D. J. Berrisford, C. Bolm, K. B. Sharpless, *Angew. Chem.* **1995**, *107*, 1159–1171; *Angew. Chem. Int. Ed. Engl.* **1995**, *34*, 1059–1070.

[21] N. Oguni, *J. Synth. Org. Chem. Jpn.* **1996**, *54*, 829–835; M. Hayashi, K. Ono, H. Hoshimi, N. Oguni, *Tetrahedron*, **1996**, *52*, 7817–7832; *idem.*, *J. Chem. Soc. Chem. Commun.* **1994**, 2699–2700.

Asymmetric Deprotonation as an Efficient Enantioselective Preparation of Functionalized Secondary Alcohols and Amino-Alcohols

Paul Knochel

The enantioselective preparation of chiral secondary alcohols has been achieved through a number of methods with great success. One pathway that is widely used is the asymmetric addition of organometallic compounds R^1M to aldehydes [1] (retrosynthetic pathway **A** of Eq. (1). A disconnection involving inversion of polarity [2] is also possible (retrosynthetic pathway **B** of [Eq. (1)].

In this case, the secondary alcohol is formed by a substitution reaction of an α-oxygen-substituted carbenoid **1**. [3] Although this alternative route has been frequently used for synthetic applications, the multistep procedures [4] required for the enantioselective preparation of the carbenoids have limited the utility of this approach. Furthermore, to be synthetically attractive, the chiral carbenoid has to be configurationally stable under the reaction conditions, and its reaction with electrophiles has to proceed with a well-defined stereochemistry (retention or inversion).

Enantioselective deprotonation reactions have been known for several years and have led to several elegant synthetic applications. [5] Hoppe found that this method allows

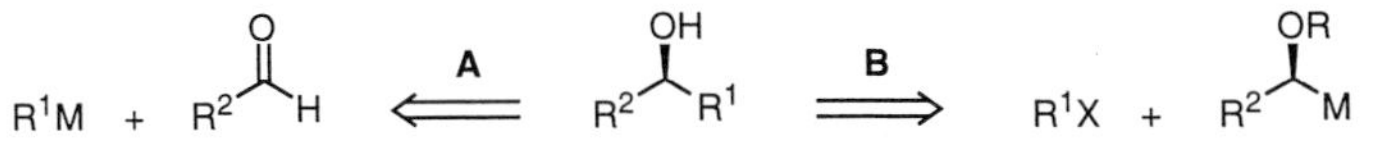

$$ R^1M \;+\; R^2\overset{O}{\underset{}{\text{-CHO}}} \quad \overset{A}{\Longleftarrow} \quad R^2\overset{OH}{\underset{}{}}R^1 \quad \overset{B}{\Longrightarrow} \quad R^1X \;+\; R^2\overset{OR}{\underset{}{}}M \qquad \textit{Eq. (1).} $$

$$ R^1\overset{OR^2}{\underset{H}{\cdots H}} \;+\; R^3Li \cdot (-)\text{-sparteine} \quad \longrightarrow \quad R^1\overset{OR^2}{\underset{Li\cdot(-)\text{-}3}{\cdots H}} $$

(S)-**1**•(−)-**3**

2: R^1 = alkyl; R^2 = Cbx, Cby; R^3 = s BuLi

R^1 = alkenyl; R^2 = Cb; R^3 = n BuLi

Cb = CON(i Pr)$_2$

Cbx = CO–N (ring with Me, Me, O)

Cby = Co–N (ring with Me Me, Me Me, O)

(−)-**3**

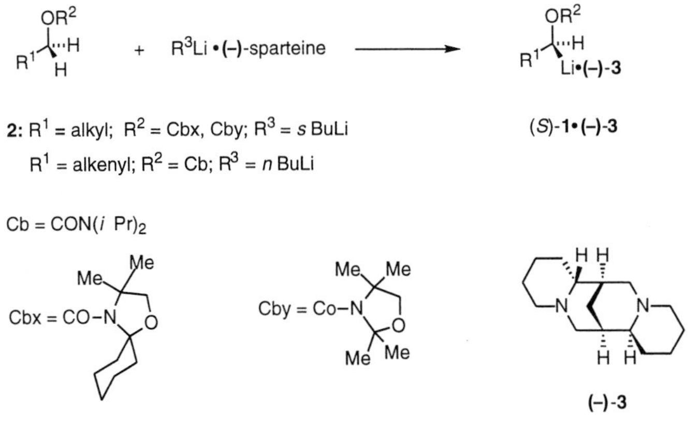

Eq. (2).

an expedient access to a wide range of chiral *a*-oxygen carbenoids **1** (M = Li, Eqs. 1 and 2) [6–12] starting from readily available achiral carbamates **2** and the complex of an alkyllithium compound and (–)-sparteine **3**. [13]

The deprotonations are complete within a few hours at –78 °C and afford the lithium carbenoid · sparteine complexes (*S*)-**1**·(–)-**3** with excellent enantioselectivities. [6–12] Whereas sparteine complexes of lithiated secondary allyl and primary alkyl carbamates are configurationally stable below –30 °C, those of primary allyl carbamates such as **4**·(–)-**3** are not configurationally stable even at –70 °C. It is, however, possible to use these reagents in synthesis, since the preferential crystallization of the *S* diastereomer in pentane/cyclohexane drives the equilibrium completely to one side. After a low-temperature transmetalation of (*S*)-**4**·(–)-**3** with an excess of tetraisopropoxytitanium, the allylic titanium reagent (*R*)-**5** is obtained with inversion of configuration. The addition of various aldehydes to (*R*)-**5** furnishes homoaldol adducts of type **6** with high *anti* diastereoselectivity and very good enantioselectivity (82–90 % *ee*) (Eq. 3). They can be further converted to lactones as shown in the synthesis of the insect pheromone (+)-eldanolide (92 % *ee*). [6d]

A crystal structure analysis of a complex between a lithiated primary carbamate and (–)-sparteine **3** unambiguously established the *S* configuration of this organometallic intermediate. [14]

Of considerable interest for synthetic applications are the deprotonations of alkyl carbamates RCH_2OCbx or RCH_2OCby with *s* butyllithium · (–)-sparteine, followed by the alkylation with an electrophile. [7] Most electrophiles such as CO_2, Me_3SnCl, Me_3SiCl, MeI, Me_2CHCHO react with retention of configuration to afford protected alcohols of type **7** (Eq. 4).

This provides a very general enantioselective synthesis of secondary alcohols. The enantioselectivities of the deprotonations are better than 95 % *ee* in most cases. These high enantiodifferentiations are kinetic in nature, and MNDO calculations show that the stabilities of the two diastereomeric ion-pairs (+)-**1**·(–)-**3** and (–)-**1**·(–)-**3** are comparable. [15] Efforts have also been made to determine the structure of the chiral base RLi·(–)-**3**, but only the structure of the complex (*i* PrLi)$_2$ · (–)-**3** in ether has been determined, by NMR techniques. [16]

The Cbx- protecting group of **7** can be removed under mild conditions. An acid treatment ($MeSO_3H$, MeOH, reflux, 16 h) afford-

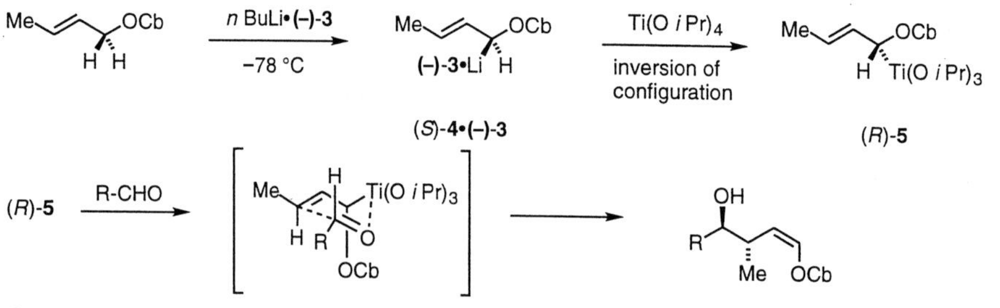

6: 90–95 %; 82–90 % *ee*   *Eq. (3).*

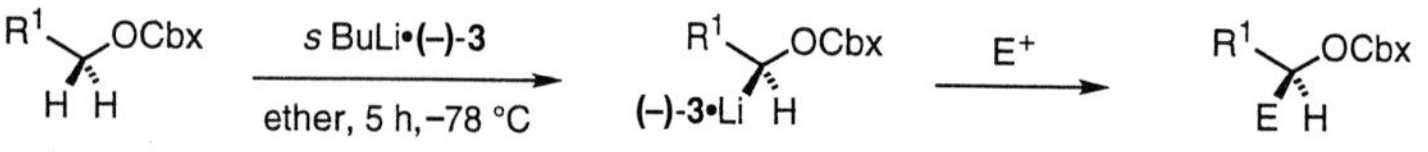

7   *Eq. (4).*

ing the intermediate carbamate **8** followed by a basic treatment (Ba(OH)$_2$ · 8H$_2$O, MeOH, reflux, 4 h) leads to the chiral alcohol **9** (Eq. 5) [7] in excellent yield.

Remarkably, this method can be applied to carbamates bearing a functional group in close proximity to the OCbx or OCby group. High enantiomeric excesses are obtained, although chelating groups such as N(CH$_2$Ph)$_2$ [8, 11] or an OR [9, 10] group are present. Thus the β-dibenzylaminocarbamate **10** can be deprotonated with high enantioselectivity with s BuLi·(–) -**3** (3 equiv; –78 °C, 3 h) and treated with various electrophiles to give almost enantiomerically pure 1,2-aminoalcohols **11** (Eq. 6). [8] Interestingly, the presence of a substituent at the a position to the nitrogen functionality strongly influences the rate and controls the enantioselectivity of the deprotonation.

This can be clearly seen in the case of the carbamate **12**, derived from (*S*)-*N,N*-dibenzyl-leucinol. After the reaction with an electrophile, substrate-controlled deprotonation yields the aminoalcohol derivative **13** (Eq. 7). [8]

Interestingly, asparagin acid derivative **14** is deprotonated enantioselectively by s BuLi·(–)-**3** in ether (–78 °C, 5 h) furnishing after carbonylation and subsequent methylation the ester **15** in 90 % yield (Eq. 8). [17]

The importance of intramolecular chelation is nicely demonstrated in the lithiation of the glutamic acid derivative **16**. With s BuLi·(–)-**3** an enantioselective deprotonation is observed at position 5 leading after the carbonylation-esterification sequence to the ester **17**, whereas with s BuLi in ether, an intramolecular complexation favors an intramolecular enantio-selective deprotonation at position 1 leading to the ester **18** (Eq. 9). [17]

The 1,3-dicarbamate **19** undergoes a stereoselective cyclopropanation after lithiation with s BuLi · (–)-**3**. Depending on the nature

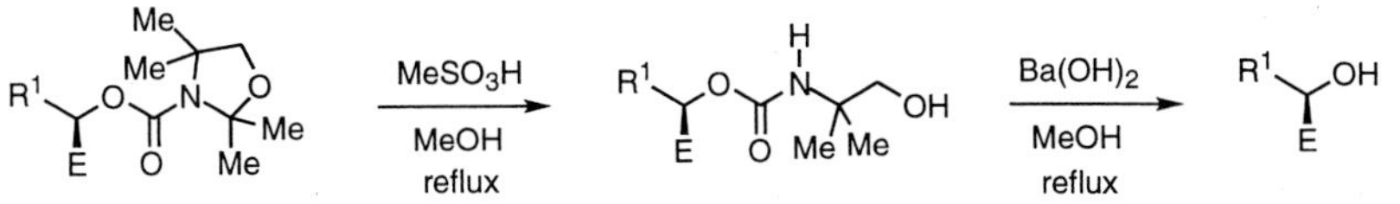

Eq. (5).

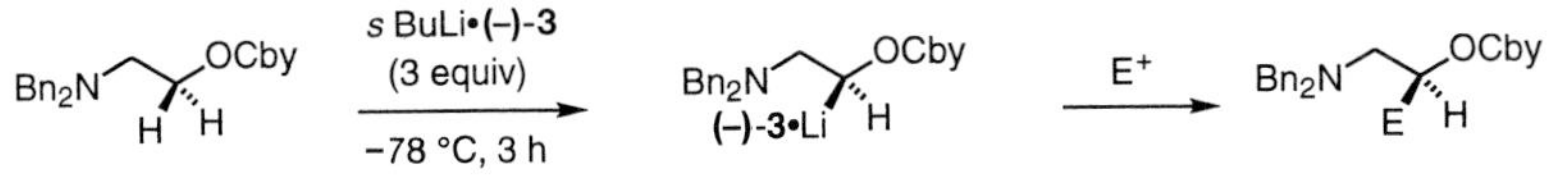

Eq. (6).

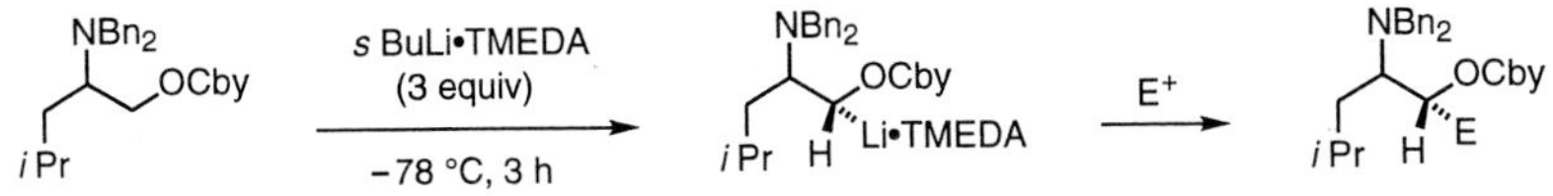

Eq. (7).

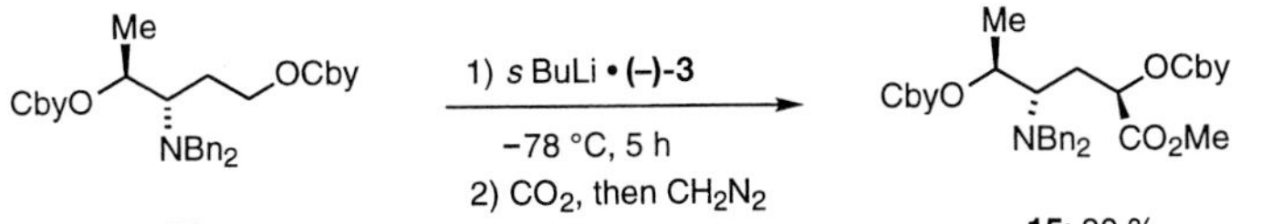

Eq. (8).

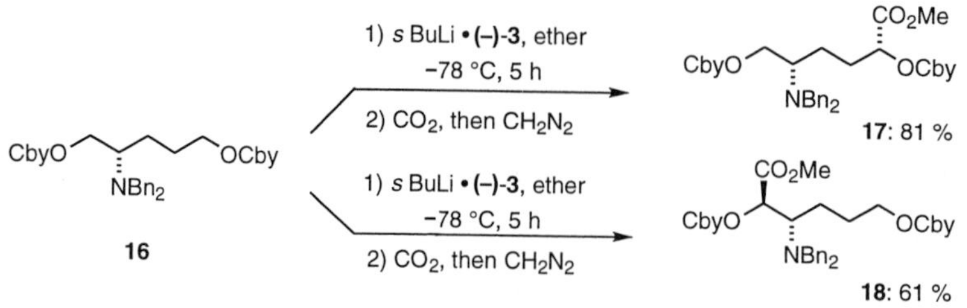

Eq. (9).

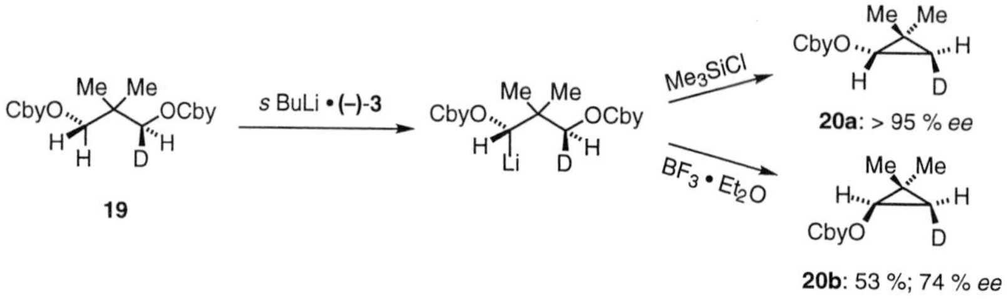

Eq. (10).

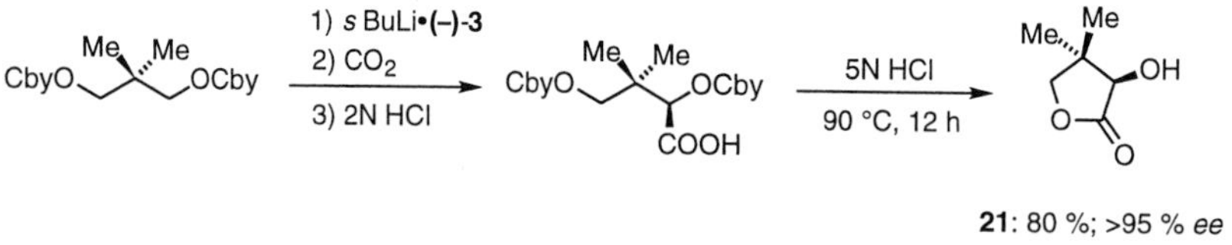

Eq. (11).

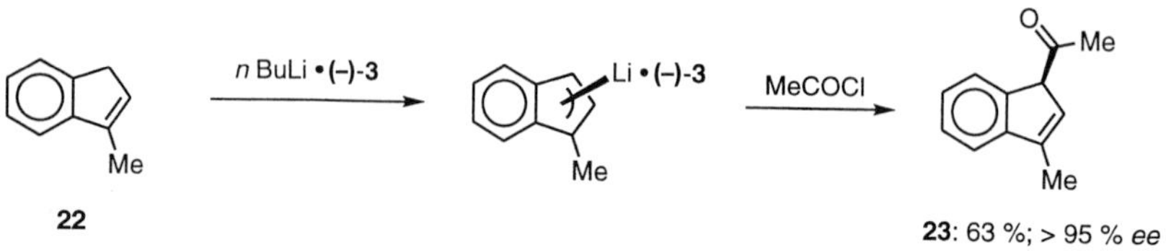

Eq. (12).

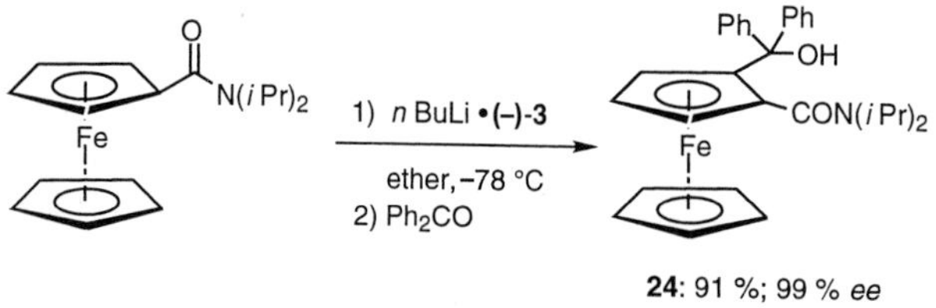

Eq. (13).

of the Lewis acid added, the cyclization occurs with retention or inversion of the carbon-lithium leading to the cyclopropanes **20a-b** (Eq. 10). [18]

Finally, the utility of the method has been demonstrated by Hoppe with a short enantioselective synthesis of (*R*)-pentolactone **21** (Eq. 11). [9]

Remarkably, the enantioselective deprotonation with (−)-3 can be extended to various systems. Thus, substituted indenes like **22** can be stereoselectively deprotonated leading to chiral allyllithiums which after reaction with electrophiles furnish chiral 1,3-disubstituted indenes such as **23** with excellent enantioselectivity (Eq. 12). [19]

25   **26**   **27**: 55–76 %; 88–99 % *ee*   *Eq. (14).*

The lithiation of ferrocenyl amides with *n* BuLi · (–)-**3** proceeds with high enantioselectivity providing after quenching with an electrophile useful new chiral ferrocene derivatives like **24** (Eq. 13). [20]

An asymmetric deprotonation of (*tert*-butoxycarbonyl)pyrrolidines **25** has been recently developed by Beak allowing a highly enantioselective synthesis of a variety of 2-substituted pyrrolidines (Eq. 14). [21] Also aminosubstituted benzylic lithiums have been generated with *s* BuLi·(–)-**3** and quenched stereoselectively with electrophiles. [22]

The enantioselective deprotonation at the *a* position to oxygen- or nitrogen-substituted carbamates with RLi·sparteine complexes allows the most practical and efficient synthesis of the corresponding chiral organolithium compounds. The high enantioselectivity of the deprotonation, the high stereoselectivity observed in the course of the reaction of these species with electrophiles, as well as the compatibility with some donating functional groups in the lithium organometallic compound make this methodology a very powerful tool for the construction of nonracemic chiral molecules.

References

[1] For recent advances in this field see: a) G. Solladié in *Asymmetric Synthesis*, Vol 2 (Ed.: J. D. Morrison), Academic Press, New York, **1983**, pp. 157–200; b) D. A. Evans, *Sience* **1988**, *240*, 420–426; c) R. Noyori, M. Kitamura, *Angew. Chem.* **1991**, *103*, 34–35; *Angew. Chem. Int. Ed. Engl.* **1991**, 30, 49–69; d) B. Schmidt, D. Seebach, *ibid.* **1991**, *103*, 1383–1385 and **1991**, *30*, 1321–1323.

[2] D. Seebach, *Angew. Chem.* **1979**, *91*, 259–278; *Angew. Chem. Int. Ed. Engl.* **1979**, *18*, 239–258.

[3] a) H. Ahlbrecht, G. Boche, K. Harms, M. Marsch, H. Sommer, *Chem. Ber.* **1990**, *123*, 1853–1858; b) G. Boche, A. Opel, M. Marsch, K. Harms, F. Haller, J.C.W. Lohrenz, C. Thümmel, W. Koch, *ibid.* **1992**, *125*, 2265–2273.

[4] a) W. C. Still, C. Sreekumar, *J. Am. Chem. Soc.* **1980**, *102*, 1201–1202; b) V. J. Jephcote, A. J. Pratt, E. J. Thomas, *J. Chem. Soc. Chem. Commun.* **1984**, 800–802; *J. Chem. Soc. Perkin Trans.1* **1989**, 1529–1535; c) J. M. Chong, E. K. Mar, *Tetrahedron* **1989**, *45*, 7709–7716; d) D. S. Matteson, P. B. Tripathy, A. Sarkar, K. N. Sadhu, *J. Am. Chem. Soc.* **1989**, *111*, 4399–4402; e) R. J. Linderman, A. Ghannam, *ibid.* **1990**, *112*, 2392–2398; f) J. A. Marshall, W. Y. Gung, *Tetrahedron* **1989**, *45*, 1043–1052.

[5] For an excellent review see: H. Waldmann, *Nachr. Chem. Tech. Lab.* **1991**, *39*, 413–418.

[6] a) D. Hoppe, O. Zschage, *Angew. Chem.* **1989**, *101*, 67–69; *Angew. Chem. Int. Ed. Engl.* **1989**, *28*, 65–67; b) O. Zschage, J.-R. Schwark, D. Hoppe, *ibid.* **1990**, *102*, 336–337 and **1990**, *29*, 296; c) O. Zschage, D. Hoppe, *Tetrahedron* **1992**, *48*, 5657–5666; d) H. Paulsen, D. Hoppe, *ibid.* **1992**, *48*, 5667–5670; e) O. Zschage, J.-R. Schwark, T. Krämer; D. Hoppe, *ibid.* **1992**, *48*, 8377–8388; f) D. Hoppe; F. Hintze; P. Tebben; M. Paetow; H. Ahrens; J. Schwerdtfeger; P. Sommerfeld; J. Haller; W. Guarnieri; S. Kolczewski, T. Hense, I. Hoppe *Pure Appl. Chem.* **1994**, *66*, 1479–1486; g) D. Hoppe; H. Ahrens; W. Guarnieri; H. Helmke; S. Kolczewski *Pure Appl. Chem.* **1996**, B 613; h) D. Hoppe, T. Hense, *Angew. Chem.* **1997**, *109*, 2376; *Angew. Chem. Int. Ed. Engl.* **1997**, *36*, 2282–2316.

[7] D. Hoppe, F. Hintze, P. Tebben, *Angew. Chem.* **1990**, *102*, 1457–1459; *Angew. Chem. Int. Ed. Engl.* **1990**, *29*, 1422–1424.

[8] J. Schwerdtfeger, D. Hoppe, *Angew. Chem.* **1992**, *104*, 1547–1549; *Angew. Chem. Int. Ed. Engl.* **1992**, *31*, 1505–1507.

[9] M. Peatow, H. Ahrens, D. Hoppe, *Tetrahedron Lett.* **1992**, *33*, 5323–5326.

[10] H. Ahrens, M. Peatow, D. Hoppe, *Tetrahedron Lett.* **1992**, *33*, 5327–5330.

[11] P. Sommerfeld, D. Hoppe, *Synlett* **1992**, 764–766.

[12] F. Hintze, D. Hoppe, *Synthesis* **1992**, 1216–1218.

[13] (–)-Sparteine has been used as additive in several types of asymmetric reactions with moderate success; see, for example, H. Nozaki, T. Aratani, T. Toraya, R. Noyori, *Tetrahedron* **1971**, *27*, 905–913.

[14] M. Marsch, K. Harms, O. Zschage, D. Hoppe, G. Boche, *Angew. Chem.* **1991**, *103*, 338–339; *Angew. Chem. Int. Ed. Engl.* **1991**, *30*, 321–323.

[15] H.-U. Würthwein, unpublished results.

[16] D. J. Gallagher, S. T. Kerrick, P. Beak, *J. Am. Chem. Soc.* **1991**, *114*, 5872–5873.

[17] W. Guarnieri; M. Grehl; D. Hoppe *Angew. Chem.* **1994**, *106*, 1815–1818; *Angew. Chem. Int. Ed. Engl.* **1994**, *33*, 1734–1737.

[18] M. Paetow; M. Kotthaus; M. Grehl; R. Fröhlich; D. Hoppe *Synlett* **1994**, 1034–1036.

[19] I. Hoppe; M. Marsch; K. Harms; G. Boche; D. Hoppe *Angew. Chem.* **1995**, *107*, 2328–2330; *Angew. Chem. Int. Ed. Engl.* **1995**, *34*, 2158–2160.

[20] M. Tsukazi; M. Trinkl; A. Rogbuo; B. J. Chapell; N. J. Taylor; V. Snieckus *J. Am. Chem. Soc.* **1996**, *118*, 685–686.

[21] a) S.T. Kerrick, P. Beak, *J. Am. Chem. Soc.* **1991**, *113*, 9708–9710; b) D. J. Gallagher, S. Wu, N. A. Nikolic, P. Beak *J. Org. Chem.* **1995**, *60*, 8148–8154; c) D. J. Gallagher, P. Beak *J. Org. Chem.* **1995**, *60*, 7092–7093.

[22] A. Basu; D. J. Gallagher; P. Beak *J. Org. Chem.* **1996**, *61*, 5718–5719.

Planar-Chiral Ferrocenes: Synthetic Methods and Applications

Antonio Togni

Since its discovery, and because of its stability and the wealth of methods for its derivatization, ferrocene has played an important role in many areas of synthetic and materials chemistry. Thus ferrocene has been incorporated into more complex structures, displaying, e. g., nonlinear optical, liquid crystalline, or ferromagnetic properties [1]. From a stereochemical point of view, however, one of the most important attributes of, e. g., 1,2- or 1,3-disubstituted ferrocenes, such as **1** and **2**, is that they are planar-chiral, and thus potentially available in enantiomerically pure form [2].

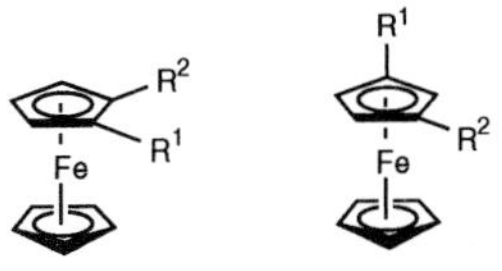

1　**2**

So far, the most relevant method for the preparation of enantiopure (or enantiomerically enriched) planar-chiral ferrocene derivatives relies upon a diastereoselective *ortho*-lithiation and subsequent reaction with an appropriate electrophile. Still the most prominent example of this methodology is due to the pioneering work of Ugi and co-workers (Scheme 1) [3].

Thus, lithiation of ferrocenylamine *(R)*-**3** with BuLi generates the two possible diastereoisomers of **4** (for simplicity formulated here as monomeric, non-solvated species) in a ratio of 96 to 4. Following the reaction with the electrophile, the diastereoisomeric side-product is usually easily separated by crystallization and/or chromatography. This strategy, involving stereogenic *ortho*-directing groups, has been further developed in more recent years by several research groups. Thus, sulfoxides (**6**) [4], acetals (**7**) [5], and oxazolines (**8**) [6] have been found to afford high diastereoselectivities. These functional groups, however, are not necessarily those

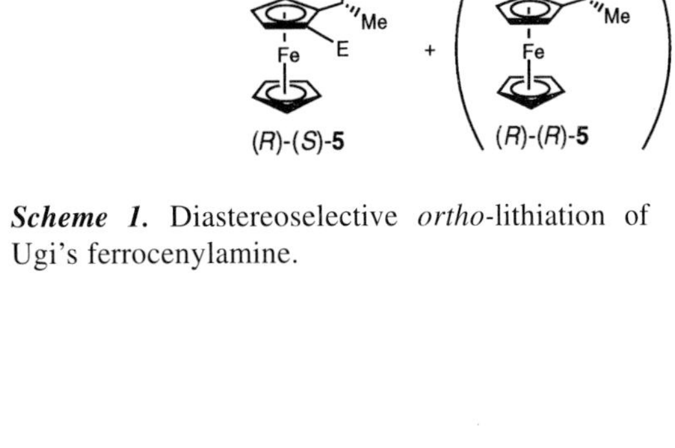

(R)-**3**　　*(R)*-*(R)*-**4**, 96%　　*(R)*-*(S)*-**4**, 4%

(R)-*(S)*-**5**　　*(R)*-*(R)*-**5**

Scheme 1. Diastereoselective *ortho*-lithiation of Ugi's ferrocenylamine.

one may want to incorporate into the final planar-chiral product, i. e., they may need removal or modification. An exception to this observation is the oxazoline fragment in compound **8** which has been demonstrated to fulfil both the purpose of a diastereoselective *ortho*-directing functionality and of a ligand for transition-metals in homogeneous catalyzed reactions.

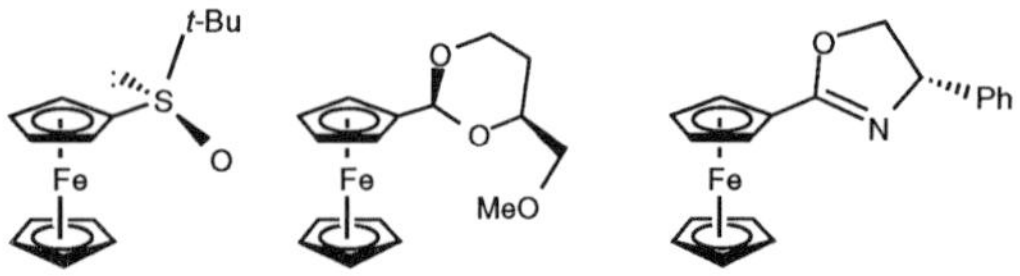

6   7   8

Since amines of type **3** (or the corresponding alcohols as synthetic equivalents) still remain the most versatile starting materials for diastereoselective lithiation processes, and because the amino group can easily be substituted in a process occuring with retention of configuration [3a], methods have been developed for their enantioselective preparation. Recent work includes the aminoalcohol-catalyzed addition of dialkylzinc reagents to ferrocene-carbaldehyde [7], and the 1,2-addition of organolithium compounds to the corresponding SAMP-hydrazone [8], as illustratetd in Scheme 2. Further methods involve the enantioselective reduction of ferrocenylketones using chiral oxazaborolidine borane [9] and lipase mediated desymmetrization of acetoxymethyl derivatives [10]. All these procedures make use of chiral auxiliaries (or catalysts) for the generation of the α-stereogenic center bearing a hetero-atom functionality. Among these methods those starting from ferrocenylketones are more straightforward and do not involve, e. g., cumbersome operations for the removal of chiral auxiliaries, or complex work-up and separation procedures. Therefore, they are anticipated to become even more significant in the near future.

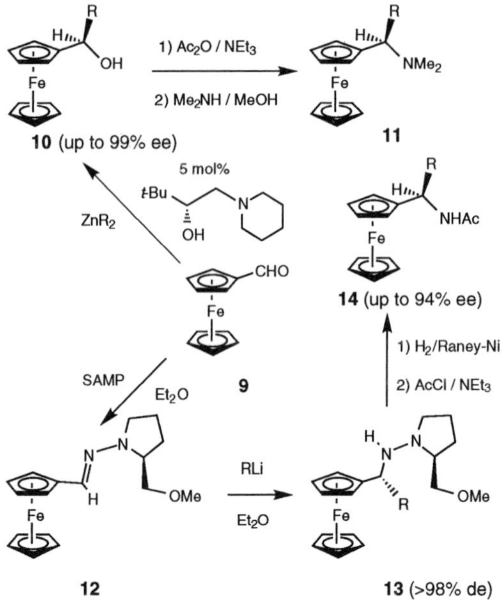

Scheme 2. New syntheses of chiral ferrocenylamines.

A completely different approach for the preparation of planar-chiral ferrocenes, representing an important development in this area was reported recently by Snieckus and co-workers [11]. Indeed it was shown that the *enantioselective ortho*-lithiation of ferrocenylamides of type **14** is effectively achieved utilizing the alkaloid (–)-sparteine as a chiral inducing agent (see example in Scheme 3). The novelty of the method is constituted by the fact that for the first time it is possible to generate planar-chiral ferrocenes highly selectively and without the need of introducing a stereogenic directing group. The application of the reagent BuLi/sparteine in ferrocene chemistry is a welcome extension of the concept of enantioselective deprotonation developed *inter alia* by Hoppe and co-workers in recent years [12]. Despite this impressive result (the phosphine **15** can be obtained in 90% *ee* and in 82% yield, whereas other electrophiles, e. g. TMSCl or benzophenone afford enantioselectivities up to 98% and

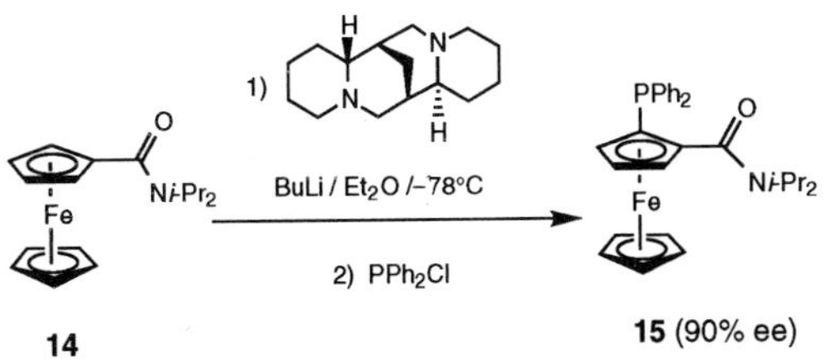

Scheme 3. Synthesis of a planar-chiral ferrocenyl-phosphine using sparteine-mediated enantioselective *ortho*-lithiation.

99 % *ee*, respectively), it remains to be seen whether or not the method may be applied so successfully also to ferrocenes bearing substituents other than amides. Furthermore, will there be a chance to overcome the disadvantage of the single enantiomeric form of (–)sparteine? [13]

Probably the most important class of planar-chiral ferrocenes accessible by the techniques delineated above is that of chelating ligands for transition-metal-catalyzed asymmetric reactions [1]. Progress in this area is characterized, among others, by ligands of type **16** [14], **17** [15], and **18** [16].

The C_2-symmetric biferrocene **16** has been reported by Ito and co-workers to behave as a *trans*-spanning chelating ligand. The combination of C_2-symmetry and *trans* coordination geometry is conceptually new in the field and bears promises for future development. TRAP (the abbreviation for this kind of compounds) affords high enantioselectivities in the Rh-catalyzed asymmetric Michael reaction of 2-cyanopropionates [17], the hydrosilylation of simple ketones [18], the hydrogenation of β,β-disubstituted *N*-acetylaminoacrylic acid

derivatives [19] 9 as well as the cycloisomerization of 1,6-enynes [20].

The *P,N*-ligand **17** represents a class of chiral auxiliaries that are amenable to easy steric and electronic tuning, an aspect that is anticipated to play an important role in the further development of asymmetric catalysis. Thus, ligands of this type have been shown to afford the highest so far reported enantioselectivities in the Rh-catalyzed hydroboration of styrenes with catecholborane (up to 98.5 % *ee*) [15b] and in the Pd-catalyzed substitution of allylic acetates and carbonates with benzylamine (up to > 99 % *ee*) [15c]. The advantage offered by these ligands is the demonstrated facility of their synthetic modification, relying upon the simple idea of a construction kit [21]. The latter is constituted by 1) the carrier of the chiral information (the amine **3**), 2) a set of chlorophosphines (for the generation of the planar-chiral phosphine-amine intermediate), and 3) a series of substituted pyrazoles to be used as nucleophiles. The assembly of these three fragment occurs in two consecutive synthetic steps, and allows the ligand to be adapted to the need of a particular reaction in an unprecedented way.

The same type of considerations applies to diphosphines of the *Josiphos* type such as **18** [16]. This specific compound, containing the bulky *t*-Bu$_2$P group, has been found to be the only ligand affording a very high diastereoface selectivity in the hydrogenation of derivative **19**, an intermediate in a new synthesis of (+)-biotin (Scheme 4A) [22]. The analogous achiral starting material containing a benzyl protecting group, instead of a stereogenic phenethyl, affords under the same reaction

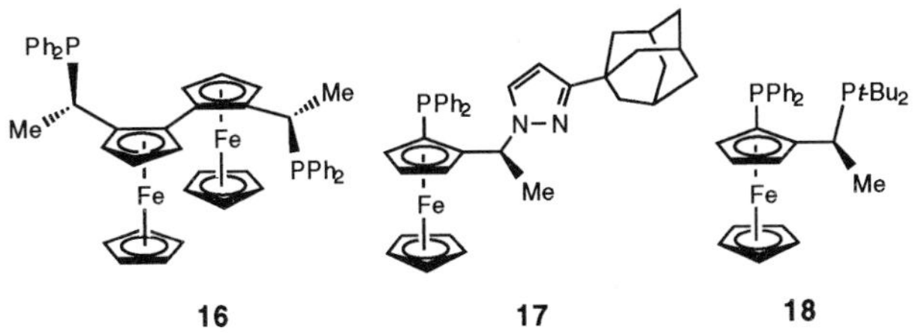

16   **17**   **18**

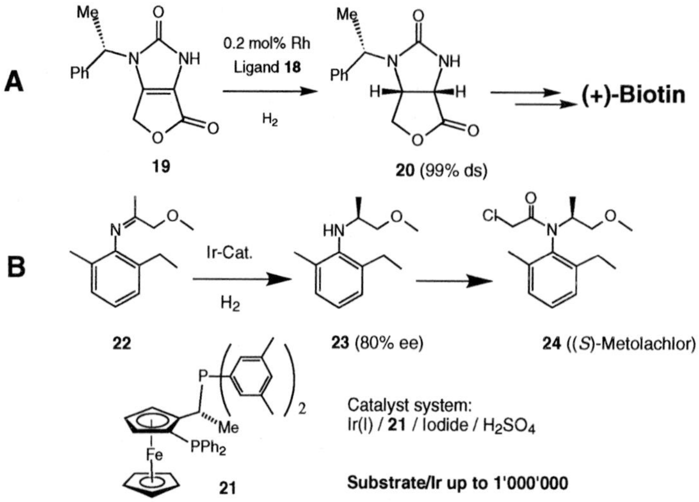

Scheme 4. Industrial applications of chiral ferrocenyl phosphines in the asymmetric hydrogenation of C=C and C=N double bonds.

conditions 90 % *ee*. This quite remarkable process, involving the hydrogenation of a fully substituted C=C bond is currently being applied on a commercial scale by LONZA Ltd.

A second and probably more important technical application of this type of ligands has recently been disclosed by Ciba-Geigy Ltd (Novartis Ltd.). Derivative **21** constitutes the chiral component of the exceptionally highly active iridium catalyst used for the enantioselective hydrogenation of imine **22**, an intermediate in the synthesis of the herbicide (*S*)-Metolachlor® (**24**) [23] (Scheme 4B). The relatively low selectivity obtained in this catalytic reaction (80 % *ee*) is tolerable for an agrochemical product. Much more important in this specific case are the activity and productivity of the catalyst. Besides a decisive co-catalytic effect of added acids (the exact role of the added iodide salt and of the acid are still unclear), one of the key feature of this large industrial process is the use of ferrocenyl ligand **21** which turns out to be superior to all commercially available chiral chelating diphosphines, with respect to a com-

bination of several important aspects. To our knowledge, the latter two examples constitute the first industrial applications of chiral ferrocenyl ligands and are clear demonstrations of the potential this class of compounds bear. Any improvement of their synthetic versatility will mirror not only their fundamental interest, but shall be welcomed also by the practitioner in the fine-chemical industry who is willing to extend the scope of homogeneous catalysis as a powerful synthetic method.

Finally, a completely new use of planar-chiral ferrocenes has been recently disclosed by Fu and co-workers [24]. Compounds of type **25** and **26** were prepared as racemic mixtures and obtained as pure enantiomers via semi-preparative HPLC. Derivatives **25**, analogues of 4-(dimethylamino)pyridine, were used as nucleophilic catalysts in the kinetic resolution of chiral secondary alcohols [24a,b]. The aminoalcohol system **26**, on the other hand, is an effective chiral ligand for the asymmetric addition of dialkylzinc reagents to aldehydes (up to 90 % *ee*) [24c].

In conclusion, this brief account summarizing recent development in the field of pla-

25   **26** (R=Me, Ph)

nar-chiral ferrocenes shows the multiplicity of activities in this area and its relevance for synthetic and organometallic chemistry at large.

References

[1] For a recent overview, see: *Ferrocenes. Homogeneous Catalysis. Organic Synthesis. Materials Science* (Eds.: A. Togni, T. Hayashi), VCH, **1995**.

[2] For a discussion of ferrocene chemistry from a stereochemical point of view, see the classical: K. Schlögl, *Top. Stereochem.* **1967**, *1*, 39.

[3] a) D. Marquarding, H. Klusacek, G. W. Gokel, P. Hoffmann, I. K. Ugi, *J. Am. Chem. Soc.* **1970**, *92*, 5389–5393; b) G. W. Gokel, I. K. Ugi, *J. Chem. Educ.* **1972**, *49*, 294–296.

[4] a) F. Rebière, O. Riant, L. Ricard, H. B. Kagan, *Angew. Chem.* **1993**, *105*, 644–646 (*Angew. Chem. Int. Ed. Engl.* **1993**, *32*, 568–570); b) H. B. Kagan, P. Diter, A. Gref, D. Guillaneux, A. Masson-Szymczak, F. Rebière, O. Riant, O. Samuel, S. Taudien, *Pure & Appl. Chem.* **1996**, *68*, 29–36.

[5] O. Riant, O. Samuel, H. B. Kagan, *J. Am. Chem. Soc.* **1993**, *115*, 5835–5836; b) A. Masson-Szymczak, O. Riant, A. Gref, H. B. Kagan, *J. Organomet. Chem.* **1996**, *511*, 193–197.

[6] a) Y. Nishibayashi, K. Segawa, K. Ohe, S. Uemura, *Organometallics* **1995**, *14*, 5486–5487; b) W. Zhang, Y. Adachi, T. Hirao, I. Ikeda, *Tetrahedron: Asymmetry* **1996**, *7*, 451–460; c) T. Sammakia, H. A. Latham, *J. Org. Chem.* **1996**, *61*, 1629–1635; d) J. Park, S. Lee, K. H. Ahn, C.-W. Cho, *Tetrahedron Lett.* **1996**, *37*, 6137–6140; e) Y. Nishibayashi, S. Uemura, *Synlett* **1995**, 79–81; f) T. Sammakia, H. A. Latham, D. R. Schaad, *J. Org.*

Chem. **1995**, *60*, 10–11; g) W. Zhang, T. Hirao, I. Ikeda, *Tetrahedron Lett.* **1996**, *37*, 4545–4548; h) C. J. Richards, T. Damalidis, D. E. Hibbs, M. B. Hursthouse, *Synlett* **1995**, 74–76; i) C. J. Richards, A. W. Mulvaney, *Tetrahedron: Asymmetry* **1996**, *7*, 1419–1430; j) K-H. Ahn, C.-W. Cho, H.-H. Baek, J. Park, S. Lee, *J. Org. Chem.* **1996**, *61*, 4937–4943.

[7] a) Y. Matsumoto, A. Ohno, S. Lu, T. Hayashi, N. Oguni, M. Hayashi, *Tetrahedron: Asymmetry* **1993**, *4*, 1763–1766. See also: b) L. Schwink, S. Vettel, P. Knochel, *Organometallics* **1995**, *14*, 5000–5001.

[8] D. Enders, R. Lochtman, G. Raabe, *Synlett* **1996**, 126–128. For a second approach utilizing SAMP-derivatives, see: C. Ganter, T. Wagner, *Chem. Ber.* **1995**, *128*, 1157–1161.

[9] a) A. Ohno, M. Yamane, T. Hayashi, N. Oguni, M. Hayashi, *Tetrahedron: Asymmetry* **1995**, *6*, 2495–2503; b) L. Schwink, P. Knochel, *Tetrahedron Lett.* **1996**, *37*, 25–28.

[10] See, e. g.: G. Nicolosi, R. Morrone, A. Patti, M. Piattelli, *Tetrahedron: Asymmetry* **1992**, *3*, 753–758.

[11] M. Tsukazaki, M. Tinkl, A. Roglans, B. J. Chapell, N. J. Taylor, V. Snieckus, *J. Am. Chem. Soc.* **1996**, *118*, 685–686.

[12] D. Hoppe, F. Hintze, P. Tebben, M. Paetow, H. Ahrens, J. Schwerdtfeger, P. Sommerfeld, J. Haller, W. Guarnieri, S. Kolczewski, T. Hense, I. Hoppe, *Pure Appl. Chem.* **1994**, *66*, 1479–1486, and refs. cited therein.

[13] Chiral amines of which both enantiomers are readily available have recently been used in analogous reactions. However, they afforded significantly lower enantioselectivities. See: a) D. Price, N. S. Simpkins, *Tetrahedron Lett.* **1995**, *36*, 6135–6138; b) Y. Nishibayashi, Y. Arikawa, K. Ohe, S. Uemura, *J. Org. Chem.* **1996**, *61*, 1172–1174.

[14] a) M. Sawamura, H. Hamashima, Y. Ito, *Tetrahedron: Asymmetry* **1991**, *2*, 593–596; b) M. Sawamura, H. Hamashima, M. Sugawara, R. Kuwano, Y. Ito, *Organometallics* **1995**, *14*, 4549–4558.

[15] a) U. Burckhardt, L. Hintermann, A. Schnyder, A. Togni, *Organometallics* **1995**, *14*, 5415–5425; b) A. Schnyder, L. Hintermann, A. Togni, *Angew. Chem.* **1995**, *107*, 996–998

(*Angew. Chem. Int. Ed. Engl.* **1995**, *34*, 931–933); c) A. Togni, U. Burckhardt, V. Gramlich, P. S. Pregosin, R. Salzmann, *J. Am. Chem. Soc.* **1996**, *118*, 1031–1037.

[16] a) A. Togni, C. Breutel, A. Schnyder, F. Spindler, H. Landert, A. Tijani, *J. Am. Chem. Soc.* **1994**, *116*, 4062–4066; b) H. C. L. Abbenhuis, U. Burckhardt, V. Gramlich, C. Köllner, R. Salzmann, P. S. Pregosin, A. Togni, *Organometallics* **1995**, *14*, 759–766.

[17] M. Sawamura, H. Hamashima, Y. Ito, *Tetrahedron* **1994**, *50*, 4439–4454.

[18] M. Sawamura, R. Kuwano, Y. Ito, *Angew. Chem.* **1994**, *106*, 92–94 (*Angew. Chem. Int. Ed. Engl.* **1994**, *33*, 111–113).

[19] R. Kuwano, M. Sawamura, Y. Ito, *Tetrahedron: Asymmetry* **1995**, *6*, 2521–2526.

[20] A. Goeke, M. Sawamura, R. Kuwano, Y. Ito, *Angew. Chem.* **1996**, *108*, 686–687 (*Angew. Chem. Int. Ed. Engl.* **1996**, *35*, 662–663).

[21] For a discussion of this aspect, see: A. Togni, *Chimia* **1996**, *50*, 86.

[22] J. McGarrity, F. Spindler, R. Fuchs, M. Eyer, Eur. Pat. Appl. EP 624 587 A2, (*LONZA AG*), *Chem. Abstr.* **1995**, *122*, P81369q.

[23] For an account about the successful development of this process, see: F. Spindler, B. Pugin, H.-P. Jalett, H.-P. Buser, U. Pittelkow, H.-U. Blaser. In: Catalysis of Organic Reactions, R. E. Malz, Jr., Ed. (Chem. Ind. Vol. *68*), Dekker, New York, **1996**, pp. 153–166.

[24] a) J. C. Ruble, G. C. Fu, *J. Org. Chem.* **1996**, *61*, 7230–7231; b) J. C. Ruble, H. A. Latham, G. C. Fu, *J. Am. Chem. Soc.* **1997**, *119*, 1492–1493; c) P. I. Dosa, J. C. Ruble, G. C. Fu, *J. Org. Chem.* **1997**, *62*, 444–445.

Asymmetric Autocatalysis with Amplification of Chirality*

Carsten Bolm, Andreas Seger, and Frank Bienewald

Asymmetric metal catalysis has been intensively studied in recent years, and some efficient methods for the synthesis of enantiomerically pure compounds have been developed. [1] With suitable metal/ligand combinations excellent enantioselectivities and high conversion rates have frequently been achieved. Normally the catalyst must remain unaffected by the continually formed new product in order to achieve a constant, high stereoselectivity. How does it behave, however, when the product itself is a catalyst and, moreover, catalyzes its own asymmetric synthesis? Wynberg recognized very early the great potential of such "asymmetric autocatalysis" for synthesis and as early as 1989 had formulated the challenge that asymmetric autocatalysis could constitute the next generation of asymmetric synthesis. [2] In spite of great efforts the discovery of the first autocatalytic system operating with high enantioselectivity has only recently been made. [3] Important contributions from Soai et al. [3–5] take pride of

place here, but before introducing them we will describe some other fundamental studies.

Seebach, Dunitz et al. first drew attention to the fact that in stereoselective reactions with organometallic reagents differing diastereo- and enantioselectivities could arise because of the formation of various mixed complexes during the course of product formation. [6, 7] This led Alberts and Wynberg to investigate the asymmetric additions of organometallic carbon nucleophiles (e.g. ethyllithium) to benzaldehyde (Scheme 1), and they showed that the stereochemical course of both stoichiometric and catalyzed reactions were influenced by metal-containing product molecules (here lithium alcoholates). [8] The product itself is not a catalyst but nevertheless ensures that the newly formed product is optically active. Alberts and Wynberg coined the term "enantioselective autoinduction" for this effect.

$$\text{PhCHO} \xrightarrow[\text{2. H}_3\text{O}^+]{\text{1. (+)-PhC*D(OLi)Et / EtLi}} \begin{array}{c} \text{PhC*H(OH)Et} \\ \text{(+ PhC*D(OH)Et)} \end{array}$$

Scheme 1. Asymmetric addition of ethyllithium to benzaldehyde. [8] Chirality centers are identified by a star.

Danda et al. established that the product powerfully influenced the catalyst in the *metal-free* catalyzed asymmetric hydrocyana-

* We thank the DFG (Graduiertenkolleg at the Philipps-Universität Marburg), BASF AG, and the Volkswagen-Stiftung for stipends and financial support. We are indebted to Professor H. Wynberg (Groningen, Netherlands) for sending the dissertation by W. M. P. B. Menge.

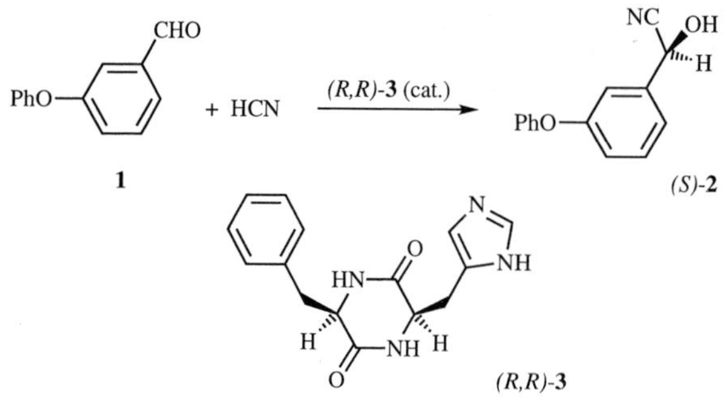

1 + HCN $\xrightarrow{(R,R)\text{-}3 \text{ (cat.)}}$ **(S)-2**

(R,R)-3

tion of 3-phenoxybenzaldehyde **1** to (*S*)-**2** by the cyclic dipeptide (*R,R*)-**3**. [9]

In the presence of the product the reaction-accelerating effect of the cyclic dipeptide was enhanced. (*R,R*)-**3** alone showed little catalytic activity at first, and with increasing reaction time the enantiomeric excess of (*S*)-**2** increased. If a small amount of (*S*)-**2** was added to the reaction mixture at the start the product was formed with an almost constant, high *ee* value. Here (*S*)-**2** itself is *not* a catalyst and only the interaction between the correct product enantiomer and the cyclic dipeptide leads to better catalysis. In addition asymmetric amplification [10] occurred only in the presence of the product. Thus, 2.2 mol% (*R,R*)-**3** with 2 % *ee* at 43 % conversion yields the cyanohydrin (*S*)-**2** with 81.6 % *ee* when 8.8 mol% (*S*)-**2** with 92 % *ee* is added at the start of the reaction. If (*S*)-**2** is not added at the start, the conversion and enantiomeric excess are 4 % and 3.4 %, respectively. It is also interesting that in this asymmetric amplification the absolute configuration of the product is not determined by the configuration of the catalyst but by that of the added cyanohydrin.

In the examples quoted the product influences the stereoselectivity or has a positive effect on an existing catalysis. [11] Things become especially interesting, however, when the chiral product itself is a catalyst for its own formation from achiral precursors. This area of asymmetric autocatalysis was hardly de-

veloped for a long time, though in 1953 Frank had already formulated a mathematical model that indicated that "spontaneous asymmetric synthesis", as he called the process, is quite possible. [12] As a natural property of life this could be of fundamental importance in the genesis of asymmetry in nature. [13–15] In the scenario described by Frank two achiral substances, **A** and **B**, react to form optically active products (*R*)-**C** and (*S*)-**C** each of which can catalyze its own synthesis (Scheme 2a). This corresponds to the working of conventional autocatalysis. If one imagines that the two enantiomeric catalysts mutually reduce (or destroy) their effects (Scheme 2b), the system would show flip-flop switching properties: even a small statistical fluctuation (for example, if the reaction catalyzed by (*R*)-**C** were preferred for a short time) would have the effect of reducing the catalytic activity of (*S*)-**C** and new (*R*)-**C** would be formed even more quickly. In such an "aggressive"

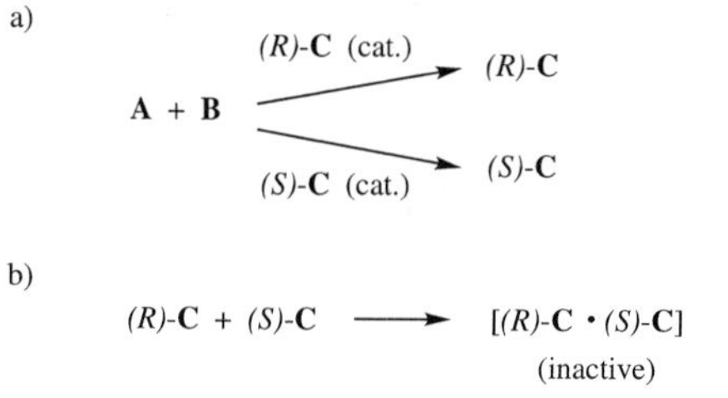

a)

$$ \mathbf{A + B} \begin{cases} \xrightarrow{(R)\text{-}\mathbf{C} \text{ (cat.)}} (R)\text{-}\mathbf{C} \\ \xrightarrow{(S)\text{-}\mathbf{C} \text{ (cat.)}} (S)\text{-}\mathbf{C} \end{cases} $$

b)

$$ (R)\text{-}\mathbf{C} + (S)\text{-}\mathbf{C} \longrightarrow [(R)\text{-}\mathbf{C} \cdot (S)\text{-}\mathbf{C}] $$
(inactive)

Scheme 2. (a) Model for "spontaneous asymmetric synthesis" according to Frank [12]. (b) Deactivation of the catalytically active chiral product.

system a trace of a chiral substance would be enough to ensure its own production in large quantities by autocatalysis. The inductor does not even need to be enantiomerically pure since the model implies the principle of asymmetric amplification. In his article Frank pleaded for the development and testing of simple autocatalytic systems. [12] The first successful reaction with a "chiral autocatalyst" was reported by Soai et al. in 1990. [16] They found that the pyridinyl alcohol **6** catalyzed its own formation from pyridine-3-carbaldehyde (**4**) and diisopropylzinc via the isopropylzinc alcoholate **5**. With 20 mol% (–)-**6** (86 % *ee*), workup afforded (–)-**6** in 67 % yield with an *ee* of 35 %.

(94.8 % *ee*) in the reaction between the corresponding aldehyde **7a** and diisopropylzinc led to the formation of (*S*)-**9a** in 48 % yield with an *ee* of 95.7 %. The product underwent automultiplication *without* a significant change in the enantiometric excess.

As if this were not enough, Soai et al. [4] also used this system to demonstrate for the first time the asymmetric amplification during autocatalysis that is inherent in the Frank model. Thus 20 mol% of (*S*)-**9b** with an *ee* of only 2 % gave, with autocatalysis, (*S*)-**9b** with an *ee* of 10 %. In further reaction cycles the enantiomeric excess rose from 10 through 57 to 81 and finally to 88 % (Fig. 1). There was a 942-fold increase in the amount of product after four cycles. The asymmetric amplification shown by this simple selfreplicating system behaved, in fact, as predicted by the simple, theoretical Frank model. [18]

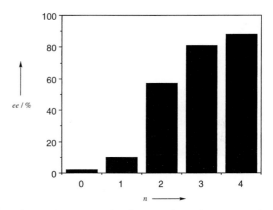

Similarly, in other enantioselective additions of organozinc compounds asymmetric autocatalyses were also demonstrated, [17] though they always gave a product with a much lower *ee* value than that of the catalyst used. Only at the end of 1995 was an important breakthrough achieved [3, 4] when Soai et al. showed that the presence of 20 mol% of the pyrimidyl alcohol (*S*)-**9a**

Figure 1. Increase in the enantiomeric excess of (*S*)-**9b** with the number of reaction cycles *n*.

As the mechanism of the reaction and possible intermediates in solution are still unknown, the phenomenon described here cannot yet be used for the targeted development of further reactions. It is, nevertheless, clear that complex formation with the organometallic reagents is essential to autocatalysis with asymmetric amplification. For this reason a deeper understanding of the behavior of chiral compounds during the formation of complexes would be very useful for the design and discovery of new autocatalytic processes.

References

[1] a) R. Noyori, *Asymmetric Catalysis in Organic Synthesis*, Wiley, New York, **1994**; b) *Catalytic Asymmetric Synthesis* (Ed.: J. Ojima), VCH, Weinheim, **1993**.

[2] a) H. Wynberg, *Chimia* **1989**, *43*, 150; b) *J. Macromol. Sci. Chem. A* **1989**, *26*, 1033; c) W. M. P. B. Menge, Dissertation, University of Groningen 1989.

[3] T. Shibata, H. Morioka, T. Hayase, K. Choji, K. Soai, *J. Am. Chem. Soc.* **1996**, *118*, 471.

[4] K. Soai, T. Shibata, H. Morioka, K. Choji, *Nature* **1995**, *378*, 767.

[5] a) T. Shibata, K. Choji, H. Morioka, T. Hayase, K. Soai, *J. Chem. Soc. Chem. Commun.* **1996**, 751; b) T. Shibata, K. Choji, T. Hayase, Y. Aizu, K. Soai, *J. Chem. Soc. Chem. Commun.* **1996**, 1235.

[6] D. Seebach, R. Amstutz, J. D. Dunitz, *Helv. Chim. Acta* **1981**, *64*, 2622.

[7] Review articles: a) D. Seebach, *Proc. Robert A. Welch Found. Conf. Chem. Res.* **1984**, *27*, 93; b) *Angew. Chem.* **1988**, *100*, 1685; *Angew. Chem. Int. Ed. Engl.* **1988**, *27*, 1624.

[8] a) A. H. Alberts, H. Wynberg, *J. Am. Chem. Soc.* **1989**, *111*, 7265; b) *J. Chem. Soc. Chem. Commun.* **1990**, 453.

[9] a) H. Danda, H. Nishikawa, K. Otaka, *J. Org. Chem.* **1991**, *56*, 6740; b) mechanistic interpretation: Y. Shvo, M. Gal, Y. Becker, A. Elgavi, *Tetrahedron: Asymmetry* **1996**, *7*, 911.

[10] a) R. Noyori, M. Kitamura, *Angew. Chem.* **1991**, *103*, 34; *Angew. Chem. Int. Ed. Engl.*

1991, *30*, 49; b) D. Guillaneux, S.-H. Zhao, O. Samuel, D. Rainford, H. B. Kagan, *J. Am. Chem. Soc.* **1994**, *116*, 9430; c) C. Bolm in *Advanced Asymmetric Synthesis* (Ed.: G. R. Stephenson), Blackie, Glasgow, **1996**, p. 9; d) H. B. Kagan, C. Girard, D. Guillaneux, D. Rainford, O. Samuel, S. Y. Zhang, S. H. Zhao, *Acta Chem. Scand.* **1996**, *50*, 345.

[11] Optically active products can also be formed with racemic metal catalysts. With a chiral additive one of the enantiomeric forms of the catalyst is "poisoned" [11 a–g] or activated [11 h] in situ; a) N. W. Alcock, J. M. Brown, P. J. Maddox, *J. Chem. Soc. Chem. Commun.* **1986**, 1532; b) J. M. Brown, P. J. Maddox, *Chirality* **1991**, *3*, 345; c) K. Maruoka, H. Yamamoto, *J. Am. Chem. Soc.* **1989**, *111*, 789; d) J. W. Faller, J. Parr, *J. Am. Chem. Soc.* **1993**, *115*, 804; e) J. W. Faller, M. Tokunaga, *Tetrahedron Lett.* **1993**, *34*, 7359; f) J. W. Faller, D. W. I. Sams, X. Liu, *J. Am. Chem. Soc.* **1996**, *118*, 1217; g) J. W. Faller, X. Liu, *Tetrahedron Lett.* **1996**, *37*, 3449; h) K. Mikami, S. Matsukawa, *Nature* **1997**, *385*, 613.

[12] a) F. C. Frank, *Biochim. Biophys. Acta* **1953**, *11*, 459. See also: b) I. Gutman, *J. Sci., Islamic Repub. Iran* **1995**, *6*, 231; *Chem. Abstr.* **1996**, *124*, 288491k; c) R. D. Murphy, T. M. El-Agez, *Indian J. Chem. Sect. A: Inorg., Bioinorg., Phys., Theor. Anal. Chem.* **1996**, *35A*, 546; *Chem. Abstr.* **1996**, *125*, 141853s.

[13] a) J. L. Bada, *Nature* **1995**, *374*, 594; b) W. A. Bonner, *Top. Stereochem.* **1988**, *18*, 1; c) W. J. Meiring, *Nature* **1987**, *329*, 712; d) P. Decker, *Nachr. Chem. Tech. Lab.* **1975**, *23*, 167; e) S. Mason, *Chem. Soc. Rev.* **1988**, *17*, 347; f) W. A. Bonner, *Chem. Ind.* **1992**, 640; g) S. Mason, *Nature* **1985**, *314*, 400; g) R. A. Hegstrom, D. K. Kondepudi, *Chem. Phys. Lett.* **1996**, *253*, 322.

[14] Nonsymmetric molecular replication and autocatalyses: a) L. E. Orgel, *Nature* **1992**, *358*, 203; b) E. A. Wintner, M. M. Conn, J. Rebek, Jr., *Acc. Chem. Res.* **1994**, *27*, 198; c) *J. Am. Chem. Soc.* **1994**, *116*, 8877; d) G. von Kiedrowski, J. Helbing, B. Wlotzka, S. Jordan, M. Mathen, T. Achilles, D. Sievers, A. Terfort, B. C. Kahrs, *Nachr. Chem. Tech. Lab.* **1992**, *40*, 578; e) T. Achilles, G. von

Kiedrowski, *Angew. Chem.* **1993**, *105*, 1225; *Angew. Chem. Int. Ed. Engl.* **1993**, *32*, 1198.

[15] Formation of homochiral crystals from solutions of optically inactive compounds: a) J. Jacques, A. Collet, S. H. Wilen, *Enantiomers, Racemates, and Resolutions*, Wiley, New York, 1981; b) D. K. Kondepudi, R. J. Kaufman, N. Singh, *Science* **1990**, *250*, 975; c) J. M. McBride, R. L. Carter, *Angew. Chem.* **1991**, *103*, 298; *Angew. Chem. Int. Ed. Engl.* **1991**, *30*, 293, and references therein.

[16] K. Soai, S. Niwa, H. Hori, *J. Chem. Soc. Chem. Commun.* **1990**, 983.

[17] a) K. Soai, T. Hayase, C. Shimada, K. Isobe, *Tetrahedron: Asymmetry* **1994**, *5*, 789; b) K. Soai, T. Hayase, K. Takai, *Tetrahedron: Asymmetry* **1995**, *6*, 637; c) C. Bolm, G. Schlingloff, K. Harms, *Chem. Ber.* **1992**, *125*, 1191; d) S. Li, Y. Jiang, A. Mi, G. Yang, *J. Chem. Soc. Perkin Trans 1* **1993**, 885; e) K. Soai, Y. Inoue, T. Takahashi, T. Shibata, *Tetrahedron* **1996**, *52*, 13555; f) for a substituted compound: T. Shibata, H. Morioka, S. Tanji, T. Hayase, Y. Kodaka, K. Soai, *Tetrahedron Lett.* **1996**, *37*, 8783.

[18] Asymmetric autocatalysis with amplification of enantiomeric excess has also been found in reactions with a 3-quinolylalkanol [5].

Resolution of Racemates by Distillation with Inclusion Compounds

Gerd Kaupp

Chiral drugs and additives are generally not permitted to be used in racemic form (most important exception: DL-methionine) so that unintended side effects and unnecessary environmental hazards are avoided. Larger amounts of enantiomerically pure compounds are economically produced by fermentation or (complete) enzymatic conversion of racemates into the desired enantiomer. Thus, since 420,000 t of L-glutamic acid and large quantities of L-aspartic acid and L-phenylalanine (for the sweetener aspartame) and D-phenylglycine are produced annually, excessive amounts of waste should and can be prevented. [1]

Absolute asymmetric synthesis, a rapidly developing field, is a simple approach to the synthesis of enantiomerically pure compounds and has low environmental impact; however, enantiomorphic crystals (with a chiral space group) are required, which are not necessarily formed by all compounds with prochiral centers. [2] Enantioselective syntheses with chiral auxiliaries and "without biology" have been fine-tuned most impressively and have a longer tradition, [3] but are still usually very costly and time-consuming. Several approaches to chemical synthesis may be outlined: the synthesis is conducted in chiral media (recently inclusion compounds have been shown to be efficient [4]) or with

chiral catalysts. Alternatively, an optically active starting material is employed in a diastereoselective synthesis, and the portion of the molecule with the initial chiral center is cleaved off afterwards. This can be done with loss of chirality in the auxiliary (e. g. with the Seebach method [3]) or without loss (e. g. with the Schöllkopf method [5]); however, recovery of the auxiliary from small-scale reactions does not appear to be very worthwhile. [3] Upscaling for industrial applications is still problematic. [6] Apparently, resolution of racemates following straightforward synthesis remains the most frequently applied method.

Fractional distillation after synthesis of volatile amides or esters met with little success, due to low differences in boiling points (4–5 K or < 1 K) and laborious transformations. [7] An early distillation approach using different volatilities of diastereomeric salts of various amines with tartaric acid or its dibenzoyl- (also di-p-toluoyl-) derivatives yielded optical purities of the distillates from 5 to 47 % with a better performance in the case of methamphetamine (56.5–66.5 %). [8] However, probably due to the ban or restrictions with this drug (and most further drugs that might be misused) for use in scientific research by many countries, that publication did not find its due recognition. A further reason might be that it did not care for the optical purities of the residues after alkaline extraction.

Fortunately supramolecular chemistry introduced new possibilities for optical resolutions that were not restricted to acids and bases; inclusion compounds were systematically examined, and versatile clathrating agents were designed and employed. [9] These findings have been used since the early 1980s to selectively and reversibly include chiral guest molecules in host lattices of chiral molecules. [4] In contrast to the formation of diastereomeric salts, compounds with almost any functional group can be treated. The enantiomerically pure host and the racemic guest (or the racemic host and the enantiomerically pure guest) are dissolved, the inclusion compound composed of the better fitting set of enantiomers is allowed to precipitate, and the crystalline material is removed by filtration. The two enantiomers of the guest (host) are obtained from the filtrate and from a solution of the crystals after the host (guest) is removed by chromatography on SiO_2, respectively. The enantiomers (atropisomers) of 2,2′-dihydroxy-1,1′-binaphthyl (1) do not interconvert at normal temperatures. Thus, if a chiral compound like (R,R)-2 (or N-benzylcinchonidinium chloride) [10] is dissolved with rac-1, only one of the diastereomeric inclusion compounds crystallizes on account of chiral recognition. (−)-1 (and (+)-1) is obtained with > 99 % *ee* (from benzene) and separated from 2 by chromatography on SiO_2. The enantiomerically pure host

(−)-1 (and (+)-1) can now be used for a variety of difficult chiral resolutions (> 99 % *ee*) by inclusion crystallization, for example the resolution of racemic sulfoxides like 3, sulfoximines like 4, phosphinates/phosphanoxides like 5, and aminoxides like 6 and 7. Hundreds of resolutions of very diverse racemic compounds have been reported with a broad range of hosts. [4]

Recently an improvement in this separation technique was reported, which seemed to indicate that enantioselective inclusion in the lattices of chiral hosts could be employed on a large scale. [11] When crystalline hosts such as (R,R)-(−)-8 (m. p. 196 °C), [12] (R,R)-(−)-9 (m. p. 165 °C), [12] and (S,S)-(−)-10 (m. p. 128 °C) are suspended in hexane or water, chiral guest molecules form the same inclusion compounds as from solution. This is by no means self-evident, since inclusion compounds have different crystal lattices than the pure host crystals. Thus crystal/liquid reactions occur, and phase rebuildings analogous to those observed in gas/solid reactions [13] must take place. Yet this suspension technique is more selective and more effective than the initially developed solution technique. Numerous racemic alcohols like 11, β-hydroxy esters like 12, epoxy esters like 13, and epoxy ketones like 14 were stirred a few hours with appropriate hosts (suspensions of 8, 9, and 10) and formed 1 : 1 complexes that could be filtered off in yields of > 85 % and with *ee* values of > 97 % (the complex of 12 and 9 formed in hexane only; 80 % *ee* in one step). Recrystallization of the inclusion compounds is not generally necessary; (–)-**11**, (+)-**12**, (+)-**13**, and (+)-**14** are released by heating under vacuum. Hosts **8**, **9**, and **10** can then be used again in further resolutions. Improvements have been made with stereo-selective gas-solid inclusions. Thus, Weber used the host (*S*)-**15** for sorptive stereoselec-tive inclusion of **16–19** and achieved mode-rate (*ee*: 8.5–71 %) resolutions. [14] Further improvements were solid-solid inclusions and slurry techniques using stoichiometric amounts of water. [15] The reasons for the success have been mechanistically revealed by AFM investigations of the gas-solid reso-lution of **11** in (*R,R*)-**8** [13, 16] and of the solid-solid resolution of (±)pantolactone in **9** and its homolog (**32**). [17] The resolution pro-cedures are simple, cheap, and environmen-tally safe. Of course, as with other resolutions, the undesired enantiomer must be disposed of, or better yet, converted into a racemic mixture and resolved again.

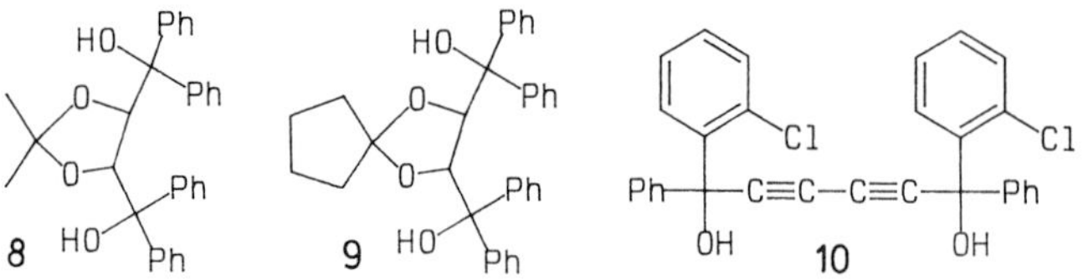

Another advantage of the new technique is the possibility of using a stoichiometric amount of the host relative to the racemate (in other words, in excess). The resolution can then be achieved by *fractional* (Kugel-rohr) *distillation*. For example, 2 mmol of *rac*-**11** and 2 mmol of crystalline **8** in 1 mL

of hexane were stirred for 1 h at room tempe-rature and the mixture subsequently distilled. At approximately 70 °C uncomplexed (+)-**11** was collected with 59 % *ee*; the distillation of the residue provided (–)-**11** in 69 % yield and with 97 % *ee* at roughly 150 °C. In a similar fashion fractional vacuum distillation of *rac*-**14** and **10** gave (–)-**14** (68 % *ee*) and 63 % (+)-**14** with 95 % *ee*. If the process with the enantiomeric host is repeated, both enan-tiomers of the guest can be isolated in pure form. Since the hosts are not consumed and do not racemize, they can be employed in sub-sequent separations. Also the amphetamine resolutions by distillation have been pursued by the Hungaryan scientists using an impro-ved acid [(+)-(*R*)-*N*-(*a*-methylbenzyl)phthalic acid monamide] or inclusion by (*R,R*)-**8**. [18]

The real breakthrough of the Toda method required the above (gas)solid-solid techniques *without any solvent* to be combined with the favorable inclusion properties that allow the non-complexed enantiomer to be distilled off at temperatures far below the melting point of the complex when the antipode is liberated. Thus, chiral recognition leads to *ee* values between 90 and 100 % in many cases by one simple distillation. Sometimes two to four consecutive enantiodiseriminating distillations

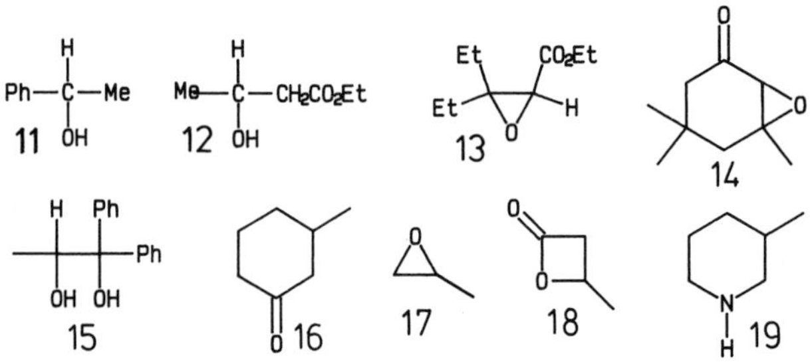

20 21 22 23

24 25 26a: X = 4-CH3 26b: X = 2-OCH3

27 28 29 30

31 32

are necessary for a complete resolution. The non-complexed enantiomer may, of course be enriched by application of the host antipode. For all systems that do not include in present chiral hosts, new ones will have to be developed. However, such endeavors might be highly rewarding, considering the ease of the procedure. At present, about 16 racemates **11–14, 20–30** have been quantitatively resolved by distillation with 5 different chiral hosts **8–10** and the homologs **31–32**. [19]

The hosts are recovered and used again. Undoubtedly, H bonds play an important role in these enantioselections: IR spectra of the complexes showed broad and shifted bands. [19] The diversity of functional groups suggests that many more racemates with other functionalities might be equally resolved. Crystallographic fit is of primary importance.

Upscaling of enantioselective distillations for industrial applications is of particular interest and initial tests cannot be far off. The heat transfer to the solid should not present major difficulties up to the kg scale, and technical solutions should be possible for larger scale separations. It can be expected that the

Toda procedure for obtaining enantiomerically pure compounds will find broad application very soon. This development could make preparative HPLC with chiral columns obsolete and be applied to distillable amino acid derivatives as well. After all, analytical resolution of amino acids was quite successful by host/guest complexation chromatography with reversed-phase packings loaded with Cram's chiral 1,1′-binaphthyl crown ethers (similar to **1**). [20]

While distillative separation of enantiomers in contact with optical selectors is the most exciting issue, the separation of isomers or mixtures with very close boiling points by distillation in the presence of structural selectors (that do not need to be chiral for that purpose) is also of high interest, because it again minimizes waste and abuse of energy. Very difficult separations, some of industrial importance, have been performed correspondingly. [21]

References

[1] Fonds der Chemischen Industrie, *Aminosäuren – Bausteine des Lebens* (Folienserie 11 des Fonds der Chemischen Industrie), Frankfurt, **1993**; R. M. Williams, *Synthesis of Optically Active α-Amino Acids*, Pergamon Press, Oxford, **1989**, Chap. 7, p. 257; I. Shiio, S. Nakamori in *Fermentation Process Development of Industrial Organisms* (Ed.: J. O. Neway), Marcel Dekker, New York, **1989**, p. 133; R. Biegelis in *Biotechnology*, Vol. 7b (Eds.: H. J. Rehm, G. Reed), VCH, Weinheim, **1989**, p. 229.

[2] Highlight: G. Kaupp, M. Haak, *Angew. Chem.* **1993**, *105*, 727; *Angew. Chem. Int. Ed. Engl.* **1993**, *32*, 694 see this book on page 99ff; L. Caswell, M. A. Garcia-Garibay, J. A. Scheffer, J. Trotter, *J. Chem. Educ.* **1993**, *70*, 785.

[3] Recent example: S. Blank, D. Seebach, *Angew. Chem.* **1993**, *105*, 1780; *Angew. Chem. Int. Ed. Engl.* **1993**, *32*, 1765, and references therein.

[4] D. Worsch, F. Vögtle, *Top. Curr. Chem.* **1987**, *140*, 22; F. Toda, *ibid.* **1987**, *140*, 43; F. Toda in *Inclusion Compounds*, Vol. 4 (Eds.: J. L. Atwood, J. E. D. Davies, D. D. MacNicol), Oxford University Press, Oxford, **1991**, p. 126; F. Toda, *Adv. Supramol. Chem.* **1992**, *2*, 141; F. Toda in *Comprehensive Supramolecular Chemistry*, Vol. 6, Chap. 15 (Eds.: D. D. MacNicol, F. Toda, R. Bishop), p. 465–516, Elsevier, Oxford, **1996**.

[5] K. Busch, U. M. Groth, W. Kühnle, U. Schöllkopf, *Tetrahedron* **1992**, *48*, 5607, and references therein.

[6] Of course, this is not true for the diastereoselective reactions of chiral natural products, for example, steroids.

[7] E. Fritz-Langhals, *Angew. Chem.* **1993**, *105*, 785; *Angew. Chem. Int. Ed. Engl.* **1993**, *32*, 753.

[8] M. Acs, T. Szili, E. Fogassy, *Tetrahedron Lett.* **1991**, *32*, 7325–8.

[9] *Comprehensive Supramolecular Chemistry*, Vol. 6 (Eds.: D. D. MacNicol, F. Toda, R. Bishop) is devoted to Solid-State Supramolecular Chemistry: Crystal Engineering; Vol. 8 (Eds.: J. E. D. Davies, J. A. Ripmeester) to Physical Methods in Supramolecular Chemistry, Elsevier, Oxford, **1996**; early review on clathrates: E. Weber, *Top. Curr. Chem.* **1987**, *140*, 2.

[10] K. Tanaka, T. Okada, F. Toda, *Angew. Chem.* **1993**, *105*, 1266; *Angew. Chem. Int. Ed. Engl.* **1993**, *32*, 1147, and references therein. Synthesis of **2** and **3** from tartaric acid: D. Seebach, H.-D. Kalinowski, B. Bastani, G. Crass, H. Daum, H. Dörr, N. P. Duprez, V. Ehsig, W. Langer, C. Nüssler, H.-A. Dei, M. Schmitt, *Helv. Chim. Acta* **1977**, *60*, 301.

[11] F. Toda, Y. Tohi, *J. Chem. Soc. Chem. Commun.* **1993**, 1238; synthesis of **8** and **9**: A. K. Beck, B. Bastani, D. A. Plattner, W. Letter, D. Seebach, H. Braunschweiger, P. Gysi, L. La Vecchia, *Chimia* **1991**, *45*, 238, and references therein; synthesis of **10**: F. Toda, K. Tanaka, K. Omata, T. Nakamura, T. Oshima, *J. Am. Chem. Soc.* **1983**, *105*, 5151. – Spontaneous inclusion of guests from suspensions of achiral components was only rarely reported: F. Toda, K. Tanaka, G. U. Daumas, M. C. Sachez, *Chem. Lett.* **1983**, 1521; H. R. Allcock

in *Inclusion Compounds*, Vol. 1 (Eds.: J. L. Atwood, J. E. D. Davies, D. D. MacNicol), Oxford University Press, Oxford, **1984**, p. 351.

[12] The molecule in ref. [11] (there **1a, b, c**) is shown with the (*S,S*)-configuration but labeled "(*R,R*)". Upon subsequent inquiry the correspondence author replied that "(*R,R*)-(–)" is correct. The optical rotations in the table apply to this configuration.

[13] G. Kaupp in *Comprehensive Supramolecular Chemistry*, Vol. 8, Chap. 9 (Eds.: J. E. D. Davies, J. A. Ripmeester) p. 381–423 + 21 color plates, Elsevier, Oxford, **1996**; G. Kaupp, J. Schmeyers, *Angew. Chem.* **1993**, *105*, 1656; *Angew. Chem. Int. Ed. Engl.* **1993**, *32*, 1587, and references therein.

[14] E. Weber, C. Wimmer, A. Llamas-Saiz, C. Foces-Foces, *J. Chem. Soc. Chem. Commun.* **1992**, 733.

[15] F. Toda, A. Sato, K. Tanaka, T. C. W. Mak, *Chem. Lett.* **1989**, 873.

[16] G. Kaupp, J. Schmeyers, M. Haak, T. Marquardt, A. Herrmann, *Mol. Cryst. Liq. Cryst.* **1996**, *276*, 315.

[17] G. Kaupp, J. Schmeyers, F. Toda, H. Takumi, H. Koshima, *J. Phys. Org. Chem.* **1996**, *9*, 795.

[18] M. Acs, A. Mravik, E. Fogassy, Z. Bocskei, *Chirality* **1994**, *6*, 314.

[19] F. Toda, H. Takumi, *Enantiomer* **1996**, *1*, 29.

[20] T. Shinbo, T. Yamaguchi, K. Nishimura, M. Sugiura, *J. Chromatogr.* **1987**, *405*, 145, and references therein.

[21] H. Takumi, M. Koga, F. Toda, *Mol. Cryst. Liq. Cryst.* **1996**, *277*, 79.

Absolute Asymmetric Synthesis by Irradiation of Chiral Crystals

Gerd Kaupp and Michael Haak

The formation of enantiopure optically active biomolecules is an important chapter in the history of evolution. The physical phenomena which can generate and augment asymmetry [1] have been known for a long time and a recent report used circularly polarized 190 nm synchrotron radiation in the enantiodifferentiating direct photoisomerization of *E*-cyclooctene and reached optical purities of 0.12 %. [2] A false claim of achieving absolute asymmetric synthesis in a static magnetic field by Breitmaier et al. [3] was highly acclaimed in *Science* and even in *Chem. Eng. News* [4] by several named researchers. That rush occurred despite the clear violation of elementary symmetry principles (an achiral field cannot create chirality) that are taught in first class courses and that were elaborated in great detail also for various field combinations [5] as a response to similarly untenable claims (magnetic *and* electrical field) in 1975. [6] Even a special seminar chaired by Feringa at Reinhoudt's Bürgenstock Conference 94 did not settle the issue. [7] Thus, the claim was only withdrawn [8] (and its immediate "confirmations" halted) after a fixed-date request to Breitmaier for response to a Correspondence telling the basic principles and the failure to reproduce the published or the "more specified" experiments. [9] Unfortunately this affair caused the cut down

of several then pending research proposals in Germany that were on their ways to apply new mechanistic knowledge for the development of preparatively valuable procedures from hitherto low-yield and low-size solid-state absolute asymmetric syntheses.

It is known since 1975 [10] that crystal chemical reactions can be used in absolute asymmetric synthesis: When achiral molecules crystallize in chiral space groups and the reaction of the crystals leads to chiral products, one speaks justly and correctly of *absolute asymmetric synthesis*. For this, the use of chiral agents and thus also the "crystal-selecting" human hand must be dispensed with. However, if autoseeding does indeed occur in a system, manipulations by seeding with crystals of a desired chirality must not detract from that term still. Most absolute asymmetric syntheses have been performed photochemically. Some short reviews have appeared. [11]

Absolute asymmetric [2 + 2] photodimerizations of achiral compounds in crystals have been known since 1982. [12] Achiral compound **1** crystallizes in the chiral space group $P2_1$. Crystals of **1** obtained from the melt or from ethanol on irradiation give either (+)- or (−)-**2** with enantiomeric excesses (*ee* values of 0 to 95 % and higher). The result varies from crystallization batch to crystallization batch; however, 100 % *ee* of the (+)- and the

(–)-enantiomer **2** were also claimed. [12] This seems to confirm that the photoreaction occurred stereospecifically (topotactic?; chemical yield and conversion not reported) and that in the batches with low optical yields the preceding crystallization led to mixtures of left- and right-handed chiral crystals. A single-crystal-to-single-crystal ("topotactic", though with disintegration of the clear crystals) photolysis was reported for electron-donor-acceptor crystals of **3** and **4** (*P2₁*), that provided the adduct **5** with 62 % (40 °C) to 95 % *ee* (–70 °C) from isolated single crystals (size and conversion not reported). [13] Unfortunately autoseeding or manual seeding were inefficient here. Most asymmetric [2 + 2]-cyclo-additions occur with considerably lower optical yields (e. g. < 7 % *ee* [12]), or they are claimed to be dependent on the conversion. [12] The former result is, however, not surprising, even when only crystals of a single chirality are present, because studies by atomic force microscopy (AFM) [14, 15] have shown that photodimerizations proceed from the surface to the interior of the crystal, during which long-range molecular movements (on the scale of the lattice constants) take place. The degree of the obtainable stereoselectivities clearly depends on the *phase rebuilding mechanisms* which are necessary for these processes. [14, 15]

The fact that relatively often spontaneous asymmetric crystallization of achiral compounds occurs with the formation of either right- or left-handed crystals is the result of autoseeding with the first crystal formed. [12, 16] Thus, in favorable cases all of the supersaturated solution yields only crystals of one chirality (analogously from melts), which, of course, cannot be predetermined. The chirality can only be influenced by manual seeding with selected crystals.

Enantioselective syntheses with achiral molecules in host crystals constructed from chiral molecules [17] belong as little to absolute asymmetric syntheses as do intramole-

cular photoreactions (di-π-methane rearrangement or γ-H transfer) following the manual selection of enantiomorphic crystals. [18a] However, Scheffer, Trotter et al. [18b] developed further their selective di-π-methane rearrangement of diisopropyl dibenzobarrelenedicarboxylate **6** to give the dibenzo-semi-bullvalene derivative **7** (claim of "quantitative *ee*" at conversions of a few percent) into an *absolute* asymmetric synthesis.

The AFM-investigation of the diethylester **6** (*P2₁2₁2₁*) yielded very pronounced surface features due to long-range molecular movements (100 nm range) by phase rebuilding. [15] These differ on all of its natural crystallographic faces ((011); (110); (010); (001)) and are nicely related to the crystal packing. [15] Thus, **6** does not undergo a topotactic reaction and the same is true for the absolute asymmetric synthesis with screwed **8** (*P2₁2₁2₁*) [19] giving equally pronounced AFM surface features, prior to becoming sticky, that again correlate with the crystal structure. [15] We suggested to increase both the chemical and the optical yield by phase-selective parallel irradiation of **6** (R = C_2H_5) on (110). The AFM features are smaller on that face of **6** (shorter mechanisms in these non-topotactic reactions for their major improvement. The same is true for asymmetric cyclobutanol formations. [18] Note, that submicromelting has to be avoided or excluded. [23]

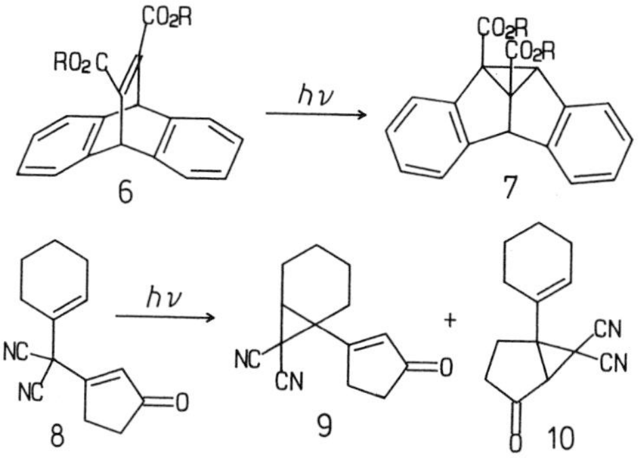

The screw/helix/spiral motive [11a] has more often been applied. Thus, irradiation of achiral **15**, which crystallizes in the chiral space group $P2_12_12_1$ yields optically active **16** without external seeding. [24] In seven of ten crystallizations from hexane the (+)-enantiomer **16** predominated, in three cases the (–)-enantiomer **16**. The optical yields (10 % *ee* at 0 °C, 75 % chemical yield) can be improved by reducing the temperature (40 % *ee* at –40 °C, 70 % yield; no reaction at –78 °C; however, formation of racemic **16** in solution also at –78 °C). This is again a reminder of the necessity of phase rebuilding and finally phase transformation to give the product lattice in solid-state reactions. Both processes are accompanied by extensive molecular movements (on the scale of the crystal lattice constants) in all known cases. [14, 15] Evidently, as a result of these, the chiral conformers **15** have – as in solution – the opportunity for racemization and further conformational alterations. Nevertheless, the

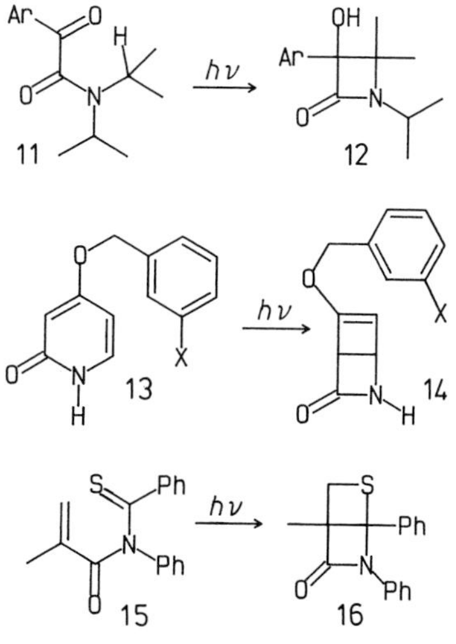

movements) than on its other natural faces and the packing is more favorable, which also decreases the disintegration tendency. [20]

"Absolute" asymmetric cyclizations (C–H to C=O additions) were performed by Toda et al. [21] Thus, **11** (Ar = Ph) and particularly *meta*-substituted derivatives of **11** crystallize in chiral space groups and form **12** with nearly quantitative *ee* at conversions of 74–95 %. As crowded molecules like **11** assume screwed/helical/spiral conformations [11a] it is understandable that the optical yields are much higher at much higher chemical conversions than in the previous cases. Clearly, the common molecular movements and phase rebuilding as well as phase transformation [14, 15] do not immediately (or easily) destroy the helices that were packed during crystallization. Thus, these reactions may gain preparative value after a mechanistic AFM study. While it has not yet been demonstrated that autoseeding is occuring in these cases, the apparently high seeding efficiencies suggest that it would work and that the attribute "absolute" may be appropriate.

Toda's ideas have been taken up and extended by Sakamoto and Scheffer. Thus, *meta*-substituted **13** give optical active **14** by 4π-cyclization although only at (very) low conversion. [22] Here the rather flat 4π-moiety is part of the chiral helix that induces the chirality. No AFM investigation has been performed in order to use the phase rebuilding

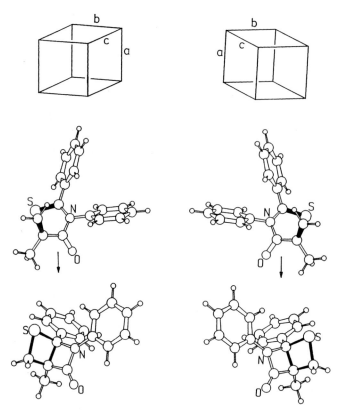

Figure 1. Top: $P2_12_12_1$-unit cell ($a = 9.535(2)$, $b = 9.782(2)$, $c = 16.428(3)$ Å) of **15**, left- and right-handed arrangement; middle: left-handed (M) and right-handed helix (P) of **15** where M and P are descriptors for helicity (M = minus and P = plus); bottom: (1R,4S)-**16** and (1S,4R)-**16**.

chirality advantage given by the crystal is partly retained. Although, in principle, it is not necessary for the action of an asymmetric crystal environment that the molecules are arranged as uniform chiral conformers, it is essential for the helical compound **15** [24] that the one enantiomorphic type of crystal contains the molecules **15** in the right-handed helix (P), the other in the left-handed helix (M).

As long as the absolute configuration of the products (e. g. **2** or **16**) and of the enantiomorphic crystals has not been determined, usually those leading to (+)-products are designated as (+)-crystals and vice versa. Without knowledge of the absolute configurations there is

also no link to the chirality of the crystal and the helicity of the lattice-frozen conformer. To elucidate the principle we arranged the unit cell ($P2_12_12_1$) of **15** with the left-handed helix arbitrarily as left handed. Compound (1R,4S)-**16** [(−) or (+)] is formed unavoidably from the left-handed helix **15**. Correspondingly, (1S,4R)-**16** must be formed from the enantiomorphic crystal arranged in the right-handed way. This is represented in Figure 1 on the basis of optimized geometries from semiempirical PM3 calculations. Regardless of whether the adoption of the configurations agrees with reality, it thus becomes clear that the spontaneous asymmetric crystallization determines the chirality of the excess enantiomer **16** exclusively. The energetically most favorable conformer of **15** (Fig. 1; *M* or *P*) is 13 kcal mol^{-1} less stable than **16** according to our calculations, [25] and thus it should show optimum reactivity. According to the X-ray structure analysis, [24] in the crystal the pitch of the helix is reduced. This is evident from the significantly shortened distances between the reactive centers: the semiempirical calculated distances S-1/C-6 and C-2/C-5 in **15** are 4.33 and 3.26 Å, respectively, (where the numbers refer to the sequence of atoms in the chain starting from S); [25] in contrast, in the crystal they were determined to be 3.59 and 3.11 Å, respectively. [24] This compression also contributes to the selectivity of the crystal photolysis.

Similarly (+)-crystals or (–)-crystals of **17** giving the (+)- and (–)-products **18** in excess have their opposite helicities. These crystals had to be manually selected even upon seeding, because two different types of racemic crystals coexisted. The two-step reactions consist of 6π-cyclization and hydrogen migration and give 100 % yield at 64 % *ee*. [26].

Further examples using rapidly equilibrating screwed/helical/spiral molecules that happen to crystallize in chiral space groups have been provided by Sakamoto. [11c] These are photochemical formations of chiral oxetanes, β-lactams, oxazolidine-2,4-diones, aziridines, oxazolines, phthalides, and β-thiolactames. [11c] In most cases the *ee* values had to be improved by going to –78 °C or to "low conversion". Whereas the latter technique is frequently preferred it is often dubious for the normal selective irradiations (only the educt absorbing the light), if the crystal reacts from its surface down into the bulk, multilayer by multilayer as these become transparent at very advanced local conversions. [14, 15] Thus, no basis exists for assumptions that the optical yield may increase to quantitative values at very low conversions, while the error limits will not allow to secure such claims. Nevertheless, there is the risk that

such claims are unduly quoted in the secondary literature. If the products absorb some of the light, extreme milling will be helpful. However, that technique does not help against long-range molecular movements that are connected to the necessary phase rebuilding and that may destroy chiral environments from the beginning.

Absolute asymmetric syntheses by irradiation of chiral crystals, now extended to intramolecular reactions, [27] support the assumption of the prebiotic origin of natural chirality. The chemical mechanisms here appear to be so conclusive and efficient that they should in any case be retained besides the complicated physical interpretations. [1, 2] It must be remembered that of the 230 space groups 65 are chiral and that of these $P2_12_12_1$ and $P2_1$ belong to the five most frequent in the organic crystals.

Regardless of these fundamental findings a wide field has opened up for asymmetric syntheses by approved photochemical reactions. However, in order to make them preparatively useful it will be essential to relate them to the crystal packing by submicroscopic AFM investigations that are easily performed and provide an unexpected wealth of kinetic information just by analysis of the far-reaching molecular movements. [14, 15] These are necessary for every non-topotactic reaction to occur. They lead to face-selective surface features regardless of topochemical allowed- or forbiddenness and can be related to the crystal packing on the reasonable premise that moving molecules will choose easy ways to do so. [15] Only with that information available many of the known absolute asymmetric syntheses will be enabled to improvement for practical use. Unduly simplifying "topochemical considerations" hold only for very rare topotactic reactions. The latter should, of course, also be checked with AFM, if they are not perfect. In short, all use of the experimental phase rebuilding mechanisms will improve the present situation. While these points are certainly more important than the collection of more and more model systems for absolute asymmetric syntheses with selected single crystals it should be pointed out that many polycrystal examples might have remained unrecognized. Far too little attention has been paid to the phenomenon of the spontaneous asymmetric crystallization to date, with the result that the step to the polarimeter or the use of chiral NMR shift reagents after the photolysis has mostly not been taken; otherwise without doubt many a supposed racemate would have turned out to be optically active.

References

[1] *Origins of Optical Activity in Nature* (Ed.: D. C. Walker), Elsevier, New York, **1979**; resolution of racemates by crystallization succeeds by Tamura's "Preferential Enrichment" in the mother liquor: T. Ushio, R. Tamura, H. Takahashi, N. Azuma, K. Yamamoto, *Angew. Chem.* **1996**, *108*, 2544–2546; *Angew. Chem. Int. Ed. Engl.* **1996**, *35*, 2372–2374; R. Tamura, T. Ushio, H. Takahashi, K. Nakamura,

N. Azuma, F. Toda, K. Endo, *Chirality*, **1997**, *9*, 220–224; H. Takahashi, R. Tamura, T. Ushio, T. Nakai, K. Hirotsu, F. Toda, *Mol. Cryst. Liq. Cryst.*, in the press and by Collet's "Entrainment" (selective crystallization): A. Collet, M.-J. Brienne, J. Jacques, *Chem. Rev.* **1980**, *80*, 215–230; A. Collet in *Comprehensive Supramolecular Chemistry, Vol. 10* (Ed. D. N. Reinhoudt), Elsevier, Oxford, **1996**, Chapter 5. Both techniques require particular phase diagram conditions.

[2] Y. Inoue, H. Tsuneishi, T. Hakushi, K. Yagi, K. Awazu, H. Onuki, *J. Chem. Soc. Chem. Commun.* **1996**, *23*, 2627–2628.

[3] G. Zadel, C. Eisenbraun, G.-J. Wolff, E. Breitmaier, *Angew. Chem.* **1994**, *106*, 460–463; *Angew. Chem. Int. Ed. Engl.* **1994**, *33*, 454–456.

[4] D. Bradley, *Science*, **1994**, *264*, 908; Anonymous, *Chem. Eng. News* **1994**, *72(9)*, 36.

[5] C. A. Mead, A. Moscowitz, H. Wynberg, F. Meuwese, *Tetrahedron Lett.* **1977**, *12*, 1063–1064; R. C. Dougherty, *J. Am. Chem. Soc.* **1980**, *102*, 380–381; L. D. Barron, *J. Am. Chem. Soc.* **1986**, *108*, 5539–5542; M. W. Evans, *Chem. Phys. Lett.* **1988**, *152*, 33–38.

[6] P. Gerike, *Naturwissenschaften*, **1975**, *62*, 38–39.

[7] W. Leitner, *Nachr. Chem. Tech. Lab.* **1994**, *42*, 716–717; B. L. Feringa, R. M. Kellogg, R. Hulst, C. Zondervan, W. H. Kruizinga, *Angew. Chem.* **1994**, *106*, 1526–1527; *Angew. Chem. Int. Ed. Engl.* **1994**, *33*, 1458–1459.

[8] E. Breitmaier, *Angew. Chem.* **1994**, *106*, 1529; *Angew. Chem. Int. Ed. Engl.* **1994**, *33*, 1461 (June 21, 1994).

[9] G. Kaupp, T. Marquardt, *Angew. Chem.* **1994**, *106*, 1527–1529; *Angew. Chem. Int. Ed. Engl.* **1994**, *33*, 1459–1461 (June 12, 1994).

[10] B. S. Green, M. Lahav, G. M. J. Schmidt, *Mol. Cryst. Liq. Cryst.* **1975**, *29*, 187–200. Bromination of manually selected enantiomorphic single crystals of 4,4'-dimethylchalcone to give optically active products (up to 6% *ee*: K. Penzien, G. M. J. Schmidt, *Angew. Chem.* **1969**, *81*, 628; *Angew. Chem. Int. Ed. Engl.* **1969**, *8*, 608) is incorrectly termed "absolute" asymmetric synthesis; similarly, no autoseeding was described in the bromination of diben-

zobarrelene with reported optical yields of 3 or 8 % using ground single crystals of 20–100 mg: M. G. Garibay, J. R. Scheffer, J. Trotter, F. Wireko, *Tetr. Letters* **1988**, *29*, 1485–1488.

[11] a) G. Kaupp, M. Haak, *Angew. Chem.* **1993**, *105*, 727–728; *Angew. Chem. Int. Ed. Engl.* **1993**, *32*, 694–695; b) L. Caswell, M. A. Garcia-Garibay, J. R. Scheffer, J. Trotter, *J. Chem. Ed.* **1993**, *70*, 785–787; c) M. Sakamoto, *Chem. Eur. J.* **1997**, *3*, 684–689.

[12] L. Addadi, J. van Mil, M. Lahav, *J. Am. Chem. Soc.* **1982**, *104*, 3422–3429; further example: M. Hasegawa, Y. Hashimoto, *Mol. Cryst. Liq. Cryst.* **1992**, *219*, 1–15; Review: V. Ramamurthy, K. Venkatesan, *Chem. Rev.* **1987**, *87*, 433–481.

[13] T. Suzuki, T. Fukushima, Y. Yamashita, T. Miyashi, *J. Am. Chem. Soc.* **1994**, *116*, 2793–2803.

[14] G. Kaupp, *Angew. Chem.* **1992**, *104*, 606–609 and 609–612; *Angew. Chem. Int. Ed. Engl.* **1992**, *31*, 592–595 and 595–598.

[15] G. Kaupp, *Adv. Photochem.* **1995**, *19*, 119–177; G. Kaupp in *Comprehensive Supramolecular Chemistry* (Ed.: J. E. D. Davies) Vol. 8, 381–423 + 21 color plates, Elsevier, Oxford, **1996**, *Chemie in unserer Zeit*, **1977**, *31*, 129–139.

[16] J. M. McBride, R. L. Carter, *Angew. Chem.* **1991**, *103*, 298–300; *Angew. Chem. Int. Ed. Engl.* **1991**, *30*, 293–295.

[17] T. Fujiwara, N. Nauba, K. Hamada, F. Toda, K. Tanaka, *J. Org. Chem.* **1990**, *55*, 4532–4537.

[18] a) S. V. Evans, M. Garcia-Garibay, N. Omkaran, J. R. Scheffer, J. Trotter, F. Wireko, *J. Am. Chem. Soc.* **1986**, *108*, 5648–5650; b) J. Chen, J. R. Scheffer, J. Trotter, *Tetrahedron* **1992**, *48*, 3251–3274; there also absolute configuration of starting conformer and of product; A. D. Gudmundsdottir, G. Rattray, J. R. Scheffer, J. Yang, *Tetrahedron Lett.* **1993**, *34*, 35; c) T. Y. Fu, Z. Liu, J. R. Scheffer, J. Trotter, *J. Am. Chem. Soc.* **1993**, *115*, 12202–12203.

[19] A. L. Roughton, M. Muneer, M. Demuth, I. Klopp, C. Krüger, *J. Am. Chem. Soc.* **1993**, *115*, 2085–2087.

[20] G. Kaupp, M. Plagmann, unpublished.

[21] F. Toda, M. Yagi, S.-I. Soda, *J. Chem. Soc. Chem. Commun.* **1987**, 1413–1414; A. Sekine, K. Hori, Y. Ohashi, M. Yagi, F. Toda, *J. Am. Chem. Soc.* **111**, *1989*, 697–699; F. Toda, H. Miyamoto, *J. Chem. Soc. Perkin. Trans. 1* **1993**, 1129–1132; D. Hashizume, H. Kogo, A. Sekine, Y. Ohashi, H. Miyamoto, F. Toda, *J. Chem. Soc. Perkin. Trans. 2* **1996**, 61–66; there also absolute configuration of starting conformer and of product; F. Toda in *Comprehensive Supramolecular Chemistry* (Ed.: D. D. MacNicol, F. Toda, R. Bishop) Vol. 6, 465–516, Elsevier, Oxford, **1996**.

[22] L.-C. Wu, C. J. Cheer, G. Olovsson, J. R. Scheffer, J. Trotter, S.-L. Wang, F.-L. Liao, *Tetrahedron Lett.*, **1997**, *38*, 3135–3138.

[23] G. Kaupp, *Mol. Cryst. Liq. Cryst.* **1992**, *211*, 1–15; **1994**, *242*, 153–169; **1994**, *252*, 259–268; *J. Vac. Sci. Technol. B* **1994**, *12*, 1952–1956.

[24] M. Sakamoto, N. Hokari, M. Takahashi, T. Fujita, S. Watanabe, I. Iida, T. Nishio, *J. Am. Chem. Soc.* **1993**, *115*, 818–820.

[25] PM3 according to J. J. P. Stewart (*J. Comput. Chem.* **1989**, *10*, 209) with Spartan Version 2.0 from Wavefunction, Inc., Irvine, CA, USA on an IBM RS6000 H32 Workstation calculated with complete geometry optimization. The best boat conformation of **15** (rotation about the C-4/C-5 bond) is about 0.6 kcal mol^{-1} less stable than the helix. Representation of the molecules with Schakal 92/AIX-UNIX version from E. Keller, Universität Freiburg, on an IBM RS6000.

[26] F. Toda, K. Tanaka, *Supramol. Chem.* **1994**, *3*, 87–88; F. Toda, K. Tanaka, Z. Stein, I. Goldberg, *Acta Cryst.* **1995**, *B51*, 856; *ibid.* **1995**, *C51*, 2722.

[27] Absolute asymmetric photodecarboxylation in a mixed crystal: H. Koshima, K. Ding, Y. Chisaka, T. Matsuura, *J. Am. Chem. Soc.* **1996**, *118*, 12059–12065.

B. Organometallic Reagents

Cyclopentadienyl Ruthenium Complexes: Valuable Assistents in the Construction of Carbon–Carbon Bonds

Holger Butenschön

Within the last few years a number of papers appeared in an area of increasing importance for organic chemistry: Organoruthenium complexes are now frequently used to form carbon–carbon bonds in a number of ways. Reagents used in this field include cyclopentadienyl complexes like [CpRu(COD)Cl] or [CpRu(PPh)$_2$Cl] as well as tetravalent cyclopentadienylruthenium derivatives and others. [1]

With respect to cyclopentadienyl metal complexes ligands other than the cyclopentadienyl (Cp) ligand are frequently regarded as less important and are considered to leave the molecule before the critical reaction step at the cyclopentadienylmetal nucleus. It is remarkable that Trost et al. found that reactions involving the 1,5-cyclooctadiene (COD) ligand for the construction of a carbon skeleton are indeed possible. [2]

Trost et al. discovered that COD can formally function as a bis-homodiene in metal-catalyzed [4+2]cycloadditions. A 0.1 M solution of the alkyne **1** reacts with 1.1 equivalents of COD in the presence of 5 mol% of chloro(η^4-cyclooctadiene)(η^5-cyclopentadienyl)-ruthenium(II) [CpRu(COD)Cl] [3] (**2**) in boiling methanol to give derivatives **3** of tricyclo[4.2.2.0^{2,5}]dec-7-ene (Scheme 1). Yields of between 78 and 100% are achieved in seven out of eight examples. Electron-defi-

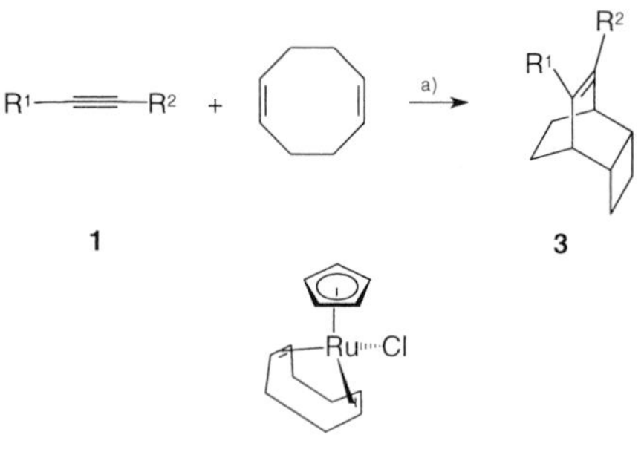

Scheme 1. a) 5% **2**, MeOH; R^1 = Me, Et, CH$_2$OSi(*i*Pr)$_3$, *p*-CH$_2$C$_6$H$_4$OCH$_3$, H; R^2 = CH$_2$OH, CH$_2$OSi(*i*Pr)$_3$, CH$_2$CH$_2$OH, *p*-CH$_2$C$_6$H$_4$OCH$_3$, CO$_2$Me, CH$_2$CH$_2$CH$_2$CO$_2$Me; 78–100%.

cient alkynes react particularly slowly, and no reaction takes place with dimethyl butynedioate. Steric hindrance slows down the reaction; **4** is obtained only in 51% yield after 80 h at reflux. This can, however, be used for the differentiation of triple bonds in alkynes with several triple bonds as shown by the selective formation of **5** in 63% (Scheme 2).

The catalytic cycle proposed by the authors starts from a cationic cyclopentadienylruthenium complex, bearing one COD ligand and one solvent molecule. The latter is displaced by the alkyne, which subsequently reacts

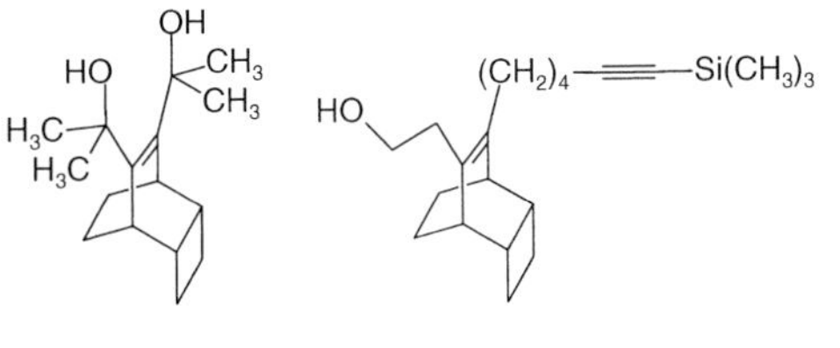

4   **5**

Scheme 2.

9   **10**

11   **12**

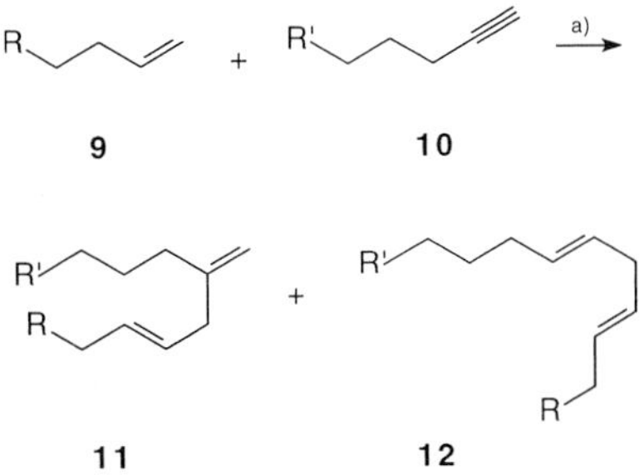

Scheme 4. a) 5 % **2**, dimethylformamide (DMF)/ H_2O 3:1, 100 °C, 2 h; R = Bu, CH_2CH_2OH, COMe, $(CH_2)_3CO_2Me$; R' = Pr, $CH_2OTBDMS$, CO_2Et; 60–90 %.

with a double bond of the COD ligand to give a metallacyclopentene. An intramolecular carbometallation of the remaining double bond of the eight-membered ring with successive reductive elimination leads to products **3**; renewed complexation of a COD molecule regenerates the catalytically active species. The authors point out that the reaction is also catalyzed by other coordinatively highly unsaturated ruthenium complexes and finally make the important remark that the ease of the reactions observed by them suggests that a range of other ruthenium-catalyzed reactions may well be possible. They assume that the coordination properties of COD cannot only be derived from the entropic effect of the spatial arrangement of the two double bonds in COD. Possibly electronic effects similar to those in norbornadiene exist in an attenuated form in COD. This is in agreement with the fact that the dienes **6–8** do not undergo the reaction found with COD (Scheme 3). [4]

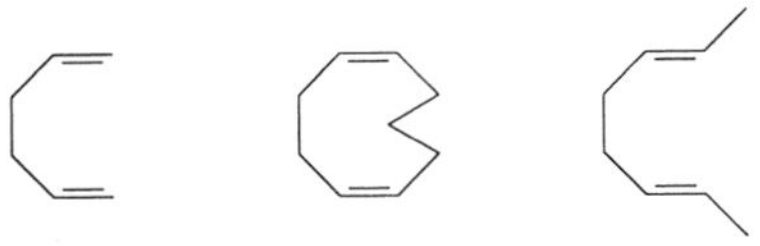

The yields lie between 60 and 90 %, and in most cases the branched isomer **11** is formed preferentially (**11** : **12** up to 6 : 1). Substituents at the propargyl position of **10** reduce the regioselectivity, and the presence of propargylic oxygen substituents even leads to an inversion of the selectivity in favor of the linear coupling products. The chemoselectivity is impressively underlined by the reaction of **13** with the α,β-unsaturated ester **14** to give exclusively **15** in 70 % yield (Scheme 5).

Also in these reaction the authors assume that a cationic cyclopentadienylruthenium complex is the catalytically active species, the only difference here being that in addition to the chloro ligand also the COD ligand is decomplexed and all three free coordination sites are filled by readily displaceable solvent molecules. After the complexation of the alkene component, the latter is transformed

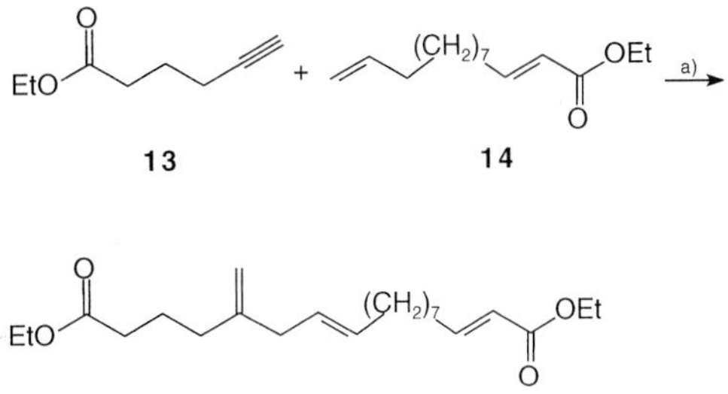

6   **7**   **8**

Scheme 3.

13   **14**

The ruthenium catalyst **2** was recently used by Trost et al. also for coupling reactions of alkenes **9** with alkynes **10**. [5] These reactions lead to the branched coupling products **11** as well as the lineaer isomers **12** (Scheme 4).

Scheme 5.   **15**

a) 5 % **2**, DMF/H_2O 3:1, 100 °C, 2 h, 70 %.

into an allyl ligand and the hydrogen atom released is bound to the metal as a hydrido ligand. After coordination, the alkyne is either intramolecularly hydrometalated or carbometalated starting from the allyl ligand. From both intermediates the reaction product, still coordinated to ruthenium, can be easily formed, which is then released with the regeneration of the above-mentioned catalytically active species.

Catalyst **2** can also be used in a butenolide synthesis on the basis of a ruthenium catalyzed Alder ene type reaction. [6–9] In the course of a reaction sequence directed to the synthesis of acetogenins, Trost et al. coupled alkene **16** with propargyl alcohol **17** in the presence of **2** to give butentolide **18** in up to 82 % yield (Scheme 6).

Several years ago Trost et al. reported on the coupling of terminal alkynes **19** with allyl alcohols **20** in the presence of ammonium hexafluorophosphate. [10] This reaction is catalyzed by [(PPh$_3$)$_2$CpRuCl] and leads to the formation of β,γ-unsaturated ketones **21** (Scheme 7). A mechanistic study reveals that vinylidene complexes are formed as intermediates from the terminal alkynes. [11] Also allyl alcohols can be isomerized with [(PPh$_3$)$_2$CpRuCl] directly to give saturated ketones. [12] Interestingly, complex **2**, which is different from [(PPh$_3$)$_2$CpRuCl] only in that instead of the two phosphane ligands a COD ligand is present, catalyzes the coupling

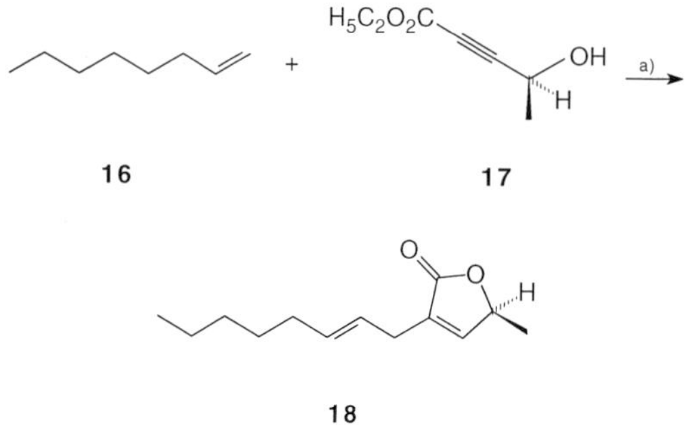

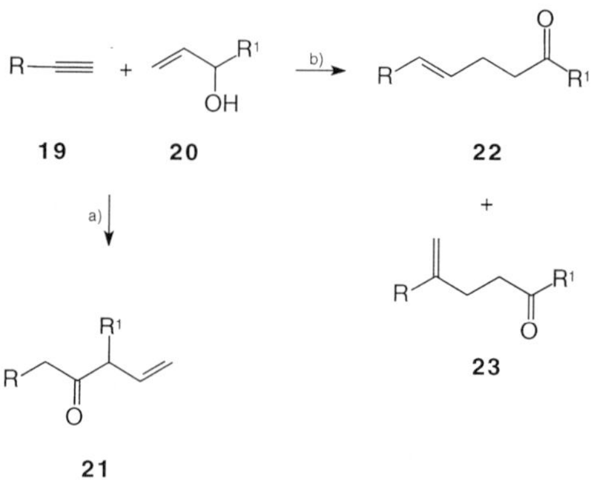

Scheme 7. a) 6 % [(PPh$_3$)$_2$CpRuCl], 10 % NH$_4$PF$_6$, 100 °C, 10 h, 44–74 %; b) 10 % **2**, 10–20 % NH$_4$PF$_6$, 100 °C, 1–2 h; R = C$_{10}$H$_{21}$, C$_6$H$_{13}$, CH(OH)C$_5$H$_{11}$, C(C$_2$H$_5$)(OH)C$_5$H$_{11}$; R^1 = Me, *i*Pr, cyclohexyl, C$_{11}$H$_{23}$; 51–85 %.

of alkynes **19** and allyl alcohols **20** in 51–85 % yields to give γ,δ-unsaturated ketones **22** and **23**. [8, 13]

From the fact that in the reaction catalyzed by **2** also internal alkynes **24** can react to give ketones **25** (Scheme 8), the authors conclude that in this case no vinylidene complex acts as an intermediate. Apparently, its formation is favored by the presence of phosphane ligands, and in their absence other reaction paths are followed. The regioselectivity of the reaction as well as its yield can be increased if a mixture of water and DMF is used as solvent. The reaction functions also with alkynes whose triple bond is conjugated to an ester group. The chemoselectivity of the reaction was demonstrated by the coupling of a steroid side chain with an allyl alcohol; one α,β-unsaturated carbonyl functionality present in the steroid part remained unaltered.

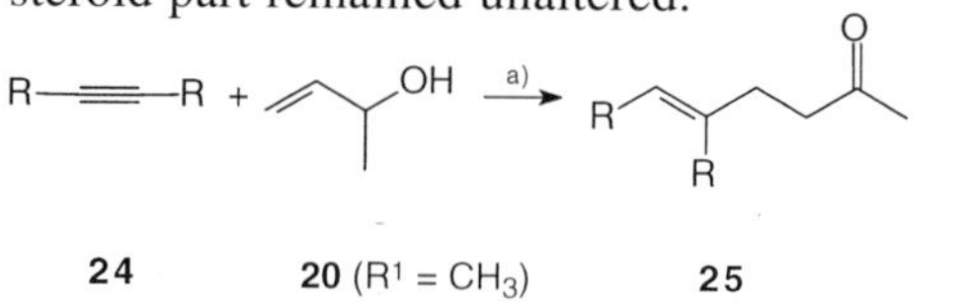

Scheme 8. a) 10 % **2**, 10–20 % NH$_4$PF$_6$, 100 °C, 1–2 h, R = Bu (45 %), Ph (50 %).

Scheme 6. a) 1.4 % **2**, CH$_3$OH, reflux, 78–82 %.

The Meyer-Schuster redox isomerization of propargyl alcohols to give α,β-unsaturated ketones or aldehydes is a valuable reaction because these compounds can be used in a variety of reactions and because many propargylic alcohols can readily be obtained by addition of acetylide anions to carbonyl compounds. [14] With ruthenium catalyst **26** in the presence of some indium trichloride the reaction becomes possible under mild conditions, allowing for example the formation of the sensitive dienal **28** from **27** (Scheme 9), which is easily prepared by a palladium catalyzed coupling reaction of 1-bromo-2-methylpropene and propargyl alcohol. [15] The authors explain the role of the indium trichloride by suggesting a bimetallic complex with the Lewis acid complexed to the carbonyl oxygen atom.

The significance of ruthenium-catalyzed reactions is also emphasized by a publication by Mitsudo et al., [16] who achieved [2+2]cycloadditions of norbornene **29** and norbornadiene **32** with a range of alkynes **30**. These reactions are catalyzed by chloro(η^4-cyclooctadiene)(η^5-pentamethylcyclopentadienyl)ruthenium(II) [Cp*Ru(COD)Cl] **(35)** and give **31** and **33**, respectively, in good yields (Scheme 10).

Ruthenium-catalyzed [2+2]cycloadditions of norbornene with butynedioates have been known for a long time. [17, 18] However, what is new is that the reaction, which was successful with other ruthenium catalysts only with butynedioates, now functions with a whole series of different alkynes, and that not only norbornene **29**, but also norbornadiene **32** can be used. In this way in the course of the reaction [Cp*RuCl(nbd)] (nbd = norbornadiene) is formed which is more stable than **2**. Thus, slightly higher reaction temperatures are required for reactions with norbornadiene. In addition to the adducts **33** to norbornadiene, *meta*-disubstituted arenes **34** are also formed. The use of phenylethyne which is deuterated at the terminal alkyne position reveals that the additional C_2 block present in **34** stems from norbornadiene. In accordance with a retro-Diels-Alder reaction, di(cyclopentadiene) was detected in the reaction mixtures.

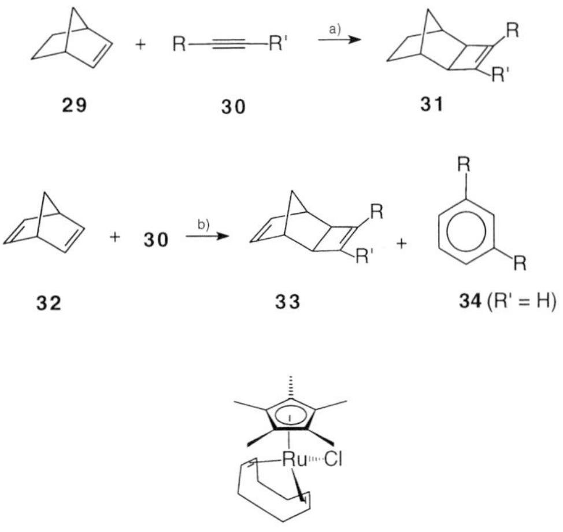

As a possible mechanism for the reaction the authors propose that **35** releases the COD ligand with the formation of the catalytically active species and the two free coordination

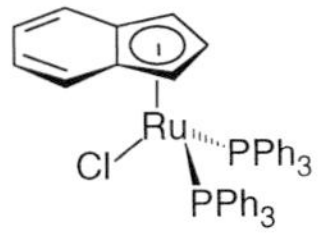

Scheme 10. a) 5 % [Cp*Ru(COD)Cl] **35**, 80 °C, 15–20 h, NEt$_3$, R, R' = Me, Et, Ph, C$_5$H$_{11}$, CH(OEt)$_2$, CO$_2$Me, CO$_2$Et, C$_8$H$_{17}$, C$_{10}$H$_{21}$, 23–87 % **33**, 12–26 % **34** (R' = H).

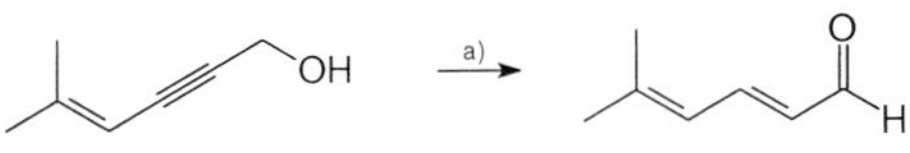

Scheme 9. a) 5 mol% **26**, 0.25 M solution of **27** in THF, 5 mol% InCl$_3$, 5 mol% Et$_3$NHPF$_6$, 1.5 h, 25 °C → reflux, 83 % of **28**.

sites are then filled by norbornadiene and the alkyne. From this a metallacyclopentene is formed which either reacts by reductive elimination to give the [2+2]cycloadduct or by insertion of a further alkyne molecule forms a metallacycloheptadiene. The latter then releases cyclopentadiene in a retro-Diels-Aler reaction and is thus transformed into the metallacycloheptatriene, the direct precursor of **34**.

Interestingly, the only difference between the catalysts used by Trost et al. (**2**) and Mitudo (**35**) is that an unsubstituted cyclopentadienyl ligand is present is **2** and a pentamethylcyclopentadienyl ligand in **35**. The electronic and steric differences of the two ligands are well known; in light of the papers presented here the question arises as to how the spectrum of ruthenium-catalyzed reactions may be extended by the use of a new ligand recently described by Gassman et al.; [19] the title of the work reads as follows: "1,2,3,4-Tetramethyl-5-(trifluoromethyl)cyclopentadienide: A Unique Ligand with the Steric Properties of Pentamethylcyclopentadienide and the Electronic Properties of Cyclopentadienide". The related pentafluorocyclopentadienyl ligand has also found its way into ruthenium chemistry, and pentafluororuthenocene has recently been reported. [20]

More recently, Dixneuf et al. reported the use of Cp*Ru(IV) as catalysts for a coupling of alkynes and allyl alcohols (Scheme 11). Phenylethyne **19** (R = Ph) and allyl alcohol **20** (R^1 = H) were transformed to isomeric acetals **36** and **37** in 20 % and 60 % yield, respectively, [21] in the presence of [RuCl$_2$(η^3-CH$_2$CMeCH$_2$)-(C$_5$H$_5$] or [RuCl$_2$(η^3-CH$_2$CMeCH$_2$)(C$_5$Me$_5$] [22]. Remarkably, not only the yield was different with the two similar catalysts, but also the regioselectivity: With the Cp complex the **36 : 37** ratio was 67 : 33, with the Cp* complex it changed to 33 : 67.

Bäckvall et al. used ruthenium complex **38** (Scheme 12) to catalyze an Oppenauer oxidation of secondary alcohols in yields between

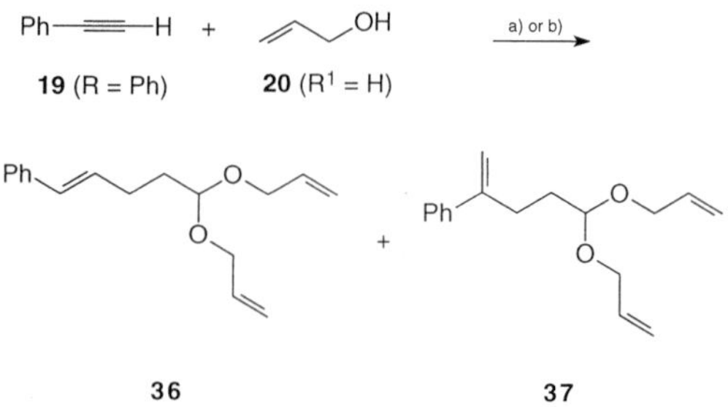

Scheme 11. a) 2.5 mmol of phenylethyne, 5 mL of allyl alcohol, 0.125 mmol of [RuCl$_2$(η^3-CH$_2$CMeCH$_2$)(C$_5$H$_5$)], 22 h, 90 °C; b) as for a) except catalyst [RuCl$_2$(η^3-CH$_2$CMeCH$_2$)(C$_5$Me$_5$)].

60 % and 96 %. The proposed mechanism includes a delicate interchange between the cyclopentadienyl ruthenium system and a corresponding cyclopentadienone complexes. [23]

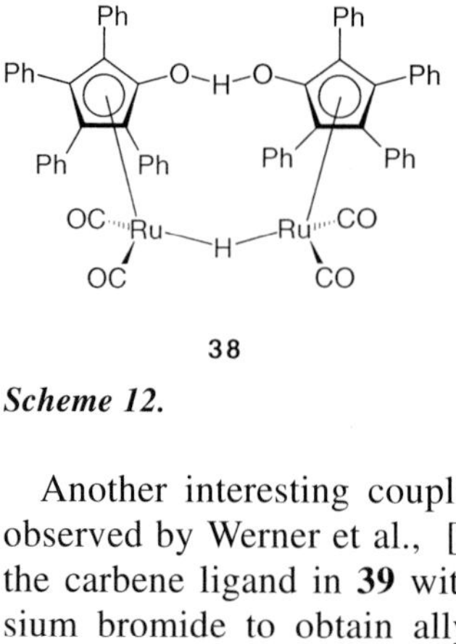

Scheme 12.

Another interesting coupling reaction was observed by Werner et al., [24] who coupled the carbene ligand in **39** with ethenylmagnesium bromide to obtain allyl complexes **40** and **41** in up to 65 % yield (Scheme 13). Subsequent decomplexation gave the corresponding trisubstituted alkenes **42**, which were obtained in quantitative yields.

Ruthenium and iron belong to the same group in the periodic system, and much of their organometallic chemistry is similar. As a special class of ferrocenylphosphines has become popular in enantioselective catalysis [25], it was quite obvious to test the corresponding ruthenium complexes for catalytic

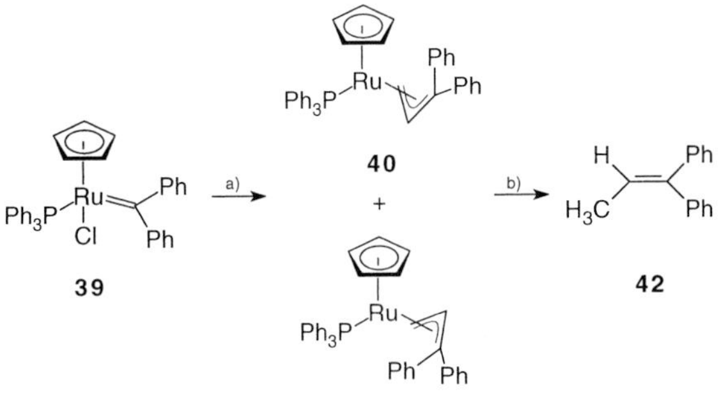

Scheme 13. a) 0.23 mmol of **39**, 5 mL C_6H_6, 0.47 mmol of $H_2C=CHMgBr$ (0.75 M sol. in THF), 45 min, room temp., 63 %, **40:41** = 1.9:1; b) CH_3COOH, quant. yield.

activity. Hayashi et al. reported in 1994 that the ring closure of **44** to **45** indeed take place with a high degree of stereoselection with catalyst ligand **43** (Scheme 14). [26]

That the Cp*Ru fragment can also be used to break down carbon frameworks was shown by Chaudret et al., [27] who, using an example from steroid chemistry, achieved the elimination of methane from a methyl substituted cyclohexadiene with consequent aromatization.

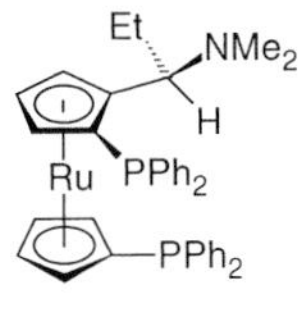

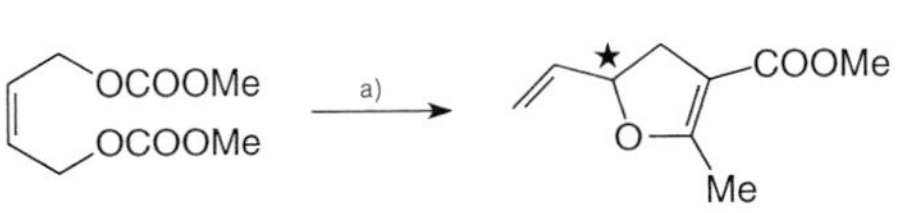

Scheme 14. a) 0.006 mmol of $Pd_2(dba)_3 \cdot CHCl_3$, 0.013 mmol of **43**, 0.9 mmol of **44**, 0.6 mmol of methyl acetylacetate in 9.0 mL of THF, −20 °C under N_2, 48 h, 83 %, *ee* 83 %.

References

[1] B. M. Trost, *Chem. Ber.* **1996**, *129*, 1313–1322.

[2] B. M. Trost, K. Imi, A. F. Indolese, *J. Am. Chem. Soc.* **1993**, *115*, 8831–8832.

[3] M. O. Albers, D. J. Robinson, A. Shaver, E. Singleton, *Organometallics* **1986**, *5*, 2199–2205.

[4] B. M. Trost, personal communication.

[5] B. M. Trost, A. Indolese, *J. Am. Chem. Soc.* **1993**, *115*, 4361–4362.

[6] B. M. Trost, T. L. Calkins, *Tetrahedron Lett.* **1995**, *36*, 6021–6024.

[7] B. M. Trost, T. J. J. Müller, J. Martinez, *J. Am. Chem. Soc.* **1995**, *117*, 1888–1899.

[8] B. M. Trost, A. F. Indolese, T. J. J. Müller, B. Treptow, *J. Am. Chem. Soc.* **1995**, *117*, 615–623.

[9] B. M. Trost, T. J. J. Müller, *J. Am. Chem. Soc.* **1994**, *116*, 4985–4986.

[10] B. M. Trost, G. Dyker, R. J. Kulawiec, *J. Am. Chem. Soc.* **1990**, *112*, 7809–7811.

[11] B. M. Trost, R. J. Kulawiec, *J. Am. Chem. Soc.* **1992**, *114*, 5579–5584.

[12] B. M. Trost, R. J. Kulawiec, *J. Am. Chem. Soc.* **1993**, *115*, 2027–2036.

[13] B. M. Trost, J. A. Martinez, R. J. Kulawiec, A. F. Indolese, *J. Am. Chem. Soc.* **1993**, *115*, 10402–10403.

[14] S. Swaminathan, K. V. Narayanan, *Chem. Rev.* **1971**, *71*, 429–438.

[15] B. M. Trost, R. C. Livingston, *J. Am. Chem. Soc.* **1995**, *117*, 9586–9587.

[16] T.-A. Mitsudo, H. Naruse, T. Kondo, Y. Ozaki, Y. Watanabe, *Angew. Chem.* **1994**, *106*, 595–597; *Angew. Chem. Int. Ed. Engl.* **1994**, *33*, 580–581.

[17] T.-A. Mitsudo, K. Kokuryo, T. Shinsugi, Y. Nakagawa, Y. Watanabe, Y. Takegami, *J. Org. Chem.* **1979**, *44*, 4492–4496.

[18] T.-A. Mitsudo, Y. Hori, Y. Watanabe, *J. Organomet. Chem.* **1987**, *334*, 157–167.

[19] P. G. Gassman, J. W. Mickelson, J. R. Sowa Jr., *J. Am. Chem. Soc.* **1992**, *114*, 6942–6944.

[20] R. P. Hughes, X. Zheng, R. L. Ostrander, A. L. Rheingold, *Organometallics* **1994**, *13*, 1567–1568.

[21] S. Dérien, P. H. Dixneuf, *J. Chem. Soc., Chem. Commun.* **1994**, 2551–2552.

[22] H. Nagashima, K. Mukai, Y. Shiota, K. Yamaguchi, K.-i. Ara, T. Fukahori, H. Suzuki, M. Akita, Y. Moro-oka, K. Itoh, *Organometallics* **1990**, *9*, 799–807.

[23] M. L. S. Almeida, M. Beller, G.-Z. Wang, J.-E. Bäckvall, *Chem. Eur. J.* **1996**, *2*, 1533–1536.

[24] T. Braun, O. Gevert, H. Werner, *J. Am. Chem. Soc.* **1995**, *117*, 7291–7292.

[25] T. Hayashi in *Ferrocenes: Homogeneous Catalysis. Organic Synthesis, Materials Science*; A. Togni, T. Hayashi, Ed.; VCH, Weinheim, **1995**; pp 105–142.

[26] T. Hayashi, A. Ohno, S.-j. Lu, Y. Matsumoto, E. Fukuyo, K. Yanagi, *J. Am. Chem. Soc.* **1994**, *116*, 4221–4226.

[27] M. A. Halcrow, F. Urbanos, B. Chaudret, *Organometallics* **1993**, *12*, 955–957.

Transition Metal Catalyzed Synthesis of Seven-Membered Carbocyclic Rings

Gerald Dyker

The Diels Alder reaction is the most important method for the preparation of functionalized cyclohexenes and is characterized by an almost inexhaustible range of possible variations: numerous functional groups and even heteroatoms can be tolerated and both diastereoselective and enantioselective syntheses are possible. Thus, there has been no shortage of attempts to develop homologous variants and to extend the range of application to the synthesis of seven-membered carbocyclic rings. In specific cases it has been possible to employ vinylcyclopropanes as "homodienes" and to carry out a cycloaddition with appropriate alkenes. [1] However, the mild reaction conditions of the example shown in Scheme 1 are an exception. This special case apparently profits from the fixed, favorable geometry of the heterocyclic-bridged vinylcyclopropane **1** and from the high reactivity of the dienophile **2**. The mechanism of this type of reaction was clarified by Klärner et al. performing kinetic measurements at high pressure. [1b, c]

For a few combinations of less reactive dienes and dienophiles, transition metal catalyzed variants of the Diels Alder reaction have been developed. An example is the cycloaddition of an unpolar diene and an unactivated alkyne; however, except when the reaction is catalyzed with iron, nickel, cobalt, or rhodium(I) complexes, the temperature required often causes competing decomposition, even for the intramolecular version. [2] Wilkinson's catalyst [3] – tris(triphenylphosphane)rhodium(I) chloride – frequently used for hydrogenations and for decarbonylations, permits the cyclization of **4** to the annelated cyclohexadiene **5** in excellent yield in only 15 minutes at 55 °C in trifluoroethanol as solvent (Scheme 2). [2c]

Recently, Wender, Takahashi, and Witulski [4] found that this rhodium complex also catalyzes the homologous Diels-Alder reaction [5] of vinylcyclopropanes with alkynes, leading to formation of cyclohepta-1,4-dienes. By this method, hitherto used exclusively as an

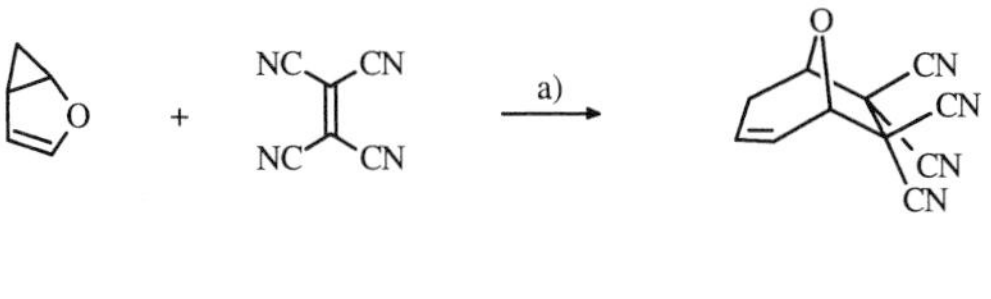

Scheme 1. a) CH$_2$Cl$_2$, 20 °C, quantitative yield.

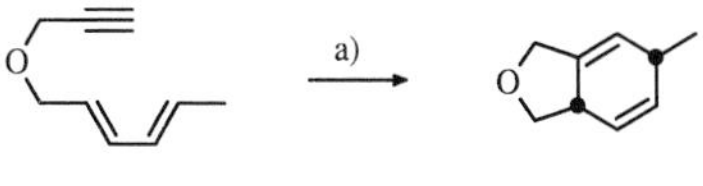

Scheme 2. a) 10 mol% [(PPh$_3$)$_3$RhCl], trifluoroethanol, 55 °C, 15 min, yield 96 %.

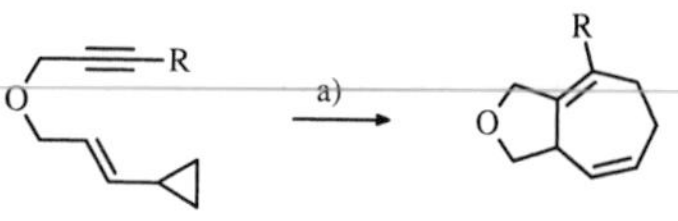

6   7

Scheme 3. R = H, Me, Ph, trimethylsilyl (TMS), CO$_2$Me: a) 10 mol% [(PPh$_3$)$_3$RhCl], toluene, 110 °C, 1.5 h, yield 50–88 %.

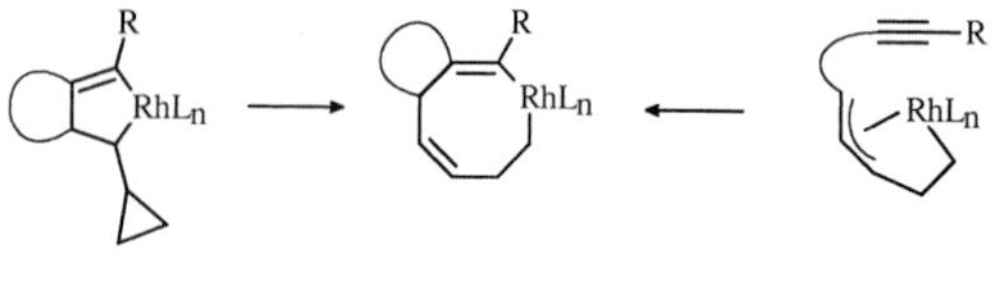

8   9   10

Scheme 4. Metallacycles as possible reactive intermediates of the reaction in Scheme 3. L = ligand.

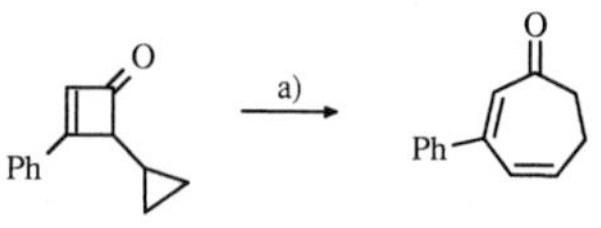

11   12

Scheme 5. a) 10 mol% [(PPh$_3$)$_3$RhCl], 60 °C, yield 84 %.

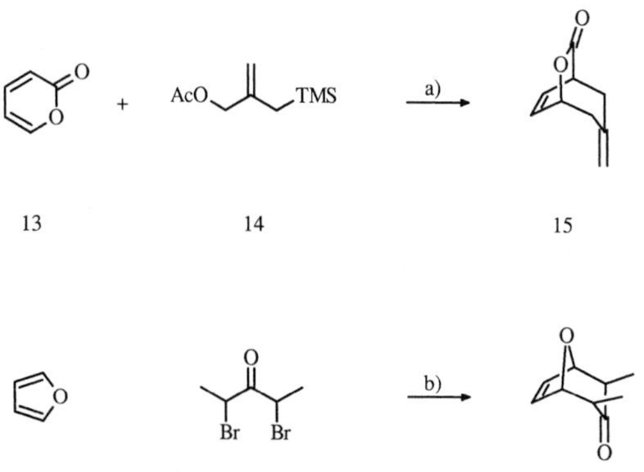

13   14   15

16   17   18

Scheme 6. a) 5 mol% Pd(OAc)$_2$, P(OiPr)$_3$, benzene 80 °C, yield 70 %; TMS: b) NaI, Cu, CH$_3$CN, 4 h, 50 °C, yield 40–48 %.

intramolecular process, annelated and functionalized seven-membered carbocycles such as **7** have been synthesized simply and efficiently (Scheme 3).

Acceleration of the reaction has been achieved by the use of the polar solvent trifluoroethanol and also by the addition of silver triflat; thus, it can be assumed that cationic rhodium complexes act as the active catalyst. Eight-membered metallacycles such as **9** are probably key intermediates. [6] Cyclopropyl-substituted five-membered metallacycles **8** and homoallyl complexes **10** can be considered as precursors of **9** [7] (Scheme 4).

Rhodium catalysis is also of crucial importance in the conceptually new type of synthesis of cyclohepta-2,4-dien-1-ones (e. g. **12**) by Huffman and Liebeskind. [8] The rearrangement of 4-cyclopropyl-2-cyclobutenones such as **11**, which are accessible in a few steps from squaric acid, [9] is similarly achieved with Wilkinson's catalyst (Scheme 5). This concept is particularly flexible in that the corresponding reaction of 4-cyclobutyl-2-cyclobutenones opens a route to eight-membered carbocyclic rings.

The new rhodium-catalyzed processes appear as competition to known and thoroughly tested methods for the synthesis of seven-membered carbocyclic rings, such as [4+3] cycloadditions catalyzed or induced by transition metals (Scheme 6).

According to Trost et al., [10] trimethylenemethane palladium complexes, which undergo cycloaddition with suitable dienes, can be prepared from the allylsilane **14** under catalytic conditions. Thus, the bridged cycloheptene **15**, containing an exocyclic methylene group, is obtained from α-pyrone **13**. Binger et al. [10c] found that methylenecyclopropane can be an advantageous reagent for the palladium-catalyzed synthesis of seven-membered rings from dienes. Cycloheptanones such as **18** can be prepared according to Hoffmann [11a–c] with α,α'-dibromoketones as coupling component. [11d, e] The preparative potential of such compounds for the diastereoselective synthesis of highly functionalized cycloheptanes and also of open-chain compounds has been demonstrated impressively quite recently. [12] It will be interesting to see whether the new rhodium-catalyzed processes also offer opportunities for enantioselective induction. [13, 14]

References

[1] a) R. Herges, I. Ugi, *Angew. Chem.* **1985**, *97*, 596–597; *Angew. Chem. Int. Ed. Engl.* **1985**, *24*, 594–596, and references therein; b) F.-G. Klärner, D. Schröer, *Chem. Ber.* **1985**, *122*, 179–185; c) T. Golz, S. Hammes, F.-G. Klärner, *Chem. Ber.* **1993**, *126*, 485–498.

[2] a) P. A. Wender, T. E. Jenkins, S. Suzuki, *J. Am. Chem. Soc.* **1995**, *117*, 1843–1844; b) L. McKinstry, T. Livinghouse, *Tetrahedron* **1994**, *50*, 6145–6154; c) R. S. Jolly, G. Luedtke, D. Sheehan, T. Livinghouse, *J. Am. Chem. Soc.* **1990**, *112*, 4965–4966; d) P. A. Wender, T. Jenkins, *ibid.* **1989**, *111*, 6432–6434; e) I. Matsuda, M. Shibata, S. Sato. Y. Izumi, *Tetrahedron Lett.* **1987**, *28*, 3361–3362; f) K. Mach, H. Antropiusova, L. Petrusova, F. Turecek, V. Hanus, P. Sedmera, J. Schraml, *J. Organomet. Chem.* **1985**, *289*, 331–339; g) H. tom Dieck, R. Diercks, *Angew. Chem.* **1983**, *95*, 801–802; *Angew. Chem. Int. Ed. Engl.* **1983**, *22*, 778–779; h) J. P. Genet, J. Ficini, *Tetrahedron Lett.* **1979**, 1499–1502; i) A. Carbonaro, A. Greco, G. Dall'Asra, *J. Org. Chem.* **1968**, *33*, 3948–3950.

[3] L. S. Hegedus, Organische Synthese mit Übergangsmetallen VCH, Weinheim, **1995**.

[4] P. A. Wender, H. Takahashi, B. Witulski, *J. Am. Chem. Soc.* **1995**, *117*, 4720–4721.

[5] The cycloaddition of norbornadiene with alkynes to give deltacyclenes has also been described as a homo-Diels-Alter reaction: M. Lautens, J. C. Lautens, A. C. Smith, *J. Am. Chem. Soc.* **1990**, *112*, 5627–5628.

[6] For an eight-membered metallacycle as reactive intermediate in the formation of a cycloheptadiene, see J. W. Herndon, G. Chatterjee,

P. P. Patel, J. J. Matasi, S. U. Turner, J. J. Harp, M. D. Reid, *J. Am. Chem. Soc.* **1991**, *113*, 7508–7509.

[7] Topical examples of ring opening of cyclopropyl-substituted transition metal complexes with formation of homoallyl metal compounds: a) R. I. Khusnutdinov, U. M. Dzhemilev, *J. Organomet. Chem.* **1994**, *471*, 1–18; b) I. Ryu, K. Ikura, Y. Tamura, J. Maenaka, A. Ogawa, N. Sonoda, *Synlett* **1994**, 941–942; c) S. Bräse, A. de Meijere, *Angew. Chem.* **1995**, 107, 2741–2743; *Angew. Chem. Int. Ed. Engl.* **1995**, *34*, 2545–2547.

[8] M. A. Huffman, L. S. Liebeskind, *J. Am. Chem. Soc.* **1993**, *115*, 4895–4896.

[9] L. S. Liebeskind, R. W. Fengl, K. R. Wirtz, T. T. Shawe, *J. Org. Chem.* **1988**, *53*, 2482–2488.

[10] a) B. M. Trost, S. Schneider, *Angew. Chem.* **1989**, 101, 215–217; ; *Angew. Chem. Int. Ed. Engl.* **1989**, *28*, 213–215; b) B. M. Trost, D. T. MacPherson, *J. Am. Chem. Soc.* **1987**, 109, 3483–3484; c) P. Binger, H. M. Büchi, *Top. Curr. Chem.* **1987**, *135*, 77–151.

[11] a) H. M. R. Hoffmann, *Angew. Chem.* **1984**, *96*, 29–48; *Angew. Chem. Int. Ed. Engl.* **1984**, *23*, 1–32; b) *ibid.* **1973**, *85*, 877–894 and **1973**, *12*, 819–835; c) M. R. Ashcroft, H. M. R. Hoffmann, *Org. Synth.* **1978**, *58*, 17–23; d) R. Noyori, *Acc. Chem. Res.* **1979**, *12*, 61–66; e) J. Mann, *Tetrahedron* **1986**, *42*, 4611–4659.

[12] a) M. Lautens, S. Kumanovic, *J. Am. Chem. Soc.* **1995**, *117*, 1954–1964; b) M. Lautens, *Pure Appl. Chem.* **1992**, *64*, 1873–1882.

[13] For other recent contributions to the synthesis of seven-membered carbocyclic rings, see: a) H. M. L. Davies, *Tetrahedron* **1993**, *49*, 5203–5223; b) A. Padwa, S. F Hornbuckle, G. E. Fryxell, Z. J. Zhang, *J. Org. Chem.* **1992**, *57*, 5747–5757; c) P. A. Wender, H. Y. Lee, R. S. Wilhelm, P. D. Williams, *J. Am. Chem. Soc.* **1989**, *111*, 8954–8957; d) K. E. Schwiebert, J. M. Stryker, *ibid.* **1995**, *117*, 8275–8276.

[14] For a detailed review on transition metal mediated cycloaddition reactions, see: M. Lautens, W. Klute, W. Tam, *Chem. Rev.* **1996**, *96*, 49–92.

[4+4]-Cycloaddition Reactions in the Total Synthesis of Naturally Occurring Eight-Membered Ring Compounds

Gerd Kaupp

Eight-membered rings can be obtained by [4+4]-cycloadditions of 1,3-dienes [1] via diradicals or other intermediates. Synthesis of such compounds has been achieved by thermal, [2] photochemical, [3] and by metal-catalyzed [4] processes: these reactions have been the subject of extensive mechanistic [5] and theoretical [5c] studies. Their strategic applications in natural product synthesis have been reviewed. [5d] The thermal version has generated little interest, except in orthoquinodimethane dimerizations and in cycloreversions; the Cope rearrangement of 1,2-divinylcyclobutanes [3] is more commonly used. [4+4]-Cycloadditions are also used with 1,3-dipoles or mesoionic heterocycles for the synthesis of six- and seven-membered rings. Sometimes also [6+4]-cycloadditions are

competing. [2b] Good yields can be obtained with the photolytic reaction when the 1,3-diene is fixed in the *s-cis* configuration. Such is the case for, amongst others, condensed arenes, α-pyrones, and 2-pyridones. Two coupled, linear [4+4]-cycloadditions lead to ten-membered rings. [2c] Catalysis by Ni^0 is the obvious method of choice for open-chain 1,3-dienes, favorable coordination (template effect) apparently solving the difficult problem of bringing the terminal groups of the bis-1,3-diene into contact. The choice of typical examples given in Scheme 1 shows that one, two, or four eight-membered rings, which are not connected by zero bridges, can be formed by [4+4]-cycloadditions.

Cyclooctanoids (eight-membered ring compounds) can also be found among biologically

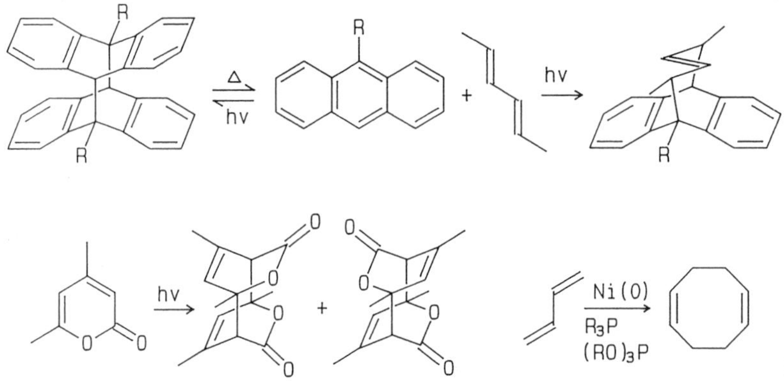

Scheme 1.

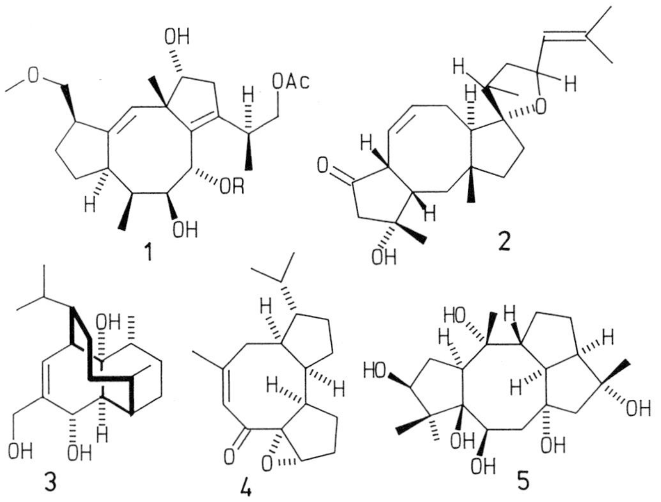

active natural products such as sesqui-, di-, and sesterterpenes, as well as non-terpenoids. The total synthesis of such compounds has been a challenge for many research groups for years. Most prominent was the more than 20 years old race for a total synthesis of the cancer chemotherapeutic taxol **20**. [6–8] All of these syntheses did not use [4+4] or other cyclovinylogous additions that could have saved steps in their long reaction sequences. Nevertheless, cycloaddition approaches in cyclooctanoid syntheses are of lasting value beyond total syntheses of taxol **20**. More practical partial syntheses of taxol have been patented. Also less highly functionalized analogues of taxol, which inhibit the depolymeri-

zation of tubulin, [9] have longer been prepared. Some of the best known naturally occuring cyclooctanoids include fusicoccin A (**1**), ophiobolin A (**2**), vinigrol (**3**), epoxybasmenone (**4**), kalmanol (**5**), (+)-asteriscanolide (**8**), crispolide, ceroplastin, vulgar olide, cotylenin, variecolin, pleuromutilin, taxusin, 7-deoxytaxol, 12,13-isobaccatin among more than 100 naturally occurring taxanes.

Strategies have been developed for the synthesis of these compounds, in some cases successfully, that do not depend on cycloadditions for formation of the eight-membered ring. The concern, unjustly, [3–4] was that the control of the regio- and stereoselectivity would not be possible and that cross-dimeriza-

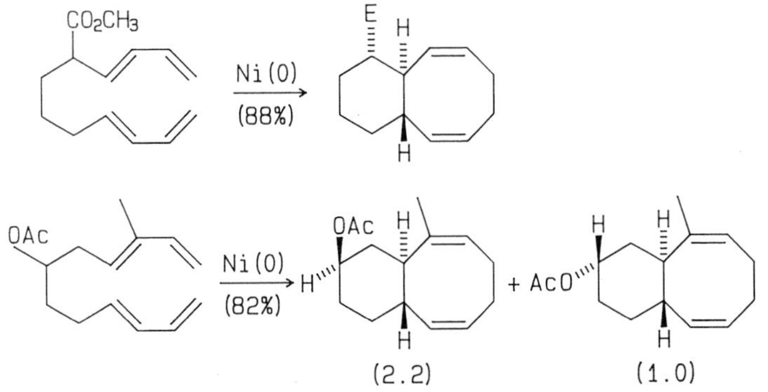

Scheme 2.

tion with substituted 1,3-dienes would lead to poor yields.

It was not until 1986 that Wender et al. showed [10] that it was possible to conduct stereoinduced Ni⁰-catalyzed [4+4]-cyclo-additions, even in the presence of oxygen-containing substituents (Scheme 2). Thereafter, the same group achieved the enantiose-lective synthesis of (+)-asteriscanolide **8**, a sesquiterpene. [11] With this approach they also planned on exploring further possibilities for the synthesis of cyclooctanoid diterpenes of the taxane series, such as **20**. [12] The key step in the synthesis of **8** was the Ni⁰-cata-lyzed intramolecular [4 + 4]-cycloaddition of **6** (accessible in 11 % yield from acrolein in nine steps) to give **7** in 67 % yield (leading to a 36 % yield of **8** in three further steps): a satisfyingly short and elegant total synthesis.

Presumably stimulated by the success of template-controlled syntheses, Sieburth et al. hoisted the flag for tether guided photochemi-cal [4+4]-cycloadditions. [13] This case re-quires, of course, that two *s-cis*-constrained dienes are connected to each other. The choice thus fell on 2-pyridones. Target molecules are fusicoccin **1** and the cancer chemothera-peutic agent taxol **20** (cf. [9]). The 3,6-bis(2-pyridone) **9**, used as the racemate but undoub-tedly separable into the enantiomers, provides favorable prerequisites for the [4+4]-photocy-cloaddition. The nitrogen atoms, the carbonyl groups, and the double bonds are so arranged that further reaction to give **1** by standard methods appears to be possible. Indeed, the epimers **10** were formed by photolysis in good yield (66 and 75 %). Moreover, suitable choice of solvent (protic or aprotic) even al-lowed preferential formation of one of the epi-

mers. Extension of this work to **11** opens the way to fusicoccines **1**. **12** is the only product in methanol and **13** is the only product in ben-zene, the difference almost certainly arising from two intramolecular hydrogen bridges in the aprotic solvent. [14] The bis-*N*-methylated derivative of **11** yields only the *trans* isomer. A systematic study of the photodimerizability and stereoselectivity in tethered pyridones has been performed [15] and extended to regiose-lection and intermolecular examples. [16] Of particular interest is, however, a thermal-pho-tochemical cycle to increase the yield of *trans*-cycloadduct, as shown with **14** to give **15**. The by-product **16** experiences thermal Cope rearrangement to give **17**, which upon photolysis provides **14**. Two cycles increase the initial 2 : 1 ratio of **15** and **16** to 18 : 1. [17]

Sieburth et al. also photolyzed the racemic 3,6-bis(2-pyridone) **18**, in which the two rings are connected by a four-membered chain, obtaining both epimers of the second-ary alcohol **19** (2 : 3, 63 %). A certain resem-blance to taxol **20** is clearly present, but there is still a long way to go. This interim result must be compared with that by Wender et al. in 1987, in which the [4+4]-cycloaddition was catalyzed with Ni⁰. [12] Using conver-sion of **21** to **22** and **23** to **24** as model reac-tions allows the eight-membered ring and each of the two six-membered rings to be con-structed, although not simultaneously. A starting compound such as **25**, in which both possibilities are combined, and from which conversion to **20** is feasible with standard reactions after treatment with Ni⁰, is not yet known. However, **25** contains the skeleton of cembrene (14-isopropyl-3,7,11-trimethyl-1,3,6,10-cyclotetradecatetraene), a

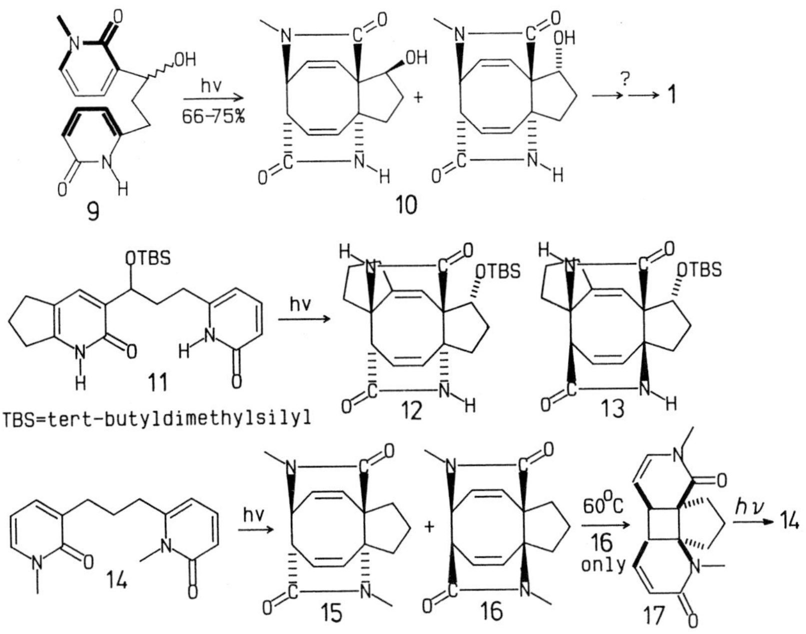

TBS=tert-butyldimethylsilyl

widespread diterpene, though with all four double bonds 1,3-shifted. Cembrene is synthesized in nature by cyclization of geranylgeraniol pyrophosphate, it is a possible precursor (or side product) in taxane biosynthesis and it has been suggested to use its isomer cembrene B for closing the taxane A- and C-rings, though separately. [18] A starting material similar to compound **23** with a methyl group at C4 and the *tert*-butyldimethylsiloxy group at C7 was used to initiate a crispolide synthesis: the [4+4]-cycloaddition succeeded with 74% yield. [19] But neither this "model" nor **23** is suitable for the synthesis of taxol **20**, of course. After the first total synthesis of taxol by Holton et al. [6] Sieburth pursued his approach with a photolysis of **26** and approached taxol more closely (an additional quaternization with a 2-hydroxypropyl group was not possible). Labile **27** was quantitatively obtained and could be selectively hydrogenated at the disubstituted double bond. The residual double bond could be stereospecifically epoxidized followed by ad-

dition of methanol. [15] However, recently it was possible to fully functionalize the eight-membered ring to give **28** with 65% yield. [14]

Thus, Sieburth et al. have come pretty close to taxol (**20**) and fusicoccin (**1**) in an elegant way. Their results have provided vital new impetus for the total synthesis of naturally occurring cyclooctanoids, many of them exhibiting mitotic inhibition that is so typical for this class of compounds. It should always be attempted, by means of stereoselective cycloaddition, to save steps in a total synthesis.

The foundation for the application of [4+4]-cycloadditions [5b] for the most varied experimental conditions already exists and has been adopted now to natural product syntheses. That event has also stimulated further synthetic work. [20] One should also not forget, that *cyclovinylogous additions* [5b] contain still further ideas for saving synthetic steps that have been used previously and adopted recently. For example, eight-membered rings can also be constructed by [2+2]-cycloaddition of cyclobutenes or of alkynes to (hetero)cyclohexenes or arenes followed by further ring expansion. [3a, 21] Diels-Alder reactions between components connected by a bridging group with six carbon atoms lead to cyclooctanone-annelated cyclohexenes, [22] homo-Diels-Alder ([4+3]-) [23] and [6+2]-additions [24] provide bridged cyclooctanoids. Further perspectives for the synthesis of natural products are opened up by the cyclovinylogous [6+4]- [2b, 5b, 24] (for ten-membered rings, see **3**, **20**, pleuromutilin), [6+6]- [25] and [4+4+4]-additions [4] (twelve-membered rings, see **5**, **20**) and are often investigated in parallel with the [4+4]-additions. Thus, synthesis strategies aimed at the largest ring present containing no zero bridges should definitely also be considered. The work of Wender [10–12, 19] and Sieburth [13–17] appears to illustrate a renaissance of cycloaddition reactions of polyenes, with very promising potential.

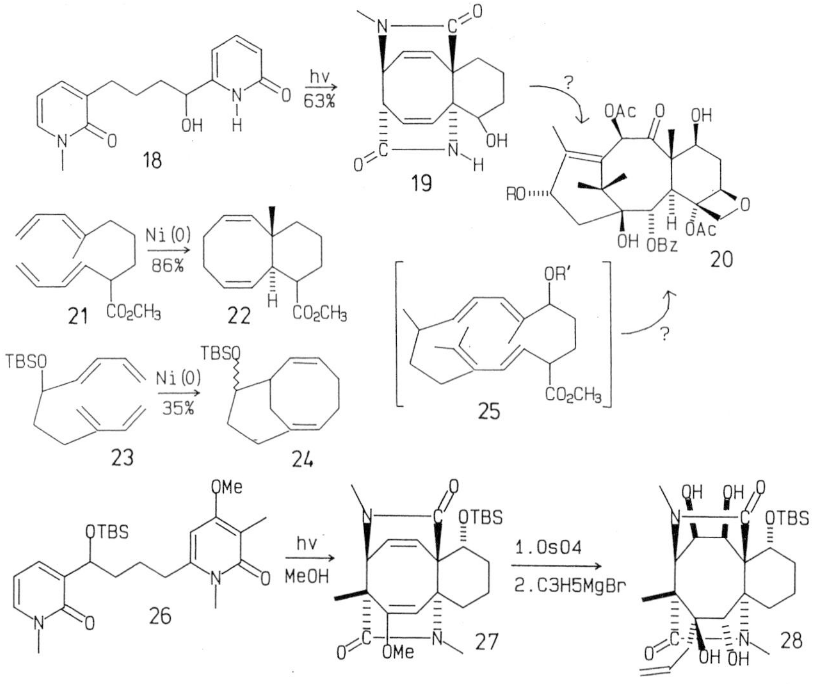

References

[1] H. Bouas-Laurent, J. P. Desvergne, *Stud. Org. Chem. (Amsterdam)* **1990**, *40*, 561–630.

[2] a) J. Wagner, J. Bendig, A. Felber, A. Sommerer, D. Kreysig, *Z. Chem.* **1985**, *25*, 64; b) W. Friedrichsen, W. D. Schröer, T. Debaerdemaeker, *Liebigs Ann. Chem.* **1980**, 1850–1858; W. Friedrichsen, H.-G. Oeser, *ibid.* **1978**, 1161–1186; c) S. A. Ali, M. I. M. Wazeer, *J. Chem. Soc. Perkin Trans. 2* **1986**, 1789–1792; J. Thesing, H. Mayer, *Chem. Ber.* **1956**, *89*, 2159–2167; J. Basan, H. Mayr, *J. Am. Chem. Soc.* **1987**, *109*, 6519–6521; H. Meier, U. Konnert, S. Graw, T. Echter, *Chem. Ber.* **1984**, *117*, 107–126; d) [6+2]-cycloadditions with 1,5-dipoles: P. A. Wender, J. L. Mascarenas, *J. Org. Chem.* **1991**, *56*, 6267–6269; e) A. Katritzky S. I. Bayyuk, N. Dennis, G. Musumarra, E.-U. Würthwein, *J. Chem. Soc. Perkin Trans 1* **1979**, 2535–2541 (classified there as an "allowed" photoreaction).

[3] a) G. Kaupp, *Methoden Org. Chem. (Houben-Weyl)* 4th Ed. **1975**, Vol. 4/5a, p. 278ff.; 360ff.; E. Leppin, *ibid.*, p. 476ff.; 484ff.; V. Zanker, *ibid.* p. 586ff.; 616ff.; b) Y. Nakamura, T. Kato, Y. Morita, *J. Chem. Soc. Perkin Trans. 1* **1982**, 1187–1191.

[4] G. Wilke, *Angew. Chem.* **1963**, *75*, 10–20; *Angew. Chem. Int. Ed. Engl.* **1963**, *2*, 105–115; P. Heimbach, P. W. Jolly, G. Wilke in *Advances in Organometallic Chemistry*, Vol. 8 (Eds.: F. G. A. Stone, R. West), *Academic Press, New York*, **1970**, p. 29ff.; P. W. Jolly, G. Wilke, *The Organic Chemistry of Nickel*, Vol. 2, Academic Press, New York, **1975**.

[5] a) J. Saltiel, R. Dabestani, K. S. Schanze, D. Trojan, D. E. Townsend, V. L. Goedken, *J. Am. Chem. Soc.* **1986**, *108*, 2674–2687, and references cited therein; b) G. Kaupp, E. Teufel, *Chem. Ber.* **1980**, *113*, 3669–3674; G. Kaupp, H.-W. Grüter, E. Teufel, *ibid.*, **1983** *116*, 630–644; G. Kaupp, D. Schmitt, *ibid.*, **1981**, *114*, 1567–1571; **1980**, *113*, 1458–1471; G. Kaupp, H.-W. Grüter, *ibid*, **1980**, *113*, 1626–1631; *Angew. Chem.* **1979**, *91*, 943–944; *Angew. Chem. Int. Ed. Engl.* **1979**, *18*, 881–882; G. Kaupp, *Liebigs Ann. Chem.* **1973**, 844–878; c) M. J. Bearpark, M. Deumal, M. A. Robb, T. Vreven, N. Yamamoto, M. Olivucci, F. Bernardi, *J. Am. Chem. Soc.* **1997**, *119*, 709–718; d) G. Kaupp, *Angew. Chem.* **1992**, *104*, 435–437; *Angew. Chem. Int. Ed. Engl.* **1992**, *31*, 422–424; S. M. Sieburth, N. T. Cunard, *Tetrahedron* **1996**, *52*, 6251–6282.

[6] R. A. Holton, C. Somoza, H. B. Kim, F. Liang, R. J. Biediger, P. D. Boatman, M. Shindo, C. C. Smith, S. Kim, H. Nadizadeh, Y. Suzuki, C. Tao, P. Vu, S. Tang, P. Zhang, K. K. Murthi, L. N. Gentile, J. H. Liu, *J. Am. Chem. Soc.* **1994**, *116*, 1597–1598; **1994**, *116*, 1599–1600 (submitted Dez. 21, 1993).

[7] K. C. Nicolaou, Z. Yang, J. J. Liu, H. Ueno, P. G. Nantermet, R. K. Guy, C. F. Claiborne, J. Renaud, E. A. Couladouros, K. Paulvannan, E. J. Sorensen, *Nature* **1994**, *367*, 630–634 (submitted Jan. 24, 1994); K. C. Nicolaou, R. K. Guy, *Angew. Chem.* **1995**, *107*, 2247–2259, *Angew. Chem. Int. Ed. Engl.* **1995**, *34*, 2079–2091.

[8] J. J. Masters, J. T. Link, L. B. Snyder, W. B. Young, S. J. Danishefsky, *Angew. Chem.* **1995**, *107*, 1886–1888, *Angew. Chem. Int. Ed. Engl.* **1995**, *34*, 1723–1725.

[9] S. Blechert, A. Kleine-Klausing, *Angew. Chem.* **1991**, *103*, 428–429; *Angew. Chem. Int. Ed. Engl.* **1991**, *30*, 412–414; F. Gueritte-Voegelein, D. Guenard, F. Lavelle, M. T. LeGoff, L. Mangatal, P. Potier, *J. Med. Chem.* **1991**, *34*, 992–998.

[10] P. A. Wender, N. C. Ihle, *J. Am. Chem. Soc.* **1986**, *108*, 4678–4679; *Tetrahedron Lett.* **1987**, *28*, 2451–2454.

[11] P. A. Wender, N. C. Ihle, C. R. D. Correia, *J. Am. Chem. Soc.* **1988**, *110*, 5904–5906.

[12] P. A. Wender, M. L. Snapper, *Tetrahedron Lett.* **1987**, *28*, 2221–2224; M. L. Snapper, Stanford University, *Diss. Abstr. Int. B* **1991**, *52*, 248–249.

[13] S. M. Sieburth, J.-L. Chen, *J. Am. Chem. Soc.* **1991**, *113*, 8163–8164.

[14] S. M. Sieburth, *Private Communication*, March 22, 1997.

[15] S. M. Sieburth, J. Chen, K. Ravindran, J.-L. Chen, *J. Am. Chem. Soc.* **1996**, *118*, 10803–10810.

[16] S. M. Sieburth, B. Siegel, *J. Chem. Commun.* **1996**, 2249–2250; S. M. Sieburth, C. H. Lin, *Tetrahedron Lett.* **1996**, *37*, 1141–1144.

[17] S. M. Sieburth, C. H. Lin, *J. Org. Chem.* **1994**, *59*, 3597–3599.

[18] T. Frejd, G. Magnusson, L. Pettersson, *Chem. Scr.* **1987**, *27*, 561–562; L. Pettersson, T. Frejd, G. Magnusson, *Tetrahedron Lett.* **1987**, *28*, 2753–2756; *Acta Chem. Scand.* **1993**, *47*, 196–207.

[19] P. A. Wender, M. J. Tebbe, *Synthesis* **1991**, 1089–1094; the reaction conditions for the conversions 6 → 7 from ref. [11] and 23 → 24 from ref. [12] are not given here for comparative purposes.

[20] F. G. West, C. E. Chase, A. M. Arif, *J. Org. Chem.* **1993**, *58*, 3794–3795.

[21] J. G. Atkinson, D. E. Ayer, G. Büchi, E. W. Robb, *J. Am. Chem. Soc.* **1963**, *85*, 2257–2263; D. Bryce-Smith, A. Gilbert, J. Grzonka, *J. Chem. Soc. Chem. Commun.* **1970**, 498–499; W. C. Agosta, W. W. Lowrance, *J. Org. Chem.* **1970**, *35*, 3851–3856; G. Kaupp, M. Stark, *Angew. Chem.* **1977**, *89*, 555–556;

Angew. Chem. Int. Ed. Engl. **1977**, *16*, 552–553, and literature cited therein; G. Kaupp, U. Pogodda, A. Atfah, H. Meier, A. Vierengel, *Angew. Chem.* **1992**, *104*, 783–785; *Angew. Chem. Int. Ed. Engl.* **1992**, *31*, 768–770; cf. also P. A. Wender, C. J. Manly, *J. Am. Chem. Soc.* **1990**, *112*, 8579–8581 (synthesis of ten-membered rings).

[22] K. Sakan, D. A. Smith, S. A. Babirad, F. R. Fronczek, K. N. Houk, *J. Org. Chem.* **1991**, *56*, 2311–2317; J. S. Yadav, R. Ravishankar, *Tetrahedron Lett.* **1991**, *23*, 2629–2632; R. V. Bonnert, P. R. Jenkins, *J. Chem. Soc. Perkin Trans. 1* **1989**, 413–418.

[23] M. Harmata, S. Elahmad, C. L. Barnes, *J. Org. Chem.* **1994**, *59*, 1241–1242.

[24] J. H. Rigby, *Acc. Chem. Res.* **1993**, *26*, 579–585.

[25] See, for example L. A. Paquette, J. H. Barrett, D. E. Kuhla, *J. Am. Chem. Soc.* **1969**, *91*, 3616–3624; G. Kaupp, E. Teufel, H. Hopf, *Angew. Chem.* **1979**, *91*, 232–234; *Angew. Chem. Int. Ed. Engl.* **1979**, *18*, 215–217, and references therein for the synthesis of para-cyclophanes via para-quinodimethanes.

"New" Reagents for the "Old" Pinacol Coupling Reaction

Thomas Wirth

An old-timer in the history of chemistry, the pinacol coupling, was first described nearly 140 years ago in a publication on the synthesis of pinacols. [1a] Today this reaction is still a versatile tool for chemists. This "longevity" can be explained by the continuous development of improved reagents and by the intrinsic elegance of the method for the preparation of 1,2-diols (pinacols) by the reductive coupling of carbonyl compounds. [1] Starting with two carbonyl functionalities a carbon–carbon bond is formed and two new adjacent stereo-centers created. 1,2-Diols are versatile inter-mediates in synthesis; for example, they can be used for the preparation of ketones by the pinacol rearrangement or alkenes with the McMurry reaction (Scheme 1).

Whether the reductive coupling of carbonyl compounds leads to 1,2-diols or to the deoxygenated or rearranged products is dependent on the oxygen affinity of the reducing agent

employed. The McMurry-reaction uses low-valent titanium compounds, which under certain reaction conditions can transform the 1,2-diol intermediates into alkenes by rapid deoxygenation (Scheme 1). When this reaction was performed on the surface of reduced titanium dioxide, the intermediate 1,2-diols could be isolated. [2]

Reducing agents suitable for the synthesis of pinacols must allow the reaction to be stopped at the 1,2-diol stage. One of the first practical reductants for pinacol coupling reactions was the Mg/MgI_2 system reported by Gomberg and Bachmann. [3] More recently these reducing agents have been extended by a magnesium-graphite system, which is competitive with other currently available reductants. [4] Different low-valent titanium compounds have proved to have similar efficiency for pinacol coupling reactions. [5] They have been used for intramolecular couplings as

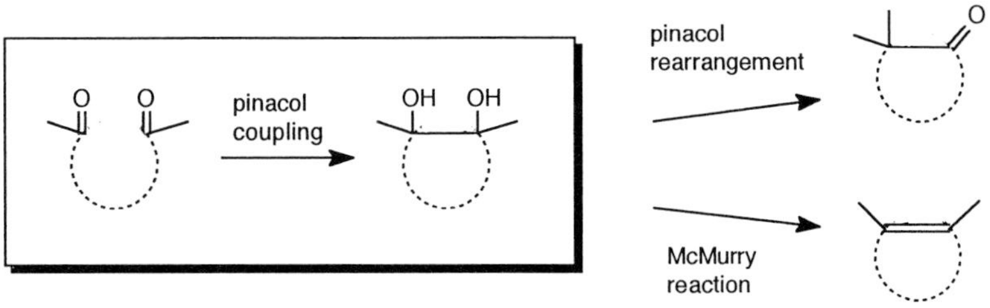

Scheme 1.

well as for the formation of unsymmetrical pinacols, and have demonstrated their potential in several key steps of natural product syntheses. [6] Recent developments show that low-valent titanium, manganese, as well as zinc reagents can be used in the pinacol coupling reaction in aqueous media. [7] Also with catalytic amounts of low-valent titanium reagents the highly diastereoselective synthesis of coupling products is possible. [8]

Typical mechanisms for pinacol couplings are shown in Scheme 2. In the first reduction step a ketyl radical is formed which can dimerize (path A) or add to a second carbonyl group, forming a C–C bond. In path B a second one-electron reduction must then follow. For pinacol coupling reactions mediated by transition metals the insertion of a carbonyl group into the metal–carbon bond of initially formed metal oxiranes has been proposed (path C).

Pinacol coupling reactions can lead to either the *syn-* or *anti-*diols. The stereochemical course of the reaction depends on the reducing agents and, of course, on the structure of the carbonyl compounds. Recent studies employing reagents that form the C–C bond according pathways B and C have met with success.

Previously, tin-ketyl radicals have been added to alkenes only in an intramolecular fashion. [9] In recent publications, however, pinacols and amino alcohols have been prepared by cyclisation of dicarbonyl compounds [10] or keto-oximes [11] with tributyltin hydride. Cyclisation of 1,5-ketoaldehydes **1** and 1,5-dialdehydes with tributyltin hydride yields *cis-*diols **2** with excellent stereoselectivities, whereas the keto-oxime **4** with four benzyloxy-substituents affords a 58 : 42 (*cis* : *trans*) mixture. The *trans-*product was transformed in two more steps to the potent glycosidase inhibitor 1-deoxynojirimycin (**6**). [11b] The reversibility of both the addition of the tributyltin-radical to the carbonyl group and the intramolecular radical C–C bond formation is believed to be responsible for the high selectivity in the formation of **2**. In the cyclisation of 1,5-pentanedial the unhydrolyzed coupling product **3** could be isolated, therefore providing evidence for a new mechanistic variant of the pinacol reaction, in which only 1.2 equivalent of the reducing agent are necessary.

1,6-Dicarbonyl compounds can also be converted into the corresponding pinacols. Tributyltin hydride can again be used as an efficient reducing agent, this time for the ste-

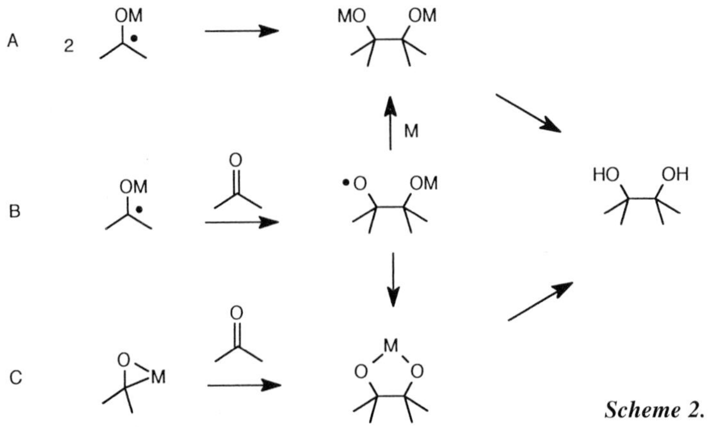

Scheme 2.

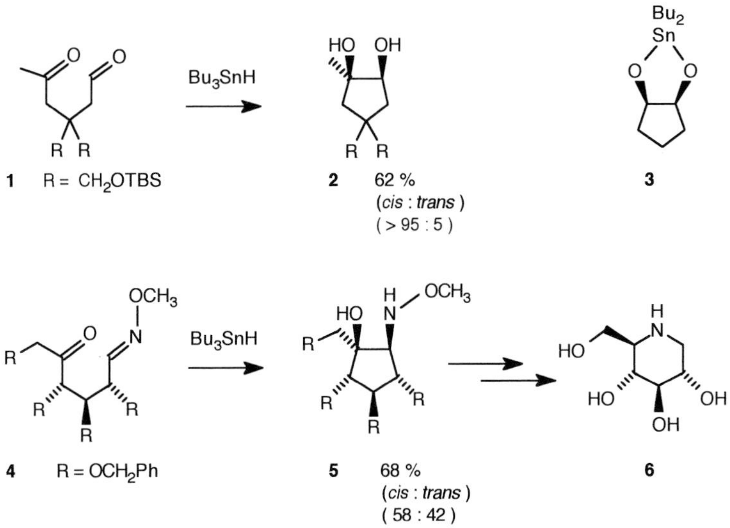

Scheme 3.

reoselective synthesis of **7**. [10] Other well established pinacol forming reagents such as low-valent titanium compounds [1b, c] or samarium diiodide [12] can cyclize 1,n-dicarbonyl compounds with comparable stereoselectivities to give 1,2-diols. X-ray crystal structures of a samarium ketyl complex and a samarium pinacolate are further evidence of the reversibility of the coupling reaction under appropriate conditions. [13] An excess of samarium diiodide was used as pinacol coupling reagent for the synthesis of the optically active inositol derivative **8** [12b] or in the highly *threo* selective coupling of tricarbonylchromium complexes of benzaldehydes. [12d] An acceleration of the pinacol coupling with samarium diiodide is possible by addition of trimethylsilyl chloride. [14] Even catalytic amounts of samarium diiodide are sufficient for the pinacol coupling reaction, when magnesium is employed as reductant for the conversion of samarium(III) to samarium(II). [15]

Low-valent compounds of other early transition metals were shown to be effective reagents for the pinacol coupling. Zirconium,

niobium, and ytterbium [16] reagents form metal oxiranes with carbonyl compounds; however, the mechanism of the recation with vanadium compounds is not clear at present. The mechanism of the pinacol coupling via metal oxiranes is supported by X-ray structure analysis of an intermediate. [17] Vanadium and niobium compounds seem to be particularly attractive reagents for intermolecular pinacol coupling reactions.

The vanadium complex [V$_2$Cl$_3$(thf)$_6$]$_2$ [Zn$_2$Cl$_6$] (Caulton's reagent) [18] is suitable for the intermolecular coupling of functionalized aldehydes which bear further coordination sites, for example amino aldehydes. [19] The formation of a bidentate or tridentate metal-aldehyde complex is proposed to be the origin of the *threo* selectivity observed in this reaction. The coupling products like **9** are interesting target structures because they are potential inhibitors of HIV-proteases. They can even be synthesized on the kg-scale with good yields. [19b] Improvements were reported in the synthesis of Caulton's reagent avoiding the highly air-sensitive VCl$_3$(thf)$_3$. [19c] Efficient pinacol cross coupling

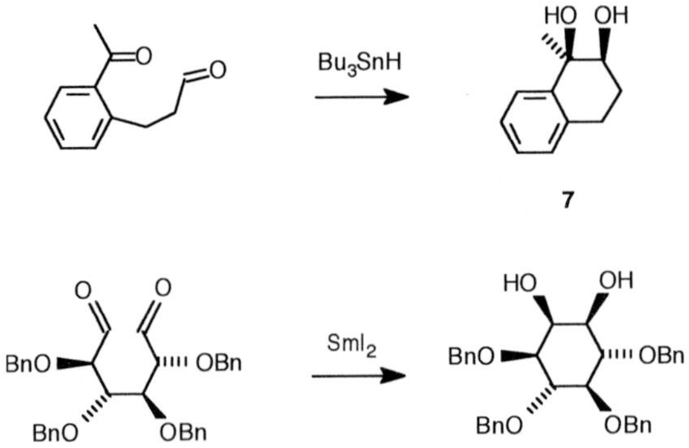

53 %
(*cis* : *trans*)
(95 : 5)

7

60 %
(*cis* : *trans*)
(93 : 7)

8   *Scheme 4.*

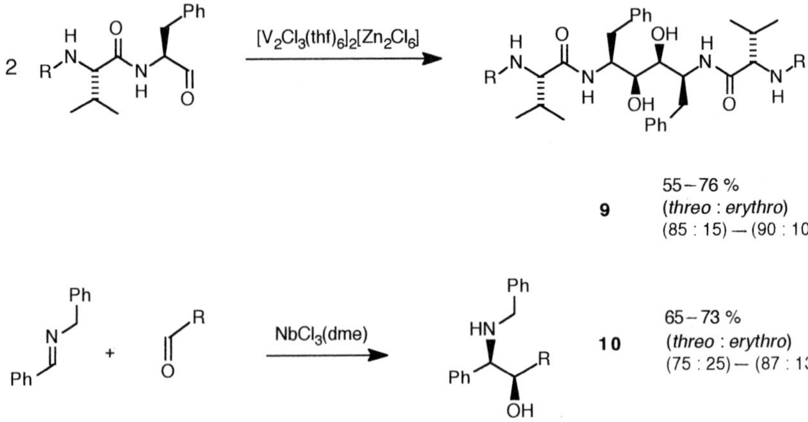

55–76 %
9   (*threo* : *erythro*)
(85 : 15) — (90 : 10)

65–73 %
10   (*threo* : *erythro*)
(75 : 25) — (87 : 13)

Scheme 5.

reactions are possible with this reagent [20] and even first catalytic systems with low-valent vanadium species have been reported. [21]

The niobium reagent [NbCl₃(dme)] can be used to couple imines with carbonyl compounds giving amino alcohols such as **10**, although generally lower stereoselectivity is observed. [6] In analogy to reactions with other early transition metals the mechanism is believed to proceed via a niobaziridine, which adds to the carbonyl functionality.

The relative and absolute stereochemistry in pinacol coupling reactions, for instance in the natural product syntheses mentioned earlier, [6] is always controlled by the stereocenters of the substrates. Very little work has concentrated on the control of the absolute stereochemistry in pinacol coupling reactions. [1b, 23] Because of the potential of the reagents presented here, the development of chiral coupling reagents will no doubt be reported in the near future.

References

[1] a) R. Fittig, *Justus Liebigs Ann. Chem.* **1859**, *110*, 23–45; b) review: G. M. Robertson in *Comprehensive Organic Synthesis*, Vol. 3 (Eds.: B. M. Trost, I. Fleming, G. Pattenden), Pergamon, Oxford, **1991**, p. 563; c) A. Fürstner, *Angew. Chem.* **1993**, *105*, 171–197; *Angew. Chem. Int. Ed. Engl.* **1993**, *32*, 164–189; d) T. Wirth, ibid. **1996**, *108*, 65–67 and **1996**, *35*, 61–63.

[2] K. G. Pierce, M. A. Barteau, *J. Org. Chem.* **1995**, *60*, 2405–2410.

[3] M.Gomberg, W. E. Bachmann, *J. Am. Chem. Soc.* **1927**, *49*, 236–257.

[4] A. Fürstner, R. Csuk, C. Rohrer, H. Weidmann, *J. Chem. Soc. Perkin Trans. 1* **1988**, 1729–1734.

[5] J. E. McMurry, *Chem. Rev.* **1989**, *89* 1513–1524.

[6] a) E. J. Corey, R. L. Danheiser, S. Chandrasekaran, P. Siret, G. E. Keck, J. Gras, *J. Am. Chem. Soc.* **1978**, *100*, 8031–8034; b) C. S. Swindell, W. Fan, P. G. Klimko, *Tetrahedron Lett.* **1994**, *35*, 4959–4962; c) K. C. Nicolaou, Y. Zang, J. J. Liu, H. Ueno, P. G. Nantermet, R. K. Guy, C. F. Claiborne, J. Renaud, E. A. Couladouros, K. Paulvannan, E. J. Sorensen, *Nature* **1994**, *367*, 630–634; d) C. S. Swindell, W. Fan, *Tetrahedron Lett.* **1996**, *37*, 2321–2324; e) C. S. Swindell, M. C. Chander, J. M. Heerding, P. G. Klimko, L. T. Rahman, J. V. Raman, H. Venkataraman, *J. Org. Chem.* **1996**, *61*, 1101–1108; f) C. S. Swindell, W. Fan, *ibid.* **1996**, *61*, 1109–1118; g) X. Yue, Y. Li, *Synthesis* **1996**, 736–740.

[7] a) M. C. Barden, J. Schwartz, *J. Am. Chem. Soc.* **1996**, *118*, 5484–5485; b) C. Li, Y. Meng, X. Yi, J. Ma, T. Chan, *J. Org. Chem.* **1997**, *62*, 8632–8633; c) T. Tsukinoki, T. Kawaji, I. Hashimoto, S. Mataka, M. Tashiro, *Chem. Lett.* **1997**, 235–236.

[8] a) A. Gansäuer, *Chem. Commun.* **1997**, 457–458; b) A. Gansäuer, *Synlett* **1997**, 363–364.

[9] B. Giese, B. Kopping, T. Göbel, J. Dickhaut, G. Thoma, K. J. Kulicke, F. Trach in *Organic Reactions*, Vol. 48 (Ed.: L. A. Paquette), Wiley, New York, **1996**, p. 301.

[10] a) D. S. Haysm G. C. Fu, *J. Am. Chem. Soc.* **1995**, *117*, 7283–7284; b) T.-H. Chuang,

J.-M. Fang, W.-T. Jiaang, Y.-M. Tsai, *J. Org. Chem.* **1996**, *61*, 1794–1805.

[11] a) T. Naito, K. Tajiri, T. Harimoto, I. Ninomiya, T. Kiguchi, *Tetrahedron Lett.* **1994**, *35*, 2205–2206; b) T. Kiguchi, K. Tajiri, I. Ninomiya, T. Naito, H. Hiramatsu, *ibid.* **1995**, *36*, 253–256; c) J. Tormo, D. S. Hays, G. C. Fu, *J. Org. Chem.* **1998**, *63*, 201–202.

[12] a) J. L. Chiara, W. Cabri, S. Hanessian, *Tetrahedron Lett.* **1991**, *32*, 1125–1128; b) J. L. Chiara, M. Martín-Lomas, *ibid.* **1994**, *35*, 2969–2972; c) J. P. Guidot, T. Le Gall, C. Mioskowski, *ibid.* **1994**, *35*, 6671–6672; d) J. L. Chiara, N. Valle, *Tetrahedron: Asymmetry* **1995**, *6*, 1895–1898; e) J. L. Chiara, J. Marco-Contelles, N. Khiar, P. Callego, C. Destabel, M. Bernabi, *J. Org. Chem.* **1995**, *60*, 6010–6011; f) N. Taniguchi, N. Kaneta, M. Uemura, *ibid.* **1996**, *61*, 6088–6089.

[13] a) Z. Hou, T. Miyano, H. Yamazaki, Y. Wakatsuki, *J. Am. Chem. Soc.* **1995**, *117*, 4421–4422; b) Z. Hou, A. Fujita, H. Yamazaki, Y. Wakatsuki, *J. Am. Chem. Soc.* **1996**, *118*, 7843–7844.

[14] T. Honda, M. Katoh, *Chem. Commun.* **1997**, 369–370.

[15] R. Nomura, T. Matsuno, T. Endo, *J. Am. Chem. Soc.* **1996**, *118*, 11666–11667.

[16] a) Y. Taniguchi, K. Nagata, T. Kitamura, Y. Fujiwara, D. Deguchi, M. Maruo, Y. Makioka, K. Takaki, *Tetrahedron Lett.* **1996**, *37*, 3465–3466; b) Y. Makioka, Y. Taniguchi, Y. Fujiwara, K. Takaki, Z. Hou, Y. Wakatsuki, *Organometallics* **1996**, *15*, 5476–5478.

[17] Z. Hou, H. Yamazaki, Y. Fujiwara, H. Taniguchi, *Organometallics* **1992**, *11*, 2711–2714.

[18] R. J. Bouma, J. H. Teuben, W. R. Beukema, R. L. Bansemer, J. C. Huffman, K. G. Caulton, *Inorg. Chem.* **1984**, *23*, 2715–2718.

[19] a) A. W. Konradi, S. J. Kemp, S. F. Pedersen, *J. Am. Chem. Soc.* **1994**, *116*, 1316–1323; b) B. Kammermeier, G. Beck, D. Jacobi, H. Jendrella, *Angew. Chem.* **1994**, *106*, 719–721; *Angew. Chem. Int. Ed. Engl.* **1994**, *33*, 685–687; c) B. Kammermeier, G. Beck, W. Holla, D. Jacobi, B. Napierski, H. Jendralla, *Chem. Eur. J.* **1996**, *2*, 307–315;

d) M. E. Pierce, G. D. Harris, Q. Islam, L. A. Radesca, L. Storace, R. E. Waltermire, E. Wat, P. K. Jadhav, G. C. Emmett, *J. Org. Chem.* **1996**, *61*, 444–450; e) M. T. Reetz, N. Griebenow, *Liebigs Ann.* **1996**, 335–348.

[20] a) Y. Kataoka, I. Makihira, M. Utsunomiya, K. Tani, *J. Org. Chem.* **1997**, *62*, 8540–8543; b) S. Torii, K. Akiyama, H. Yamashita, T. Inokuchi, *Bull. Chem. Soc. Jpn.* **1995**, *68*, 2917–2922; c) M. Kang, J. Park, S. F. Pedersen, *Synlett* **1997**, 41–43.

[21] T. Hirao, T. Hasegawa, Y. Muguruma, I. Ikeda, *J. Org. Chem.* **1996**, *61*, 366–367.

[22] E. J. Roskamp, S. F. Pedersen, *J. Am. Chem. Soc.* **1987**, *109*, 6551–6553.

[23] a) R. Annunziata, M. Benaglia, M. Cinquini, F. Cozzi, P. Giaroni, *J. Org. Chem.* **1992**, *57*, 782–784; b) M. Shimizu, T. Iida, T. Fujisawa, *Chem. Lett.* **1995**, 609–610.

Exciting Results from the Field of Homogeneous Two-Phase Catalysis

Boy Cornils

With the advent of two-phase catalysis, the last and in fact inherent disadvantage of homogeneous catalysis relative to its heterogeneous counterpart appears to have been swept away. This break-through has been achieved by arranging the process such that, at least when reaction is complete, the organometallic complex serving as a catalyst on one hand, and the reaction product (and residual starting material) on the other, are located in different phases, so a simple phase separation is sufficient for isolating the product from the catalyst (which may then be introduced immediately into another catalyst cycle). [1] The catalyst is situated in a mobile phase which, because it is confined whithin the reactor, also serves as an immobilization medium. Interestingly, quite atypically, the principle itself was implemented on a large (approx. 10^6 tonnes per annum) scale in industry in the Shell higher olefins process (SHOP), as well as the Ruhrchemie/Rhône-Poulenc oxo process even prior to the onset of comprehensive scientific investigation, initiated later by a number of academic research groups. [2] Some of these later studies have been pursuing unusual course: The work undertaken by Chaudhari et al. [3], which deserves much credit (for introducing a systematic approach) assumes, for instance, that it is possible to manage without quotations from the pioneers of the biphase technique (Joó and Ruhrchemie).

An indication of the significance of the process as far as the scientific world is concerned is the fact that the "standard ligand" for aqueous two-phase catalysis, triphenylphosphine trisulfonate (TPPTS), now appears in the Aldrich Catalogue of Fine Chemicals. The elegant approach to catalyst separation has in turn led to intense preoccupation with possible laboratory and industrial applications of two-phase catalysis, as well as extension of the process into new areas. A series of important contributions to the literature has recently cast a rather bright spotlight in this direction.

Not surprisingly the ideal form of the process is *aqueous* biphase catalysis, in which the organometallic two-phase catalyst resides in a stationary aqueous solution in the reaction system. This is not only the most convenient arrangement on both the laboratory and industrial scale, but also the optimal modification wich respect to cost and environmental considerations. Use of water as the second phase has its limitations however, especially when the water solubility of the starting materials proves too low, preventing adequate transfer of organic substrate into the aqueous phase or at the phase boundary, and consequently reducing the reaction rate to such an extent that it becomes unacceptable. Cases like this can be dealt with by introducing a surfactant (or by using ligands that confer surfactant properties), but other alternatives include addition of a solvating agent that produces a quasi "mechanical" effect, or use perhaps of a cosolvent, two measures that presumably increase either the mutual solubility of the components or mobility across the phase boundaries. The effectiveness of additives of this type has been the subject of much controversy in the literature, and it depends heavily on such specific factors as the ligands and their purity, the nature and behavior of the substrate, the presence of miscibility gaps, and precise reaction conditions. From an engineering and economic standpoint it is also important to remember that any "foreign" additive will inevitably increase the difficulty and cost of purification, thereby mitigating the clear advantages associated wich two-phase catalysis in its purest form.

Chaudhari et al. [4] were the first to suggest introducing "promoter ligands" soluble exclusively in the organic phase, thus altering the solubility of the complet *internally.* These authors were able to show through hydroformylation of the extremely water-insoluble 1-octene (equilibrium solubility in water at 298 K: 2.408×10^{-5} kmol m^{-3}; cf. propene, with a solubility of approx. 7×10^{-3} kmol m^{-3}) with [HRh(CO)(TPPTS)$_3$] as catalyst that addition of triphenylphosphine to the 1-octene increases the reaction rate by a factor of 10–50. Ligand exchange during the course of the reaction leads to the formation of isolable mixed-ligand complexes of the type [HRh(CO)(TPPTS)$_{3-x}$(TPP)$_x$] ($x = 0$–3). These display activities somewhat greater than is observed for catalysts modified only with TPPTS, though still inferior to the activity obtained by simple physical addition of TPP alone, a result that indicates the importance of as yet unexplained effects at the boundary surfaces of the two reaction phases (cf. Fig. 1).

With respect to the development of an industrial process for the hydroformylation

of higher olefins $>C_6$ (a challenge not yet met) it must of course be noted that the suggestion by Chaudhari et al. of adding a promoter ligand once again raises the problem of a "foreign substance", with the resulting need for a costly separation step. Nevertheless, these experiments do suggest that a search for ligands with carefully matched hydrophilic and hydrophobic characteristics may lead to a solution to the problem, provided the electronic characteristics of the oxo catalyst can be optimized simultaneously. From among the large number of relevant papers, reference may be made here to suggestions by Fell on ligands of matched hydrophilicity and lipophilicity for converting raw materials used in fat chemistry [5], the paper by Purwanto and Delmas on hydroformylation in the presence of externally added solvents [6], that by Hanson on surface-active ligands or special spectator cations [7] (supplementing earlier studies by Ruhrchemie AG [8]) or that by Montflier and Mortreux on the two-phase hydroformyla-

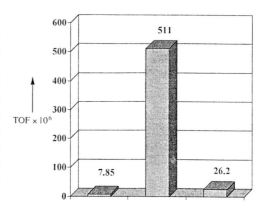

Figure 1. Two-Phase hydroformylation of 1-octene. Turnover frequencies (TOF) obtained under three different operating conditions: Column 1: Hydroformylation with [HRh(CO)(TPPTS)$_3$]. Column 2: Addition of 3.33×10^3 kmol TPP per m^3 of organic phase to TPPTS. Column 3: [HRh(CO)(TPPTS)$_{3-x}$ (TPP)$_x$]. The TOF values indicate how many kmol of product per kg rhodium are formed per second [4].

tion of water-insoluble olefins with the aid of molecular recognition with functional β-dextrines [9]. The various possibilities from a process engineering viewpoint for catalyst recycling in two-phase processes and hence also on the necessary circulation of any auxiliaries or solvents have been dealt with by Behr [10].

The foreign additive problem is avoided in an approach suggested by Jin, Fell, et al. [11, 12] who have proposod a two-phase hydroformylation of higher olefins (up to 1-dodecene) based on replacing the ligands in [HRh(CO)₄] with ethoxalated tris(p-hydroxyphenyl)phoshines of the structure indicated in **1**.

1, *n* > 6–8

The corresponding Chemical Abstracts entry fails to provide any indication of the real novelty involved in the Chinese work, but the German version of the paper makes it clear that this particular oxo catalyst takes advantage of a temperature dependent "cloud point" associated wich the phosphorus bound poly (alkylene glycol ether) ligands. Thus, above the cloud point the ligand (and thus the complex catalyst) loses its hydration shell, just as in the case of other compounds of this type, causing the two-phase reaction mixture normally obtained when olefin is added to the catalyst solution to merge into a single phase, thereby initiating a rapid transformation that is no longer impeded by material-transport problems. Subsequent lowering of the temperature causes the hydration shell to be reversibly restored, inducing the complex catalyst solution once again to separate this time from the reaction product (the de-

sired higher aldehydes) as an independent phase. Since the agent responsible for the merger and subsequent separation of the phases is the appropriately custom-designed ligand itself, there is no call for investing extra effort in the removal and recycling of a foreign additive, and this must therefore be regarded as a promising avenue for further exploration on a commercially realistic scale.

Results reported in two other papers by Bergbreiter et al. [13] and Wan and Davis [14] take advantage of similar effects. Bergbreiter et al. applied the designation "smart ligands" in describing certain phosphorus bound poly(alkylene oxide) oligomers that together with rhodium precursors form complexes with the structure **2**, complexes that display both hydrogenation and oxo activity. Here again, an inverse temperature dependency of the complex catalyst's solubility leads at high temperature to single-phase behavior and correspondingly increased reaction rates.

$$[Rh(Cl)\{Ph_2P(CH_2CH_2O)_nCH_2CH_2PPh_2\}_{1.5}]$$
2

Wan and Davis combined the possibilities inherent in the supported liquid phase catalysis (SLPC) and supported aqueous phase catalysis (SAPC) techniques with a modified water-soluble (and thus hydrophilic) ruthenium catalyst dissolved in ethylene glycol in their demonstration of an enantioselective hydrogenation of 2-(6'-methoxy-2'-naphthyl) acrylic acid (**3**). The reaction leads to (*S*)-naproxen (**4**) with *ee* values as high as 96 %. Ligand modification was accomplished with tetrasulfonated BINAP [2,2'-bis(diphenylphosphino)-1,1'-binaphthyl], which in conjunction with ruthenium chloride generates the active hydrogenation complex [Ru(BINAP-4-SO₃Na)(C₆H₆)Cl]Cl (Scheme 1).

The substrate is dissolved in the hydrophobic solvent mixture cyclohexane/chloroform. This combination forms a second phase in

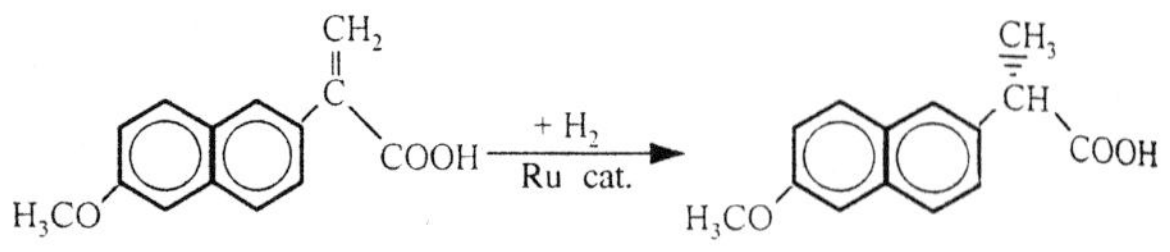

Scheme 1. Homogeneous hydrogenation of the acrylic acid derivate **3** to give (S)-naproxene (**4**).

the presence of the highly polar supported-catalyst solution, which is in turn immiscible with organic liquids. Wan and Davis attempted in this way to achieve a double "heterogenization" or "immobilization" of a complex catalyst that would otherwise be homogeneous: Deposition on the selected support CPG (controlled-pore glass) causes the catalyst to become *heterogeneous*, and the two-phase technique results in its further *immobilization*. The success reported in conjunction with this remarkable experiment is quite astounding. It will certainly be worth waiting a bit to see if the success proclaimed [15] can be confirmed with other substrates, and whether the indirect "proofs" cited for localized retention of the ruthenium catalyst in fact withstand closer scrutiny. Given all the previous experience with liquid-solid phase catalysts and their susceptibility to "leaching" this would seem somewhat unlikely.

Another approach to two-phase catalysis and the immobilization of homogeneons catalysts has been suggested by Horváth and Rábai at Exxon. Following preliminary conference reports [16] and a paper in *Science* [17] the corresponding European Patent application has now been published as well [18]. Horváth and Rábai discuss their new "fluorous biphase system" (or "fluorous multi-phase system" as it is characterized in the broader set of claims constituting the patent application) also in the context of the hydrofonmylation reaction as an illustrative example. Thus, a modified oxo catalyst bearing partially fluorinated ligands bound via phosphorus, such as that illustrated in structure **5**, is introduced into certain perfluoro (or partially fluorinated)

solvents with which ordinary organic substrates and solvents are immiscible

$$[HRh(CO)\{P[CH_2CH_2(CF_2)_5CF_3)]_3\}_3]$$
5

Proper "tailoring" of the ligands is said to be extraordinarily important. For example, the methylene group located between the phosphorus atom and the perfluoro end group in **5** is alleged to be responsible for controlling the influence of the electron withdrawing end groups ("fluorous ponytails" [19]) and finetuning their effect to the appropriate level. The two-phase mixture consisting of catalyst phase and olefin becomes homogeneous at higher temperature, causing the catalytic process itself to occur with a correspondingly high rate of reaction in a single phase. The system is subsequently cooled, permitting one to take advantage of a thermally defined miscibility gap in order to separate the reaction product from the catalyst, which then becomes available for immediate reuse.

The practical applicability on an industrial scale of this rather exotic two-phase system remains to be demonstrated. Doing so will require a clarification of such basic issues as activity, cost, catalyst lifetimes (and thus catalyst life and economic feasibility), toxicity, concerns regarding the ozone depletion potential (ODP) and greenhouse warming potential (GWP) values of the corresponding fluorinated compounds, etc. There is also the possibility of competitive extraction of fluorinated hydrocarbons by the aldehyde phase in the oxo reaction (leading to potential problems in subsequent hydrogenation to the plasticizer alcohols that are in fact the desired products). Some other work describe the environmental advantages of the "classical" aqueous biphase systems [20].

The professional world will also follow with some interest the future course of Exxon's patent application itself and the possibility that there may be appeal, since at least part of the basic idea underlying the Exxon process was described previously in a 1991 dissertation by M. Vogt (under Keim's supervision) entitled "Zur Anwendung perfluorierter Polyether bei der Immobilisierung homogener Katalysatoren" (On the Application of Perfluorinated Polyethers in the Immobilization of Homogeneous Catalysts). [21] Vogt was attempting to develop ligands based on hexafluoropropene oxide (HFPO) oligomers for subsequent complexation with appropriate cobalt or nickel precursors, and to exploit the latter for oligomerization or polymerization reactions. It did indeed prove possible in principle to bind HFPO oligomers to homogeneous transition metal catalysts, conferring upon the modified complexes the solvophobic properties characteristic of this class of compounds. The result was a two-phase reaction procedure conducted in a perfluoropolyether medium facilitating separation of the product from the catalyst.

The current enthusiasm for biphase catalysts will also benefit work being undertaken with ionic liquids. According to Chauvin et al. [22] olefins in particular can be dimerized with this special form of biphasic reaction in molten salt media (e. g. the Dimersol® process of dimethyl butenes).

There now exists evidence for the extension of two-phase catalysis into a new area, suggesting interesting possibilities in an entirely different field of application. Gassner and Leitner [23] have described a hydrogenation of carbon dioxide to formic acid [Eq. (1)] that is accompanied by astonishingly high turnover values in aqueous solution in the presence of TPPTS-modified rhodium catalysts. This potentially relevant process will certainly generate considerable interest if it proves feasible to carry out a selective transformation of "technical grade" carbon dioxide (i.e. without prior purification) at "industrial concentrations" (meaning a CO_2 content in the range of a few percent). The high specificity associated with two-phase processes makes rapid progress in this regard appear likely as well.

$$H_{2(aq)} + CO_2(aq) = HCOOH_{(aq)} \qquad (1)$$

The specificity of biphase catalyst systems (in conjunction with activities comparable to those anticipated with single-phase systems) has been illustrated in an elegant way by Mortreux et al. [24] In a regioselective hydroformylation of methyl acrylate to give alpha-formylpropionic acid with the catalyst [HRh(CO)(TPPTS)₃], these authors observed an aldehyde yield comparable to that achieved with TPP, as well as an alpha/beta product ratio of the expected order of magnitude and an even higher turnover frequency (TOF) (Scheme 2).

A [HRh(CO)(TPPTS)₃] catalyst of the SAPC type on SiO_2 led to TOF values that were higher still, exceeding those obtainable with

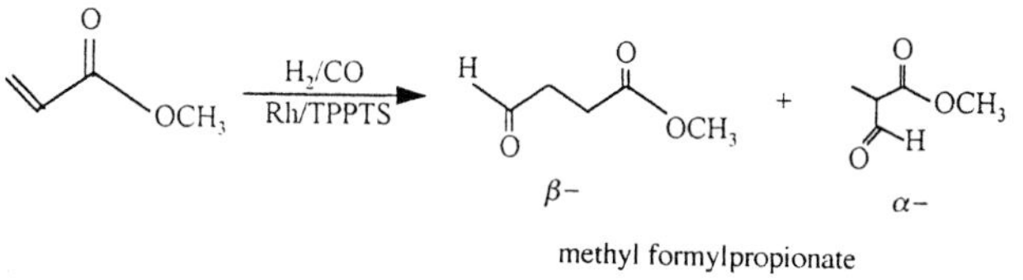

Scheme 2. Hydroformylation of methyl acrylate.

the homogeneous single-phase [HRh(CO)(TPP)₃] system by a factor of 200 depending on experimental conditions. This is impressive evidence of the fact that it is dangerous to make predictions about reaction rates with biphasic catalyst systems, and that the nature of the associated substrate must always be taken into account. The same applies to results of Chaudhari et al. [25], who proved that, in the hydroformylation of allyl alcohol, the two-phase form of reaction affords special advantages through the reduced level of catalyst inactivation and a more effective circulation.

What are also exciting are the many and varied methods by which the field of homogeneous two-phase catalysis can be expanded. Following on from biphasic carbonylation and the Heck reaction, two highly promising approaches pioneered by Beller, new results now suggest that Wacker oxidation [26] and Suzuki coupling [27] also have great potential.

References

[1] Introduction to the topic: (a) B. Cornils, W. A. Herrmann, *Aqueous-Phase Organometallic Catalysis – Concepts and Applications*, VCH, Weinheim, **1998**; (b) W.A. Herrmann, C. W. Kohlpaintner, Angew. Chem. **1993**, *105*, 1588; Angew. Chem. Int. Ed. Engl. **1993**, *32*, 1524.

[2] A *NATO Advanced Research Workshop* was held in August/September 1994 on the subject "Aqueous Organometallic Chemistry and Catalysis" (Debrecen, Hungary); followed by other workshops on this topic, such as the symposium *"Catalysis in Multiphase Reactors"*, Dec. 1994 in Lyon (France), Spring ACS Meeting in Anaheim (CA) in April 1995 and the Fall Meeting in Las Vegas, Scp. 1997.

[3] R. V. Chaudhari, A. Bhattacharya, B. M. Bhanage, *Catalysis Today* **1995**, *24*, 123

[4] R. V. Chaudhari, B. M. Bhanage, R. M. Deshpande, H. Delmas, *Nature* **1995**, *373*, 501.

[5] B. Fell, C. Schobben, G. Papadogianakis, *J. Mol. Catal. A:* **1995**, *101*, 179; S. Kanagasabapathy, Z. Xia, G. Papadogianakis, B. Fell *J.Prakt. Chem./Chem.-Ztg.* **1995**, *337*, 446.

[6] P. Purwanto, H. Delmas, *Catalysis Today* **1995**, *24*, 135.

[7] H. Ding, B. E. Hanson, *J. Mol. Catal. A* **1995**, *99*, 131; T. Bartik, H. Ding, B. Bartik, B. E. Hanson, *J. Mol. Catal. A* **1995**, *98*, 117.

[8] The appropriate patents of Ruhrchemie AG are described in detail in *J. Organomet. Chem.* **1995**, *502*, 177.

[9] E. Montflier, G. Fremy, Y. Castanet, A. Mortreux, *Angew. Chem.* **1995**, *107*, 2450.

[10] A. Behr, *Henkel-Referate* **1995**, *31*, 31.

[11] Y. Yan, H. Zhuo, Z. Jin, *Fenzi Cuihua* **1994**, *8*(2), 147, Chem. Abstr. **1994**, *121*, 111.875a.

[12] Z. Jin, Y. Yan, H. Zhuo, B. Fell, *J. Prakt. Chem./Chem. Ztg.* **1996**, *338*, 124.

[13] D. E. Bergbreiter, L. Zhang, V. M. Mariagnanam, *J. Amer. Chem. Soc.* **1993**, *115*, 9295.

[14] K. T. Wan, M.E. Davis, *Nature* **1994**, *370*, 449.

[15] J. M. Brown, S. G. Davies, *Nature* **1994**, *370*, 418.

[16] *NATO Advanced Research Workshop* (cf. Ref. [2], p. 35).

[17] I. T. Horváth, J. Rábai, *Science* **1994**, *266*, 72.

[18] Exxon Res. Eng. Comp. (I. T. Horváth, J. Rábai), EP-Appl. 0.633.062 (1994).

[19] J. A. Gladysz, *Science* **1994**, *266*, 55.

[20] a) B. Cornils, E. Wiebus, *Recl. Trav. Chim. Pays-Bas* **1996**, *115*, 211; (b) G. Papadogianakis, R. A. Sheldon, *New J. Chem.* **1996**, *20*, 175.

[21] M. Vogt, Dissertation, Technische Hochschule Aachen, 26. August 1991.

[22] Y. Chauvin, H. Olivier in *Applied Homogeneous Catalysis with Organometallic Compounds* (Eds. B. Cornils, W. A. Herrmann), Vol. 1, p. 258, VCH, Weinheim/Germany, **1996**; Y. Chauvin, H. Olivier-Bourbigou, *CHEMTECH* **1995**, (9), 26; Y. Chauvin, S. Einloft, H. Olivier, *Ind. Eng. Chem. Res.* **1995**, *34*, 1149.

[23] F. Gassner, W. Leitner, *J. Chem. Soc. Chem. Commun.* **1993**, 1465; W. Leitner, *Angew. Chem.* **1995**, *107*, 2391.

[24] G. Fremy, E. Montflier, J.-F. Carpentier, Y. Castanet, A. Mortreux, *Angew. Chem.* **1995**, *107*, 1608; *Angew. Chem. Int. Ed. Engl.* **1995**, *34*, 1474.

[25] R. M. Deshpande, S. S. Divekar, B. M. Bhanage, R. V. Chauhari, *J. Mol. Catal. Letter* **1992**, *75*, L19.

[26] E. Montflier, S. Tilloy, E. Blouet, Y. Barbaux, A. Mortreux, *J. Mol. Catal. A* **1996**, *109*, 27.

[27] Hoechst AG (S. Haber), DE-Appl. 19527118 A1 (1997)

Palladium-Catalyzed Amination of Aryl Halides

Matthias Beller and Thomas H. Riermeier

"The discovery of truely *new* reactions is likely to be limited to the realm of transition-metal organic chemistry, which will almost certainly provide us with additional 'miracle reagents' in the years to come", was predicted by D. Seebach in 1990. [1] No other area of classic synthetic chemistry today still offers such innovative possibilities as metal-catalyzed processes. Transition metal-catalyzed reactions are becoming increasingly more significant not only for the synthesis of complex building blocks, but also for structurally simpler, industrially important intermediates. Since the inception of the Wacker-Hoechst oxidation [2] and the Heck reaction, [3] palladium-catalyzed processes, in particular, have aroused the interests both of the synthetic chemist in the laboratory and also of the industrial chemist. Based on the almost unique range of catalytic transformations of palladium complexes, whose use however is often limited by low selectivities and deactivations during the reactions, a multitude of palladium supported syntheses were discovered or developed further in the last few years. [4] So far, most of the research has focussed on methods for C–C coupling. Because of the significance of unsymmetrically substituted ethers and amines in organic synthesis, the transferability of catalysis by palladium complexes of C–C couplings to the equally impor-tant C–O and C–N couplings is of general interest.

In this respect, new methodic developments for hetero Heck reactions leading to aromatic amines, which are important substructures in natural products as well as industrial chemicals are noteworthy. Palladium-catalyzed C–N bond-forming coupling processes were first reported by Migita and coworkers [5]. Unfortunately the original procedure applies toxic and air sensitive tributyl-N,N-diethyl-aminostannane as transamination reagent (Scheme 1).

According to investigations by Hartwig et al. [6a] the actual catalytically active species in the amination reaction seems to be a palladium(0)-bis(tri-o-tolylphosphine) complex. The catalytic cycle starts with an oxidative addition of the palladium(0) complex into the aryl-halogen bond. Then the resulting arylpalladium(II) complex reacts with the tin amide under transmetalation; this step is postulated

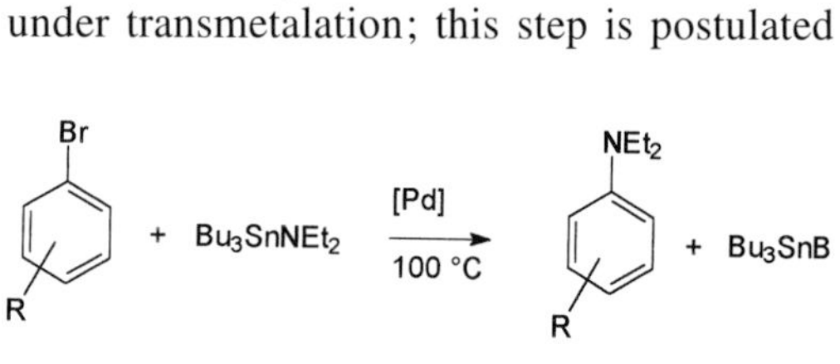

Scheme 1. [Pd] = PdCl$_2$[P(o-tolyl)$_3$]$_2$. R = H, 2-CH$_3$, 3-CH$_3$, 4-CH$_3$, 4-OCH$_3$, 4-COCH$_3$, 4-NO$_2$, 4-N(CH$_3$)$_2$.

to be rate determining. The subsequent reductive elimination leads back to the palladium(0) catalyst. Recent investigations show that in addition to tri-*o*-tolylphosphine complexes, simpler arylpalladium(II)-bis(triphenylphosphine) complexes also react with alkali metal amides to form amido complexes, which give access to arylamines by reductive elimination. [7]

By clever coupling of a transamination reaction of tributyl-*N,N*-diethylaminostannane with higher boiling amines with palladium catalysis, Buchwald and Guram succeeded in extending the method (Scheme 2). [6b] It was shown that secondary aliphatic and aromatic amines react with substituted aryl bromides to afford the corresponding arylamines in good yields. As in the classic Heck reaction, depending on the substitution pattern of the arene, electron acceptor substituents promote the reactivity of the aryl bromide.

Clearly, the use of stoichiometric amounts of organo tin compounds is the main disadvantage of this type of C–N coupling reaction both for ecological reasons and with regard to practicability. Thus, from an industrial point of view the aim was to replace the tin amides by simpler amino sources, ideally the amines themselves. Again independently, Buchwald et al. [8] and Hartwig et al. [9] reported

shortly afterwards the first catalytic aminations of aryl bromides with free amines. The palladium-catalyzed reactions occur in the presence of stoichiometric amounts of a sterically hindered base such as NaO*t*Bu in toluene or tetrahydrofuran at temperatures of 65–100 °C (Scheme 3). More recently Buchwald et al. have been able to expand this reaction protocol to aryl iodides. [10]

The arylamines are generally formed in good yields (Table 1). Dehalogenation products are the only by-products observed, which probaly arise from base-induced β-hydride elimination of the amido arylpalladium complex and subsequent reduction. Interestingly, the base employed has a decisive influence on the course of the reaction. In the amination of 1-bromo-4-*n*-butylbenzene with free amines in the presence of silyl amides as base – in contrast to the coupling with tin amides – the rate-determining step in the catalytic cycle is the oxidative addition of bis(tri-*o*-tolylphosphine)palladium(0) to the aryl halide. However, when LiO*t*Bu is used as base, the formation and reductive elimination of the amido arylpalladium complex is decisive for the rate of the reaction. In the presence of NaO*t*Bu both reaction steps seem to take place at similar rates. [9]

Scheme 2. [Pd] = PdCl$_2$[P(*o*-tolyl)$_3$]$_2$. R = 4-CH$_3$, 3-CH$_3$, 4-CF$_3$, 3-OCH$_3$, 4-CO$_2$Et; R′ = CH$_2$C$_6$H$_5$, C$_6$H$_5$; R″ = H, CH$_3$.

Scheme 3. [Pd] = PdCl$_2$[P(*o*-tolyl)$_3$]$_2$. X = Br, I; R = 4-C$_4$H$_9$, 4-CF$_3$, 4-OCH$_3$, 4-C$_6$H$_5$, 4-(CH$_3$)$_2$N; R′ = C$_6$H$_5$, C$_6$H$_{13}$; R″ = H, CH$_3$; R′–R″ = H$_2$C(CH$_2$)$_3$CH$_2$, H$_2$CCH$_2$N(CH$_3$)CH$_2$CH$_2$.

As can be seen from Table 1 sodium *tert*-butoxide gives similar results compared to lithium amides as base (entries 1 and 2). Both are superior to lithium *tert*-butoxide (entries 3 and 4). Since NaOtBu is easier to handle it seems to be the base of choice for this reaction. With tri-*o*-tolylphosphine as ligand the procedure is limited to secondary amines. Nevertheless, some application of this new method – mainly from industrial research groups – appeared in the literature: Zhao et al. (Roche Bioscience) demonstrated elegantly the synthetic potential of the amination reaction for the synthesis of various *N*-arylpiperazines. [11] In the case of piperazine itself, the appropiate choice of reaction stoichiometry leads to either a symmetrical *N,N'*-bisarylpiperazine or *N*-monoarylpiperazine in good yield. Amination reactions with C-substituted unsymmetrical piperazines pro-

Table 1. Palladium-catalyzed reactions of aryl bromides and iodides with (*o*-tolyl)₃P palladium complexes.

Entry	Aryl halide	Amine	Base	Arylamine	mol% Pd	Yield [%]	Lit.
1	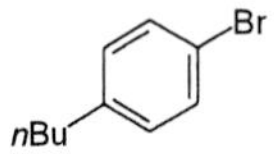	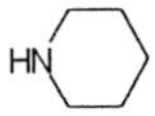 HN⟨⟩	LiN(SiMe₃)₂	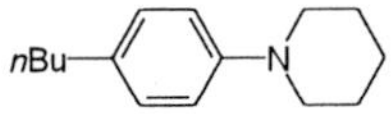	5	89	[9]
2	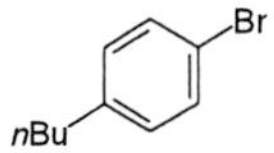	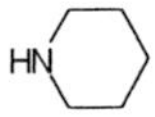 HN⟨⟩	NaOtBu	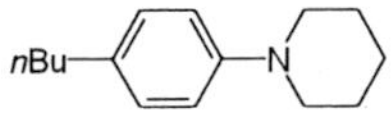	5	89	[9]
3	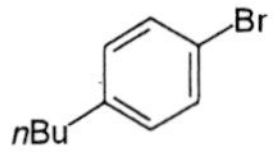	HNEt₂	LiN(SiMe₃)₂	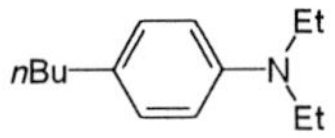	5	40	[9]
4	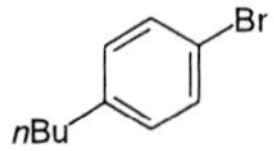	HNEt₂	LiOtBu	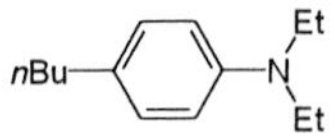	5	<2	[9]
5	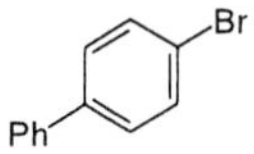	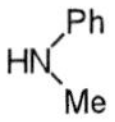 HN(Ph)(Me)	NaOtBu	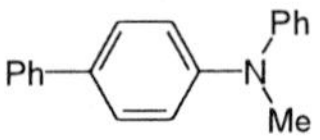	2	88	[8]
6	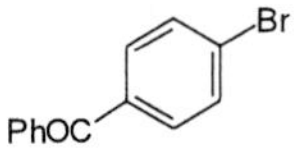	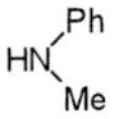 HN(Ph)(Me)	NaOtBu	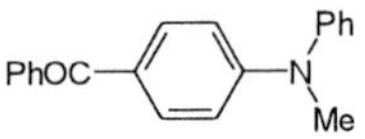	2	89	[8]
7	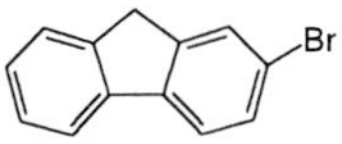	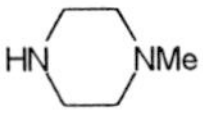 HN⟨N⟩NMe	NaOtBu	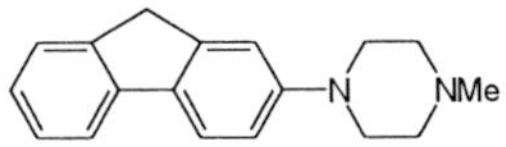	2	79	[8]
8	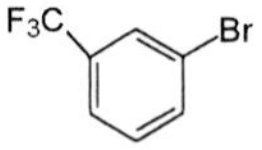	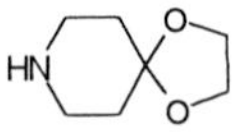 HN⟨⟩	NaOtBu	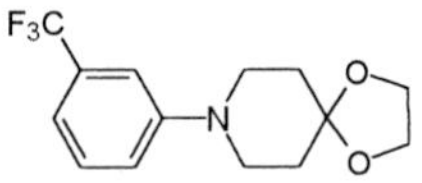	2	67	[8]
9	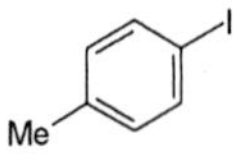	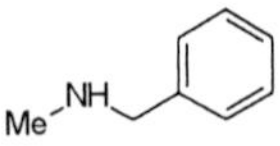 Me-NH-	NaOtBu	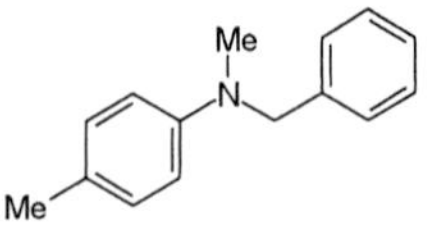	1	79	[10]
10	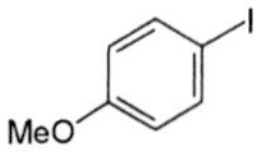	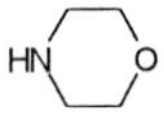 HN⟨O⟩	NaOtBu	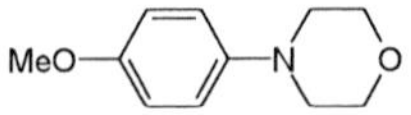	1	66	[10]

ceeded with high regioselectivity, allowing facile preperation of several novel arylpiperazines.

Two other research groups, Ward et al. (Boehringer Ingelheim Pharmaceuticals) and Willoughby et al. (Merck Research Laboratories) succeded in expanding the amination protocol to the solid phase synthesis of aryl amines. [12] Thus, this new reaction was already added to the tools of combinatorial synthesis.

Extending the intermolecular coupling, dihydroindoles (Scheme 4), dihydroquinolines, and other N-heterocycles were successfully synthesized by simple intramolecular trapping reactions, starting from alkylamino substituted aryl bromides. [8] Intramolecular amination can be achieved in the presence of tetrakis(triphenylphosphine)palladium as catalyst and stoichiometric amounts of base in toluene. Here, best results are obtained with mixtures of NaOtBu and potassium carbonate.

The catalysts used for the aforementioned aminations – usually 1 to 5 mol% palladium – contained monodentate triarylphosphines especially tri-o-tolylphosphine as ligands. In case of tri-o-tolylphosphine the steric bulk leads to superior reactivity by favoring low coordination numbers at the metal center. Studies from the groups of Buchwald [13] and Hartwig [14] of late-transition metal amido complexes led to new second-generation aryl halide amination catalysts based on chelating bisphosphine ligands. While Hartwig used a 1,1'-bis(diphenylphosphino)ferrocene ligand Buchwald employed BINAP as ligand. As shown in Table 2 the new catalyst system works efficiently for the cross coupling of a variety of primary amines and secondary amines with both aryl bromides as well as aryl iodides. More recently, the aryl amination methodology was extended to the use of aryl triflates using chelating ligands. [15a, b] Because of the diversity of available phenols this extension of the reaction protocol is of synthetic value on laboratory scale.

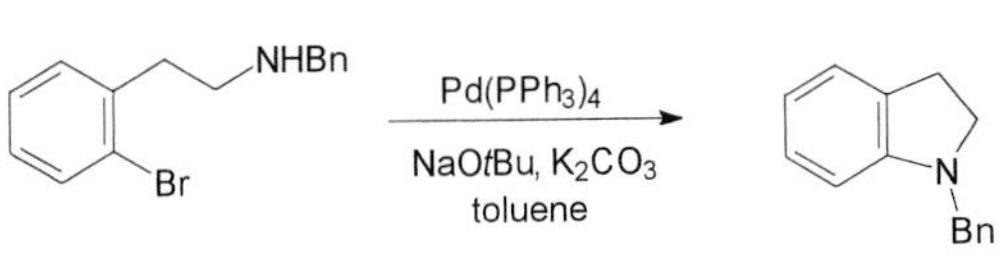

Noteworthy, are the superior yields for coupling of amines with *ortho*-substituted halides and halopyridines. [15c] In case of the synthesis of aminopyridines the improved catalyst activity is explained by the ability of chelating ligands to prevent formation of bispyridyl palladium complexes that terminate the catalytic cycle. Interestingly, electron rich aryl bromides (entry 2) gave similar high yields as electron poor aryl halides (entry 1). The sterically hindered aryl bromide 1-bromo-2,5-dimethylbenzene can be coupled with *N*-methylpiperazine even in the presence of just 0.05 mol % palladium (entry 4). Thus, catalyst turnover numbers up to 2000 were realized for the first time. When primary amines are used, just small ammounts of double arylated products were detected.

Although the reported method give high yields for aminations of aryl bromides and aryl iodides, clearly, an extension of the methodology towards chloroarenes as starting materials is of high interest due to their availability and low cost. Thus, we studied the coupling reaction of 4-trifluoromethyl-1-chlorobenzene with piperidine in the presence of new palladium catalysts (palladacycles, e. g. *trans*-di(μ-acetato)-bis[*o*-(di-*o*-tolylphosphino)-benzyl]dipalladium(II)) and additional bromide ions as co-catalysts [16] as model reaction (Scheme 5) [17]. Crucial for the success of the C–N bond forming reaction is the use of potassium *tert*-butoxide as base and reaction temperatures > 120 °C. Turnover numbers up to 900 and yields up to 80 % have been obtained for the amination of 4-trifluoromethyl-1-chlorobenzene. Small amounts of the *meta*-regioisomere are observed. This is explained by aryne

Scheme 4. Intramolecular palladium-catalyzed amination of aryl halides.

Table 2. Aminations of aryl bromides and iodides catalyzed by palladium complexes with chelating ligands ('second-generation catalysts').

Entry	Aryl halide	Amine	Base	Arylamine	mol% Pd	Yield [%]	Lit.
1	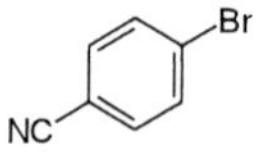 NC-C6H4-Br (para)	nHexNH$_2$	NaOtBu	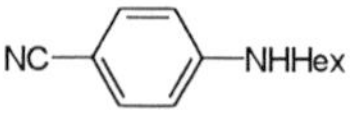 NC-C6H4-NHHex (para)	0.5	98	[13]
2	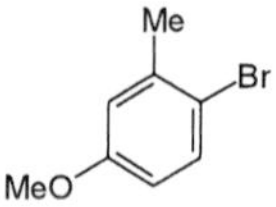 2-Me-4-MeO-C6H3-Br	nHexNH$_2$	NaOtBu	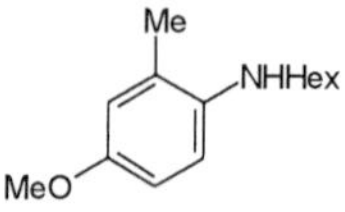 2-Me-4-MeO-C6H3-NHHex	0.5	95	[13]
3	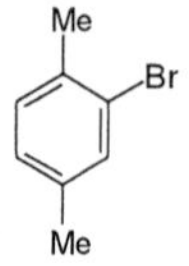 2,5-Me$_2$-C6H3-Br	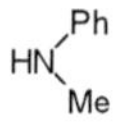 HN(Me)Ph	NaOtBu	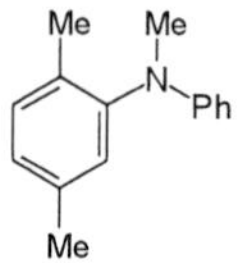 2,5-Me$_2$-C6H3-N(Me)Ph	0.5	94	[13]
4	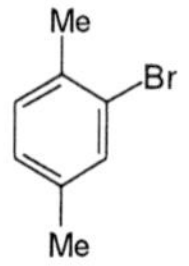 2,5-Me$_2$-C6H3-Br	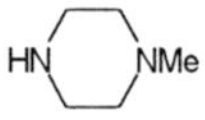 HN(CH$_2$CH$_2$)$_2$NMe (N-methylpiperazine)	NaOtBu	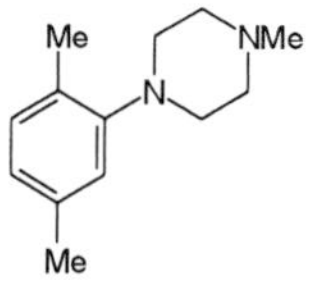 2,5-Me$_2$-C6H3-(4-Me-piperazin-1-yl)	0.05	94	[13]
5	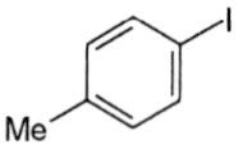 4-Me-C6H4-I	H$_2$N-Ph	NaOtBu	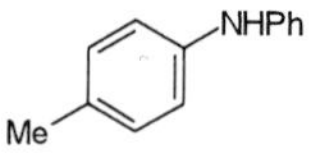 4-Me-C6H4-NHPh	5	92	[14]
6	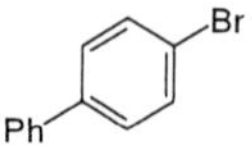 4-Ph-C6H4-Br	H$_2$N-Ph	NaOtBu	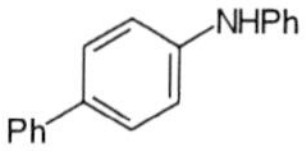 4-Ph-C6H4-NHPh	5	94	[14]
7	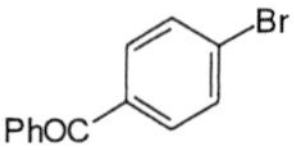 4-PhOC-C6H4-Br	nBuNH$_2$	NaOtBu	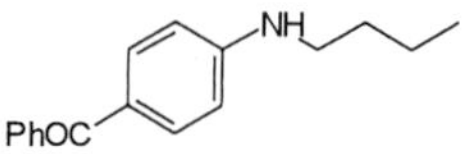 4-PhOC-C6H4-NH-nBu	5	96	[14]
8	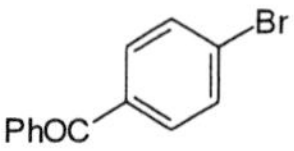 4-PhOC-C6H4-Br	iBuNH$_2$	NaOtBu	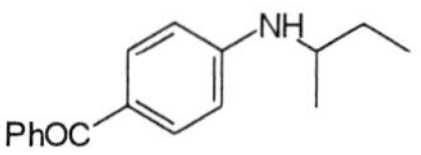 4-PhOC-C6H4-NH-iBu	5	84	[14]
9	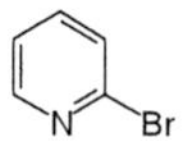 2-bromopyridine	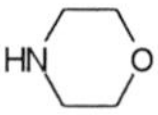 HN(CH$_2$CH$_2$)$_2$O (morpholine)	NaOtBu	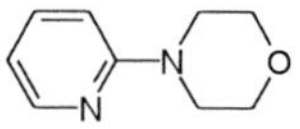 2-(morpholin-4-yl)pyridine	1	87	[15]
10	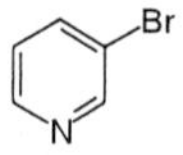 3-bromopyridine	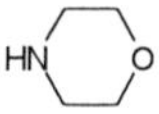 HN(CH$_2$CH$_2$)$_2$O (morpholine)	NaOtBu	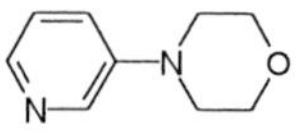 3-(morpholin-4-yl)pyridine	1	75	[15]

intermediates which can be formed under the conditions.

Apart from the aforementioned results the usefulness of palladium-catalyzed aryl aminations is shown by applications in natural product synthesis. [18] In this regard the total synthesis of the toad poison dehydrobufetenine is of particular interest. Here, the key step of the synthesis is the intramolecular amination of an aryl iodide (Scheme 6).

In conclusion, the palladium-catalyzed amination of aryl halides, discovered by Migita et al. and developed by Buchwald at al. and Hartwig et al. is now a powerfull tool for or-

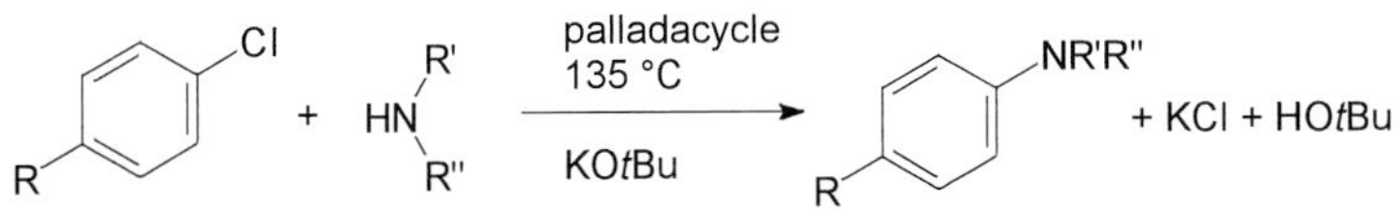

Scheme 5. Palladacycle: *trans*-di(μ-acetato)-bis[*o*-(di-*o*-tolylphosphino)benzyl]dipalladium(II); R = CF$_3$, COPh; R′ = C$_4$H$_9$, C$_6$H$_5$; R″ = C$_4$H$_9$, CH$_3$; R′-R″ = H$_2$C(CH$_2$)$_3$CH$_2$, H$_2$CCH$_2$OCH$_2$CH$_2$.

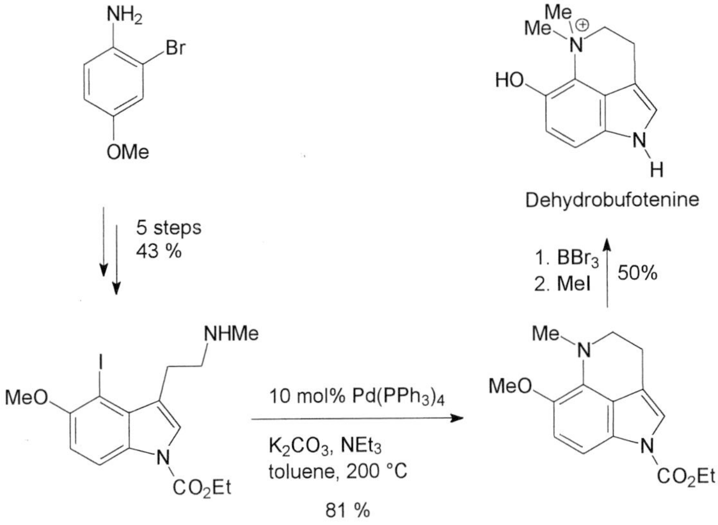

Scheme 6. Total synthesis of dehydrobufotenine by Buchwald et al. [18]

ganic synthesis. The availibility of a variety of secondary and primary amines together with the broad range of aryl halides – even aryl chlorides – allows the synthesis of complex arylamines. First applications of this method, both by research groups from industry as well as universities, showed the importance such C–N coupling reactions.

Interestingly, a first successful extension of the amination to oxygen nucleophils has been reported very recently. [19] Still at the beginning, this discovery opens the door to C–O coupling reactions for the synthesis of arylethers, including oxygen heterocycles.

It becomes evident that there is still a need for practical catalytic methods also for apparently simple molecules such as arylamines. The future will show that in this respect orga-

nometallic chemistry and catalysis offers manifold possibilities.

References

[1] D. Seebach, *Angew. Chem.* **1990**, *102*, 1363; Angew. Chem. Int. Ed. Engl. **1990**, *29*, 1320.
[2] J. Tsuji in *Comprehensive Organic Synthesis,* Vol. 7 (Eds.: B. M. Trost, J. Fleming, S. V. Ley), Pergamon, Oxford, **1991**, p. 449.
[3] Recent reviews: a) R. F. Heck, *Palladium Reagents in Organic Synthesis*, Academic Press, London, **1985**; b) R. F. Heck in B. M. Trost and I. Flemming (Eds.), *Comprehensive Organic Synthesis*, Vol. 4, Pergamon Press, Oxford, **1991**, p. 833; c) W. Cabri and I. Candiani, Acc. Chem. Res **1995**, *28*, 2; d)

J. Tsuji, *Palladium Reagents and Catalysts – Innovations in Organic Synthesis*, Wiley, Chichester, UK, **1995**; e) W. A. Herrmann, in B. Cornils and W. A. Herrmann (Eds.), *Applied Homogeneous Catalysis*, VCH, Weinheim, **1996**, p. 712; f) A. de Meijere, F. Meyer, *Angew. Chem* **1994**, *106*, 2437; *Angew. Chem. Int. Ed. Engl.* **1994**, *33*, 2379.

[4] Recent examples: a) H. M. R. Hoffmann, A. R. Otte, A. Wilde, *Angew. Chem.* **1992**, *104*, 224; *Angew. Chem. Int. Ed. Engl.* **1992**, *31*, 2379; b) T. I. Wallow, B. M. Novak, *J. Org. Chem.* **1994**, *59*, 5034; c) T. Ohishi, J. Yamada, Y. Inui, T. Sakaguchi, M. Yamashita, *J. Org. Chem.* **1994**, *59*, 7521; d) G. Dyker, *Angew. Chem.* **1994**, *106*, 117; *Angew. Chem. Int. Ed. Engl.* **1994**, *33*, 117; e) E. Drent, J. A. M. van Broekhooven, M. J. Doyle, *J. Organomet. Chem.* **1991**, *417*, 235; f) L. F. Tietze, R. Schimpf, *Angew. Chem.* **1994**, *106*, 1138; *Angew. Chem. Int. Ed. Engl.* **1994**, *33,* 1089; g) Y. Sato, S. Nukui, M. Sodeoka, M. Shibasaki, *Tetrahedron* **1994**, *50,* 371; h) Y. Ben-David, M. Portnoy, D. Milstein, *J. Am. Chem. Soc.* **1989**, *111*, 8742; i) W. Oppolzer, *Pure Appl. Chem.* **1990**, 62, 1941; j) S. C. A. Nefkens, M. Sperrle, G. Consiglio, *Angew. Chem.* **1993**, *105*, 1837; *Angew. Chem. Int. Ed. Engl.* **1993**, *32*, 1719; k) B. M. Trost, *Angew. Chem.* **1995**, *107*, 285; *Angew. Chem. Int. Ed. Engl.* **1995**, *34*, 259; l) C. Copéret, S. Ma, E. Negishi, *Angew. Chem.* **1996**, *108*, 2255; *Angew. Chem. Int. Ed. Engl.* **1996**, *35*, 2125; m) L. F. Tietze, T. Nöbel, M. Spescha, *Angew. Chem.* **1996**, *108*, 2385; *Angew. Chem. Int. Ed. Engl.* **1996**, *35*, 2259; n) M. Brenner, G. Mayer, A. Terpin. W. Steglich, *Chem. Eu. J.* **1997**, *3*, 70; o) A. Heim, A. Terpin, W. Steglich, *Angew. Chem.* **1997**, *109*, 158; *Angew. Chem. Int. Ed. Engl.* **1997**, *36*, 155.

[5] M. Kosugi, M. Kameyama, T. Migita, *Chem. Lett.* **1983**, 927.

[6] a) F. Paul, J. Patt, J. F. Hartwig, *J. Am. Chem. Soc.* **1994**, *116*, 5969; b) A. S. Guram, S. L. Buchwald, *J. Am. Chem. Soc.* **1994**, *116*, 7901.

[7] M. S. Driver, J. F. Hartwig, *J. Am. Chem. Soc.* **1995**, *117*, 4708.

[8] A. S. Guram. R. A. Rennels, S. L. Buchwald, *Angew. Chem.* **1995**, *107*, 1456; *Angew. Chem. Int. Ed. Engl.* **1995**, *34*, 1348.

[9] J. Louie, J. F. Hartwig, *Tetrahedron Lett.* **1995**, *36*, 3609.

[10] J. P. Wolfe, S. L. Buchwald, *J. Org. Chem.* **1996**, *61*, 1133.

[11] S. Zhao, A. K. Miller, J. Berger, L. A. Flippin, *Tetrahedron Lett.* **1996**, *37*, 4463.

[12] a) Y. D. Ward, V. Farina, *Tetrahedron Lett.* **1996**, *37*, 6993; b) C. A. Willoughby, K. T. Chapman, *Tetrahedron Lett.* **1996**, *37*, 7181.

[13] J. P. Wolfe, S. Wagaw, S. L. Buchwald, *J. Am. Chem. Soc.* **1996**, *118*, 7215.

[14] M. S. Driver, J. F. Hartwig, *J. Am. Chem. Soc.* **1996**, *118*, 7217.

[15] a) J. P. Wolfe, S. L. Buchwald, *J. Org. Chem.* **1997**, *62*, 1264; b) J. Louie, M. S. Driver, B. C. Hamann, J. F. Hartwig, *J. Org. Chem.* **1997**, *62*, 1268; c) S. Wagaw, S. L. Buchwald, *J. Org. Chem.* **1996**, *61*, 7240.

[16] a) W. A. Herrmann, C. Broßmer, K. Öfele, C.-P. Reisinger, T. Priermeier, M. Beller, H. Fischer, *Angew. Chem.* **1995**, *107*, 1989; *Angew. Chem. Int. Ed. Engl.* **1995**, *34*, 1844; b) W. A. Herrmann, C. Broßmer, C.-P. Reisinger, T. H. Riermeier, K. Öfele, M. Beller, *Chem. Eur. J.* **1997**, in press.

[17] M. Beller, T. H. Riermeier, C.-P. Reisinger, W. A. Herrmann, *Tetrahedron Lett.* **1997**, *38*, 2073.

[18] A. J. Peat, S. L. Buchwald, *J. Am. Chem. Soc.* **1996**, *118*, 1028.

[19] a) M. Palucki, J. P. Wolfe, S. L. Buchwald, *J. Am. Chem. Soc.* **1996**, *118*, 10333; b) G. Mann, J. F. Hartwig, *J. Am. Chem. Soc.* **1996**, *118*, 13109.

The Metal-mediated Oxidation of Organic Substrates via Organometallic Intermediates: Recent Developments and Questions of Dispute

Jörg Sundermeyer

The complex-catalyzed oxyfunctionalization of organic substrates is of fundamental importance, for example in chemical processes of nature, in the synthesis of fine chemicals, and in the production of commodity chemicals. [1] Regardless of the results achieved in this interdisciplinary area of research, there is still a great lack of experimental evidence that could help to settle very controversial debate [2] about the mechanism of the metal-mediated C–O bond formation. The catalytically active complexes are usually differentiated on the basis of their mechanisms. [1e] Thus, there is a group of metal complexes that catalyzes the homolytic cleavage the O–O bond of peroxides and triplet oxygen, a second group that induces its heterolytic cleavage, and a third that has at its disposal polarizable and transferable oxo functions such as M–O$^{\delta-}$, M=O, and M≡O$^{\delta+}$. The first group includes low- and high-spin complexes of Fe, Co, Ni, Mn, Cu, and other biologically relevant metals, which, for example, form "oxen" complexes [M=O] with transferable oxygen atom [1] and control the radical chain initiations as well as the propagation and inhibition of autoxidation reactions. In contrast, d^0 complexes of the second category (e. g. those of Ti, V, Mo, W, and Re) are diamagnetic and strong Lewis acids. These compounds activate H$_2$O$_2$ or alkyl hydroperoxides by formation of reactive d-block metal peracids [M–OOH], peroxides [M(η^2-O$_2$)], or peracid esters [M–OOR], that is, they polarize the O–O bond for nucleophilic attack of the substrate at the peroxidic oxygen atom. In contrast, electron-rich transition metals (Rh, Ir, Pd, Pt) cause an inverse polarization, which makes an electrophilic attack at the peroxy function more favourable. The group of reactive d^0 oxo complexes includes, for example, [MnO$_4$]$^-$, RuO$_4$ and OsO$_4$ well-established for hydroxylations.

Not so long ago the suggestion that organometallic intermediates might participate in polar oxygen-transfer reactions would have been refuted and considered as unrealistic. The knowledge accumulated over the last few years about oxo and peroxo complexes functionalized with organometallic co-ligands demonstrates that, in particular, the relatively nonpolar M–C bonds in complexes of molybdenum(VI), tungsten(VI), and rhenium(VII) can be stable under catalytic conditions (protic medium, hydroperoxides, O$_2$), and that they are sometimes essential for the activity and lifetime of a catalyst. In this respect the oxidation catalyst methyltrioxorhenium (MTO)/H$_2$O$_2$/tBuOH developed and intensely studied by Herrmann et al. should be considered a breakthrough in homogeneous oxidation catalysis. [3] They have succeeded in bridging the gap between the Mimoun reagents of the type

1 (L = HMPA, DMF, etc.) [4] well established for epoxidations, and the classic Milas reagent $OsO_4/H_2O_2/tBuOH$ [5] for the catalytic *syn* hydroxylation of olefins. The isolation and complete characterization of the catalytically active species **2**, R = CH_3, [3b] in the catalyst system $RReO_3/H_2O_2/tBuOH$ is noteworthy and stimulates comment.

Compounds **1** and **2** have the same pentagonal-bipyramidal molecular structure; [3b, 4c] however, the equatorial [d^0-Mo-L] building block in **1** is replaced by the isoelectronic [d^0-Re-R] fragment in **2**. The symmetry and relative arrangement of all frontier orbitals should be similar in the two compounds. This relationship between **1** and **2** is reflected in the similar reactivity of the two compounds, particularly in the *anti* hydroxylation of alkenes by means of the opening of epoxides formed as intermediates. However, it is already apparent that the Herrmann reagent excels as a result of higher reactivity (oxyfunctionalization of certain alkynes, similar as OsO_4 [5d]) and high catalytic activity, whereas the readily prepared Mimoun reagent has proved successful primarily for stoichiometric reactions. [2a, 4]

under catalytic conditions (weakly acidic, protic medium). The hydrolytic instability of the peroxidic function [Re(η^2-O_2)] in absence of an organyl group at the rhenium center is clearly demonstrated by the peroxo perrhenic acid {$ReO(O_2)_2(H_2O)$}$_2$(μ-O) [3c] which is structurally related with **2**. This remarkable peroxo complex catalyses the same oxidation reactions as $ReO(O_2)_2(H_2O)(CH_3)$, but water, the reaction product from H_2O_2 strongly inhibits the catalytic cycle due to the formation of the catalytically inactive hydrolysis product perrhenic acid $ReO_3(OH)$. The MTO/H_2O_2 system has been successfully used not only for epoxidations [3a] but for many other oxidations, the most prominent being the arene oxidation [3d], the Baeyer-Villiger oxidation [3e] and the oxidation of carbonyl complexes [3c].

Other organometallic compounds have also proven their value as catalyst precursors in the activation of hydroperoxides. For example, Trost and Bergmann have reported [6] that molybdenum complex **3** catalyzes the epoxidation of olefins with alkyl hydroperoxides ROOH. In sharp contrast to **2**, **3** is not capable of activating H_2O_2, conversely **2** cannot activate alkyl hydroperoxides. The catalytically active species in the ROOH/**3**/olefin system has not been characterized unambiguously as yet; the authors report, however, that the surprisingly catalytic inactive peroxo complex **4** forms from **3** and ROOH or H_2O_2 in a *deactivating* side reaction. Complexes with a [(η^5-C_5Me_5)Mo(η^2-OOR)] building block could be considered as a plausible organometallic intermediate for the oxygen-transfer step.

Why has it taken a period of more than twenty years since the discovery of Mimoun type complexes that the quality of isoelectronic rhenium oxo complexes for catalytic O-transfer reactions has been recognized and proven? The key to success with **2** lies in its organometallic functionality. More polar Re–X groups (X = OR, NR_2, Cl, etc.) are generally more rapidly protolyzed to give catalytically inactive compounds than the relatively nonpolar Re–R unit in **2**, which is stable

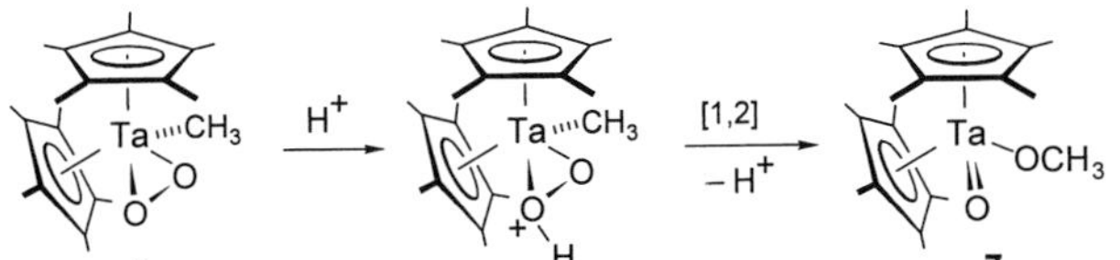

The stability of M–C bonds in the presence of hydroperoxides could also be confirmed for the structurally related alkyl complexes **5** (M = Mo, W; R = CH$_3$, CH$_2$SiMe$_3$). [7] An intramolecular [1,2] shift of the alkyl groups to one of the peroxidic oxygen atoms similar to that which occurs in the tantalocene complexes, for example **6**, investigated in detail by Bercaw et al. [8] does not appear to be a preferred reaction pathway in analogous complexes of Mo, W, and Re. However, Bercaw et al. were able to show that the rearrangement of **6** into the alkoxy complex **7** does not take place either spontaneously or according to a first-order rate law, but is initiated by the attack of an electrophile (e. g. H$^+$) at the peroxo group which leads to substantial polarization of the O–O bond.

In spite of intensive efforts, the knowledge accumulated to date about polar oxygen-transfer reactions on peroxo, hydroperoxo, and alkylperoxo complexes is not yet sufficient to decide whether to agree with the direct attack of the olefin at the positively polarized oxygen center [9] ("butterfly mechanism" without an organometallic intermediate, step A in Scheme 1) discussed by Chong and Sharpless, or with the mechanism postulated by Mimoun based on an olefin coordination,

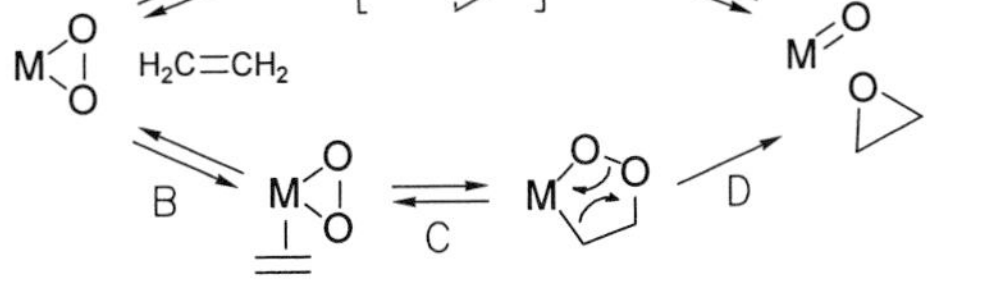

followed by a 1,3-dipolar cycloinsertion with the formation of a metalladioxacyclopentane and its cycloreversion (steps B–D in Scheme 1). [2a]

The question of whether the metal center adopts only the activating role of a π-acidic substituent, analogous so that of a –I, –M substituent in percarboxylic acids or persulfonic acids, or whether in an equilibrium on account of unoccupied metal acceptor orbitals it forms weak interactions [10] with the olefin to be oxidized, remains a point of controversal debate. However an increasing number of recent experimental and theoretical studies support the idea of direct nucleophilic attack of the olefin HOMO into a low-lying σ^*_{O-O} orbital (LUMO) of an η^2-metal-bonded peroxidic function [11]. The oxenoid [12] character of the two coordinate oxygen atom in three membered rings [M(η^2-OOE)] is strongly enhanced by an electron withdrawing substituent E$^+$ on the adjacent peroxidic oxygen atom. This substituent is polarizing the nonpolar O–O bond of the less reactive oxenoid [M(η^2-O$_2$], thus facilating a reaction of S$_N$2 type centred at the less substituted oxygen the [M(η^2-OOE)] moiety [11]. The polarizing substituent E$^+$ may be a proton [8], an alkyl group R$^+$ [12], an acyl group or the Lewis acidic metal center of a second coordinatively unsaturated transition metal peroxide as found in complexes with [M(μ,η^2-O$_2$)M] functionality. [13]

How easily well-established mechanistic ideas about the metal-mediated oxyfunctionalization of organic substrates can begin to falter and consolidate again is also illustrated in a kinetic study by Göbel and Sharpless on the influence of the reaction temperature on the enantioselectivity of the asymmetric *syn*

Scheme 1. Mechanistic alternatives for polar oxygen-transfer reactions of peroxo complexes.

hydroxylation of olefins with chirally modified alkaloid-OsO$_4$ complexes. [14] The Eyring plots for several olefin/alkaloid combinations provide indications of a multistep mechanism with at least two diastereoselective steps. [14a] Thus, a concerted [2 + 3] addition mechanism (pathway A in Scheme 2) should be ruled out, a fact which can be considered as evidence for but not proof of the two diastereomeric oxametallacyclobutanes discussed as intermediates by Sharpless (step B in Scheme 2) [14b] and their stereoselective [1,2] rearrangement into the glycolate complexes **8** (step C in Scheme 2).

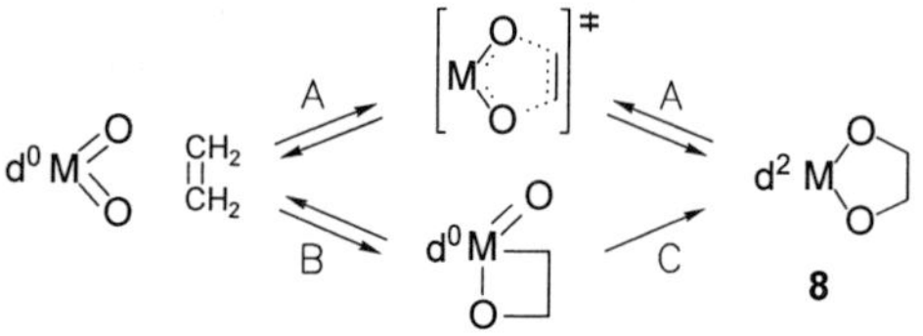

Scheme 2. Mechanistic alternatives for *syn* hydroxylations of olefins with OsO$_4$ complexes.

A few model reactions known from the literature also support the idea of an oxametallacyclobutane intermediate such as the formation of the irida(III)oxetane **10** modeling the metal-mediated transfer of oxygen to a coordinated olefin (step B). This reaction is explained by autoxidation of the iridium(I)-cyclooctadiene complex **9** via the plausible dinuclear oxoiridium(III) intermediate. [15]

A model reaction related to step C in Scheme 2 is the photochemically induced [1,2] migratory insertion of metal bonded aryl groups to a metal oxo group in hydrido trispyrazolylborato rhenium(V) complexes. [16]

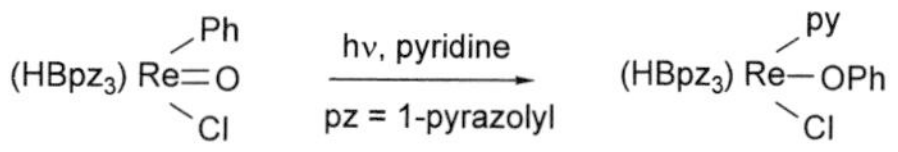

There is also theoretical evidence supporting oxametallacyclobutanes as intermediates in a two step (B and C) mechanism as they are minima and not too high in energy on the potential energy surface. [17] Very recently however, three more detailed and independently performed theoretical studies [18] clearly demonstrate, that activation barriers for a two step [2 + 2] addition/[1,2] rearrangement mechanism are definitively too high in energy. This again together with experimentally kinetic isotope effects observed by Corey et al. [19] suggests that osmylation reactions proceed by a more or less concerted [3 + 2] pathway as originally suggested by Böseken [20a] and Criegee. [20b] This point is still a topic of ongoing debate [21].

A third class of mechanistically not well understood oxidation reactions, that might involve organometallic intermediates is the allylic oxidation of olefins, technically realized in the important catalytic oxidation of propene to acrolein by molecular oxygen on a heterogeneous bismuth molybdate contact (SOHIO process). According to Grasselli et al. [22] a Bi(V)-O functionality is believed to be responsible for the allylic hydrogen abstraction from propene while a high valent molybdenum oxide site is believed to be responsible for the redox O-transfer reaction with the allyl radicals initially formed. The key step of C–O bond formation could be rationalized by direct attack of the allyl radical at the oxygen atom of a d^1 metal oxo species (step A in Scheme 3).

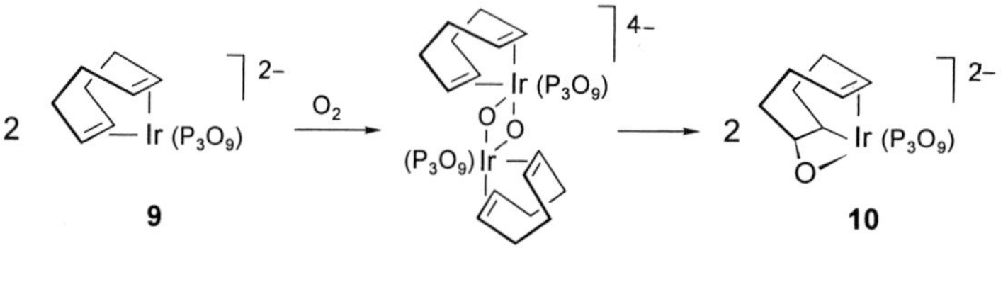

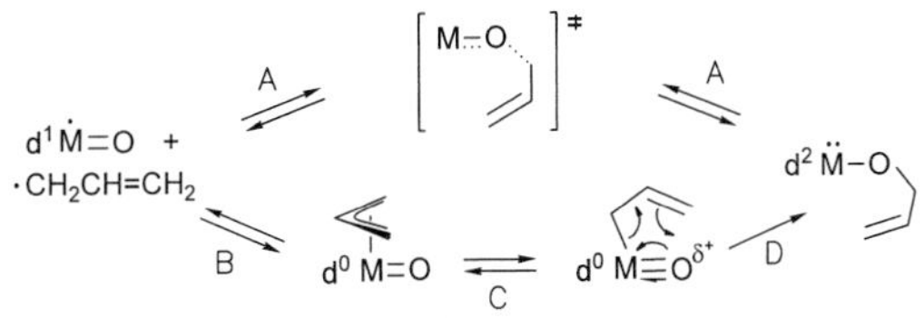

Scheme 3. Mechanistic alternatives for allylic oxidation of olefins, e. g. propene, at a MoO_3/Bi_2O_3 catalyst site.

On the other hand a coordinatively unsaturated paramagnetic d^1 metal center could oxidatively add the allyl radical at the metal center to form a labile d^0 π-allyl oxo species as organometallic intermediate (step B). The latter may rearrange via a σ-allyl intermediate to a d^2 allyloxy species (step D). The latter is believed to convert to acrolein and a reduced metal center via an β-hydrogen abstraction by the metal center. In an attempt to gain insight into fundamental reactions involved in this catalytic cycle, some authentic d^0 allyl oxo complexes of rhenium [23] and tungsten [24] have been prepared. So far reactivity studies of these could not yet accumulate any proof for an [1,2] allyl migration to terminally bonded oxo group (step D). However these and other unsaturated organometallic compounds prooved to be interesting substrates in selective photooxygenation reactions with singlet oxygen. [24b]

Whereas there is still an immense lack of knowledge whether organometallic intermediates play a crucial role in O-transfer reactions of metal oxo and peroxo complexes, our understanding for the mechanisms that control oxidation reactions of authentic organometallic compounds, e. g. intermediates in organic synthesis, was put into more precise terms in recent years. In this respect a combined theoretical and experimental study of Boche et al. [12] provided good evidence that lithiation [12, 25] and titanation [26] of an alkyl hydroperoxide leads to an increase of the electrophilic character of the formally anionic oxygen atom thus making it a better oxenoid than the protonated parent. It is known for a long time and even of technical importance [27] that alkylhydroperoxides and alcohols are products from autoxidations of reactive organometallic intermediates. It was suggested that alkylperoxy species, the formal insertion products of O_2 into the M–C bond, are preferably formed on radical chain pathways and that they may act as oxygen transfer reagents to the remaining part of polar M–C bonds within the reaction mixture. Progress in this field arises from a better understanding for this second oxygen transfer step. It can be envisaged as polar S_N2 type of reaction centred at the formally anionic oxygen atom of the oxenoid $[M(\eta^2\text{-OOR})]$ (M = Li, Al, Mg, Ti etc.) rather than as an electron transfer/recombination sequence. The configuration of the attacking carbon nucleophile is retained. The attack is directed into the low-lying σ^*_{O-O} orbital of the oxenoid. [11, 12, 25]

Progress from a synthetic point of view has been achieved in the selective oxidation of alkyl zinc reagents with oxygen in perfluorohexane solutions to give good yields of either hydroperoxides or alcohols. [28] As many zinc organometallics are easily obtained from olefins via a hydroboration/boron-zinc exchange sequence or by a nickel catalyzed hydro- or carbozincation reaction, this method may proof its synthetic potential in the future development.

References

[1] See also the Highlight from C. Bolm, *Angew. Chem.* **1991**, *103*, 414–415; *Angew. Chem. Int. Ed. Engl.* **1991**, *30*, 403–404. Reviews: a) R. H. Holm, *Chem. Rev.* **1987**, *87*, 1401–1449; b) K. A. Jørgensen, *ibid.* **1989**, *89*, 431–458; c) B. Meunier, *ibid.* **1992**, *92*, 1411–81456; d) H. Mimoun in *Comprehensive Coordination Chemistry*, Vol. 6 (Eds. G. Wilkinson, R. D. Gillard, J. A. McCleverty),

Pergamon, Oxford, **1987**, 317–410; e) R. A. Sheldon, J. K. Kochi, *Metal-Catalyzed Oxidation of Organic Compounds*, Plenum, New York, **1981**; f) G. Strukul, *Catalytic Oxidations with Hydrogen Peroxide as Oxidant*, Kluwer Academic Publishers, Dordrecht, **1992**.

[2] a) H. Mimoun, *Angew. Chem.* **1982**, *94*, 750–766; *Angew. Chem. Int. Ed. Engl.* **1982**, *21*, 734–750; b) K. A. Jørgensen, B. Schiøtt, *Chem. Rev.* **1990**, *90*, 1483–1506.

[3] a) W. A. Herrmann, R. W. Fischer, D. W. Marz, *Angew. Chem.* **1991**, *103*, 1706–1709; *Angew. Chem. Int. Ed. Engl.* **1991**, *30*, 1638–1641; b) W. A. Herrmann, R. W. Fischer, W. Scherer, M. U. Rauch, *Angew. Chem.* **1993**, *105*, 1209–1212; *Angew. Chem. Int. Ed. Engl.* **1993**, *32*, 1157–1160; c) W. A. Herrmann, J. D. G. Correia, F. E. Kühn, G. R. J. Artus, C. C. Romão, *Chem. Eur. J.* **1996**, *2*, 168–173; d) W. Adam, W. A. Herrmann, J. Lin, C. R. Saha-Möller, R. W. Fischer, J. D. G. Correia, *Angew. Chem.* **1994**, *106*, 2545–2546; *Angew. Chem. Int. Ed. Engl.* **1994**, *33*, 2475–2477; e) W. A. Herrmann, R. W. Fischer, J. D. G. Correia, *J. Mol. Cat.* **1994**, *94*, 213–223.

[4] a) H. Mimoun, I. Seree de Roch, L. Sajus, *Bull. Soc. Chim.* **1969**, 1481–1492; b) H. Mimoun, I. Seree de Roch, L. Sajus, *Tetrahedron* **1970**, *26*, 37–50; c) P. J.-M. Le Carpentier, R. Schlupp, R. Weiss, *Acta Crystallogr. Sect. B* **1972**, *28*, 1278–1288; d) Lit. [1e], p. 93 f.

[5] N. A. Milas, S. Sussman, *J. Am. Chem. Soc.* **1936**, *58*, 1302–1304; *ibid.* **1937**, *59*, 2345–2347; b) N. A. Milas, S. Sussman, H. S. Mason, *ibid.* **1939**, *61*, 1844–1847; c) R. Criegee, *Angew. Chem.* **1937**, *50*, 153–155; *ibid.* **1938**, *51*, 519–520; d) Review: M. Schröder, *Chem. Rev.* **1980**, *80*, 187–213.

[6] M. K. Trost, R. G. Bergman, *Organometallics* **1991**, *10*, 1172–1178.

[7] a) P. Legzdins, E. C. Phillips, S. J. Rettig, L. Sánchez, J. Trotter, V. C. Yee, *Organometallics* **1988**, *7*, 1877–1878; b) P. Legzdins, E. C. Phillips, L. Sánchez, *ibid.* **1989**, *8*, 940–949; c) J. W. Faller, Y. Ma, *ibid.* **1988**, *7*, 559–561; d) *J. Organomet. Chem.* **1989**,

368, 45–56; e) G. Parkin, J. E. Bercaw, *Polyhedron* **1988**, *7*, 2053–2082.

[8] a) A. van Asselt, M. S. Trimmer, L. M. Henling, J. E. Bercaw, *J. Am. Chem. Soc.* **1988**, *110*, 8254–8255.

[9] A. O. Chong, K. B. Sharpless, *J. Org. Chem.* **1977**, *42*, 1587–1590; for η^2-alkyl peroxides both reaction pathways can also be formulated.

[10] $p_\pi \rightarrow d_\pi$ Charge transfer interactions between ligand and metal center in d^0 electron configuration, without participation of metal $d_\pi \rightarrow p_\pi$ backbonding; spectroscopic indications: W. A. Nugent, *J. Org. Chem.* **1980**, *45*, 4533–4534.

[11] Review: R. Curci, O. E. Edwards in ref. [1f], 45–95.

[12] G. Boche, F. Bosold, J. C. W. Lohrenz, *Angew. Chem.* **1994**, *106*, 1228–1230; *Angew. Chem. Int. Ed. Engl.* **1994**, *33*, 1161–1163, and references cited therein.

[13] L. Salles, J.-Y. Piquemal, R. Thouvenot, C. Minot, J.-M. Brégeault, *J. Mol. Catal.* **1997**, *117*, 375–387.

[14] a) T. Göbel, K. B. Sharpless, *Angew. Chem.* **1993**, *105*, 1417–1418; *Angew. Chem. Int. Ed. Engl.* **1993**, *32*, 1329–1330; b) K. B. Sharpless, A. Y. Teranishi, J.-E. Bäckvall, *J. Am. Chem. Soc.* **1977**, *99*, 3120–3128.

[15] V. W. Day, W. G. Klemperer, S. P. Lockledge, D. J. Maine, *J. Am. Chem. Soc.* **1990**, *112*, 2031–2033.

[16] S. N. Brown, J. M. Mayer, *Organometallics* **1995**, *14*, 2951–2960.

[17] A. Veldkamp, G. Frenking, *J. Am. Chem. Soc.* **1994**, *116*, 4937–4946.

[18] a) U. Pidun, C. Boehme, G. Frenking, *Angew. Chem.* **1996**, *108*, 3008–3011; *Angew. Chem. Int. Ed. Engl.* **1996**, *35*, 2817–2820; b) S. Dapprich, G. Ujaque, F. Maseras, A. Lledós, D. G. Musaev, K. Morokuma, *J. Am. Chem. Soc.* **1996**, *118*, 11660–11661; c) M. Torrent, L. Deng, M. Duran, M. Sola, T. Ziegler, *Organometallics* **1997**, *16*, 13–19.

[19] E. J. Corey, M. C. Noe, M. J. Grogan, *Tetrahedron Lett.* **1996**, *37*, 4899–4902.

[20] a) J. Böseken, *Recl. Trav. Chim.* **1922**, *41*, 199–207; b) R. Criegee, B. Marchand, H. Wannowius, *Justus Liebigs Ann. Chem.* **1942**, *550*, 99–133.

[21] a) P.-O. Norrby, H. Becker, K. B. Sharpless, *J. Am. Chem. Soc.* **1996**, *118*, 35–42; b) The controversy about the mechanism of the Jacobsen-Katsuki epoxidation has recently been highlighted, while this manuscript was in print: T. Linker, *Angew. Chem.* **1997**, *109*, 2150–2151; *Angew. Chem. Int. Ed. Engl.* **1997**, *36*, 2060–2062.

[22] a) R. K. Grasselli, *J. Chem. Edu.* **1986**, *63*, 216–221; b) J. D. Burrington, C. T. Kartisek, R. K. Grasselli, *J. Catal.* **1984**, *87*, 363–380; c) L. C. Glaeser, J. F. Brazdil, M. A. Hazle, M. Mehicic, R. K. Grasselli, *J. Chem. Soc., Faraday Trans. 1* **1985**, *81*, 2903–2912.

[23] W. A. Herrmann, F. E. Kühn, C. C. Romão, H. T. Huy, *J. Organomet. Chem.* **1994**, *481*, 227–234.

[24] a) J. Sundermeyer, J. Putterlik, H. Pritzkow, *Chem. Ber.* **1993**, *126*, 289–296; b) W. Adam, J. Putterlik, R. Schuhmann, J. Sundermeyer, *Organometallics* **1996**, *15*, 4586–4596.

[25] G. Boche, K. Möbus, K. Harms, J. C. W. Lohrenz, M. Marsch, *Chem. Eur. J.* **1996**, *2*, 604–607.

[26] G. Boche, K. Möbus, K. Harms, M. Marsch, *J. Am. Chem. Soc.* **1996**, *118*, 2770–2771.

[27] The AlEt$_3$ catalyzed oligomerization of ethene followed by an stoichiometric autoxidation with molecular oxygen leads – via Al(OR)$_3$ – to even-numbered primary alcohols used in production of PVC plasticizers and biodegradable detergents.

[28] I. Klement, H. Lütjens, P. Knochel, *Tetrahedron Lett.* **1995**, *36*, 3161–3164; b) I. Klement, P. Knochel, *Synlett* **1995**, 1113–1114.

The Oxofunctionalization of Alkanes

Oliver Reiser

The selective oxidation of alkanes remains a challenge in organic synthesis. The development of economical processes for the functionalization of hydrocarbons is of tremendous technical importance. In nature such reactions are mediated efficiently by various enzymes. The systems related to cytochrome P-450, which for example detoxify lipid-soluble compounds in the human liver, [1] have received the most attention.

The fundamental problem in the functionalization of saturated hydrocarbons is that their components, carbon and hydrogen, do not have lone electron pairs, and the molecules do not have orbitals of sufficient energy that are easily accessible. Thus, very reactive reagents and/or extreme reaction conditions are typically required, for example for the oxidation of alkanes. However, the initial products are almost always more reactive than the starting compounds, and undesired side reactions may occur.

Other complications arise if the molecule to be oxidized contains different types of C–H bonds. Since formation of tertiary radicals and carbenium ions is favored and they are also more stable than their secondary and primary analogues, functionalization processes that proceed via these intermediates generally have the selectivity for C atoms in the order tertiary > secondary > primary. For steric reasons, however, the attack of bulky reagents at primary C atoms may be preferred; the best example known is the oxidative addition of transition metal complexes.

The metal-mediated oxyfunctionalization of organic compounds, of olefins in particular, is becoming increasingly important. [2] Research on metal catalysts that activate elemental oxygen and make it usable as a selective oxidizing agent is therefore especially worthwhile. Inspired by the enzyme cytochrome P-450, in which the active center is an oxo-iron(IV) unit, researchers have examined numerous iron-containing reagents for the oxidation of alkanes. For example, the so-called Gif systems [3] developed by Barton et al., which consist essentially of air, catalytic amounts of iron and/or zinc, acetic acid (HOAc), pyridine (py), and water, oxidize adamantane (**1**) to give adamantanone (**2**) as the major product along with 2- (**3**) and 1-adamantanol (**4**) as side products.

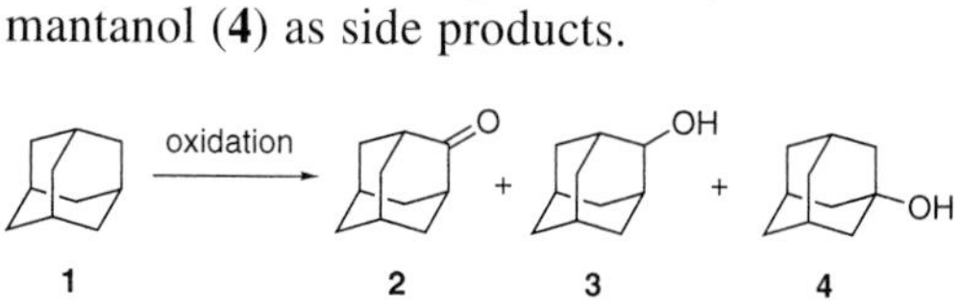

The oxidizing species are postulated to be $[Fe^{II}Fe^{III}{}_2O(OAc)_6(py)_{3.5}]$ [4] and heteroaromatic *N*-oxide radical cations, [5] in keeping with the necessity of a nitrogen base component such as pyridine. Apparently these species attack C–H bonds on secondary C atoms in marked preference to those on primary and tertiary C atoms. Mechanistic studies suggest that the oxidation yielding ketones like **2** does not proceed via the corresponding alcohols **3** and **4**. Turnover rates of over 3000 and almost quantitative reactions have been achieved with several Gif reagents. One disadvantage is that the reactions can be conducted only up to 10–15 % conversion to avoid considerable amounts of side products.

The allotropic form of oxygen, ozone, can also be employed for the oxidation of saturated hydrocarbons. The reactivity of ozone without additional reagents is not sufficient for the preparative functionalization of alkanes in solution; however, its reactivity is increased substantially by the addition of iron(III) chloride [6] or antimony pentafluoride. [7] The dry ozonation variant [8] of Mazur et al. [9] by which alkanes are hydroxylated at tertiary C atoms with high selectivities and yields, was shown to be especially useful. According to this method, silica gel is coated with roughly 1 wt% of the substrate, cooled to −78 °C, saturated with ozone, and subsequently allowed to warm to room temperature within 0.5–2 h. Adamantane (**1**) is converted almost quantitatively into 1-adamantanol (**4**) in this way (Table 1), and this method of oxyfunctionalization has been applied successfully even on certain steroids. [10]

The selective oxidation of hydrocarbons by dry ozonation is not restricted to reaction at tertiary C atoms: the CH_2 groups adjacent to a cyclopropane ring are smoothly converted into carbonyl groups. This "cyclopropyl activation" can be explained by the well-known ability of the cyclopropyl group to effectively stabilize a neighboring R_3C^+ center. Although in the oxidation of alkanes with ozone the first intermediate, the hydrotrioxide, [11] does not arise from carbenium ions, the effect of polarity upon attack of the electronegative ozone molecule, either as a radical or by 1,3-dipolar

addition, at the C–H bond certainly leads to development of a positive charge on the pertinent C atom. [11] Dry ozonation was the first efficient method for the *a*-functionalization of cyclopropylidene hydrocarbons and served as the key step in a synthesis of [6]rotane [12] (**7**) via the tetraspiro hydrocarbon 5. [13]

The ketonation of cyclopropylidene hydrocarbons with ruthenium tetraoxide (generated *in situ* from $RuCl_3$ and $NaIO_4$ in CH_3CN/CCl_4/aqueous phosphate buffer) [14] proved to be more convenient and better suited for large-scale reactions. This oxidation of organic substrates originally developed by K. B. Sharpless et al. [15] for the oxidative cleavage of alkenes can also be applied for the selective hydroxylation of tertiary C–H groups; [14c, d] for example, adamantane (**1**) is converted into 1-adamantanol (**4**) in 75 % yield. [14d]

Since the development of a handy, even large-scale method for the generation of dimethyldioxirane (**9a**), [16] it has become known best as a reagent for the epoxidation of double bonds. [17] Its application for the selective hydroxylation of C–H bonds at tertiary C atoms is less well known. For example, the reaction of **1** with **9a** as the oxidizing agent provides **4** with only minor amounts (< 3 %) of side products. [18]

8a: R = CH3
8b: R = CF3

9a: R = CH3
9b: R = CF3

Oxidations with **9a** proceed stereoselectively with retention, as the reactions with *cis*-decalin (**10**) and *trans*-decalin (**12**) demonstrate. Remarkably, the oxidation of **10** is faster than that of **12**: after a reaction time of 17 h **11** had been formed in 84 % yield, **13** in only 20 %. This indicates that the oxidation of equatorial positions is preferred to that of axial positions (Table 1).

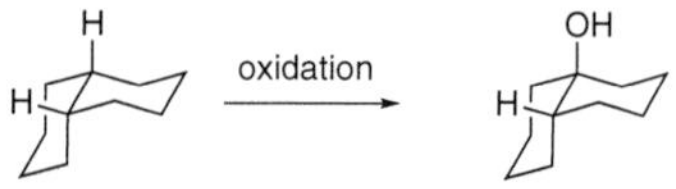

An oxidizing agent up to 7000 times more reactive than **9a** is methyl(trifluoromethyl)-dioxirane (**9b**), which not only hydroxylates tertiary C–H positions (Table 1) but also converts secondary C–H groups into carbonyl functions. As an illustration, the oxidation of cyclohexane at –22 °C afforded cyclohexa-none in yields of over 98 % in only 18 min. [19] With this reagent not only the monohy-droxylation of adamantane (**1**) providing **4** proceeds in excellent yield, but also the two-, three-, and even four-fold hydroxylation of the bridgehead positions of adamantane. [20]

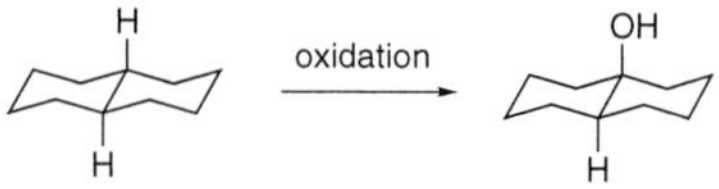

Remarkably, **9b** is stable towards acids like trifluoroacetic acid, trifluoromethanesulfonic acid, and sulfuric acid. Based on this observa-tion, Asensio et al. employed **9b** for the oxida-tion of tertiary C–H positions in the presence of amino groups. [21] Their trick was protec-ting the amino group as an ammonium tetra-fluoroborate by treatment with tetrafluorobo-ric acid. In the subsequent reaction with **9b**, the C–H bond at a tertiary C atom, which must be at least two C atoms removed from the ammonium group, is converted into a hydroxyl or acetamido function in excellent yield. C–H bonds on secondary C atoms in acyclic amines such as **14** (R = Me, Et, Pr) may also be oxidized in this way to afford the corresponding ketones **15**; the oxygen insertion takes place exclusively at the ε-

Table 1. Comparison of reagents for the oxidation of adamantane (**1**) and decalins **10** and **12**.

Substrate	Oxidation agent	Product	t [h] T [°C]	Yield [%]
1	O$_3$/SiO$_2$	4	0.5/–78 to 25	99
1	RuCl$_3$/NaIO$_4$	4	3–7/90	75
1	9a	4	18/22	87
1	9b	4	1/–22	92
1	F$_2$/H$_2$O/CH$_3$CN	4	–/0	80
1	hν/TCB/O$_2$	4/2	–/–	60/18
1	19a	4	< 1/25	90
1	19b	4	< 1/25	90
10	O$_3$/SiO$_2$	11	0.5/–78 to 25	99
10	9a	11	17/22	84
10	9b	11	0.2/–22	91
10	19a	11	few/25	85
10	19b	11	few/25	88
12	O$_3$/SiO$_2$	13	0.5/–78 to 25	72
12	9a	13	17/22	20
12	9b	13	0.1/–22	61
12	19a	13	many/25	73

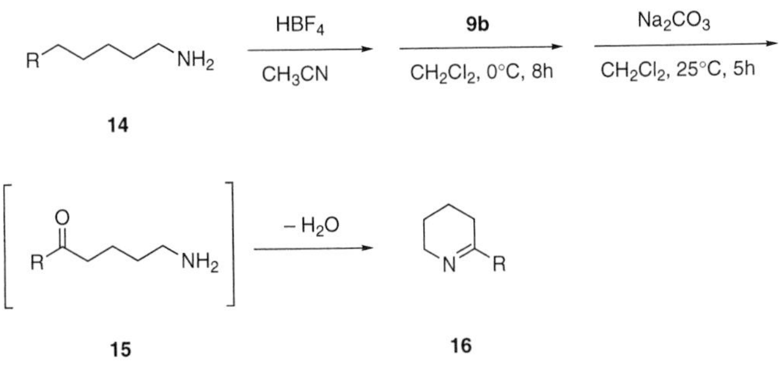

R⌒⌒NH₂ →(HBF₄, CH₃CN) →(9b, CH₂Cl₂, 0°C, 8h) →(Na₂CO₃, CH₂Cl₂, 25°C, 5h)

14

[R-C(O)⌒⌒NH₂] →(− H₂O) 16

15   **16**

methylene group. The authors explain the excellent selectivity by the coordination of **9b** to the ammonium group by hydrogen bonding.

Recently, the scope of the reagent **9b** could be considerably broadened in that hydroxylations of secondary and even primary C–H bonds were achieved. [22] The innovation is to carry out the oxidation in the presence of trifluoroacetic anhydride, which acylates the alcohols being the primary reaction products. For example, cyclohexane is converted in quantitative yield to the trifluoroacetate of cyclohexanol.

Very exciting results for the oxyfunctionalization of alkanes were also obtained using photoinduced SET reactions as the key step. [23] The principle applied here is that after electron tranfer oxidation of the alkane **18** the aciditiy of the C–H bonds of the resulting radical cation **20** is dramatically increased. Thus, deprotonation to the radical **23** occurs, which then can be trapped by oxygen. For example, if **1** is irradiated in an oxygen saturated acetonitrile solution in the presence of 1,2,4,5-tetracyanobenzene (TCB), **4** is obtained as the main product. Changing from TCB to 2,3,5,6-tetrachloro-*p*-benzoquinone (Chl) the radical **23** is further oxidized to the cation **26**, which in turn can now also be trapped by acetonitrile leading to amidated products **24**.

Hydroxylation of C–H bonds on tertiary C atoms can be conducted with solutions of fluorine in aqueous acetonitrile. [24] This

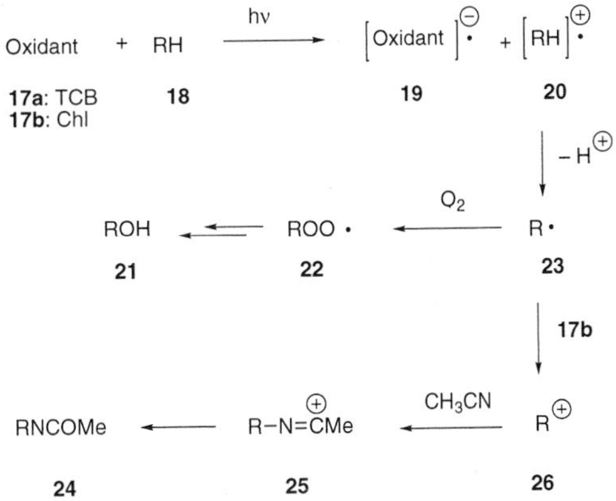

Oxidant + RH →(hν) [Oxidant]⁻• + [RH]⁺•

17a: TCB **18**   **19**   **20**

17a: TCB
17b: Chl

| −H⁺

ROH ← ROO• ← R•
21   **22**   **23**

| O₂ (at 22←23)

| **17b**

RNCOMe ← R–N=CMe⁺ ← R⁺ (CH₃CN)
24   **25**   **26**

reagent appears particularly attractive for the synthesis of ¹⁸O-labeled compounds; for example, [¹⁸O]-1-adamantanol ([¹⁸O]-**4**) is prepared in over 80 % yield by using a solution of F₂ in acetonitrile and H₂¹⁸O.

Another promising reagent for the oxyfunctionalization of hydrocarbons are the di(perfluoroalkyl)oxaziridines **29**, [25] which can be synthesized from commercially available tri(perfluoroalkyl)amines **27** on a large scale [25b] in two steps. The reaction of **27** with antimony pentafluoride produces α-fluoroketimines **28**, which are then epoxidized with *m*-chloroperbenzoic acid (MCPBA) to give

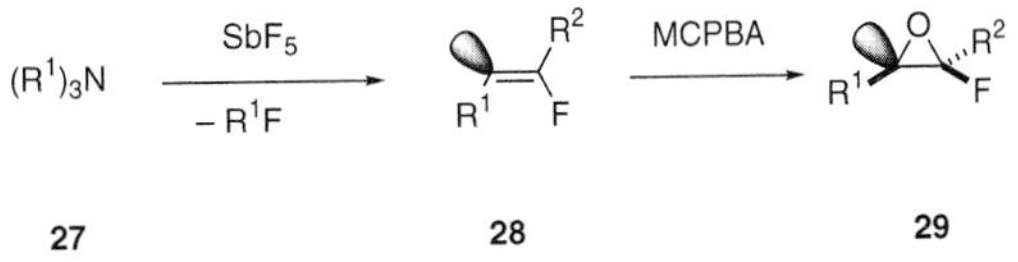

(R¹)₃N →(SbF₅, − R¹F) R¹R²C=N–F →(MCPBA) epoxide R¹R²C–N(F) O

27   **28**   **29**

29. Compound **29** can be purified by vacuum distillation and stored indefinitely at room temperature.

Solutions of hydrocarbons such as **1**, **10**, and **12** in chloroform, carbon tetrachloride, or trichlorofluoromethane are readily mono-hydroxylated at their tertiary C atoms with only minimal amounts of side products. In addition to the high selectivity for tertiary C atoms over secondary, **29** shows a marked preference for the oxidation of equatorial over axial positions and a high stereospecificity for retention of configuration of the C atoms oxidized (Table 1). The authors propose the concerted O insertion into a C–H bond as the mechanism, since in the reaction of **19** with **1** in CCl₄ no chlorine-containing trapping products are formed from **1** via intermediate radicals, as is observed for the oxidation of **1** in CCl₄ with ozone. **29** has proven to be useful also for the epoxidation of alkenes, [26] and for the oxidation of sulfides to sulfoxides or sulfones, [27] pyrdines to the corresponding *N*-oxides, [28] or trialkylsilanes to trialkylsilanols, the latter occurs with retention of configuration. [29]

As these examples show there has been considerable progress in achieving the goal of finding efficient methods to selectively oxidize alkanes. There have been also several promising reports on transition metal complexes which are able to insert into non activated C–H bonds at very low temperatures. [30] All these developments give rise to optimism that in the near future general methods for the functionalization of non activated molecules will become available.

Acknowledgement: This work was supported by the Winnacker foundation and the Fonds der Chemischen Industrie.

References

[1] H. L. Holland, *Organic Synthesis with Oxidative Enzymes*, VCH, New York, **1991**.

[2] See also J. Sundermeyer, *Angew. Chem* **1993**, *105*, 1195–1197; *Angew. Chem. Int. Ed. Engl.* **1993**, *32*, 1144–1146, and references therein.

[3] D. H. R. Barton, E. Csuhai, N. Ozbalik, *Tetrahedron* **1990**, *46*, 3743–3752, and references therein.

[4] a) D. H. R. Barton, M. J. Gastiger, W. B. Motherwell, *J. Chem. Soc. Chem. Commun.* **1983**, 731–733. b) G. Balavoine, D. H. R. Barton, Y. V. Geletii, D. R. Hill in *The Activation of Dioxygen and Homogeneous Catalytic Oxidation*; D. H. R. B. et. al., Ed.; Plenum Press: New York, **1993**; pp. 225–242.

[5] a) Y. V. Geletii, A. E. Shilov in *The Role of Oxygen in Chemistry and Biochemistry*; W. Ando Y. Moro-oka, Ed.; Elsevier: Amsterdam, **1988**; 33, pp. 293–300. b) E. M. Koldasheva, Y. V. Geletii, A. F. Shestakov, A. V. Kulikov, A. E. Shilov, *New. J. Chem.* **1993**, 421–24.

[6] T. M. Hellman, G. A. Hamilton, *J. Am. Chem. Soc.* **1974**, *96*, 1530–1535.

[7] a) G. A. Olah, G. K. S. Prakash, J. Sommer, *Superacids*, Wiley, New York, **1985**, p. 315ff, and references therein; b) N. Yoneda, T. Kiuchi, T. Fukuhara, A. Suzuki, G. A. Olah, *Chem. Lett.* **1984**, 1617.

[8] Reviews: a) A. de Meijere, *Nachr. Chem. Tech. Lab.* **1979**, *27*, 177–182; b) E. Keinan, T. H. Varkony in *The Chemistry of Peroxides* (Eds.: S. Patai), Wiley, New York, **1983**, p. 649ff.

[9] Z. Cohen, E. Keinan, Y. Mazur, T. H. Varkony, *J. Org. Chem.* **1975**, *40*, 2141–2142.

[10] Z. Cohen, E. Keinan, Y. Mazur, A. Ulman, *J. Org. Chem.* **1976**, *41*, 2651–2652.

[11] a) M. Zarth, A. de Meijere, *Chem. Ber.* **1985**, *118*, 2429–2449; b) A. de Meijere, F. Wolf, *Methoden Org. Chem.* (Houben-Weyl) 4th ed., Vol. E13, **1988**, pp. 971–990, and references therein.

[12] L. Fitjer, *Angew. Chem.* **1976**, *88*, 804–5; *Angew. Chem. Int. Ed. Engl.* **1976**, *15*, 762.

[13] a) E. Proksch, A. de Meijere, *Angew. Chem* **1976**, *88*, 802–803; *Angew. Chem. Int. Ed. Engl.* **1976**, *15*, 761–762; b) *Tetrahedron Lett.* **1976**, 4851–4854.

[14] a) T. Hasegawa, H. Niwa, K. Yamada, Chem. *Lett.* **1985**, 1385–86, and references therein; b) B. Waegell, J.-L. Coudret, S. Zöllner, A. de Meijere, unpublished; J.-L. Coudret, Ph. D. thesis **1995**, Universite d'Aix Marseille III; c) A. Tenaglia, E. Terranova, B. Waegell, *Tetrahedron Lett.* **1989**, *30*, 5271–5274; d) H. J. Carlsen, *Synth. Commun.* **1987**, *17*, 19–23.

[15] P. H. J. Carlsen, T. Katsuki, V. S. Martin, K. B. Sharpless, *J. Org. Chem.* **1981**, *46*, 3936–3938.

[16] W. Adam, J. Bialas, L. Hadjiarapoglou, *Chem. Ber.* **1991**, *124*, 2377.

[17] a) W. Adam, L. P. Hadjiarapoglou, R. Curci, R. Mello in *Organic Peroxides* (Ed.: W. Ando), Wiley, New York, **1992**, pp. 195–219ff; b) R. Curcin in *Adv. Oxygenated Processes*, (Ed.: A. L. Baumstark), JAI, Greenwich, CT, USA, Vol. 2, **1990**, pp. 1–59.

[18] R. W. Murray, R. Jeyaraman, L. Mohon, *J. Am. Chem. Soc.* **1986**, *108*, 2470–2472.

[19] R. Mello, M. Fiorentino, C. Fusco, R. Curci, *J. Am. Chem. Soc.* **1989**, *111*, 6749–6757.

[20] R. Mello, L. Cassidei, M. Fiorentino, R. Curci, *Tetrahedron Lett.* **1990**, *31*, 3067–3070.

[21] G. Asensio, M. E. Gonzalez-Nunez, C. B. Bernadini, R. Mello, W. Adam, *J. Am. Chem. Soc.* **1993**, *115*, 7250–7253.

[22] G. Asensio, R. Mello, M. Elena, M. E. Gonzaleznunez, G. Castellano, J. Corral, *Angew. Chem.* **1996**, *108*, 196–8; *Angew. Chem. Int. Ed. Engl.* **1996**, *35*, 217–218.

[23] M. Mella, M. Freccero, A. Albini, *J. Chem. Soc. Chem. Commun.* **1995**, 41–42.

[24] S. Rozen, M. Brand, M. Kol, *J. Am. Chem. Soc.*. **1989**, *111*, 8325–8326.

[25] a) D. D. Desmarteau, A. Donadelli, V. Montanari, V. A. Petrov, G. Resnati, *J. Am. Chem. Soc.* **1993**, *115*, 4897–4898; b) V. A. Petrov, D. D. DesMarteau, *J.Org. Chem.* **1993**, *58*, 4754–4755. c) A. Arnone, M. Cavicchioli, V. Montanari, G. Resnati, *J. Org. Chem.* **1994**, *59*, 5511–5513. (d) V. A. Petrov, G. Resnati, *Chem Rev* **1996**, *96*, 1809–1823.

[26] A. Arnone, D. D. Desmarteau, B. Novo, V. A. Petrov, M. Pregnolato, G. Resnati, *J. Org. Chem.* **1996**, *61*, 8805–8810.

[27] a) D. D. Desmarteau, V. A. Petrov, V. Montanari, M. Pregnolato, G. Resnati, *J. Org. Chem.* **1994**, *59*, 2762–2765. (b) M. Terreni, M. Pregnolato, G. Resnati, E. Benfenati, *Tetrahedron* **1995**, *51*, 7981–7992. (c) J. P. Begue, A. Mbida, D. Bonnetdelpon, B. Novo, G. Resnati, *Synthesis Stuttgart* **1996**, 399.

[28] a) C. Balsarini, B. Novo, G. Resnati, *J. Fluorine Chem.* **1996**, *80*, 31–34. b) R. Bernardi, B. Novo, G. Resnati, *J. Chem. Soc. Perkin Trans 1* **1996**, 2517–2521.

[29] M. Cavicchioli, V. Montanari, G. Resnati, *Tetrahedron Lett.* **1994**, *35*, 6329–6330.

[30] Review: W. A. Herrmann, B. Cornils, *Angew. Chem.* **1997**, *109*, 1074–1095.

Cooperativity in Rh$_2$ Complexes: High Catalytic Activity and Selectivity in the Hydroformylation of Olefins

Georg Süss-Fink

Several important steps forward in the search for catalytically active polynuclear metal complexes have been made over the last years by George G. Stanley's group: By reaction of a novel tetraphosporus ligand L, which exist in the R,R, S,S, and R,S forms and which can be separated into the racemate (R,R/S,S) and the meso form (R,S) [1], with the norbornadiene (C$_7$H$_8$) complex [Rh(η^4-C$_7$H$_8$)$_2$][BF$_4$], the new dinuclear complex cations rac-[Rh$_2$(η^4-C$_7$H$_8$)$_2$(η^2:η^2-L)$_2$]$^{2+}$ (rac-1) and meso-[Rh$_2$(η^4-C$_7$H$_8$)$_2$(η^2:η^2-L)$_2$]$^{2+}$ (meso-1) were obtained and isolated as the tetrafluoroborate salts (Scheme 1). [2] The racemic complex rac-1, whose structure was determined by a single-crystal X-ray structure analysis, is an excellent hydroformylation catalyst for olefins such as propylene, 1-hexene and 1-octene, combining high activity with high regioselectivity. Up to 12000 turnovers (10 cycles per minute) were achieved in a single batch run without observing any ligand degradation reactions. The conversion of olefin to aldehyde for 1-hexene is 85 %, the n/i ratio being 96.5 : 3.5. [2] The combination of high activity with high selectivity can be explained by cooperativity between both metal centers in the catalytically active dinuclear complex.

The hydroformylation of olefins is the world's most important homogeneous catalytic industrial process with more than six million tonnes of oxo products produced per

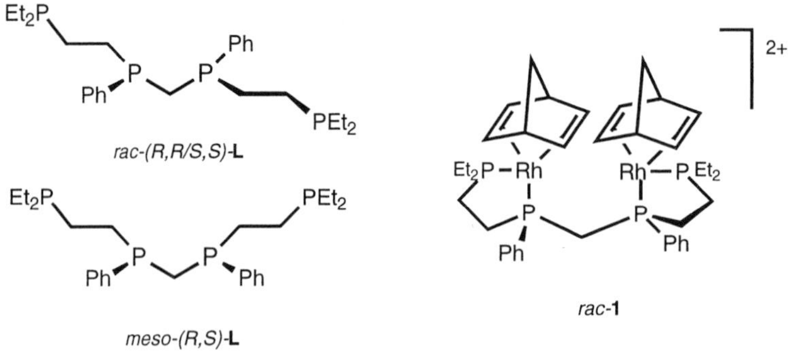

Scheme 1. The tetraphosphorus ligands *rac*-**L** and *meso*-**L**, and the hydroformylation active dirhodium complex *rac*-**1.**

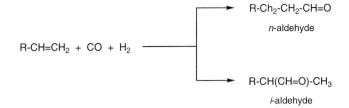

R-CH=CH₂ + CO + H₂

R-Ch₂-CH₂-CH=O

n-aldehyde

R-CH(CH=O)-CH₃

i-aldehyde

Scheme 2. The hydroformylation of terminal olefins and the regio-selectivity problem.

year. [3] Besides the problem of chemical selectivity, the hydroformylation of olefins raises the question of regioselectivity depending on whether the formyl group is added to the terminal or to the internal carbon atom of the double bond. This results in the ratio of linear to branched product (*n/i* ratio of the aldehyde formed) (Scheme 2). The industrial processes are all based on mononuclear cobalt or rhodium complexes.

Ever since transition metal clusters have been discussed as catalysts, [4] there have been many attempts to develop catalytically active polynuclear complexes in which the metal centers interact during the formation of the target molecule ("cooperativity") to control activity and selectivity of the catalytic process. In this way, a highly selective hydroformylation catalyst for propene was found in the cluster anion $[HRu_3(CO)_{11}]^-$ (linear to branched product ration of butyraldehyde 98.6 : 1.4), however, its activity is very low with only 57 cycles in 66 h. [5] The high selectivity and low activity suggest the catalytic process to occur *via* a sterically demanding, intact trinuclear structure, for which there is experimental evidence. [6] In contrast to $[HRu_3(CO)_{11}]^-$, the tetranuclear neutral cluster $[Rh_4(CO)_{12}]$ was reported to yield high catalytic conversions (84000 cycles) for the hydroformylation of 1-hexene, the selectivity, however, is unfavorable (*n/i* 54 : 46). [7] Accordingly, evidence for a catalytic route *via* mononuclear rhodium fragments was found. [8] The higher hydroformylation activity of the mixed-metal cluster $[Co_2Rh_2(CO)_{12}]$ as compared to those of either the parent clusters

$[Co_4(CO)_{12}]$ and $[Rh_4(CO)_{12}]$, originally interpreted by cooperativity between Co and Rh centers in the bimetallic catalyst, [9] was later shown to be simply due to the more facile fragmentation of the mixed-metal cluster into reactive $[HRh(CO)_4]$ species. [10] The catalyst *rac*-**1** described by Stanley et al. is the only polynuclear system as yet which compares favorably with the hydroformylation catalysts used industrially.

Stanley et al. explain the combination of high selectivity with high activity of their catalyst by bimetallic cooperativity in the form of an *intramolecular* hydride transfer which assists in the elimination of the aldehyde from an intermediate containing the RCH_2CH_2CO fragment as a ligand at one Rh centre and the H ligand at the other Rh center. This interpretation is supported by the lack of catalyst fragmentation and the fact that mononuclear model complexes $[Rh(\eta^4\text{-}C_7H_8)\text{-}(\eta^2\text{-diphos})][BF_4]$ (diphos = $Et_2PCH_2\text{-}CH_2\text{-}PEt_2$, $Et_2PCH_2CH_2PMePh$, $Et_2PCH_2CH_2PPh_2$, $Ph_2PCH_2CH_2PPh_2$) generate, under the same conditions as for *rac*-**1**, extremely poor hydroformylation catalysts for 1-hexene from both activity and selectivity viewpoints. [2]

Further persuasive evidence for this interpretation came from a systematic investigation of well-designed dinuclear model systems in which the central methylene group (with respect to complex **1**) has been replaced by propylene (**2**) or *para*-xylylene (**3**) groups (Scheme 3). These "spaced" precursors **2** and **3** (although comparable to *rac*-**1** both electronically and sterically) are only poor hydroformylation catalysts as compared to *rac*-**1**,

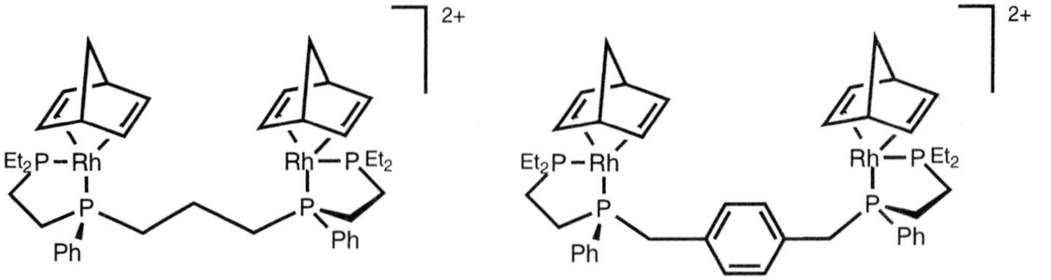

Scheme 3. The model "spacer" complexes **2** and **3**.

being in line with the cooperativity concept: Unlike in *rac*-**1**, an intramolecular hydride transfer is not possible in **2** or **3**, since both Rh atoms are kept apart by the spacers. [2]

The most internally self-consistent check, however, is the dramatic decrease in both activity and selectivity by using the *R,S* diastereomer *meso*-**1** as hydroformylation catalyst (as compared to the racemic *R,R/S,S* diasteromers *rac*-**1**). The higher rate of the *racemic* system was thought to arise from its ability to form a double-bridged hydrido-carbonyl intermediate (or transition-state) which favors the intramolecular hydride transfer. The *meso* catalyst can do an intramolecular hydride transfer, but cannot form a double-bridged H/CO situation facilitating the hydride transfer, since the terminal CO ligand at the acyl-bound Rh center is unfavorably oriented with respect to the other Rh atom. [2]

The mechanism proposed in the original paper, [2] however, proved to be inconsistent with further mechanistic studies by Stanley et al.: The original catalytic cycle was based on neutral dinuclear rhodium intermediates, the first of which, the neutral hydrido-carbonyl complex *rac*-[Rh$_2$(CO)$_2$(η^2:η^2-L)], was believed to form from the dicationic *rac*-**1** under syngas (CO/H$_2$) conditions with liberation of H$^+$ and norbornadiene. In order to model the intermediacy of this neutral species, Stanley et al. designed the neutral allyl (C$_3$H$_5$) analogue *rac*-[Rh$_2$(η^3-C$_3$H$_5$)$_2$(η^2:η^2-L)] (*rac*-

4), accessible from *rac*-**1** and allylmagnesium bromide. Surprisingly, *rac*-**4** turned out to be a rather slow and poorly selective hydroformylation catalyst (with respect to *rac*-**1**). [12]

In the face of these confusing results, Stanley and co-workers decided to initiate an extensive series of *in situ* spectroscopic studies to understand what was occurring under hydroformylation conditions with the neutral and dicationic precursor complexes *rac*-**1** and *rac*-**4**, why they give such dramatically different catalytic results, in marked contrast to mononuclear hydroformylation catalyst. In a recent book on catalysis by di- and poly-nuclear metal complexes, edited by R. D. Adams and F. A. Cotton, George G. Stanley gives a detailed account of this fascinating story which reads as exciting as a detective novel. [12]

In situ FT-IR studies showed that the neutral complex *rac*-**4** reacts under CO atmosphere with carbon monoxide through the intermediacy of *rac*-[Rh$_2$(CO)$_2$(η^3-C$_3$H$_5$)$_2$(η^2:η^2-L)] (*rac*-**5**) to give *rac*-[Rh$_2$\{η^1-C(O)C$_3$H$_5$\}$_2$(CO)$_4$(η^2:η^2-L)] (*rac*-**6**). Under syngas conditions (CO/H$_2$), *rac*-**6** is converted into a zwitterionic Rh(−I)–(Rh+I) species *rac*-[Rh$_2$(CO)$_2$(μ_2-CO)(η^3:η^1-L)] (*rac*-**7**), for which the crystal structure analysis revealed the tetraphosphine ligand L to be unsymmetrically coordinated (with 3 P atoms to one Rh center and with one P to the other Rh center). [11]

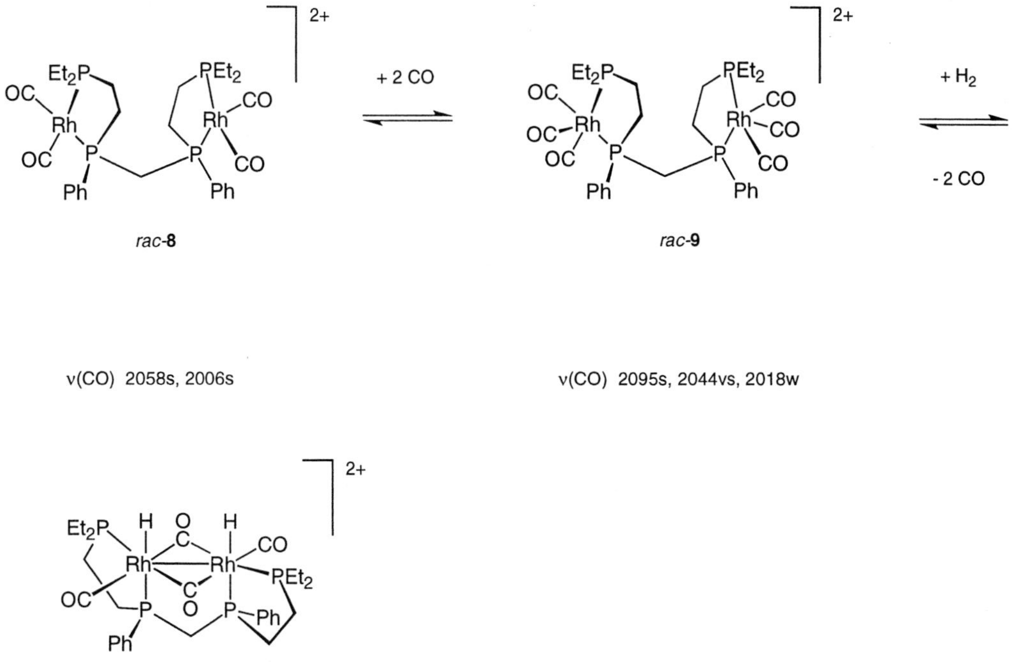

rac-8

v(CO) 2058s, 2006s

rac-9

v(CO) 2095s, 2044vs, 2018w

rac-12

v(CO) 2076m, 2038s, 1832m, 1818w

δ(^{1}H) - 9.0, ^{1}J(Rh-H) 164 Hz, ^{2}J(Rh-H) 15 Hz)

Scheme 4. Interpretation of the *in situ* spectroscopic studies on the reactivity of the dicationic complex *rac-8* towards CO and H$_2$.

In situ FT-IR and NMR studies on *rac-1*, however, revealed the catalytically active species to be a dicationic Rh(+II)–Rh(+II) complex: The dicationic precursor *rac-*[Rh$_2$(CO)$_4$(η^2:η^2-L)]$^{2+}$ (*rac-8*), which generates under hydroformylation conditions the same active species and gives identical spectroscopic results as *rac-1*, was shown to react, under CO atmosphere, with carbon monoxide to give *rac-*[Rh$_2$(CO)$_6$(η^2:η^2-L)]$^{2+}$ (*rac-9*). Under syn gas (CO/H$_2$) atmosphere, *rac-9* takes up molecular hydrogen to give, with elimination of CO (and presumably through the intermediacy of uncharacterized species *rac-10* and *rac-11*), a complex which could be identified by its IR, ^{1}H and ^{31}P NMR data as being *rac-*[Rh$_2$H$_2$(CO)$_2$(μ_2-CO)$_2$(η^2:η^2-L)]$^{2+}$ (*rac-12*) (Scheme 4). [13]

Complex *rac-12* is now proposed, based on the *in situ* spectroscopic studies, to be the catalytically active species in the olefin hydroformylation using *rac-1* as catalyst precursor. In contrast to the original hypothesis [2], the active species now is dicationic and contains a Rh–Rh bond: *rac-12* is believed to be a electron-precise 34e Rh(+II)–Rh(+II) complex with an edge-sharing bioctahedral structure. [13] The current proposed hydroformylation mechanism is depicted in Scheme 5.

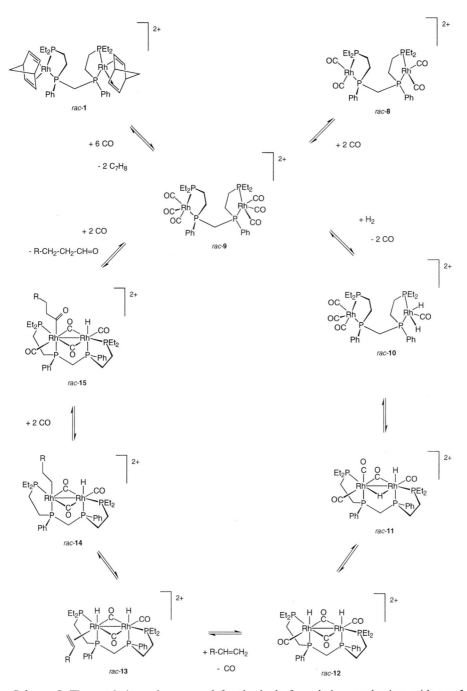

Scheme 5. The catalytic cycle proposed for the hydroformylation mechanism with *rac*-**1** as catalyst precursor. The two *intramolecular* hydride transfer processes reflect the "cooperativity" of the two rhodium centers.

The facile loss of CO from *rac*-**9**, formed from the precursor *rac*-**1** (or *rac*-**8**), opens up a vacant coordination site allowing the oxidative addition of H_2 to give *rac*-**10** which is considered to be a Rh(+I)–Rh(+III) 18e/16e species. Here the first cooperative step takes place with the rearrangement of *rac*-**10** by an *intramolecular* hydride transfer *via* the bridged intermediate *rac*-**11** to the metal-metal bonded Rh(+II)–Rh(+II) 34e complex *rac*-**12**. Dissociation of CO in *rac*-**12** provides entry for an olefin ligand to give *rac*-**13** which forms, by migratory insertion, the alkyl species *rac*-**14**. CO addition gives *rac*-**15** and finally, with reductive elimination of the aldehyde *via* another *intramolecular* hydride transfer, *rac*-**9** with which the cycle started. Despite the modification of the original hypothesis, [2] the cooperativity concept still holds in the current proposed cycle. [12]

Furthermore, Stanley and co-workers have also been able to resolve the racemic tetraphosphorus ligand into the enantiomerically pure compounds *R,R*-**L** and *S,S*-**L** by preparatory HPLC using a Chiralcel OD column. From these, they prepared the chiral catalyst precursors *R,R*-[Rh₂(η^4-C₇H₈)₂(η^2:η^2-L)₂]²⁺ (*R,R*-**1**) and *S,S*-[Rh₂(η^4-C₇H₈)₂(η^2:η^2-L)₂]²⁺ (*S,S*-**1**). These turned out be excellent asymmetric hydroformylation catalysts for non-sterically hindered substrates such as vinyl esters (Scheme 6). Enantioselectivities of 85 % (*ee* 85 %) and regioselectivities of 80 % (*n/i* ratio 4 : 1) have been achieved for vinyl acetate at 85 °C and 6.2 bars of CO/H_2 (1 : 1), the turnover frequency being 476 cycles per hour. [14]

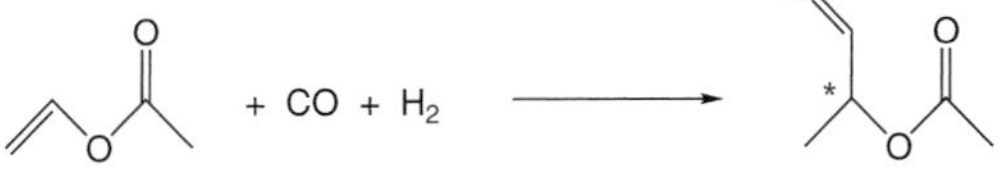

Scheme 6. The asymmetric hydroformylation of vinyl acetate, a key step in the synthesis of *D*- or *L*-threonine.

In a brilliant systematic study, Stanley et al. have demonstrated that their dinuclear rhodium hydroformylation catalyst is radically different from anything previously observed for this industrially important process: It combines high activity with high chemo- and regioselectivity and, provided the pure enantiomers are employed, with excellent enantioselectivities. It is not exaggerated to consider these results as a breakthrough in the development of tailor-made homogeneous catalysts. Fixing two or more cooperating metal centers in the geometry required for a particular catalytic process by means of a suitable ligand matrix, with the aim of controlling activity and selectivity of the catalyst, is now in reach.

References

[1] S. A. Laneman, F. R. Fronczek, G. G. Stanley, *J. Am. Chem. Soc.* **1988**, *110*, 5585.
[2] M. E. Broussard, B. Juma, S. G. Train, W.-J. Peng, S. A. Laneman, G. G. Stanley, *Science* **1993**, *260*, 1784.
[3] K. Weissermel, H.-J. Arpe, *Chimie organique industrielle*, 1ère dition, Masson, Paris, **1981**, p. 113; *Industrielle Organische Chemie*, 2. Auflage, Verlag Chemie, Weinheim, **1978**, p. 120.
[4] B. F. G. Johnson, J. Lewis, *Pure Appl. Chem.* **1975**, *44*, 43; E. L. Muetterties, *Bull. Soc. Chim. Belg.* **1975**, *84*, 959; *Science* **1977**, *196*, 839.
[5] G. Süss-Fink, G. F. Schmidt, *J. Mol. Catal.* **1987**, *42*, 361.
[6] G. Süss-Fink, G. Herrmann, *J. Chem. Soc. Chem. Commun.* **1985**, 735.
[7] R. Lazzaroni, P. Pertici, S. Bertozzi, G. Fabrici, *J. Mol. Catal.* **1990**, *58*, 75.
[8] C. Fyhr, M. Garland, *Organometallics* **1993**, *12*, 1735.
[9] A. Ceriotti, L. Garlaschelli, G. Longoni, M. C. Malatesta, D. Strumolo, A. Fumagalli, S. Martinengo, *J. Mol. Catal.* **1984**, *24*, 309.
[10] M. Garland, *Organometallics* **1993**, *12*, 535.
[11] W.-J. Peng S. G. Train, D. K. Howell, F. R. Fronczek, G. G. Stanley, *J. Chem. Soc., Chem. Commun.* **1996**, 2607.

[12] G. G. Stanley, "Bimetallic Homogeneous Hydroformylation", in: R. D. Adams, F. A. Cotton (editors), *Catalysis by Di- and Polynuclear Metal Complexes*, VCH, New York, to be published in **1997**.

[13] R. C. Matthews, D. K. Kowell, W.-J. Peng, S. G. Train, W. D. Treleaven, G. G. Stanley, *Angew. Chem.* **1996**, *108*, 2402; *Angew. Chem. Int. Ed. Engl.* **1996**, *35*, 2253.

[14] G. G. Stanley, "Bimetallic Hydroformylation Catalysis: The Uses of Homobimetallic Cooperativity in Organic Synthesis", in: M. G. Scaros, M. L. Prunier (editors), *Catalysis of Organic Reactions*, Marcel Dekker, New York, **1995**, p. 363.

Synthesis of Optically Active Macromolecules Using Metallocene Catalysts

Jun Okuda

Although Nature is capable of efficiently producing monodisperse, optically active macromolecules from functional monomers, the rational synthesis of high molecular weight polymers having a well-defined microstructure, a narrow molecular weight distribution, and variable architectures remains a significant challenge. One of the goals in synthetic chemistry therefore is the development of structurally characterized complexes which can efficiently catalyze the polymerization of unsaturated monomers in a controlled or living fashion and, in the case of prochiral monomers, in a stereoselective manner.

Several years ago, the cocatalyst methylalumoxane was found to increase the activity of Ziegler catalysts derived from group 4 metallocenes dramatically. [1] Shortly thereafter the fundamental possibility of controlling the tacticity of polypropylene through the molecular structure of the transition metal catalyst was reported. [2] Since the pioneering work by Turner et al. [3a], Jordan [3b], and others it has been established in a plethora of studies that cationic 14-electron alkyl $[Cp_2MR]^+$ is the active site for chain propagation. [3] Elegant kinetic studies recently performed by the teams of Brintzinger et al. [4a] and of Bercaw et al. [4b] have proved the decisive role of α-agostic inter-

action of the growing chain with the transition metal center. At present intense effort worldwide is concentrating on evaluating the critical parameters for the strereoregularity of α-olefin polymerization and finding a transparent relationship between ligand properties of the catalyst and polymer structure. [5] Thus, the use of racemic C_2-symmetrical zirconocene rac-[1,1'-ethylenebis(4,5,6,7-tetrahydro-1-indenyl)]zirconium dichloride (**1**) developed by Brintzinger leads to highly isotactic polypropylene. Here an enantiomorphic site control is proposed to be operative on the basis of defects within the polymer chain. [2b] On the other hand, the achiral titanocene derivative Cp_2TiPh_2 gives isotactic block polymer that is formed via chain end control. [2a] Finally the C_S-symmetrical metallocene complex **2** with a bridged cyclopentadienyl-fluorenyl ligand system yields syndiotactic polypropylene [6].

Recently Waymouth and Coates have demonstrated that it is also possible to produce an optically active polyolefin using a chiral non-racemic metallocene catalyst of the Brintzinger type. [7] Although optically active oligomers can be obtained in the presence of a resolved metallocene complex according to the above mentioned procedure [8], high molecular weight isotactic as well as syndiotactic polypropylene generally con-

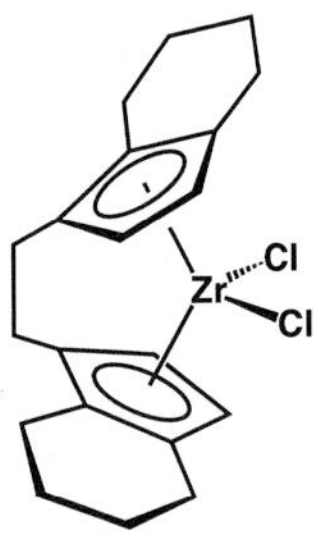

(S)-1 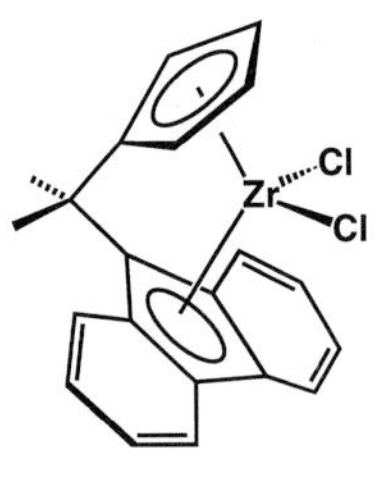   2

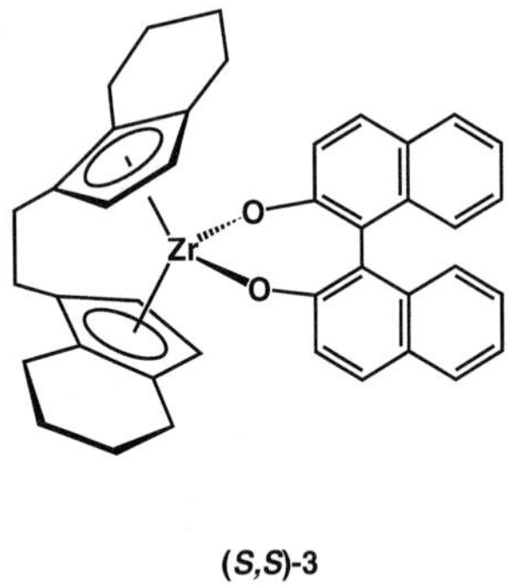

(S,S)-3

tain mirror planes and are therefore achiral ("cryptochiral"). [9]

Waymouth and Coates employed the homogeneously catalyzed cyclopolymerization of 1,5-hexadiene giving poly(methylene-1,3-cyclopentane) as previously developed in his group in order to utilize the stereoselectivity of the monomer insertion for the construction of a polymer with main-chain chirality. The cyclopolymerization is a remarkable chain growth reaction during which a conventional

insertion of a vinylic function into the transition metal–carbon bond is followed by an intramolecular insertion resulting in a cyclization. [10] Thus from 1,5-hexadiene, a polymer chain is obtained in which methylene and 1,3-cyclopentanediyl fragments are arranged in a strictly alternating sequence (Scheme 1). By modifying the catalyst's ligand sphere, control over the diastereoselectivity could be achieved ([Cp$_2$ZrCl$_2$] leads to trans, [Cp*$_2$ZrCl$_2$] to cis connection of the cyclopentane fragments). An analysis of possible stereoisomers shows four structures of maximum order of which all but the racemo-diisotactic polymer are achiral. The latter does not contain a mirror plane and is chiral due to configurationally determined main-chain stereochemistry. The Stanford group recognized that by using a chiral catalyst the enantioselective cyclopolymerization of 1,5-hexadiene would be possible. The catalyst used was chiral [1,1'-ethylene-bis(4,5,6,7-tetrahydro-1-indenyl)]zirconium 1,1'-binaph-tholate (3) which had been prepared according to *Brintzinger* [11] by reacting the racemic dichloro derivative 1 with optically active (R)- or (S)-1,1'-binaphthol.

Poly(methylene-1,3-cyclopentane) synthesized in the presence of (−)-(R,R)-3 and methylalumoxane revealed a molar optical rotation of [Φ]$^{28}_{405}$ = +51.0° (c 8.0, CHCl$_3$), whereas the polymer analogously prepared using the enantiomer (+)-(S,S)-3 showed a value of [Φ]$^{28}_{405}$ = −51.2° (c 8.0, CHCl$_3$). These findings are consistent with the formation of polymers with main-chain chirality. [9a]

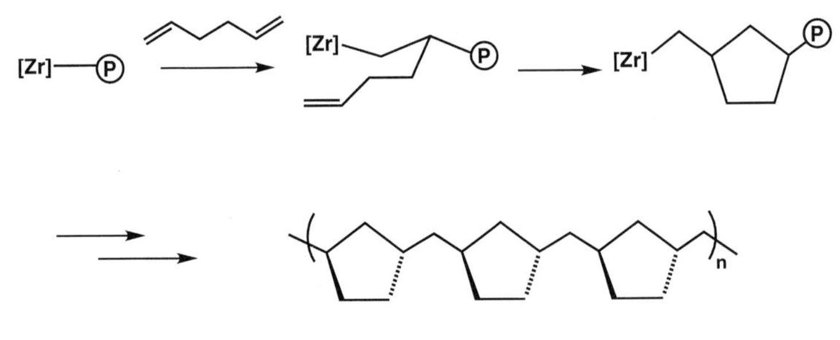

Scheme 1.
[Zr] = zirconocene fragment,
-Ⓟ = polymer chain.

In summary, Waymouth and Coates have for the first time realized the enantioselective synthesis of an optically active polymer starting with an achiral monomer by using an optically active, structurally well-defined metallocene catalyst. Their results are promising, since the synthesis of organic polymers in a more rational way through modification of the transition metal catalyst seems within reach. In addition, this study emphasizes that knowledge of well-characterized organotransition metal complexes offers an unprecedented opportunity for the design of novel polymerization reactions.

References

[1] H. Sinn, W. Kaminsky, *Adv. Organomet. Chem.* **1980**, *18*, 99.

[2] a) J. A. Ewen, *J. Am. Chem. Soc.* **1984**, *106*, 6355; b) W. Kaminsky, K. Kulper, H. H. Brintzinger, F. R. W. P. Wild, *Angew. Chem.* **1985**, *97*, 507; *Angew. Chem. Int. Ed. Engl.* **1985**, *24*, 507.

[3] a) G. G. Hlatky, H. W. Turner, R. R. Eckman, *J. Am. Chem. Soc.* **1989**, *111*, 2728; b) R. F. Jordan, *J. Chem. Educ.* **1988**, *65*, 285; *Adv. Organomet. Chem.* **1991**, *32*, 325.

[4] a) H. Krauledat, H. H. Brintzinger, *Angew. Chem.* **1990**, *102*, 1459; *Angew. Chem. Int. Ed. Engl.* **1990**, *29*, 1412. H. H. Brintzinger in *Organic Synthesis via Organometallics* (Eds.: K. H. Dötz, R. W. Hoffmann), Vieweg, Braunschweig, **1991**, p. 33; b) W. E. Piers, J. E. Bercaw, *J. Am. Chem. Soc.* **1990**, *112*, 9406.

[5] a) For some recent reviews, see: P. C. Möhring, N. J. Coville, *J. Organomet. Chem.* **1994**, *479*, 1. H. H. Brintzinger, D. Fischer, R. Mülhaupt, B. Rieger, R. M. Waymouth, *Angew. Chem.* **1995**, *107*, 1255; *Angew. Chem. Int. Ed. Engl.* **1995**, *34*, 1143. *Ziegler Catalyst, Recent Scientifc Innovations and Technological Improvements* (Eds.: G. Fink, R. Mülhaupt, H. H. Brintzinger), Springer, Berlin, **1995**; b) G. Erker, R. Nolte, Y.-H. Tsay, C. Krüger, *Angew. Chem.* **1989**, *101*, 642; *Angew. Chem. Int. Ed. Engl.* **1989**, *28*, 628. G. Erker, *Pure Appl. Chem.* **1991**, *63*, 797. G. Erker, C. Fritze, *Angew. Chem.*, **1992**, *104*, 204.; *Angew. Chem. Int. Ed. Engl.* **1992**, *31*, 619728; c) W. Röll, H. H. Brintzinger, B. Rieger, R. Zolk, *Angew. Chem.* **1990**, *102*, 339; *Angew. Chem. Int. Ed. Engl.* **1990**, *29*, 279; d) W. Spaleck, M. Antberg, V. Dolle, R. Klein, J. Rohrmann, A. Winter, *New. J. Chem.* **1990**, *14*, 499; e) T. Mise, S. Miya, H. Yamazaki, *Chem. Lett.* **1989**, 1853; f) J. A. Bandy, M. L. H. Green, I. M. Gardiner, K. Prout, *J. Chem. Soc, Dalton Trans.* **1991**, 2207.

[6] J. A. Ewen, R. L. Jones, A. Razavi, J. D. Ferrara, *J. Am. Chem. Soc.* **1988**, *110*, 6255.

[7] G. W. Coates, R. M. Waymouth, *J. Am. Chem. Soc.* **1991**, *113*, 6270; *idem, ibid.* **1993**, *115*, 91. G. W. Coates, A. Mogstad, R. M. Waymouth, *J. Macromol. Sci., Chem.* **1992**, *31*, 47.

[8] P. Pino, P. Cioni, J. Wei, *J. Am. Chem. Soc.* **1987**, *109*, 6189. W. Kaminsky, A. Ahlers, N. Möller-Lindenhof, *Angew. Chem.* **1989**, *101*, 1304; *Angew. Chem. Int. Ed. Engl.* **1989**, *28*, 1216.

[9] a) G. Wulff, *Angew. Chem.* **1989**, *101*, 22; *Angew. Chem. Int. Ed. Engl.* **1989**, *28*, 21; b) M. Farina, *Top. Stereochem.* **1987**, *17*, 1.

[10] a) L. Resconi, R. M. Waymouth, *J. Am. Chem. Soc.* **1990**, *112*, 4953; b) G. B. Butler, *Acc. Chem. Res.* **1982**, *15*, 370.

[11] F. R. W. P. Wild, M. Wasiucionek, G. Huttner, H. H. Brintzinger, *J. Organomet. Chem.* **1985**, *288*, 63.

C. Biological and Biomimetic Methods

Discovering Biosynthetic Pathways – A Never Ending Story

Sabine Laschat and Oliver Temme

There is an increasing interest in the study of biosynthetic reactions due to the fact that a detailed understanding of biosynthetic and metabolic pathways eventually enables us to control malfunctions which often lead to severe health disorders and diseases. In addition, biosynthetic reactions have largely improved synthetic organic methodology. This is especially true for the biomimetic reactions, which proceed highly chemo-, regio- and stereoselective under extremely mild (and often neutral) conditions in an aqueous environment without the requirement of protecting groups. The following chapter deals with several biosyntheses, which have been investigated recently.

Pericyclic reactions, that is one-step processes that proceed through cyclic transition states under control of orbital symmetry, represent an important tool for the synthetic organic chemist. A large number of pericyclic key steps occur also in biological systems [1]. The classical example, which led Woodward and Hoffmann to establish the rules of conservation of orbital symmetry, is the formation of vitamin D_3 from 7-dehydrocholesterol by a photochemically induced, conrotatory cycloreversion of the steroid B-ring, followed by a thermally allowed [1,7]-sigmatropic H-shift. Claisen rearrangements are found as well in natural systems. The [3,3]-sigmatropic

Claisen rearrangement of chorismic acid **1** to prephenic acid **2** (Scheme 1), which is catalyzed by the enzyme chorismate mutase, can be considered as the key step in the biosynthesis of aromatic compounds, that is the so-called shikimic acid pathway. The chair-like transition state geometry **3** was proved by double isotope-labeling experiments [2]. However, in the laboratory this particular reaction can be accelerated not only by enzymes but also by catalytic antibodies [3]. For the generation of such antibodies haptenes such as **4** were used, that is, molecules whose structure is very similar to the transition state of the particular reaction and which are tightly bound by the antibody.

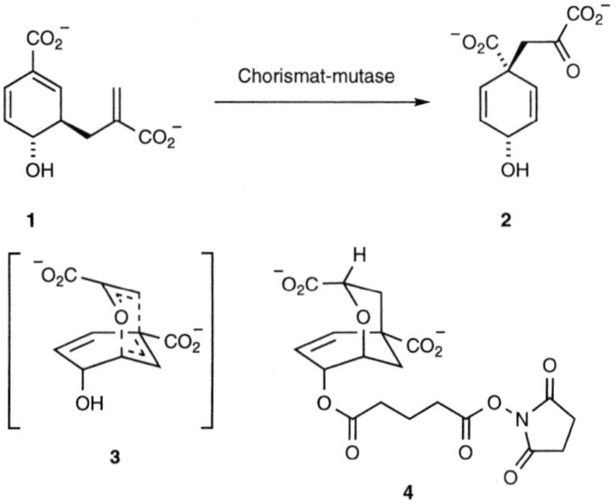

Scheme 1.

Besides Claisen rearrangements many other types of reactions can be catalyzed by antibodies. Reduction of diketone **5** with NaBH$_4$ yielded the hydroxyketone **6** in 95 % (96 % *ee*) (Scheme 2) [4]. This remarkable regioselectivity was achieved, because the reduction of the *p*-nitrobenzyl-substituted carbonyl group is 75 times faster than the reduction of the corresponding *m*-methoxybenzyl-substituted carbonyl group in the presence of the catalytic antibody 37B.39.3, which was raised against the haptene **7**. In the absence of the catalytic antibody no kinetic preference for one of the carbonyl group was observed. In the case of antibody-catalyzed Diels-Alder reactions an additional problem has to be tackled with, that is the product inhibition resulting from the close structural similarity between the transition state (or haptene respectively) and the product. In order to avoid the product inhibition, the cycloaddition must be either followed by a cheleotropic reaction or the haptene must be bulkier than the corresponding transition state [5]. However, Janda et al. reported a completely different approach to this problem [6]. They used the conforma-

Scheme 2.

tional highly flexible ferrocene **9** as haptene instead of the conformational rigid bicyclo-[2.2.2]octane system **8** (Scheme 3). Due to the low rotational barrier of **8** along the metallocene axis this haptene mimics both the *ortho-endo* transition state **10a** as well as the *ortho-exo* transition state **10b** of the

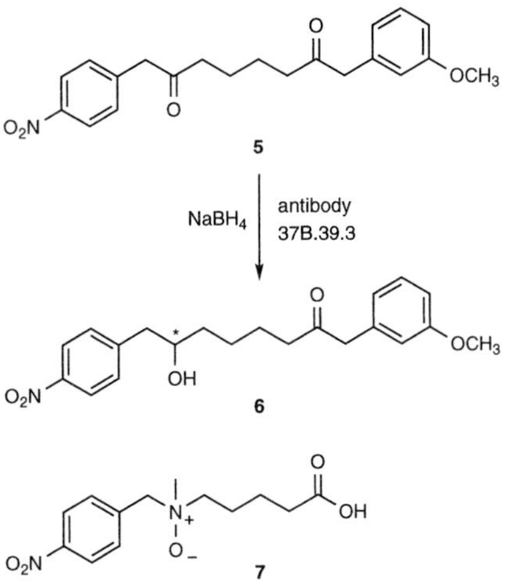

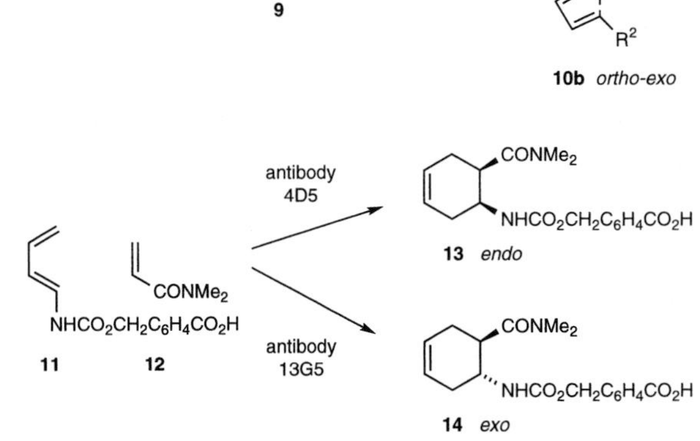

Scheme 3.

cycloaddition of diene **11** and dienophile **12**. Thus two different antibodies 4D5 and 13G5 respectively were generated from **9**, which catalyze highly diastereo- and enantioselectively the formation of **13** and **14** respectively.

In a more application-oriented approach an antibody was developed, which is able to deactivate cocaine **15** by conversion to (–)-ecgonine **16**, before the drug is delivered to the central nervous system (Scheme 4) [7]. The antibody, which was raised against haptene **17**, indeed shows similar enzyme kinetics compared to butyrylcholinesterase, the natural metabolic enzyme for compounds like **15**, although the catalytic activity of the antibody still needs to be improved in order to successfully fight drug abuse. Despite the progress, which has been achieved by using catalytic antibodies, there still remain severe problems to be solved especially concerning the production of antibodies. The immunization process is rather time- and material-consuming. However, the tools of molecular biology should allow to generate whole antibody libraries without the requirement of animals [8].

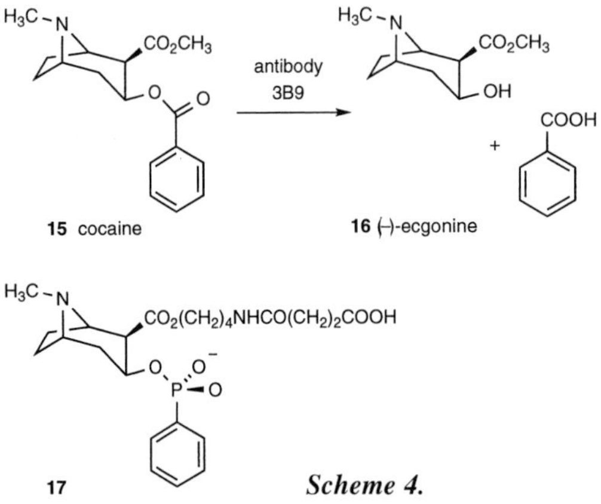

15 cocaine   **16** (–)-ecgonine

17   *Scheme 4.*

However, pericyclic reactions in natural systems do not necessarily require the presence of an enzyme. Several examples proceed spontaneously, e. g. the above mentioned

1,7-H shift in the biosynthesis of vitamin D$_3$. Recently Boland et al. reported, that the ectocarpenes **19** and related substances, which were isolated from the brown algae *Ectocarpus siliculosus*, are not the sexual pheromones of the algae, but that these cycloheptadienes **19** are formed from the actual pheromones, namely the thermolabile *cis*-divinylcyclopropanes **18**, by a spontaneous Cope rearrangement (Scheme 5) [9]. Thus the Cope rearrangement of **18** resembles a deactivation pathway of the pheromones.

18   R = H, Et, CH=CH$_2$ **19**

Scheme 5.

Diels-Alder reactions have been postulated as key steps in a number of biosynthetic conversions. The biosynthesis of the endiandric acids **26** and **27** is discussed below as an example for a spontaneous [4+2]-cycloaddition. This class of compounds which is produced by the Australian plant *Endiandra introsa (Lauraceae)* is remarkable, because different constitutional isomers were formed simultaneously and in all cases racemic mixtures were synthesized by the plant. It was first postulated by Black et al. [10] and later confirmed via biomimetic total syntheses by Nicolaou et al. [11] that the endiandric acids A, B and C (**26a**, **26b** and **27**) are formed by a cascade of electrocyclic reactions (Scheme 6). The cascade is initiated by a conrotatory 8π-electron ring closure of the polyene carboxylic acids **20**, **21** to the isomeric cyclooctatrienes **22** and **23**, respectively, which subsequently undergo a disrotatory 6π-electron cyclization to **24** and **25**, respectively. Termination of the cascade by an intramolecular Diels-Alder reaction yields either the tetracyclic endiandric acid **26a, b** or the bridged derivative **27**.

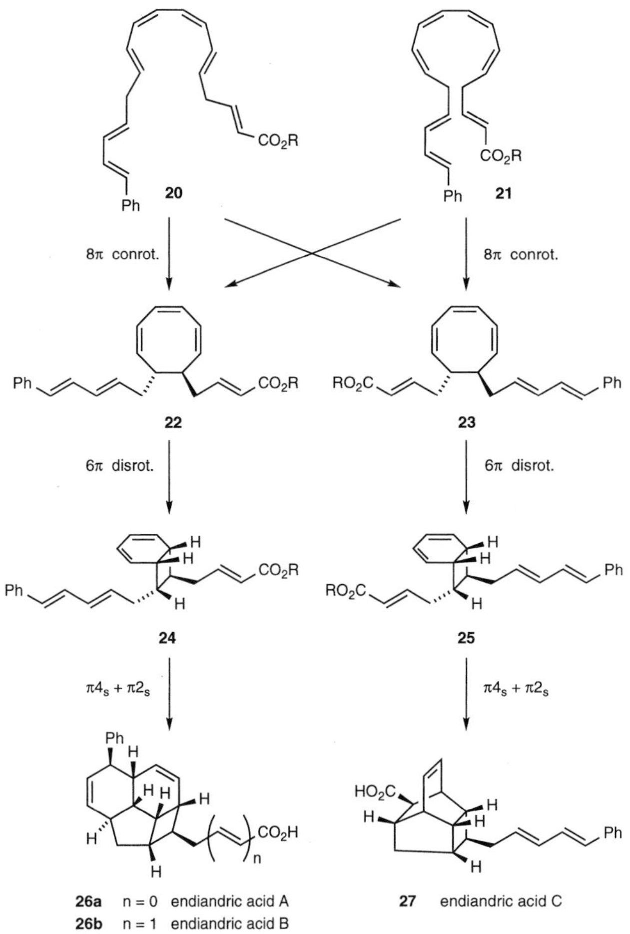

CO_2R
Ph
20
CO_2R
Ph
21
8π conrot.
8π conrot.
Ph
CO_2R
RO_2C
Ph
22
23
6π disrot.
6π disrot.
24
25
$\pi 4_s + \pi 2_s$
$\pi 4_s + \pi 2_s$
CO_2H
HO_2C
26a n = 0 endiandric acid A
26b n = 1 endiandric acid B
27 endiandric acid C

Scheme 6.

Besides those spontaneous processes a variety of [4+2]-cycloadditions exists for which it still remains unclear whether they are Diels-Alder reactions and if so, whether they proceed spontaneously or only in the presence of an enzyme. In this respect the stereospecific synthesis of the sesterpenoid heliocide H_2 **28**, an insecticide isolated from cotton wool, from hemigossypolon **29** and myrcene **30** (Scheme 7) [12, 13] should be mentioned. Other examples of biomolecules which are probably formed by Diels-Alder reactions contain the chalcone kuwanon J **31** from the mulberry tree *Morus alba* L. [14], the mycotoxins chaetoglobosin A **33** [15] and brevianamide A **35** [16], the polyketide nargenicine **37** [17] and mevinoline **39**, a metabolite of the fungus *Aspergillus terreus* MF4845, which is used as a drug for decreasing the blood cholesterol level [18].

However, until now there is no case known, where the corresponding enzyme system, that is the Diels-Alder-ase, could be detected. Even in the biosynthesis of the iboga and aspidosperma alkaloids the final proof for a Diels-Alder reaction is still missing (Scheme 8) [19]. In 1970 Scott et al. were able to elaborate nearly the whole biosynthetic pathway by feeding experiments with plant seedlings

[20]. According to Scott isovincoside **41** is converted to stemmadenine **42** via several steps. Heterolytic ring opening and concomitant dehydration leads to the postulated intermediate dehydrosecodine **43**. Starting from the acrylic ester **43** two [4+2]-cycloadditions are possible. If the 2-dehydropyridine system of **43** reacts as dienophile, the aspidosperma alkaloid tabersonine **44** is formed, whereas participation of the 2-dihydropyridine moiety as a diene leads to the iboga alkaloid catharanthine **45**. The reversibility of the reactions lead the authors to assume stepwise Michael additions.

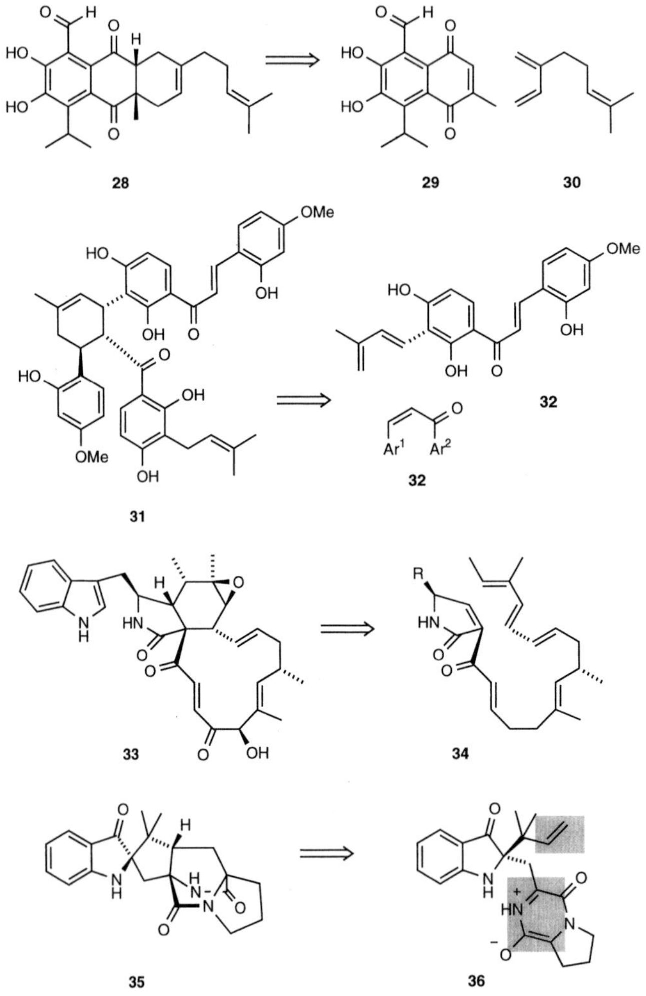

Scheme 7.

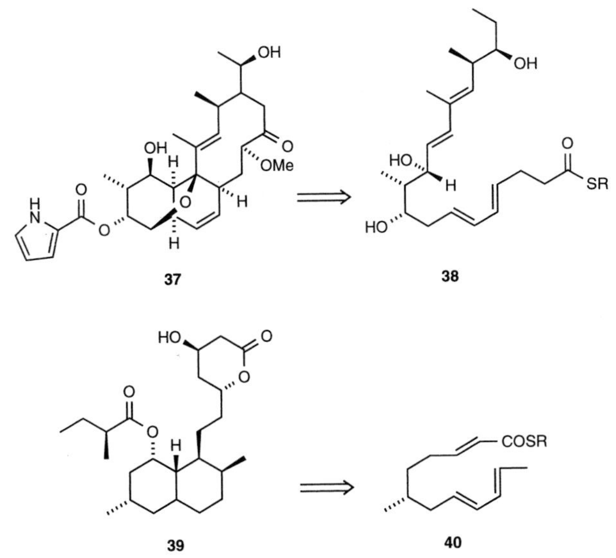

Scheme 7 (Forts.).

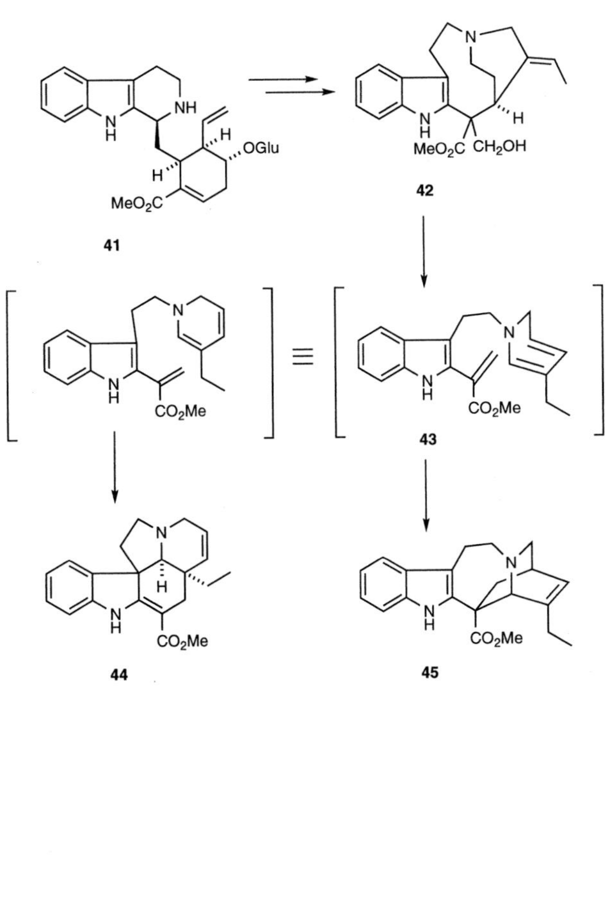

Scheme 8.

With regard to such difficulties in detecting and isolating Diels-Alder-ases, one might assume, that there are no enzyme-catalyzed [4+2]-cycloadditions involved in natural systems. However, this assumption seems to be incorrect. Recently Oikawa, Ishihara et al. published experimental evidence, that the solanapyrones **48** and **49**, two phytotoxines produced by the pathogenic fungus *Alternaria solani*, are probably formed by enzyme-catalyzed [4+2]-cycloadditions (Scheme 9) [21]. Treatment of prosolanapyrone III **47** with a cell-free extract of *A. solani* at 30 °C for 10 min yielded a mixture of **48** and **49** (25 % conversion) with an *exo/endo* ratio of 53 : 47. In the absence of the cell-free extract under the same conditions the aldehyde **47** undergoes an uncatalyzed Diels-Alder reaction to **48** and **49** (15 % conversion) with an *exo/endo* ratio of 3 : 97 [22]. Treatment of **47** with the denatured cell-free extract yielded likewise an *exo/endo* ratio of 3 : 97 (10 % conversion). Thus, the enzyme-related conversion is about 15 % and the *exo/endo* ratio 87 : 13 (>92 % *ee* for **48**). The high *exo* selectivity is remarkable, because it can not be achieved by chemical methods. It should be emphasized, that in the above mentioned investigations no isolated enzymes were used but a cell-free extract. Thus the postulated Diels-Alder mechanism needs to be further confirmed by characterization of the enzyme or enzyme complex. The biosynthesis of the solanapyrones **48**, **49** shows an additional feature worthy of mention. Whereas the alcohol prosolanapyrone II **46** does not form any cycloaddition products in aqueous medium without any catalyst, the presence of the cell-free extract induces the production of 19 % of the solanapyrones **48** and **49** (*exo/endo* 85 : 15) and 6 % of the aldehyde **47**. Apparently, the Diels-Alder-ase is accompanied by a dehydrogenase, which catalyzes the oxidation of **46** to **47**. Further studies with isolated enzymes are required in order to investigate, whether these are separated enzymes or coupled in an enzyme complex.

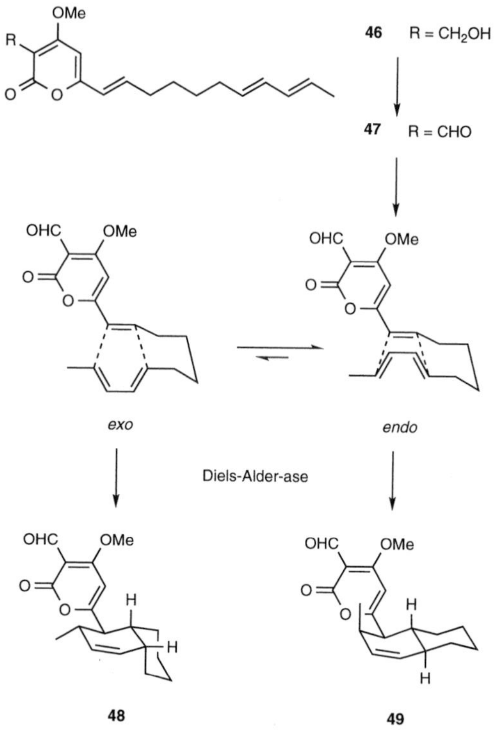

Scheme 9.

The conversion of oxidosqualene **50** to lanosterol **52**, the so-called squalene folding in the biosynthesis of steroids has initiated much research efforts (Scheme 10) [23]. This process is catalyzed by the enzyme lanosterol synthase, which controls precisely the formation of four rings and six new stereocenters. According to the pioneering work by Eschenmoser and Stork the cyclization proceeds in a concerted fashion due to favorable orbital overlap [24, 25]. In contrast Nishizawa et al. were able to trap several cationic intermediates **55–57** from a related model system **54** [26]. The most recent mechanistic studies of lanosterol biosynthesis, which have been conducted with purified enzyme from recombinant sources, have revealed that the cyclization involves discrete carbocations during C-ring formation and specifically that a five-membered C-ring cationic structure **58** is first

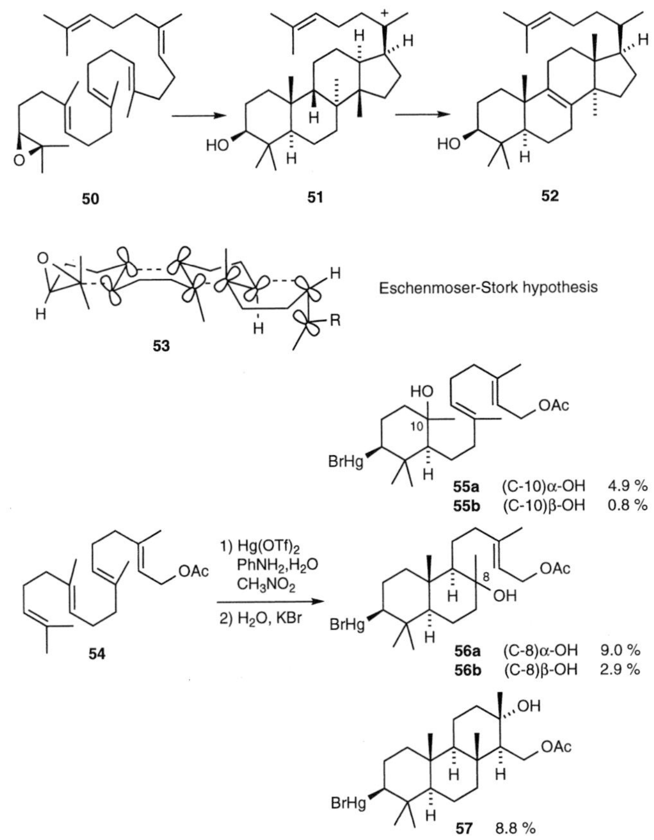
50
51
52
53
Eschenmoser-Stork hypothesis
55a (C-10)α-OH 4.9 %
55b (C-10)β-OH 0.8 %
1) Hg(OTf)2
PhNH2,H2O
CH3NO2
2) H2O, KBr
54
56a (C-8)α-OH 9.0 %
56b (C-8)β-OH 2.9 %
57 8.8 %

Scheme 10.

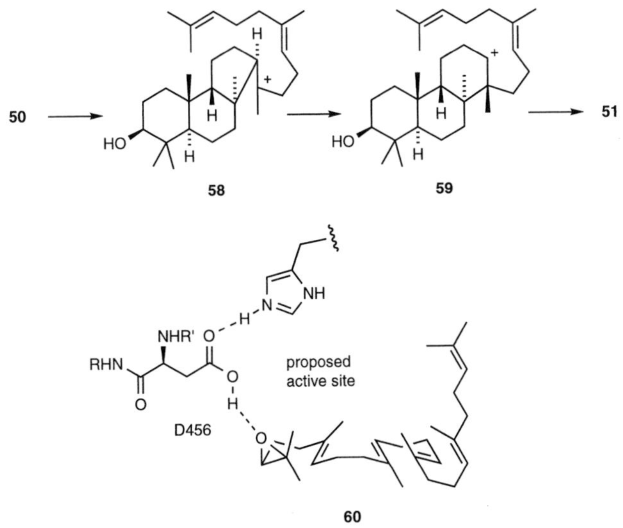
50
58
59
51
proposed
active site
D456
60

Scheme 11.

formed and then enlarged by expansion to a six-membered structure **59** (Scheme 11) [27]. Corey et al. overexpressed lanosterol synthase from *Saccharomyces cerevisiae* in baculovirus-infected cells [27]. With the purified enzyme the kinetics of the cyclization were determined using Michaelis-Menten analysis for **50** and two analogs, in which the methyl group at C6 was replaced by H or Cl. The measured V_{max}/K_M ratios indicated that oxirane cleavage and cyclization to form the A-ring are concerted, since the nucleophilicity of the proximate double bond influences the rate of oxirane cleavage. In addition, atomic absorption analysis revealed that activation of the oxirane **60** is effected by an acidic group of the enzyme (presumably aspartic acid D456) rather than a Lewis acidic metal ion.

Complex polyketides like erythromycin A **61** not only have stimulated many synthetic organic chemists due to the complexity of their molecular backbone [28], there is also a growing interest in their biosynthesis, because they exhibit prominent antibiotic activity (Scheme 12) [29]. Despite the structural diversity of these polyketides, it is proposed that their biological origin is related to the biosynthesis of fatty acids. Experiments with mutant strains of *Saccharopolyspora erythaea* yielded 6-desoxyethronolide B **63** as the first intermediate, which is not bound to an enzyme. Thus it was concluded that the biosynthesis of **61** occurs in two different sequences. First the carbon skeleton is built up in a chain reaction involving several acyl transfer, reduction and dehydration steps by the enzyme polyketide synthase (PKS) and then cyclized to 6-desoxyerythronolide B **63**, which is further modified and glycosylated to finally yield **61**. The following mode of action of PKS was established via isotope labeling experiments and feeding studies with larger fragments of **63** [30]. As shown in Scheme 13 the methyl groups in the backbone of **63**

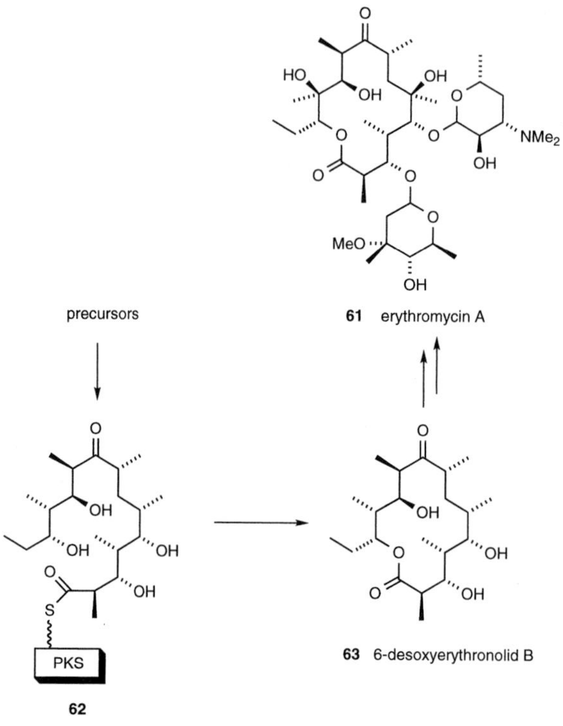

precursors   **61** erythromycin A

63 6-desoxyerythronolid B

62

Scheme 12.

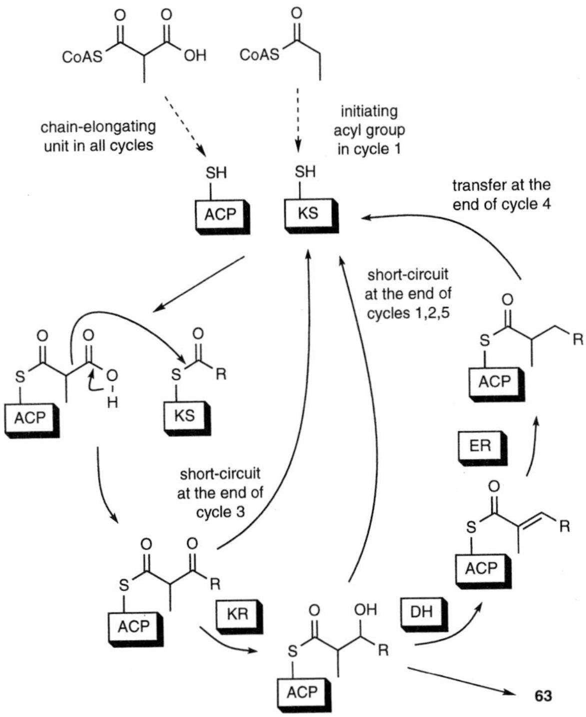

Scheme 13. Proposed mode of action of erythromycin-PKS. KS = ketoacyl synthase; KR = keto reductase; DH = deydratase; ER = enol reductase; ACP = acyl carrier protein; CoA = coenzyme A.

can be introduced, if the chain initiator acetate in the fatty acid synthesis is replaced by propionate and if 2-methyl-malonate is used instead of malonate for the chain elongating steps. Hydroxy- and keto groups can be obtained by a short-circuit of the usual fatty acid synthase cycle, that is the final dehydration and hydrogenation (enol reduction) steps are skipped out. In 1991 Katz et al. reported the complete sequence of the PKS gene, which encodes a large multifunctional protein [31]. From these findings it was proposed that PKS is organized as an array of homologous enzymes, each of them catalyzing a certain chain elongation step in the synthesis of **63**. However, the exact mode of coordination and cooperation between these enzymes as well as the three-dimensional folding of the protein are not known until now.

In conclusion, although our knowledge about many biosynthetic pathways has improved much during the last few years, there still remains a lot of work to be done, especially concerning a detailed investigation of structural and functional aspects of the corresponding enzymes.

References

[1] U. Pindur, G. H. Schneider *Chem. Soc. Rev.* **1994**, 409–415; S. Laschat *Angew. Chem.* **1996**, *108*, 313–315; *Angew. Chem. Int. Ed. Engl.* **1996**, *35*, 289–291.

[2] Despite various mechanistic investigations, it is still a matter of debate whether the enzymatic Claisen rearrangement proceeds through a concerted mechanism. See: Y. Asano, J. J. Lee, T. L. Shieh, F. Spreafico, C. Kowal, H. G. Floss *J. Am. Chem. Soc.* **1985**, *107*, 4314–4320. J. J. Delany, R. E. Padykula, G. A. Berchtold *J. Am. Chem. Soc.* **1992**, *114*, 1394–1397 and refs. cited therein.

[3] Reviews: *Acc. Chem. Res.* **1994**, *26*, 391–411. P. G. Schultz *Angew. Chem.* **1989**, *101*, 1336–1348; *Angew. Chem. Int. Ed. Engl.* **1989**, *28*, 1283–1295. R. A. Lerner, S. J. Benkovic, P. G. Schultz *Science* **1991**, *252*, 659–667. C. Leumann *Angew. Chem.* **1993**, *105*, 1352–1354; *Angew. Chem. Int. Ed. Engl.* **1993**, *32*, 1291–1293.

[4] L. C. Hsieh, S. Yonkovich, L. Kochersperger, P. G. Schultz *Science 1993*, **259**, 490–493.

[5] D. Hilvert, K. W. Hill, K. D. Nared, M. T. M. Auditor *J. Am. Chem. Soc.* **1989**, *111*, 9261–9262. A. C. Braisted, P. G. Schultz *J. Am. Chem. Soc.* **1990**, *112*, 7430–7431.

[6] J. T. Yli-Kauhaluoma, J. A. Ashley, C.-H. Lo, L. Tucker, M. M. Wolfe, K. D. Janda *J. Am. Chem. Soc.* **1995**, *117*, 7041–7047.

[7] D. W. Landry, K. Zhao, G. X.-Q. Yang, M. Glickman and T. M. Georgiadis *Science* **1993**, *259*, 1899–1901.

[8] C. F. Barbas III, J. D. Bain, D. M. Hoekstra, R. A. Lerner *Proc. Natl. Acad. Sci. USA* **1992**, *89*, 4457–4461.

[9] W. Boland, G. Pohnert, I. Maier *Angew. Chem.* **1995**, *117*, 1717–1719; *Angew. Chem. Int. Ed. Engl.* **1995**, *34*, 1602–1604.

[10] W. M. Bandaranayake, J. E. Banfield, D. S. C. Black *J. Chem. Soc. Chem. Commun.* **1980**, 902–903.

[11] K. C. Nicolaou, N. A. Petasis, R. E. Zipkin, J. Uenishi *J. Am. Chem. Soc.* **1982**, *104*, 5555–5564.

[12] R. D. Stipanovic, A. A. Bell, D. H. O'Brien, M. J. Lukefahr *Tetrahedron Lett.* **1977**, *18*, 567–570.

[13] In addition an enzymatic prenylation of the sesquiterpene hydroquinone **29** to the corresponding sesterpene **28** might be a possible alternative to the Diels-Alder mechanism.

[14] Y. Hano, A. Ayukawa, T. Nomura *Naturwissenschaften* **1992**, *79*, 180–182.

[15] H. Oikawa, Y. Murakami, A. Ichihara *J. Chem. Soc. Perkin Trans. I* **1992**, 2955–2959.

[16] J. F. Sanz-Cervera, T. Glinka, R. M. Williams *J. Am. Chem. Soc.* **1993**, *115*, 347–348.

[17] D. E. Cane, W. Tan, W. R. Ott *J. Am. Chem. Soc.* **1993**, *115*, 527–535.

[18] Y. Yoshizawa, D. J. Witter, Y. Liu, J. C. Vederas *J. Am. Chem. Soc.* **1994**, *116*, 2693–2694.

[19] W. A. Carrol, P. A. Grieco *J. Am. Chem. Soc.* **1993**, *115*, 1164–1165. W. G. Bornmann, M. E. Kuehne *J. Org. Chem.* **1992**, *57*, 1752–1760.

[20] A. I. Scott *Acc. Chem. Res.* **1970**, *3*, 151–157. E. Wenkert *J. Am. Chem. Soc.* **1962**, *84*, 98–102.

[21] H. Oikawa, K. Katayama, Y. Suzuki, A. Ichihara *J. Chem. Soc. Chem. Commun.* **1995**, 1321–1322.

[22] This ratio is comparable to the *endo* ratio of other Diels-Alder reactions in aqueous media.

[23] R. Bohlmann, *Angew. Chem.* **1992**, *104*, 596–598; *Angew. Chem. Int. Ed. Engl.* **1992**, *31*, 582–584.

[24] A. Eschenmoser, L. Ruzicka, O. Jeger, D. Arigoni *Helv. Chim. Acta* **1955**, *38*, 1890–1904.

[25] G. Stork, A. W. Burgstahler *J. Am. Chem. Soc.* **1955**, *77*, 5068–5077.

[26] M. Nishizawa, H. Takenaka, Y. Hayashi *J. Am. Chem. Soc.* **1985**, *107*, 522–523.

[27] E. J. Corey, H. Cheng, C. H. Baker, S. P. T. Matsuda, D. Li, X. Song *J. Am. Chem. Soc.* **1997**, *119*, 1277–1288, 1289–1296, and references cited therein.

[28] J. Mulzer *Angew. Chem.* **1991**, *103*, 1484–1486; *Angew. Chem. Int. Ed. Engl.* **1991**, *30*, 1452–1454.

[29] J. Staunton *Angew. Chem.* **1991**, *103*, 1331–1335; *Angew. Chem. Int. Ed. Engl.* **1991**, *30*, 1302–1307.

[30] D. E. Cane, C. Yang *J. Am. Chem. Soc.* **1987**, *109*, 1255–1257.

[31] S. Donadio, M. J. Staber, J. B. McAlpine, J. B. Swanson and L. Katz *Science* **1991**, *252*, 675–678.

Peptide Ligases –
Tools for Peptide Synthesis

Hans-Dieter Jakubke

Introduction

Following the first chemical formation of a peptide bond by Theodor Curtius in 1881 in the laboratory of Hermann Kolbe in Leipzig, chemical peptide synthesis [1] developed over the next ninety years culminating in the first total synthesis of an enzyme. Ribonuclease A (RNase A), composed of 124 amino acid residues, was constructed by condensation of fragments in solution [2] as well as by solid-phase synthesis [3]. Roughly ten years later Yajima and Fujii [4] synthesized RNase A according to an improved conventional synthetic strategy and obtained the first crystalline synthetic enzyme having full enzymatic activity. But an ideal, universally applicable method for chemical formation of the peptide bond has still not been found. Since chemosynthetic reactions are typically conducted with residues having chiral centers, racemization is a constant concern. Even stepwise chain extension using urethane type protecting groups, which are supposed to prevent racemization, may be affected [5]. Recently, the rate of racemization during solid-phase peptide synthesis was studied using capillary electrophoresis [6]. Using this separation method with a limit of detection of 0.05 % the formation of stereoisomers could be verified and was around 0.4 % per synthesis cycle. There is no doubt, that chemical linkage of peptide fragments is yet unreliable.

Principles of Enzyme-Catalyzed Synthesis

Owing to the stereo- and regiospecificity of enzymes, their applicaton for the formation of peptide bonds certainly offers a list of advantages over chemical procedures, for example mild reaction conditions, no need for permanent protection of functional groups in side chains, racemization-free course of reaction, and simple scale-up after optimization of the process. To make full use of these advantages, however, a close to universally applicable peptide ligase is needed which in principle would display high catalytic efficiency for all theoretically possible combinations of amino acids. In deviation from EC nomenclature, the term peptide ligase here refers to a biocatalyst for the formation of the peptide bond. In other words, such a peptide ligase should be nonspecific and not affected by the side chain functions of the amino acids to be coupled. Since an ideal biocatalyst like the ribosomal peptidyl transferase is not available, and multienzyme complexes in bacterial peptide synthesis are limited to

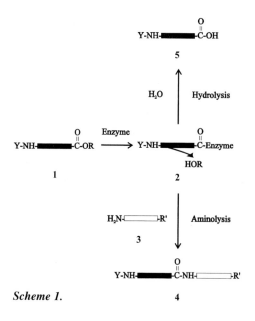

Scheme 1.

specific purposes, only proteases can be used for formation of the peptide bond. The idea of using enzymes to make peptide bonds is about as old as chemical peptide synthesis itself. In 1898, evoking the reversibility of chemical reactions, van't Hoff predicted that enzymes could be used to catalyze the formation of peptide bonds. This idea was first taken up in 1938 by Max Bergmann and confirmed experimentally. In principle, two different mechanistic strategies for protease-catalyzed peptide synthesis are distinguished [7]. In kinetically controlled enzymatic peptide synthesis [8], the protease acts as a transferase using a weakly activated acyl donor to rapidly acylate a serine or cysteine protease, whereas the equilibrium-controlled approach represents the direct reversal of the proteolysis and ends with a true equilibrium. The kinetic approach can be more efficiently manipulated than the equilibrium approach. However, proteases are not perfect acyl transferases; owing to their limited specificity, other undesired reactions may take place parallel to acyl transfer, e. g. hydrolysis of the acyl enzyme 2, secondary hydrolysis of the peptide

product 4, and other undesired cleavages of possible protease-labile bonds in 1, 3 and 4 (Scheme 1).

To eliminate or minimize these disadvantages, the enzyme can be engineered and/or the reaction medium and the mechanistic features of the process can be adjusted as shown schematically in Figure 1 [9]. Undesired hydrolytic side reactions may be eliminated by adapting the medium. Protease-catalyzed syntheses in monophasic organic solvents or in biphasic aqueous-organic systems proceed with minimal proteolytic side reactions. Since the reactants are better soluble in organic media, the chemical and enzymatic steps can be accomodated. Another option in modifying a kinetically controlled reaction is manipulating the leaving group so that the enzyme reacts exclusively with the acyl donor 1 and not with the peptide product 4 or the amino component 3. The even protease-labile products can be obtained in good yields. Two recent reviews give an overview of the current possibilities for manipulation [9, 10].

Enzyme engineering

The term enzyme engineering [11] (see Fig. 1) describes a range of techniques from deliberate chemical modification to remodeling a wild-type enzyme by gene technology. "Subtiligase", a mutant of subtilisin BNP', was prepared by Jackson et al. [12] by protein design and used in a further total synthesis of RNase A. This work, which combines solid-phase synthesis of oligopeptides and enzymatic coupling of these fragments, demonstrates the state of the art in peptide synthesis and proves impressively the potential of enzymes for the formation of peptide bonds. Average yields of roughly 75 % were obtained in the fragment condensations. The excellent leaving group of the Phe-NH$_2$-modified carboxamido methyl ester, which even in unmodified

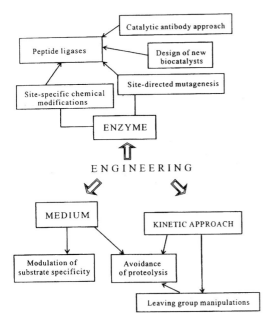

Figure 1. Possibilities for engineering enzymatic peptide synthesis at different levels.

Non-conventional Enzymatic Synthesis Strategies

Some new approaches to suppress competitive reactions in protease-catalyzed peptide synthesis have been developed in our group [14], namely leaving group manipulations at the acyl donor in kinetically controlled reactions, enzymatic synthesis in organic solvent-free microaqueous systems, cryoenzymatic peptide synthesis, and biotransformations in frozen aqueous systems using the "reverse hydrolysis potential" of proteases and other hydrolases [15]. Limitations for general application of the enzymatic approach result from the restrictive specificity of proteases and the permanent danger of undesired proteolytic side reactions.

While investigating irreversible strategies in protease-catalyzed peptide synthesis, we have focussed our interest in using 'substrate mimetics' as acyl donors in the kinetic approach. We have introduced the term substrate mimetics for acyl donor esters in which a cationic center is included in the leaving group instead of being in the acyl moiety, i. e. the appropriate protease should be predominantly specific for the leaving group. The term "inverse esters" was firstly used by Wagner and Horn [16] for p-amidinophenyl esters of aromatic carboxylic acids instead of p-amidinobenzoic acid aryl esters in inhibition studies of trypsin and other serine proteases. In 1991 we described the first application of inverse substrates in trypsin-catalyzed peptide synthesis [17]. Using acyl amino acid p-guanidinophenyl ester the leaving group acts as a mimetic for the basic side chain function of arginine or lysine and this type of inverse substrate interacts readily with trypsin forming the acyl enzyme. The latter reacts with the amino component yielding the desired peptide. Since the formed peptide bond does not correspond to trypsin specificity the enzyme cannot split it. This irreversible synthesis approach allows even the synthesis of Pro-

form was found to be a very good acyl donor in other syntheses [13], and the use of a considerable excess of acyl donor ensured that most of the side reactions were suppressed. The total yield after the five fragment condensations was 15%, and after the folding of the final product, 8%. Thus starting with 100 mg of the initial peptide (residues 98–124), RNase A can be obtained on a 10-mg scale. The fragments (77–97, 64–76, 52–63, 21–51, 1–20) were chosen such that the C-terminal residues, Tyr[97], Tyr[76], Val[63], and Ala[20], were the closest to matching the substrate specificity of the subtilisin mutant. A similar strategy was also used in the synthesis of three variants of RNase A, in which the two histidine residues His[12] and His[119] at the active center were exchanged individually and simultaneously for L-4-fluorohistidine; the affect of these substitutions in the three analogues of Rnase A was studied in detail.

Xaa bonds catalyzed by trypsin. The improvements in the syntheses of acyl amino acid p-guanidino as well as p-amidinophenyl esters [18] promote the application of this new irreversible strategy. Furthermore, the potential of substrate mimetics mediated synthesis could be extended to other suitable proteases [19].

Zymogen-catalyzed Synthesis

In 1994, for the first time we were able to show that zymogens, which are known as catalytically inactive precursors of proteases, can be used as biocatalysts for practically irreversible peptide bond formation [20]. In order to confirm that the results obtained with zymogens can be actually attributed to proenzyme catalysis, it was essential to establish active enzyme-free zymogen preparations [14]. Firstly, we made use of the significantly different affinity of both enzyme and zymogen to the basic pancreatic trypsin inhibitor (BPTI). Doing so it was possible to analyze the esterase activity of the zymogen which is an efficiency parameter in estimating their peptide bond forming potential. Due to the fact that differences in the specificity constant k_{cat}/K_M cover a range of about 5 orders of magnitude, for a general use of zymogen catalysis it is essential to improve the acylation rate (k_2). Using the carboxamido methyl ester leaving group [13] the acylation efficiency can be improved. Furthermore, guanidiated trypsinogen (which cannot be activated by trypsin) could be used for successful acyl transfer experiments. In addition, no secondary hydrolysis of the formed peptide bond was observed in long-time experiments. Our studies show that zymogen catalysis can be regarded as an unexpected synthesis tool which allows practically irreversible peptide bond formation. Starting with first sucessful experiments, the improvement of this new strategy necessitates simultaneous develop-

ment of mutants based on proteolytically inactive zymogens with higher efficiency combined with the application of acyl donor esters with an improved acylation potential.

Abzyme-catalyzed Synthesis

Substrate binding at the active site [8] plays a crucial role in protease-catalyzed peptide bond formation. Unfortunately, a simple C–N ligase is not capable of developing different substrate binding regions, for example by induced fit, for the structurally diverse amino acid side chain functionalities. According to Linus Pauling the action of an enzyme depends on the complementarity of the active site to the transition state structure of a reaction, as shown in Scheme 2 for peptide bond formation.

Given the fact that the enzyme should not bind the substrate very strongly but must stabilize the transition state considerably, it is unlikely that a relatively simple enzyme can act as a universal peptide ligase. This premise was recently confirmed by reports by Hirschmann et al. [21] and by Jacobsen and Schultz [23] on peptide bond formation with catalytic antibodies (abzymes). Analogues of the transition state were used as haptens to induce antibodies with the correct arrangement of catalytic groups. The idea behind both strategies is illustrated in Scheme 2 by the transition state analogue **6** synthesized by Hirschmann et al. [21]. When these haptens are injected in animals, their immune systems generate antibodies against them. Since the structure of the tetrahedral intermediate **5** is also greatly affected by the amino acid side chains R^1 and R^2, it is doubtful whether a universally applicable catalytic antibody can be generated for peptide bond formation. The structures of the transition state analogues **7**, **8a,** and **8b** were changed deliberately, in particular by incorporation of a cyclohexyl group, such that the

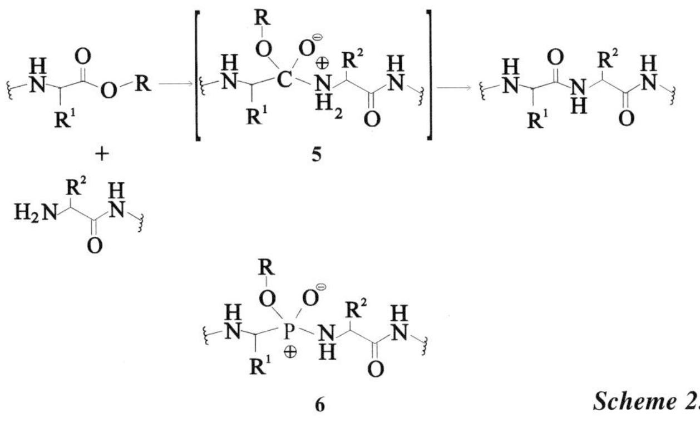

Scheme 2.

abzyme that catalyzed the reaction of **9** and **10** would have broad substrate specificity (Scheme 3).

The abzyme-catalyzed dipeptide syntheses gave all possible stereoisomers in yields of 44 to 94 %. The abzymes did not catalyze the hydrolysis of either the dipeptide products or the activated esters employed. Further studies have revealed that the monoclonal antibody not only couples activated amino acids to form dipeptides with high turnover rates but also couples an activated amino acid with a dipeptide to form a tripeptide, as well as an activated dipeptide with another dipeptide to give a tetrapeptide [22]. Jacobsen and Schultz [23] elicited antibodies against a neutral phosphonate diester transition state analogue, which significantly accelerated the coupling

of an N^{α}-acylalanine azide with a phenylalanine derivative relative to the uncatalyzed reaction. The application of catalytic antibodies exploits the fundamental property of the immune system, generating binding sites in the folds of the antibodies which are analogous to the active sites in enzymes. However, extensive development is needed before this approach is applied as routinely as enzymatic peptide synthesis currently is, and this perspective emphasizes the importance of irreversible strategies in protease-catalyzed peptide synthesis.

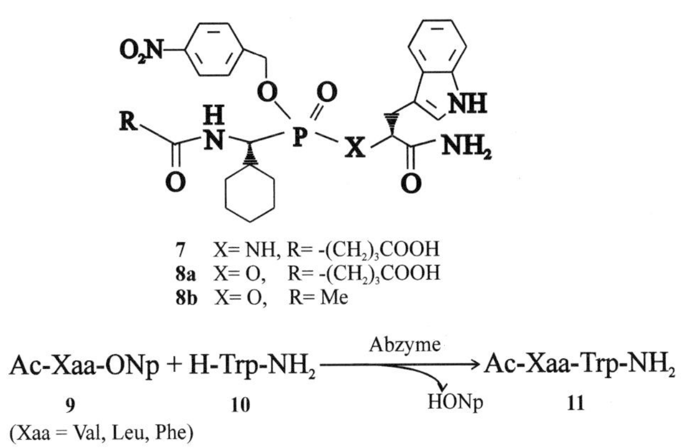

7   X= NH, R= -(CH₂)₃COOH
8a   X= O,   R= -(CH₂)₃COOH
8b   X= O,   R= Me

Ac-Xaa-ONp + H-Trp-NH₂ $\xrightarrow[\text{HONp}]{\text{Abzyme}}$ Ac-Xaa-Trp-NH₂

9   10   11

(Xaa = Val, Leu, Phe)

Scheme 3.

References

[1] H.-D. Jakubke, *Peptide: Chemie und Biologie*, Spektrum Akademischer Verlag, Heidelberg, **1996**; B. Gutte (Ed.), *Peptides: synthesis, structure, and applications*, **1995**, Academic Press, San Diego.

[2] R. Hirschmann, R. F. Nutt, D. F. Veber, R. A. Vitali, S. L.Varga, T. A. Jacob, F. W. Holly, R. G. Denkewalter, *J. Am. Chem. Soc.* **1969**, *91*, 507–508.

[3] B. Gutte, R. B. Merrifield, *J. Am. Chem. Soc.* **1969**, *91*, 501–502.

[4] H. Yajima, N. Fujii, *J. Am. Chem. Soc.* **1981**, *103*, 5867–5871.

[5] N. L. Benoiton, *Int. J. Peptide Protein Res.* **1994**, *44*, 399–400.

[6] D. Riester, K.-H. Wiesmüller, D. Scholl, R. Kuhn, *Anal. Chem.* **1996**, *68*, 2361–2365.

[7] Selected reviews: a) H.-D. Jakubke, P. Kuhl, A. Könnecke, *Angew. Chem.* **1985**, *97*, 79–87; *Angew. Chem. Int. Ed. Engl.* **1985**, *24*, 85–93; b) W. Kullmann, *Enzymatic Peptide Synthesis*, **1987**, CRC, Boca Raton, **1987**; c) H.-D. Jakubke in *The Peptides: Analysis, Biology*, (Eds.: S. Udenfriend, J. Meienhofer), Vol. 9, pp. 103–165, Academic Press, New York, **1987**; d) P. Sears, C. H. Wong, *Biotechnol. Prog.*, **1996**, *12*, 423–433.

[8] V. Schellenberger, H.-D. Jakubke, *Angew. Chem.* **1991**, *103*, 1440–1452; *Angew. Chem. Int. Ed. Engl.* **1991**, *30*, 1437–1449.

[9] H.-D. Jakubke, *J. Chin. Chem. Soc.* **1994**, *41*, 355–370.

[10] J. Bongers, E. P. Heimer, *Peptides* **1994**, *15*, 183–193.

[11] C.-H. Wong, *Chimia* **1993**, *47*, 127–132.

[12] D. Y. Jackson, J. Vurnier, C. Quan, M. Stanley, J. Tom, J. A. Wells, *Science* **1994**, *266*, 243–247.

[13] P. Kuhl, U. Zacharias, H. Burckhardt, H.-D. Jakubke, *Monatsh. Chem.* **1986**, *117*, 1195–1204.

[14] H.-D. Jakubke, U. Eichhorn, M. Hänsler, D. Ullmann, *Biol. Chem.* **1996**, *377*, 455–464.

[15] M. Hänsler, H.-D. Jakubke, *J. Pept. Sci.* **1996**, *2*, 279–289; *Amino acids,* **1996**, *11*, 379–395.

[16] G. Wagner, H. Horn, *Pharmazie* **1973**, *28*, 428–431; K. Tanizawa, A. Kasaba, Y. Kanaoka, *J. Am. Chem. Soc.* **1977**, *99*, 4485–4488

[17] V. Schellenberger, H.-D. Jakubke, N. P. Zapevalova, Y. V. Mitin, *Biotechnol. Bioeng.* **1991**, *38*, 104–108; *38*, 319–321; K. Itoh, H. Sekizaki, E. Toyota, K. Tanizawa, *Chem. Pharm. Bull.* **1995**, *43*, 2082–2087

[18] H. Sekizaki, K. Itoh, E. Toyota, K. Tanizawa, *Chem. Pharm. Bull.* **1996**, *44*, 1577–1579; 1585–1587; D. R. Kent, W. L. Cody, A. M. Doherty, *Tetrahedron Lett.* **1996**, *37*, 8711–8714.

[19] F. Bordusa, D. Ullmann, C. Elsner, H.-D. Jakubke, *Angew. Chem. Int. Ed. Engl.* **1997**, *36*, 2473–2475.

[20] D. Ullmann, K. Salchert, F. Bordusa, R. Schaaf, H.-D. Jakubke in *Proc. 5th Akabori-Conference, Max-Planck-Gesellschaft*, (Ed. E. Wünsch), pp. 70–75, R. & J. Blank, München, **1994**.

[21] R. Hirschmann, A. B. Smith, C. M. Taylor, P. A. Benkovic, S. D. Taylor, K. M. Yager, P. A. Sprengeler; S. J. Benkovic, *Science* **1994**, *265*, 234–237.

[22] D. B. Smithrud, P. A. Benkovic, S. J. Benkovic, C. M. Taylor, K. M. Yager, J. Witherington, B. W. Philips, P. A. Sprengeler, A. B. Smith, R. Hirschmann, *J. Am. Chem. Soc.* **1997**, *119*, 278–282.

[23] J. R. Jacobsen, P. G. Schulz, *Proc. Natl. Acad. Sci. USA* **1994**, *91*, 5888–5892.

Synthetic Ribozymes and Deoxyribozymes

Petra Burgstaller and Michael Famulok

The principal aims of the early stages of protein engineering were limited to understanding how enzymes work. Today's greatest challenge is the *de novo* design of synthetic enzymes with novel activities. Since the discovery of natural catalytic RNAs [1], however, progress in this field is no longer limited just to proteins. The development of methods for the screening of combinatorial nucleic acid libraries have made it possible to identify scarce functional molecules in pools of up to 10^{15} different sequences. [2, 3] In a typical *in vitro* selection or *in vitro* evolution experiment, a pool of RNA or DNA molecules consisting of a randomized region flanked by defined primer binding sites is subjected to a selection step to isolate those molecules capable of performing the desired task. Active molecules are then amplified and applied to another round of selection and amplification. This iterative procedure allows the enrichment of functional molecules until they dominate the library even though they initially may have been represented in a very low proportion.

Using the methods of *in vitro* selection and *in vitro* evolution not only the functionality and specificity of natural ribozymes could be altered or improved, but also nucleic acid based catalysts with new catalytic activities could be evolved. The latest results in this field demonstrate that ribozymes or deoxyribozymes are able to catalyze a wide range of chemical reactions.

Aptamer-based Libraries

The isolation of new ribozymes can be achieved by either direct selection of active sequences from completely randomized pools, or by first selecting for binding to a cofactor needed in the reaction and using the aptamer as a basis for the selection of functional molecules. The more challenging method of direct selection allows a more complete search of the available sequence space, while an aptamer-based selection strategy might facilitate the isolation of functional molecules.

Lorsch and Szostak started [5] from a specific ATP-binding RNA, which had previously been selected by affinity chromatography on ATP agarose, [4] and used it to evolve an ATP-dependent oligonucleotide kinase. [5] An RNA pool in which the central sequence of the ATP binder had been partly randomized was synthesized and was surrounded by three completely randomized sequences of a total of 100 bases (Fig. 1). This pool was incubated with ATP-γ-S. RNAs onto which the γ-thiophosphate group of the ATP-γ-S had been transferred could be separated from the rest

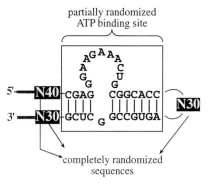

Figure 1. Design of the RNA library for the ATP-dependent kinase selection. The grey lines flanking the 40 and 30 nt randomized region at the 3′- and 5′-end represent primer binding sites for PCR amplification.

of the library on activated thiopropyl agarose, since they specifically formed a disulfide bond between the thiophosphate group and the agarose. These covalently bound RNAs were then eluted by washing with an excess of 2-mercaptoethanol, amplified and reselected. After thirteen cycles of selection, seven classes of ribozymes were characterized which catalyzed various reactions. Five of these RNA classes catalyzed the transfer of the γ-thiophosphate onto their own 5′-hydroxyl group. The other two classes phosphorylated the 2′-hydroxyl groups of specific internal base positions. Starting from one of these sequences, a ribozyme was developed which can phosphorylate the 5′-end of oligonucleotides in an intermolecular reaction (Fig. 2). In this way, multiple turnover of ATP-dependent phosphorylation was demonstrated with enzyme kinetics according to the Michaelis-Menten equation. Comparison of the sequence region corresponding to the original randomized ATP binding site in the selected ribozymes with the sequence of the ATP aptamer revealed that in the seven different ATP-dependent kinase

Intramolecular thiophosphorylation

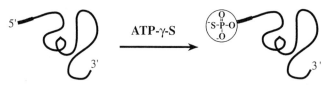

Intermolecular thiophosphorylation

Figure 2. Schematic for the inter and intramolecular 5′-thiophosphorylation reaction catalyzed by one class of ribozymes. Other ribozyme classes catalyzed the 2′-thiophosphorylation.

ribozymes only four retained the ATP motif to some extent. The other three classes had completely lost the aptamer motif.

An analogous strategy was used by Wilson and Szostak [6] to isolate self-alkylating ribozymes using an iodoacetyl derivative of the cofactor biotin. After isolating biotin-binding RNA aptamers by repeated rounds of affinity chromatography and amplification, a second library was generated which consisted of the mutagenized aptamer sequence flanked on either side by 20 random nucleotides. Molecules from this library which were able to self-alkylate with the biotin derivative were separated from inactive sequences by binding to streptavidin. By the seventh round of selection, more than 50 % of the RNA performed the self-biotinylation reaction. The sequencing of individual clones revealed that a majority of the ribozymes was derived from a single ancestral sequence. To optimize the activity, a third selection was carried out in which the incubation time as well as the concentration of the cofactor were progressively lowered. The resulting ribozyme alkylated itself at the N7 position of a specific guanosine residue within a conserved region (Fig. 3). It shows a rate acceleration of 2×10^7 compared to the uncatalyzed reaction. This rate enhancement is comparable to that of highly active catalytic antibodies.

The self-biotinylating ribozymes which originated from a biotin-binding aptamer show an astounding structural change compared to their ancestor. A highly conserved nucleotide stretch of the biotin binder which seems to directly mediate the interaction between biotin and the aptamers as well as the catalytically active molecules is retained in the self alkylating ribozymes with only a single point mutation. Yet, the secondary structural context of this consensus sequence is highly unrelated in the two classes of molecules. While the biotin aptamer contains a pseudoknot motif, the ribozyme forms a cloverleaf which resembles the structure of tRNAs.

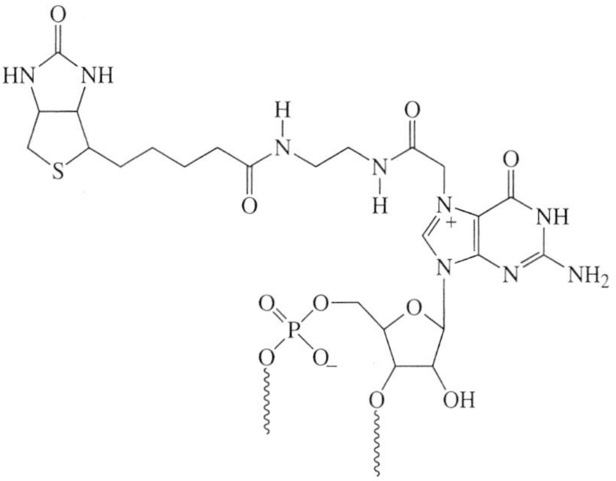

Figure 3. Mode of alkylation of the N7 position of an internal guanosine residue catalyzed by the alkylation ribozyme.

Random Libraries

Bartel and Szostak [7] succeeded in isolating ribozymes from a synthetic RNA library which were capable of ligating an oligonucleotide to their 5'-end with cleavage of the 5'-terminal pyrophosphate. The mechanism of this reaction corresponds to the mode of action of RNA-dependent RNA polymerases. The selection started with a pool containing a very long randomized synthetic sequence of 220 nucleotides. This length was chosen to increase the probability of isolating catalytically active sequences.

To bring the 3'-hydroxyl group of the oligonucleotide in proximity to the triphosphate at the 5'-end of the RNA pool, a sequence in the constant region of the 5'-primer was constructed such that the substrate oligonucleotide could hybridize to the 5'-end of the pool. Catalytically active molecules were enriched in two steps. Following incubation with the oligonucleotide, the pool was purified by passage through an oligonucleotide affinity column, to which only successfully ligated sequences could bind. These were then eluted, reverse transcribed, and amplified in a PCR reaction with a primer that was complementary to the sequence of the ligated oligonucleotide. In a second PCR reaction the original, unligated structure was then regenerated, from which an enriched RNA pool was transcribed for the next round of selection. Three of the ten selection rounds contained a mutagenic PCR step, so that "worse" catalysts had the chance to develop into "better" ones. This strategy led to an increase in the ligation rate from 3×10^{-6} h^{-1}, for the uncatalyzed rate, to 8 h^{-1}.

The characterization of individual sequences from this selection experiment revealed three different structural classes of ligases from which only one, the class I ligase, generates a $3'-5'$ phosphodiester bond at the ligation site. [8] An optimized version of this complex ribozyme [9] that differs at 10 positions from the original sequence has a k_{cat} of 100 min^{-1}, a value comparable to that of the protein enzyme ligase (Fig. 4). [9]

Ekland and Bartel [10] took one step further in creating a plausible scenario for an "RNA world" in which ribozymes catalyzing the replication of RNA are postulated. They demonstrated that the class I ribozyme is capable of extending a separate RNA primer by one nucleotide in the presence of a template oligonucleotide and nucleoside triphos-

phates. The polymerization reaction exhibits a remarkable template directed fidelity. Mismatched nucleotides are added 1000-fold less efficiently. By linking the primer covalently to the ribozyme, the reaction could be expanded to the addition of three nucleotides. By designing the primer in a way that it was able to slip onto the template, even six nucleotides could be added to the primer in a template directed way (Fig. 5).

In another attempt to provide a starting point for the evolution of self-replicating ribozymes, Chapman and Szostak [11] designed a selection experiment for the generation of RNA molecules that ligate their 3'-end to a hexanucleotide with a 5'-phosphate activated as phosphorimidazolide. However, the isolated ribozyme catalyzed the attack of the 5'-terminal γ-phosphate group on the 5'-phosphorimidazolide of the substrate oligonucleotide forming a 5'-5' tetraphosphate linkage. Depending on whether a 5'- mono-, di-, or tri-

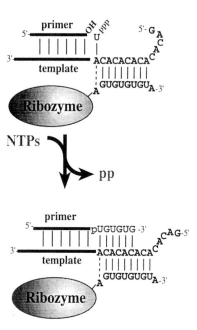

Figure 5. RNA polymerization of up to six nucleotides catalyzed by the class I ligase (Ribozyme).

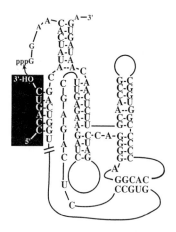

Figure 4. Secondary structure of the class I ligase.

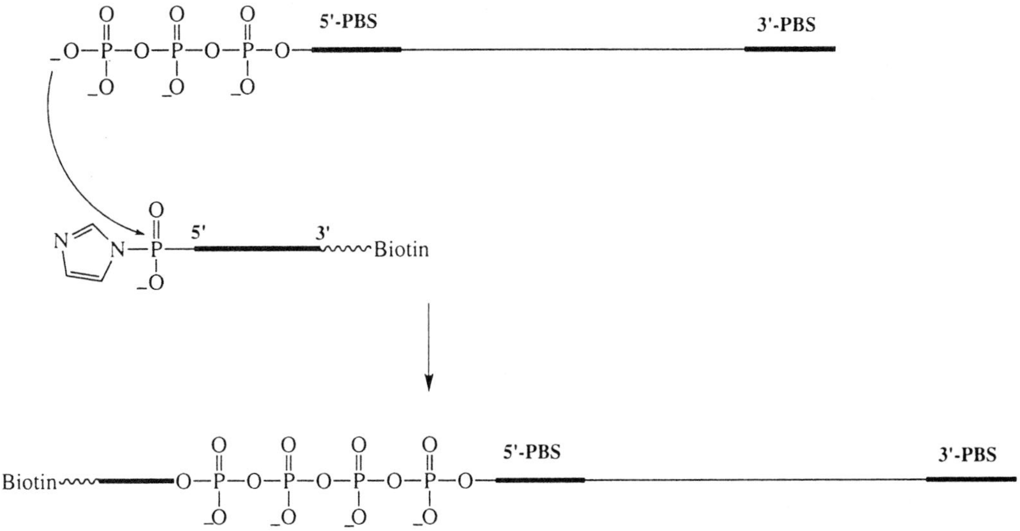

Figure 6. Reaction catalyzed by the RNA to generate a 5′-5′-tetraphosphate linkage.

phosphate was present, the 54 nucleotide long pseudoknot motif was also capable of generating di- or triphosphate linkages (Fig. 6).

To make the existence of an "RNA world" as a step in prebiotic evolution feasible, not only self-replicating ribozymes are essential, but also a second type of catalytically active molecules. The transition from a world based on RNA catalysis to a world of protein enzymes demands ribozymes which are able to synthezise peptides from activated amino acids.

A prerequisite for peptide and protein synthesis in all modern-day life forms is the formation of aminoacyl-tRNA carried out by aminoacyl-tRNA synthetases. These enzymes first activate the carbonyl group of the amino acid by forming an aminoacyl-adenylate containing a highly activated mixed anhydride group which is then used to transfer the amino acid to the 3′(2′)-hydroxy terminus of the cognate tRNA. Illangasekare et al. [12] used an *in vitro* selection strategy to obtain an RNA that catalyzes the esterification of an activated phenylalanine to its own 3′(2′)-end.

An RNA library consisting of $\sim 10^{14}$ different molecules was incubated with synthetic phenylalanyl-5′-adenylate. Those RNA molecules which had catalyzed their own aminoacylation contained a free a-NH_3^+-group from the amino acid. This nucleophilic a-NH_3^+-group was selectively reacted with the *N*-hydroxysuccinimide of naphthoxyacetic acid. Thus, only those RNAs with the amino acid covalently attached to themselves contained the naphthoxy residue and therefore differed significantly in their hydrophobicity from inactive molecules. Consequently, separation from inactive sequences could be achieved by reversed phase HPLC (Fig. 7). Eleven cycles of selection resulted in a variety of self-aminoacylating ribozymes of which the most active showed a rate enhancement of 10^5-fold compared to the background rate.

Lohse and Szostak [13] used a selection scheme that mimicks the transfer of formyl-methionine from a fragment of fMet-tRNA onto the hydroxy group of hydroxypuromycin, a simplified version of the ribosomal peptidyltransferase reaction. An RNA library

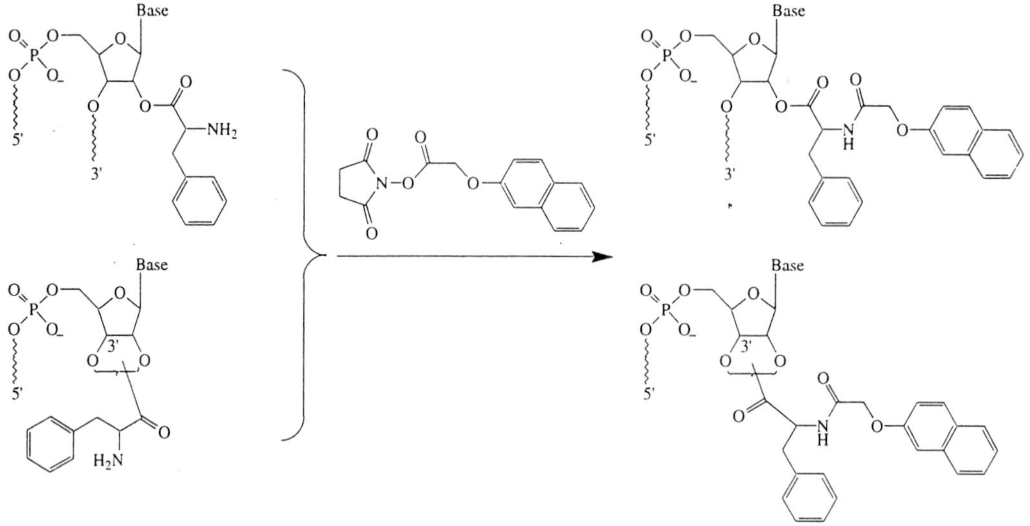

Figure 7. 2'- and 2',3'-aminoacylation of catalytic RNAs. Catalytic RNAs that can self-aminoacylate are made more hydrophobic with a naphthoxyacetyl label.

comprising a 90 nucleotide random region was treated with alkaline phosphatase to expose the 5'-hydroxyl group and was incubated with a hexanucleotide charged at its 3'-end with biotinylated methionine acting as an amino acid donor. Functionally active RNA molecules were separated from the inactive part of the library on streptavidin agarose via the biotin tag. After eleven cycles of selection, including two rounds of mutagenic PCR, the enriched library showed a 10^4-fold increase in acyl transferase activity. Cloning and sequencing revealed that the pool was dominated by one class of sequences with a highly conserved internal region. The 3'-end of this consensus sequence is complementary to the hexa-oligonucleotide substrate, the 5'-part is able to basepair with the 5'-end of the RNA, thus bringing the 3'-end of the donor and the 5'-end of the acceptor into close proximity. The ribozyme, however, does not act merely as a template, since it shows a rate enhancement of 10^3 compared to the rate of template-directed acyl transfer. Interestingly, two

G·U wobble base pairs at the donor/acceptor junction are important for the reaction. When replaced by G:C pairs, the value of k_{cat} decreased 14-fold, while the template-only directed reaction is optimal with a completely Watson-Crick base paired duplex. The described ribozyme is also capable of forming amide bonds which was demonstrated by replacing the attacking 5'-hydroxyl group by an amino group.

TSA-based Selections

All ribozymes described so far were isolated by the method of direct selection, whereby the partly or completely randomized pool of nucleic acids is subjected to a competitive situation in which only those molecules survive that can catalyze a particular reaction. A different strategy by which catalysis can be achieved is the indirect selection for binding to transition state analogs (TSAs), a tech-

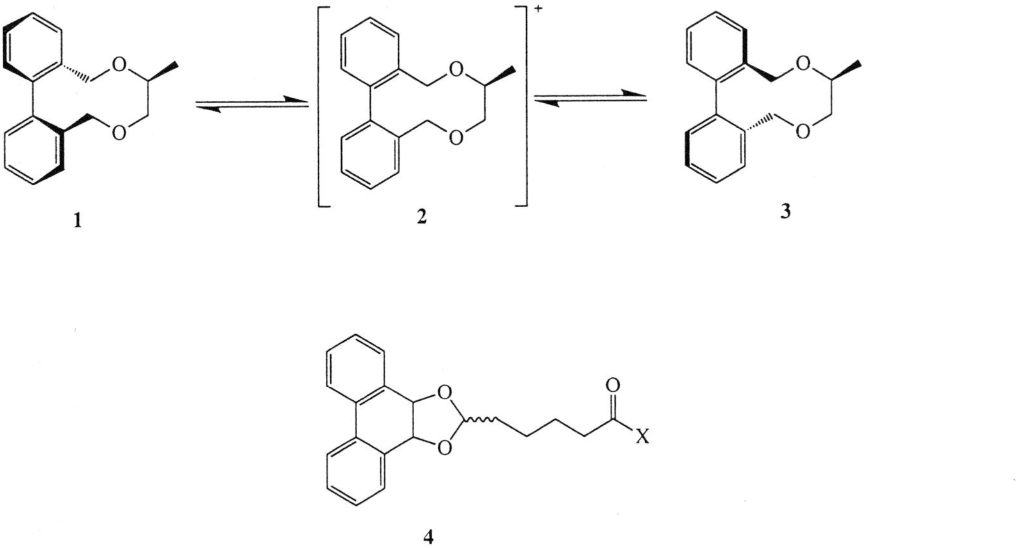

1   2   3

4

Figure 8. Isomerization of the bridged biphenyl derivative **1** to its diastereomer **3** catalyzed by the rotamase ribozyme. The reaction proceeds through the transition state **2**, mimicked by the transition state analog **4**.

nique used for the isolation of catalytic antibodies.

Following this strategy, Prudent et al. [14] have isolated RNA aptamers that could bind specifically to the TSA of the isomerization of an asymmetrically substituted biphenyl derivative (Fig. 8). The selection was performed by affinity chromatography of a randomized pool on the TSA immobilized on agarose.

After seven rounds of selection, the RNA pool was able to accelerate the reaction by a factor of 100 over the uncatalyzed reaction. A kinetic study showed that a Michaelis-Menten complex is formed initially, followed by the isomerization reaction and release of the reaction product. The reaction was completely inhibited by the planar TSA.

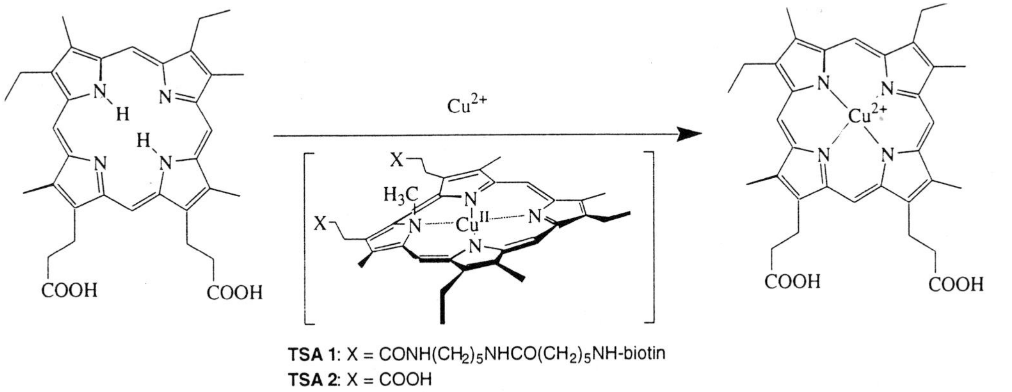

TSA 1: X = CONH(CH$_2$)$_5$NHCO(CH$_2$)$_5$NH-biotin
TSA 2: X = COOH

Figure 9. Metalation of Mesoporphyrin and transition state analog *N*-methyl mesoporphyrin **TSA1** and **TSA2**. **TSA1** was used for the selection by streptavidin immobilization.

The second example of a catalytic RNA obtained by isolating an RNA aptamer for TSA binding was reported from the same laboratory [15]. Conn et al. recently reported the isolation of a 35 nucleotide RNA molecule which binds mesoporphyrin IX and catalyzes the insertion of Cu^{2+} into the porphyrin with a value of k_{cat}/K_m of 2100 M^{-1} s^{-1}. The k_{cat}/K_m value achieved by the RNA was close to that of the Fe^{2+}-metalation of mesoporphyrin catalyzed by the protein enzyme ferrochelatase (Fig. 9).

The isolation of ribozymes by selection of TSA-binding nucleic acids may consitute an efficient method for investigating the diveristy of available chemical reactions that can be catalyzed by nucleic acids.

Deoxyribozymes

A new departure in the development of nucleic acid based catalysts was made three years ago by Breaker and Joyce, who reported the isolation of the first deoxyribozyme. The proof that DNA can also catalyze chemical reactions is not completely unexpected, especially since it was shown shortly after the development of the *in vitro* selection technique that single-stranded DNA molecules can also be selected to bind to a variety of ligands. Meanwhile, several catalytically active DNAs have been described, expanding the range of nucleic acid catalyzed reactions even further.

Breaker and Joyce [16] accomplished the selection of a DNA enzyme that could specifically cleave the phosphodiester bond of a ribonucleotide. The DNA pool synthesized for this purpose contained a single ribonucleotide at a specific position within a primer binding site, in order to avoid the possible effect of RNA on the catalysis. In addition, they assumed that a DNA-dependent cleavage at pH 7 would require a cofactor, possibly a

metal ion. All ribozymes whose reaction mechanisms have been studied are metalloenzymes. Metal ions such as Mg^{2+}, or even Pb^{2+} in the case of the self-cleaving yeast $tRNA^{Phe}$, were always involved in the catalytic mechanism and also served to maintain the correct folding of the RNA. Pan and Uhlenbeck [17] have shown that new ribozymes with Pb^{2+}-dependent phosphodiesterase activity could be isolated from a randomized pool of $tRNA^{Phe}$ molecules. The specificity of the reaction depends on the coordination of a Pb^{2+} ion to a defined position within the RNA.

Breaker and Joyce also used Pb^{2+} as the cofactor in their selection. They generated a pool of about 10^{14} double-stranded DNAs with a randomized region of 50 nucleotides. One of the strands contained a biotin group at the 5'-end, followed by a defined 43-mer base sequence containing a single ribose adenosine unit. The double-stranded DNA was immobilized on streptavidin agarose, and the non-biotinylated strand was removed by increasing the pH. The immobilized single-stranded DNA was then incubated in Pb^{2+}-containing buffer. In active sequences this led to cleavage of the phosphodiester bond at the ribose phosphate group, and hence to loss of the covalent attachment to the biotin-bearing 5'-end. The catalytically active sequences were thereby specifically eluted from the column, and were subsequently amplified and selected once more (Fig. 10). Five rounds of selection afforded a population of single-stranded DNAs that could catalyze the Pb^{2+}-dependent cleavage of the ribose residue. These DNAs also show similarities in their sequences. Like the hammerhead and hairpin ribozymes, they contain two conserved, unpaired regions between two sequences which can pair with bases upstream and downstream of the splice site. This structural motif served as the basis for the construction of a shortened version of the catalytic and substrate domains. In this way, it was shown that the 38 nucleotide long catalytic domain could cleave the 21-mer substrate specifically and with high turnover rates. The deoxyribozyme is, however, not capable of cleaving a pure RNA substrate, although the substrate binding to the enzyme is ensured with two longer base-paired regions.

Using the same selection strategy, other experiments have been carried out to isolate DNA molecules with either RNA or DNA phosphoesterase activity. Since Mg^{2+}-dependent rather than Pb^{2+}-dependent cleavage is compatible with intracellular conditions and thus, more suitable for possible medical applications, Breaker and Joyce [18] also developed deoxyribozymes that used Mg^{2+} as a cofactor instead of Pb^{2+}. The deoxyribozyme showed a cleavage rate of 0.01 min^{-1} and was also capable of intermolecular cleavage.

Faulhammer and Famulok [19] attempted to develop deoxyribozymes that use a non-metal cofactor rather than divalent metal ions for the cleavage of a ribonucleotide residue. They performed a selection experiment under conditions of low magnesium concentration, or even without any divalent metal ions, by incubating the immobilized single-stranded DNA library in a large excess of histidine. Surprisingly, non of the resulting eight classes of deoxyribozymes utilized the histidine as a cofactor for the strand scission. Instead, some of them were dependent on the presence of divalent metal ions while others were able to accelerate the cleavage reaction even in the absence of divalent metal ions. Remarkably, one of the catalysts, showed higher cleavage activity in the presence of Ca^{2+} than of Mg^{2+}, even though magnesium was part of the selection buffer. A possible explanation

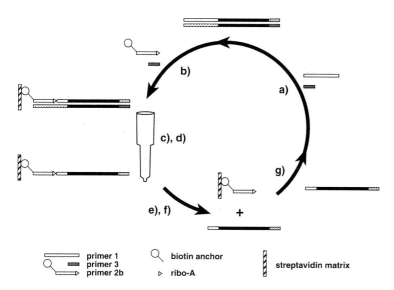

Figure 10. Selection scheme for deoxyribozymes which cleave RNA/DNA chimeric oligonucleotides. In a first PCR (a), the starting pool was amplified using **primer 1** and **primer 3**. (b) In a second PCR, the 5'-end of the pool was biotinylated and the ribonucleotide serving as cleavage site was introduced using **primer 2b** and **primer 3**. The double-stranded pool was then loaded on a streptavidin column (c). The antisense strand is removed by raising the pH of the solution (d) and the remaining pool of single-stranded DNA allowed to fold (e) by rinsing the column with equilibration buffer. The equilibration was followed by addition of cleavage buffer (f). Cleavage products were eluted from the column (g) and amplified by PCR using the same primers as above.

might be that in this special case Ca^{2+} can be more suitably positioned at the cleavage site. This suggestion is supported by the observation that two Ca^{2+}-ions are bound in a co-operative fashion by the deoxyribozyme. [20] Thus, the cleavage mechanism performed by the seelected deoxyribozyme might be highly similar to the "Two-metal ion"-cleavage mechanism recently suggested by Pontius et al. [21] for the hammerhead ribozyme (Fig. 11). From the unexpected result of this selection, one might conclude that in nucleic acids the number of potential binding sites for metal-ions is by far larger than for non-metal cofactors.

To further examine the catalytic potential of DNA, Carmi et al. [22] employed the *in vitro* selection protocol described above to generate deoxyribozymes that facilitate self-cleavage by a redox-dependent mechanism. The design of this selection was based on the fact that DNA is more sensitive to cleavage via depurination followed by β-elimination or via oxidative mechanisms than by hydrolysis. Therefore, single-stranded DNA (ssDNA) bound to streptavidin by its biotin-tag was incubated with $CuCl_2$ and ascorbate. The pool isolated after seven rounds of selection consisted of two distinct classes of self-cleaving ssDNA molecules. While "class I" deoxyribozymes performed strand scission in the presence of both Cu^{2+} and ascorbate, "class II" molecules only required the copper ion as a cofactor. An optimized version of the "class II" deoxyribozyme shows a rate enhancement of more than 10^6-fold compared to the background reaction.

However, the catalytic potential of DNA is not only limited to phosphoresterase activity. Cuenoud and Szostak [23] designed a selection scheme to isolate DNA molecules that catalyze the ligation of their free 5'-hydroxyl group to the 3'-phosphate group of a substrate oligonucleotide activated by imidazolide. The resulting deoxyribozyme is dependent on the presence of Zn^{2+} or Cu^{2+}. It contains two conserved domains which position the 5'-hydroxyl group and the 3'-phosphorimidazolide of the substrate oligonucleotide in close proximity. However, the consensus sequences are flanked by highly variable regions, suggesting that several independent ways of arranging the consensus sequences resulted in active deoxyribozymes. Based on the selected sequences, a truncated version of the DNA ligase was designed that is able to ligate two DNA sub-

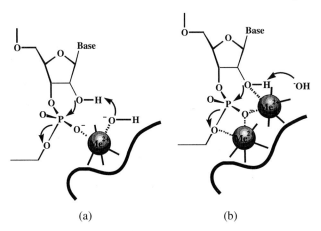

(a)   (b)

Figure 11. Proposed cleavage mechanisms for the hammerhead ribozyme. (a) One metal ion mechanism. (b) Two metal ion mechanism. This mechanism is also proposed for the calcium-dependent deoxyribozyme which binds at least two Ca^{2+} ions in a cooperative manner.

strates in a multiple turnover reaction with a 3400-fold rate enhancement compared to the template-only directed ligation.

Recently, in a similar approach as reported for an RNA selection [17], a DNA molecule able to catalyze porphyrin metallation was isolated [24]. An oligomer consisting of the 33 nucleotide long guanine-rich binding site of a randomly chosen aptamer that binds to the TSA *N*-methylmesoporphyrin IX showed a k_{cat} of 13 h^{-1} for the insertion of Cu^{2+} into mesoporphyrin IX. This corresponds to a rate enhancement of 1400 compared to the uncatalyzed reaction. A catalytic antibody for the same reaction shows a rate acceleration in the same range. [25] The deoxyribozyme is strongly dependent on the presence of potassium in the reaction buffer, which suggests the formation of guanine quartets.

Despite the general acceptance that the 2'-hydroxyl group of RNA is essential for its capability to form complex secondary and tertiary structures and therefore for its catalytic activity, the results of these *in vitro* selection experiments make clear that DNA also can adopt a variety of complex structures enabling it to catalyze a wide range of chemical reactions. [26] The man-made ribo- and deoxyribozymes will certainly be applicable in some diagnostic, technical or medical purposes, where it is advantagous that they are based not on proteins but on nucleic acids.

We expect that we will soon see examples of ribo- and deoxyribozymes evolved for the catalysis of complex chemical transformations. There is enough reason to assume that such synthetic enzymes will be used as catalysts in organic syntheses. The novel catalysts not only support theories of an "RNA world", [27] in which the metabolism and replication of primitive organisms were controlled by RNA enzymes. [28] There is also the potential for biotechnological, synthetical and diagnostical applications. One of the most obvious applications of synthetic nucleic acid-based enzymes is their use *in vivo*, for example as

reporter systems in eucaryotic transcription assays which would greatly facilitate the development of highly sensitive screening assays.

References

[1] Kruger, K., Grabowski, P. J., Zaug, A. J., Sands, J., Gottschling, D. E., Cech, T. R., *Cell* **1982**, *31*, 147–157; Guerrier-Takada, C., Gardiner, K., Marsh, T., Pace, N., Altman, S., *ibid.* **1983**, *35*, 849–857.

[2] Ellington, A. D., Szostak, J. W., *Nature* **1990**, *346*, 818–822; Tuerk, C., Gold, L., *Science* **1990**, *149*, 505–510; Robertson, D. L., Joyce, G. F., *Nature* **1990**, *344*, 467–468.

[3] Gold, L., Polisky, B., Uhlenbeck, O. C., Yarus, M. *Annu. Rev. Biochem.* **1995**, *64*, 763–797; Joyce, G. F. *Curr. Opin. Struct. Biol.* **1994**, *4*, 331–336; Ellington, A. D. *Curr. Biol.* **1994**, *4*, 427–429; Klug, S. J. Famulok, M. *Mol. Biol. Reports* **1994**, *20*, 97–107; Famulok, M. Szostak, J. W. *Angew. Chem. Int. Ed. Engl.* **1992**, *31*, 979–988.

[4] Sassanfar, M., Szostak, J. W., *Nature* **1993**, *364*, 550–553.

[5] Lorsch, J. R., Szostak, J.W., *Nature* **1994**, *371*, 31–36.

[6] Wilson, C., Szostak, J. W., *Nature* **1995**, *374*, 777–782.

[7] Bartel, D., Szostak, J. W., *Science* **1993**, *261*, 1411–1418.

[8] Ekland, E., Szostak, J. W., Bartel, D., *Science* **1995**, *269*, 364–370.

[9] Ekland, E., Szostak, J. W., *Nucl. Acids Res.* **1995**, *23*, 3231–3238.

[10] Ekland, E., Bartel, D., *Nature* **1996**, *382*, 373–376.

[11] Chapman, K. B., Szostak, J. W., *Chem. Biol.* **1995**, *2*, 325–333.

[12] Illangasekare, M., Sanchez, G., Nickles, T., Yarus, M., *Science* **1995**, *267*, 643–647.

[13] Lohse, P. A., Szostak, J. W., *Nature* **1996**, *381*, 442–444.

[14] Prudent, J. R., Uno, T., Schultz, P. G., *Science* **1994**, *264*, 1924–1927.

[15] Conn, M. M., Prudent, J. R., Schultz, P. G. *J. Am. Chem. Soc.* **1996**, *118*, 7012–7013.

[16] Breaker, R. R., Joyce, G. F. , *Chem. Biol.* **1994**, *1*, 223–229.

[17] Pan, T. Uhlenbeck, O. C., *Nature* **1992**, *358*, 560–563.

[18] Breaker, R.R., Joyce, G. F., *Chem. Biol.* **1995**, *2*, 655–660.

[19] Faulhammer, D., Famulok, M., *Angew. Chem. Int. Ed. Engl.* **1996**, *35*, 2837–2841.

[20] Faulhammer, D., Famulok, M., *J. Mol. Biol.* **1997**, in press.

[21] Pontius, B. W., Lott, W. B., von Hippel, P. H., *Proc. Natl. Acad. Sci. USA* **1997**, *94*, 2290–2294.

[22] Carmi, N., Shultz, L. A., Breaker, R. R., *Chem. Biol.* **1996**, *3*, 1039–1046.

[23] Cuenoud, B., Szostak, J. W., *Nature* **1995**, *375*, 611–615.

[24] Li, Y., Sen, D., *Nature Struct. Biol.* 1996, *3*, 743–747.

[25] Cochran, A. G., Schultz, P. G., *Science* **1990**, *249*, 781–783.

[26] Berger, I., Egli, M. *Chem. Europ. J.* **1997**, in press.

[27] Gilbert, W., *Nature* **1986**, *319*, 618.

[28] Benner, S. A., Ellington, A. D., Tauer, A., *Proc. Natl. Acad. Sci. USA* **1989**, *86*, 7054–7058; Joyce, G. F., Orgel, L. E., in *The RNA World* (Eds.: Gesteland, R. F., Atkins, J. F.), Cold Spring Harbor Laboratory Pess, New York, **1993**, 1–25.

Enzyme Mimics

Anthony J. Kirby

Introduction

This *Highlight* is part of an extraordinary story (also a cautionary tale) in the area of biocatalysis. The point of particular interest was the incredible catalytic activity claimed for so-called 'pepzymes' – small synthetic peptides modelled to mimic the active site structures of trypsin and chymotrypsin. One was claimed to hydrolyse a simple peptide (a trypsin substrate) with efficiency comparable to that of the native enzyme. This extraordinary result provoked at least as much scepticism as excitement, and in the following months several groups tried to reproduce the results. They failed, comprehensively. [1, 2] Some reasons why this failure came as no surprise were subsequently summarised by Matthews, Craik and Neurath, [3] and by Corey and Corey [4]. The background has been discussed in an Angewandte Review on Enzyme Mechanisms, Models and Mimics. [5]

The *Highlight* noted a second example of remarkably effective catalysis of amide hydrolysis, this time in a surfactant system. This work involved activated amides, and accelerations – though exceptional – less spectacular than those claimed for the pepzymes. Nevertheless, this result too was not what it seemed: the apparent release of *p*-nitroaniline turned out to be an unusually slow physical change,

of chromophore intensity wiht aggregation state. [6]

The pepzymes debacle has never been satisfactorily explained, though the published results point unmistakeably to enzyme contamination. The surfactant story on the other hand has been properly sorted out, to the satisfaction of the original authors. Yet the lessons to be learned from these two case histories are basically the same – and by no means new. Mechanisms generally cannot be proved, only ruled out. So the design of control experiments to test alternative explanations is at least as important as the key experiment which appears to confirm a favourite or new and exciting theory. It is an inconvenient but inescapable fact of life that the more remarkable the result, the more important are those control experiments.

One of the great intellectual challenges presented to Science by Nature is a proper understanding of how enzymes work. At one level we can ‚explain' enzyme catalysis – what an enzyme does is bind, and thus stabilise, selectively the transition state for a particular reaction. [7] But our current level of understanding fails the more severe, practical test – that of designing and making artificial enzyme systems with catalytic efficiencies which rival those of natural enzymes.

Enzyme mimics have long been high-profile targets for bioorganic chemists. The picture to date has been the familiar one of steady progress, with occasional flashes of inspiration; and a heightening awareness of just how complex the problem is. But two recent reports [8, 9] – approaching the problem from completely different directions, claim catalytic efficiencies in artificial systems which are extraordinary in terms of conventional wisdom. First we should set the scene.

The 'conventional wisdom' is based on the one hand on mechanistic work on enzymes, and on the other hand on (mostly separate) studies of binding and catalysis in simpler, artificial systems. Enzymes are of course more than just highly evolved catalysts: they also recognise and respond to molecules other than their specific substrate and product, as part of the control mechanisms of the cell. But the evolution of enzyme mimics is at a stage where the efficient combination of binding and catalysis is the main objective. A starting point may be a system which shows efficient binding; or the mechanism of the reaction concerned, since it is the rate-determining transition state which is the most important target for the binding process. Eventually the two approaches must converge if a genuine enzyme mimic is to emerge. This sort of work holds out the promise of artificial catalysts, which may be more robust than proteins, for unnatural reactions of interest. On the other hand, the most natural basis for an enzyme mimic is inevitably a real enzyme. All these approaches are currently producing interesting results.

Enzyme-based Mimics

It is possible to modify a natural enzyme, chemically or more commonly by the methods of protein engineering, in such a way that its specificity is altered, even to the point that the modified system will catalyse a new reaction. Such systems are modified enzymes rather than enzyme mimics, which to qualify in this context should have been constructed artificially.

When it became possible to identify the functional groups in enzyme active sites it was natural to look at small peptides containing active-site sequences of amino-acids as possible catalysts. The results of this sort of work were uniformly negative. We know now that a working enzyme active site has its functional groups disposed in a specific, dynamic three-dimensional array, with significant interactions with the rest of the protein, and this cannot successfully be modelled in two dimensions.

A true enzyme-based mimic might try to reproduce the three-dimensional arrangement of the functional groups of the active site in a synthetic framework. The reasoning is simple; turning the idea into real molecules less so. However, what appears to be a major success for this approach is the report of Atassi and Manshouri [8] of the preparation of two so-called 'pepzymes', modelled from the active site structures of trypsin and chymotrypsin by 'surface-simulation'. This involved the design and synthesis of a series of relatively small (29 residue) peptides containing the key catalytic and binding amino-acid sequences of the enzymes. These were connected by glycine spacers so as to model the 3D arrangement known from the X-ray structures of the enzyme and its complexes with substrate analogues. An early version which showed some trypsin-like binding activity was modified systematically to the point where the peptide shown (native active-site residues indicated in bold) shows extraordinary catalytic activity, specifically in the cyclic (disulphide) form.

Not only does this molecule hydrolyse the simple trypsin 'substrate' N-tosyl-L-arginine methyl ester with k_{cat} and K_m comparable to those of the native enzyme, but it also hydro-

Cys-Gly-**Tyr-His-Phe**-Gly$_2$-**Ser-Asp-Gly-Gln**-Gly-**Ser-Ser**-Gly$_2$-**Val-Ser-Trp**-Gly-**Leu**-Gly$_2$-**Asp**-Gly-**Ala-Ala-His-Cys**

lyses test proteins to give similar peptide profiles. A closely related peptide based on the chymotrypsin active site had similar activity and the expected, different specificity.

It must be said that this level of activity is surprising, especially against amide bonds, and intensive efforts to repeat the results are under way. If confirmed this work will be seen as an important advance: the design stage may be complex, but with modern synthetic methods peptides of this sort of size are quite reasonable synthetic targets. A practical limitation, as always with enzyme-based mimics, is that – at this stage of development at least – complete success simply means doing a reaction as well as an available natural catalyst.

Mechanism-based Mimics

At the other extreme, it is possible to achieve enormous rate accelerations in quite simple systems by by-passing the binding process and making reactions intramolecular – that is, by bringing the functional groups concerned together on the same molecule. [10] Typically, making the reaction of interest part of a thermodynamically favourable cyclisation can produce systems in which the extraordinarily stable groups of structural biology (amides, glycosides and phosphate esters have half-lives of many years under physiological conditions near pH 7) can be cleaved in a fraction of a second. Detailed chemical mechanisms of catalysis can then be worked out for specific reactions, studied under the same conditions and going at similar rates as the same reactions between the same two (or more) functional groups in enzyme active sites. Two measures of catalytic efficiency are

relevant: the Effective Molarity (EM: the effective concentration of the catalytic group, that would needed to make an intermolecular reaction go at the rate of the intramolecular one). [10] And – of course – the absolute rate of the reaction. Because reactions in enzyme active sites are very fast: fast enough for many enzymes to have reached evolutionary perfection, defined by Albery and Knowles [11] as catalytic efficiency so high that the rate determining step of the reaction concerned is diffusion away of the products.

EM's as high as 10^{13-14} – meaning half-lives of the order of a second – can be attained in systems where an ordinary aliphatic amide is forced into close proximity with a COOH group, [11, 12] or a phosphate diester with a neighbouring OH, [13] and it is possible to define detailed mechanisms for such model reactions. These serve as an essential basis for the discussion of the mechanisms of the same reactions in enzyme active sites, or for the design of enzyme mimics. Because it is not possible – so far at least – to attain rate-enhancements of anything like this magnitude when the reacting groups are brought together by non-covalent binding.

Binding-step Based Mimics

A minimum requirement for a true enzyme mimic is a binding interaction between two molecules preliminary to the catalytic reaction, indicated by Michaelis-Menten kinetics. Intramolecular systems can support very rapid reactions because we can use synthesis to bring groups together into close and unavoidable proximity. But an enzyme must select and bind its substrate non-covalently in a dynamic equilibrium. The chemistry of the Molecular Recognition processes involved is one of the most active areas of current research, and a popular topic for meetings [14, 15] highlights [16] and reviews. [17] As with chemical catalysis, much of our understanding of non-covalent interactions comes from studies of simple systems designed to answer specific questions about the basic process. More directly relevant to the development of enzyme mimics are systems designed to achieve catalysis by binding, i. e. without specific catalytic groups built in. These fall into two important classes: synthetic hosts designed to bring two reactants into close and productive proximity, and most catalytic antibodies.

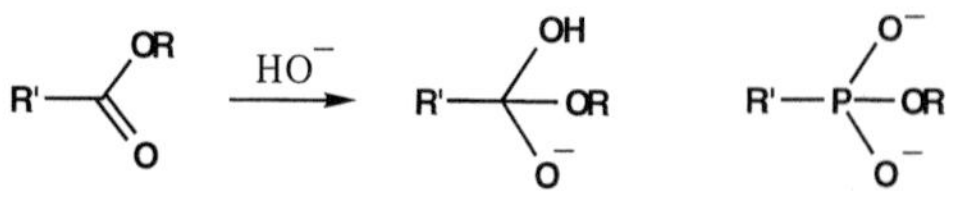

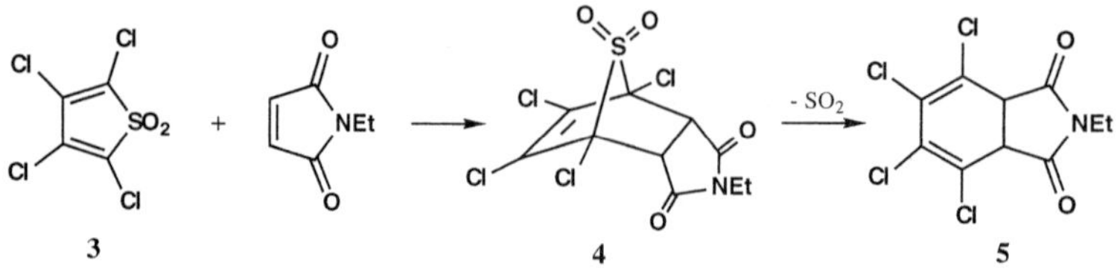

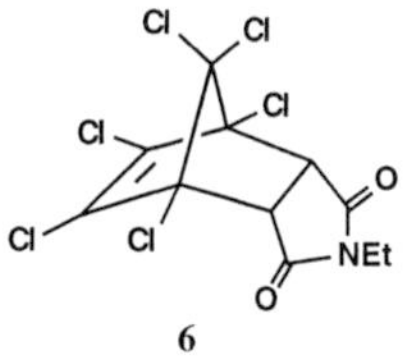

The majority of catalytic antibodies [18, 19] so far known have been designed to catalyse the hydrolysis of carboxylic acid derivatives, and have been raised against phosphonate haptens **2**, which models the structure of the tetrahedral transition states involved (**1**).

Catalysis of ester hydrolysis is rather reliably obtained with suitable haptens (and substrates), with the nucleophile coming from the solvent. More ambitious systems with catalytic groups built-in can in principle be obtained by careful hapten design (coupled with a large slice of luck); or from an existing catalytic antibody by protein engineering. This is an area of definite promise, and much current activity.

Careful hapten design also allowed the preparation of an antibody that catalysed the Diels-Alder reaction of tetrachlorothiophene dioxide **3** and *N*-ethylmaleimide. [20] Again, catalysis results simply from productive binding (an EM of > 110 M is estimated): turnover depends on the instability of the initial adduct **4**, which loses SO_2 very rapidly to give the aromatic product **5**. This avoids product inhibition, which is a common problem with such potential catalytic systems: the hapten **6** is a reasonable transition state analogue, but geometrically very different from the final product.

The Diels-Alder reaction can also be catalysed by simple artificial systems. A recent example is the reversible reaction between **7** and **8**, which is accelerated by (stoicheiometric amounts of) a cyclic zinc-porphyrin trimer host which binds pyridine derivatives inside the cavity. [22] The product is the *exo*-adduct, produced up to 1000 times faster than the cor-

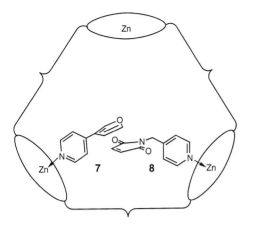

7 **8**

responding *endo*-isomer (which is obtained as the kinetic product in the absence of the macrocycle). This corresponds to an EM of about 20 M. The system is not catalytic because of product inhibition (as was an early example using the cavity of a cyclodextrin as host [23]).

Enzyme Mimics Showing Binding *and* Catalysis

Many of the most successful enzyme mimics have involved functionalised cyclodextrins, and the work of Breslow in particular is familiar to anyone who has followed the field. [24] These hosts bind aromatic rings within a hydrophobic cavity. In another seminal contribution Lehn [25, 26] has used polyammonium macrocycles to catalyse phosphate transfer reactions of ATP, demonstrating that multiple hydrogen-bonds can also be an effective source of binding between flexible systems in aqueous solution.

A different approach has been reported by Benner and his group, [26] who based the design of a synthetic decarboxylase on the known properties of proteins and the mechanism of amine-catalysed decarboxylation of β-ketoacids. Their enzyme mimics are 14-

residue peptides (called oxaldie 1 and 2) based on leucine and lysine, in a sequence known to favour α-helix formation, and thus expected to adopt protein-like conformations, with a hydrophobic core and a hydrophilic exterior. They catalyse the decarboxylation of oxaloacetate by the expected mechanism shown, with the cationic lysine side-chains presumably involved in binding the two carboxylates of both substrate and transition state (acetoacetate, with a single CO_2^-, is not a substrate). Michaelis-Menten kinetics are observed and imine formation is 10^{3-4} times faster than with simple amine catalysts. And activity does indeed seem to depend on the degree of α-helix formation.

The simplest, and one of the most remarkable, new enzyme mimics has emerged from a piece of 'lateral thinking' by Menger and Fei. [9] No synthesis is involved. These authors simply mixed long-chain carboxylic acids, amines, alcohols and alkylimidazoles, of the sort known to form aggregates, and eventually micelles, in aqueous solution: then screened large numbers of such mixtures for catalytic activity. The test reaction was the hydrolysis of the reactive ester **9** (X = O), which is easily followed above pH 7 by the release of the *p*-nitrophenolate chromophore. Some of the mixtures used effected the hydrolysis of **9** (X = O) at rates too fast to measure manually. Remarkably this was also true in the presence of a single component when this was the hexadecanoate anion, and this system also effects the hydrolysis of the p-nitroanilide (**9**, X = NH).

This is an activated amide, but one that is not hydrolysed detectably in the presence of 0.2 M acetate at pH 7 (25 °C). But under similar conditions, in the presence of just

$C_{12}H_{25}$ — N$^+$(Me)(Me) — CH_2 — C(=O) — X — $C_6H_4NO_2$

9

10

2×10^{-5} M hexadecanoate its half-life is only 3 minutes.

Only nucleophilic catalysis could account for such an efficient process. [10] No EM can be calculated as no data are available for a suitable comparison, but hexadecanoate is at least 10^8 times more effective than acetate. Interestingly the reaction is stoicheiometric: though a mixed anhydride is almost certainly an intermediate its hydrolysis must be rate determining in the overall hydrolysis of the anilide. The authors suggest that the reaction takes place in sub-micellar aggregates or 'clumps,' in which hydrophobic association of the long alkyl chains brings anilide C=O and hexadecanoate CO_2^- groups into close and remarkably productive proximity (see 10). (Simply adding three methylene groups to the substrate [9, X = $(CH_2)_3NH$] eliminates the observed reaction.)

Two points are of special interest here: the high reactivity, which is much greater than previously observed for such apparently loosely-associated systems: and the principle, of screening large numbers of simple systems, rather than actually synthesising complex, carefully designed ones. The newer approach supplements existing ways of thinking, and practical applications could result.

References

[1] D. R. Corey, M. A. Philips, *Proc. Natl. Acad. Sci. U.S.* **1994**, *91*, 4106–4109.

[2] J. A. Wells, W. J. Fairbrother, J. Otlewski, M. Lagowski, J. Burnier, *Proc. Natl. Acad. Sci. U.S.* **1994**, *91*, 4110–4114.

[3] B. W. Matthews, C. S. Craik, H. Neurath, *Proc. Natl. Acad. Sci. U.S.* **1994**, *91*, 4103–4105.

[4] M. J. Corey and E. Corey, *Proc. Natl. Acad. Sci. U.S.* **1996**, *93*, 11428–11434.

[5] A. J. Kirby, *Angew. Chem., Intl. Ed. Engl.,* **1996**, *35*, 707–724.

[6] W. K. Fife and S. Liu, *Angew. Chem., Intl. Ed. Engl.,* **1995**, *34*, 2718–20.

[7] Fersht, A. R., *Enzyme Structure and Mechanism*, second edition, W. H. Freeman, New York, **1985**.

[8] M. Z. Atassi and T. Manshouri, *Proc. Natl. Acad. Sci. U.S.*, **1993**, *90*, 8282–8286.

[9] F. Menger and Z. X. Fei, *Angew. Chem., Intl. Ed. Engl.,* **1994**, *33*, 346–348.

[10] A. J. Kirby, Effective molarities for intramolecular reactions. *Adv. Phys. Org. Chem.,* **1980**, *17*, 183–278.

[11] W. J. Albery and J. R. Knowles, *Biochemistry,* **1976**, *15*, 5631.

[12] F. M. Menger and M. Ladika, *J. Am. Chem. Soc.,* **1988**, *110*, 6794.

[13] K. N. Dalby, A. J. Kirby and F. Hollfelder, *J. Chem. Soc., Perkin Trans. 2,* **1993**, 1269.

[14] Host-Guest Interactions: from Chemistry to Biology, D. J. Chadwick and K. Widdows, eds., *CIBA Foundation Symposium No. 158,* Wiley, 1991.

[15] The Chemistry of Biological Molecular Recognition, A. J. Kirby and D. H. Williams, eds., *Phil Trans. Roy. Soc. Lond. A*, **1993**, *345*, pp. 1–164.

[16] H.-J. Schneider, *Angew. Chem., Intl. Ed. Engl.*, **1993**, *32*, 848.

[17] R. J. Pieters and J. Rebek, *Rec. Trav. Chim.*, **1993**, *112*, 330. I. Chao and F. Diederich, *Rec. Trav. Chim.*, **1993**, *112*, 335.

[18] P. G. Schultz, *Angewandte Chemie*, **1989**, *28*, 1283.

[19] U. K. Pandit, *Rec. Trav. Chim.*, **1993**, *112*, 431.

[20] D. Hilvert, K. W Hill, K. D. Narel & M.-T. M. Auditor, *J. Am. Chem. Soc.*, **1989**, *111*, 9261. See also reference 15.

[21] A. C. Braisted and P. G. Schultz, *J. Am. Chem. Soc.*, **1992**, *112*, 7431.

[22] C. J. Walter, H. L. Anderson and J. K. M. Sanders, *J. Chem. Soc., Chem. Commun.*, **1993**, 458. See this paper for references to recent related work involving reactions between two bound molecules.

[23] D. Rideout and R. Breslow, *J. Am. Chem. Soc.*, **1980**, *102*, 7816.

[24] For recent references and a review see R. Breslow, P. J. Duggan and J. P. Light, *J. Am. Chem. Soc.*, **1992**, *114*, 3982; and R. Breslow, reference 8, p. 115.

[25] M. W. Hosseini, J.-M. Lehn, K. C. Jones, K. E. Plute, K. B. Mertes and M. P. Mertes, *J. Am. Chem. Soc.*, **1989**, *111*, 6330–6335. M. P. Mertes and K. B. Mertes, *Accts. Chem. Res.*, **1990**, *23*, 413–418.

[26] K. Johnsson, R. K. Allemann, H. Widmer and S. Benner, *Nature (London)*, **1993**, *365*, 530–532.

Metal-Assisted Peptide Organization: From Coordination Chemistry to De Novo Metalloproteins

Heinz-Bernhard Kraatz

Contemporary research in chemistry often crosses the lines that previously divided the classical disciplines of chemistry and biology. Coordination chemistry has certainly been one of the areas of chemistry that has spanned the old boundaries. An inclusive definition of what constitutes coordination chemistry is emerging which requires "only that distinct molecular species be formed by the binding interaction." [1] The modeling of active sites of metalloenzymes has always been a great stimulus to coordination chemists. An excellent example for this is the "nitrogenase problem", which has played a pivotal role in the rapid development of the coordination chemistry of molybdenum and sulfur-based ligands. [2] Recent advances in the *de novo* design of nonnatural proteins have allowed the combination of classical coordination chemistry and protein biochemistry. The main idea of the design is to reduce the complexity of naturally occurring proteins to a set of minimal structural features necessary for a certain function. The model protein is prepared by standard synthetic methodologies, and can be used to evaluate the structural assumptions that led to its design and thereby lead to an understanding of the intricate interplay between protein structure and function. Using this general approach, research in this area has led to the successful design of a range of artificial proteins and protein mimics. [3]

Two major routes to the structural design of a *de novo* protein can be distinguished: (a) the synthesis of a larger peptide with subdomains of known secondary structure, which will adopt a particular tertiary structure, and (b) the design of smaller peptide subunits of known secondary structure and their assembly on a template to yield the final protein of a particular tertiary structure, such as described by Mutter et al. [4] The construction of specific and topologically predetermined protein structures from smaller amphiphilic peptide precursors by simple self-assembly is hampered by the tendency of peptides to establish an equilibrium between monomers and aggregates in aqueous solution. This makes controlling the number of participating peptides and their relative stereochemistry a formidable task indeed. There are however examples in the literature where control over the number the participating peptide subunits can be obtained, such as in Ghadiri's peptide nanotubes [5], which is are formed by a self-assembly process starting from cyclic peptide subunits.

In an novel approach, a metal ion mediated self-assembly process [6] is used in the spirit of the new inclusive definition of coordination chemistry. Peptides or amino acids can be covalently linked to a metal-binding ligand.

The coordination of chemically modified peptides to a metal center through the ligating group leads to the formation of a metal–peptide complex, thereby allowing control of the stereochemistry of the complex and the number of peptide subunits participating in the final de novo protein. To a large extent, the art of successfully applying this strategy lies in the choice of the metal center and the chemical modification of a specific peptide. This surprisingly simple approach has been applied to the synthesis of several helical metalloproteins with well-defined stereochemistry such as **1** (py = pyridine) [6a] and **2**. [6b]

***trans*-[RuCl$_2$(py-peptide)$_4$]** **1**

***cis*-[Ru(NH$_3$)$_4$(His-peptide)$_2$]$_3^+$** **2**

The final coordination geometry and stereochemistry of the complex is determined by the particular electronic and steric requirements of the metal atom and its ligands. In addition, interactions of the peptide subunits with each other and with the metal atom are of paramount importance in determining the overall structure of the complex. This makes it possible to create small artificial metallopeptides and metalloproteins with well-defined and stable structural peptide motifs, such as the β-sheet structure and helical bundle proteins consisting of helical peptide subunits. Recently a number of *de novo* metallo–peptide and protein designs based on coordination chemistry have appeared in the literature. [7] The general aim is to utilize metal coordination to a ligating part of the peptide to influence the overall structure of the peptide. Recent examples by Kelly and Fairlie demonstrate this simple design strategy.

Using a bipyridine-based template, Kelly and coworkers were able to stabilize a β-sheet like structure by the coordination of the peptide-modified bipyridine (bpy) to a Cu(II) center. [8] The 6,6′-bis-(acylamino)-2,2′-bipyridine-based peptide **3** is designed to promote a β-sheet structure upon Cu(II) binding. This template was chosen because the distance between the chains in the cisoid conformation is ca. 5.2 Å, which is similar to the distance between the two strands in β-sheet structures. In the uncoordinated state, the peptide-modified bpy adopts a *transoid* configuration to minimize steric interactions. The *cisoid* conformation has to be adopted in order for efficient Cu(II) binding (**4**) and hence the peptide residues will be forced in a spacial arrangement favoring β-sheet formation (Fig. 1).

The coordination geometry around the copper atom in **4** is similar to simple Cu complexes of 6,6′-bis-(acylamino)-2,2′-bipyridine amino acids in that the Cu(II) is coordinated by the two bpy nitrogen atoms and the amide oxygens. Amide oxygen coordination to Cu(II) is observed also in Fairlie's complex. In their work, Fairlie and coworkers have

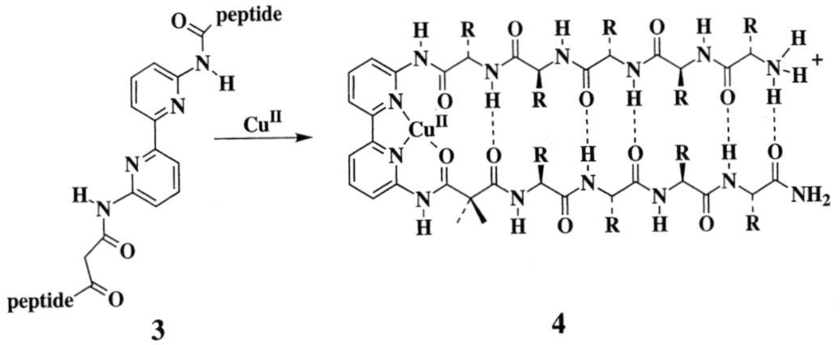

3 **4**

Figure 1. Conformational changes mediated by Cu(II) coordination to the bipyridine group of the *transoid* 6,6′-bis-(acylamino)-2,2′-bipyridine-peptide **3** to complex **4** having a *cisoid* conformation; R = α-amino acid side chain (adapted from [8]).

used the tridentate macrocycle 1,4,7-triaza-cyclononane modified by phenyl alanine methylester (**5**) and prepared a Cu(II) complex **6** (Fig. 2). [9] Like its unsubstituted analogue, **6** coordinates to the copper center through the three nitrogen atoms. The X-ray crystallographic study clearly shows that the additional coordination of the amide oxygens to the copper atom forces the amino acid residues to one face of the molecule. Attachment of helical amphiphilic peptide chains to this metallo-template are very likely to form bundle-like structures, similar to **1** and **2**. [6a, 6c]

Ghadiri and coworkers have refined this idea to such an extent that it has been possible to synthesize a Ru^{II} metalloprotein with a well-defined metal-binding site. [10] This was achieved with peptide **7**, which had two potential metal-binding sites – an N-terminal bpy and an imidazole nitrogen atom of a histidine residue close to the C-terminus of **7** – and should favor a helical structure. Incorporation of the bpy functionality at the N-terminus of each peptide allowed direction of the self-assembly of three peptides with a Ru^{II} ion as

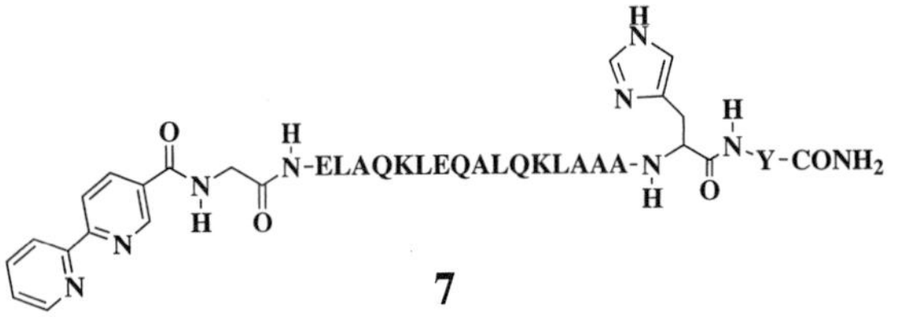

a template to form a three-helix bundle protein. Kinetically inert **8** is produced exclusively (Fig. 3). The higher kinetic stability of Ru(bpy)$_3{}^{2+}$ compared to Ru(imH)$_6{}^{2+}$ is seen as the determining factor for the formation of **8**. This leaves the three His18-residues unligated.

Since three peptide subunits are in close proximity, short-range intramolecular hydrophobic interactions amongst the three metal-bound peptides exceed intermolecular interactions which thereby facilitates the formation of the overall helical tertiary structure of **8** (circular dichroism (CD) spectrum: θ_{222} = $-23\,000\,^{\circ}\,cm^2\,dmol^{-1}$). UV/VIS spectroscopy ($\lambda_{max}$ = 289 and 475 nm), electrospray mass specrometry (found: M = 6887 ± 2; calcd.: M = 6887), and gel permeation chromatography all provide unequivocal evidence for the identity of **8**. It is important to point out that because of this helical arrangement, the three His18-residues close to the C-termini of each of the three peptide subunits are close together, thereby forming an effective metal binding site. Coordination of Cu^{2+} to this binding site does not significantly alter the structure of the protein, which indicates the presence of this binding after the self-assembly of **8** (CD spectrum: θ_{222} = $-24\,000\,^{\circ}\,cm^2\,dmol^{-1}$). Based on the marked decrease in the fluorescence emisson of the adjacent tryrosine chromophore and analysis of the EPR (g$_{||}$= 2.27 and A$_{||}$ = 173 cm^{-1} x 10^4) and UV/VIS (λ_{max} = 375 (charge transfer) and 495 nm) spectra,

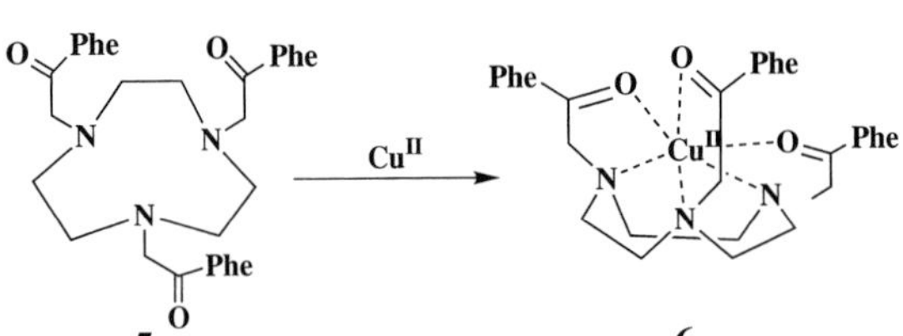

7

fac-[RuII(bpy-peptide)$_3$]$_2{}^+$ **8**

5   **6**

Figure 2. Coordination of Cu(II) to the amino acid N-functionalized 1,4,7-triazacyclononane **5** (adapted from [9]).

Ghaderi and Case [10] demonstrated that Cu^{II} is able to bind to this metal-binding site and leads to the formation of the Ru^{II}Cu^{II} protein 9 (Fig. 3). Cu^{II} binding increases the overall stability of the protein towards denaturation by 1.5 Kcal mol^{-1}.

Relevant in this context are the recent reports of the design and syntheses of cytochrome (cyt) models by Diederich et al. [11] based on polyether-amide dendrites encapsulating a central Zn porphyrin core, and the approach by DeGrado and Dutton's for the synthesis of artificial proteins. [12] In DeGrado and Dutton's approach, several peptides have been synthesized consiting of two identical 31-residue subunits linked by a disulfide bridge. The peptides contain histidine residues which allow the coordination of Fe^{II} heme through two *trans* imidazole groups. The peptides associate to dimeric four-helix bundle structures. Model protein 11 was designed from the two identical subunits 10 (by oxidation of the cystein sulfhydryl groups), which each possess two histidine units (His10 and His24), to mimic the cyt b subunit of cyt bc$_1$.

Compound 11 has two possible heme binding sites per peptide (based on two histidine residues) and an estimated separation between the Fe atoms of about 20 Å. Due to dimerization of the model protein 11, addition of Fe^{II} heme leads to the incorporation of four redox centers at the four predetermined binding sites of the protein dimer. CD spectroscopy also confirms in tis case that well-defined binding sites are present prior to heme coordination ($\theta_{222} = -26\,000\,°\,cm^2\,dmol^{-1}$). As expected the heme-protein complex exhibits a rhombic EPR spectrum ($g_z = 2.89$, $g_y = 2.24$, $g_x = 1.54$) confirming the coordination of low-spin Fe^{2+} through two *trans* imidazole groups.

These are only a few examples indicating the scope of this emerging area of research at the interface of peptide and protein chemistry and coordination chemistry. The synthesis of metal-binding metalloproteins and the introduction of redox-active metal centers into artificial proteins by self-assembly processes is surely only the beginning. One can certainly think of extending this approach to the incorporation of substrate-binding enzymatic func-

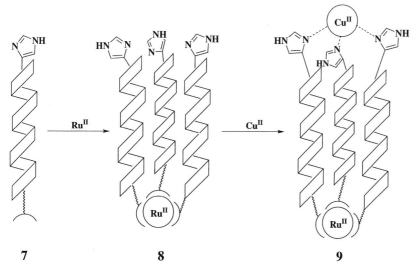

Figure 3. Assembly of Ru(II) metallopeptide 8 from bpy functionalized peptide 7. 8 possess a well-formed His$_3$ metal-binding site to which Cu(II) binds to form 9.

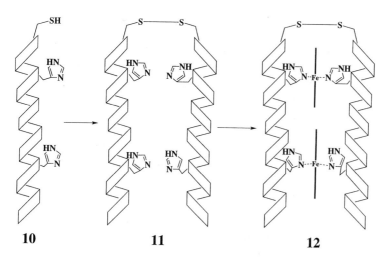

10   **11**   **12**

Figure 4. Formation of a Fe(II) heme binding protein **11** by dimerization of helical peptide subunits **10**.

tions, as already demonstrated by Sasaki and Kaiser, [13] and photosensitive metal centers in proteins which can then act as light-harvesting analogues of the photosystem of green plants. The rational synthesis of enzyme analogues allowing efficient catalytic transformations of industrial relevance is perhaps one of the key driving forces. In addition, *de novo* design of metalloproteins allows for a systematic study of several key biological functions, such as electron transfer [14] and other biological transformations and allows to develope an understanding of the intricate interplay between protein structure and its function.

References

[1] D. H. Busch *Chem. Rev.* **1993**, *93*, 847–860.

[2] a) D. Sellmann and J. Sutter in: Transition Metal Sulfur Chemistry (Eds.: E. I. Stiefel, and K. Matsumoto) pp. 101–116, ACS Symposia Series 653, Washington D.C. 1996; b) D. Coucouvanis, K. D. Demadis, S. M. Malinak, P. E. Mosier, M. A. Tyson, L. J. Laughlin,

ibid. pp. 117–134; c) Ian Dance, *ibid*, pp. 135–152.

[3] a) W. F. DeGrado, Z. R. Wasserman, J. D. Lear *Science (Washington D.C. 1883)* **1989**, *243*, 622–628; b) K. T. O'Neil, R. H. Hoess, W. F. DeGrado *ibid.* **1989**, *243*, 622–628; c) K. T. O'Neil, W. F. DeGrado *Trends Biochem. Sci.* **1990**, *15*, 59–64; d) K. S. Åkerfeldt, J. D. Lear, Z. R. Wasserman, L. A. Chung, W. F. DeGrado *Acc. Chem. Res.* **1993**, *26*, 191–197; e) K. W. Hahn, W. A. Klis and J. M. Stewart *Science (Washington D.C. 1883)* **1990**, *248*, 1544–1547; f) A. Grove, M. Mutter, J. E. Rivier, M. Montal *J. Am. Chem. Soc.* **1993**, *115*, 5919–5924; g) M. R. Ghadiri, J. R. Granja, L. K. Buehler *Nature (London)* **1990**, *369*, 301–304; h) N. Voyer, M. Robitaille *J. Am. Chem. Soc.* **1995**, *117*, 6599–6600.

[4] a) M. Mutter, S. Vuilleumier *Angew. Chem.* **1989**, *101*, 551–571; b) M. Mutter, G. G. Tuchscherer, C. Miller, K.-H. Altman, R. I. Carey, D. F. Wyss, A. M. Labhardt, J. E. Rivier *J. Am. Chem. Soc.* **1992**, *14*, 1463–1470.

[5] a) M. R. Ghadiri, K. Kobayashi, J. R. Granja, R. K. Chadha, D. W. McRee *Angew. Chem.* **1995**, *107*, 76; *Angew. Chem. Int. Ed. Engl.* **1995**, *34*, 93–95; b) M. Engels, D. Bashford, M. R. Ghadiri *J. Am. Chem. Soc.* **1995**, *117*, 9151–9158.

[6] a) M. R. Ghadiri, A. M. Fernholz *J. Am. Chem. Soc.* **1990**, *112*, 9633–9635; b) M. R. Ghadiri, C. Soares, C. Choi *ibid*, **1992**, *114*, 825–831; c) *ibid.* **1992**, *114*, 4000–4002.

[7] J. P. Schneider, J. W. Kelly *Chem. Rev.* **1995**, *95*, 2169–2187.

[8] J. P. Schneider, J. W. Kelly *J. Am. Chem. Soc.* **1995**, *117*, 2533–2546.

[9] A. A. Watson, A. C. Willis, D. P. Fairlie *Inorg. Chem.* **1997**, *36*, 752–753.

[10] M. R. Ghadiri, M. A. Case *Angew. Chem.* **1993**, *105*, 1663–1667; *Angew. Chem. Int. Ed. Engl.* **1993**, *32*, 1594–1597.

[11] P. J. Dandliker, F. Diederich, M. Gross, C. B. Knobler, A. Louati, E. M. Sandford *Angew. Chem.* **1994**, *106*, 1821–1824; *Angew. Chem. Int. Ed. Engl.* **1994**, *33*, 1739–1742.

[12] D. E. Robertson, R. S. Farid, C. C. Moser, J. L. Urbauer, S. E. Mulholland, R. Pidikiti, J. D. Lear, W. F. DeGrado, P. L. Dutton *Nature (London)* **1994**, *368*, 425–432.

[13] T. Sasaki, E. T. Kaiser *J. Am. Chem. Soc.* **1989**, *111*, 380–381.

[14] H.-B. Kraatz, J. Lusztyk, G. D. Enright *Inorg. Chem.* **1997**, *36*, 2400–2405.

Artificial Replication Systems

Siegfried Hoffmann

Introduction

"It has not escaped our notice that the specific pairing we have postulated immediately suggests a possible copying mechanism for the genetic material" – rarely has anything fascinated science as much as Watson and Crick's "holy" structure of DNA [1a, b], in which the prophecy of Schrödinger for an "aperiodic crystal" [1c] to be the genetic material [1d, e] became reality.

The discovery of the DNA-structure altered our view of life. The prototype of a matrix became the beacon that enlighted research into the fields of molecular biology and redirected organic chemistry back to its early native pretension. The reduplication of the holy structure rendered the key mechanisms for prebiotic systems, when passing the borderline from inanimated to animated world.

Molecular Matrices

Todd sought to answer for the outruled chemistry in a landmark appeal: "The use of one molecule as a template to guide and facilitate the synthesis of another ... has not hitherto been attempted in laboratory synthesis, although it seems probable that it is common in living systems. It represents a challenge, which must, and surely can, be met by Organic Chemistry" [2a].

And the grand concept appeared to be portentous. Quite soon Schramm et al. [2b] aroused enthusiasm with the "nonenzymatic synthesis of nucleosides and nucleic acids, and the origin of self-replicating systems": the polycondensation of uridylic acid onto the orientating $(A)_n$-matrix – and the first attempt at an artificial matrix reaction [2]. Was this the breakthrough that would lead chemistry not only into the wonderland of biology, but also rapidly to ordered and instructed macromolecular organizations in its own dominion [2c–f]?

Schramm's dramatic experiments had to push the native standards back into the high error phase of a prebiotic beginning. Purely artificial template experiments [2c–e], however, were devoid of even this possible relationship to evolutionary development [3]. The first great "chemical" departure into the temptations of such matrix reactions ran aground on innumerable difficulties. The dominant element in scientific progress came to be not the variety of chemical matrices and the distant aim of building systems capable of self-replication. Rather, fascination with the elegance of the natural prototypes stimulated the vanguards of chemistry and biology in their campaign of molecular penetration of biological systems. The golden age of molecular biology had begun [2f, 3].

Oparin's [3a] and Haldane's [3b] heirs, Eigen [3d, e] and Kuhn [3f], gave these events a time perspective, and with their "information" also described the vector of the "Grand Process". Self-replication, mutation, and metabolism (as prerequisites for selection) made up the list of criteria; through these, information and its origin, evaluation, processing, and optimization had governed the evolutionary history of prebiotic and biotic systems [2f, 3].

Self-Reproduction Models

In connection with Spiegelman's [4a] initial experiments, Eigen and Schuster [3d, e], Joyce [3g, 4b], and others attempted to "bring to life" the theoretical premises behind the laboratory realities of enzyme-catalyzed RNA replication and evolution experiments, an approach later on forwarded into the rapidly broadening creative fields of artificially directed evolution [4]. Orgel's groups [3h, 4c], by contrast, exerted themselves to transform diverse matrix relationships into artificial, enzyme-free nucleic acid formation. When Altman [4d] and Cech [4e] raised RNA to the throne of an archaic informational and functional omnipotence, it amounted to a late justification of the toil and trouble of this "nucleic acids first" route. "A tRNA looks like a nucleic acid doing the job of a protein", Crick once observed. Now, self-splicing RNA complexes, polymerase activities, ribozymes and hypothetical RNA-somes afforded unusual insights into the genotypical and(!) phenotypical complex behavior of a single nucleic acid species. And with the supposedly greater understanding of this "RNA-world", fresh impetus was given even to the purely chemical approaches, once so promising and

now almost forgotten. A "Kuhn" period of divergent evolution of matrix systems between molecular biology and classical chemistry had begun.

"Minimal Self-Replicating Systems"

A rapidly growing number of artificial self-reproduction [2, 3l, 5] models (Fig. 1), covering nucleic acids [5a–c] as well as their near and distant analogs [5a–d], but also followers of nearly forgotten Fox-microspheres [3c] in the field of peptide [5e] and membrane [5f, g] components, endeavor to gain insights into the transition stages between chemical and biological evolution.

After innumerable attempts to elucidate the matrix relationships of mono- and oligo-nucleotides to orientating and catalytic oligo- and polynucleotide templates, two self-replicating nucleic acid models, a DNA-analogous hexamer system by von Kiedrowski et al. [6a, b] and an RNA-analogous tetrameric assembly from the Orgel group [6c], opened new ways. Von Kiedrowski's first successful model-hexadeoxyribonucleotide duplex became a leitmotif in the detailed treatment of growth kinetics and anticipated later self-replicating "minimal systems" in many ways [6]. The ligation of cooperative oligonucleotides increased the stability of the matrix duplex and the ternary formation complex. The choice of palindromic systems reduced the complexity of biological replication experiments to a simplified kinetic measurement of its identical (since self-complementary) matrix components, demonstrating self-replication by autocatalytic behavior. The surprising square root law of matrix growth kinetics [6a, b, d–h] (curiously even computer viruses seem to be in love with it) with its ideal case of a parabolic reaction course – derived from the hexadeoxynucleotide duplex, confirmed slightly later by the Zielinski/Orgel system [6c], and, finally, also verified in the Rebek-

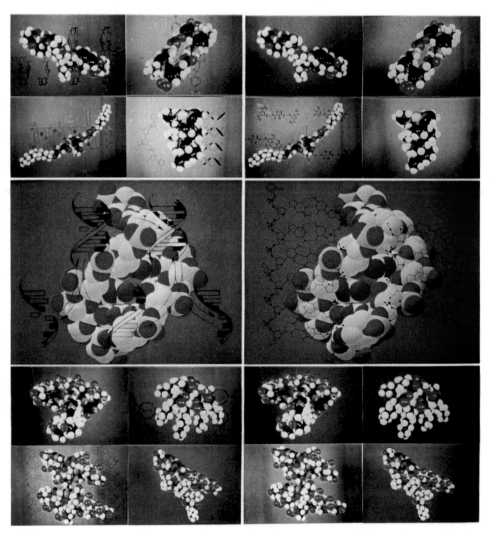

Figure 1. Artificial replicators and matrices (left to right and top to bottom): Rebek's distant nucleoside [6k–m] and v. Kiedrowski's amidinium-carboxylate [7f] replicators; Lehn's chiral organic [7g] and elementorganic [7h] helicates; v. Kiedrowski's [6a, b, d–g] and Orgel's [6c] "minimal self-replication systems"; Hoffmann's alienated nucleic acids [2f, 5c, 7b, 10e] and Luisi's "minima-vita" approach [5f, g, 8a–d]; Eschenmoser's hexose-nucleic-acids [15], compared with Olson-type R/DNAs and hypothetical nucleation dynamics of nucleoprotein systems [2f, 3k, 5c].

dimer assembly [5d, 6i–l] – was recognized as autocatalytic system behavior of self-replicating oligonucleotide templates under the constraints of isothermal conditions, where stability relationships of matrix reaction partners might exclude the expected exponential growth kinetics. A hexameric system, constructed from two trimeric blocks by phosphoamidate coupling, revealed sigmoidal growth behavior and extraordinary autocatalytic efficiency [6d–g]. Continuations in more complex matrix block offers enriched the informational content of the model systems and invited for the first time studies of selection behavior [5d, 6].

The patterns of autocatalysis with respect to parabolic and exponential reaction courses, that closely affect the conclusions of Eigen's evolution experiments concerning the decision criteria for mutant selection and coexistence [3d, 1, 6h], can by now be derived from the thermodynamic and kinetic data of the matrix partners and offer quite new views with autocatalytic cooperation between competitive species [5b, 6g, h]. Separate from "enzyme-catalyzed" evolution experiments with RNA- and DNA-systems, basic questions of prebiotic behavior can for the first time become the object of detailed experimental research. While continuing their studies on complex autocatalysis patterns, von Kiedrowski et al. diagnosed modulation of molecular recognition as an operational deficit of earlier artificial self-replicational nucleic acid systems with regard to exponential reaction courses and identified it as an ideal aim for future models [6f]. On its way to the nucleoprotein system, evolution must have had a similar view of the problem, when it endowed nucleic acids with proteins, experienced in phase and domain regulation strategies, and thus achieved an ideal milieu for directing modulations of recognition. While, presumably, the both informational and functional RNAs allowed for first successful self-replications, they seem to have been outclassed in future developments by the cooperative efforts of nucleic acids and proteins. The urgent demands to establish suitable and reliable regimes of strand-recognition, annealing, separation and reannealing – so far having only been brought about by drastic variations of reaction temperature – seem to have been accounted for under stringent isothermal conditions only by the complexity patterns of nucleoprotein systems. It had been the impressive, both functional and informational potencies of proteins that could like some "dei ex

machina" provide, by a permanently renewing coherency continuation of transiently acting complex order-disorder patterns along desirable trajectory bundles, suitable system-inherent isothermal conditions. Only the integrative efforts of nucleic acids and proteins, that submerged their structural individualities into the biomesogenic unifications of functionally and informationally completely new characteristics of dynamic nucleoprotein systems, reached the evolutionary breakthrough in self-organizational, self-replicational, and general information-processing abilities [2f, 3l, 5c].

Matrix growth kinetics as known for oligonucleotides are followed even by a drastically abstracted artificial replication system developed by Rebek's group [5d, 6i–l]. The dimer assembly of a distant peptide-nucleic-acid analog developed from host-guest relationships combined the interactive and cognitive possibilities of native nucleic acid matrices with the more general biopolymer relationships of the amide bond formation involved in the matrix reaction. When Rebek et al., who have in the meantime sought to stimulate the competition and selection behavior of their self-replicating species by "chemical mutations," stated that one of their main aims is artificial peptide synthesis on a nucleic acid matrix [5d, 6, i–l], it is slightly reminiscent of Todd's template vision [2a] in the very beginning.

The dualities of supramolecular chemistry [7a] and biomesogen systems [7b] generate a tremendous number of new matrix and replication variations [7]. 2-Pyridone versions, already foreseen in mesogenic mono- and polymeric nucleic acid analog approaches [7b–d], reoccur in new appearances [7e]. In the attempt to reduce generalizing principles as far as possible, desirable operation modes astoundingly become accessible. Dramatically abstracted amidinium-carboxylate systems, which, nevertheless, cover certain essentials of complex nucleic acid-protein interactions, prove to be susceptible to molecular recognition modulation and seem even to delight their examiners with exponential growth kinetics [7f]. The beauties of some sort of chiral main-chain LC-polymers – built up as impressive supramolecular helical arrangements from bifunctional recognition units – convincingly confirm the selection and discrimination facilities of supramolecular organizations in transitions from hetero- to homochirality [7g]. Detailed aspects, such as the possibilities of coordination matrices in native nucleic acid assemblies, make themselves independent in helicates [7h–l], whose structures reflect also relationships between our life process and its basic matrices [2f]. Matrix studies of complex duplex-triplex systems model regulation strategies of nucleic acid organization and by this detect not only protein-like behavior of RNA-Hoogsteen strands in reading informational DNA-duplex patterns, but bridge also the gap to basic hysteretic mechanisms of information processings in highly condensed systems [2f, 5c, 7b, m–q].

"Minima Vita Models"

It is fully within this context when, in addition to the nucleic acid/nucleic acid analog pioneering triad of self-replication systems, a fourth forwarding approach adds to the "minimal systems" of preferentially informational replicators the new view of a "minima vita" challenge [3l, 5f, g]. It appears somehow as a reincarnation of Fox's microspheres [5d], when micelles advance in the hands of Luisi et al. [5f, g, 8] as first examples of "minimal life" models, where the chemical autopoiesis is taken as a minimum criterium for not only self-reproductive, but, moreover, in some way life-bearing systems. An intriguing approach, forwarded to stages where core [8e] and shell [8a–d] reproduction has been achieved by spherically bounded micelle systems that host inside replicative nucleic acids – demonstrating by this "minimal-cell-models" [3l, 5f, g, 8e].

Playing the Game of Artificial Evolution

And then there is, finally, the "evolution" of an individual scientific life's work [3i, 9], which itself follows decisive stages of the Grand Process: the exploration of early chemical requirements, the development of prebiotic ligand systems, the fixation into the ordered structures of informational inorganic matrix patterns and, finally, the liberation of their inherent wealth of design and information into the order-disorder dialectics of today's nucleoprotein system. Using basic hexoses – somewhat the successor molecules of evolution – a never attempted, or perhaps only forgotten, "evolutionary step" is now taking place, once again in the area of replicative (homo)nucleic acid systems [9a–d].

But this is another whole story – and is another great game. A game that is representative in all its loveful utilizations and impressive manifestations of today's chemistry standards and facilities for our future ways of modelling of what has created us – without any chance, however, to renew artificially the whole on our own.

It is, indeed, just this native complexity which for our today's chemistry provides provocation and stimulation, intimidation and temptation, love and hate and fate together. The present artificial systems still remain utterly outclassed by even the most primitive life forms such as RNA-viruses. The possibilities of describing natural selection behavior according to quasi-species distributions in the extreme multidimensionalities of sequence spaces [3d, e] are, for artificial systems, at best a very distant utopia. With all its early primitivity, but also with its promising inherent potential of "minimal models" of self-replication [6, 7, 9] and – just to follow – "minima vita models" [8], chemistry, nevertheless, is gaining new qualities by retracing transitions to life.

References

[1] Replicative double helix: a) J. D. Watson, F. H. C. Crick, *Nature* **1953**, *171*, 737; b) F. H. C. Crick, J. D. Watson, *Proc. R. Soc. London [Ser. A]* **1954**, *223*, 80; c) E. Schrödinger, *What is life?*, Cambridge Univ. Press, New York, **1944**; d) F. Miescher, "On the chemical composition of pyocytes", *Hoppe-Seyler's med. Untersuchungen,* **1871**; *Die histochemischen und physiologischen Arbeiten* (Ed.: F. C. W. Vogel), Leipzig, **1897**; e) R. Altmann, *Arch. Anat. Phys. Phys. Abt.* **1889**; *Die Elementarorganismen,* Veit, Leipzig, **1890**.

[2] Molecular matrices: a) A. Todd, in *Perspectives in Organic Chemistry* (Ed.: A. Todd), Interscience, New York, **1956**, p. 245; b) G. Schramm, H. Grötsch, W, Pollmann, *Angew. Chem.* **1962**, *74*, 53; *Angew. Chem. Int. Ed. Engl.* **1962**, *1*, 1 c) W. Kern, H. Kämmerer, *Chem. Ztg.* **1967**, *91*, 73; d) H. Kämmerer, *ibid.* **1972**, *96*, 7; e) J. H. Winter, *Angew. Chem.* **1966**, *78*, 887; *Angew. Chem. Int. Ed. Engl.* **1966**, *5*, 862; f) S. Hoffmann, *Molekulare Matrizen (I Evolution, II Proteine, III Nucleinsäuren, IV Membranen)*, Akademie-Verlag, Berlin, **1978**.

[3] Evolution views: a) A. I. Oparin, *Origin of Life*, Moscow, **1924**; b) J. B. S. Haldane, *The Course of Evolution*, Longman, New York, **1932**; c) S. W. Fox (Ed.), *The Origin of Prebiological Systems and of Their Molecular Matrices*, New York, Academic Press, **1965**; *Science* **1960**, *132*, 200; d) M. Eigen, *Naturwissenschaften* **1971**, *58*, 465; *Stufen zum Leben*, Pieper, München, **1987**; *Cold Spring Harbor Symp. Quant. Biol.* **1987**, *LII*, 307; e) M. Eigen, P. Schuster, *Naturwissenschaften* **1977**, *64*, 541; *ibid.* **1978**, *65*, 7; f) H. Kuhn, *Angew. Chem.* **1972**, *84*, 838; *Angew. Chem. Int. Ed. Engl.* **1972**, *11*, 798; g) G. F. Joyce, *Nature* **1989**, *338*, 217; *Cold Spring Harbor Symp. Quant. Biol.* **1987**, *LII*, 41; h) L. Miller, L. E. Orgel, *The Origins of Life on Earth*, Prentice Hall, Englewood Cliffs, New York, **1974**; i) A. Eschenmoser, in *Origins Life* **1994**, *24*, 389; k) S. Hoffmann, in *Chirality – from Weak Bosons to the α-Helix* (Ed.: R.Janoschek), Springer, Berlin-Heidelberg-New York, **1992**, p. 205; l) *Self-Production of Supramolecular Structures* (Eds.: R. Fleischaker, S. Colonna, P. L. Luisi), *NATO ASI Ser. 446*, Kluwer Acad. Publ., Dordrecht-Boston-London, **1994**; cf. also [2f].

[4] Artificially directed evolution: a) S. Spiegelman, *Quart. Rev. Biophys.* **1971**, *4*, 213; b) G. F. Joyce, *Curr. Biol.* **1996**, *6*, 965, and preceding communications (a.p.c.); in [31], 127; *Scientific American* **1992**, *269*, 48; c) L. E. Orgel, *ibid.* 9, a.p.c.; d) S. Altmann, *Angew. Chem.* **1990**, *102*, 735; *Angew. Chem. Int. Ed. Engl.* **1990**, *29*, 707; e) T. R. Cech, *ibid.* 745 and 716; cf also [3d, g, h, l].

[5] Self-reproduction models: a) L. E. Orgel, *Nature* **1992**, *358*, 203; b) D. Sievers, T. Achilles, J. Burmeister, S. Jordan, A. Terfort, G. von Kiedrowski, in [31], *45*; c) S. Hoffmann, *Angew. Chem.* **1992**, *103*, 1032; *Angew. Chem. Int. Ed. Engl.* **1992**, *31*, 1013; in [31], 3; d) J. Rebek, *Acta. Chem. Scand.* **1996**, *50*, 469, a.p.c.; *Scientific American* **1994**, *271*, 48; e) R. Ghadiri, K. Kobayashi, J. R. Granja, R. K. Chadha, D. E. McRee, *Angew. Chem.* **1995**, *107*, 76; *Angew. Chem. Int. Ed. Engl.* **1995**, *34*, 93; f) P. L. Luisi, in [31], 179; g) P. L. Luisi, P. Walde, T. Oberholzer, *Ber. Bunsenges. Phys. Chem.* **1994**, *98*, 1160; h) D. Philp, J. F. Stoddart, *Angew. Chem.* **1996**, *108*, 1242; *Angew. Chem. Int. Ed. Engl.* **1995**, *35*, 1154.

[6] "Minimal models" of self-replication: a) G. von Kiedrowski, *Angew. Chem.* **1986**, *93*, 932; *Angew. Chem. Int. Ed. Engl.* **1986**, *25*, 932; b) G. von Kiedrowski, B. Wlotzka, J. Helbing, *Angew. Chem.* **1989**, *102*, 1259; *Angew. Chem. Int. Ed. Engl.* **1989**, *28*, 1235; c) W. S. Zielinski, L. E. Orgel, *Nature* **1987**, *327*, 346; d) G. von Kiedrowski, B. Wlotzka, J. Helbing, M. Matzen, S. Jordan, *Angew. Chem.* **1991**, *103*, 456, 1066; *Angew. Chem. Int. Ed. Engl.* **1991**, *30*, 423, 892; e) G. von Kiedrowski, J. Helbing, B. Wlotzka, S. Jordan, M. Mathen, T. Achilles, D. Sievers, A. Terfort, B. C. Kahrs, *Nachr. Chem. Tech. Lab.* **1992**, *40*, 578; f) T. Achilles, G. von Kiedrowski, *Angew. Chem.* **1993**, *105*, 1225; *Angew.*

Chem. Int. Ed. Engl. **1993**, *32*, 1189; g) D. Sievers, G. von Kiedrowski, *Nature* **1994**, *369*, 221; h) E. Szathmáry, in [3l], 65; i) T. Tjivikua, P. Ballester, J. Rebek jr., *J. Am. Chem. Soc.* **1990**, *112*, 1249; k) J. Rebek jr., *Angew. Chem.* **1990**, *102*, 261; *Angew. Chem. Int. Ed. Engl.* **1990**, *29*, 245; l) T. K. Park, Q. Feng, J. Rebek jr., *J. Am. Chem. Soc.* **1992**, *114*, 4529, a.p.c.; cf. also [5a–c].

[7] Surrounding matrix and replicator patterns: a) J.-M. Lehn, *Science* **1985**, *227*, 849; b) S. Hoffmann, in: *Polymeric Liquid Crystals* (Ed.: A. Blumstein), New York, Plenum, **1985**, p. 423; *Z. Chem.* **1987**, *27*, 395; c) R. Heinz, J. P. Rabe, W.-V. Meister, S. Hoffmann, *Thin Solid Films* **1995**, *264*, 246; a.p.c.; d) S. Hoffmann, *Z. Chem.* **1979**, *19*, 241; e) F. Persico, J. D. Wuest, *J. Org. Chem.* **1993**, *58*, 95; f) A. Terfort, G. von Kiedrowski, *Angew. Chem.* **1992**, *104*, 626; *Angew. Chem. Int. Ed. Engl.* **1992**, *31*, 654; g) T. Gulik-Krzywicki, C. Fouquey, J.-M. Lehn, *Proc. Nat. Acad. Sci. USA* **1993**, *90*, 163; h) B. Hasenknopf, J.-M. Lehn, B. O. Kneisel, G. Baum, J. Femske, *Angew. Chem.* **1996**, *108*, 1987; *Angew. Chem. Int. Ed. Engl.* **1996**, *35*, 1838; a.p.c.; i) C. R. Woods, M. Benaglia, F. Cozzi, J. S. Siegel, *ibid.* 1977, 1830; a.p.c.; k) A. F. Williams, C. Piguet, G. Bernardinelli, *Angew. Chem.* **1991**, *103*, 1530; *Angew. Chem. Int. Ed. Engl.* **1991**, *30*, 1490; l) E. C. Constable, *ibid.* **1991**, *103*, 1482 and **1991**, *30*, 1450; m) K. J. Luebke, P. B. Dervan, *J. Am. Chem. Soc.* **1989**, *111*, 8733; n) E. Neumann, A. Katchalsky, *Proc. Natl. Acad. Sci. USA* **1972**, *69*, 993; o) E. Neumann, *Angew. Chem.* **1973**, *85*, 430; *Angew. Chem. Int. Ed. Engl.* **1973**, *12*, 356; p) W. Guschlbauer, *Encycl. Polymer Sci. Eng.* **1988**, *12*, 699; in *Dynamic Aspects of Conformation Changes in Biological Macromolecules* (Ed.: C. Sadron), Reidel, Dordrecht, **1973**; q) S. Hoffmann, in *2nd Swedish-German Workshop on Modern Aspects of Chemistry and Biochemistry of Nucleic Acids* (Ed.: H. Seliger), *Nucleosides & Nucleotides* **1988**, *7*, 555; cf. also [2f; 3k].

[8] "Minima-vita" models: a) P. A. Bachmann, P. Walde, P. L. Luisi, J. Lang, *J. Am. Chem. Soc.* **1990**, *112*, 8200; *J. Am. Chem. Soc.* **1991**, *113*, 8204; b) P. A. Bachmann, P. L. Luisi, J. Lang, *Nature* **1992**, *357*, 57; c) P. L. Luisi, F. J. Varela, *Origins Life* **1990**, *19*, 633; d) K. Morigaki, S. Dallavalle, P. Walde, S. Colonna, P. L. Luisi, *J. Am. Chem. Soc.* **1997**, *119*, 292, a.p.c.; e) T. Oberholzer, R. Wick, P. L. Luisi, C. K. Biebricher, *Biochem. Biophys. Res. Commun.* **1995**, *207*, 250, a.p.c.; cf. also [3l, 5f, g].

[9] Evolutionary hexose nucleic acids: a) A. Eschenmoser, *Angew. Chem.* **1988**, *100*, 5; *Angew. Chem. Int. Ed. Engl.* **1988**, *27*, 5; *Nachr. Chem. Tech. Lab.* **1991**, *39*, 795; *Nova Acta Leopold.* **1992**, *NF 67/281*, 201; b) A. Eschenmoser, M. Dobler, *Helv. Chim. Acta* **1992**, *75*, 218; c) A. Eschenmoser, M. V. Kisakurek, *Helv. Chim. Acta* **1996**, *79*, 1249; d) R. Krishnamurthy, S. Pitsch, M. Minto, C. Miculka, N. Windhab, A. Eschenmoser, *Angew. Chem.* **1996**, *108*, 1619; *Angew. Chem. Int. Ed. Engl.* **1996**, *35*, 1537, a.p.c..

[10] Supramolecular chemistry and biomesogen systems: a) J.-M. Lehn, *Angew. Chem.* **1988**, *100*, 91; Angew. Chem. Int. Ed. Engl. **1988**, *27*, 89; *ibid.* **1990**, *102*, 1347 and **1990**, *29*, 1304; b) J.-M. Lehn, *Supramolecular Chemistry*, VCH, Weinheim, **1995**; c) H. Ringsdorf, B. Schlarb, J. Venzmer, *Angew. Chem.* **1988**, *100*, 117; *Angew. Chem. Int. Ed. Engl.* **1988**, *27*, 113; d) M. Ahlers, W. Müller, A. Reichert, H. Ringsdorf, H. Venzmer, *ibid.* **1990**, *102*, 1310 resp. **1990**, *29*, 1269; e) S. Hoffmann, W. Witkowski, in *Mesomorphic Order in Polymers and Polymerization in Liquid Crystalline Media* (Ed.: A. Blumstein), *Am. Chem. Soc. Symp.-Ser.* **1978**, *74*, 178; f) S. Hoffmann, *Living systems*, in *Handbook of Liquid Crystals*, VCH, Weinheim, in press.

D. General Methods and Reagents

LiClO$_4$ and Organic Solvents – Unusual Reaction Media

Uschi Schmid and Herbert Waldmann

Many reactions can be influenced in a variety of ways by the solvent employed. This is especially the case when polarized transition states or ionic intermediates are involved and when the solvent is nucleophilic or electrophilic. A case to the contrary is the Diels-Alder reaction, which remains largely unaffected by the surrounding organic medium. In the mid-eighties, however, Breslow et al. [1] and Grieco et al. [2] demonstrated that Diels-Alder reactions proceed with increased reaction rate and with improved *endo/exo* selectivity when they are carried out not in organic solvents but in aqueous solutions. The effect is further enhanced by salts such as LiCl (salting-in effect), whereas the addition of guanidinium chloride has the opposite influence (salting-out effect). The use of water as solvent for such cycloadditions had already been described earlier by Alder et al. [3a] and later by Koch et al. [3b] The accelerating effect of this reaction medium is also manifested in many other reactions, [4] e. g. asymmetric hetero-Diels-Alder reactions [4b] and asymmetric nonhetero-Diels-Alder reactions, [4c, d] nucleophilic additions to iminium ions [4e] and carbonyl compounds, [4f] Claisen rearrangements, [4g] the benzoin condensation, [1b] and aldol reactions. [4h] It is attributed to the fact that a suitable aggregation is generated by hydrophobic interactions between the reaction partners (hydrophobic effect), and so exercises an "internal pressure" on the reactants encapsulated in "solvent cavities" whose effects are, in turn, comparable with a high external pressure, at least in the case of the Diels-Alder reaction.

A solvent system described by Grieco et al., [5] namely a 5 M solution of LiClO$_4$ in diethyl ether, has a comparable, if not greater accelerating effect on Diels-Alder reactions. Already in 1959 Winstein et al. [6] noted that dissolving LiClO$_4$ in an organic solvent greatly enhances the polarity of the solution, i. e. a solution of LiClO$_4$ in diethyl ether was found to be more polar than glacial acetic acid. This remarkable effect was used for the first time by Sauer et al. [7] to influence the steric course of a Diels-Alder reaction. Based on the finding that the cycloaddition of methacrylic acid and cyclopentadiene proceeds with 88 % *endo* selectivity in a 4 M solution of LiClO$_4$ in ether, the authors proposed that this reaction medium might be an advantageous solvent for different organic reactions. Mainly through the pioneering investigations of Grieco et al. was this expectation brought to reality. [8]

The reaction of cyclopentadiene with ethyl acrylate to give the diastereomeric bicycloheptenes proceeds more rapidly and with higher *endo*-selectivity in 5 M LiClO$_4$/diethyl ether (*endo* : *exo* = 8 : 1) than in water. [8] Particularly impressive is the reaction of furan **1**, which, owing to its aromaticity does not react with the thiophene derivative **2** under normal conditions (Scheme 1). In LiClO$_4$/ether, however, the cycloadducts **3** and **4** are formed in 70 % yield in the ratio 85 : 15 after 9.5 h at room temperature and under atmospheric pressure. These values must be compared with those reported by Dauben et al., [9] who required 6 h and 15 kbar in order to achieve the same product ratio in their synthesis of cantharidin by the same reaction in CH$_2$Cl$_2$.

Already in the early eighties B. Föhlisch et al. [10] exploited the advantageous effects of the medium presented here in their detailed investigations of the [4+3] cycloadditions of α-halogeno- or α-sulfonyloxy-substituted ketones to 1,3-dienes, in particular to furan. The 8-oxabicyclo[3.2.1]oct-6-en-3-ones formed thereby are of interest, inter alia, as starting materials for the synthesis of tropones and other natural products. For the cycloadditions with the α-halogeno- or the α-mesyloxyketones, room temperature suffices, and inter- and intramolecular reactions can be carried out in satisfactory to good yields. A possible use of this method for the synthesis of natural terpenes, e. g. of the guajanolide, azulene, and hydroazulene type, seems very promising.

The thermal reaction between α,β-unsaturated carbonyl compounds like **5** and phenyl vinyl sulfide **6** usually proceeds to give the hetero-Diels-Alder adducts **7**. However, according to Hall et al., [11] in 5 M LiClO$_4$/ether exclusively the [2+2] cycloadducts **8** are formed (Scheme 2).

Tandem [2+2] cycloaddition-cycloreversion processes to form substituted alkenes and dienes have been described separately with highly reactive ketenes and are usually conducted at relatively high temperatures. Cossio et al. [12] have reported that even non-activated ketenes such as dimethylketene **9**, generated in situ from the acyl chloride **10**, react at room-temperature in 5 M LiClO$_4$/ether with the aromatic aldehyde **11** to form the alkene **12**. In this case, the reaction does not take place in the absence of lithium perchlorate (Scheme 3).

These unusual results raise the question how LiClO$_4$ influences the course of these and further reactions. The similarity to the accelerating effect of water as solvent for carbo- and hetero Diels-Alder reactions suggests that an "internal pressure" might be operative, [5, 13] but also an electrostatic catalysis by ion pairs [14] might be involved if polar transition states are passed. Finally, the lithium cation may serve as Lewis acid in an essentially non-acidic medium, a notion

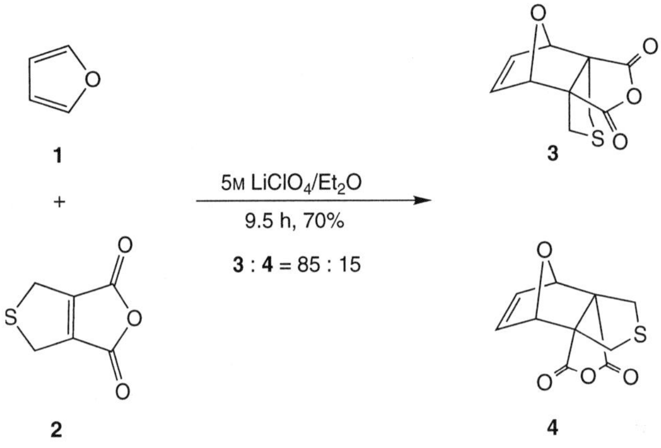

1

+

2

5M LiClO$_4$/Et$_2$O

9.5 h, 70%

3 : 4 = 85 : 15

3

4

Scheme 1.

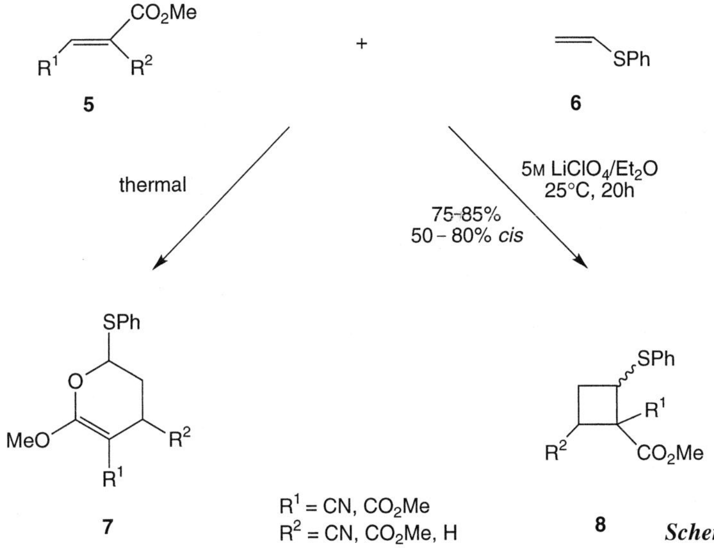

R¹ = CN, CO₂Me
R² = CN, CO₂Me, H

Scheme 2.

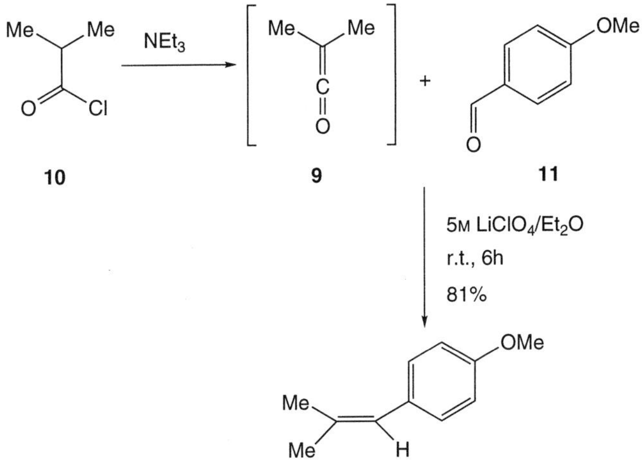

Scheme 3.

which is supported by kinetic measurements, [15] NMR experiments and MNDO calculations. [16] Whereas Li⁺ in the gas phase is a very strong Lewis acid in solution its acidity is weakened by solvation and by interaction with the accompanying anion.

Not only cycloadditions are accelerated in solutions of LiClO₄ in organic solvents, but, in particular, these reaction media activate carbonyl compounds towards attack by C-nucleophiles like TMS-CN and silyl ketene acetals. For instance, Grieco et al. [17] showed that **14** adds to the sterically hindered enones **13** exclusively in a 1,4-fashion (Scheme 4). In the presence of titanium Lewis acids **15a** was obtained in only 10% yield, and in the case of **13b** strong Lewis acids can not be employed at all. Reetz et al.

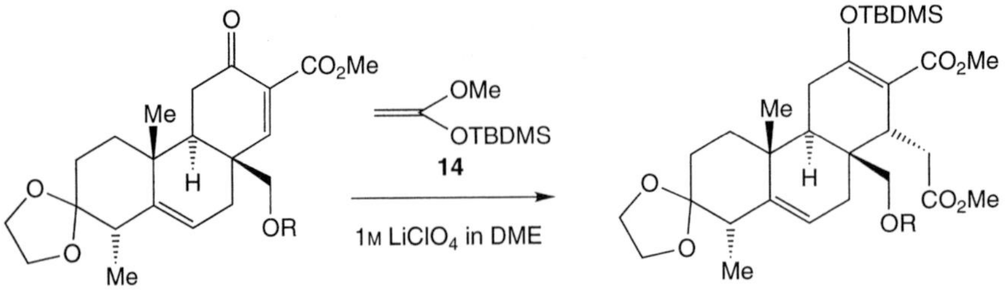

13a R = Me
13b R = TBDMS

15a R = Me   93%
15b R = TBDMS   quant.   *Scheme 4.*

[18] described the use of LiClO$_4$ in CH$_2$Cl$_2$ as solvent. In this solvent the reaction mixture is heterogeneous but the additions of e.g. silyl ketene acetals to carbonyl groups proceed faster than in ether.

Furthermore the addition of allyl tin and -silicon nucleophiles to aldehydes is accelerated in LiClO$_4$/ether. Ipaktschi et al. [19] found that for instance allyl tributyl tin **17a** adds to the a-epoxy aldehyde **16** with high *syn*-selectivity (Scheme 5). Typically Lewis

acids like BF$_3 \cdot$OEt$_2$ in these cases give *anti*-products with similar selectivity.

A chemo- and regioselective conversion of epoxides to carbonyl compounds in 5 M LiClO$_4$/ether was reported by Sankararaman et al. [20] The stereoselectivity in the case of limonene oxide **20** can be explained by invoking the rule of diaxial ring opening (Scheme 6). The small differences in the activation barriers of the two diastereomers is manifested in LiClO$_4$/ether.

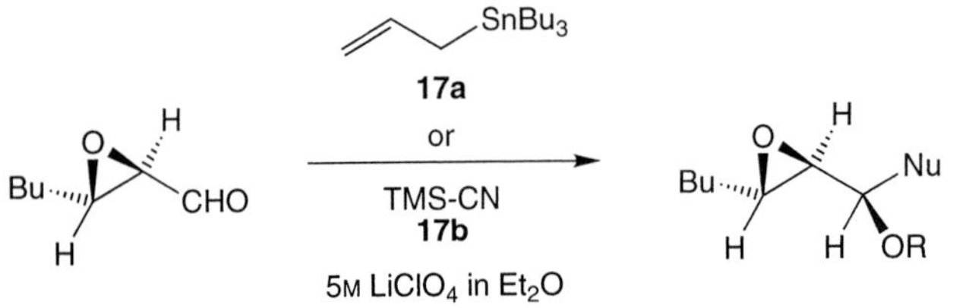

16

	R	Nu	yield	*syn : anti*
18a	H	allyl	95%,	90 : 10
18b	SiMe$_3$	CN	90%,	85 : 15

Scheme 5.

cis: trans
1:1

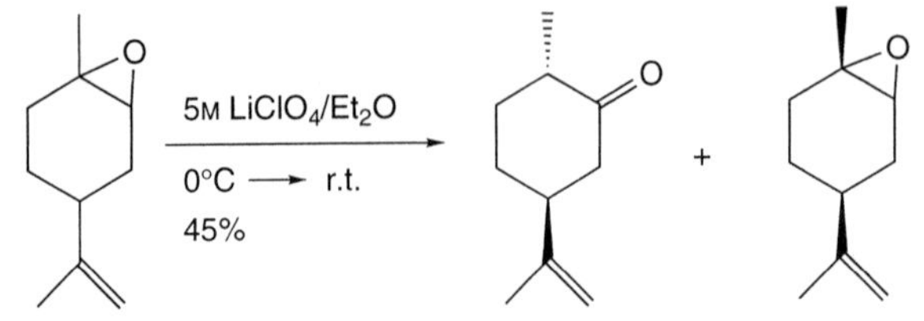

19   **20**   **19** *cis*   *Scheme 6.*

Pearson et al. [21] described that allyl alcohols and their acetic acid esters (21) are subject to a nucleophilic substitution by silyl ketene acetals and other C- and N-nucleophiles (Scheme 7). This process offers an advantageous alternative to transition metal catalysed processes.

Recently solutions of LiClO₄ in organic solvents were employed as media for the activation of various glycosyl donors. [22] In these solvents, glycosyl phosphates, glycosyl trichloroacetimidates and even the usually very stable glycosyl fluorides like the fucosyl donor 23 could be converted to glycosides like the trisaccharide 25 under neutral conditions and without the use of any further promotor (Scheme 8).

Finally, the use of LiClO₄/ether for the elimination of acetic acid from serine derivatives [23] and for the synthesis of aromatic amines [24] was reported.

Grieco et al. and Ghosez et al. [25] have recently reported that lithium trifluoromethanesulfonimide (LiNTf₂) in acetone or diethyl ether is a safe alternative to lithium perchlorate solutions for effecting Diels-Alder and hetero-Diels-Alder reactions. In addition, the lithium salt of tetrakis(polyfluoroalkoxy)aluminate (LiAl(OC(Ph)(CF₃)₂)₄) was described, a new hydrocarbon-soluble catalyst, for carbon–carbon bond-forming reactions like the 1,4 conjugate addition of silyl ketene acetals to α,β-unsaturated carbonyl compounds by Grieco et al. and Strauss et al. [26]

In conclusion, the use of solutions of LiClO₄ in organic solvents opens up new opportunities to direct the course of various reactions and to carry out transformations under exceptionally mild conditions.

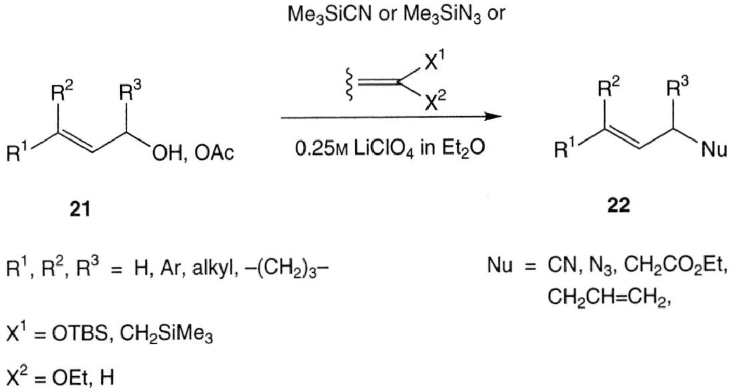

R^1, R^2, R^3 = H, Ar, alkyl, –(CH₂)₃–

X^1 = OTBS, CH₂SiMe₃

X^2 = OEt, H

Nu = CN, N₃, CH₂CO₂Et, CH₂CH=CH₂,

Scheme 7.

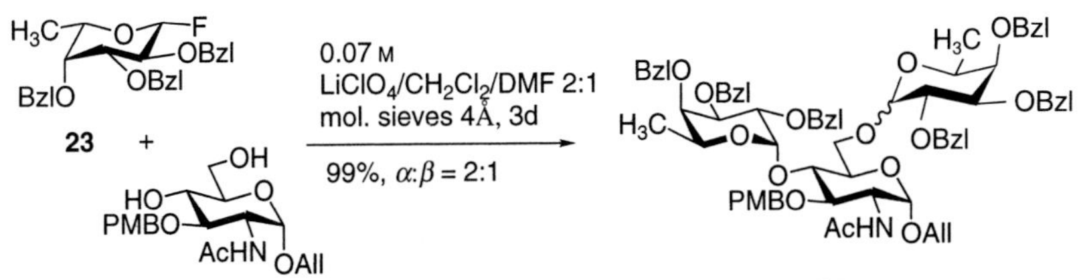

Scheme 8.

References

[1] a) R. Breslow, U. Maitra, *Tetrahedron Lett.* **1984**, *25*, 1239–1240, and references cited therein; b) review: R. Breslow, *Acc. Chem. Res.* **1991**, *24*, 150–164.

[2] P. A. Grieco, P. Galatsis, R. F. Spohn, *Tetrahedron* **1986**, *42*, 2847–2853, and references cited therein.

[3] a) O. Diels, K. Alder, *Justus Liebigs Ann. Chem.* **1931**, *490*, 243–257; b) H. Koch, J. Kotlan, H. Markert, *Monatsh. Chem.* **1965**, *96*, 1646–1657.

[4] a) Review: H. U. Reissig, *Nachr. Chem. Tech. Lab.* **1986**, *34*, 1169–1171; b) H. Waldmann, *Liebigs Ann. Chem.* **1989**, 231–238, and references cited therein; c) H. Waldmann, M. Dräger, *ibid.* **1990**, 681–685; d) A. Lubineau, Y. Queneau, *Tetrahedron* **1989**, *45*, 6697–6712; e) S. D. Larsen, P. A. Grieco, W. F. Fobare, *J. Am. Chem. Soc.* **1986**, *108*, 3512–3513; f) H. Waldmann, *Synlett* **1990**, 627–628, and references cited therein; g) P. A. Grieco, E. B. Brandes, S. McCann, J. D. Clark, *J. Org. Chem.* **1989**, *54*, 5849–5851; h) A. Lubineau, *ibid.* **1986**, *51*, 2142–2144.

[5] P. A. Grieco, J. J. Nunes, M. D. Gaul, *J. Am. Chem. Soc.* **1990**, *112*, 4595–4596.

[6] S. Winstein, S. Smith, D. Darwish, *J. Am. Chem. Soc.* **1959**, *81*, 5511–5512.

[7] R. Braun, J. Sauer, *Chem. Ber.* **1986**, *119*, 1269–1274.

[8] Reviews: a) P. A. Grieco, *Aldricimica Acta* **1991**, *24*, 59–66; b) H. Waldmann, *Angew. Chem.* **1991**, *103*, 1335–1337; *Angew. Chem. Int. Ed. Engl.* **1991**, *30*, 1306–1308; c) A. Flohr, H. Waldmann, *J. Prakt. Chem.* **1995**, *337*, 609–611.

[9] a) W. G. Dauben, C. R. Kessal, K. H. Takemura, *J. Am. Chem. Soc.* **1980**, *102*, 6893- 6894; b) for further details see: W. G. Dauben, J. Y. L. Lam, Z. R. Guo, *J. Org. Chem.* **1996**, *61*, 4816–4819.

[10] a) R. Herter, B. Föhlisch, *Synthesis* **1982**, 976–979; b) B. Föhlisch, D. Krimmer, E. Gerlach, D. Käshammer, *Chem. Ber.* **1988**, *121*, 1585–1593, and references cited therein.

[11] W. Srisiri, A. B. Padias, H. K. Hall, Jr., *J. Org. Chem.* **1993**, *58*, 4185–4186.

[12] I. Arrastia, F. P. Cossío, *Tetrahedron Lett.* **1996**, *37*, 7143–7146.

[13] P. A. Grieco, J. P. Beck, S. T. Handy, N. Saito, J. F. Daeuble, *Tetrahedron Lett.* **1994**, *35*, 6783–6786.

[14] Y. Pocker, J. C. Ciula, *J. Am. Chem. Soc.* **1989**, *111*, 4728–4735, and cited literature.

[15] a) M. A. Forman, W. P. Dailey, *J. Am. Chem. Soc.* **1991**, *113*, 2761–2762; b) G. Desimoni, G. Faita, P. P. Righetti, G. Tacconi, *Tetrahedron* **1991**, *47*, 8399–8406.

[16] R. M. Pagni, G. W. Kabalka, S. Bains, M. Plesco, J. Wilson, J. Bartmess, *J. Org. Chem.* **1993**, *58*, 3130–3133.

[17] P. A. Grieco, R. J. Cooke, K. J. Henry, J. M. VanderRoest, *Tetrahedron Lett.* **1991**, *32*, 4665–4668.

[18] a) M. T. Reetz, D. N. A. Fox, *Tetrahedron Lett.* **1993**, *34*, 1119–1122; b) M. T. Reetz, A. Gansäuer, *Tetrahedron Lett.* **1993**, *34*, 6025–6030.

[19] J. Ipaktschi, A. Heydari, H.-O. Kalinowski, *Chem. Ber.* **1994**, *127*, 905–909.

[20] R. Sudha, K. M. Narasimhan, V. G. Saraswathy, S. Sankararaman, *J. Org. Chem.* **1996**, *61*, 1877–1879.

[21] W. H. Pearson, J. M. Schkeryantz, *J. Org. Chem.* **1992**, *57*, 2986–2987.

[22] a) H. Waldmann, G. Böhm, U. Schmid, H. Röttele, *Angew. Chem.* **1994**, *106*, 2024–2025; *Angew. Chem. Int. Ed. Engl.* **1994**, *33*, 1936–1938; b) G. Böhm, H. Waldmann, *Tetrahedron Lett.* **1995**, *36*, 3843–3847; c) G. Böhm, H. Waldmann, *Liebigs Ann. Chem.* **1996**, 613–619 and 621–625; d) U. Schmid, H. Waldmann, *Tetrahedron Lett.* **1996**, *37*, 3837–3840; e) U. Schmid, H. Waldmann *Chem. Eur. J.* **1998**, *4*, 494–501.

[23] T. L. Sommerfeld, D. Seebach, *Helv. Chim. Acta* **1993**, *76*, 1702–1714.

[24] I. Zaltsgendler, Y. Leblanc, M. A. Bernstein, *Tetrahedron Lett.* **1993**, *34*, 2441–2444.

[25] a) S. T. Handy, P. A. Grieco, C. Mineur, L. Ghosez, *Synlett* **1995**, 565–567; b) R. Tamion, C. Mineur, L. Ghosez, *Tetrahedron Lett.* **1995**, *36*, 8977–8980.

[26] T. J. Barbarich, S. T. Handy, S. M. Miller, O. P. Anderson, P. A. Grieco, S. H. Strauss, *Organometallics* **1996**, *15*, 3776–3778.

Reactions in Supercritical Carbon Dioxide

Gerd Kaupp

The increasing global stress on the environment due to atmospheric pollutants from anthropogenic sources, in addition to those stemming from natural emissions, necessitates certain restrictions, for example on the use of organic solvents. For instance, lacquering is more and more frequently performed in a solvent-free fashion, and a great deal of effort is being made to avoid organic solvents in chemical syntheses, through the use of crystal photolyses, [1] gas/solid reactions, [2] and solid/solid reactions. [3] An additional environmentally friendly approach is the use of supercritical carbon dioxide sc-CO_2 as solvent for chemical syntheses. It is nontoxic, cheap, and nonflammable. It has already proven useful in some large-scale extraction processes in the food industry (e. g., decaffeination of coffee or extraction of hops). [4] In these processes, pressure or density changes lead to solubility variations which can also be exploited for chemical reactions in sc-CO_2.

The favorable ecological properties of CO_2, especially when it is used in a closed loop system, are counterbalanced by the effort and expense associated with the use of high-pressure installations. High-pressure operations are more readily realized in large-scale production processes than in university or research laboratories. These suffer not only from an increasing lack of funding, but also from the continually changing safety directions and lack of clearly assigned responsibilities. This has deterred academic researchers from undertaking and teaching high-pressure experiments. This misdirected development must be halted because of the detrimental effects on the environment. Many syntheses that cannot be performed as solid-state reactions rely on the use of inert solvents of medium polarity, which are frequently delicate. Therefore sc-CO_2, with the critical temperature, pressure, and density values of $T_c = 31.06\,°C$, $p_c = 73.83$ bar, $D_c = 0.467$ g mL^{-1}, ought to be utilized. [5] In terms of polarity, sc-CO_2 [Dimroth-Reichardt $E_T(30)$ value of 32 [6]] is roughly comparable to carbon tetrachloride. However, differences remain when the Kamlet-Taft π^* solvatochromicity parameter ($\pi^* = -0.2$) [7] is taken into account. Thus, sc-CO_2 could be used as a replacement for commonly used organic solvents. Until 1994 it appears that most experiments have been primarily governed by academic curiosity, however, interest has largely increased since.

Diels-Alder reactions have shown no discontinuities in reaction rates when liquid CO_2 is replaced by sc-CO_2, as long as the density is kept constant. [8] The reaction rates were roughly equal to those in solvents at normal pressures. [9, 10] The reactions of isoprene and methylacrylate [11] or cyclopentadiene and methylacrylate [12] have been studied in mechanistic detail. It appears that the more involved cycloadditions of acetylenedicarboxylate to azulene derivatives in sc-CO_2 are of some preparative value. One example is shown in Scheme 1 (yields are 30% and 22%). [13] Also [2+2] photocycloadditions of N-acylindoles with alkenes have been investigated in sc-CO_2. [14]

Addition/condensation reactions have been performed in liquid and sc-CO_2 when phytol or isophytol and trimethylhydroquinone gave α-tocopherol (vitamine E; Scheme 1). [15]

Mechanistic laser-flash photolysis studies of benzophenone with H donors in sc-CO_2 have shown that the local substrate clusters, which are considered crucial for the reactivity, exchange rapidly (in picoseconds) with the supercritical medium. [16] The photochemical α-cleavage of asymmetric benzyl ketones also indicated that there is no cage effect for radical recombinations in sc-CO_2. [17] The products are formed in a statistical ratio (Scheme 2).

The formation of diphenylcarbene by laser flash photolysis of diphenyldiazomethane was studied in sc-CO_2 and other supercritical media. [18] A laser-flash induced ring-closure reaction of a bipyridyl complex (Scheme 2) revealed solvation properties. [19] Laser-flash impact to metal carbonyl complexes activated hydrogen and simple alkanes like CH_4, C_2H_4, C_2H_6, and further inorganic reactions in supercritical fluids have been reviewed. [20]

The carboxy inversion of diacyl peroxides has been interpreted as being an ionic reac-

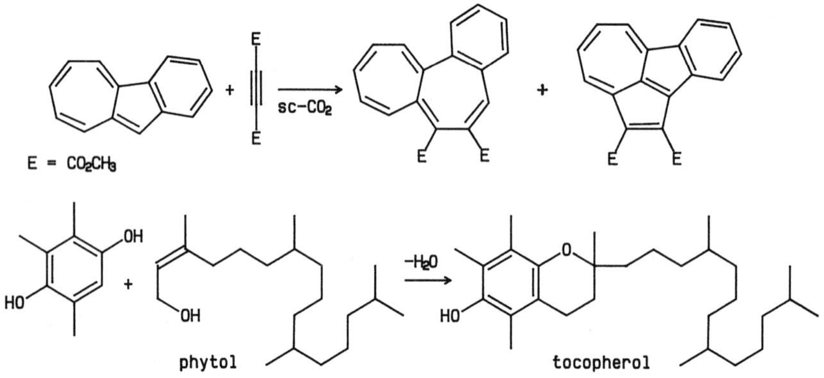
E
E
sc-CO2
E = CO2CH3
E E
E E
OH
HO
OH
phytol
-H2O
O
HO
tocopherol

Scheme 1.

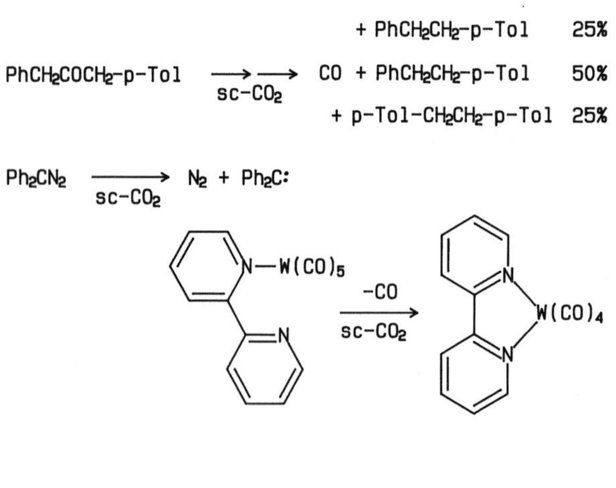
+ PhCH2CH2-p-Tol 25%
PhCH2COCH2-p-Tol
sc-CO2
CO + PhCH2CH2-p-Tol 50%
+ p-Tol-CH2CH2-p-Tol 25%
Ph2CN2
sc-CO2
N2 + Ph2C:
N—W(CO)5
N
-CO
sc-CO2
N
W(CO)4
N

Scheme 2.

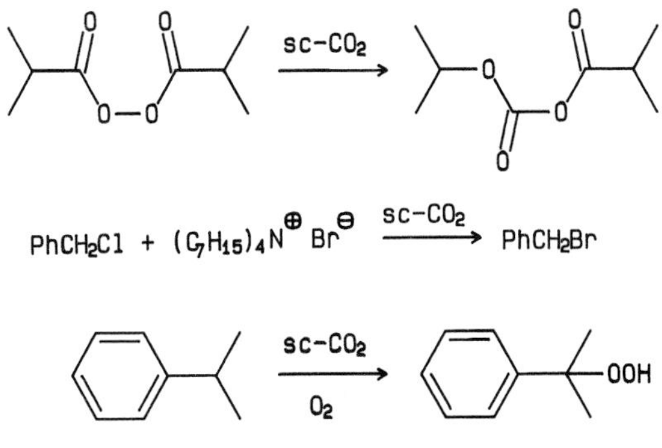

$$\text{(diisobutyryl peroxide)} \xrightarrow{\text{sc-CO}_2} \text{(product)}$$

$$\text{PhCH}_2\text{Cl} + (\text{C}_7\text{H}_{15})_4\overset{\oplus}{\text{N}} \ \text{Br}^{\ominus} \xrightarrow{\text{sc-CO}_2} \text{PhCH}_2\text{Br}$$

$$\text{(cumene)} \xrightarrow[\text{O}_2]{\text{sc-CO}_2} \text{(cumene—OOH)}$$

Scheme 3.

tion. In sc-CO$_2$ it proceeds with a 17 % yield (40 °C, $d = 0.93$ g mL^{-1}), but slower than in CCl$_4$ or CHCl$_3$ (Scheme 3). [7] However, ionic reactions in sc-CO$_2$ can be facilitated by phase-transfer catalysis. This is clearly shown by the substitution reaction of benzyl chloride in Scheme 3. [21] Acetone as a cosolvent helps. Radical chain reactions like the well-known oxidation of cumene to cumene hydroperoxide led to lower yields in sc-CO$_2$ (110 °C, 200–414 bar), probably because it was not possible to prevent chain termination reactions due to the metal container (Scheme 3). [22]

Recently, however, a number of studies were published which demonstrated that radical brominations [23] and polymerizations, [24, 25] can proceed providing better (or roughly equal) results than the corresponding reactions in conventional solvents. Furthermore, hydroformylations, [26] CO$_2$ hydrogenations, [27] catalytic additions/cycloadditions on CO$_2$, [28, 29] and enzymic reactions [30, 31] in sc-CO$_2$ were successful. In many cases both reaction and extraction of products can profit from the supercritical phase.

Radical reactions are nicely performed in sc-CO$_2$. Thus, the homolytic cleavage of bitropenyl to give cycloheptatrienyl radical and its trapping by oxygen have been used to evaluate activation, stabilization and bond energies. [32]

The bromination of toluene with bromine in sc-CO$_2$, photochemically initiated through a sapphire window, led to benzyl bromide (74 %) and 4-bromotoluene (11 %). Ethylbenzene afforded 1-bromo-1-phenylethane in 95 % yield (Scheme 4). [23]

In order to demonstrate that uncomplexed bromine atoms act as chain propagators, toluene and ethylbenzene were photobrominated in a competition study at pressures of 75 to 423 bar and at 40 °C. Over the entire pressure range, the reactivity of the benzylic secondary C–H bond in ethylbenzene was found to be about 30 times greater than that of the corresponding primary C–H bond in toluene. The analogous value for the reactivity in CCl$_4$ at 40 °C is 36. The bromine atoms in sc-CO$_2$ are therefore particularly free. It would be important to determine quantum yields (chain lengths) at various pressures to learn more about mechanistic aspects and other details of the reaction. Local solvent structures on model free-radical reactions in sc-CO$_2$ have been analyzed in some detail. [33]

The heterogeneous N-bromosuccinimide (NBS) bromination (Ziegler bromination; Scheme 4) of toluene in sc-CO$_2$ was initiated photochemically with azobis(isobutyronitrile) (AIBN) (40 °C, 170 bar, 4 h). [23] In CCl$_4$, this reaction proceeds without succinimidyl radicals, with the same selectivity as the direct bromination. The yield is quantitative on the 4 mmol scale; no 4-bromotoluene is observed.

Homogeneous fluoroalkene polymerizations (and copolymerizations) also proceed in sc-CO$_2$ after AIBN initiation. [24, 25] In this case it is advantageous that highly fluorinated polymers (> 250 000 g mol^{-1}) and copolymers are very soluble (up to 25 %) in sc-CO$_2$ at high pressures. This means that no chlorofluorocarbons need to be employed as solvents. The homopolymer (270 000 g mol^{-1}) shown in Scheme 5 can be obtained from FOA in sc-CO$_2$ (59.4 °C, 207 bar, 48 h, no Trommsdorff effect [24]) in 65 % yield. The product can be precipitated from the homogeneous solu-

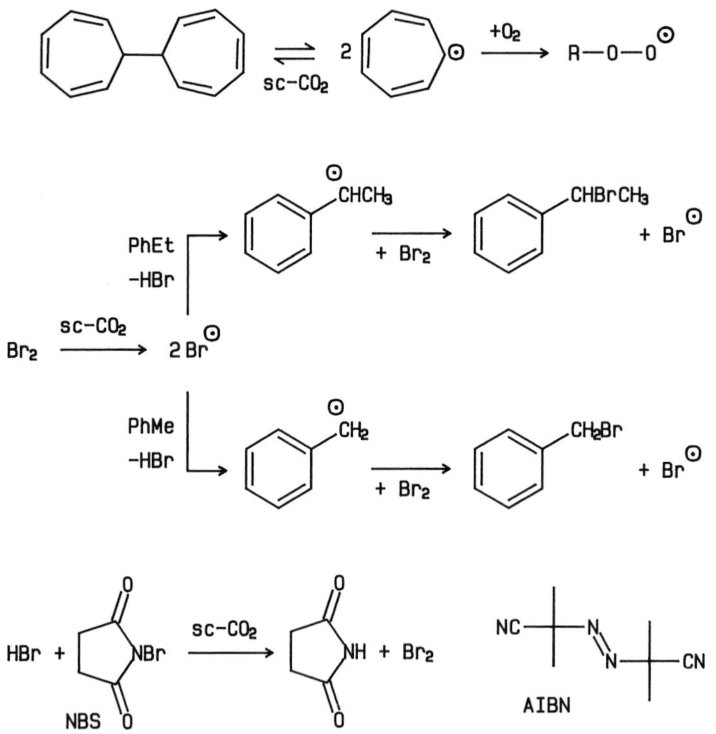

Scheme 4.

tion by pressure release. It would be useful, however, to develop a workup procedure that does not offset the use of sc-CO$_2$ by the necessity to dissolve the raw product in 1,1,2-trifluorotrichloroethane.

Statistical copolymers of FOA are also surprisingly soluble in sc-CO$_2$. The kinetic parameters of the thermal AIBN decomposition in sc-CO$_2$ have been determined accurately, and 1,1-difluoroethylene was telomerized correspondingly. [22]

After these pioneering investigations a large number of radical and cationic polymerizations have been performed in sc-CO$_2$. These

FOA

Scheme 5.

include *polymethylmethacrylate, polystyrene, polyesters, polyvinylalcohols, poly(2-hydroxypropylmethacrylate), polyisobutylene,* and *polyacrylamide.* [34] Spherical latex particles are usually formed and modified in (inverse) emulsion polymerizations. Organic aerogels with large inner surfaces (600–1000 m^2 g^{-1}) and ultrafine pores were prepared by condensation polymerization of resorcinol and formaldehyde in water with subsequent supercritical drying. [35] Finally, crosslinked and natural *rubber* could be controllably *depolymerized* to obtain useful feedstocks in supercritical H$_2$O and CO$_2$. [36]

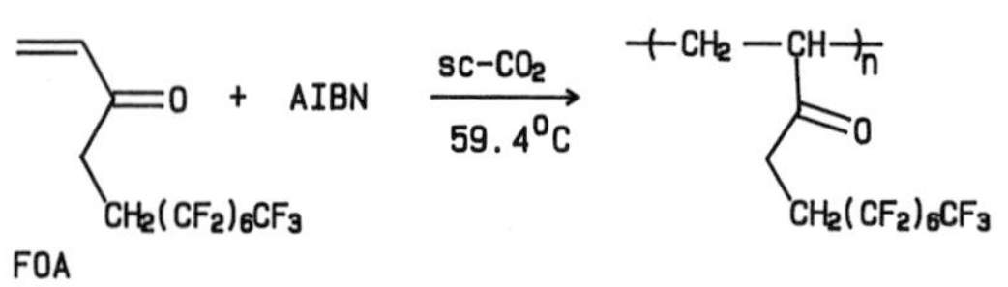

The inertness of sc-CO$_2$ is also useful for metal-catalyzed *hydroformylations* and *hydrogenations* of alkenes to the corresponding aldehydes. Selective hydroformylations were obtained with Co catalysts. [26] They profit from the good miscibility in sc-CO$_2$ (Scheme 6). The reaction mixture is less viscous, which leads to sharp ^{59}Co NMR signals (quadrupole nucleus), which in turn allows the identification of the catalytically active species by high-pressure NMR spectroscopy. The hydrogenation of double bonds does compete if MnH(CO)$_5$ is the reagent (Scheme 6). [37]

The solubility of Rh(I)phosphane complexes has been increased by substitution with four *n*-tridecafluorooctane and two trifluoromethyl side groups and the selectivity of the hydroformylation of 1-octene in sc-CO$_2$ was *n/iso* 3.7 at 100 % conversion. The same catalyst was used to hydrogenate isoprene. [38] Asymmetric hydrogenations in sc-CO$_2$ have been obtained heterogeneously and homogeneously. Thus, ethyl pyruvate gave (*R*)-ethyl lactate enantioselectively on a Pt/Al$_2$O$_3$ catalyst with cinchonidine. However, CO$_2$ deactivates the catalyst due to its reduction. [39] More general are soluble chiral Rh catalysts {Rh(I)(cod)(*R,R*)-Et-DuPHOS trifluoromethane-sulfonate} or { -tetrakis-(3,5-bis(trifluoromethyl)phenyl) borate} that hydrogenate the prochiral enamides (Scheme 6). Four β-monosubstituted compounds gave similar, two β,β-disubstituted ones considerably better *ee* values as those found in liquid organic media (methanol or hexane). [40]

Hydrogenations of carbon dioxide to give formic acid (derivatives) are of particular concern. Noyori summarized his contributions, patents and improvements in that field in two full papers. [27] Simple soluble catalysts like RuCl$_2$[P(CH$_3$)$_3$]$_4$, RuH$_2$[P(CH$_3$)$_3$]$_4$ or others are active in homogeneous hydrogenation of CO$_2$ to formic acid as long as this reaction is coupled to salt formation. The reaction in sc-CO$_2$ (50 °C, 86 bar H$_2$, 207 bar CO$_2$, dissolved NEt$_3$) proceeds with the highest rate of the numerous variants of this reaction, namely 1400 mole of formic acid per mole of catalyst per hour. This turnover frequency (TOF) is reduced to 1.3 h^{-1} for the same composition when liquid CO$_2$ (at 15 °C) is employed. If cosolvents are added (water, or DMSO) a TOF of 4000 h^{-1} is achieved. If methanol is added, the product is methyl formate, the presence of NEt$_3$ still being necessary. Turnover numbers (TON) up to 3500 have been reached. All other high-pressure variants of this reaction in liquid solvents yield much lower TOF values. [27] The formic acid synthesis can also be coupled with subsequent reactions of formic acid with secondary amines, to even higher TON values up to 420 000 for the synthesis of DMF (TOF 8000 h^{-1}), when in contact with a liquid phase starting with solid

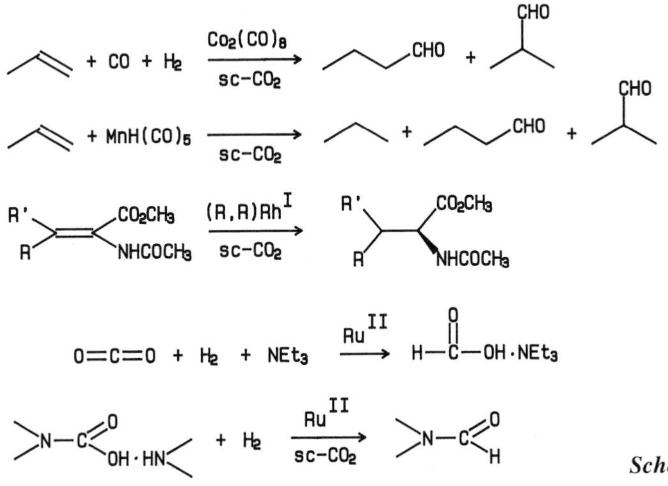

Scheme 6.

dimethyl-ammonium dimethylcarbamate as the dimethylamine source at 50 °C (Scheme 6). [27] This behavior indicates the favorable effect of the high solubility of H_2 in sc-CO_2 as well as the high mass transport rate and recommends the system for continuous operation in industrial plants. It should be pointed out that the concentration of H_2 in a supercritical mixture of H_2 (85 bar) and CO_2 (120 bar) at 50 °C is 3.2 M, while the concentration of H_2 in THF under the same pressure is merely 0.4 M. [27] Importantly, the nonpolar catalyst could be recycled by extraction with sc-CO_2.

Oxidations with O_2 in sc-CO_2 (Scheme 7) also profit from miscibility of gases and high diffusivity. Oxirane and acetaldehyde were obtained by KrF excimer laser irradiation of ethylene and O_2 in CO_2 under sub- and supercritical conditions. Also ethane and cyclohexane oxidized. [41] The cyclohexane oxidation giving cyclohexanone and cyclohexanol may be greatly manipulated in sc-CO_2. [42] Total oxidations in sc-CO_2 (to give CO_2 and H_2O) have been obtained with ethanol,

toluene and tetraline at > 300 °C. [43] Such reactions might have a bearing for waste treatment similar to related use of sc-H_2O. Finally, it is also possible to perform Rh(I) catalyzed oxidation of THF in supercritical CO_2/O_2 to give butyrolactone (TON 100) (Scheme 7). [38]

Polycarbonates and cyclic carbonates are obtained by the well-known and well-studied reactions of CO_2 and oxiranes (Scheme 7). These reactions have also been successfully performed in supercritical mixtures. [28] It turned out, however, that the industrial production of ethylene carbonate (similar to propylene carbonate) in a liquid (product) phase (190–200 °C, total pressure 80 bar) is more economical for capacities of 4000 t per year and installation when it is run nearly stoichiometrically. [44] In the cases of substituted or polycyclic oxiranes, solvents are usually added and 1 to 40 bar of CO_2 are introduced, depending on the catalyst employed. [28] However, the development of a CO_2-soluble Zn catalyst formed the polycarbonate from

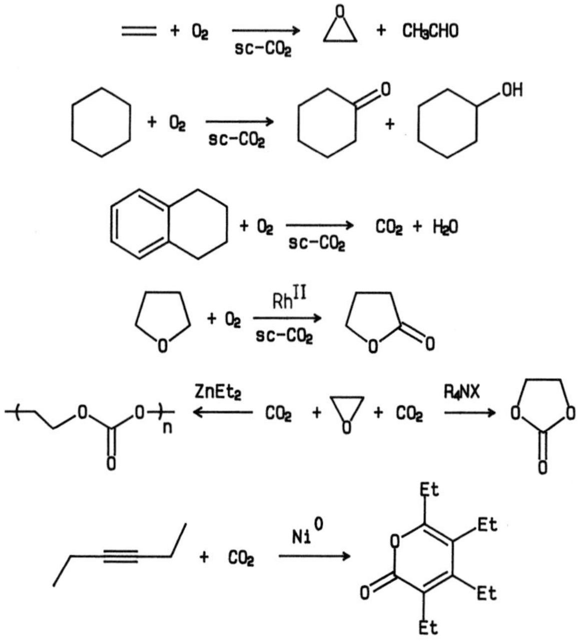

Scheme 7.

cyclohexene oxide in the absence of any additional organic solvent. [45] Homologous cyclic ethers (oxetanes, tetrahydrofurans, etc.) are also expected to undergo cycloaddition to CO_2 and should be tried in sc-CO_2. A [2+2+2] cycloaddition of two moles of 3-hexyne to one mole of CO_2 in the presence of the catalyst [{Ni(cod)$_2$}Ph$_2$P(CH$_2$)$_4$PPh$_2$] and of benzene (25 °C, 50 bar, 20 h at 120 °C), led to tetraethylpyranone (57 %) (Scheme 7). [46] A preliminary attempt to improve this yield by using the same catalyst in sc-CO_2 failed (102 °C, 93 g CO_2 in 200 mL, 69 h, 35 %). [29] It appears however, that attempts to improve this yield are still in progress in spite of a number of experiments that have reproduced it. [29]

Enzymic reactions in sc-CO_2 cover oxidations and solvolyses. Good yields (75 % at a residence time of only 13 s) were reported for the enzymic oxidation (immobilized cholesterol oxidase from *G. chrysocreas*) of cholesterol in supercritical CO_2/O_2 (9 : 1) (Scheme 8). Cosolvents, like *tert*-butyl alcohol, that increase the solubility and, to an even larger extent, those that assist aggregate formation, increase the rate of the reaction (fourfold in this case). [31] However, it appears that this line has not been pursued any further. Horseradish peroxidase was used in the oxidative polymerization of *p*-cresol by H_2O_2 in sc-CO_2. Cosolvents were useful. The method was evaluated for manufacture of phenolic resins without incorporating formaldehyde. [47]

One of the first enzymic esterifications (immobilized lipase MY) of *rac*-citronellol with oleic acid in sc-CO_2 close to the critical point led stereoselectively to 1.2–5.8 % of citronellol oleate (Scheme 9). Only at the critical temperature (31 °C, 84 bar) does the (*S*)-(−)-citronellol oleate form with 99 % optical purity. Even small temperature increases (4–9 °C) led to a drastic reduction of the optical purity; pressure increases (up to 190 bar) had the same effect. This result was explained by the formation of clusters near the critical point, where pressure and temperature changes have particularly strong effects. No esterification took place in water-saturated cyclohexane. [30] Similarly, *rac*-ibuprofen was enantioselectively esterified in sc-CO_2 with 70 % ee, by immobilized Mucor miehei lipase. [48] After these pioneering results much work has been put into enzymic hydrolyses, esterifications, transesterifications in sc-CO_2 (Scheme 9), due to high demands of the food industry. Numerous groups in Austria, Canada, Finland, France, Germany, Italy, Japan, Slovenia, Sweden, and USA have published on such research and several reviews are available. [49] Major topics are continuous processes, technology of fats, fatty acids and alcohols. Frequently, sc-CO_2 is used both for reaction and extraction. Also demands of fuel industry are covered (e. g. bio-Diesel). However, for the latter applications of sc-CO_2 non-catalyzed processes at higher temperatures may be more important, even though methanolysis of seed oils or soy

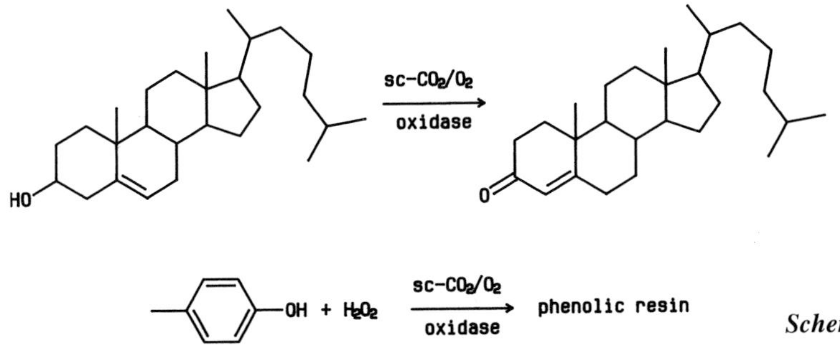

Scheme 8.

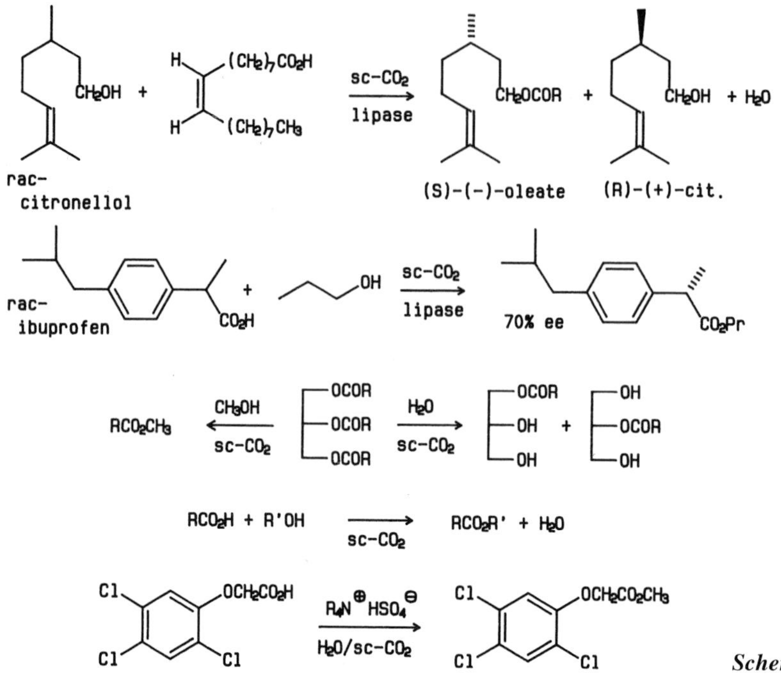

Scheme 9.

flakes have enzymicly succeeded. [50] A non-enzymic simultaneous extraction and methylation of herbicides like 2,4,5-T with CH_3I and the phase transfer catalyst $(C_6H_{13})_4N^+HSO_4^-$ was complete from 1 ppm solutions (Scheme 9). [51]

Apparently, unnoticed by the recent studies, the photolysis of hops extracts in sc-CO_2 has been developed into an established industrial process. [52] The sc-CO_2 extract contains the so-called "α-acids" (vinylogous carboxylic acids), for instance humulone, which are primary bittering agents with limited stability. These are isomerized into α-isoacids, such as *trans*-isohumulone (Scheme 10), stable and more soluble bittering agents, which give beer its characteristic taste. The conventional brewing method achieved this transformation in low yields by boiling the hops in the wort. It is more efficient to add the ingredients stemming from hops to the cold finished beer. This, however, requires the isomerization of the air-sensitive α-acids to be performed pho-

tochemically in the sc-CO_2 extract before the addition to the beer.

The quantum yield ϕ is 0.03. The mechanism of this [1,2,3,4]–rearrangement [53] is unknown. Formally, the reaction can be considered an H migration from the OH group to the end of a four-membered chain with a concomitant acyl group migration from position 2 to 3 within that chain. According to the current canon of photochemistry, an oxa-di-π-methane rearrangement with subsequent [1,3]-H-migration and ring opening was formulated. [52] It is said that this procedure affords special protection to the flavors of the hops ingredients.

The current state of knowledge [54] indicates the versatility of sc-CO_2 as a reaction medium. The last three years saw a tremendous increase in material giving reason to be optimistic that in the future even more reactions will be carried out in the ecologically sound sc-CO_2 (closed loop systems) rather than in organic solvents. The theoretical and

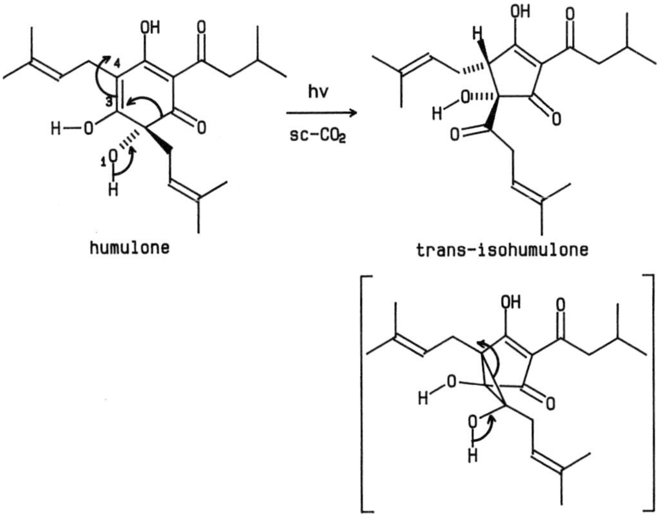

Scheme 10.

mechanistic basics have been complemented by a great deal of practical and technical knowledge now exceeding the widespread knowledge from extraction processes utilizing sc-CO₂. Industrial applications have become reality. However, teaching deficiencies persist. Textbooks on experimental techniques are still not covering unusual reaction media. Academia seems to be largely caught in thermodynamic measurements, although reactions are at work and thus, the more exciting creative aspects of the issue with new techniques including AFM and SNOM [2,55] should find due public support for the sake of an environmentally benign future.

References

[1] V. Ramamurthy, *Tetrahedron 1986, 42*, 5753; G. Kaupp, *Adv. Photochem.* **1994**, *19*, 119; G. Kaupp in *Handbook of Organic Photochemistry and Photobiology*, (Ed.: W. Horspool), CRC, Cleveland OH, **1995**, p. 50–63.

[2] G. Kaupp, D. Matthies, *Chem. Ber.* **1986**, *119*, 2387; *Mol. Cryst. Liq. Cryst.* **1988**, *161*, 119; G. Kaupp, D. Lübben, O. Sauerland, *Phosphorus Sulfur Silicon Relat. Elem.* **1990**, *53*, 109; G. Kaupp, *Mol. Cryst. Liq. Cryst.* **1992**, *211*, 1; G. Kaupp, J. Schmeyers, *Angew. Chem.* **1993**, *105*, 1656; *Angew. Chem. Int. Ed. Engl.* **1993**, *32*, 1587, and references therein; G. Kaupp in *Comprehensive Supramolecular Chemistry*, Vol. 8, Chap. 9 (Eds.: J. E. D. Davies, J. A. Ripmeester) p. 381–423 + 21 color plates, Elsevier, Oxford, **1996**.

[3] F. Toda, K. Kiyoshige, M. Yagi, *Angew. Chem.* **1989**, *101*, 329; *Angew. Chem. Int. Ed. Engl.* **1989**, *28*, 320; Review: F. Toda in *Reactivity in Molecular Crystals* (Ed.: Y. Ohashi), VCH/Kodansha Weinheim/Tokyo, **1993**, p. 177; G. Kaupp, M. Haak, F. Toda, *J. Phys. Org. Chem.* **1995**, *8*, 545; G. Kaupp, J. Schmeyers, F. Toda, H. Takumi, H. Koshima, *ibid.* **1996**, *9*, 795.

[4] R. Eggers, *Angew. Chem.* **1978**, *90*, 799; *Angew. Chem. Int. Ed. Engl.* **1978**, *17*, 751; K. Zosel, *ibid.* **1978**, *90*, 748 and **1978**, *17*, 702; P. Hubert, O. G. Vitzthum, *ibid.* **1978**, *90*, 756 and **1978**, *17*, 710; E. Stahl, W. Schilz, E. Schütz, E. Willing, *ibid.* **1978**, *90*, 778 and **1978**, *17*, 731; K. Zosel (Studiengesellschaft Kohle), US-A 4260639, **1981**; A. B. Caragay, *Perfum Flavor* **1981**, *6*, 43; H. Brogle, *Chem.*

Ind. (London) **1982**, 385; R. Vollbrecht, *ibid.* **1982**, 397; liquid: D. S. Gardner, *ibid.* **1982**, 402.

[5] According to its phase diagram, CO_2 can only be liquefied by compression between the triple point at $-57\,°C/5.2$ bar and the critical point. Above both T_c and p_c, CO_2 is in the supercritical state.

[6] Y. Ikushima, N. Saito, M. Arai, *J. Phys. Chem.* **1992**, *96*, 2293; Y. Ikushima, N. Saito, M. Arai, K. Arai, *Bull. Chem. Soc. Jpn.* **1991**, *64*, 2224; J. A. Hyatt, *J. Org. Chem.* **1984**, *49*, 5097.

[7] M. E. Sigman, J. T. Barbas, J. E. Leffler, *J. Org. Chem.* **1987**, *52*, 1754.

[8] For example, reaction of 1.5 g cyclopentadiene with 0.5 g *p*-benzoquinone in 20 g CO_2: $25–40\,°C$, $60–240$ bar, $k^I = 0.597 \times 10^{-3}$ to 1.147×10^{-3} s^{-1}.

[9] N. S. Isaacs, N. Keating, *J. Chem. Soc. Chem. Commun.* **1992**, 876.

[10] M. E. Paulaitis, G. C. Alexander, *Pure Appl. Chem.* **1987**, *59*, 61.

[11] Y. Ikushima, N. Saito, O. Sato, M. Arai, *Bull. Chem. Soc. Jpn.* **1994**, *67*, 1734.

[12] R. D. Weinstein, A. R. Renslo, R. L. Danheiser, J. G. Harris, J. W. Tester, *J. Phys. Chem.* **1996**, *100*, 12337.

[13] R. Hunziker, D. Sperandio, H.-J. Hansen, *Helv. Chim. Acta* **1995**, *78*, 772.

[14] B. T. Des Islet, Univ. of Western Ontario, Diss. Abstr. Int., **B 1996**, *57*, 334. Avail.: Univ. Microfilms Int. Order No. DANN03447.

[15] R. Lowack, J. Meyer, M. Eggersdorfer, P. Grafen (BASF A.G.) US 5523420, **1995**; DE 4243464 A1, **1992**.

[16] C. D. Roberts, J. E. Chateauneuf, J. F. Brennecke, *J. Am. Chem. Soc.* **1992**, *114*, 8455.

[17] K. E. O'Shea, J. R. Combes, M. A. Fox, K. P. Johnston, *Photochem. Photobiol.* **1991**, *54*, 571; C. B. Roberts, J. Zhang, J. F. Brennecke, J. E. Chateauneuf, *J. Phys. Chem.* **1993**, *97*, 5618; for high-resolution reference spectra see G. Kaupp, E. Teufel, H. Hopf, *Angew. Chem.* **1979**, *91*, 232; *Angew. Chem. Int. Ed. Engl.* **1979**, *18*, 215.

[18] J. E. Chateauneuf, *Res. Chem. Intermed.* **1994**, *20*, 159.

[19] Q. Ji, C. R. Lloyd, E. M. Eyring, R. van Eldik, *J. Phys. Chem. A* **1997**, *101*, 243.

[20] J. A. Banister, A. I. Cooper, S. M. Howdle, M. Jobling, M. Poliakoff, *Organometallics* **1996**, *15*, 1804; M. Poliakoff, M. W. George, S. M. Howdle, *Chem. Extreme Non-Classical Cond.* p. 189–218 (Ed.: R. Van Eldick, C. D. Hubbard), Wiley, New York, **1997**.

[21] A. K. Dillow, S. L. J. Yun, D. Suleiman, D. L. Boatright, C. L. Liotta, C. A. Eckert, *Ind. Eng. Chem. Res.* **1996**, *35*, 1801.

[22] G. J. Suppes, R. N. Occhiogrosso, M. A. McHugh, *Ind. Eng. Chem. Res.* **1989**, *28*, 1152.

[23] J. M. Tanko, J. F. Blackert, *ACS Symp. Ser.* **1994**, *577(Benign by Design)*, 98.

[24] J. R. Combes, Z. Guan, J. M. DeSimone, *Macromolecules* **1994**, *27*, 865.

[25] Z. Guan, J. R. Combes, Y. Z. Menceloglu, J. M. DeSimone, *Macromolecules*, **1993**, *26*, 2663.

[26] J. W. Rathke, R. J. Klingler, T. R. Krause, *Organometallics* **1991**, *10*, 1350.

[27] P. G. Jessop, Y. Hsiao, T. Ikariya, R. Noyori, *J. Am. Chem. Soc.* **1995**, *117*, 8277; **1994**, *116*, 8851.

[28] U. Petersen, *Methoden Org. Chem. (Houben Weyl)* 4th. ed. 1952, Chap. E4, **1983**, p. 95.

[29] M. T. Reetz, W. Könen, T. Strack, *Chimia* **1993**, *47*, 493.

[30] Y. Ikushima, N. Saito, T. Yokoyama, K. Hatakeda, S. Ito, M. Arai, H. M. Blanch, *Chem. Lett.* **1993**, 109; Y. Ikushima, N. Saito, M. Arai, *J. Chem. Eng. Jpn.* **1996**, *29*, 551.

[31] H. W. Blanch, T. Randolph, C. R. Wilke (University of California, Berkeley) US 4925790 A, **1985**; T. Randolph, D. S. Clark, H. W. Blanch, J. M. Prausnitz, *Science* **1988**, *239*, 387.

[32] W. R. Roth, F. Hunold, M. Neumann, F. Bauer, *Liebigs Ann.* **1996**, 1679.

[33] S. Ganapathy, C. Carlier, T. W. Randolph, J. A. O'Brien, *Ind. Eng. Chem. Res.* **1996**, *35*, 19.

[34] Review of the contributions of J. M. DeSimone: K. A. Schaffer, J. M. DeSimone, *Trends Polym. Sci.* **1995**, *3*, 146; J. L. Kerschner, S. H. Jureller, R. Harris, *Polym. Mater. Sci. Eng.* **1996**, *74*, 246; G. Deak, T. Pernecker, J. P. Kennedy, *Macromol. Rep.* **1995**, *A32*, 979; F. A. Adamsky, E. J. Beckman,

Macromolecules **1994**, *27*, 5238; 312; T. Pernecker, J. P. Kennedy, *Polym. Bull. (Berlin)* **1994**, *33*, 13.

[35] J.-H. Song, H.-J. Lee, J.-H. Kim, *Han'guk Chaelyo Hakhoechi* **1996**, *6*, 1082; *Chem. Abstr.* **1996**, *126*, 104500.

[36] D. T. Chen, C. A. Perman, M. E. Riechert, J. Hoven, *J. Hazard. Mater.* **1995**, *44*, 53.

[37] P. G. Jessop, T. Ikariya, R. Noyori, *Organometallics* **1995**, *14*, 1510.

[38] W. Leitner, S. Kainz, D. Koch, C. Six, K. Wittmann, *Lecture at the Chemiedozententagung*, March 16–19, **1997**, Berlin; S. Kainz, D. Koch, W. Baumann, W. Leitner, *Angew. Chem.* **1997**, *109*, 1699; *Angew. Chem. Int. Ed. Engl.* **1997**, *36*, 1628.

[39] B. Minder, T. Mallat, K. H. Pickel, K. Steiner, A. Baiker, *Catal. Lett.* **1995**, *34*, 1.

[40] M. J. Burk, S. Feng, M. F. Gross, W. Tumas, *J. Am. Chem. Soc.* **1995**, *117*, 8277.

[41] S. Koda, Y. Oshima, J. Otomo, T. Ebukuro, *Process Technol. Proc.* **1996**, *12*, 97.

[42] P. Srinivas, M. Mukhopadhyay, *Ind. Eng. Chem. Res.* **1994**, *33*, 3118.

[43] L. Zhou, A. Akgerman, *Ind. Eng. Chem. Res.* **1995**, *34*, 1588; *AIChE J.* **1995**, *41*, 2122.

[44] G. Hechler, *Chem. Ing. Tech.* **1971**, *43*, 903.

[45] C. A. Costello, E. Berluche, S. J. Han, D. A. Sysyn, M. S. Super, E. J. Beckman, *Polym. Mater. Sci. Eng.* **1996**, *74*, 430.

[46] Y. Inoue, Y. Itoh, H. Kazama, H. Hashimoto, *Bull. Chem. Soc. Jpn.* **1980**, *53*, 3329.

[47] K. Ryu, S. Kim, *Korean J. Chem. Eng.* **1996**, *36*, 415.

[48] M. Rantakylae, O. Aaltonen, *Biotechnol. Lett.* **1994**, *16*, 825; enantioselective acetylations: E. Cernia, C. Palocci, F. Gasparrini, D. Misiti, N. Fagnano, *J. Mol. Catal.* **1994**, *89*, L11-L18.

[49] O. Aaltonen, M. Rantakylä, *CHEMTECH* **1991**, 240; S. V. Kamat, E. J. Beckman, A. J. Russell, *Crit. Rev. Biotechnol.* **1995**, *15*, 41–71; K. Nakamura, *Supercrit. Fluid Technol. Oil Lipid Chem.* (Ed.: J. W. King, G. R. List), AOCS Press, Champaign, Ill, **1996**, 306–320.

[50] M. A. Jackson, J. W. King, *J. Am. Oil Chem. Soc.* **1996**, *73*, 353.

[51] M. Y. Croft, E. J. Murby, R. J. Wells, *Anal. Chem.* **1994**, *66*, 4459.

[52] J. C. Andre, A. Said, M. L. Viriot (Centre National de la Recherche Scientifique), FR-A1 2590589, **1987**; *J. Photochem. Photobiol. A* **1988**, *42*, 383; M. L. Viriot, J. C. Andre, M. Niclause, D. Bazard, R. Flayeux, M. Moll, *J. Inst. Brew.* **1980**, *86*, 21; *J. Am. Soc. Brew. Chem.* **1980**, *38*, 61; J. C. Andre, M. L. Viriot, J. Villermaux, *Pure Appl. Chem.* **1986**, *58*, 907.

[53] G. Kaupp, *Top. Curr. Chem.* **1988**, *146*, 57.

[54] Recently, interesting high-yield olefin metathesis cyclizations have been performed in sc-CO$_2$ under the influence of Ru and Mo catalysts: A. Fürstner, D. Koch, K. Langemann, W. Leitner, *Angew. Chem.* **1997**, *109*, 2562; *Angew. Chem. Int. Ed. Engl.* **1997**, *36*, 2466

[55] G. Kaupp, *Chemie in unserer Zeit* **1977**, *31*, 129; English translation is available in WWW under http://Kaupp.chemie.uni-oldenburg.de.

The Selective Blocking of *trans*-Diequatorial, Vicinal Diols; Applications in the Synthesis of Chiral Building Blocks and Complex Sugars

Thomas Ziegler

For efficient chemical synthesis of complex oligosaccharides, protecting group strategies and blocking techniques are of overriding significance. It is usually unavoidable first to prepare protected smaller saccharide units that enable a directed selective formation of the glycosidic bond and the sequential construction of larger saccharides. To this end all functional groups with the exception of the projected reaction center in the saccharide building block to be glycosylated (the glycosyl acceptor) must, as a rule, be protected, and only in particularly favorable cases can partially protected glycosyl acceptors be used for regioselective glycosylation. In the glycosyl donor, the saccharide building block that must be linked to the glycosyl acceptor, the protecting groups must be very precisely tuned to the planned glycosylation method. Both the reactivity of glycosyl donor and glycosyl acceptor and the anomeric selectivity of the glycosylation step is dominated by the protecting groups. The search for novel protecting group strategies is therefore as urgent for the carbohydrate chemist as the need for useful glycosylation methods.

Of the classical protecting groups of esters, ethers, and acetals that are favored in saccharide synthesis, the benzylidene and isopropylidene protecting groups have particular significance, because both open the possibility of blocking two hydroxyl functions in a monosaccharide simultaneously and selectively. As the example of methyl α-D-galactopyranoside in Scheme 1 illustrates, benzylidene acetals can be the major product from 1,3-diols, so that in the protected galactose derivative **1** the hydroxyl functions of positions 2 and 3 are available for further reaction. In contrast, isopropylidene acetals can be selectively prepared from vicinal *cis*-diols, with the result that in the example **2** chosen here the hydroxyl functions of positions 2 and 6 remain free. [1] Although isopropylidene acetals can also be synthesized from monosaccharides with vicinal *trans*-diequatorial diols, these acetals are usually so acid-labile that they are difficult to handle and are therefore less suitable for glycosylation reactions.

In the search for acetal protecting groups for vicinal *trans*-diequatorial diols that can be selectively introduced, the research group of S. V. Ley in Cambridge, England, recently developed the dispiroketal (Dispoke) [2] and cyclohexane-1,2-diacetal (CDA) protecting groups. [3] The former can be introduced by acetalization of a diol with the readily accessible 3,3′,4,4′-tetrahydro-6,6′-bis-2*H*-pyran [4, 5] (bis-DHP, **3a**), the latter by acetalization with the just as easily obtainable 1,1,2,2,-tetramethoxycyclohexane **4**. [3] In the case of the methyl α-D-galactopyranoside (Scheme

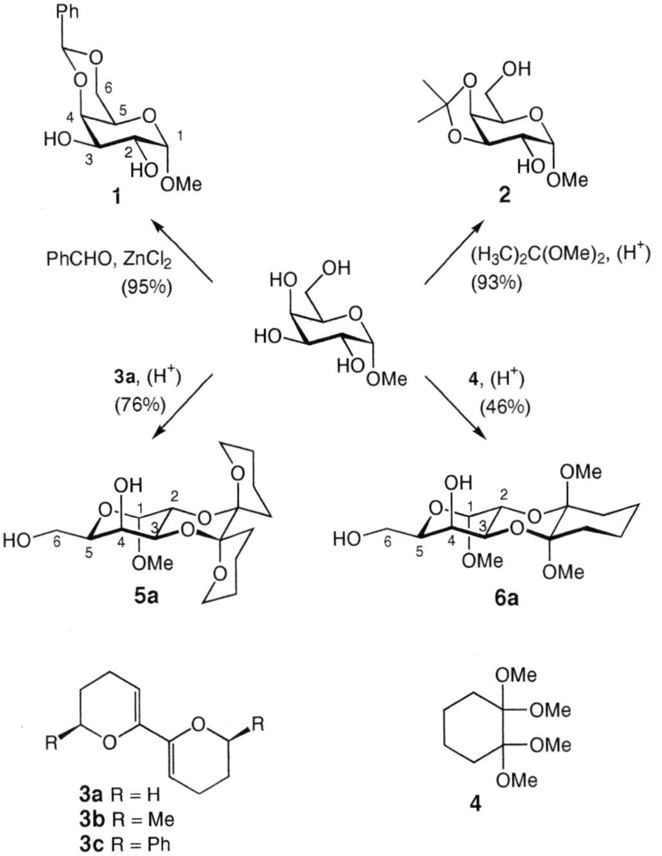

Scheme 1.

1), the *trans*-diequatorial hydroxyl functions of positions 2 and 3 can be highly selectively blocked as the Dispoke-protected galactoside **5a** (76%) [5] or the CDA-protected galactoside **6a** (46%) [3] in this way. Dispoke and CDA derivatives of monosaccharides are therefore useful complements to benzylidene and isopropylidene derivatives and to the 1,1,3,3-tetraisopropyl-1,3-disiloxane-1,3-diyl group which can be used as well for the protection of *trans*-diequatorial, vivinal diols in monosaccharides. [6] Like the classical acetal protecting groups, the Dispoke group can be removed hydrolytically under acidic conditions, preferably by transacetalization with ethylene glycol. [5]

The exceptionally high selectivity of the Dispoke and CDA groups can be explained by a steric interaction of the neighboring spiro centers and the equatorial arrangement of the alkyl residue as well as a strong anomeric effect of the two acetal functions. [2, 3] This high selectivity is particularly impressively reflected in the reaction of (S)-1,2,4-butanetriol with bis-DHP (**3a**) and a catalytic amount of camphorsulphonic acid (CSA), in which the dispiroketal **7** (96%) is obtained as exclusive stereoisomer [2] (Scheme 2). In the case of the symmetric glycerol, a "desymmetrization" of glycerol can even be achieved during the Dispoke formation by the use of chiral C_2-symmetric bis-DHP derivative **3b**.

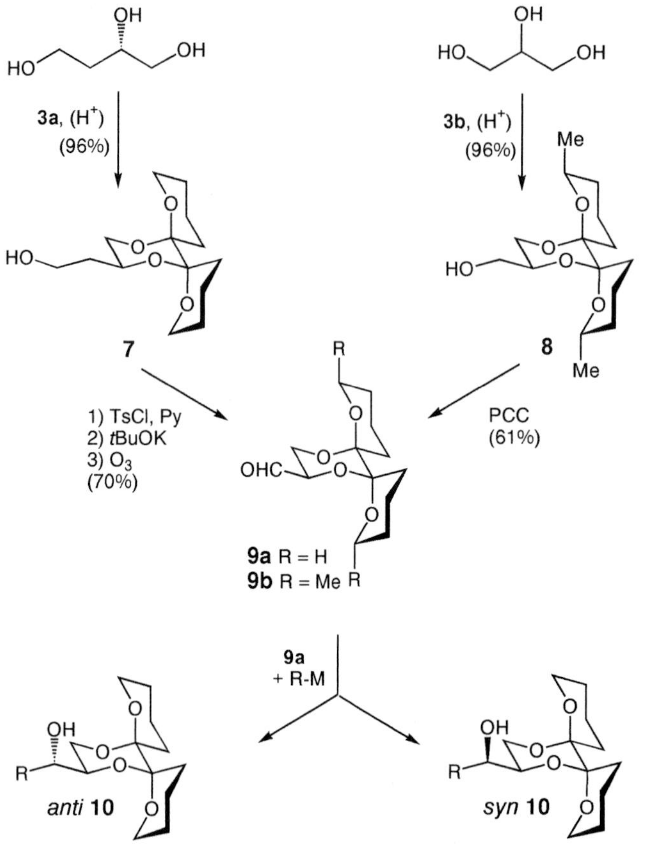

Scheme 2.

Here the only product is the enantiomerically pure compound **8** (96%). The compounds **7** and **8** can be transformed into the corresponding Dispoke-protected glyceraldehydes **9**, which not only are thermally more stable than the isopropylidene glycerinaldehyde, [2, 7] but also have a distinctly higher *anti*-selectivity in the 1,2-addition of carbon nucleophiles to the aldehyde function [2] (Table 1).

In the same way, glycolic acid can be stereoselectively transformed into the Dispoke derivative **10** (61%) with enantiomerically pure **3b**. [8] Formation of the ester enolate, alkylation with methyl iodide, renewed enolate formation and treatment with benzyl bro-

mide then give highly diastereoselectively in 51% yield the protected α-hydroxycarbonic acid derivative **11** (Scheme 3). [8] Similarly, other *meso*-diols like *myo*-inositols are easily desymmetrisized by Dispoke derivatives **3b** and **3c**, respectively. [9, 10] It can be predicted that the C_2-symmetric Dispoke group will rapidly become popular as auxilliary in the stereoselective synthesis of small building blocks.

Besides the 2,3-protected galactosides **5a** and **6a** (Scheme 1), other monosaccharides from the *gluco, manno, rhamno, fuco, xylo, lyxo,* and *arabino* series can be converted into the corresponding Dispoke- and CDA-blocked derivatives **5** and **6**. [3, 10, 11] How-

Table 1. Reaction of **9a** with various organometallics R–M [2].

R–M	Conditions	Yield **10**	anti/syn **10**	anti/syn Lit.[a]
MeLi	Et$_2$O/THF, –78 °C, 22 h	82 %	82 : 18	60 : 40
MeMgCl	THF, –78 °C, 24 h	92 %	81 : 19	–
Me$_2$CuLiMe$_2$S	Et$_2$O, –78 °C, 20 h	69 %	12 : 88	18 : 82
EtMgBr	THF, –78 °C, 6 h	62 %	73 : 27	–
H$_2$C=CHMgBr	THF, –78 °C, 4 h	56 %	91 : 9	60 : 40
(H$_2$C=CH)$_2$Zn	THF, 25 °C, 48 h	84 %	67 : 33	–
HC≡CMgBr	THF, –78 °C, 5 h	65 %	89 : 11	44 : 56
AllylMgBr	THF, –78 °C, 18 h	89 %	68 : 32	60 : 40

[a] Reaction of isopropyliden-glyceraldehyd with the corresponding organometallic.

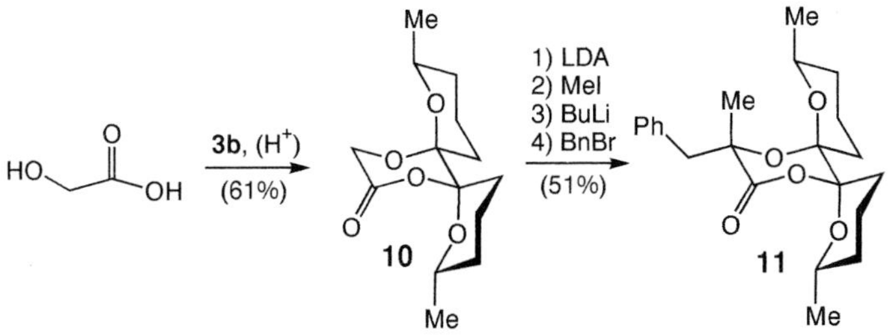

Scheme 3.

ever, the regioselectivity of the acetal formation of Dispoke derivatives can in some cases (for instance, methyl α-L-rhamnopyranoside) lead to mixtures of the 3,4- and 2,3-derivatives **5b** and **5b'**, respectively. [11] For the CDA analogues **6**, the ratio is most often shifted so far to the side of the *trans*-diequatorially blocked compounds **6b** that a preparative application seems meaningful [3] (Scheme 4).

On the other hand the Dispoke and CDA protection for D-glucopyranosides presents problems, because here all secondary OH groups are *trans*-diequatorially arranged. Both with bis-DHP **3a** and with 1,1,2,2-tetramethoxycyclohexane **4**, mixtures of the 2,3- and 3,4-protected glucosides **5c,6c** and **5c',6c'**, respectively, are obtained. [3, 11, 12] If, however, analogous to the "desymmetrization" of glycerol, phenyl-substituted, chiral

bis-DHP **3c** is used, for glucose derivatives the two regioisomers **12** and **13** can be prepared in a highly selective synthesis through double diastereoselection [13] (Scheme 5). [14]

For strategies with the Dispoke or CDA protecting groups in oligosaccharide synthesis the concept of armed and disarmed glycosyl donors [15] is especially fruitful. According to this principle the reactivity of a glycosyl donor is diminished by deactivating protecting groups (e. g. acyl groups) or by restricting the flexibility of the pyranose ring by bridging two hydroxyl functions (e. g. acetal protecting groups). Such a disarmed donor can then function a glycosyl acceptor and participate in a selective reaction with a reactive armed (most often benzyl-protected) glycosyl donor. The conformation-stabilizing Dispoke and CDA groups are plainly predestined to disarm a glycosyl donor. For example, the

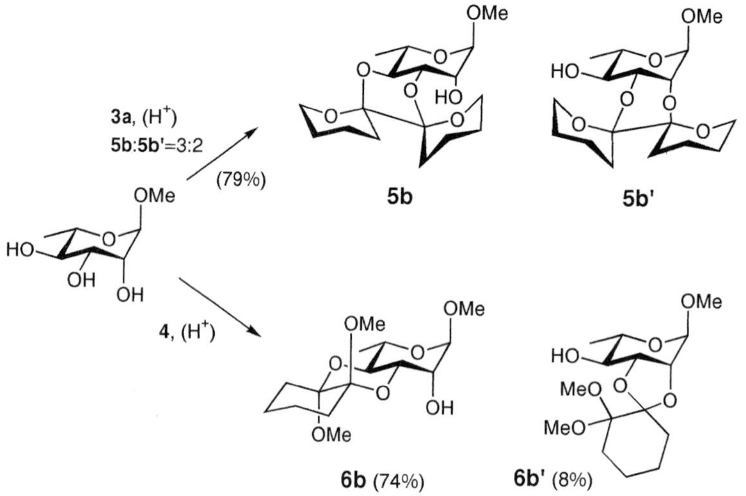

Scheme 4.

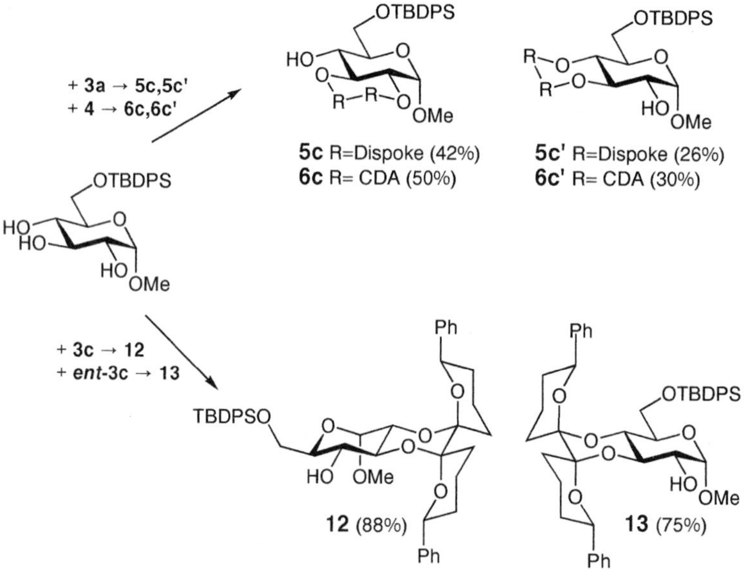

Scheme 5.

ethyl 1-thio-β-D-galactoside **14**, conformationally fixed by a Dispoke group and thus disarmed, undergoes a smooth reaction with thiogalactoside **15** – reactive as a result of benzyl protection – and the mild activator iodonium dicollidine perchlorate (IDCP) to form disaccharide **16** [16] (Scheme 6). Through the subsequent use of the more reac-

tive activator N-iodosuccinimide (NIS), disaccharide **16** is armed and reacts with the 2-*O*-benzoyl-protected and thus disarmed thiomannoside **17** as glycosyl acceptor in attractive yield to afford trisaccharide **18**.

The transformation of **18** as trisaccharide donor without further protecting group manipulation is achieved with the pseudo disaccha-

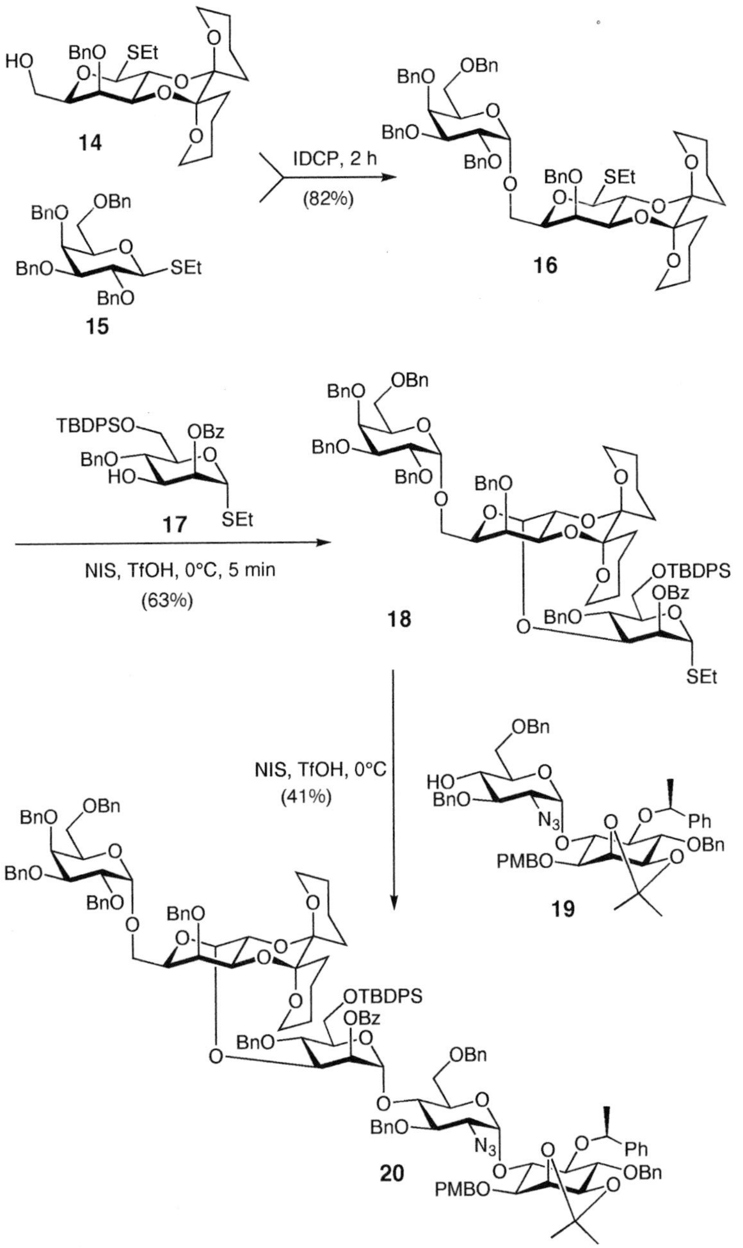

Scheme 6.

ride acceptor **19** and NIS after prolonged reaction times. The pentasaccharide **20**, a fragment of the GPI anchor of *Trypanosoma brucei*, is thus obtained in 41 % yield. [16]

Based on the concept of armed and disarmed glycosyl donors, the trisaccharide **23** can be constructed very elegantly from the two CDA-protected rhamnosides **6b** and **6d** in a one-pot reaction similar to the ciclamycin trisaccharide synthesis [17] in two steps as follows: [18] The IDCP-catalyzed condensation of armed thiorhamnoside **21** and disarmed glycosyl acceptor **6d** yields as intermediate a disaccharide that reacts immediately without further purification with the CDA derivative **6b** under NIS activation to form **23** (62 %). The disaccharide **22** is also found as by-product (10 %), which arises in the second glycosylation step from unconverted **21** and added **6b.** The final deblocking of saccharide **23** furnishes the *Streptococcus* antigen trisaccharide **24** (Scheme 7).

In particular, the examples given in Schemes 6 and 7 for the application of Dispoke and CDA groups in oligosaccharide synthesis demonstrate magnificently that with these novel protecting groups complex sugars can be synthesized very efficiently. This was also recently demonstrated for the elegant synthesis of nonasaccharide fragments related to glycoproteins. [19] The Dispoke and CDA acetals should therefore rapidly become standard practice in carbohydrate chemistry.

References

[1] T. W. Greene, P. G. M. Wuts, *Protective Groups in Organic Chemistry*, 2nd Ed., J. Wiley, New York, **1991**; P. J. Kocienski, *Protecting Groups*, Thieme, Stuttgart, **1994**.

[2] S. V. Ley, M. Woods, A. Zanotti-Gerosa, *Synthesis* **1992**, 52–54.

[3] S. V. Ley, H. W. M. Priepke, S. L. Warriner, *Angew. Chemie* **1994**, *106*, 2410–2412.

[4] S. Gohsal, G. P. Luke, K. S. Kyler, *J. Org. Chem.* **1987**, *52*, 4296–4298.

[5] S. V. Ley, R. Leslie, P. D. Tiffin, M. Woods, *Tetrahedron Lett.* **1992**, *33*, 4767–4770.

[6] J. J. Oltvoort, M. Klosterman, J. H. van Boom, *Recl. Trav. Chim. Pays-Bas* **1983**, *102*, 501–505.

[7] G.-J. Boons, D. A. Entwistle, S. V. Ley, M. Woods, *Tetrahedron Lett.* **1993**, *34*, 5649–5652.

[8] R. Downham, K. S. Kim, S. V. Ley, M. Woods, *Tetrahedron Lett.* **1994**, *35*, 769–772.

[9] P. J. Edwards, D. A. Entwistle, S. V. Ley, D. Owen, G. Visentin, *Tetrahedron Lett.* **1994**, *35*, 777–780.

[10] S. V. Ley, R. Downham, P. J. Edwards, J. E. Innes, M. Woods, *Contemp. Org. Synth.* **1995**, *2*, 365–392; and references cited therein.

[11] S. V. Ley, G.-J. Boons, R. Leslie, M. Woods, D. M. Hollinshead, *Synthesis* **1993**, 689–692.

[12] A. B. Hughes, S. V. Ley, H. W. M. Priepke, M. Woods, *Tetrahedron Lett.* **1994**, *35*, 773–776.

[13] S. Masamune, W. Choy, J. S. Petersen, L. R. Sita, *Angew. Chem.* **1985**, *97*, 1–31.

[14] C. Genicot, S. V. Ley, *Synthesis* **1994**, 1275.

[15] B. Fraser-Reid, U. E. Udodong, Z. Wu, H. Ottoson, R. Merritt, C. S. Rao, C. Roberts, *Synlett* **1992**, 927–942; G. H. Veeneman, J. H. van Boom, *Tetrahedron Lett.*, **1990**, *31*, 275–278.

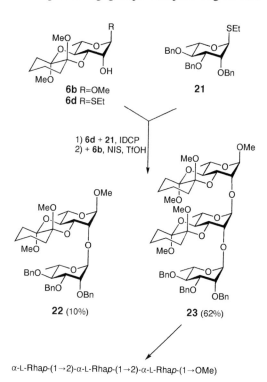

1) **6d** + **21**, IDCP
2) + **6b**, NIS, TfOH

22 (10%)   **23** (62%)

α-L-Rhap-(1→2)-α-L-Rhap-(1→2)-α-L-Rhap-(1→OMe)

24 (53%)

Scheme 7.

[16] G.-J. Boons, P. Grice, R. Leslie, S. V. Ley, L. L. Yeung, *Tetraheron Lett.* **1993**, *34*, 8523–8526.

[17] S. Raghavan, D. Kahne, *J. Am. Chem. Soc.* **1993**, *115*, 1580–1581.

[18] S. V. Ley, H. W. M. Priepke, *Angew. Chem.* **1994**, *106*, 2412–2414.

[19] P. Grice, S. V. Ley, J. Pietruszka, H. M. I. Osborn, H. W. M. Priepke, S. L. Warriner, *Chem. Eur. J.* **1997**, *3*, 431–440; and references cited therein.

Oxidative Polycyclization Versus the "Polyepoxide cascade": New Pathways in Polyether (Bio)synthesis?

Ulrich Koert

In 1983, Cane et al. postulated that polyether biosynthesis takes place as a two step process. The first of these steps was proposed to be the enzymatically polyepoxidation of an acyclic hydroxypolyene precursor, and the second comprised a cascade of intramolecular epoxide ring openings with formation of the polyether frame work. [1] For monensin A (**3**), for example, this is shown in Scheme 1: the (*E,E,E*)-triene **1** is transformed biosynthetically into the tris-epoxide **2**, which reacts further in a cascade of epoxide ring openings to yield the natural product **3**. [2]

This biosynthetic scheme gained rapid acceptance, above all because of the elegance of the second step, the "polyepoxide cascade". Polyepoxide cascades of the type **2** → **3** were reproduced in the laboratory [3] and used as key steps in the synthesis of polyethers. [4] The first step in the Cane-Celmer-Westley hypothesis, the stereoselective epoxidation, was subject to far less attention, even though it may be regarded as a weak point in the hypothesis. In feeding studies with *Streptomyces cinnamonensis*, no conversion of the (*E,E,E*)-triene **1** to monensin A (**3**) could be detected. [5] Earlier labeling studies had shown merely that the oxygen atoms in two tetrahydrofuran rings (O-7,8) and in the tetrahydropyran hemiacetal (O-9) were derived from molecular oxygen, and it was unclear

whether these were incorporated by epoxidation or by another oxidation process. [2] Although the second step of the the Cane-Celmer-Westley hypothesis, the polyepoxide cascade, is compelling, the initial, polyepoxidation step is still not sufficiently proven experimentally, and hence the entire hypothesis is uncertain.

McDonald and Towne [6] extended studies of Townsend and Basak, [7] and presented an alternative model to the Cane-Celmer-Westley hypothesis: *syn*-oxidative polycyclization. The Townsend-Basak-McDonald hypothesis for the biosynthesis of monensin A (**3**) is depicted in Scheme 2.

The starting point is the alkoxy-bound oxo metal derivative of a (*Z,Z,Z*)-triene **4**. Intramolecular [2+2] cycloaddition affords a metallaoxetane **5**, [8] and the first tetrahydrofuran ring is then closed by reductive elimination of the metal to form compound **6**. An oxidation step from **6** to **7** activates the alkoxy bound metal for the next [2+2] cycloaddition (**7** → **8**). Reductive elimination of the metal with closure of the second tetrahydrofuran ring leads to the compound **9**. The third tetrahydrofuran ring of monensin A is closed in a further, analogous oxidative cyclization sequence (**9** → **10** → **11** → **12**). From compound **12** the natural product can be obtained *via* the intermediate **13**. The stereochemical

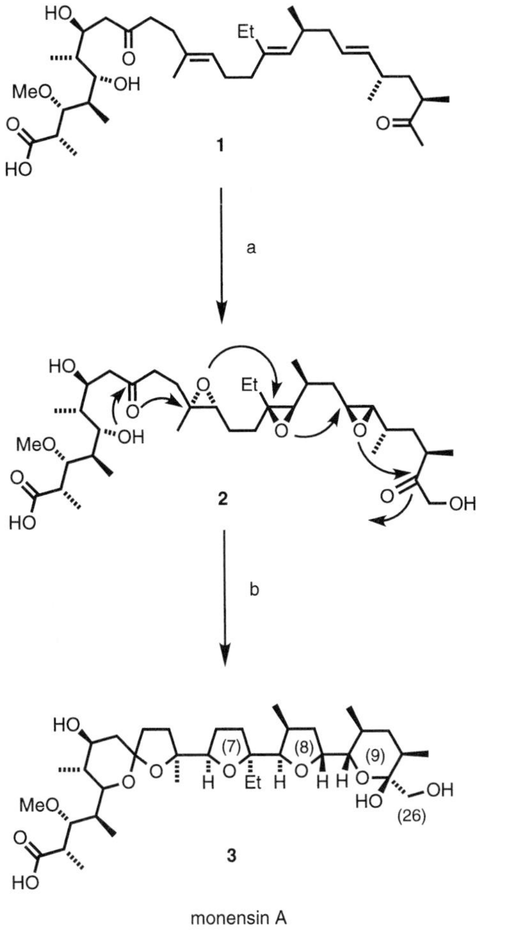

Scheme 1. The Cane-Celmer-Westley hypothesis for the biosynthesis of monensin A (**3**); (a) stereoselective epoxidation; (b) polyepoxide cascade; it is assumed that the OH group at C(26) is also introduced during the epoxidation step.

course of the oxidative cyclization should be emphasized: because a [2+2] cycloaddition is involved, the two oxygen atoms are added to the double bond in a *syn* configuration. In order to construct the correct stereochemistry for the polyether framework of monensin A (**3**), the Townsend-Basak-McDonald hypothesis therefore postulates a (Z,Z,Z)-triene **4** as a biosynthetic precursor.

Using more simple systems, McDonald and Towne were able to demonstrate that pyridinium chlorochromate (PCC) is a suitable reagent for carrying out *syn*-oxidative polycyclizations (Scheme 3). [6] Thus, transformation of the (Z)-hydroxydiene **14** yielded the bis-tetrahydrofurans **15** and **16**. Starting with the (E)-hydroxydiene **17**, the bis-tetrahydrofurans **18** and **19** were obtained.

The formation of **15** and **16** from the (Z)-diene and of **18** and **19** from the (E)-diene support the assumption of a *syn*-oxidative mechanism. The authors present conformational-analysis arguments to explain the preferred formation of the *trans*-tetrahydrofurans **15** and **18**. [6]

The low yields attained for the reactions shown in Scheme 3 call into doubt the usefulness of the *syn*-oxidative polycyclization under the published PCC-conditions for preparative purposes. However, the palette of possible oxidation reagents is by no means exhausted by PCC. Valuable ideas for a solution of the problem come from the work of Kennedy et al. with Re_2O_7 as an oxidizing agent for oxidative olefin cyclization. [6] As shown in Scheme 4 Keinan et al. [10] applied Re_2O_7 successfully in an iterative oxidative-olefin-cyclization sequence leading to oligo-tetrahydrofurans (**20** → **21** → **22**).

A fully satisfying answer to the question of the biosynthesis of polyether compounds such as monensin A (**3**) cannot be given from the results described here. However, it is clear that the new the Townsend-Basak-McDonald hypothesis has gained ground on the older Cane-Celmer-Westley hypothesis. A key experiment to test the Townsend-Basak-McDonald hypothesis will be the use of labeled (Z,Z,Z)-triens of type **4** in feeding experiments with *Streptomyces cinnamonensis*.

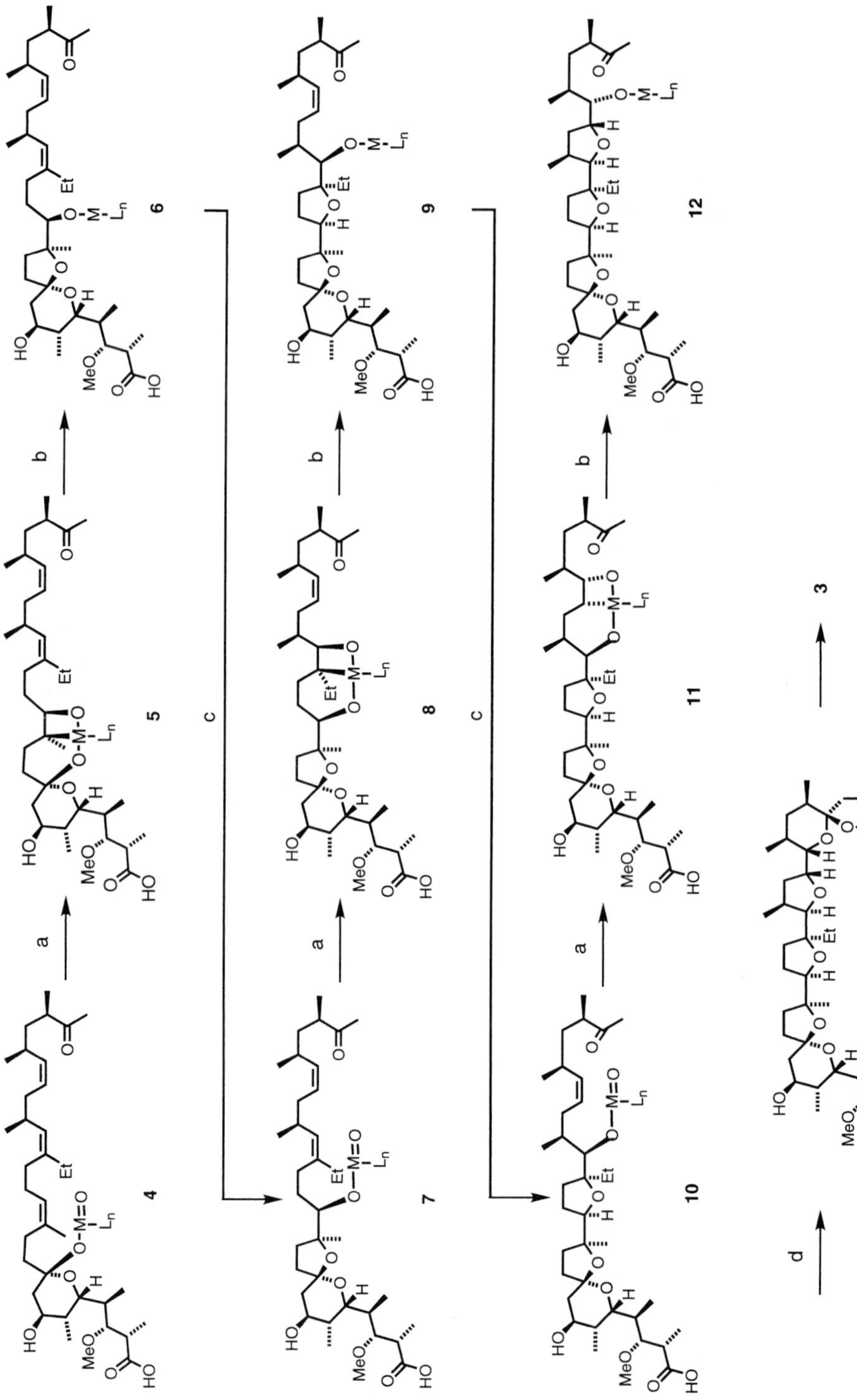

Scheme 2. Townsend-Basak-McDonald hypothesis for the biosynthesis of monensin A (**3**); (a) [2+2] cycloaddition; (b) reductive elimination of the metal with formation of the tetrahydrofuran ring; (c) reoxidation of the metal; (d) hemiacetal formation and oxidative introduction of the OH group at C(26).

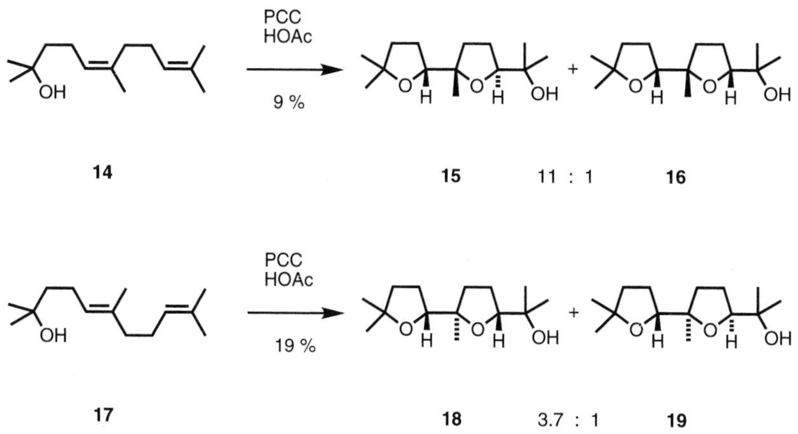

Scheme 3. Model reactions of the Townsend-Basak-McDonald hypothesis using PCC as the oxidizing agent.

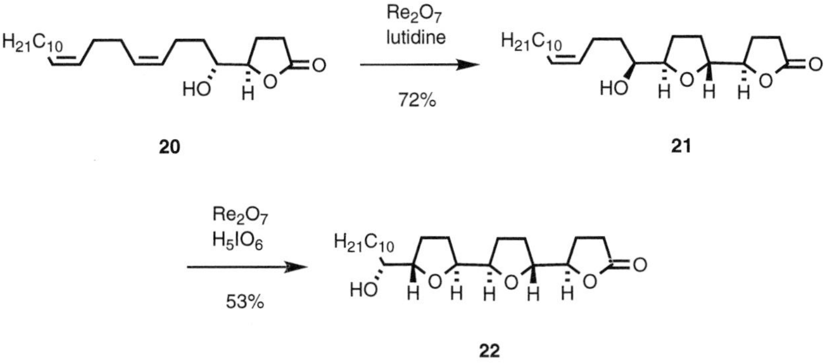

Scheme 4. Use of Re$_2$O$_7$ in an oxidative-olefin-cyclization sequence.

References

[1] D. E. Cane, W. D. Celmer, J. W. Westley, *J. Am. Chem. Soc.* **1983**, *105*, 3594–3600.

[2] J. A. Robinson, *Prog. Chem. Org. Nat. Prod.* **1991**, *58*, 1–81.

[3] a) W. C. Still, A. G. Romero, *J. Am. Chem. Soc.* **1986**, *108*, 2105–2106; b) S. L. Schreiber, T. Sammakia, B. Hulin, G. Schulte, *ibid.* **1986**, *108*, 2106–2108; c) S. A. Russell, J. A. Robinson, D. J. Williams, *J. Chem. Soc. Chem. Commun.* **1987**, 351–352.

[4] For representative examples see: a) T. R. Hoye, J. C. Suhadolnik, *J. Am. Chem. Soc.* **1985**, *107*, 5312–5313; b) K. Nozaki, H. Shirama, *Chem. Lett.* **1988**, 1847; c) I. Paterson, R. D. Tillyer, J. B. Smaill, *Tetrahedron. Lett.* **1993**, *34*, 7137–7140; d) U. Koert, H. Wagner, M. Stein, *ibid.* **1994**, *35*, 7629–7632.

[5] D. S. Holmes, J. A. Sherringham, U. C. Dyer, S. T. Russell, J. A. Robinson, *Helv. Chim. Acta* **1990**, *73*, 239–259: the authors point

out that solubility problems may have played a role in the failure of the labeling experiment.

[6] F. E. McDonald, T. B. Towne, *J. Am. Chem. Soc.* **1994**, *116*, 7921–7922.

[7] C. A. Townsend, A. Basak, *Tetrahedron*, **1991**, *47*, 2591–2602.

[8] Metallaoxetanes have not yet been directly identified. Their role as intermediates in oxidative processes is however the subject of intensive discussion: K. A. Jorgensen, B. Schiott, *Chem. Rev.* **1990**, *90*, 1483–1506.

[9] S. Tang, R. M. Kennedy, *Tetrahedron Lett.* **1992**, *33*, 5303–5306.

[10] S. C. Sinha, A. Sinha-Bagchi, E. Keinan, *J. Am. Chem. Soc.* **1995**, *117*, 1447–1448.

Radical Reactions as Key Steps in Natural Product Synthesis

Ulrich Koert

Radical reactions have developed into indispensable methods in organic synthesis [1] and are often used as key steps in the construction of complex natural products. Impressive demonstrations are found in the following examples taken from the current literature: the dactomelyne synthesis by Lee et al., [2] the camptothecin synthesis by Curran et al., [3] and of (7)-deoxypancratistatin by Keck et al., [4] and the approaches towards aspidospermidine by Murphy et al. [5] and pseudopterosine A by Schmalz et al. [6].

The total synthesis of the marine natural product (*3Z*)-dactomelyne (**1**) requires the elaboration of the *cis*-linked pyranopyran skeleton and the stereoselective introduction of halogen atoms on the tetrahydropyran rings. Lee et al. [2] solved this problem by using of β-alkoxy-acrylates twice as radical acceptors (Scheme 1).

A radical cyclization of the trichloro compound **2** afforded the bicyclic product **3**. The selective monodehalogenation of the dichloro compound **3** giving **4** was accomplished with a radical reaction as well. Important for the desired stereoselectivity was the use of the silane $(Me_3Si)_3SiH$. Stannanes like *n*-Bu_3SnH gave the other epimer **5** as the major product. The highlight of the synthesis was the construction of the second tetrahydropyran ring and simultaneous stereoselective

introduction of the bromo-substituted stereocenter: the radical reaction of β-alkoxy-acrylate **6** with *n*-Bu_3SnH gave exclusively the desired bicyclic product **7**. The authors propose a chairlike transition state **8** to explain the observed stereoselectivity. The bicyclic product **7** was converted in few steps into the target compound (*3Z*)-dactomelyne (**1**).

The potent antitumor agents camptothecin (**11**) and its derivatives are targets of extensive synthetic activities. [7] A major contribution to this research area came from the laboratory of D. P. Curran [3] in Pittsburgh with a total synthesis of camptothecin relying on radical chemistry. In the key step of his synthesis (Scheme 2) the iodoalkyne **9** was allowed to react with phenyl isonitrile to provide the target compound **11**.

In this radical cascade reaction, an excellent example of a "sequential-reaction", [8] the simultaneous construction of the B- and C-rings of camptothecin was accomplished. The mechanism for this radical reaction is best explained by the transformation of the structurally less complex bromoalkyne **12** into the tetracyclic compound **17**. First, a trimethylstannyl radical produced from hexamethyldistannane attacks the C–Br bond of compound **12**. The resulting pyridone radical **13** reacts intermolecularly with the isonitrile **10** to yield the radical intermediate **14**. An intra-

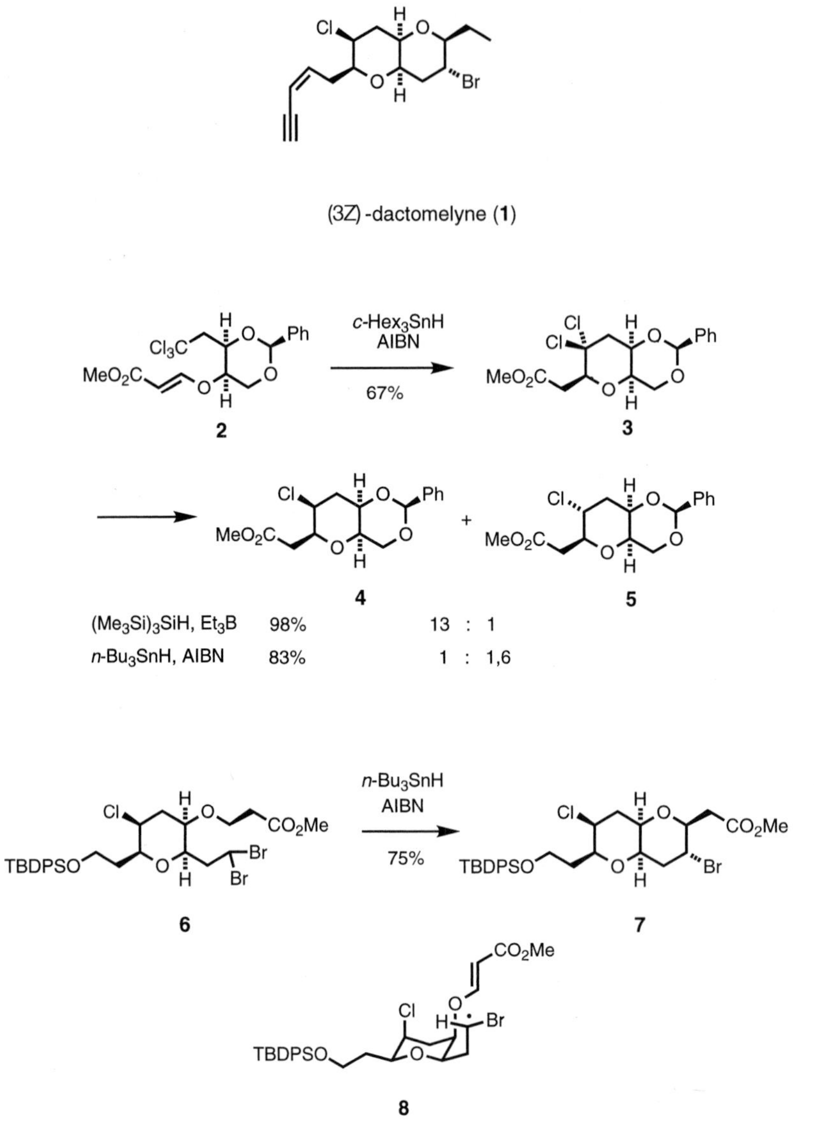
(3Z)-dactomelyne (1)
c-Hex3SnH
AIBN
67%
2
3
4
5
(Me3Si)3SiH, Et3B 98% 13 : 1
n-Bu3SnH, AIBN 83% 1 : 1,6
n-Bu3SnH
AIBN
75%
6
7
8

Scheme 1.

molecular attack of the radical center in **14** on the alkyne functionality leads to radical **15**. Finally, the A-ring is attacked by the radical center in **15** leading to **16**, which rearomatizes to the desired compound **17**.

The alkaloid pancratistatin **18** isolated from *Amaryllidaceous* plants displays promising antineoplastic and antiviral activity. [9a] The (7)-deoxy compound **19** exhibits both better therapeutic properties and decreased toxicity.

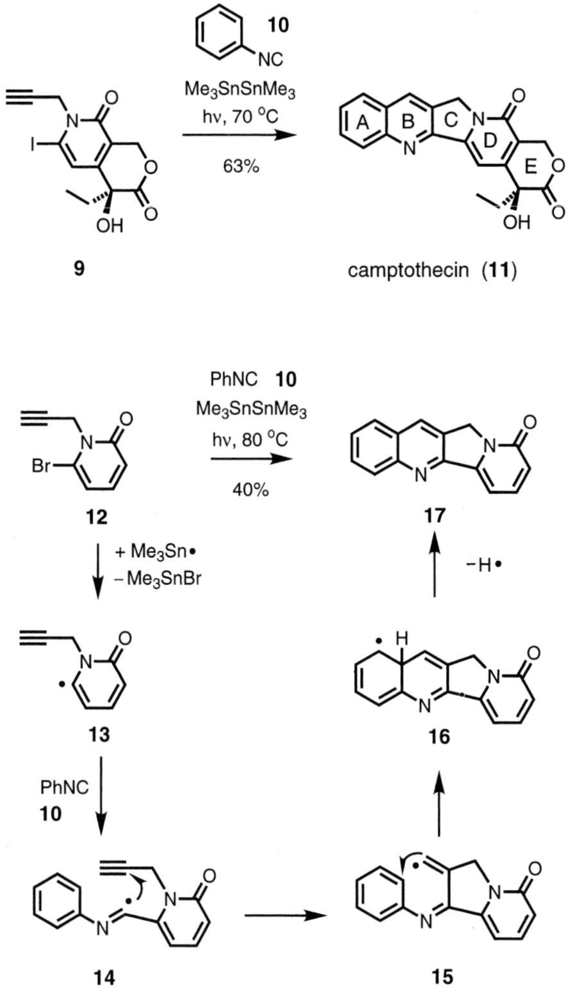

Scheme 2.

[9b] The group of G. E. Keck [4] successfully completed the total synthesis of (7)-deoxy-pancratistatin (**19**) using a radical reaction as a key step (Scheme 3).

Here, the radical precursor **20** was treated with nBu$_3$SnH to yield the benzyl radical **21**, which reacted further by a 6-*exo*-cyclization to the cyclohexylamine **22**. Important for the excellent stereoselectivity of this step was

the conformational control dictated by the *tert*-butyldimethylsilyl (TBDMS) protected six-membered lactol ring. Compound **22** was then transformed in few standard steps into the target compound **19**.

J. A. Murphy et al. [5] were able to show the synthetic potential of radical reactions in a short and efficient synthesis of the ABCE ring system of aspidospermidine (**23**)

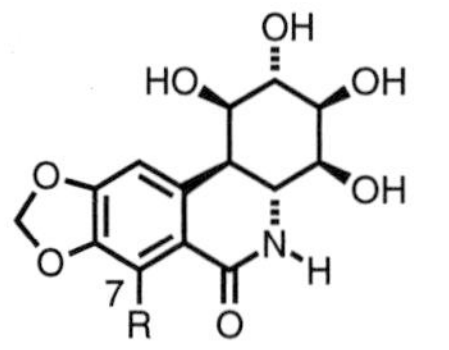

pancratistatin (**18**, R = OH)

7-deoxypancratistatin (**19**, R = H)

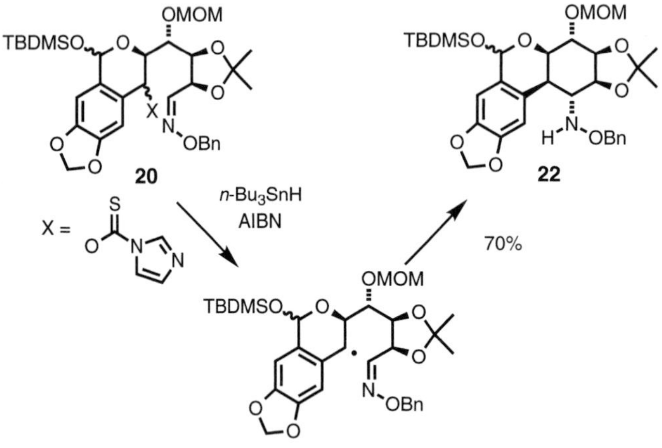

Scheme 3.

(Scheme 4). They used a tandem radical cyclization to directly convert the aryl iodide **24** into the tetracycle **25**.

The mechanism of this reaction sequence was rationalized as follows: A silyl radical produced from (Me₃Si)₃SiH attacks first the aryl iodide bond in **24** forming the aryl radical **26**. Intramolecular attack of the radical center in **26** on the double bond then affords the alkyl radical **27**. Lastly, addition of this alkyl radical to the azide function followed by loss of N₂ gives the observed product **25** via the N-radical intermediate **28**. The high stereoselectivity and the excellent yield of this tandem radical cyclization is noteworthy.

New perspectives are opened by the combination of radical reactions with organometallic chemistry, as was demonstrated by the group of H.-G. Schmalz from the TU Berlin. [6] In the course of synthetic studies towards Pseudopterosin A (**29**) and related compounds they succeeded in the SmI₂- induced cyclization of the arenetricarbonylchromium-complex **30** to the tricyclic pseudopterosin scaffold **31** (Scheme 5).

The authors propose the following mechanism for this reaction: In the first step the ketyl radical **32** is produced, which adds to the arene ring to give **33**. A second transfer of one electron by SmI₂ affords the anionic

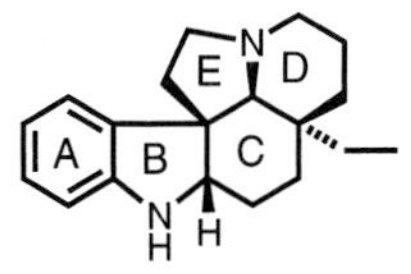

aspidospermidine (**23**)

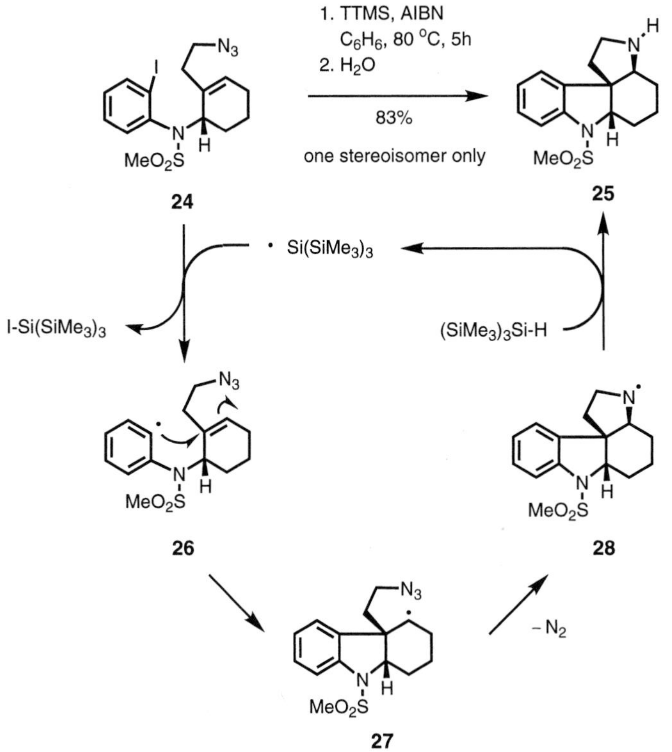

Scheme 4.

η^5-complex **34**. Subsequent protonation yileds η^4-complex **35**, from which the final product **31** is obtained after elimination of methanol and hydrolytic workup. It should be noted that the pendant trisubstituted double bond does not give rise to any side reactions.

The present selection of natural product syntheses with radical reactions as key steps demonstrates the extraordinary potential applications of modern radical chemistry.

However, one limitation is evident: four out of the five reaction sequences presented involved an intramolecular cyclization reaction. [10] Intermolecular radical bond formations with companally high yields and stereoselectivities are still very rare in the total synthesis of bioactive compounds. One exception is Curran's camptothecin synthesis. However, progress in acyclic stereoselection of radical reactions [11] should soon help to formulate new solutions for these synthetic challenges.

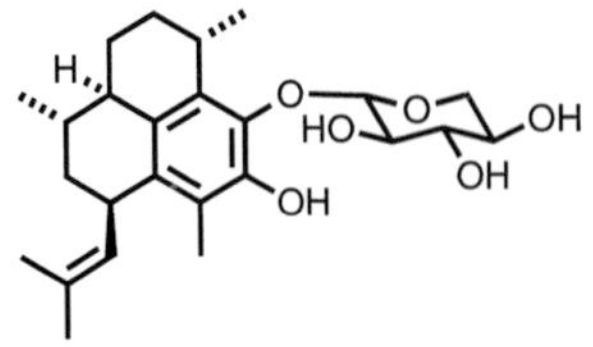

pseudopterosin A (**29**)

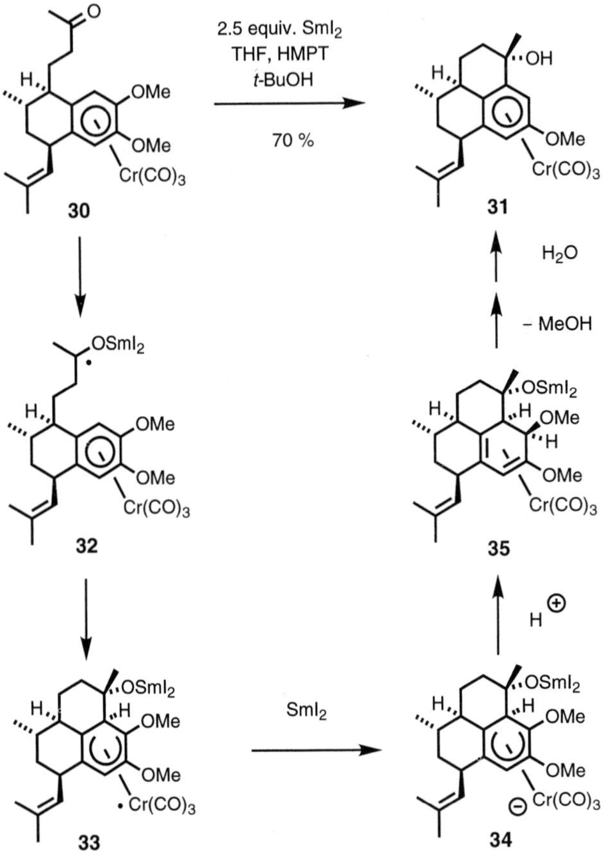

Scheme 5.

References

[1] a) B. Giese, *Radicals in Organic Synthesis: Formation of Carbon-Carbon-Bonds*, Pergamon, Oxford, **1986**; b) D. P. Curran in *Comprehensive Organic Synthesis*, Vol. 4 (Eds.: B. M. Trost, I. Fleming), Pergamon, Oxford, **1991**, p. 779; c) D. P. Curran, N. A. Porter, B. Giese, *Stereochemistry of Radical Reactions*, VCH, Weinheim, **1996**.

[2] E. Lee, C. M. Park, J. S. Yun, *J. Am. Chem. Soc.* **1995**, *117*, 8017.

[3] a) D. P. Curran, S.-B. Ko, H. Josien, *Angew. Chem.* **1995**, *107*, 2948; *Angew. Chem. Int. Ed. Engl.* **1995**, *34*, 2683; b) D. P. Curran, H. Liu, *J. Am. Chem. Soc.* **1992**, *114*, 5863.

[4] G. E. Keck, S. F. McHardy, J. A. Murry, *J. Am. Chem. Soc.* **1995**, *117*, 7289.

[5] M. Kizil, J. A. Murphy, *J. Chem. Soc., Chem. Commun.* **1995**, 1409.

[6] a) H.-G. Schmalz, S. Siegel, J. W. Bats, *Angew. Chem.* **1995**, *107*, 2597; *Angew. Chem. Int. Ed. Engl.* **1995**, *34*, 2383.

[7] a) D. P. Curran, *J. Chin. Chem. Soc.* **1993**, *40*, 1; b) U. Koert, *Nachr. Chem. Tech. Lab.* **1995**, *43*, 686.

[8] L. F. Tietze, U. Beifuss, *Angew. Chem.* **1993**, *105*, 137; *Angew. Chem. Int. Ed. Engl.* **1993**, *32*, 131.

[9] a) G. R. Pettit, V. Gaddamidi, G. M. Cragg, *J. Nat. Prod.* **1984**, *47*, 1018; b) S. Ghosal, S. Singh, Y. Kumar, R. S. Srivastava, *Phytochemistry* **1989**, *28*, 611.

[10] D. P. Curran, J. Xu, E. Lazzarini, *J. Am. Chem. Soc.* **1995**, *117*, 6603.

[11] a) W. Smadja, *Synlett* **1994**, 1; b) W. Damm, J. Dickhaut, F. Wetterich, B. Giese, *Tetrahedron Lett.* **1993**, *34*, 431; c) N. A. Porter, B. Giese, D. P. Curran, *Acc. Chem. Res.* **1991**, *24*, 296; d) P. Renaud, M. Gerster, *J. Am. Chem. Soc.* **1995**, *117*, 6607.

Light-Directed Parallel Synthesis of Up to 250 000 Different Oligopeptides and Oligonucleotides

Günter von Kiedrowski

The development of enzyme inhibitors based on oligopeptides makes it imperative to prepare a daunting number of oligopeptides with different sequences and to test their binding properties in order to find an optimally binding ligand.

Even if we restrict ourselves to the twenty naturally occurring amino acids, $20^2 = 400$ dipeptides, $20^3 = 800$ tripeptides, and in general 20^N oligopeptides of length N are possible. A systematic testing of all sequence alternatives with conventional techniques is already out of the question with tetrapeptides (160 000 compounds). Molecular modeling and quantitative structure-activity relationships rule out many sequence alternatives based on a known guide sequence, but for a "rational drug design" [1] in the true sense, a three-dimensional structure of the receptor must be known. The prerequisite for "irrational drug design" [2] (i. e., the filtering out of binding sequences from the pool of statistical sequences) is the ability to isolate the best ligand in pure form and in high enough quantities for a structure determination – or it must be possible to raise the concentration to the required level. This last criterion restricts irrational drug design largely to biopolymers that can be cloned (e. g., antibodies) or replicated *in vitro* (nucleic acids). Such attempts to select and amplify biomolecules possessing the desired binding properties have already been honored in a Highlight [3].

Recently Fodor et al. described a chemical method for irrational drug design. [4] A combination of three processes – solid-phase synthesis, photolithography, and affinity labeling – makes it possible to lay down next to each other up to 250 000 different oligopeptides or oligonucleotides per square centimeter on a glass slide and to test their binding to a given receptor molecule. The principle of photolithographic solid-phase synthesis is shown in Figure 1 for the example of the parallel construction of two dipeptides from two amino acids.

The glass surface is first treated with 3-(aminopropyl)triethoxysilane and then the free amino groups, now bound to the glass surface by a C_3 spacer, are protected with a nitroveratryloxycarbonyl (NVOC) group (Step **I**), which can be photochemically removed. On illumination (Step **II**) only those NVOC groups that are not covered by an opaque mask are removed. The irradiated amino groups on the surface (field **A**) are now acylated with the first amino acid, which is protected by NVOC and is activitated as its 1-hydroxybenzotriazol (HOBt) ester (Step **III**). Field A is now masked, and field B exposed to light and thereafter acylated with the second amino acid, thus completing the first phase

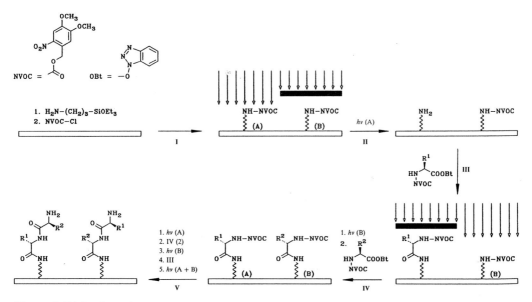

Figure 1. Light-directed parallel synthesis of two dipeptides from two amino acids in two irradiation fields.

of the synthesis (Step **IV**). Further analogous irradiation and coupling steps construct the dipeptides in the second phase (Step **V**).

In practice strips of the surface are exposed in each photolysis and coupling round of the synthesis. The strips irradiated in the next round lie perpendicular to those of the preceding round. This corresponds to oligopeptide construction in the rows and columns of a square matrix and enables a considerably shorter synthesis than were possible with successive illumination of each field. The shortest synthesis is achieved with the binary masking technique, in which the direction of the strip is changed in every cycle and its width is reduced to half every second cycle, while the number of strips in the mask is doubled. In this way, with n synthesis cycles 2^n different oligomers can be constructed, whose lengths are described by a binomial distribution (combinatorial synthesis). Ten synthesis cycles, for example, yield 1024 (minus 1) different peptides with "1 of length 0", and successively 10, 45, 120, 210, 252, 210, 120, 45, and 1 with lengths 1–10. Each peptide sequence is situated at one field of the (32×32) matrix.

The last irradiation step removes the N-terminal NVOC groups of all peptides; other deprotecting steps may be necessary to free the funtional groups in side chains. For the affinity test the glass slide is simply held in a solution of the fluorescence-labeled receptor, rinsed, and scanned under the fluorescence microscope. The affinity of a peptide for a given receptor is determined from the fluorescence intensity of the corresponding field in the array. The authors demonstrated the whole process on a combinatorial synthesis of 1024 peptides, whose binding to the antibody 3E7 was to be tested. The 15 peptides with clearly recognizable affinity all begin with the sequence 5'-H$_2$N-Tyr-Gly. This result agrees with those by Cwirla et al., who claim an N-terminal tyrosine residue as the key determinant in the binding to 3E7. [5]

As light source, the authors used a mercury arc lamp, that required an exposure of 20 min at $\lambda = 365$ nm and 12 mWcm^{-2}. Shorter exposure times should be possible with an argon laser. To prevent excitation at aromatic side chains and, in particular, a photochemical decomposition of tryptophan, shorter wavelengths are not recommended. The total yield per synthesis cycle (irradiation and coupling) lies between 85 and 95 %, so that for oligopeptides with more than ten amino acids the process only makes sense in exceptional cases. With mask technology a resolution of > 20 μm^2 per synthesis field can be achieved; thus with 50 μm^2 elements, 250 000 syntheses can be addressed on a glass cover slide for microscopy. The authors show that their process is not only restricted to oligopeptides, but can also be applied to light-directed oligonucleotide syntheses, in which case 5′-NVOC-protected nucleoside-3′-cyanoethylphosphoramidites are suitable monomeric building blocks.

Three critical questions must, however, be answered experimentally before a general application to affinity studies is recommended:

1. Does the carrier-bound oligopeptide have the same conformation as in solution?
2. Which spacer can be used as alternative to a C$_3$ chain, so that a large proportion of biological receptors can bind unhindered?
3. Can the yields of the synthesis cycles be improved, for instance, by the use of other protecting groups?

We wait in suspense for a report in the near future that the authors can suggest a solution to these problems. If this is the case it is foreseeable that light-driven parallel synthesizer machines, similar to the present peptide and DNA synthesis machines with soon revolutionize many university and pharmacetical research laboratories. The site of the Affymax Research Institute of the authors in "Silicon Valley" will, of course, not hinder the development of the first prototype of the commercial machines.

References

[1] H. Frühbeis, R. Klein, H. Wallmeier, *Angew. Chem.* **1987,** *99,* 413; *Angew. Chem. Int. Ed. Engl.* **1987,** *26,* 403.
[2] The term "irrational drug design" was recently proposed by S. Brenner at a symposium on natural and artificial selection processes (Max-Planck-Institut für Biophysikalische Chemie, Göttingen, 17th–19th April 1991). In contrast to "rational drug design", in which more or less detailed knowledge about the receptor, the ligands, and their interaction is available and does influence the development of better ligands, "irrational drug design" embodies attempts in the face of a void of knowledge in which a large repertoire of ligands are tested blindly.
[3] A. Plückthun, L. Ge, *Angew. Chem.,* **1991** *103,* 301; *Angew. Chem. Int. Ed. Engl,* **1991,** *30,* 296.
[4] S. P. A. Fodor, J. L. Read, M. C. Pirrung, L. Stryer, A. T. Lu, D. Solas, *Science* **1991,** *251,* 767.
[5] S. E. Cwirla, E. A. Peters, R. W. Barret, W. J. Dower, *Proc. Natl. Acad. Sci. USA* **1990,** *87,* 6378.

Opportunities for New Chemical Libraries: Unnatural Biopolymers and Diversomers

Rob M. J. Liskamp

The solid-phase methodology for the synthesis of biopolymers is nowadays indispensable in chemistry, pharmacology, immunology, biology, physiology and biophysics. Using this fantastic tool, [1] it is not only possible to prepare (large) peptides in a (semi)automatic manner in which amino acids and other reagents are added, and excess of reagents and waste are removed (without the tedious purification of intermediates), but this methodology has also been successfully applied to the synthesis of nucleic acids and recently carbohydrates. [2, 3] Once developed the next logical step was the adaptation of the efficient solid-phase methodology to the *simultaneous* synthesis of *several* peptides. This led to the development of several multiple peptide synthesis strategies with a large range of applications. [4] A significant step further were the peptide libraries, for example, aimed at the systematic synthesis of many – if not all – possible peptides of a certain lenth. When each amino acid residue is one of the 20 proteinogenic L-amino acids, there are 400 possible dipeptides, 8000 possible tripeptides etc. in a library: with each amino acid increment the number of peptides increases 20-fold. However, there will be intrinsic difficulties in preparing and handling such large libraries, [4] in addition to difficulties in identification, selection, and enrichment of promising compounds from a library. In order to facilitate identification and enrichment, encoded libraries might be very interesting. In one type of these libraries a chemically synthesized entity, for example a peptide, is linked to a particular oligonucleotide sequence. It is proposed that the use of the encoding genetic tag should serve to identify and to enrich to promising compounds from the library. [5, 6] Peptide libraries are undeniably useful in the discovery of chemical lead compounds. However, the chemical lead discoveries from these libraries still require extensive modifications before suitable drug candidates are produced. This is due to the disadvantages of peptides, such as their water solubility and in many cases their facile degradation by proteases, which limit their use in biological systems. Therefore the results described by Cho et al. and Hobbs DeWitt et al. were the next significant steps on the way to new libraries, [7, 8] which first may – to a certain extent – circumvent these disadvantages, second provide lead compounds, possibly requiring less extensive modifications, and third may provide new frameworks for generating macromolecules with novel properties.

Cho et al. described the principles of the solid-phase synthesis of an oligocarbamate as well as the generation of an oligocarbamate library. [7] This oligocarbamate was referred to as "an unnatural biopolymer", [9] "unnatural" because it is obviously not found in nature and "biopolymer" to hint at the origin of the monomers which were derived from proteinogenic amino acids. The term "biopolymer mimetic" is perhaps a better alternative, as "unnatural biopolymer" is a contradiction in itself. The monomeric, *N*-protected aminoalkylcarbonates were conveniently prepared from the corresponding *N*-protected amino acids or from amino alcohols derived from amino acids (Scheme 1). By using the base-labile Fmoc-protecting group, an oligocarbamate was synthesized by standard solid-phase methods; the coupling yields were greater than 99 % per step. Even more interesting was the use of the photolabile nitroveratryloxycarbonyl (Nvoc) amino-protecting group (Scheme 1). This enabled the authors to prepare a library containing 256 oligocarbamates, by employing a binary masking strategy as was described by Fodor et al. in their light-directed parallel synthesis of libraries of oligopeptides and oligonucleotides. [10] The library contains all of the compounds that can be formed by deleting one or more carbamate units from the parent sequence Ac-Tyr^c-Phe^c-Ala^c-Ser^c-Lys^c-Ile^c-Phe^c-Leu^c (the superscript "c" indicates the presence of a carbamate linkage). The library was attached to

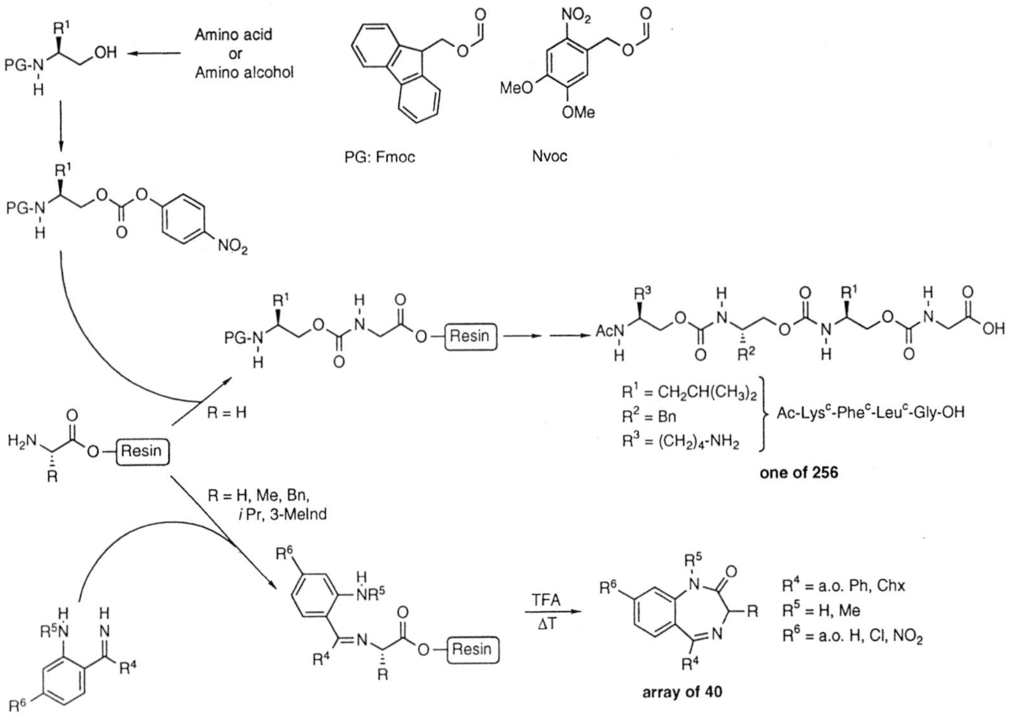

Scheme 1. Solid-phase synthesis of oligocarbamates and benzodiazepines starting with a resin-bound amino acid. The Fmoc-protected 4-nitrophenyl carbonate monomers derived from amino acids or amino alcohols were used in a normal solid-phase synthesis of an oligocarbamate, whereas the corresponding Nvoc-protected monomers were used in the generation of a library of 256 oligocarbamates of which one representative (Ac-Lys^c-Phe^c-Leu^c-Gly-OH) is shown. An array of 40 diazepines was synthesized by reacting the resin-bound amino acid with eight different 2-aminobenzophenone imines followed by cyclization. Ind = indolyl, Chx = cyclohexyl.

a glass surface of approximately 1.3 cm by 1.3 cm and screened for its ability to bind a monoclonal antibody against Ac-Tyr^c-Lys^c-Phe^c-Leu^c. It was found that the oligocarbamates Ac-Lys^c-Phe^c-Leu^c-Gly-OH (Gly is the linker in the keyhole limpet hemocyanin conjugate of this oligocarbamate), Ac-Phe^c-Lys^c-Phe^c-Leu^c-Gly-OH, Ac-Tyr^c-Lys^c-Phe^c-Leu^c-Gly-OH, Ac-Ala^c-Lys^c-Phe^c-Leu^c-Gly-OH, and Ac-Ile^c-Phe^c-Leu^c-Gly-OH were among the ten highest affinity ligands based on fluorescence intensities. Competitive enzyme-linked immunosorbent assay experiments in solution using the free oligocarbamates resulted in IC$_{50}$ values between 60 and 180 nM. It was therefore suggested that the dominant epitope of the antibody was Phe^c-Leu^c. Although, Ac-Tyr^c-Phe^c-Leu^c-Gly-OH containing this epitope did bind in solution with an IC$_{50}$ value of about 160 nM, the fluorescence signal associated with this oligocarbamate on the solid support, that is the glass surface in the library, ranked in the bottom 30%. This suggests that the conformation of the oligocarbamate on the solid support may be different from that in solution. [11]

Although the oligocarbamate differs significantly from the corresponding polypeptide, for example, it is more extended (that is there are four atoms located between the side chains instead of two atoms as in a peptide), some inherent disadvantages of peptides seem to be less pronounced in oligocarbamates: a) the oligocarbamates are significantly more hydrophobic, as was determined from a comparison of the water/octanol partitioning coefficients; b) the oligocarbamates are resistant to degradation by trypsin and pepsin.

Contrary to the Berkeley and Affymax report, the diversomer paper by Hobbs DeWitt et al. placed no emphasis on the preparation of an oligomer library but focussed entirely on the generation of a library of diverse, albeit related, organic compounds called "diversomers". [8] An important reason for the creation of these libraries was to discover lead compounds, which will be structurally more related to the ultimate drug, since lead compounds discovered from peptide or nucleotide libraries still require extensive modifications (vide supra). Another equally important reason is to rapidly obtain a large library of organic compounds (comparable to an in-house sample collection of a pharmaceutical company), which can be offered to the often automated, fast high throughput biological screening assays.

In contrast to the well-known procedure in which polyethylene rods or pins were used as a solid support for multiple peptide synthesis in the "PEPSCAN" method, [4, 12] hollow pins, ending in a glass frit were used, each containing approximately 100 mg resin to which a Fmoc amino acid was attached by a linker. The removal of the Fmoc-group from eight amino acid resins and treatment of each resin with five different isocyanates followed by cyclization led to 39 of the 40 desired hydantoins in 4–81% yield. Similarly, by starting from Boc-protected amino acid Merrifield resins (five), 40 different benzodiazepines were synthesized by reacting with eight different 2-aminobenzophenone imines followed by cyclization (Scheme 1). The 40 products were obtained in 9–63% yield and their purity was typically >90% (determined by ^{1}H-NMR spectroscopy). These compounds were then used in an assay for inhibition of fluoronitrazepam. A report by Bunin and Ellman had earlier described the solid-phase synthesis of ten benzodiazepine derivatives using an alternative approach. [13]

An important message from the reports is that the authors were able – by carrying out synthesis on a solid support – to prepare an array of unnatural compounds, which can be used for screening purposes to discover new lead compounds. These lead compounds may then be closer to ultimate drugs, because they are obtained from an array of compounds further removed from natural systems such as peptides and nucleotides.

Using the solid-phase strategy also libraries of other biopolymer mimetics have become accessible, for example, oligoureas, oligosulfones, peptidosulfonamides, [14] peptidophosphoramidates, and peptoids (Scheme 2). [15] These compounds will undoubtedly provide new biologically active compounds and also opportunities for the construction of biopolymers with novel properties.

The diversomer library showed that an array of organic compounds can be synthesized, and, although the number was much smaller than that which can be achieved by most of the current methods for generating peptide and other libraries, a significant number of structural analogues was obtained in a faster way than was the case when each of the compounds was synthesized separately.

Another challenge is still the development of "conformationally restricted" libraries in which the conformation of an oligomer is more accurately defined, for example, by introducing rigid monomers at different locations in an oligomer. This is noteworthy, since there is ample evidence in the literature that in many cases merely the presence of functional groups delivered, for example, by an amino acid sequence is not sufficient for an optimal biological activity, but that a proper orientation of functional groups is essential too. [16] In addition, the conformation on the solid support of the library may be different from that in solution (vide supra), wich may diminish the reliability of a library for the selection process. This underlines the importance of creating library compounds with defined conformations that are similar if not identical in different environments.

The scientist has an innate desire to be able to *rationally* design a compound with a predicted (biological) activity. Computer-assisted molecular modeling techniques, X-ray crys-

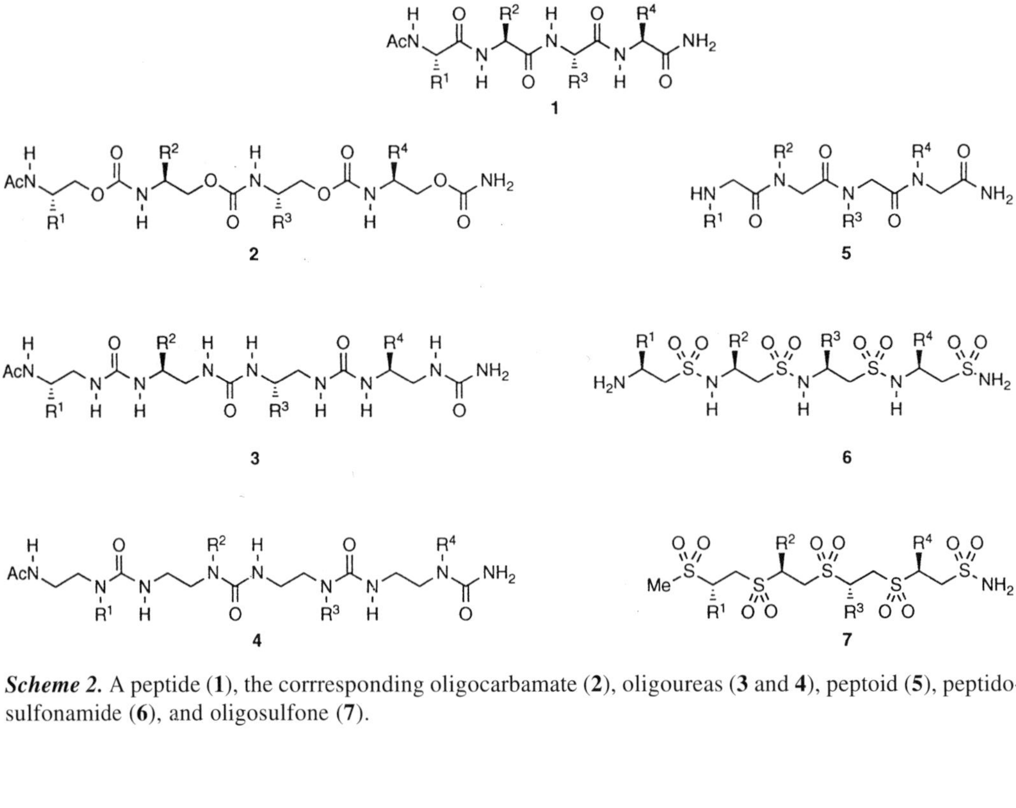

Scheme 2. A peptide (**1**), the corresponding oligocarbamate (**2**), oligoureas (**3** and **4**), peptoid (**5**), peptidosulfonamide (**6**), and oligosulfone (**7**).

tallographic analysis, and NMR methods have intensified this desire and belief that this should ultimately be possible for all drugs. Nevertheless, many of the current drugs or at least the lead compounds which initiated their development have been obtained by screening approaches, that is, by an *"irrational"* approach. Although it is expected that the rational design of drugs will remain important, screening approaches have gained momentum, since they contribute to a faster and more direct access to information. Therefore the rational design of irrational screening procedures holds promises for the future, and as a result there will be plenty of opportunities for the creation and applications of new libraries. [17]

References

[1] R. B. Merrifield, *J. Am. Chem. Soc.* **1963**, *85*, 2149; *Angew. Chem.* **1985**, *97*, 801; *Angew. Chem. Int. Ed. Engl.* **1985**, *24*, 799.

[2] For a review: S. L. Beaucage, R. P. Iyer, *Tetrahedron* **1993**, *48*, 2223.

[3] R. Verduyn, P. A. M. van der Klein, M. Douwes, G. A. van der Marel, J. H. van Boom, *Recl. Trav. Chim. Pays-Bas* **1993**, *112*, 464; S. J. Danishefsky, K. F. McClure, J. T. Randolph, R. B. Ruggeri, *Science* **1993**, *260*, 1307; S. P. Douglas, D. M. Whitfield, J. J. Krepinsky, *J. Am. Chem. Soc.* **1991**, *113*, 5095; G. H. Veeneman, R. M. Liskamp, G. A. van der Marel, J. H. van Boom, *Tetrahedron Lett.* **1987**, *28*, 6695.

[4] For a review on multiple peptide synthesis: G. Jung, A. G. Beck-Sickinger, *Angew. Chem.* **1992**, *104*, 375; *Angew. Chem. Int. Ed. Engl.* **1992**, *31*, 367.

[5] S. Brenner, R. A. Lerner, *Proc. Natl. Acad. Sci USA* **1992**, *89*, 5381; I. Amato, *Science* **1992**, *257*, 330; J. Nielsen, S. Brenner, K. D. Janda, *J. Am. Chem. Soc.* **1993**, *115*, 9812.

[6] For use of halogenated aromatic compounds as encoding tags, see: M. H. J. Ohlmeyer, R. N. Swanson, L. W. Dillard, J. C. Reader, G. Asouline, R. Kobayashi, M. Wigler, W. C. Still, *Proc. Natl. Acad. Sci. USA* **1993**, *90*, 10922–10926.

[7] C. Y. Cho, E. J. Moran, S. R. Cherry, J. C. Stephans, S. P. A. Fodor, C. L. Adams, A. Sundaram, J. W. Jacobs, P. G. Schultz, *Science* **1993**, *261*, 1303.

[8] S. Hobbs DeWitt, J. S. Kiely, C. J. Stankovic, M. C. Schroeder, D. M. Reynolds Cody, M. R Pavia, *Proc. Natl. Acad. Sci. USA* **1993**, *90*, 6909.

[9] The antisense oligonucleotides are an earlier category of "unnatural biopolymers": A. Peyman, *Chem. Rev.* **1990**, *90*, 543.

[10] S. P. A. Fodor, J. L. Leighton Read, M. C. Pirrung, L. Stryer, A. T. Lu, D. Solas, *Science* **1991**, *251*, 767; a highlight was devoted to this topic: G. von Kiederowski, *Angew. Chem.* **1991**, *103*, 839; *Angew. Chem. Int. Ed. Engl.* **1991**, *30*, 822.

[11] This was one of the critical questions raised in the highlight by von Kiederowski [10].

[12] H. M. Geysen, R. H. Meloen, S. J. Barteling, *Proc. Natl. Acad. Sci. USA* **1984**, *81*, 3998.

[13] B. A. Bunin, J. A. Ellman, *J. Am. Chem. Soc.* **1992**, *114*, 10997.

[14] For example: W. J. Moree, G. A. van der Marel, R. M. J. Liskamp. *J. Org. Chem.* **1995**, *6*, *5157*; D. B. A. de Bont, G. D. H. Bykstra, J. A. J. den Hartog, R. M. J. Liskamp, *Bioorganic. Med. Chem. Lett.* **1996**, *6*, 3035.

[15] Peptoids were highlighted by H. Kessler, *Angew. Chem.* **1993**, *105*, 572; *Angew. Chem. Int. Ed. Engl.* **1993**, *32*, 543.

[16] See for example the review on peptidomimetics: A. Giannis, T. Kolter, *Angew. Chem.* **1993**, *105*, 1303; *Angew. Chem. Int. Ed. Engl.* **1993**, *32*, 1244.

[17] A highlight was devoted to the "rationality of random screening": A. Plückthun, L. Ge, *Angew. Chem.* **1991**, *103*, 301; *Angew. Chem. Int. Ed. Engl.* **1991**, *30*, 296.

Reactive Intermediates

Henning Hopf

The study of reactive intermediates will always be of prime importance in organic chemistry whether its synthetic, mechanistic or theoretical aspects are involved. And with organic chemistry being the chemistry of carbon compounds the prototype intermediates will always be carbon centered radicals, carbenes, carbanions, and carbocations as well as the numerous small acyclic and cyclic π-systems which form the very basis of the structural and reactive richness of this branch of chemistry.

During the last decades the methods for studying reactive intermediates have changed: purely chemical approaches like the isolation or independent synthesis of reactive intermediates or their study by classical kinetic methods – although still of importance – have more and more been replaced by physical and theoretical/computational methods. Techniques reaching from the extremely "slow" ones encountered under e. g. matrix conditions at liquid helium temperatures to ultrafast femtosecond spectroscopic methods are having an enormous – and ever growing – impact on the study and characterization of reactive intermediates. In fact, the latter techniques make the distinction between reactive intermediates and transition states – so fundamental in the introductory teaching of chemistry – more and more meaningless.

Keeping with the title of this volume no attempt is made to present more than a small number of highlights which have appeared during the first half of the present decade. Highlights, though, which illustate some modern trends in "reactive intermediate chemistry" particularly well.

Matrix isolation coupled with matrix spectroscopy has been used on numerous occasions to prepare and characterize highly reactive molecules which would be "non-existent" under normal laboratory conditions. A selection from the vast literature includes C_4O_2 **1**, the first dioxide of carbon with an even number of carbon atoms [1] and dicarbondisulfide C_2S_2 **2** [2], the longsought diisocyan **3**, [3] an isomer of cyanogen first prepared by *Gay-Lussac* in 1815, and various halogen derivatives of cyclopropylidene **4**. [4] When the parent system **4** is irradiated it isomerizes to the allenediradical **5** ("propargylene"), [5] which – incidentally – is one of the most abundant organic compounds in interstellar

$$O{=}C{=}C{=}C{=}C{=}O \qquad\qquad S{=}C{=}C{=}S$$

1   **2**

$$|C{=}\overline{N}{-}\overline{N}{=}C| \qquad\qquad \overset{h\nu}{\longrightarrow} \qquad H\dot{C}{=}C{=}\dot{C}H$$

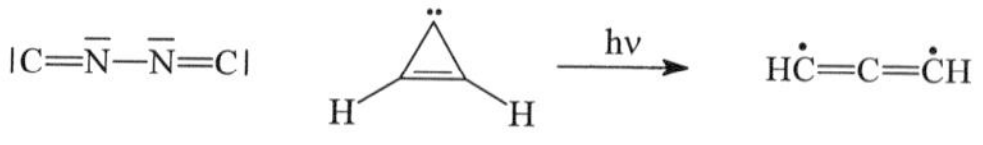

3   **4**   **5**

space, itself a more and more important "laboratory" for highly reactive compounds. Among the structurally more complex reactive intermediates which have been prepared and spectroscopically characterized in matrix, 2-adamantylidene **6**, a singlet carbene, [6] and 2,4-dehydrophenol **7**, the first derivative of *m*-dehydrobenzene [7] should be mentioned.

Unfortunately, matrix-isolated species normally cannot be studied by NMR spectroscopy, and hence the chemical shifts of the parent systems, which are of particular importance for comparison with the results obtained by theoretical calculations are unavailable. Substitution by bulky substituents or complexation by (transitions) metals, often used tricks to stabilize highly reactive species, lead to more or less pronounced disturbances of the parent molecules. An elegant esacpe from this dilemma has been accomplished by Cram and co-workers who were able to record the ^{1}H-NMR spectrum of 1,3-cyclobutadiene **8**, [8] the epitomy of the unstable monocyclic, fully-conjugated π-electron systems, *at room-temperature*, the vibrational spectra having been recorded in a low-temperature matrix already many years ago. When a-pyrone **9** is refluxed in chlorobenzene in the presence of the "hemicarcerand" **10** the lactone can slip into the hollow container molecule thus forming the encapsulated system **11**. At room temperature, however, the "exits" of the latter are too narrow and do not allow **9** to leave the complex again. Irradiation of **11** with a xenon lamp first causes electrocyclization to **12** which then splits off carbon dioxide to provide cyclobutadiene in a "molecular prison" **13**. According to the ^{1}H-NMR spectrum the cyclobutadiene in **13** exists as a stable singlet.

The so-called cryptophanes which are structurally related to **10** have shown to be capable of complexing stable molecules like methane or various fluorocarbons [9] and the number of "fullerenes with a content" – helium, neon, numerous metal ions [10, 11] is rapidly increasing.

Despite all the successes to prepare highly reactive intermediates in matrices, no "hard" structural information, i. e. X-ray structural data can be recorded under these conditions. The time-honored approach to stabilize highly reactive compounds by attaching sterically shielding or electronically stabilizing substituents to them has therefore not lost its attractiveness. This is borne out by the first stable and crystalline carbene **15** which has been prepared by Arduengo et al. by treating the precursor chloride **14** with sodium hydride in tetrahydrofuran in the presence of catalytic amounts of the DMSO-anion [12].

Carbene **15** melts at 240 °C; the bonding angle at the carbene carbon center, determined by X-ray structural analysis to 102°, agrees well with the value calculated for π-donor substituted singlet carbenes. The extreme stability of **15** has several origins: the compound is stabilized thermodynamically by the

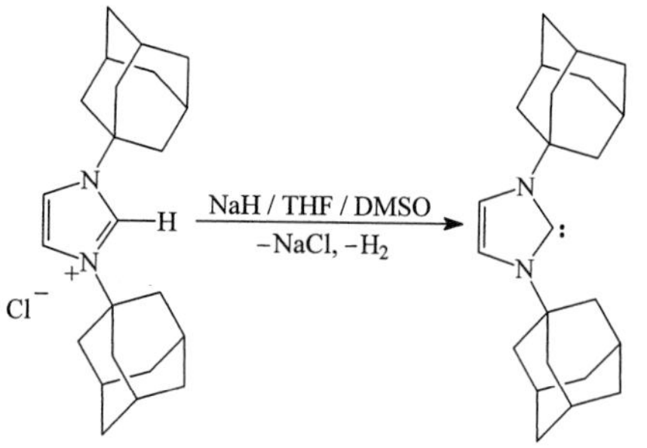

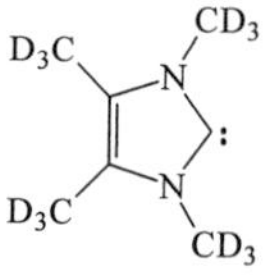

14   **15**   **16**

N–C=C–N fragment which serves as an electron donor and the σ-electronegativity of the two heteroatoms. And it profits from the kinetic stabilization provided by the two bulky adamantanyl substituents. Still, these groups are not mandatory as illustrated by the perdeuterio derivative **16** which was shown to be a true carbene with negligible ylidic character by X-ray and neutron diffraction studies. [13] A quantitative study of the influence of sterically demanding substituents demonstrated that dimesitylcarbene **17** is ca. 160 times more stable than diphenylcarbene at room temperature; [14] and for didurylcarbene a still further increase of life-time has been observed because of the buttressing effect of the additional methyl substituents.

In fact, **17** is so stable that it can be trapped by oxygen to yield the dioxirane **19** in 55 % yield, the first derivative of this three-membered heterocycle which is stable in condensed phase. The **17**→**19** oxidation process proceeds

by way of the carbonyl oxide **18**, itself stable for several hours in solution at –78 °C. [15] Carbonyloxides have been discussed as intermediates in the ozonization of alkenes for a long time; apparently **18** is the first spectroscopically oberservable representative of these highly reactive intermediates.

Whereas carbenes of type **17** are singlets the hexabromide **20** is a stable triplet both in solution and in the crystalline state as shown by ESR spectroscopy. Prepared by photolysis of a diazomethane precursor, crystalline **20** is stable for months at room temperature. [16]

The question of singlet and triplet stability is not only of importance for carbenes but in many other diradicaloid systems as well. Especially triplet states are increasingly studied by material scientists because of the possible ferromagnetism of the appropriate compounds. In photobiology triplets may be involved in the primary steps of the light-induced charge separation which takes place

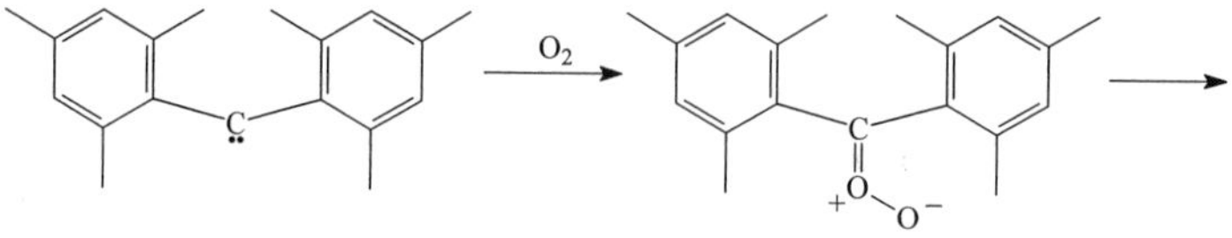

17   **18**

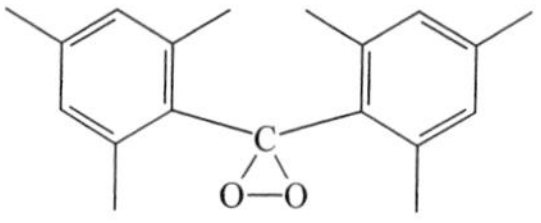

19 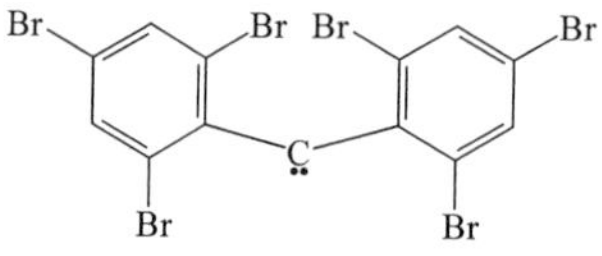   **20**

in the reactions centers of the photosynthesis system.

Two of the classical diradicals are 1,3-propanediyl **21** and trimethylenemethane **22** which have often been invoked as intermediates in photochemical and thermal processes. A recent surprising observation by Maier et al. makes these species available under the controlled conditions of a matrix experiment (at 10 K as compared to several hundreds of degrees in e. g. pyrolysis experiments [17, 18]. Rather then using a pure noble gas matrix these authors employed a halogen-doped matrix. Working under these conditions led to clean C–C bond cleavage and to the generation of **21** and **22**, allowing the registration of the IR spectrum of the latter species for the first time. [18] Diradical **22**, the parent system of the non-Kekulé hydrocarbons, is a ground state triplet which is 110 kJ/mol more stable than its singlet state [19].

Another classical non-Kekulé molecule is triangulene **23** ("*Clar's* hydrocabon"); which so far has escaped detection. With two half-filled degenerate molecular orbitals **23** should possess a triplet ground state and be paramagnetic. As a first derivative of triangulene, the trianion **24** has recently been prepared. According to its ESR spectrum it is a triplet molecule with threefold symmetry [20].

Non-Kekulé structures are of importance for the design of organic ferromagnets. Having prepared a hexacarbene with a tridecet ground state [21] Iwamura and co-workers have now been able to obtain a branched nonacarbene, **25**, which possesses the record number of 18 unpaired electrons [22].

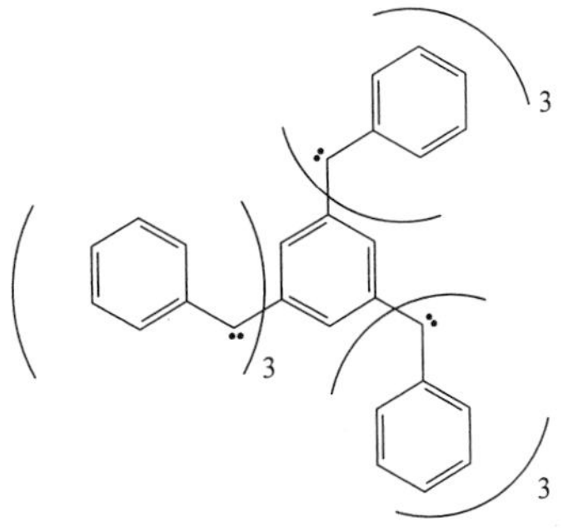

25

Turning to carbene related reactive species, alkylidene carbenoids like **26** (X = halogen, OR, NR$_2$) are particularly valuable for preparative purposes since they can undergo cycloaddition reactions with olefins (to methylenecyclopropanes), isomerizations (to alkynes by the so-called Fritsch-Buttenberg-Wiechell rearrangement), and dimerization (to [3]cumulenes). Although carbenoids have been studied extensively by NMR spectroscopy [23], the first X-ray structural analysis of a stable carbenoid, **27**, as a TMEDA · 2THF complex has been reported only recently [24].

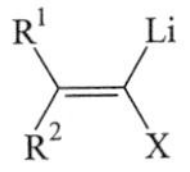

26

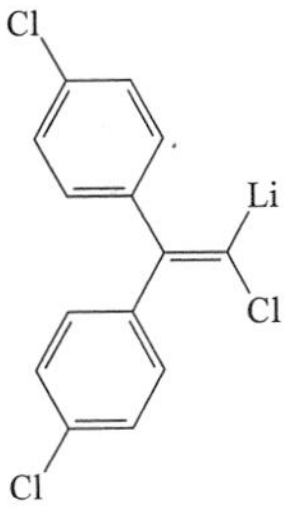

27

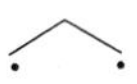

21    **22**

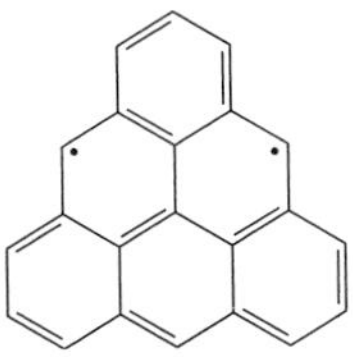

23

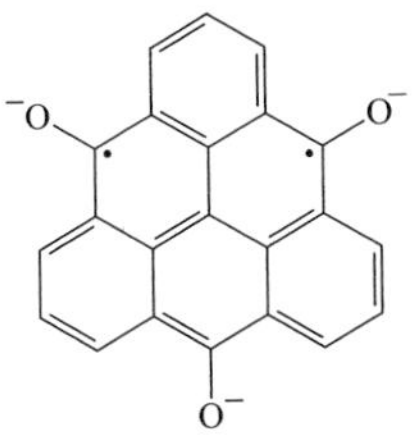

24

The carbenoid carbon atom shows interesting rehybridization phenomena which are caused by an increase in both the *s*-character of the C–Li- and the *p*-character of the C–X bond.

Another class of reactive intermediates that in recent years has enormously profited from the symbiosis of theoretical chemistry and modern physical methods, is carbanion chemistry. A famous case in point is provided by CLi_6 or hyperlithiated methane, calculated to possess considerable stability by Schleyer, Pople and co-workers in 1983 [25]. Since these calculations are usually carried out for isolated molecules in the gas phase it is obvious to search for them by mass spectrometric measurements. In fact, when crystalline dilithioacetylene is subjected to radio frequency heating in a Knudsen cell and the gas mixture formed is analyzed by quadrupol mass spectrometry besides CLi_3 and CLi_4 CLi_6 could be detected [26]. Whereas this approach clearly does not allow to prepare isolable amounts of poly- or perlithiated hydrocarbons, several preparative methods have been described for the prepartion of perlithiomethane (CLi_4), [27] a grey extremely pyrophoric solid. As shown by extensive quantummechanical calculations hyperlithiation is not restricted to "saturated" systems. For example, several isomers of C_2Li_4 have been calculated whose most stable structure is not ethylenic but shows the typical bonding characteristics of hypermetalated species [28].

X-ray crystallography is by now one of the most important methods for the determination of carbanion structures. From the very large number of recently characterized carbanions only a few can be mentioned here. They include preparatively important species such as the first lithiated carbamide [29], the lithium and tetra-*n*-butyl ammonium salts of a *a*-sulfonyl substituted carbanion [30], and a protoype of the synthetically very useful Lochmann bases. [31] On the other hand structural work – both by X-ray crystallography and by various modern NMR spectros-

28

29

copic techniques – has also been carried out on the theoretically interesting dilithium salt of acepentalene **28** [32] and the tetralithium salt of corannulene **29**. [33]

That charged species form various types of ion-pairs in solution (solvent separated, solvent shared, and contact ion pairs) has been known for a long time in carbanion (and carbocation) chemistry. That cations and anions of sufficient structural complexity can even penetrate each other has, however, been established only recently. ^{1}H,^{1}H- and ^{11}B,^{1}H- NOE measurements on tetra-*n*-butyl ammonium borohydride show for example that the BH_4^- anion and the quarternary nitrogen atom of the cation must be in very close contact, i. e. that the anion is positioned between the alkyl chains of the cation ("anion within the cation"). [34] Interpenetration of ions has also been established for **30**. The (chiral) cation of this complex has been employed in phase transfer catalysis. It carries the BH_4^- anion into the organic phase where the former can cause optical induction in reduction reactions. [35]

Whereas carbanion chemistry is of large current interest, carbocation chemistry is

30

nowadays associated largely with the sixties and seventies. Still, in this area of physical organic chemistry interesting results keep to be published. These concern classical carbocations or those with novel structures on the one hand and carbocations as reactive intermediates in important organic reactions on the other other.

Among the long sought for carbocations, the unsubstituted allyl cation **31** should be mentioned, which has finally been prepared in condensed phase by reaction of cyclopropyl bromide with SbF_5 at 140 K in a matrix where it also could be studied by IR spectroscopy.

All efforts to study **31** in solution by NMR spectroscopy have met with failure so far, though [36]. The facile ring-opening of cyclopropyl cations can be prevented, when a ferrocenyl substituent is introduced into the three-membered ring system: cation **32** is so stable that its NMR spectrum can be recorded at −60 °C. [37] The central eight-membered in the biscation **33** which should be planar according to Hückel's rule, in fact prefers a tub-like conformation. It appears that the gain in delocalization energy for a planar structure is overbalanced by an increase in angle strain and growing non-bonding interactions between the bicyclic substituents. [38]

Among the novel carbocations discovered recently the "internally H-bridged" bicyclic species **35** is noteworthy.

When the precursor *in-out*-hydrocarbon **34** is treated with trifluoromethanesulfonic acid at room temperature in dichloromethane **35** is produced in quantitative yield while a stoi-

chiometric amout of hydrogen is released. [39] This example is not only of interest in its own right but also sheds light on the mechanism of carbocation formation by protonolysis of alkanes discovered independently in the late sixties by Olah and Hogeveen.

How far carbocation chemistry has evolved from the old solvolysis days is demonstrated by two recent structure determinations: the IR spectrum of the nonclassical CH_5^+ cation has been measured by "solvating" this unusual species with molecular hydrogen in the gas phase. This slows down the ultrafast fluxional process which so far prevented the recording of vibrational spectra. The cluster ions $CH_5^+(H_2)_n$ ($n = 1, 2, 3$), after mass selection by an ion trap, were then subjected to IR laser spectroscopy/quadrupol mass spectrometry which ultimately yielded the IR absorption. [40] And the benzene cation, formed by removal of one electron from the parent hydrocarbon was shown to possess D_{6h} symmetry by rotation resolved ZEKE-photoelectron spectroscopy (*Zero Kinetic Energy*-PES). [41]

Turning to reaction mechanisms involving carbocations, important results have been published concerning one of the oldest organic reactions: the electrophilic addition of halogens to olefins, which for a long time has been formulated as a three step process: after the initial π complex formation, a halonium-ion complex (σ complex) is formed which in the last step is attacked in *trans*-fashion to provide the 1,2-halogen adduct. In principle this process is reversible, and in fact for a sterically strongly shielded alkene such as adamantylidenadamantane **36** an equilibrium has been shown to exist between the free olefin and bromine, the π complex, and the bromoniumion **37**. NMR studies have furthermore revealed a fast degenerate equilibration between **36** and **37** by Br$^+$ transfer. [42–45]

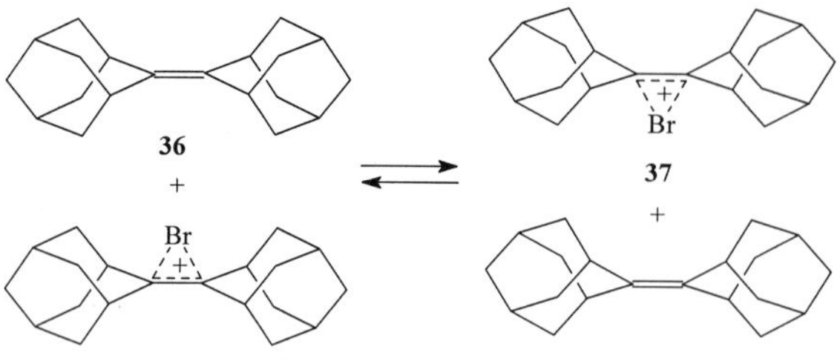

Whereas crystal structures of bromonium and iodonium salts could be obtained, [43, 46] experimental data on π complexes were scarce until recently when Legon and co-workers introduced a new method for studying the structures of labile molecular complexes. [47] This technique involves the rapid mixing of the components (halogen and olefin) while they expand into the evacuated resonator of a FT microwave spectrometer. Because of the collisonfree ultrafast expansion the molecular complexes formed cool down rapidly and hence possess only a very small internal energy. This prevents addition to the normally observed products and allows measurement of the rotational spectra of the π complexes and determination of their structures. The π complexes of ethylene and acetylene with chlorine were shown to have structures **38** [48] and **39**, [49] respectively, i. e. arrangements in which the halogen molecule stands perpendicularly above the middle of the respective carbon–carbon multiple bond:

The structural parameters of both the halogen and the hydrocarbon component are hardly influenced by the complex formation attesting to the weak interaction between the π bonds and the halogen molecule; and the latter is not polarized to any significant extent. All in all these experimental results beautifully confirm theoretical calculations and structural properties derived from matrix isolation studies.

In a more complex case – which nevertheless demonstrates how new observations can be made for an old reaction – the highly hindered butadiene derivative **40** – which possesses an orthogonal structure – yields the substitution product **42** on reaction with bromine, not the *a priori* expected 1,2- and 1,4-dibromides. [50] In this case the sterically hindered cationic intermediate **41** prefers a-proton elimination over the usually observed interception by bromide ion.

Propelled by recent developments in natural products chemistry (catchword: "enediyne antibiotics") and material science ("novel carbon allotropes, carbon networks") a veritable renaissance of acetylene chemistry has taken place during the last few years. [51] As far as reactive intermediates are concerned the emphasis has been on reducing the ring size of cyclic acetylenes to the ultimate limit: the three-membered ring.

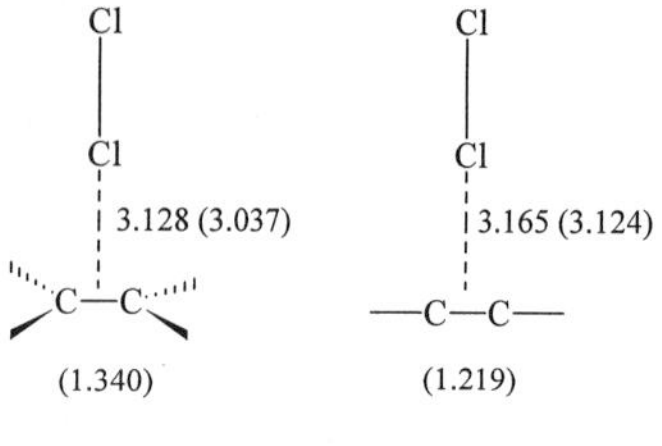

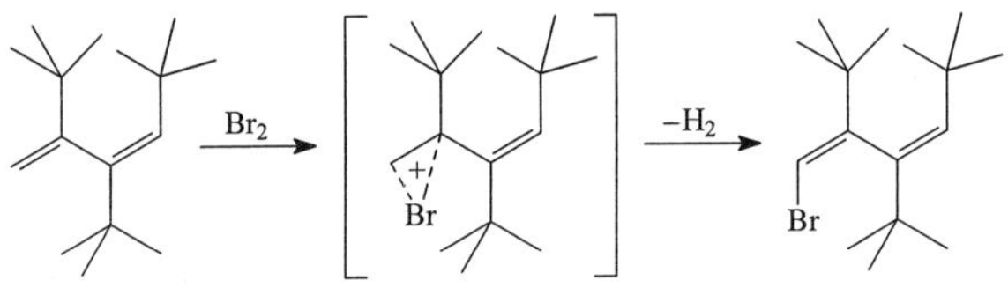

The small-ring acetylene project – originally initiated by Wittig – illustrates very nicely how an "isolable system" can be gradually transformed into a "reactive intermediate". Or to put it another way: When discussing reactive intermediates *vs.* isolable molecules the question of reaction conditions and methods used – both chemical and physical – must always be asked and answered. Under "normal laboratory conditions" cyclooctyne is a stable molecule, although the deformation angle between the triple bond and the adjoining methylene group is ca. 22°, and the reactivity is increased as compared to a corresponding acyclic alkyne. In cycloheptyne this angle is 30–35°, and the laboratory conditions have changed to −76 °C and a half-life of one hour. And cyclohexyne is already so unstable that it can only be studied in a low-temperature matrix. [52] However, exploiting the trick of introducing heteroatoms (especially silicon has often been used) into the ring and shielding the reactive triple bond by voluminous groups, [53] Ando and co-workers have been able to prepare the cycloalkyne **43** – a compound with a half-life of 8 hours at 174 °C ! [54]

When subjecting the disilane **48** to a newly developed pyrolysis technique, trimethylsilane is split off and 1-silacyclopropylidene **49** is formed as the most stable C_2H_2Si isomer. Photolysis of this "silyiene" with monochromatic light yields **50** which was identified by comparing its IR spectrum with a calculated spectrum. Whether **50** is a "true" cycloalkyne or a diradical, a singlet or a triplet must await further investigation as must the study of its chemical behavior.

Highly strained compounds also result when allenic or cumulenic units are introduced into a cyclic or bicyclic molecule. A particularly interesting example is provided by 1,2,4-hexatriene or isobenzene **52**, an allenic isomer of benzene.

51   52   53

54

43   44   45   46   47

Cyclopentyne **44** could not be generated so far; during an attempted photochemical synthesis the allene **45** was obtained, indicating that **44** – if formed at all – is very likely photolabile. [55] Extensive recent abinitio calculations suggest that cyclobutyne **46** should be observable in a matrix, whereas for the "end point" of the series, cyclopropyne **47**, no minimum could be found on the potential energy surface. [56] It is thus all the more surprising that Maier and co-workers have been able to produce the silicon-containing **50**, as the first observable cyclopropyne derivative: [57]

This cycloallene was first prepared by Christl and co-workers from the bromofluorohydrocarbon **51** by treatment with methyl lithium. [58] The C_6H_6 isomer does not survive the conditions of its generation but can be trapped with e. g. styrene to yield the bicyclic adduct **54**. In the meantime a further route to **52** has been found, consisting in the thermal cyclization of 1,3-hexadien-5-yne **53**. [59]

Hydrocarbon **53** is a close relative of 3-hexen-1,5-diyne **55**, the parent enediyne system for which Bergman first described the cyclization **55**→**56** which later should carry his name.

48   49   50

The remainder of this brief overview will be dedicated to some methods for the direct observation of transitions states, clearly a question of utmost importance for the determination of reaction mechanisms.

One of classical and well studied examples of a molecule undergoing dynamic changes is cyclooctatetraene **60**, which, of course, is also an important representative of a $4n\pi$-electron system; it avoids antiaromaticity by bond fixation and molecular deformation in the ground state (tub-like structure with D_{2d} symmetry). Cyclooctatetraene participates in two types of dynamic processes: ring inversion *via* a planar eight-membered ring with alternating single and double bonds **61** (D_{4h} symmetry) and a double bond shift which takes place *via* antiaromatic transition state **62** (*D*8h symmetry) in which all C–C bonds are of the same length.

The delocalized structure **62** has now been observed directly by, Borden and co-workers [65] using photo electron spectroscopy. As starting material for **62** the authors used the radical anion of cyclooctatetraene **63** which can be prepared by electron capture in an ion source, and which is known to be planar from experimental and theoretical studies. When this species is irradiated with monochromatic light the weakly bonded excess

Because this process forms the basis for the biological action (DNA cleavage) of several highly potent cancerostatica [51] a lot of mechanistic work has been carried out on it including the exact determination of its activation parameters, [59] the measurement of the heat of formation of the diradical involved, **56** (*p*-dehydrobenzene), [60] and the incorporation of the enediyne unit into a large number of more complex molecules. [51, 61]

Another interesting diradical – also formed by interaction between two triple bonds – is the bicyclic 1,4-didehydro-1,3-butadiene **58**. When strained cyclic dialkynes, symbolized by the general formula **57** are heated to temperatures between 80 and 150 °C they cyclize to **58**, a reactive intermediate which can be trapped by various hydrogen donors to provide the bicyclic dienes **59**. [62]

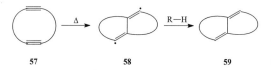

As already shown by many of the above examples, physical methods, especially spectroscopic ones, are having a very strong impact on the development of physical organic chemistry, and it is only for lack of space that such important techniques as laser flash spectroscopy can only be mentioned here in passing. Having established itself as a highly efficient method for the study of carbenes and carbocations [63] laser flash spectroscopy has also been used to generate carbanions like the trityl anion recently, [64] and promises to become a useful method for the determination of carbanion reactivity as well.

electron is split off and neutral cycloocta-tetraene is formed. Since this latter process is faster than the movement of the nuclei of cyclooctatetraene the neutral hydrocarbon produced retains the planar structur of the pre-scursor ion. The kinetic energy of the liberated electrons – which is measured by the PE-spec-trum – contains the spectroscopic information of planar cyclooctatetraene (i.e. **62**). As the analysis of the fine structure of the PE spec-trum shows, both D_{4h}- and D_{8h} cycloocta-tetraene are generated in this experiment; even more: it could also be observed that D_{8h} cyclooctatetraene occurs as a singlet (the tran-sition sate **62**) and as a triplet species **64**, with the latter being a true, though very short-lived intermediate.

Whether bond-alternation can be induced in $4n+2$, i.e. aromatic π systems has been discus-sed in organic chemistry long before Hückel formulated his famous rule. After many attempts, hydrocarbon **65** has now been shown to be the first mononuclear benzenoid hydrocarbon with a true cyclohexatriene structure. [66] Its internal ring is planar, all internal angles amount to 120°, yet the two nonequivalent C–C bonds in the benzene ring differ by 0.089 Å, corresponding to Pauling bond orders of 1.86 and 1.39 for the shorter and the longer bond, respectively.

65

If there is an ultimate method for the direct observation of transition states femtosecond spectroscopy could well qualify for it. This method, largely developed by Zewail and co-workers, is now more and more used to study – even complex – uni- and bimolecular processes. It employs an extremely short pulse from a femtosecond laser to initiate a reaction, for example the dissociation or an isomerization of a molecule. By a second laser pulse from the same pump laser, triggered a few femto (10^{-15}) or picoseconds (10^{-12} s) later, the products formed and their internal states are then registered. As a particularly important example for organic reaction mechanisms Zewail et al. recently reported a study of the real-time dynamics of the retro-Diels-Alder reaction of norbornene and nor-bornadiene, [67] which, of course, also sheds light on the reverse process, the Diels-Alder addition. After starting the decomposition, which was carried out in a molecular beam to isolate the elementary process, the authors "watched" the excited species by studying the molecular beam with pulses at femtose-cond intervals from the second laser. These pulses ionize the reactive intermediates thus making them amenable for mass spectral anal-ysis. The mass spectra show peaks corres-ponding to the starting materials ($m/z = 94$ and 92, respectively), but also a peak at $m/z = 66$ corresponding to one of the cleav-age products, cyclopentadiene. Interestingly, however, this signal builds up in the course of the decomposition and eventually it disap-pears. It can hence not be due to cyclopenta-diene itself but must represent a reactive inter-mediate. For this signal a decay time of 220 fs is registered. The decay time of the $m/z = 94$ signal (for norbornene) amounts to 160 fs, i.e. it takes this time for *two* C–C bonds to rupture. A non-concerted and a concerted route are thus both available for the retro-Diels-Alder reaction, and by implication for the forward (addition) process as well. The reaction times measured are also much shorter than the 1 to 10 picoseconds it takes for a C–C bond to rotate. There is hence no contradiction betwen the stereospecificity of the process – which, of course, has been known for a long time from classical mechanistic studies – and the presence of an intermediate. The old problem of concertedness *vs.* nonconcerted-ness which has dominated the discussion on cycloadditions for decades, may hence finally have been solved: both symmetric and non-symmetric motions of the two C–C bonds in the above examples are possible, leading eventually to (competing) concerted and non-concerted reaction paths! The route which a particular cycloaddition or its reverse process prefers will hence depend on the asymmetry of the structures involved, the different energy barriers, and the energy available.

The impact which femtosecond spectroscopy is also having on the study of the reaction mechanisms of biological systems is finally illustrated by an investigation of the *cis/trans* isomerization of rhodopsin **66** → **67**, [68] the primary step of the vision process. The reaction lasts 200 fs and is thus one of the fastest photochemical processes known so far. The formation of the photoproduct **67** occurs without an activation barrier.

References

[1] G. Maier, H. P. Reisenauer, H. Balli, W. Brandt, R. Janoschek, *Angew. Chem.* **1990**, *102*, 920–923; *Angew. Chem. Int. Ed. Engl.* **1990**, *29*, 905; cf. D. Sülzle, H. Schwarz, *Angew. Chem.* **1990**, *102*, 923–925; *Angew. Chem. Int. Ed. Engl.* **1990**, *29*, 908.

[2] G. Maier, H.P. Reisenauer, J. Schrot, R. Janoschek, *Angew. Chem.* **1990**, *102*, 1475–1477; *Angew. Chem. Int. Ed. Engl.* **1990**, *29*, 1464.

[3] G. Maier, H. P. Reisenauer, J. Eckwert, C. Sierakowski, Th. Stumpf, *Angew. Chem.* **1992**, *104*, 1287–1289; *Angew. Chem. Int. Ed. Engl.* **1992**, *31*, 1218.

[4] G. Maier, T. Preiss, H. P. Reisenauer, B. A. Hess, Jr., L. J. Schaad, *J. Am. Chem. Soc.* **1994**, *116*, 2014–2018 ; cf. G. Maier, T. Preiss, H. P. Reisenauer, *Chem. Ber.* **1994**, 127, 779–782.

[5] R. Herges, A. Mebel, *J. Am. Chem. Soc.* **1994**, *116*, 8229–8237; cf. R. A. Seburg, R. J. McMahon, *Angew. Chem.* **1995**, *107*, 2198–2201; *Angew. Chem. Int. Ed. Engl.* **1995**, *34*, 2009.

[6] T. Bally, S. Matzinger, L. Truttmann, M. S. Platz, S. Morgan, *Angew. Chem.* **1994**, *106*, 2048–2051; *Angew. Chem. Int. Ed. Engl.* **1994**, *33*, 1964.

[7] G. Bucher, W. Sander, E. Kraka, D. Cremer, *Angew. Chem.* **1992**, *104*, 1225–1228; *Angew. Chem. Int. Ed. Engl.* **1992**, *31*, 1230. In the meantime the parent system of **7**,1,3–didehydrobenzene (*m*-benzyne) has also been generated and spectroscopically characterized in matrix: R. Marquardt, W. Sander, E. Kraka, *Angew. Chem.* **1996**, *108*, 825–827; *Angew. Chem. Int. Ed. Engl.* **1996**, *35*, 746–748.

[8] D. J. Cram, M. E. Tanner, R. Thomas, *Angew. Chem.* **1991**, *103*, 1048–1051; *Angew. Chem. Int. Ed. Engl.* **1991**, *30*, 1024; cf. H. Hopf, *Angew. Chem.* **1991**, *103*, 1137–1139; *Angew. Chem. Int. Ed. Engl.* **1991**, *30*, 1117. For the activation energy of the valence isomerization of cyclobutadiene see G. Maier, R. Wolf, H.-O. Kalinowski, *Angew. Chem.* **1992**, *104*, 764–766; *Angew. Chem. Int. Ed. Engl.* **1992**, *31*, 738. Very recently dehydrobenzene has been prepared and studied by the scene approach: R. Warmuth, *Angew. Chem.* **1997**, *109*, 1406–1409; *Angew. Chem. Int. Ed.* **1997**, *36*, 1347–1350.

[9] L. Garel, J.-P. Dutasta, A. Collet, *Angew. Chem.* **1993**, *105*, 1249–1251; *Angew. Chem. Int. Ed. Engl.* **1993**, *32*, 1169.

[10] Review: H. Schwarz, *Angew. Chem.* **1992**, *104*, 301–305; *Angew. Chem. Int. Ed. Engl.* **1992**, *31*, 293.

[11] Review: F. T. Edelmann, *Angew. Chem.* **1995**, *107*, 1071–1075; *Angew. Chem. Int. Ed. Engl.* **1995**, *34*, 981.

[12] A. J. Arduengo III, R. L. Halow, M. Kline, J. Am. Chem. Soc. **1991**, *113*, 361–363.

[13] A. J. Arduengo III, H. V. Rasika Dias, D. A. Dixon, R. L. Harlow, W. T. Klooster, T. F. Koetzle, *J. Am. Chem. Soc.* **1994**, *116*, 6812–6822.

[14] H. Tomioka, H. Okada, T. Watanabe, K. Hirai, *Angew. Chem.* **1994**, *106*, 944–946; *Angew. Chem. Int. Ed. Engl.* **1994**, *33*, 873.

[15] A. Kirschfeld, S. Muthusamy, W. Sander, *Angew. Chem.* **1994**, *106*, 2261–2263; *Angew. Chem. Int. Ed. Engl.* **1994**, *33*, 2212.

[16] H. Tomioka, T. Watanabe, K. Hirai, K. Furukawa, T. Takui, K. Itoh, *J. Am. Chem. Soc.* **1995**, *117*, 6376–6377.

[17] G. Maier, St. Senger, *Angew. Chem.* **1994**, *106*, 605–606; *Angew. Chem. Int. Ed. Engl.* **1994**, *33*, 558.

[18] G. Maier, H. P. Reisenauer, K. Lanz, R. Troß, D. Jürgen, B. Andes Hess, Jr., L. J. Schaad, *Angew. Chem.* **1993**, *105*, 119–121; *Angew. Chem. Int. Ed. Engl.* **1993**, *32*, 74.

[19] O. Claesson, A. Lund, T. Gillbro, T. Ichikawa, O. Edlund, H. Yoshida, *J. Chem. Phys.* **1980**, *72*, 1463–1470.

[20] G. Allinson, R. J. Bushby, J.-L. Paillaud, D. Oduwole, K. Sales, *J. Am. Chem. Soc.* **1993**, 115, 2062–2064.

[21] N. Nakamura, K. Inoue, H. Iwamura, T. Fujioka, Y. Sawaki, *J. Am. Chem. Soc.* **1992**, *114*, 1484–1485.

[22] N. Nakamura, K. Inoue, H. Iwamura, *Angew. Chem.* **1993**, *105*, 900–901; *Angew. Chem. Int. Ed. Engl.* **1993**, 32, 872.

[23] D. Seebach, R. Hässig, J. Gabriel, *Helv. Chim. Acta.* **1983**, *66*, 308–337.

[24] G. Boche, M. Marsch, A. Müller, K. Harms, *Angew. Chem.* **1993**, *105*, 1081–1082; *Angew. Chem. Int. Ed. Engl.* **1993**, *32*, 1279.

[25] P. v. R. Schleyer, E.-U. Würthwein, E. Kaufmann, T. Clark, J. A. Pople, *J. Am. Chem. Soc.* **1983**, *105*, 5930–5932.

[26] H. Kudo, *Nature* **1992**, *355*, 432–434; see also A. E. Reed, P. v. R. Schleyer, R. Janoschek, *J. Am. Chem. Soc.* **1991**, *113*, 1885–1892.

[27] Review: A. Maercker, M. Theis, *Top. Curr. Chem.* **1987**, *138*, 1–61.

[28] A. E. Dorigo, N. J. R. van Eikema Hommes, K. Krogh-Jespersen, P. v. R. Schleyer, *Angew.*

[29] Th. Maetzke, C. P. Hidber, D. Seebach, *J. Am. Chem. Soc.* **1990**, *113*, 8248–8250.

[30] H.-J. Gais, G. Hellmann, H. J. Lindner, *Angew. Chem.* **1990**, *102*, 96–99; *Angew. Chem. Int. Ed. Engl.* **1990**, *29*, 100; cf. H. J. Gais, G. Hellmann, J. Am. Chem. Soc. **1992**, *114*, 4439–4440. The increasing importance of heteroatom substituted organolithium compounds for enantioselective synthesis has been demonstrated by several authors e. g. T. Ruhland, R. Dress, R. W. Hoffmann, *Angew. Chem.* **1993**, *105*, 1487–1489; *Angew. Chem. Int Ed. Engl.* **1993**, *32*, 1467; H. J. Reich, R. R. Dykstra, *Angew. Chem.* **1993**, *105*, 1489–1491; *Angew. Chem. Int. Ed. Engl.* **1993**, *32*, 1469.

[31] M. Marsch, K. Harms, L. Lochmann, G. Boche, *Angew. Chem.* **1990**, *102*, 334–336; *Angew. Chem. Int. Ed. Engl.* 1990, *29*, 308.

[32] R. Haag, R. Fleischer, D. Stalke, A. de Meijere, *Angew. Chem.* **1995**, *107*, 1642–1644; *Angew. Chem. Int. Ed. Engl.* **1995**, *34*, 1492.

[33] M. Baumgarten, L. Gherghel, M. Wagner, A. Weitz, M. Rabinovitz, P.-C. Cheng, L. T. Scott, *J. Am. Chem. Soc.* **1995**, *117*, 6254–6257.

[34] T. C. Pochapsky, P. M. Stone, *J. Am. Chem. Soc.* **1990**, *112*, 6714–6715.

[35] T. C. Pochapsky, P. M. Stone, S. S. Pochapsky, *J. Am. Chem. Soc.* **1991**, *113*, 1460–1462.

[36] P. Buzek, P. v. R. Schleyer, H. Vancik, Z. Mihalic, J. Gauss, *Angew. Chem.* **1994**, *106*, 470–473; *Angew. Chem. Int. Ed. Engl.* **1994**, *33*, 448.

[37] G. K. S. Prakash, H. Buchholz, V. P. Reddy, A. de Meijere, G. A. Olah, *J. Am. Chem. Soc.* **1992**, *114*, 1097–1098.

[38] T. Nishinaga, K. Komatsu, N. Sugita, *J. Chem. Soc. Chem. Commun.* **1994**, 2319–2320.

[39] J. E. McMurry, T. Lectka, *J. Am. Chem. Soc.* **1990**, *112*, 869–870.

[40] D. W. Boo, Z. F. Liu, A. G. Suits, J. S. Tse, Y. T. Lee, *Science* **1995**, *269*, 57–59.

[41] R. Lindner, H. Sekiya, B. Beyl, K. Müller-Dethlefs, *Angew. Chem.* **1993**, *105*, 631–634; *Angew. Chem. Int. Ed. Engl.* **1993**, *32*, 603.

[42] R. S. Brown, R. W. Nagorski, A. J. Bennet, R. E. D. McClung, G. H. M. Aarts, M. Klobukowski, R. McDonald, B. D. Santasiero, *J. Am. Chem. Soc.* **1994**, *116*, 2448–2456.

[43] G. Belluci, R. Bianchini, C. Chiappe, V. R. Gadgil, A. P. Marchand, *J. Org. Chem.* **1993**, *58*, 3575–3577.

[44] G. Bellucci, R. Bianchini, C. Chiappe, F. Marioni, R. Ambrosetti, R. S. Brown, H. Slebocka-Tilk, *J. Am. Chem. Soc.* **1989**, *111*, 2640–2647.

[45] R. S. Brown, R. Geyde, H. Slebocka-Tilk, J. M. Buschek, K. R. Kopecky, *J. Am. Chem. Soc.* **1984**, *106*, 4515–4521.

[46] H. Slebocka–Tilk, R. G. Ball, R. S. Brown, *J. Am. Chem. Soc.* **1985**, *107*, 4504–4508.

[47] A. C. Legon, C. A. Rego, *J. Chem. Soc. Faraday Trans.* **1990**, *86*, 1915–1921.

[48] H. I. Bloemink, K. Hinds, A. C. Legon, J. C. Thorn, *J. Chem. Soc. Chem. Commun.* **1994**, 1321–1322.

[49] H. I. Bloemink, K. Hinds, A. C. Legon, J. C. Thorn, *Chem. Phys. Letters* **1994**, *223*, 162.

[50] H. Hopf, R. Hänel, P. G. Jones, P. Bubenitschek, *Angew. Chem.* **1994**, *106*, 1444–1445; *Angew. Chem. Int. Ed. Engl.* **1994**, *33*, 1369.

[51] P. J. Stang, F. Diederich (Eds.), *Modern Acetylene Chemistry*, VCH, Weinheim, **1995**.

[52] C. Wentrup, R. Blanch, H. Briehl, G. Gross, *J. Am. Chem. Soc.* **1989**, *110*, 1874–1880; cf. W. Sander, O. L. Chapman, *Angew. Chem.* **1988**, *100*, 402–403; *Angew. Chem. Int. Ed. Engl.* **1988**, *27*, 398–399.

[53] A. Krebs, J. Wilke, *Top. Cur. Chem.* **1983**, *109*, 189–233.

[54] W. Ando, F. Hojo, S. Sekigawa, N. Nakayama, T. Shimizu, *Organometallics* **1992**, *11*, 1009–1011; cf. F. Hojo, S. Sekigawa, N. Nakayama, T. Shimzu, W. Ando, *Organometallics* **1993**, *12*, 803–810.

[55] O. L. Chapman, J. Gato, P. R. West, M. Regitz, G. Maas, *J. Am. Chem. Soc.* **1981**, *103*, 7033–7036.

[56] H. A. Carlson, G. E. Quelch, H. F. Schaefer III, *J. Am. Chem. Soc.* **1992**, *114*, 5344–5348.

[57] G. Maier, H. P. Reisenauer, H. Pacl, *Angew. Chem.* **1994**, *106*, 1347–1349; *Angew. Chem. Int. Ed. Engl.* **1994**, *33*, 1248–1250.

[58] M. Christl, M. Braun, G. Müller, *Angew Chem.* **1992**, *104*, 471–473; *Angew. Chem. Int. Ed. Engl.* **1992**, *31*, 473; cf. R. Janoschek, *Angew. Chem.* **1992**, *104*, 473–475; *Angew. Chem. Int. Ed. Engl.* **1992**, *31*, 476. For the generation of two other benzene isomers, 1,2,3-cyclohexatriene and 1-cyclohexen-3-yne see W. C. Skakespeare, R. P. Johnson, *J. Am. Chem. Soc.* **1990**, *112*, 8578–8579.

[59] W. R. Roth, H. Hopf, C. Horn, *Chem. Ber.* **1994**, 127, 1765–1779; cf. H. Hopf, H. Berger, G. Zimmermann, U. Nüchter, P. G. Jones, I. Dix, *Angew. Chem.* **1997**, *109*, 1236–1238; *Angew. Chem. Int. Ed.* **1997**, *36*, 1187–1190.

[60] P. G. Wenthold, R. R. Squires, *J. Am. Chem. Soc.* **1994**, *116*, 6401–6412.

[61] Review: R. Gleiter, D. Kratz, *Angew. Chem.* **1993**, *105*, 884–887; *Angew. Chem. Int. Ed. Engl.* **1993**, *32*, 842.

[62] R. Gleiter, J. Ritter, *Angew. Chem.* **1994**, *106*, 2550–2552; *Angew. Chem. Int. Ed. Engl.* **1994**, 33, 842.

[63] See for example: F. Cozens, J. Li, R. A. McClelland, S. Steenken, *Angew. Chem.* **1992**, *104*, 753–755; *Angew. Chem. Int. Ed. Engl.* **1992**, *31*, 743.

[64] M. Shi, Y. Okamoto, S. Takamuku, *Bull. Chem. Soc. Jpn.* **1990**, *63*, 453–460.

[65] P. G. Wenthold, D. A. Hrovat, W. T. Borden, W. C. Lineberger, *Science*, **1996**, *272*, 1456–1459; cf. W. T. Borden, H. Iwamura, J. A. Berson, *Acc. Chem. Res.* **1994**, *27*, 109–116.

[66] H.–B. Bürgi, K. K. Baldrige, K. Hardcastle, N. L. Frank, P. Gantzel, J. S. Siegel, J. Ziller, *Angew. Chem.* **1995**, *107*, 1575–1577; *Angew. Chem. Int. Ed. Engl.* **1995**, *34*, 1454. Traditionally bond alternation is discussed in terms of the so-called Mills-Nixon effect, see J. S. Siegel, *Angew. Chem.* **1994**, *106*, 1808–1810; *Angew. Chem. Int. Ed. Engl.* **1994**, *33*, 1721.

[67] B. A. Horn, J. I. Herek, A. H. Zewail, *J. Am. Chem. Soc.* **1996**, *118*, 8755–8756.

[68] R. W. Schoenlein, L. A. Peteanu, R. A. Mathies, C. V. Shank, *Science*, **1991**, *254*, 412–415.

Part II. Applications in Total Synthesis Synthesis of Natural and Non-natural Products

Pentazole and Other Nitrogen Rings

Rudolf Janoschek

Although the road of the pentazoles through chemistry has not only just begun, its end is not yet in sight; they have proved that they can stay the distance. Their history began in 1903 when Hantzsch [1] tried in vain to form phenylpentazole (**2**) by rearrangement of benzene-diazonium azide (**1**); (formerly the azide structure was assumed to be a three-membered ring. This is documented on the Curtius monument in Heidelberg).

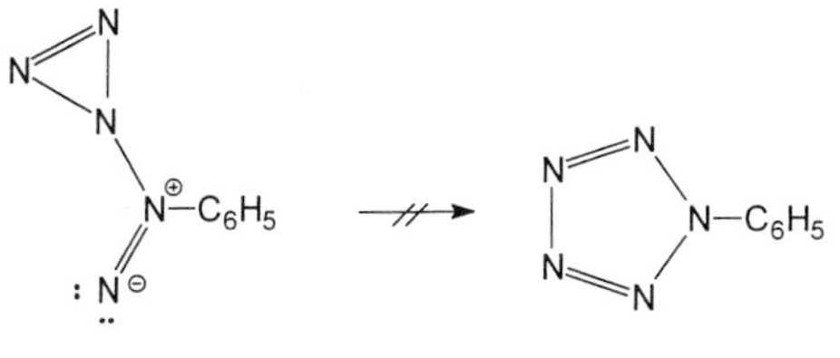

Dimroth and de Montmollin [2] also had no luck when they planned to prepare phenylpentazole from a chain of six nitrogen atoms (**3**). Then in 1915 success seemed to be at hand when Lifschitz [3] believed he had synthesized pentazolyl acetic amidrazone (**5**) from tetrazolecarbo-nitrile (**4**). But in the same year came a rebuttal from Curtius et al. [4] with the title "Die sogenannten Pentazol-Verbindungen von Lifschitz" ("The so-called pentazole compounds of Lifschitz"). They did not mince their words: "Ein derartiger Reaktionsverlauf wäre im höchsten Grade überraschend ... Lifschitz glaubt die Richtigkeit seiner Auffassung 'mit vollkommener Sicherheit' folgern zu müssen ... ohne weitere Prüfung in das Reich des Unmöglichen zu verweisen ... dass alle seine Beobachtungen und Folgerungen auf Irrtum beruhen ... von

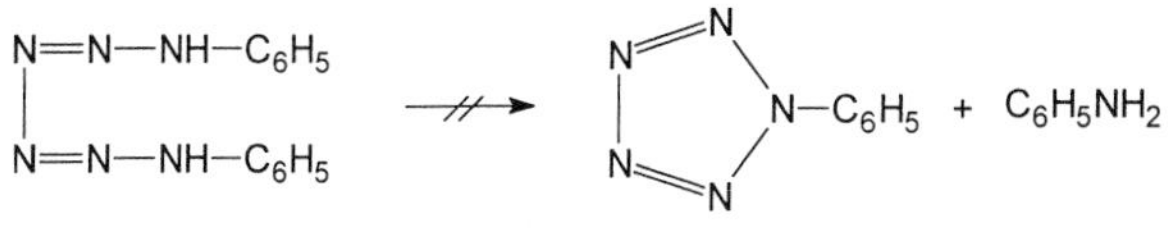

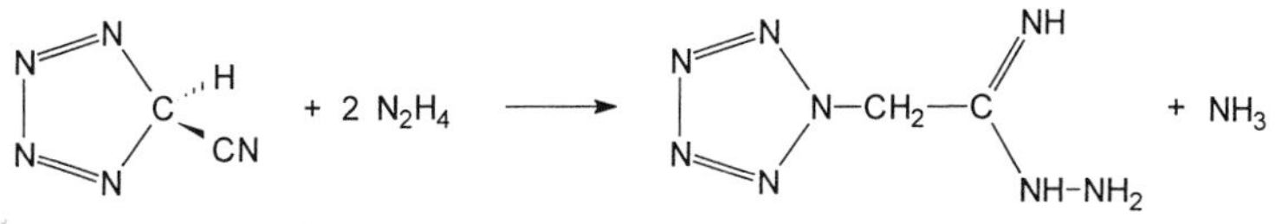

$$1 \quad\longrightarrow\!\!\!/\!\!\!/\quad 2$$

$$3 \quad\longrightarrow\!\!\!/\!\!\!/\quad 2 \;+\; C_6H_5NH_2$$

$$4 \;+\; 2\,N_2H_4 \quad\longrightarrow\quad 5 \;+\; NH_3$$

Pentazol keine Rede sein kann." ("Such a course for the reaction would be extremely surprising ... Lifschitz believes he must conclude that his interpretation is correct 'in perfect confidence' ... without further proof it must be relegated to the realm of the impossible ... all his observations and conclusions are based on error ... there can be no question of pentazoles in this case").

Dispute, misunderstanding, and failure dogged the pentazoles from the beginning. Curtius's verdict hit hard. Almost half a century passed before anyone again ventured a search for pentazole. Huisgen and Ugi [5] thus solved a classic problem in 1956 when they proved that they could link the benzenediazonium ion (**6**) with the azide ion to form benzene-diazoazide (**7**), which on ring-closure yielded phenylpentazole (**2**).

The first structure determination by X-ray diffraction on a pentazole was performed in 1983 by Dunitz and Wallis. [6] The N–N bond lengths in the five-membered ring were all similar (1.31–1.35 Å) and lie between the standard values for N–N (1.449 Å) and N=N (1.252 Å), which can be interpreted as a clear sign of an aromatic ring with six π electrons. Indeed, the recently performed natural bond orbital analysis (NBO) of the parent pentazole HN_5 (**8**) has shown that formula **8** is not a good Lewis structure. The $2p\pi$ lone pair at N(H) donates 25 % of its charge into the antibonding NN π^*-orbitals. This information does not, however, close the chapter on the pentazoles. In fact quite to the contrary, it is now all the more puzzling why the parent compound of the pentazoles, HN_5 (**8**), is still unknown. After all, it would be the final member of a well-known series that begins with pyrrole and breaks off prematurely at tetrazole. Perhaps theory can help solve the enigma.

8

The interest of the quantum chemists in pentazoles was aroused by Ferris and Bartlett. [7] They employed ab initio calculations such as the *n*th order *many-body perturbation theory*, abbreviated to MBPT(n), whose better known variant is the *n*th order Møller-Plesset perturbation theory, MPn. The MBPT(n) procedure takes account of *all* the terms to a predetermined final order *n* for the electron correlation. In the *coupled cluster* (CC) theory, on the other hand, certain contributions from the MBPT formalism are summated for *all* orders *n*. Use of the DZP basis set with these procedures yielded the anticipated structural parameters for cyclic HN_5 (**8**) with C_{2v} symmetry and proof that it represents a minimum on the energy hypersurface. The energy barrier for potential decomposition into HN_3 and N_2 was given on the MBPT(2) level as 19.8 kcal mol^{-1}. The question in the title of reference [7], "Does it exist?", can, of course, not yet be answered on these grounds alone. Experience shows that the MBPT procedure yields energy barriers that are too high. This will be demonstrated later in this report for the N_4 molecule. To estimate the kinetic stability of **8** more accurately, other calculation methods such as MP2, MP4, CC, or MCSCF-CI should also be employed. Meanwhile, density functional theory (DFT) has been developed and the available basis sets are extended to higher angular momentum quantum numbers. Therefore, the author of this report applied himself

$$N{=}N{=}N^{\ominus} \;+\; N{\equiv}N{\overset{\oplus}{-}}C_6H_5 \longrightarrow N{=}N{=}N{-}N{=}N{-}C_6H_5 \longrightarrow$$

6   **7**   **2**

to the problem of the uncertainty in this decomposition barrier, and in addition, also to other problems of nitrogen rings. The computational method is Becke's density functionals (B3LYP) using the *c*orrelation *c*onsistent *p*olarized *v*alence *t*riple *z*eta (cc-pVTZ) basis sets which are in standard notation: N(4s,3p,2d,1f), H(3s,2p,1d). Table 1 collates some of the results for the decomposition barrier of **8**. The height of the barrier varies between 13.7 and 20.9 kcal mol^{-1}, which might cast doubt on the kinetic stability. Furthermore, it was stressed repeatedly in the past that the energy hypersurface for the triplet state may not be ignored, because it might cross energy barriers leading to spin-forbidden processes (intersystem crossing) and thus to other decomposition pathways. However, recent B3LYP/cc-pVTZ calculations could not confirm a low-lying triplet state at the transition structure. Moreover, the triplet energy is significantly above the ground state singlet energy at any point of the energy profile of the ring opening process. Only one issue is definite: theory has pointed the way; the next step must be left to experiment. Matrix isolation has provided many a surprise in the past.

As dramatic as the tale of the pentazoles is the story of the unsuccessful attempt to detect a six-membered nitrogen ring. The first calculations were performed in the early seventies. In 1980 Vogler et al. [8] found indications that N_6 is formed on UV irradiation from *cis*-[Pt(N$_3$)$_2$(PPh$_3$)$_2$] embedded in a matrix, but it is only stable at low temperature. At that time hexaazabenzene (**9**) was considered the only feasible structure for N_6. This report unleashed an avalanche of quantum chemical calculations, in particular because **9** is isoelectronic with benzene and thus should likewise display aromatic stabilization. The calculations over the decade produced every conceivable answer: hexaazabenzene has D_{6h} symmetry ... has D_{3h} symmetry ... does not exist, because **9** is the transition structure for nitro-

Table 1. The activation barrier ΔE [kcal mol^{-1}] incorporating the zero-point vibrational correction for the decomposition HN$_5$ → HN$_3$ + N$_2$, calculated with several methods.

Method[a]	ΔE
MBPT(2)/DZP [7]	19.8
MP2/6-31G*	13.7
MP4/6-31G*	18.9
CCSD/6-31G*	20.9
B3LYP/cc-pVTZ	16.4

[a] The calculations with the 6-31G* basis set were performed with the program GAUSSIAN 92 [1993]; for the B3LYP calculations GAUSSIAN 94 was used [1997].

gen exchange for 3 N$_2$. The latest MBPT calculations of Bartlett et al. [9] nullified all previous efforts. According to them, hexaazabenzene cannot exist because this arrangement of atoms (D_{6h}) gives two imaginary frequencies for the out-of-plane vibrations. Recently performed B3LYP/cc-pVTZ calculations yielded three imaginary frequencies. The naive view and the conclusions from it that *all* cyclic compounds with six π electrons must be aromatic had already been convincingly disproved by Shaik et al. [10] The idea that Vogler's N$_6$ could be diazide (**10**), which was shown by calculations [11] to be 35 kcal mol^{-1} more stable than **9**, was supported repeatedly during the past ten years.

Now that the birth of the five-membered nitrogen rings has been announced only after a long labor, and a six-membered nitrogen ring has even been disputed by theory, it is

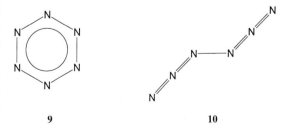

     9   **10**

the turn of three- and four-membered nitrogen rings. In 1977 triaziridine N_3H_3 was discovered in small amounts bound as a complex to a zeolite crystal. X-ray diffraction confirmed the ring structure. [12] More recent ab initio calculations suggest that this compound can also exist in uncomplexed form, since the three-membered nitrogen ring represents a minimum on the energy hypersurface. [13] The authors of this paper, however, are even self-critical and issue a warning that applies to all the theoretical investigations considered here: "... but we must do more work to understand the kinetic lability of these compounds".

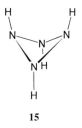

15

11   12

The localization of an energy minimum is certainly not sufficient for the stability of a chemical structure. In addition, heights of activation barriers for ring opening processes are necessary for the discussion of kinetic lability. New computational results for three-, four-, and five-membered nitrogen rings are presented in Table 2. These cyclic systems exhibit low activation barriers to ring opening and high product stability. Curtius's cyclic azide structure **11** and pentazole **8** are seen to be in competition with respect to their instability. Triaziridine **13** is the most stable one among this series, according to its lowest product stability (triimide **14**) and highest activation barrier of 33.0 kcal mol^{-1}. No singlet transition structure could be localized for

the concerted decomposition of the hypothetic tetraaziridine **15** to diimide. For this case, a two-step mechanism is assumed for the decomposition where a low-lying triplet diradical indicates the instability of the four-membered ring.

In the form of **16**, a structure isoelectronic with cyclobutadiene, N_4 exhibits a similar stability than tetraazatetrahedrane (**17**). Because of the homology to the known tetrahedral P_4, more attention has been paid to structure **17** than to **16**. After the most thorough theoretical study yet undertaken, Lee and Rice [14] calculate a considerable energy barrier of 61 kcal mol^{-1} for the decomposition $N_4 \rightarrow N_2$. The authors exploited all possibilities both with the basis set (DZP, [4s, 3p, 2d, 1f]) and

Table 2. Calculated energy profiles for the decomposition of nitrogen rings. B3LYP/cc-pVTZ calculated relative energies [kcal mol^{-1}] including zero-point energy correction. TS: transition structure; T: triplet minimum structure.

11	TS	**12**
45.6	65.4	0.0

13	TS	**14**
28.1	61.1	0.0

15	T(HN·-(NH)$_2$-N·H)	2 N$_2$H$_2$
32.4	33.2	0.0

8	TS	**12** + N$_2$
41.7	58.1	0.0

8	TS	Diazoazide
41.7	71.1	66.4

13   14

with the methods of calculation (CASSCF, MP2, CC) to separate the less appropriate methods from the better ones by comparing the different results. In the process, the energy barrier for the decomposition was lowered from 80 to 61 kcal mol⁻¹. Bartlett et al. [9] report a quite high value of 79 kcal mol⁻¹ from only one MBPT calculation, which leads to the conclusion that energy barriers calculated with MBPT are too high as a result of a systematic error, and the kinetic stability of all the compounds covered here is therefore overemphasized. The lower, but nevertheless still substantial barrier for N_4 decomposition must not lead to a premature euphoria, [9] because the triplet energy hypersurface intersects this barrier low down. [14] A fast spin-forbidden process (double intersystem crossing) can therefore annihilate N_4. Most importantly, the relative energy of N_4 with respect to 2 N_2 turned out to be as high as 186 kcal mol⁻¹.

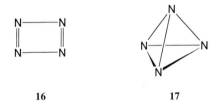

16   17

This trail through the territory of the nitrogen rings should not end without mention of the latest candidates for experimental proof. Minima have been located on the energy hypersurface for bipentazole N_5–N_5 [7] and octaazacubane N_8, [9] and structural data have been reported.

Have nitrogen rings or cages any interesting property that makes the effort to synthesize them worthwhile? A glance at the reaction energies on decomposition to N_2 gives an answer. The energy released per N_2 unit on decomposition of N_4 is 93 kcal mol⁻¹ and of N_8, 112 kcal mol⁻¹. Such high energy density makes these compounds promising rocket fuels. In fact, theory predicts a better effici-

ency for N_8 than for the conventional fuel, which is based on H_2 and O_2.

The intention of this short report was to show the advance of theory through quantum chemical calculations in the field of nitrogen rings. Although the calculated molecular properties can still be extended and refined, already a quite clear picture emerges. Nitrogen rings, and in particular nitrogen cages, are extremely energy-rich and kinetically labile systems. In comparison with the results of theoretical studies, the yield of the experimental investigations is as yet small. N_3H_3 is an unintentional success. Well-targeted photochemical studies under matrix conditions will probably still expose several surprises.

References

[1] A. Hantzsch, *Ber. Dtsch. Chem. Ges.* **1903**, *36*, 2056.
[2] O. Dimroth, G. de Montmollin, *Ber. Dtsch. Chem. Ges.* **1910**, *43*, 2904.
[3] J. Lifschitz, *Ber. Dtsch. Chem. Ges.* **1915**, *48*, 410.
[4] T. Curtius, A. Darapsky, E. Müller, *Ber. Dtsch. Chem. Ges.* **1915**, *48*, 1614.
[5] R. Huisgen, I. Ugi, *Angew. Chem.* **1956**, *68*, 705; *Chem. Ber.* **1957**, *90*, 2914.
[6] J. D. Wallis, J. D. Dunitz, *J. Chem. Soc. Chem. Commun.* **1983**, *16*, 910.
[7] K. F. Ferris, R. J. Bartlett, *J. Am. Chem. Soc.* **1992**, *114*, 8302.
[8] A. Vogler, R. E. Wright, H. Kunkely, *Angew. Chem.* **1980**, *92*, 745; *Angew. Chem. Int. Ed. Engl.* **1980**, *19*, 717.
[9] W. J. Lauderdale, J. F. Stanton, R. J. Bartlett, *J. Phys. Chem.* **1992**, *96*, 1173.
[10] S. S. Shaik, P. C. Hiberty, J. M. Lefour, G. Ohanessian, *J. Am. Chem. Soc.* **1987**, *109*, 363.
[11] M. T. Nguyen, *J. Phys. Chem.* **1990**, *94*, 6923.
[12] Y. Kim, J. W. Gilje, K. Seff, *J. Am. Chem. Soc.* **1977**, *99*, 7057.
[13] D. H. Magers, E. A. Salter, R. J. Bartlett, C. Salter, B. A. Hess, Jr., L. J. Schaad, *J. Am. Chem. Soc.* **1988**, *110*, 3435.
[14] T. J. Lee, J. E. Rice, *J. Chem. Phys.* **1991**, *94*, 1215.

New Total Syntheses of Strychnine

Uwe Beifuss

Strychnine (**1**), the active component of a notorious arrow poison in Southeast Asia, has a mysterious history. [1] It is a convulsant blocking synaptic inhibition in the spinal cord by acting as an antagonist of the inhibitory neurotransmitter, glycine. [2] In therapeutic doses strychnine has a mildly analeptic effect; in toxic doses it leads to uncoordinated tonic convulsions induced by acoustic, tactile, or optical stimuli. Paralysis of the respiratory organ results in death; 100–300 mg is the lethal dose für an adult human. [2c]

In 1818 strychnine (**1**), which in larger amounts occurs in the poison nut (*Strychnos nux vomica L.*) and the St. Ignatius' bean (*Strychnos ignatii Bergius*), was isolated by Pelletier and Caventou [3] and was one of the first alkaloids obtained in pure form. The structure determination by chemical degradation of the natural product, a complex structure with seven rings and six stereogenic centres, proved to be difficult and tedious. Many

1	2
Strychnine	Wieland-Gumlich aldehyde

structural formulae were proposed and rejected before Robinson et al. (1946) and Woodward et al. (1947) presented the correct formula. [4] A few years later the relative and absolute configuration was determined by X-ray crystallography. [5]

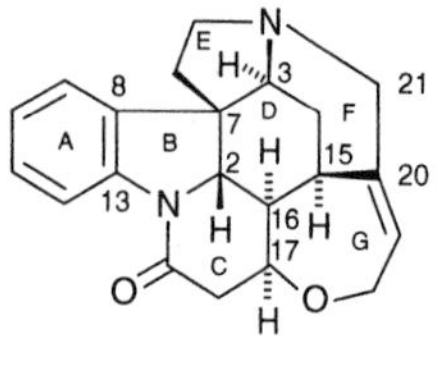

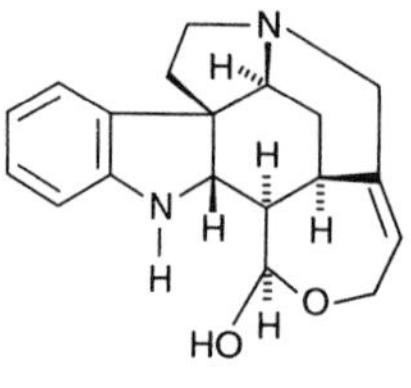

The structure determination culminated in the total synthesis of **1** by Woodward et al., which was the only total synthesis of this natural product for nearly forty years. [6] Even today it is still considered a highlight in the development of modern organic synthesis, for the deliberate design and execution of the synthesis of a such a complex molecule was unprecedented at that time. Despite the rapid development of organic chemistry since then, strychnine has remained an attractive and challenging synthetic target. A number of novel approaches to the strychnos alkaloids have been devised, [7, 8] but the molecules prepared frequently lack the functionalities required for the construction of the seven-membered allylic ether G ring of strychnine. Recently, almost forty years after Woodward, four research groups achieved the total synthesis of strychnine. In 1992 Magnus et al. reported on the successful conclusion of the second total synthesis of strychnine and, at the same time, the first total synthesis of the so-called Wieland-Gumlich aldehyde **2**. [9] In 1993 Overman et al. published the first

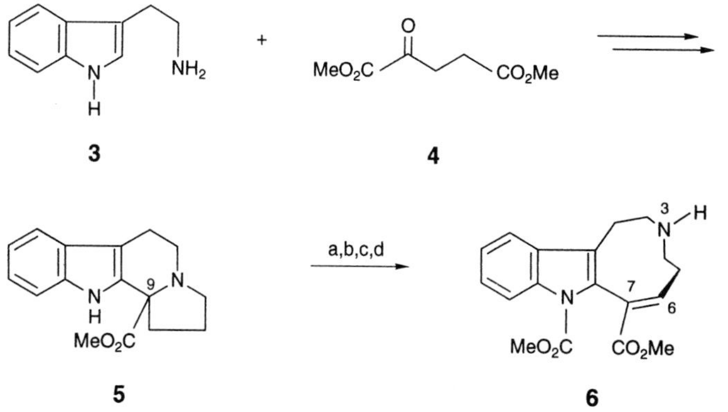

3 + **4** →

5 →a,b,c,d→ **6**

Scheme 1. (a) ClCO$_2$CH$_2$CCl$_3$, CH$_2$Cl$_2$; (b) NaOMe, MeOH; (c) 50 % aq NaOH, CH$_2$Cl$_2$, ClCO$_2$Me, benzyltriethylammonium chloride; (d) Zn, CH$_3$COOH, THF.

and so far only enantioselective route to strychnine (**1**). [10] In an extension of this work Overman et al. recently reported on the first synthesis of unnatural strychnine (*ent*-**1**). [10b] In the meantime Kuehne et al. [11] and Rawal et al. [12] have also completed their total syntheses of *rac*-**1**. This certainly provides ample grounds for discussing these new syntheses.

Magnus' synthesis [9] follows the retrosynthetic analysis of the strychnos framework successfully employed by Harley-Mason in the syntheses of several natural products in the 1960s and 1970s. [13] The key step of this strategy is the transannular iminium ion cyclization of a nine-membered ring for the stereoselective construction of rings D and E of the strychnos alkaloids. Magnus follows the classical approach right from the start with the multistep conversion of the tetracyclic amine *rac*-**5**, which is readily accessible from **3** and **4**, into **6** (Scheme 1). A crucial step is the β,β,β-trichloroethyl chloroformate induced fragmentation of the tertiary amine *rac*-**5** to provide the expanded nine-membered ring system. One of the most remarkable steps in the remainder of the synthesis is the construction of the F ring by intramolecular conjugate addition. For this purpose amide **7a** is prepared and treated with sodium hydride in THF, which results in the facile diastereoselective transformation into *rac*-**8**. The stereo-

selectivity of this 1,4-addition is attributed to the protonation of the intermediate ester enolate from the top face. Oxidation of *rac*-**8** to give the corresponding mixture of sulfoxides is then followed by a Pummerer reaction and Hg^{2+}-mediated hydrolysis to furnish *rac*-**9** (Scheme 2).

Magnus et al. obtained both enantiomers of **9** with considerable effort by acylation of **6** with (+)-(*R*)-*p*-toluenesulfinylacetic acid and cyclization of the resulting sulfoxide **7b**, separation of the four diastereomers formed this way, combination of the pairs with the same absolute configurations at C6 and C7, and subsequent conversion into the two enantiomers **9** and *ent*-**9**. [9a] The cyclization of sulfoxide **7b** yields the two products enantiomeric at C6 and C7 in a 55:45 ratio. This is why this route is not only laborious but then only insignificantly more efficient than resolution.

After transformation of *rac*-**9** into acetal *rac*-**10** (Scheme 2) rings D and E are constructed in the most critical step of the entire synthesis. A transannular iminium ion cyclization [7, 13] provides predominantly *rac*-**12** in 65 % yield with remarkable regio- and stereoselectivity. [14] It is assumed that treatment of *rac*-**10** with mercury(II) acetate in acetic acid leads primarily to the cyclic iminium ion *rac*-**11**, which then gives *rac*-**12**.

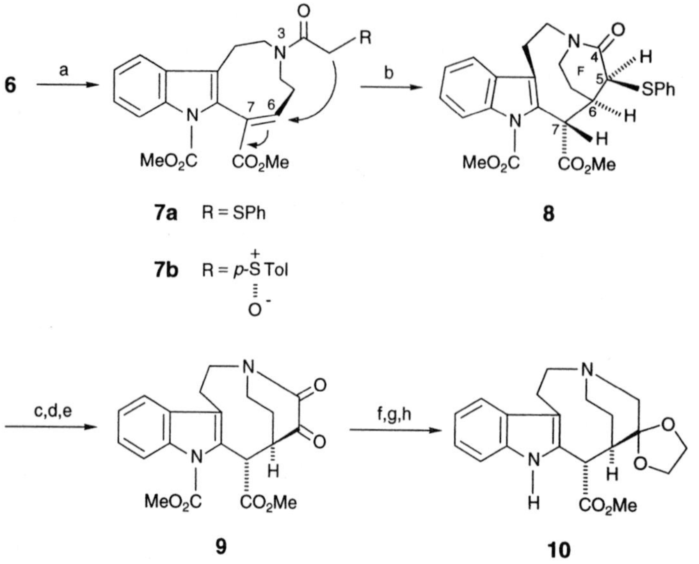

Scheme 2. (a) PHSCH$_2$CO$_2$H, bis(2-oxo-3-oxazolidinyl)phosphinic acid (BOPCl), Et$_3$N, CH$_2$Cl$_2$; (b) NaH, THF; (c) *m*-chloroperbenzoic acid (MCPBA), CH$_2$Cl$_2$, 0 °C; (d) trifluoroacetic anhydride (TFAA), 2,6-di-*tert*-butyl-4-methylpyridine; (e) HgO, CdCO$_3$, THF, H$_2$O; (f) BrCH$_2$CH$_2$OH, 1,8-diazobicyclo[5.4.0]undec-7-ene (DBU), C$_7$H$_8$; (g) BH$_3$ · THF; (h) Na$_2$CO$_3$, MeOH, 65 °C.

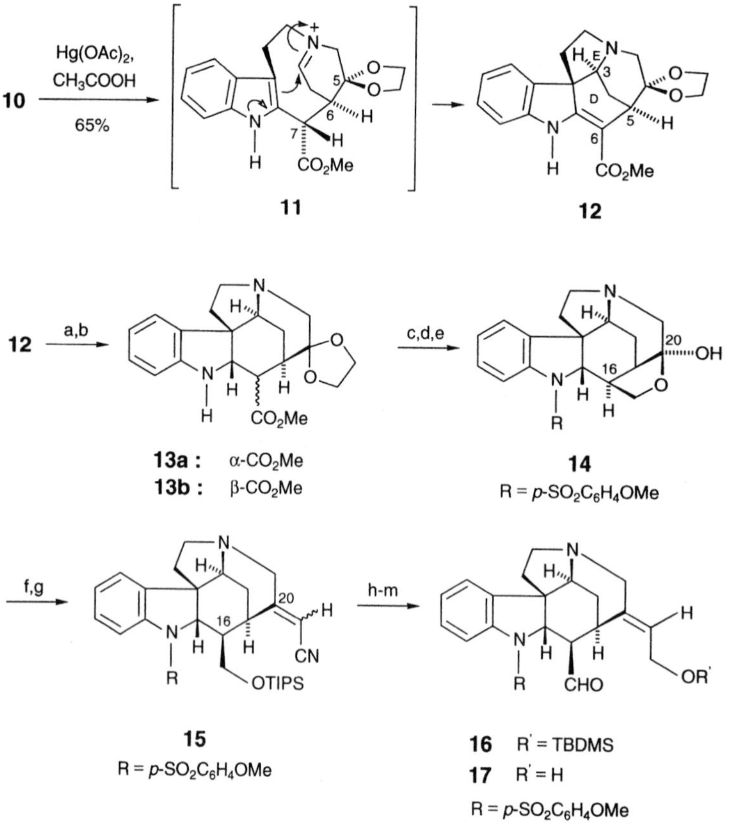

Scheme 3. (a) Zn, H$_2$SO$_4$, MeOH; (b) MeONa, MeOH; (c) *p*-MeOC$_6$H$_4$SO$_2$Cl, EtN*i*Pr$_2$, 4-dimethylaminopyridine (DMAP), CH$_2$Cl$_2$; (d) LiBH$_4$, THF, HN(CH$_2$CH$_2$OH)$_2$; (e) HClO$_4$; (f) triisopropylsilyltrifluoromethanesulfonate (TIPSOTf), DBU, CH$_2$Cl$_2$; (g) (EtO)$_2$-P(O)CH$_2$CN, potassium hexamethyldisilazide (KHMDS), THF, 25 °C; (h) diisobutylaluminium hydride (DIBAL), CH$_2$Cl$_2$, H$_3$O$^+$; (i) NaBH$_4$, MeOH; (j) 2 N HCl, MeOH; (k) *tert*-butyldimethylsilyltrifluoromethanesulfonate (TBDMSOTf), DBU, CH$_2$Cl$_2$, –20 °C; (l) SO$_3$ · C$_5$H$_5$N, DMSO, Et$_3$N; (m) pyridine, HF.

A number of steps are required for the conversion of *rac*-**12** into the cyclic hemiacetal *rac*-**14** (Scheme 3). This compound is important to the rest of the synthesis, since it cannot only be prepared from tryptamine (**3**) and dimethyl 2-ketoglutarate (**4**), but can also be obtained readily in substantial amounts and in enantiomerically pure form by degradation of strychnine. As soon as enough of this relay hemiacetal has been secured, assembly of the G ring can be tackled.

Since just a compound with an (*E*)-configuration double bond as in **17** can be used for the synthesis of the G ring, the construction of the hydroxyethylidene double bond must be diastereoselective. This very problem had challenged Woodward et al., and like those pioneers Magnus et al. were also foiled: Wittig-Horner reaction furnishes the *α,β*-unsaturated cyanide **15** as a 2:3 mixture of the (*Z*) and (*E*) isomers. Although the undesired (*Z*) isomer can be isolated and isomerized photochemically to give a mixture of both isomers, this step is far from being optimal.

The next steps in the synthesis include the multistep conversion of (*E*)-**15** into **16**, subsequent cleavage of the silyl ether protecting group, and cyclization of **17**, which cannot be isolated, to provide hexacyclic **18**. With the reductive removal of the sulfonamide Magnus et al. arrived at their first goal, the total synthesis of the Wieland-Gumlich aldehyde (**2**). The conclusion of the total synthesis of strychnine is also within reach; the authors succeed in converting **2** into the natural product in a one-step reaction with malonic acid, following the method of Robinson et al. [15] With 27 steps this synthesis is only negligibly shorter than Woodward's 28-step synthesis from 1954. But Magnus et al. achieved an overall yield of roughly 0.03 %, which is more than 1000 times greater than Woodward's overall yield of 0.00006 %. This result would have been impossible without powerful new synthetic methods. Magnus' synthetic strategy has a classical form, and he prepared enantiomerically pure compounds very effectively by traditional means, namely via a relay compound obtained by degradation of the natural product itself.

The third total synthesis of strychnine (**1**), so far the only enantioselective route to the natural product, was accomplished by Overman et al. [10] The key to their approach to **1** is the sequential cationic aza-Cope rearrangement/Mannich cyclization, which is frequently employed with success in alkaloid synthesis. With the synthesis of akuammicine *rac*-**19** the authors proved that this strategy offers an efficient route to the strychnos alkaloids. [8g]

The crucial compound in this strychnine synthesis is azabicyclo[3.2.1]octane **31**, which is the substrate for the aza-Cope-Mannich sequence (Scheme 4). In the preparation of **31**, *meso*-diester **20** is subjected to acetylcholine esterase catalyzed hydrolysis to yield **21** with high enantiomeric purity. Eighteen ensuing steps then provide **31** in 14 % yield. The allylic carbonate obtained from **21** is

18 R = *p*-SO$_2$C$_6$H$_4$OMe
2 R = H
Wieland-Gumlich aldehyde

19 Akuammicine

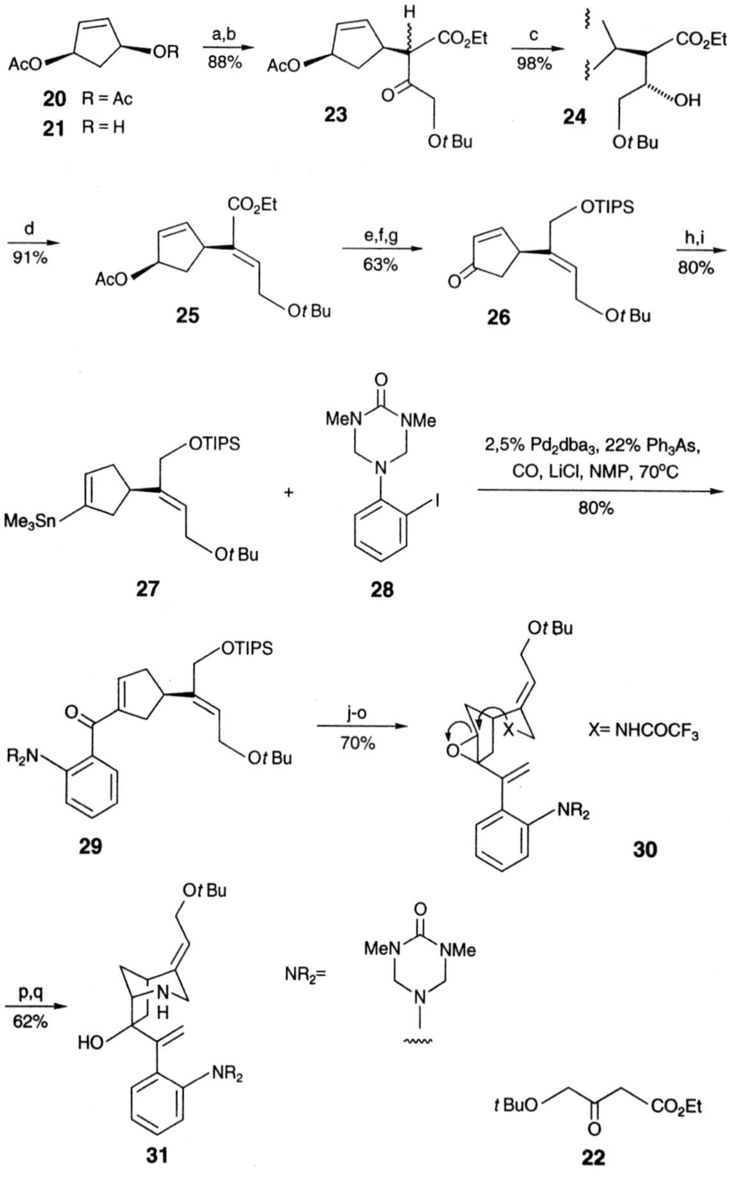

Scheme 4. (a) ClCO₂Me, pyridine, CH₂Cl₂, 23 °C; (b) *t*BuOCH₂COCH₂CO₂Et (**22**), NaH, 1 % [Pd₂(dba)₃], 15 % PPh₃, THF, 23 °C; (c) NaCNBH₃, TiCl₄, −78 °C; (d) dicyclohexylcarbodiimide (DCC), CuCl, C₆H₆, 80 °C; (e) DIBAL, CH₂Cl₂, −78 °C; (f) TIPSCl, tetramethylguanidine, *N*-methyl-2-pyrrolidone (NMP), −10 °C; (g) Jones oxidation, −5 °C; (h) L-Selectride, PhNTf₂, THF, −78 °C → 0 °C; (i) Me₆Sn₂, 10 % [Pd(PPh₃)₄], LiCl, THF, 60 °C; (j) *t*BuO₂H, Triton-B, THF, −15 °C; (k) Ph₃P=CH₂, THF, 0 → 23 °C; (l) tetrabutylammonium fluoride (TBAF), THF, −15 °C; (m) methanesulfonyl chloride (MsCl), *i*Pr₂NEt, CH₂Cl₂, −23 °C; (n) LiCl, DMF, 23 °C; (o) NH₂COCF₃, NaH, DMF, 23 °C; (p) NaH, C₆H₆, 100 °C; (q) KOH, EtOH–H₂O, 60 °C.

used in a palladium-catalyzed allylic substitution with **22** to furnish the *cis* products **23**, which are reduced with high diastereoselectivity (> 20 : 1) in accord with the Felkin-Anh model to provide a mixture of the *trans-β*-hydroxy esters **24**. Subsequent *syn* elimination affords the (*E*) isomer **25** almost exclusively (97 : 3). In this way Overman et al. succeeded in solving the problem of the stereoselective construction of what will become the allylic ether double bond at C20 of the natural product early in the synthesis. The next important intermediate is the *α,β*-unsaturated ketone **29**, which is accessible by the palladium-catalyzed, carbonylative cross-coupling of vinyl stannane **27** with the triazone-protected *o*-iodoaniline **28**. In turn, **27** is prepared by the palladium-catalyzed Stille coupling of the enol triflate obtained regioselectively from **26**. The key step in the transformation of **29** into **31** is the stereoselective, intramolecular aminolysis of epoxide **30**, which is the product of the substrate-controlled stereoselective epoxidation and subsequent Wittig methylenation of **29**.

Now the aza-Cope-Mannich cascade must be triggered, in other words, amine **31** must be converted into the corresponding formaldiminium ion **32** (Scheme 5). This is achieved by reaction with paraformaldehyde without added acid. Intermediate **32** undergoes cationic [3, 3] sigmatropic rearrangement under these reaction conditions to give **33**, which contains all the structural features that are required for the ensuing intramolecular Mannich cyclization. At the end of the sequence **34** is obtained stereoselectively and almost quantitatively (98 % yield) – an outstanding example of the efficiency and flexibility of this synthetic strategy. The cyclization product **34** is then acylated with methyl cyanoformate, and cleavage of the triazone protecting group affords pentacycle **35**, which contains all of the C atoms required for the synthesis of the Wieland-Gumlich aldehyde **2**. The conversion of **36** into strychnine (**1**) by conventional means is the concluding step in the first synthesis of this natural product proceeding without the resolution of racemates and without relay compounds. Overman et al. accomplish this in a total of 25 steps and with an overall yield of approximately 3 %. This first enantioselective total synthesis is achieved in excellent yield. But just as impressive is the underlying retrosynthetic analysis, which is then followed by combining several modern palladium-catalyzed reactions and the elegant aza-Cope-Mannich sequence, a distinctive feature of Overman's work.

In an extension of this work Overman and his group have achieved the total synthesis of unnatural strychnine (*ent*-**1**) as well. [10b] Their approach is based on the palladium catalyzed coupling of the hydroxy acetate **21** with the sodium salt of **22** followed by acetylation to yield the cyclopentenyl keto ester *ent*-**23**. This intermediate was readily converted to unnatural strychnine (*ent*-**1**) following the chemistry developed in the natural series.

The beginning of Kuehne's total synthesis of racemic strychnine *rac*-**1** [11] is promising, since the highly diastereoselective construction of the tetracycle *rac*-**42** from tryptamine **38** and butenal **39** proceeds in just one synthetic operation (Scheme 6). The yield for this novel and efficient sequential reaction is 51 % after cleavage of the acetal to provide aldehyde *rac*-**43**. Presumably the cyclizing Mannich reaction leading to *rac*-**40** is followed by a [3, 3] sigmatropic rearrangement giving *rac*-**41**, which in turn undergoes acid-catalyzed cyclization to afford *rac*-**42**. This new method also provides access to the indolenine *rac*-**44**, [8f] a compound containing all but two carbon atoms of ring C of the target molecule. However, it has not been possible so far to employ the furan ring in **44** to construct rings F and G in strychnine. This is why Kuehne et al. relied on *rac*-**42** in their approach to the natural product.

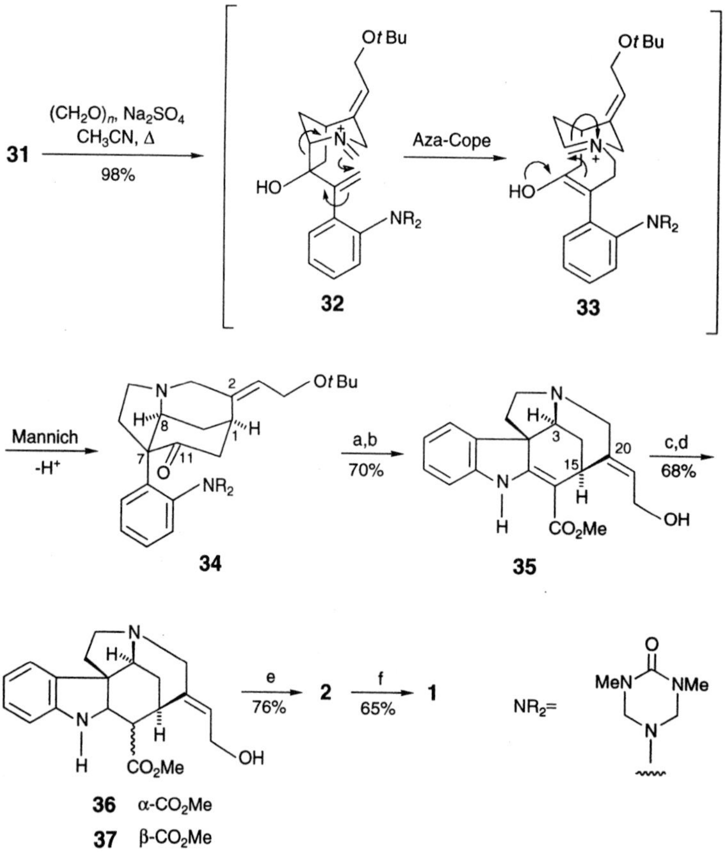

Scheme 5. (a) Lithium diisopropylamide (LDA), NCCO$_2$Me, THF, $-78\,°C$; (b) 5 % HCl–MeOH, reflux; (c) Zn, 10 % H$_2$SO$_4$-MeOH, reflux; (d) NaOMe, MeOH, 23 °C; (e) DIBAL, CH$_2$Cl$_2$, $-78\,°C$; (f) CH$_2$(CO$_2$H)$_2$, Ac$_2$O, NaOAc, HOAc, 110 °C.

This entailed the lengthy and difficult successive construction of the three missing rings F, G, and C. The synthesis of the F ring is the least troublesome, as it succeeds in only three steps and with 67 % yield (*rac*-**45** → *rac*-**47**) (Scheme 7). The key step is the intramolecular nucleophilic ring-opening of the unisolated epoxide *rac*-**45**; the thermodynamically controlled reaction yields *rac*-**46** exclusively.

In contrast to Magnus and Overman, who directed their syntheses towards the Wieland-Gumlich aldehyde *rac*-**2** and its straightforward conversion into strychnine *rac*-**1**,

Kuehne focussed on the synthesis of isostrychnine *rac*-**53** (Scheme 8) and had to rely on the isostrychnine – strychnine transformation, which Woodward had already recognized as being exceptionally difficult. First, ring C is assembled. The key step is the intramolecular Claisen reaction of the acetamido ester *rac*-**48** to give ketolactam *rac*-**49**. Although the overall yield for the construction of the C ring is good (*rac*-**47** → *rac*-**51**), the eight steps are relatively laborious. In the next part of the synthesis Kuehne, like Magnus, learned by experience that the (*E*)-selective construction of the allylic ether group at C20, which

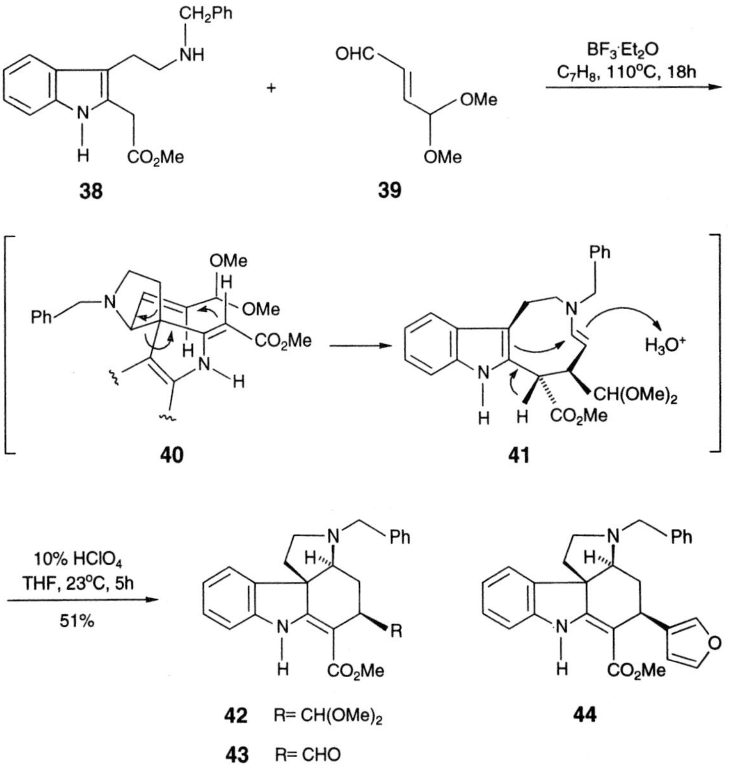

Scheme 6.

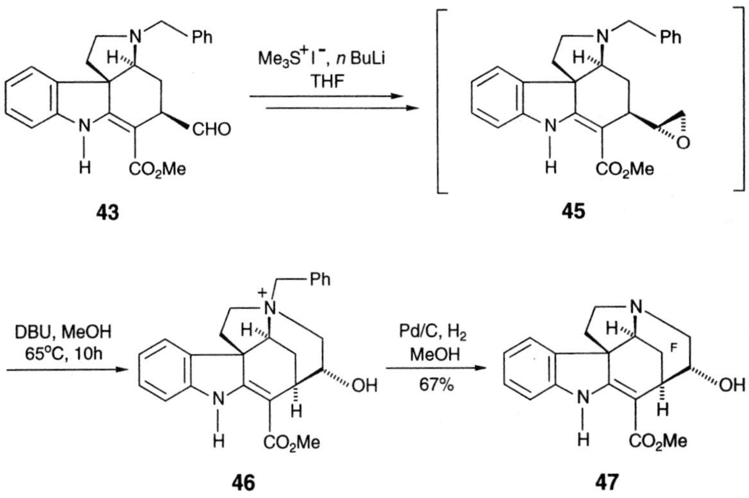

Scheme 7.

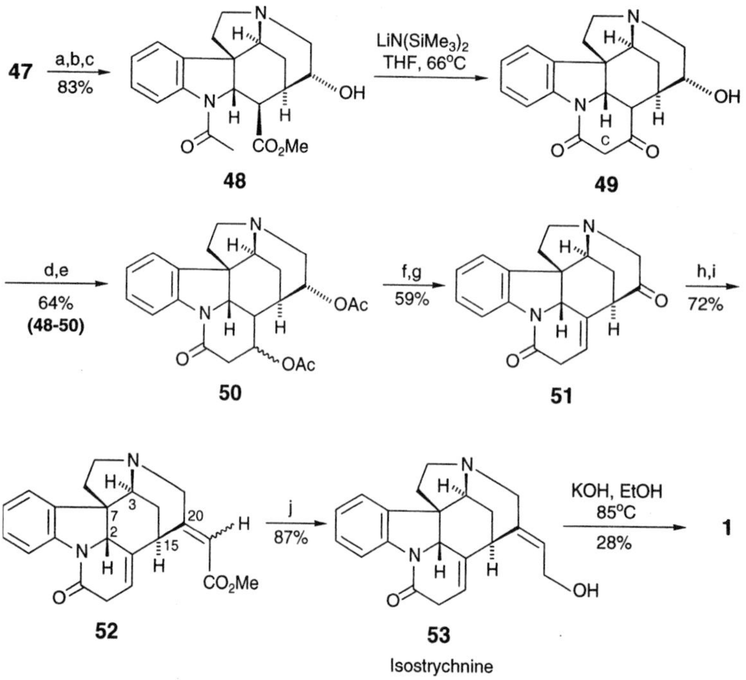

Scheme 8. (a) NaBH₃CN, HOAc, 23 °C; (b) Ac₂O, pyridine; (c) NaOMe, MeOH, 0 °C; (d) NaBH₄, MeOH; (e) Ac₂O, pyridine; (f) DBU, dioxane-H₂O, 100 °C; (g) Swern oxidation; (h) (EtO)₂P(O)CH₂CO₂Me, KN(SiMe₃)₂, THF, 23 °C, 2 h; (i) *hv*; (j) DIBAL, BF₃ · Et₂O, –78 °C.

is critical for closure of the G ring, is not possible at such a late stage of the synthesis. Wittig olefination of *rac*-**51** provides a 1 : 1 mixture of the (*E*) and (*Z*) isomers of *rac*-**52**, which can be enriched by irradiation in favour of the required (*E*)-acrylic ester (8 : 1). The reduction of (*E*)-*rac*-**52** to isostrychnine *rac*-**53** is straightforward; as expected, however, the last step of the synthesis, the problematic conversion of isostrychnine *rac*-**53** into strychnine *rac*-**1**, could not be solved in a satisfactory manner: strychnine *rac*-**1** was isolated in only 28 % yield in addition to 61 % unreacted starting material. One could argue that not every step in Kuehne's synthesis proceeds with the desired selectivity and yield and that the synthesis is not enantioselective. Yet it should be stressed that this synthesis, designed around a new and efficient key sequence, is one of the shortest routes to

strychnine *rac*-**1** with 17 steps and an overall yield of roughly 2 %. The efficiency of the synthesis could also be improved by avoiding the isostrychnine – strychnine conversion and the consequent time-consuming assembly of the C ring.

A strategy combining the advantages of both the intramolecular Diels-Alder reaction and the intramolecular Heck reaction was successfully tested in Rawal's synthesis of *rac*-**1**. [12] Heating precursor *rac*-**59**, which is prepared from **54** in eight steps, [8i] to 185 °C provides tetracycle *rac*-**60** in 99 % yield as the sole product of the Diels-Alder reaction (Scheme 9). The rapid construction of the C ring by intramolecular amide formation in only two steps is also remarkable. Allylation of *rac*-**61** with **62** furnishes *rac*-**63**. Subsequent intramolecular Heck reaction leads to the diastereoselective ring closure providing

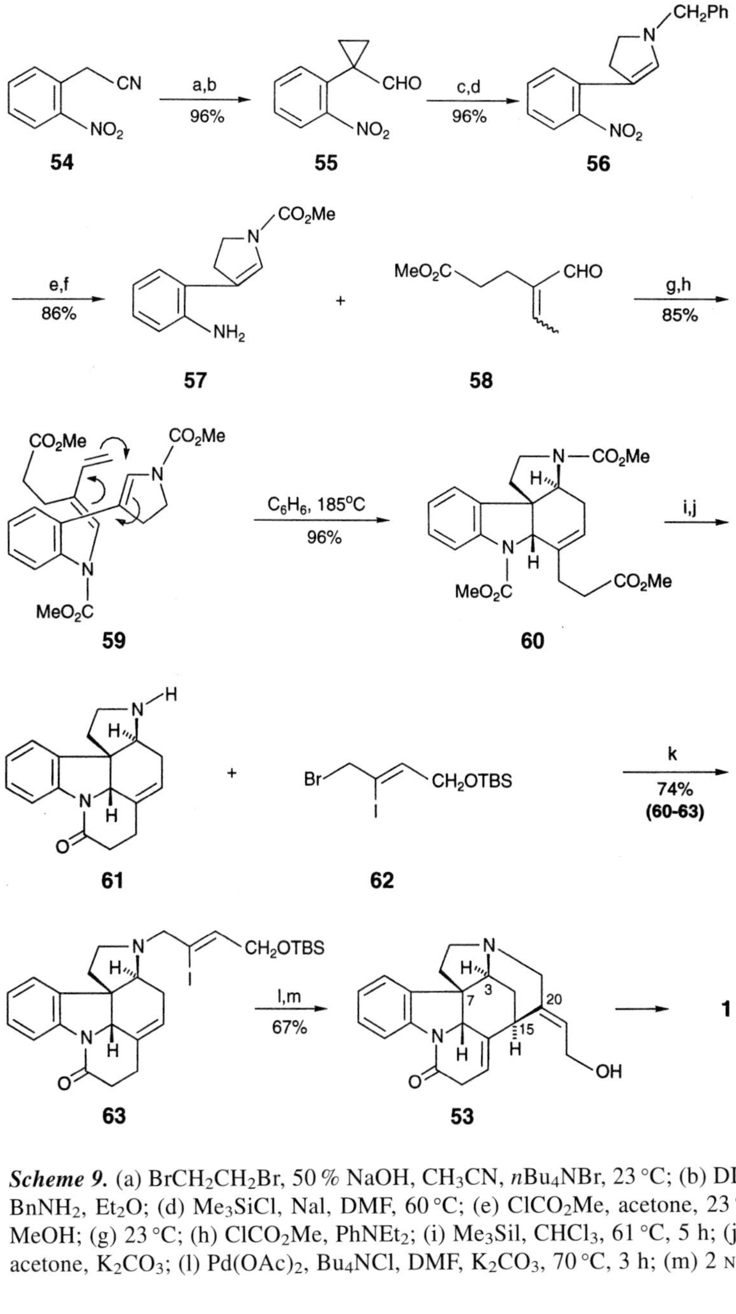

Scheme 9. (a) BrCH$_2$CH$_2$Br, 50 % NaOH, CH$_3$CN, nBu$_4$NBr, 23 °C; (b) DIBAL, C$_7$H$_8$, −78 °C, H$_3$O$^+$; (c) BnNH$_2$, Et$_2$O; (d) Me$_3$SiCl, NaI, DMF, 60 °C; (e) ClCO$_2$Me, acetone, 23 °C; (f) 10 % Pd/C, HCO$_2$NH$_4$, MeOH; (g) 23 °C; (h) ClCO$_2$Me, PhNEt$_2$; (i) Me$_3$SiI, CHCl$_3$, 61 °C, 5 h; (j) MeOH, 65 °C, 6 h; (k) DMF, acetone, K$_2$CO$_3$; (l) Pd(OAc)$_2$, Bu$_4$NCl, DMF, K$_2$CO$_3$, 70 °C, 3 h; (m) 2 N HCl, THF.

the bridged piperidine system with retention of stereochemistry at the double bond of the vinyl iodide. Deprotection concludes this hitherto shortest synthesis of racemic iso-strychnine *rac*-**53** which has 14 steps and a 35 % yield. But Rawal et al. are confronted with the problem of the inefficient final transformation to give *rac*-**1**. If one assumes Kuehne's reported yield of 28 % for this step, [16] then Rawal formally achieves the synthesis of the natural product – in racemic form, though – in only 15 steps and with almost 10 % yield, which is in the range of Kuehne's and Overman's results.

Whether the four new total syntheses represent a fundamental improvement over Woodward's strychnine synthesis can certainly be debated, as well as the extent of this improvement. It cannot, however, be contested that Overman et al. accomplished the first and only enantioselective synthesis of the natural product, and that Kuehne and Rawal with their respective 17- and 15-step syntheses devised approaches with markedly fewer reaction steps than Woodward's 28-, Magnus' 27- and Overman's 25-step syntheses. The considerable improvement in the overall yields relative to that of the first total synthesis is also noteworthy. Whereas Magnus improved the yield by a factor 1000, Overman, Kuehne and Rawal upped the overall yield by a factor of 100 000! These impressive numbers cannot be attributed solely to improved synthetic methods and modern reagents, but emphasize the importance that sequential reactions [17] have achieved in the construction of complex natural products.

The four new total syntheses demonstrate that some of the difficulties in the synthesis of strychnine, such as the diastereoselective construction of the double bond at C20, can now be solved elegantly, but that others like the isostrychnine – strychnine conversion remain unsolved. Thus it can be expected that the search for improved solutions for the

efficient synthesis of strychnine and strychnos alkaloids will continue.

References

[1] L. Lewin, *Die Pfeilgifte*, J. A. Barth, Leipzig, **1923**.

[2] a) Betz, *Angew. Chem.* **1985**, *97*, 363; *Angew. Chem. Int. Ed. Engl.* **1985**, *24*, 365; b) E. Teuscher, U. Lindequist, *Biogene Gifte*, Gustav Fischer, Stuttgart, **1987**; c) *Allgemeine und spezielle Pharmakologie und Toxikologie* (Ed.: W. Forth), 5th ed., Bibliographisches Institut & F. A. Brockhaus, Mannheim, **1987**.

[3] P. J. Pelletier, J. B. Caventou, *Ann. Chim. Phys.* **1818**, *8*, 323.

[4] a) H. T. Openshaw, R. Robinson, *Nature* **1946**, *157*, 438; b) R. B. Woodward, W. J. Brehm, A. L. Nelson, *J. Am. Chem. Soc.* **1947**, *69*, 2250.

[5] a) J. H. Robertson, C. A. Beevers, *Acta Crystallogr.* **1951**, *4*, 270; b) C. Bokhoven, J. C. Schoone, J. M. Bijvoet, *Acta Crystallogr.* **1951**, *4*, 275; c) A. F. Peerdeman, *Acta Crystallogr.* **1956**, *9*, 824.

[6] a) R. B. Woodward, M. P. Cava, W. D. Ollis, A. Hunger, H. U. Daeniker, K. Schenker, *J. Am. Chem. Soc.* **1954**, *76*, 4749; b) *Tetrahedron* **1963**, *19*, 247.

[7] Reviews: a) J. Bosch, J. Bonjoch in *Studies in Natural Product Chemistry. Vol. 1. Stereoselective Synthesis* (Part A) (Ed.: Atta-ur-Rahman), Elsevier, Amsterdam, **1988**, p. 31; b) G. Massiot, C. Delaude in *The Alkaloids*, Vol. 34 (Ed.: A. Brossi), Academic Press, New York, **1988**, p. 211.

[8] a) J. Bonjoch, D. Solé, J. Bosch, *J. Am. Chem. Soc.* **1995**, *117*, 11017; b) J. Bonjoch, D. Solé, J. Bosch, *J. Am. Chem. Soc.* **1993**, *115*, 2064; c) M. Amat, J. Bosch, *J. Org. Chem.* **1992**, *57*, 5792; d) G. A. Kraus, D. Bougie, *Synlett* **1992**, 279; e) M. E. Kuehne, C. S. Brook, D. A. Frasier, F. Xu, *J. Org. Chem.* **1995**, *60*, 1864; f) R. L. Parsons, J. D. Berk, M. E. Kuehne, *J. Org. Chem.* **1993**, *58*, 7482; g) S. R. Angle, J. M. Fevig, S. D. Knight, R. W. Marquis, Jr., L. E. Overman, *J. Am. Chem. Soc.* **1993**, *115*, 3966; h) J. M. Fevig, R. W. Marquis, Jr., L. E. Overman, *J. Am. Chem.*

Soc. **1991**, *113*, 5085; i) V. H. Rawal, C. Michoud, R. F. Monestel, *J. Am. Chem. Soc.* **1993**, *115*, 3030; j) H.-J. Teuber, C. Tsaklakidis, J. W. Bats, *Liebigs Ann. Chem.* **1992**, 461; k) J. Nkiliza, J. Vercauteren, J.-M. Léger, *Tetrahedron Lett.* **1991**, *32*, 1787; l) D. B. Grotjahn, K. P. C. Vollhardt, *J. Am. Chem. Soc.* **1986**, *108*, 2091; m) D. B. Grotjahn, K. P. C. Vollhardt, *J. Am. Chem. Soc.* **1990**, *112*, 5653; n) S. F. Martin, C. W. Clark, M. Ito, M. Mortimore, *J. Am. Chem. Soc.* **1996**, *118*, 9804.

[9] a) P. Magnus, M. Giles, R. Bonnert, C. S. Kim, L. McQuire, A. Merritt, N. Vicker, *J. Am. Chem. Soc.* **1992**, *114*, 4403; b) P. Magnus, M. Giles, R. Bonnert, G. Johnson, L. McQuire, M. Deluca, A. Merritt, C. S. Kim, N. Vicker, *J. Am. Chem. Soc.* **1993**, *115*, 8116.

[10] a) S. D. Knight, L. E. Overman, G. Pairaudeau, *J. Am. Chem. Soc.* **1993**, *115*, 9293; b) S. D. Knight, L. E. Overman, G. Pairaudeau, *J. Am. Chem. Soc.* **1995**, *117*, 5776.

[11] M. E. Kuehne, F. Xu, *J. Org. Chem.* **1993**, *58*, 7490.

[12] V. H. Rawal, S. Iwasa, *J. Org. Chem.* **1994**, *59*, 2685.

[13] J. Harley-Mason, *Pure Appl. Chem.* **1975**, *41*, 167.

[14] Typically *rac*-**12** was obtained in 50 % yield [9b].

[15] F. A. L. Anet, R. Robinson, *Chem. Ind. (London)* **1953**, 245.

[16] The authors do not give a yield for the conversion of isostrychnine *rac*-**53** into strychnine *rac*-**1**.

[17] a) L. F. Tietze, U. Beifuss, *Angew. Chem.* **1993**, *105*, 137; *Angew. Chem. Int. Ed. Engl.* **1993**, *32*, 131; b) L. F. Tietze, *Chem. Rev.* **1996**, *96*, 115.

Total Syntheses of Zaragozic Acid

Ulrich Koert

Who is nowadays not aware of the negative effects on health of high cholesterol levels in blood? Atherosclerosis, hypertension, and myocardiac infarction can be the consequences of this metabolic disturbance (Scheme 1). [1] On the one hand the diet influences the cholesterol level exogenously, on the other hand the body builds up cholesterol endogen-

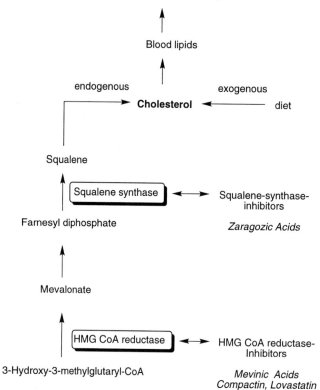

Atherosclerosis, hypertension, myocardial infarction

Blood lipids

endogenous — Cholesterol ← exogenous — diet

Squalene

Squalene synthase ←→ Squalene-synthase-inhibitors

Farnesyl diphosphate

Zaragozic Acids

Mevalonate

HMG CoA reductase ←→ HMG CoA reductase-Inhibitors

3-Hydroxy-3-methylglutaryl-CoA

Mevinic Acids
Compactin, Lovastatin

Scheme 1. Cholesterol in the human body: endogeneous build up, exogeneous introduction, and pathological consequences.

ously. The biosynthetic pathway of cholesterol begins with a branching of the citric acid cycle. Starting from 3-hydroxy-3-methyl-glutaryl-(HMG)-CoA, farnesyl-diphosphate is formed via mevalonate. The coupling of two molecules of farnesyldiphosphate yields squalene, which is then converted into cholesterol.

If this endogenous production of cholesterol could be reduced, it would be of significant therapeutic use. With the discovery of the mevinic acids almost 15 years ago, an enzyme inhibitor for an enzyme participating in cholesterol biosynthesis, namely HMG-CoA reductase, was found for the first time. Today lovastatin, simvastatin, and pravastatin are standard components of the battery of pharmaceuticals.

Now a group of recently discovered natural products that inhibit another enzyme, squalene synthase, has become the focal point of attention: the zaragozic acids. In 1992 researchers at Merck, Sharp and Dohme, [2] and Glaxo [3] independently reported the discovery of a new family of natural products with squalene synthase inhibiting properties. The Merck group gave the new compounds the name zaragozic acids after the Spanish city of Zaragoza in which they had been found for the first time in fungal cultures. The Glaxo researchers named the new compounds squalestatins after their squalene synthase inhibiting effect.

Representative compounds of this new group of natural products are zaragozic acid C (**1**) and zaragozic acid A (**2**) (= squalestatin S1). All zaragozic acids share a common 2,8-dioxabicyclo[3.2.1]octane skeleton (Fig. 1a), which is extremely polar as a result of three carboxyl groups in positions 3, 4, and 5 as well as three hydroxyl groups at C4, C6, and C7. An alkyl side chain (CH$_2$-C1-SK) is found at C1 and a fatty acid side chain at C6-O (C6-O-SK-CO). The spatial arrangement of the three COOH groups and the OH group at C4 is remarkable. This arrangement of functionalities corresponds to that of a

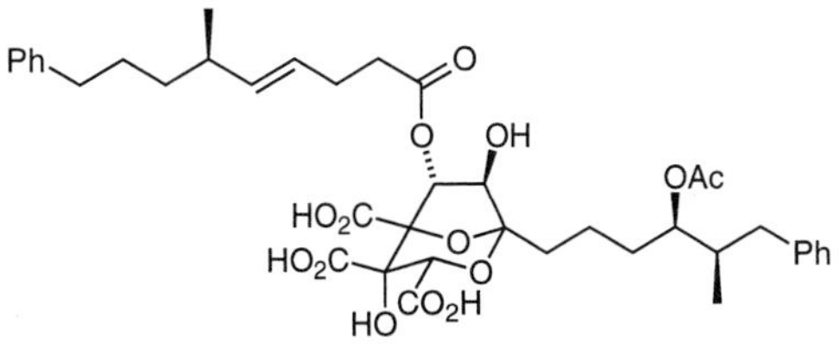

Zaragozic Acid C **1**

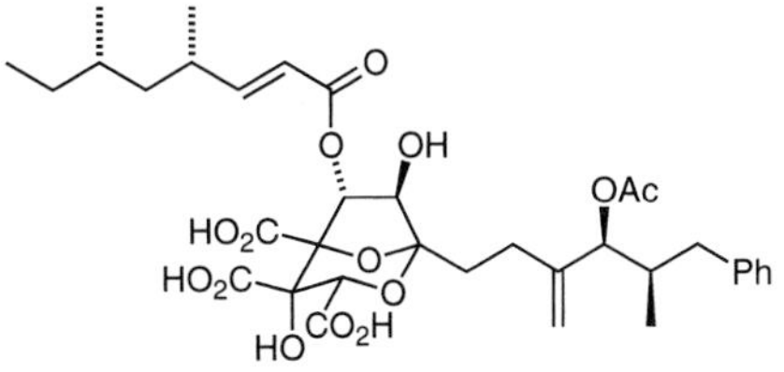

Zaragozic Acid A **2**
(Squalestatin S1)

citric acid molecule (Fig. 1b) in a frozen conformation. Thus, the binding site of squalene synthase for zaragozic acids could be one for citric acid.

The biological activity of zaragozic acids and the complexity of their structures have attracted the attention of numerous synthetic

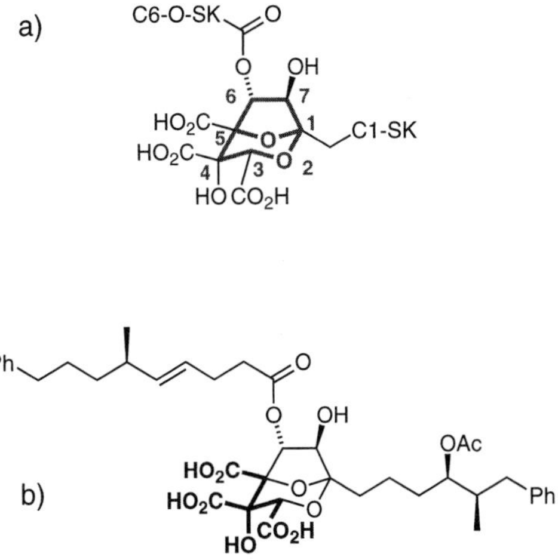

Figure 1. (a) Structure of the zaragozic acids with the 2,8-dioxabicyclo[3.2.1]-octane skeleton (red); (b) conformationally frozen citric acid partial structure (blue) in zaragozic acid C.

chemists within a very short space of time. [4] At the end of 1994, three total syntheses were published: one by Carreira et al. (zaragozic acid C), [5] one by Nicolaou et al. (zaragozic acid A), [6] and one by Evans et al., which was carried out in collaboration with researchers at Merck, Sharp and Dohme (zaragozic acid C). [7] These three syntheses are compared in this highlight.

Scheme 2 shows the three retrosyntheses of zaragozic acids **3**. All three research groups introduce the acyl side chain at C 6-O as the last step. Thus, the alcohol **4** is obtained as the precursor of the complete molecular skeleton **3**.

Apart from that, the synthetic pathways differ. Nicolaou derives the dioxabicyclo[3.2.1]-octane skeleton from the ketone **5**, which can be split into the two building blocks **6** and **7** by cleavage of the C1–C7 bond. The synthetic strategy of Evans regards the ketone **8** as the

precursor of the dioxabicyclo[3.2.1]octane **4**. The retrosynthetic cut between C1 and the adjacent C atom of the C1 side chain yields the building blocks **9** and **10**. These two syntheses have the common feature that the dioxabicyclo[3.2.1]octane is constructed inclusive of the carboxyl functions. A different route was followed by Carreira. In this the three carboxyl groups are introduced after the completion of the dioxabicyclo[3.2.1]-octane. Thus, compound **4** is traced back to the tetraol **11**, which should be accessible from the acyclic precursor **12**. The cleavage of the C1–C7 bond leads to the building blocks **13** and **14**. Compared with the strategies of Nicolaou and Evans, Carreira thus has the chance to rapidly assemble the dioxabicyclo[3.2.1]octane. However, as it turned out the construction of the three carboxyl functions requires a considerable number of steps at the end of the synthesis. In all three

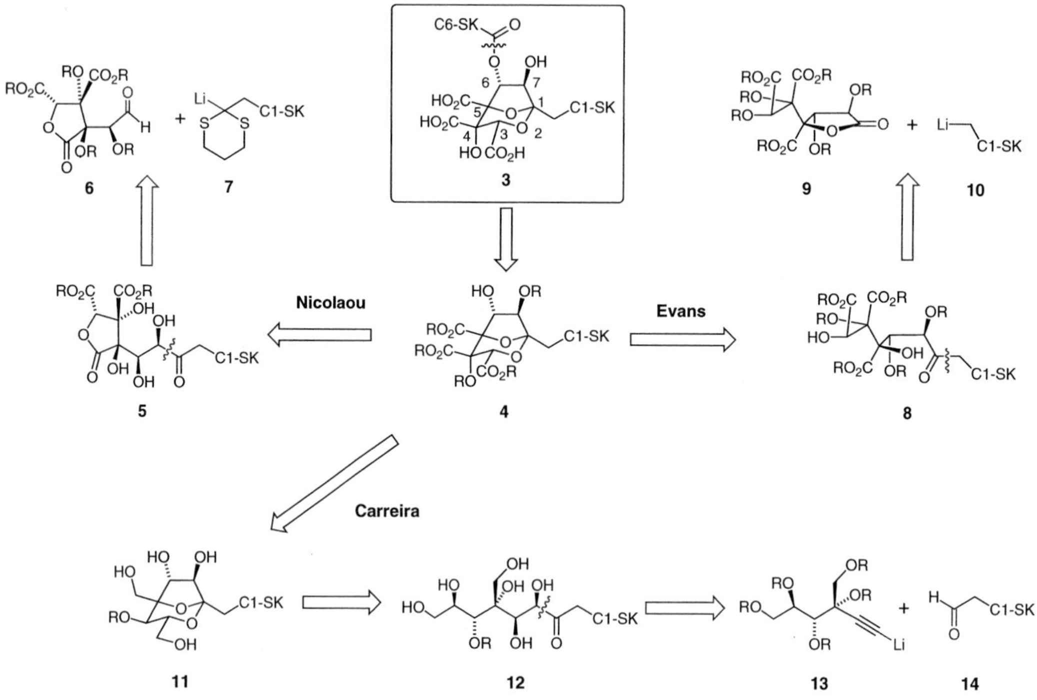

Scheme 2. Comparison of the retrosyntheses by Nicolaou, Carreira und Evans.

syntheses, the dioxabicyclo[3.2.1]octane ske-leton is obtained by acetal formation of a 4,6-dihydroxyketone by using standard proce-dures. The efficiency with which the acyclic precursors **5**, **8**, and **12** with their five stereo-genic centers can be assembled will be im-portant in the assessment of the syntheses.

The synthesis of zaragozic acid A [5] by Nicolaou is summarized in Schemes 3–7. Four of the five stereogenic centers are con-structed by two stereo-selective Sharpless dihydroxylations (Scheme 3). In the first, the diene **20**, which is synthesized in a few steps from the simple building blocks **15–18**, is

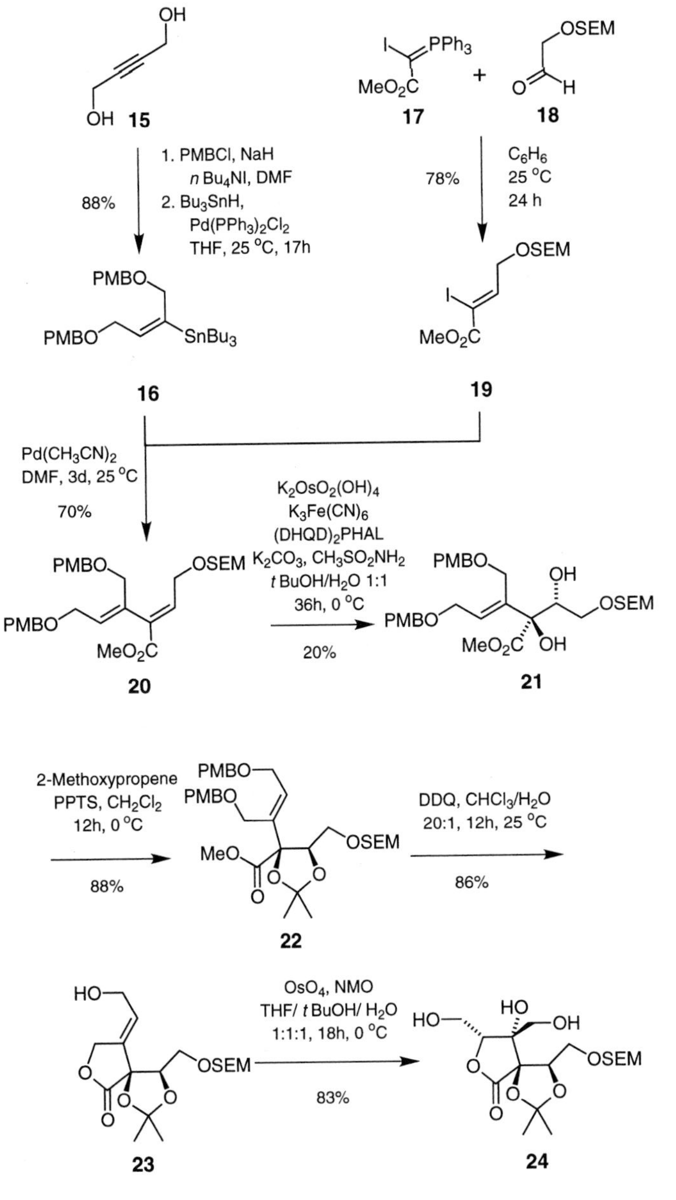

Scheme 3. Total synthesis of zara-gozic acid A (**2**) by Nicolaou, part 1.

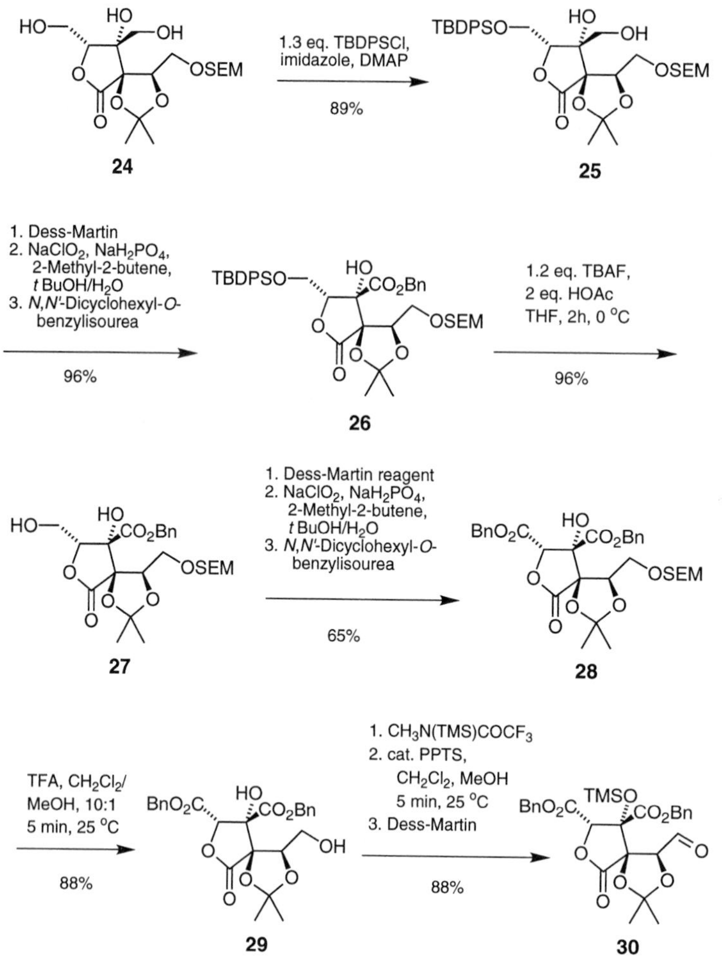

Scheme 4. Total synthesis of zaragozic acid A (**2**) by Nicolaou, part 2.

converted enantioselectively to give the diol **21** (20 % yield, 78 % *ee*); in the second the olefin **23** is diastereoselectively dihydroxylated, and the product undergoes spontaneous lactonization to give **24** (83 % yield, as a single diastereomer after crystallization).

In the subsequent 12 steps (Scheme 4) the two primary alcohol functions in **24** are oxidized to carboxyl functions and protected as benzyl esters. After trimethylsilyl(TMS)-protection of the OH group at C4 Nicolaou then obtains the aldehyde **30**. Addition of the lithium compound prepared from the dithiane **31** to the C1 side chain (Scheme 5) of **30** deli-

vers, with the formation of the alcohol **32**, the missing fifth stereogenic center. This addition is however unselective: two epimeric alcohols are formed (total yield 75 % in a 1:1 ratio) and must be separated by chromatography. Removal of the thioacetal function in **33** leads to the hemiacetal **34**.

Under carefully worked out acidic conditions the lactone hemiacetal **34** rearranges into the dioxabicyclo[3.2.1]octane **35** (Scheme 6). By changing the alcohol protecting group and exchanging the methyl ester for a benzyl ester the diol **38a** is obtained, whose C6-OH function is esterified with the

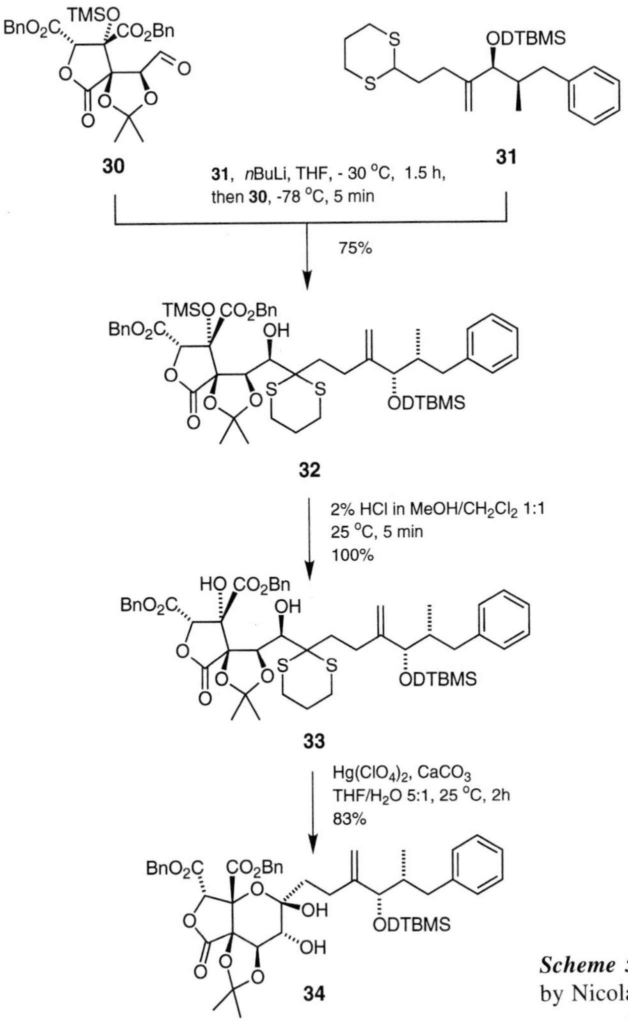

Scheme 5. Total synthesis of zaragozic acid A (**2**) by Nicolaou, part 3.

carboxylic acid **38b** leading to the formation of the tetraester **39**. The latter already contains the complete skeleton of zaragozic acid A. In this esterification the C6- and the C7-OH group can react. The selectivity of 3:2 in favor of the desired C 6-OH group established by Nicolaou is rather low. Starting from **39** protecting group manipulations (5 steps, Scheme 7) gave the target compound zaragozic acid A (**2**).

The synthesis of zaragozic acid C (**1**) by Carreira [6] is given in Schemes 8–10. This synthesis starts from D-erythronic-*γ*-lactone (**42**), a compound from the chiral pool. An impressive feature of the synthesis is the conversion of the amide **43** into the alcohol **44** (Scheme 8): The reaction of ethoxyvinyllithium with the dimethylamide **43** can be stopped at the ketone intermediate. The subsequent reaction of the carbonyl function with the organomagnesium compound prepared from TMS–C≡CH proceeds stereoselectively (20 : 1). The vinyl ether group in **44** is a latent ester function: the ozonolysis of **44** in ethanol

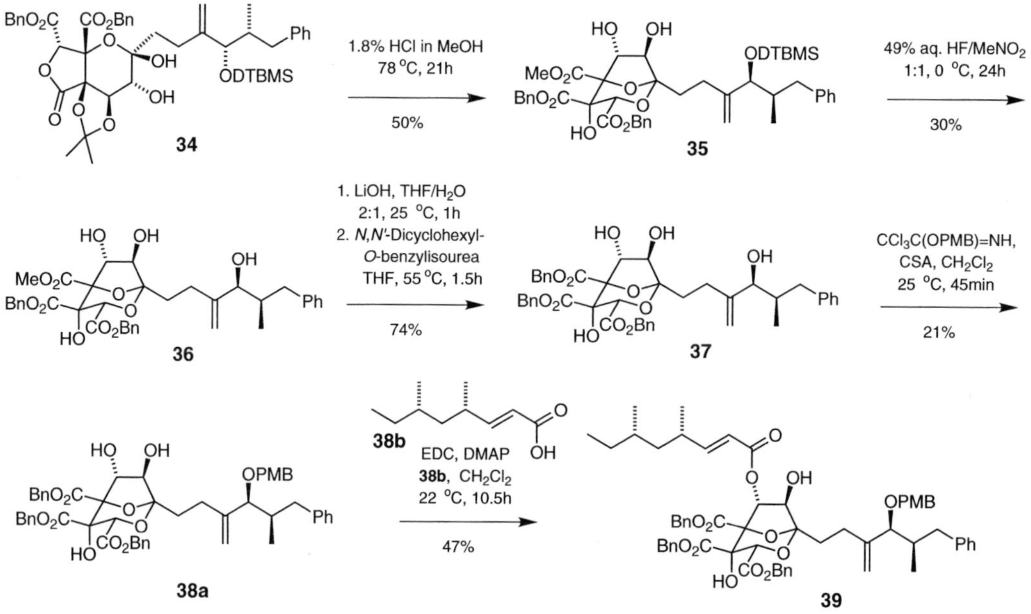

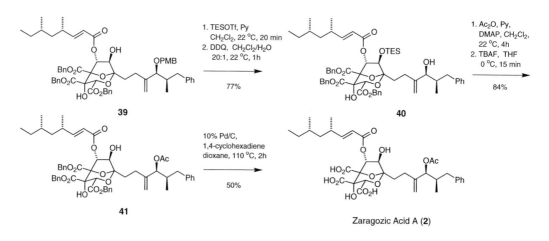

Scheme 6. Total synthesis of zaragozic acid A (**2**) by Nicolaou, part 4.

Scheme 7. Total synthesis of zaragozic acid A (**2**) by Nicolaou, part 5.

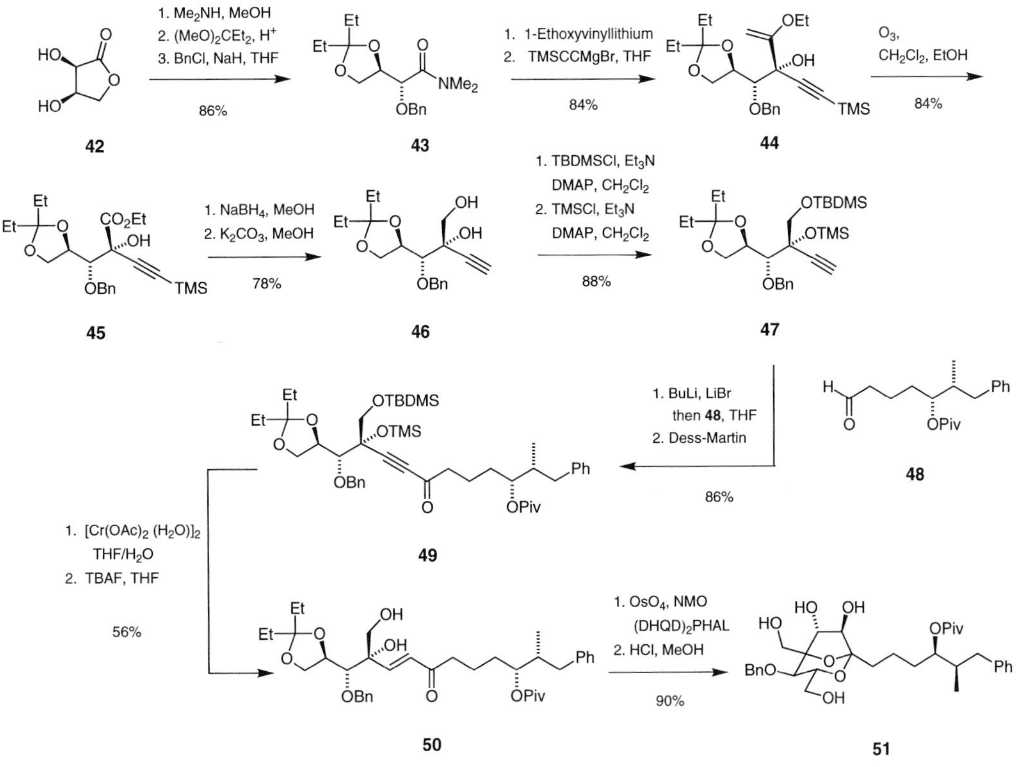

Scheme 8. Total synthesis of zaragozic acid C (**1**) by Carreira, part 1.

affords the ethyl ester **45**. Subsequently the terminal acetylene **47** is obtained via **46** by a standard sequence. The C1 side chain is attached through the coupling of **47** with the aldehyde **48** to give the ketone **49**. The two stereogenic centers still missing are produced by Sharpless dihydroxylation of the olefin **50** obtained after the oxidation and removal of the silyl protecting groups. Without a chiral ligand stereoselectivity cannot be achieved in this bishydroxylation step. With a chiral ligand, the diol is formed in a reagent-controlled reaction in the ratio 64 : 36 in favor of the desired stereoisomer. In the subsequent synthetic step, the diol is cyclized under acidic conditions to give the dioxabicyclo-[3.2.1]octane acetal (**51**).

To construct the missing quaternary center C4 Carreira transforms the compound **51** into the ketone **53** (Scheme 9). The addition of lithiated TMS–C≡CH to the carbonyl function of **53** to give the alcohol **54** is achieved with a stereoselectivity of 86 : 14. At this stage the pivaloyl protecting groups are replaced by acetyl protecting groups and the triple bond is reduced to a double bond to give compound **56**

Carreira now oxidizes the two primary alcohol groups sequentially to give the carboxyl functions (**56 → 57 → 58**, Scheme 10). The third carboxyl function is formed by ozonolysis and subsequent oxidation of the aldehyde (**58 → 59**). This is followed by a remarkably selective hydrolysis of the triacetate **59** to

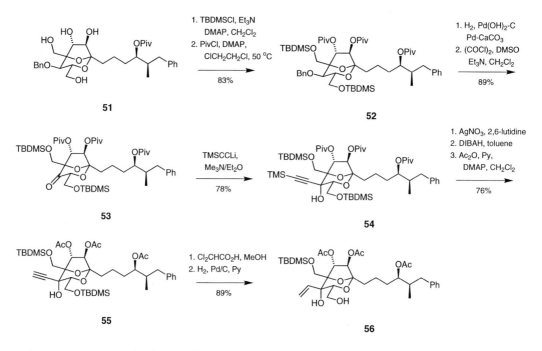

Scheme 9. Total synthesis of zaragozic acid C (**1**) by Carreira, part 2.

give the monoacetate **60** in 92 % yield (!). With regard to his final step, the introduction of the acyl side chain at C6-O, Carreira faces the same problem as Nicolaou: how can the C6-OH group be esterified selectively in the presence of the free C7-OH group? Carreira is unable to provide a satisfactory solution to the selectivity problem. From the reaction of the diol **60** with the acid chloride **61** he obtains the regioisomeric esters in the ratio 3 : 1 in favor of the undesired isomer. The desired isomer can be isolated by chromatography and after acidic hydrolysis of the *tert*-butyl esters the target compound zaragozic acid C (**1**) is obtained.

The total synthesis of zaragozic acid C (**1**) by Evans [7] is shown in Schemes 11 and 12. Evans identified a tartaric acid unit in the C3–C4 partial structure of zaragozic acid. Accordingly, his synthesis commences from a tartaric acid derivative (Scheme 11). The

enantiomerically pure acetal **61** is readily accessible from di-*tert*-butyl D-tartrate. This is transformed into the silylketene acetal **62**, which reacts in a Lewis acid catalyzed aldol addition with the aldehyde **66** to give the adduct **67**. The aldehyde **66** was prepared stereoselectively with the oxazolidinone method developed by Evans (**64** + **65**→**63**→**66**). The Dess-Martin oxidation of **67** yields the ketone **68**. Addition of vinylmagnesium bromide to the keto-carbonyl function in **68** leads to the stereo-controlled construction of the quaternary center C5 with the formation of **69**. The stereochemical course of this addition can be explained with a chelate-controlled transition state controlled by the adjacent benzyl ether. Next, Evans transforms the styryl group in **69** oxidatively into a carboxyl function and closes the butyric lactone ring to give **70**. By the sequence **70**→**71** the vinyl groups is transformed into the missing

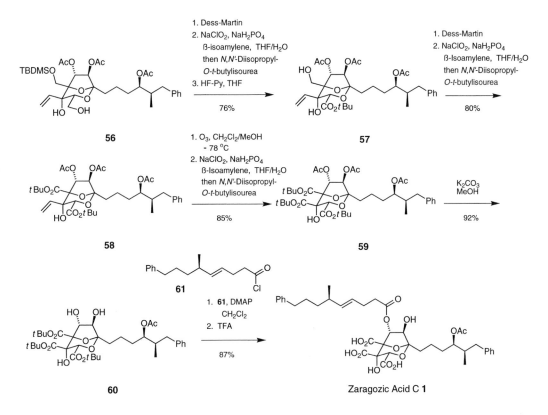

Scheme 10. Total synthesis of zaragozic acid C (**1**) by Carreira, part 3.

third carboxyl function of the zaragozic acid skeleton.

In the further course of the synthesis (Scheme 12) Evans attaches the side chain to C1 by the reaction of the organolithium compound accessible from the iodide **72** with the lactone **71** under the formation of **73**. The replacement of the *para*-meth-oxy-benzyl(PMB) protecting group with an acetyl group yields the hemiacetal **74**. Subsequently, the hemiacetal **74** is converted under acidic conditions into the dioxabicyclo[3.2.1]octane **75**. With wise foresight the C7-OH group has been protected since the beginning of the synthesis as a silyl ether with the result that the problem of selectivity between C6-OH and C7-OH does not arise in the following acyla-

tion step. The acyl side chain can thus be attached to C6-O without any problems: the reaction of the alcohol **75** with the carboxylic acid **76** affords the desired ester **77** in 82 % yield. The deprotection of **77** to give the target compound zaragozic acid C (**1**) is almost quantitative.

A discussion of the preparation of the C1 side chains and the C6-O carboxylic acid, which was also effected by the three research groups by different methods, is not within the scope of this highlight. The interested reader should refer to the information in the original literature. [4–7]

All three total syntheses of zaragozic acid are examples of modern, efficient syntheses of natural products. The comparison of the

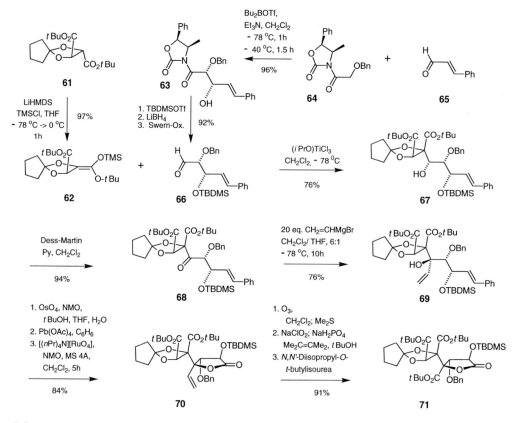

Scheme 11. Total synthesis of zaragozic acid C (**1**) by Evans, part 1.

number of steps and the overall yields of different synthetic routes is always difficult and of only limited value. The longest linear sequence as a measure for the number of steps for the presented syntheses is 33 (Nicolaou), 36 (Carreira), and 21 steps (Evans). The overall yields are calculated to be 1 (Nicolaou), 1 (Carreira), and 15 % (Evans). Noteworthy in the synthesis by Evans is the highly selective formation of each new stereogenic center. The stereo- and regioselectivities of some of the steps in the syntheses by Nicolaou and Carreira will surely be improved. The two research groups of Carreira and Nicolaou have been the first cross the finish line in the race for the total synthesis of zaragozic acids.

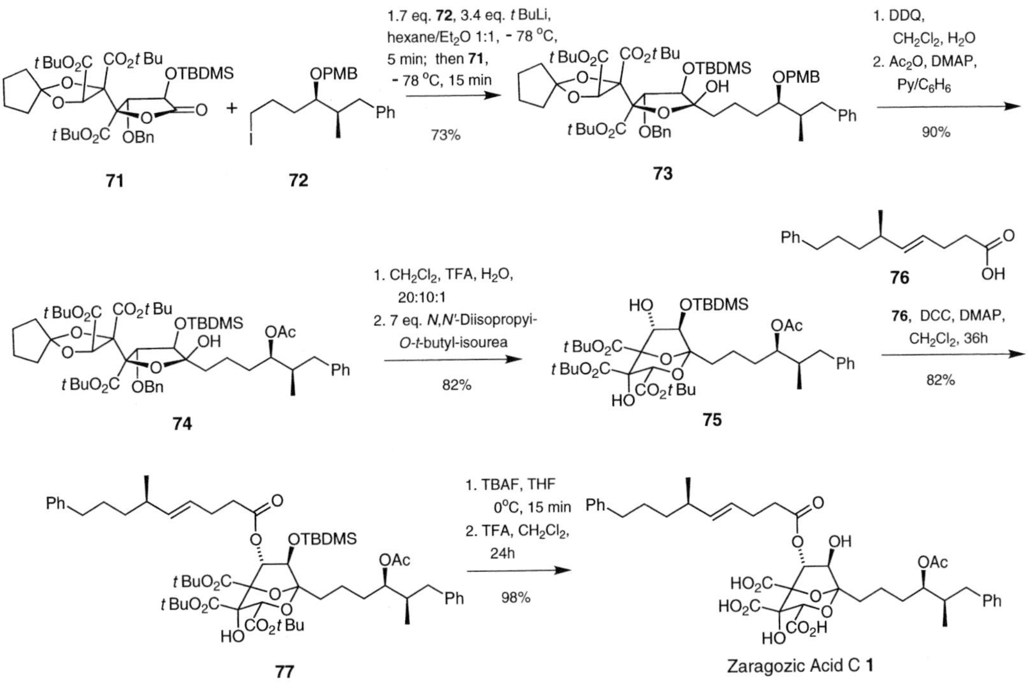

Scheme 12. Total synthesis of zaragozic acid C (**1**) by Evans, part 2.

References

[1] Goodman and Gilman's *The Pharmacological Basis of Therapeutics* 8th ed. (Eds.: A. Goodman Gilman, T. W. Rall, A. S. Nies, P. Taylor) Mc Graw-Hill, New York, **1993**, p. 874.

[2] K. E. Wilson, R. M. Burk, T. Biftu, R. G. Ball, K. Hoogsteen, *J. Org. Chem.* **1992**, *57*, 7151–7158; J. D. Bergstrom, M. M. Kurtz, D. J. Rew, A. M. Amend, J. D. Karkas, R. G. Bostedor, V. S. Bansal, C. Dufresne, F. L. Van-Middlesworth, O. D. Hensens, J. M. Liesch, D. L. Zink, K. E. Wilson, J. Onishi, J. A. Milligan, G. Bills, L. Kaplan, M. Nallin Omstead, R. G. Jenkins, L. Huang, M. S. Meinz, L. Quinn, R. W. Burg, Y. L. Kong, S. Mochales, M. Mojena, I. Martin, F. Pelaez, M. T. Diez, A. W. Alberts, *Proc. Natl. Acad. Sci. USA* **1993**, *90*, 80; C. Dufresne, K. E. Wilson,

S. B. Singh, D. L. Zink, J. D. Bergstrom, D. Rew, J. D. Polishook, M. Meinz, L. Huang, K. C. Silverman, R. B. Lingham, M. Mojena, C. Cascales, F. Pelaez, J. B. Gibbs, *J. Nat. Prod.* **1993**, *56*, 1923.

[3] J. M. Dawson, J. E. Farthing, P. S. Marshall, R. F. Middleton, M. J. O'Neill, A. Shuttleworth, C. Stylli, R. M. Tait, P. M. Taylor, H. G. Wildman, A. D. Buss, D. Langley, M. V. Hayes, *J. Antibiot.* **1992**, *45*, 639–647; P. J. Sidebottom, R. M. Highcock, S. J. Lane, P. A. Procopiou, N. S. Watson, *ibid.* **1992**, *45*, 648; W. M. Blows, G. Foster, S. J. Lane, D. Noble, J. E. Piercey, P. J. Sidebottom, G. J. Webb, *ibid.* **1994**, *47*, 740.

[4] a) V. K. Aggarwal, M. F. Wang, A. Zaparucha, *J. Chem. Soc. Chem. Commun.* **1994**, 87;

b) R. W. Gable, L. M. McVinish, M. A. Rizzacasa, *Aust. J. Chem.* **1994**, *47*, 1537; c) H. Abdel-Rahman, J. P. Adams, A. L. Boyes, M. J. Kelly, D. J. Mansfield, P. A. Procopiou, S. M. Roberts, D. H. Slee, N. S. Watson, *J. Chem. Soc. Chem. Commun.* **1993**, 1839; synthesis of the C1 side chains: d) A. J. Robichaud, G. D. Berger, D. A. Evans, *Tetrahedron Lett.* **1993**, *34*, 8403; synthesis of the C6 acyl-side chain: e) C. Santini, R. G. Ball, G. D. Berger, *J. Org. Chem.* **1994**, *59*, 2261; for a recent synthesis of the core of zaragozic acids see: f) I. Paterson, K. Fessner, M. R. V. Finlay, M. F. Jacobs, *Tetrahedron Lett.* **1996**, *37*, 8803; g) Y. Xu, C. R. Johnson, *Tetrahedron Lett.* **1997**, *38*, 1117.

[5] E. M. Carreira, J. DuBois, *J. Am. Chem. Soc.* **1994**, *116*, 10825–10826 E. M. Carreira, J. DuBois, *ibid.* **1995**, *117*, 8106–8125.

[6] a) K. C. Nicolaou, E. W. Yue, Y. Naniwa, F. De Riccardis, A. Nadin, J. E. Leresche, S. La Greca, Z. Yang, *Angew. Chem.* **1994**, *106*, 2306; *Angew. Chem. Int. Ed. Engl.* **1994**, *33*, 2184; b) K. C. Nicolaou, A. Nadin, J. E. Leresche, S. La Greca, T. Tsuri, E. W. Yue, Z. Yang, *Angew. Chem.* **1994**, *106*, 2309; *Angew. Chem. Int. Ed. Engl.* **1994**, *33*, 2187; c) K. C. Nicolaou, A. Nadin, J. E. Leresche, E. W. Yue , S. La Greca *Angew. Chem.* **1994**, *106*, 2312; *Angew. Chem. Int. Ed. Engl.* **1994**, *33*, 2190; d) K. C. Nicolaou, E. W. Yue, S. La Greca, A. Nadin, Z. Yang, J. E. Leresche, T. Tsuri, Y. Naniwa, F. D. Riccardis, *Chem. Eur. J.* **1995**, *1*, 467; e) for a review of the chemistry and biology of the zaragozic acids by Nicolaou see: K. C. Nicolaou, A. Nadin, *Angew. Chem.* **1996**, *108*, 1732; *Angew. Chem. Int. Ed. Engl.* **1996**, *35*, 1622.

[7] D. A. Evans, J. C. Barrow, J. L. Leighton, A. J. Robichaud, M. Sefkov, *J. Am. Chem. Soc.* **1994**, *116*, 12111–12112.

The First Total Syntheses of Taxol

Ludger A. Wessjohann

Introduction

In modern chemistry it is a considerable rarity for the total synthesis of a medicinally and commercially significant natural product not to appear till 22 years after its complete structural elucidation in 1971. Though taxol (paclitaxel, **3a**), [1] a terpenoid first isolated in 1962 from the pacific yew tree (*Taxus brevifolia* Nutt), is one of the most promising agents against breast cancer and other types of cancer, it took some time before synthetic chemists became interested in it, for several reasons. Thus, nearly a decade elapsed before taxol's likely biological mode of action – an unusual stabilization of the microtubuli – and its significance for a novel cancer therapy were recognized (cf. [1]). [2]

However, from the middle of the 1980s onwards, the number of synthetic attempts increased in an almost explosive manner. Initial milestones and still the most important synthetic processes for the industrial production of taxoids are semisynthetic methods in which a suitable side chain is attached to baccatin III (**1**) to give either taxol (**3a**) or taxotere (**3b**). [1a–d] Baccatin III is available from renewable parts of a variety of yews in larger amounts. [1a–d, f, g]

In the early 90's an intense competition developed to complete the first total synthesis of this unusual and extremely demanding molecule, a race which involved (and involves) over forty of the best known research groups in the field of natural product synthesis. Taxol and the efforts to complete its synthesis gave rise – at least in the USA and amongst natural product chemists – to at least as much excitement as did fullerenes in physical organic chemistry, though with the decided difference that taxol (paclitaxel) and its derivative taxotere rapidly became real market products for the treatment of breast and ovarian cancer and useful tools in cell biology. Indeed, because of problems with supply and limited possibilities to synthesize derivatives, its full medicinal potential has still only been investigated to a limited extent. [1g–i, 3, 4]

Most studies of structure activity relationships (SAR) rely on degradations of baccatin and semisynthesis. [1, 3, 4] First insights into the SAR of taxol derivatives have been summerized in a recent article on C ring aryl derivatives, [4] and more can be expected in the near future as patented information becomes released. Total synthesis, especially in preparing the way for the synthesis of derivatives, will be of great importance to elucidate the relationship between structure, microtubuli stabilization and anticancer activity more thoroughly. This impact of total synthesis is exemplified by SAR-studies done on rearrangement products of intermediates of Wender's taxol synthesis (v. i. and [3]), which could be developed into compounds with excellent microtubuli stabilization, partly exceeding that of taxol, but with a dramatically decreased cancerostatic effect, i. e. microtubuli and cytotoxic effect could be separated. [3]

An initial peak of the taxol excitement was reached in 1994 when the first two total syntheses by R. A. Holton et al. [5] (Florida State University) and K. C. Nicolaou et al. [6] (University of California, San Diego and Scripps Research Institute) were published almost simultaneously. These highlights of organic synthesis will be discussed in more detail in the second part of this review, followed in a third part by a brief survey of the two other syntheses known to date, accomplished by S. J. Danishefsky et al. [7] (Sloan-Kettering Cancer Center, New York) and by Paul A. Wender and coworkers [8] (Stanford University).

What was it that made the synthesis of taxol so difficult? For a while, the assertion that every molecule that can exist can today be made, given enough manpower, [9] was almost called into question. To analyse the synthetic problem, it is useful to concentrate on the tricyclic A/B/C ring system of baccatin

III (1), which can be treated with 2, for example, by the method of Holton and Ojima [1d] (Scheme 1) to give taxol (3a). [1a–d] The oxetane ring (D) may be treated as a functional group. Apart from this group, the most remarkable feature is the unusual anti-Bredt double bond of the A ring. However, in contrast to normal doctrine, in this case it leads to a reduction in the total strain (ca. -1.5 kcal mol^{-1}), whereas the very rigid, arched structure of the entire molecule is highly strained from the steric effect of the bridges dimethyl group at C15 (strain energy $\sim +10$ kcal mol^{-1}), in cooperation with the transannular C8-methyl group, which also projects outwards.

The main problem in taxol synthesis is, then, the construction of the highly functionalized and strained eight-membered ring (B). Although the methyl groups (C16/C19) projecting into this ring are not too close to each other in taxol (convex face), most reactions with prefabricated A/C rings bring them into too close a proximity during B ring closure reactions. In addition, the possibilities for subsequent stereoselective functionalization of the central eight-membered ring are limited, because of its rigidity and steric shielding, and because of the lability of the oxetane ring. Accordingly, in most early syntheses of the taxane skeleton (A/B/C

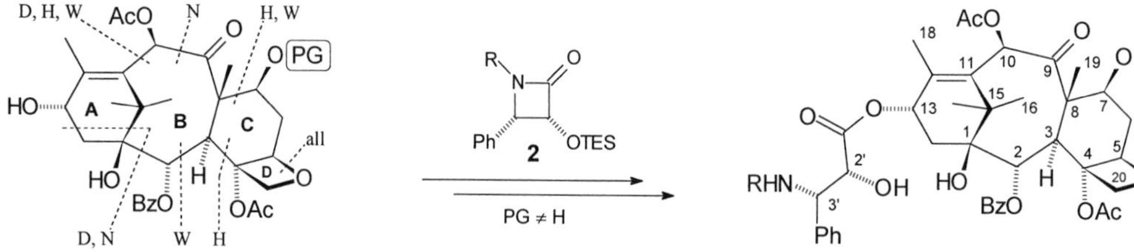

Baccatin III **1**, PG = H

Taxol   **3a**, R = Bz
Taxotere **3b**, R = BOC

Scheme 1. Semisynthesis of taxol (paclitaxel) and taxotere from protected baccatin III according to Holton and Ojima (cf. [1a–d]); strategic bond cleavages for the syntheses of Holton et al., [5] Nicolaou et al., [6] Danishefsky et al. [7], and Wender et al. [8] are indicated by the last names initials. PG = protecting group. [10]

rings) one of the two (or three) central methyl groups have been missing. If all of them were indeed present, it has not been possible to introduce the necessary functional groups afterwards. Alternative routes with the annulation and functionalization of a lateral ring after the B ring has been completed must be less convergent in principal. C ring annulations also will have to address the base sensitivity of the C7–C8 bond to retroaldol/aldol reactions equilibrating the stereocenter at C7 in favor of the more stable unnatural epimer. The successful research groups have solved these problems very differently, but in part also with almost identical reactions. [10]

The Pioneer Syntheses of Holton and Nicolaou

Holton et al. [5] first constructed the A and B rings, in a linear strategy using the elegant fragmentation of the [3.3.0] system **5** derived from β-patchoulene oxide (Scheme 2), which they had previously used for the synthesis of *ent*-taxusine. However, to obtain the correct enantiomeric series, they had to start extravagantly, using (–)-camphor (**4**). The resulting A/B fragment **6** contains the complete, homochiral A ring and all the methyl groups, as well as one oxygen functional group in both

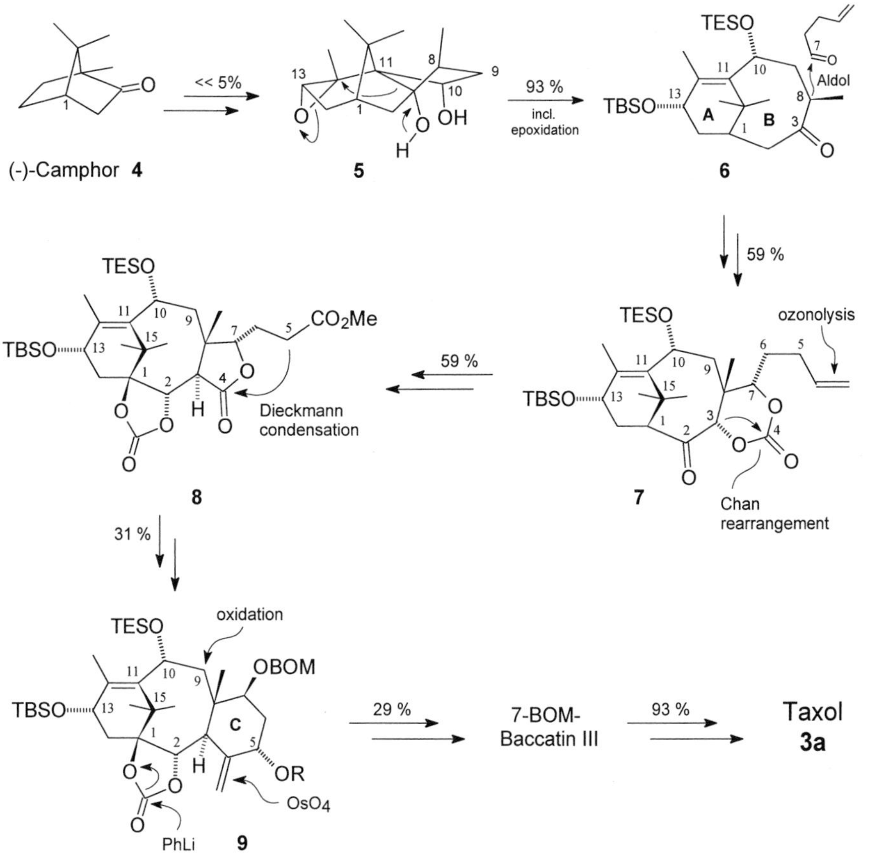

Scheme 2. Total synthesis of taxol according to Holton et al. [5] Taxol numbering (Scheme 1) is used. [10]

the upper and lower region of the B ring for further modifications.

The problematic stereocenter at C7 was constructed early, by using a diastereoselective aldol reaction with magnesium diisopropylamide and 4-pentenal. Since three of four missing C ring atoms could be introduced here in a simple manner, derivatizations will later presumably be easy at this step. C4 was provided by a carboxylic ester – initially introduced as a carbonate protecting group in intermediate **7** – through a Chan rearrangement. This was followed by a few less elegant steps for deoxygenation at C3 (SmI_2) and oxidation at C1 (cf. Wender's findings [8b]). The latter was only possible with the B/C-*cis* compound, by an unusual, selective enolate formation to the bridgehead C1 rather than to the doubly activated C3 – a further proof for the unusual steric and conformational characteristics of taxoids. After reduction and isomerization to the *trans* compound, the resulting C1/C2 diol was quantitatively protected with phosgene (cf. **8**) and, at the same time, prepared for the later formation of the benzoate at C2 by addition of phenyllithium and selective fragmentation [11] (cf. **9**). First, however, the vinyl group had to be degraded oxidatively to give ester **8**. The C ring was then closed by an effective Dieckmann condensation, and decarbalkoxylated.

The remaining keto group at C4 enables the oxidation of C5, and the conversion to methylene compound **9**. This last step is sterically problematic and was achieved in moderate yield by a methyl Grignard addition in dichloromethane (instead of ether) and elimination with Burgess' reagent. Dihydroxylation afforded the precursor to the oxetane, which was obtained by nucleophilic substitution of a secondary mesyl or tosyl group (OR at C5), followed by acetylation of the remaining tertiary OH group. These steps, with substitution at the secondary carbon rather than at the primary one (as suggested by the textbooks), were similar to those used by the

other groups, [6–8a] albeit with considerable variation of the leaving groups and yields. In principle they correspond to the biosynthesis proposed by Halsall and Potier (see [1]). Finally, the upper bridge in the B ring had to be oxidized, for which benzeneseleninic anhydride was very cleverly employed. On the route to 7–BOM-Baccatin III, the unusual release of the TBS-protected C13-OH group with TASF [tris-(dimethylamino)sulfur (difluorotrimethyl-silanide)] should be mentioned.

Nicolaou et al., who only worked on the synthesis of taxol for about two years, [6] follow a much more convergent route (Scheme 3), which begins, however, with achiral precursors. The A-ring precursor was constructed by a Diels-Alder reaction (**14** + **15**) and refunctionalized to the sulfonylhydrazone **16**. This reacted according to Shapiro to give the corresponding alkenyllithium nucleophile, a procedure which is reminiscent of Danishefsky's [7] and of Funk's synthetic routes (cf. [1b]). The C2 aldehyde group of **13** served as the electrophile and reacted diastereoselectively in excellent 82 % yield to give the A-C coupled alcohol. Chelate-directed epoxidation of this alcohol then led to **18**. Compound **13** was accessible from racemic **12**, the product of the Diels-Alder reaction between 3-hydroxypyrone **11** (from mucic acid) and the dienophile **10**. The epoxide **18** was selectively hydrogenated at C14, and the resulting C1/C2 diol protected as the carbonate (see above). [11]

After oxidation of C9 and C10 to the aldehyde level, the key step followed. A well-established method for the construction of strained structures, the McMurry pinacol coupling, was chosen to form the top B ring bridge, which afforded the diol **17** in 23 % yield. The only other successful eight-membered ring closure starting from an "advanced" A/C system with all methyl groups in place, which was known at that time, was also based on a McMurry reaction at these positi-

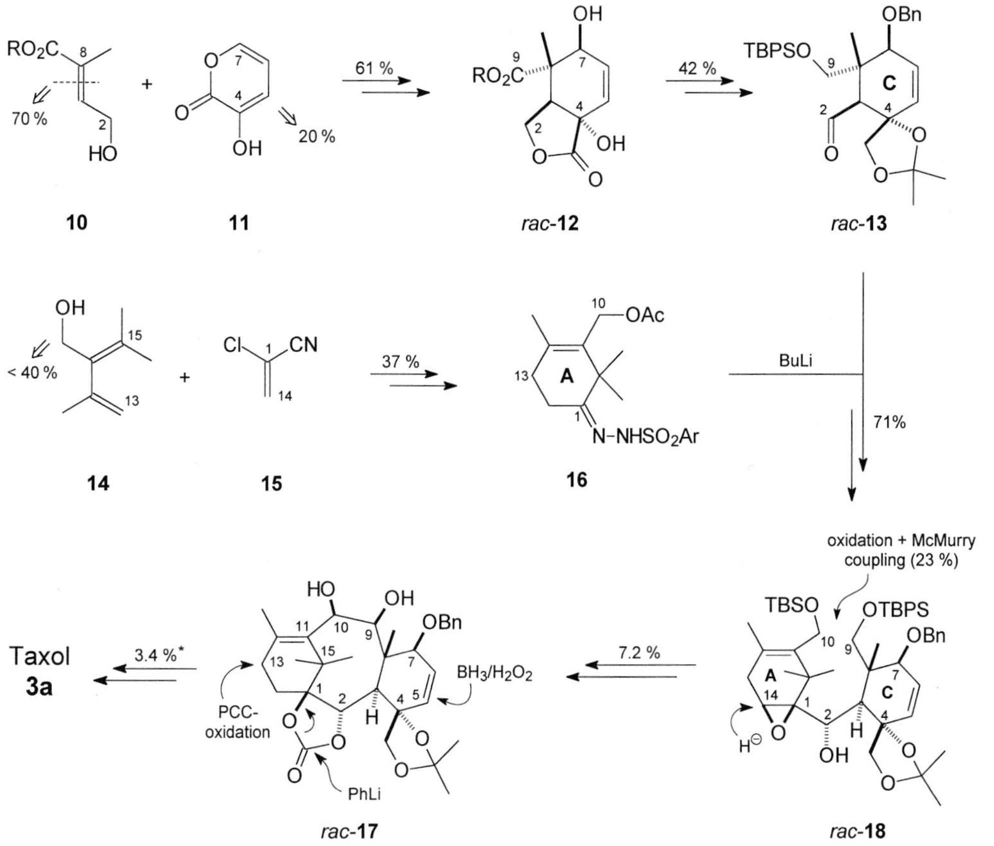

Scheme 3. Total synthesis of taxol according to Nicolaou et al. [6] Taxol numbering (Scheme 1) is used. [10] * = yield obtained without enantiomer separation. The percentages below the retrosynthetic arrows for the compounds **10**, **11** and **14** relate to their yields from readily available precursors according to the literature given in [6].

ons. It also gave yields in the lower 20 % region (Kende, cf. also Pattenden). [1b] Nicolaou et al., however, stopped this reaction at the diol level, permitting further functionalization and enantiomer separation. With Kende's alkene, this was not possible.

In order to construct the oxetane ring, the C5–C6 double bond was hydroborated and oxidized with moderate regioselectivity and yield. Oxetane ring closure and benzoate formation at C2 [11] followed as already discussed. The late, regioselective oxidation at C13,

the linkage point to the taxol side chain, is unusual and was accomplished with a very large excess of pyridinium chlorochromate (PCC, cf. also Danishefsky et al. [7]). The ketone formed was then reduced stereoselectively with borohydride. This introduction of the C13-hydroxy group into an otherwise fully functionalized baccatin derivative and the beforementioned transformation of the C1/C2 carbonate to the C2 benzoate probably were the most valuable contributions of this route with an impact on later approaches.

The Danishefsky and Wender Syntheses of Taxol

Danishefsky's synthesis (Scheme 4) [7] started from the readily available Wieland-Miescher ketone (**19**) which, by a series of mainly protection and oxidation reactions, was transformed to the fully functionalized C ring precursor **21**. The oxetane moiety was introduced very early on in the synthesis, from a corresponding triol, again by nucleophilic substitution at C5. Noteworthy is the selective protection or modification of primary versus secondary versus tertiary hydroxy groups for this purpose. The benzyl protected enolized form **20** then could be oxidized, cleaved oxidatively and processed to compound **21** which, apart from complete C/D rings, possesses the necessary handles (C2 and C9/10) to bind to the A ring precursor **22** and thus form the B ring.

Counterpart **22** was available from the corresponding 2,2,4-trimethylcyclohexan-1,3-dione by the same strategy discussed for compound **16** (cf. Scheme 3). Similar to Nicolaou's approach are the C1-C2 coupling of the two building blocks, the oxidation at C1, and the formation of the C1-C2 carbonate (cf. **23**). Finally formation of an enol triflate at C11 and extension at C10 to form a vinyl group set the stage for a Heck coupling reaction of **23** to close ring B. The Heck reaction gave up to 49 % of tetracyclus **24**, but required more than equimolar amounts of palladium(0) and long reaction times. This is no surprise considering the strain and steric impediment which has to be overcome during bond formation and may inhibit not only the addition reactions but also the subsequent elimination of palladium. At this stage it should be noted that other attempts for ringclosure at this or other positions (e. g. C1–C2 closures as final step) failed or were thwarted.

Unfortunately it proved to be very difficult to functionalize and cleave the highly shielded *exo*-methylene group at C10 in taxoid **24**. Only after intermediate transformation of the

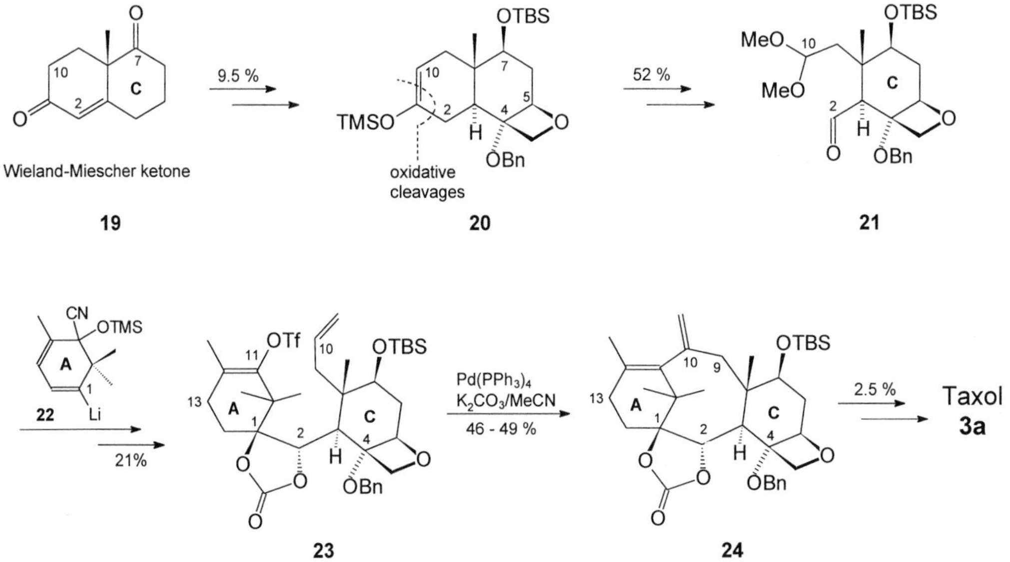

Scheme 4. Total synthesis of taxol according to Danishefsky et al. [7] Taxol numbering (Scheme 1) is used. [10]

other (bridgehead) double bond to an epoxide and reaction of the carbonate to form the C2-benzoate, it became accessible to oxidation [equimolar OsO$_4$, then Pb(OAc)$_4$]. Oxidation at C9 and C13 were performed as discussed for the Holton and Nicolaou syntheses, respectively, and completion of the molecule followed accordingly.

It is noticable that the strategy which has probably been chosen most often in successful constructions of basic taxane tricycles, the fragmentation of bicyclo[4.2.0] systems (cf. e. g. **28**) to yield the B ring, [1a–c] was not involved in one of the first successes for a complete synthesis. The advantage of the 6/4/6 to 6/8 strategy in A/B ring or B/C ring construction are evident: usually easily acces-

sible six-membered rings can be used (or formed), instead of difficult eight-membered ones; the strain energy hidden in the four-membered ring can inherently accomodate for the strain energy of the B ring, thus eliminating thermodynamic problems of linear ring closures; and the steric interaction and pre-organisation in the intermediates is better. Finally the Wender group succeeded with such an approach, [8a] which also not only proved to be the most efficient to date, but also undoubtly is the most elegant one, at least regarding the construction of the A/B fragment (Scheme 5).

Wender's construction of the A/B ring system [8a] in principle resembles the Holton synthesis, but makes use of the 4/6 precursor

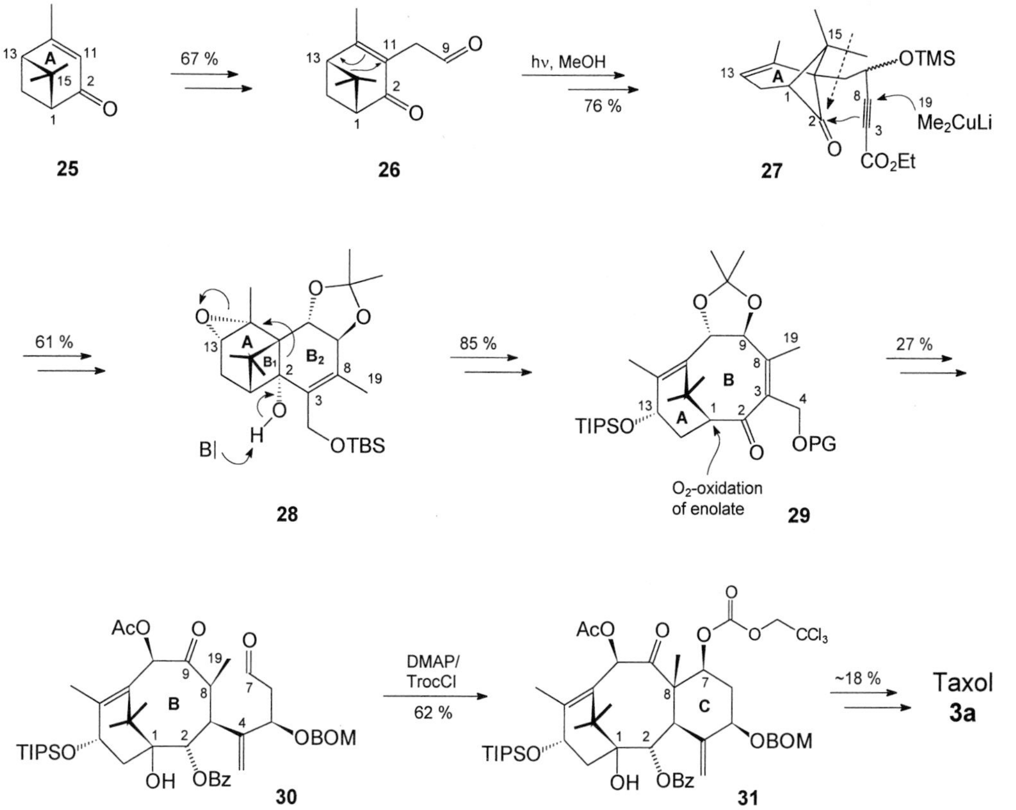

Scheme 5. Total synthesis of taxol according to Wender et al. [8a] Taxol numbering (Scheme 1) used. [10]

28 of the B ring instead of Holtons 5/5 precursor (**5**). This allows the use of readily available verbenone **25** (an oxidation product from *α*-pinene) to produce enantiomerically pure aryltaxanes in only six steps. [8b, d] In order to set the stage for the 4/6 ringopening, verbenone **25** was premodified at C11 to give **26**. It then was rearranged to the corresponding chrysanthenone derivative by a photochemically induced allylshift (cf. arrows in verbenone derivative **26**), [8b–d] and further functionalized at C9 to yield Michael acceptor **27**. In order to prepare the 6/4/6 system **28**, the 6–membered ring B_2 has to be formed by addition of a suitably tethered C3 nucleophile to the chrysanthenone carbonyl group (C2) in **27**. This is a very difficult task. Not only is the hindered chrysanthenone keto group difficult to attack in general without initiating an uncontrolled cleavage of the cyclobutanone moiety, it also does not allow free access from the *β*-face which is shielded by the C15-methyl group, definately not along a Bürgi-Dunitz trajectory (dashed arrow in **27**). [8c, d] A four atom tethered nucleophile certainly has an entropic advantage but can access C2 exclusively from the *β*-face and under a disfavored angle. Wender et al. achieved such attacks with sp^2-nucleophiles, which do not have steric bulk orthogonal to the attacking electron pair and thus can pass by the C15-methyl group (for attempts with larger groups like sp^3-C ring precursors see ref. [8d]). The small alkinyl precursor **27** is extremely suitable, and a domino 1,4/1,2 addition reaction with methyl cuprate gave the ring B_2 together with the introduction of C19 in fantastic 97 % yield (cf. arrows in **27**).

After some standard transformations the base induced double ringopening of **28** to A/B ring compound **29** was achieved in 85 % yield, including protection at C13. For the oxidation at C1 in **29** the proton was abstracted with potassium *t*-butoxide at the tertiary carbon and not at C9 to give once again the unusual bridgehead enolate, which was oxi-

dized with oxygen or air as previously reported by the same group. [8b] The introduction of a side chain at C3 and its functionalization to prepare for the construction of the C- and D ring followed, as well as oxidation at the upper rim and benzoylation at O2 via the cyclic carbonate as described previously. The resulting ketoaldehyde **30** could be closed by an aldol reaction with DMAP or 4-pyrrolidinopyridine and immediate protection of the kinetic aldolate with trichloroethoxycarbonyl chloride (TrocCl) to give **31** with the desired stereochemistry at C7 (11 : 1 ratio), thus providing a solution to the epimerization problem encountered with other bases, without selectively scavanging the desired aldolate (see introduction). Interestingly the aldol reaction did not proceed with a C1-C2 carbonate. Further transformations were achieved analogously to previously described methods to yield deacetylbaccatin III, baccatin III (**1**) and finally taxol (**3a**).

Résumé

The more convergent routes of Nicolaou et al. and Danishefsky et al. with B ring closure from a tethered A–C precursor are clearly the least efficient with respect to total yield (about 0.01 % starting from commercially available materials) if directly compared to the other two routes. Additionally the Nicolaou route needs enantiomer separation or a – presumably less favorable – asymmetric synthesis, whereas the Danishefsky route addresses absolute stereochemistry but involves more steps. Holton's synthesis shows quite good yields throughout with a total one of approximately 0.1 %, but has less room for improvements and is based on a not readily available starting material. The best synthesis in all respects to date is the one of Wender and coworkers, which is about 1–2 orders of magnitudes better in total yield (ca. 0.8 %), utilizes cheap chiral pool material, and offers possibilities for derivatization which are at least equal to those of the other approaches. In the end, however, a comparison of number of steps and yields in this case should not be overestimated. More important factors are practicability and versatility of the syntheses, their usefulness to prepare derivatives for SAR studies and their impact on the solution of other synthetic problems.

Since none of the total syntheses yet can compete with the semisynthesis from baccatin III, the potential for the preparation of derivatives is crucial. Convergent syntheses are principally better if varied building blocks shall be combined. This requirement is met by Nicolaou's and Danishefsky's routes. They also permit easy mutations in the A ring building blocks. With Nicolaou's procedure, however, increased steric hindrance with more complex derivatives may lead to problematic yields for the ring closure. B ring and especially C ring derivatives can, in principle, be provided by all procedures. Ideally derivatizations should take place as late as possible in the reaction sequence, in order to minimize the additional labor required to produce them. With respect to C ring versatility the Wender and Holton routes provide probably the best basis. However, for any alteration desired it must be kept in mind that experience with taxanes has taught that even the tiniest change in the substituents or the skeleton can cause great problems with the synthesis. Most published fully synthetic analogues have been C ring aromatics, [4] but also spin-offs of the total synthesis have been reported (see introduction [3]).

In spite of the fantastic successes from the groups of Holton, Nicolaou, Danishefsky and Wender much excitement remains in the total syntheses of taxoids [12]. Compounds and procedures, superior to the known ones with respect to variability, yield, and commercial feasibility are more than ever a challenge for the inventive chemist. Apart from the more

practical viewpoint, all approaches demonstrate state of the art organic synthesis and the high intellectual standard reached in this area, be it through earlier efforts of the developers of selective synthetic methods and tools (cf. e. g. the McMurry reaction or the interplay of protective groups), the proper combination and selection of such tools (e. g. leaving an oxetane ring untouched through multiple transformations), or the beauty of the concept as found in the Holton and especially the Wender synthesis with their multiple rearrangements not immediately obvious from standard retrosynthetic considerations.

References

[1] Several excellent review articles on the chemistry and biology of taxol have been published, as well as its history and even politics related to the compound. For this reason the preliminary work and synthetic routes of other authors are not quoted individually in the text; they can be found in the following reviews on chemistry and related subjects: a) G. I. Georg, T. T. Chen, I. Ojima, D. M. Vyas (Eds.), *Taxane Anticancer Agents – Basic Science and Current Status*, Vol. 583, American Chemical Society, Washington, **1995**; b) K. C. Nicolaou, W.-M. Dai, R. K. Guy, *Angew. Chem.* **1994**, *106*, 38–69; *Angew. Chem. Int. Ed. Engl.* **1994**, *33*, 45; c) D. Guenard, F. Gueritte-Voegelein, P. Potier, *Acc. Chem. Res.* **1993**, *26*, 160–167; d) I. Ojima, *Acc. Chem. Res.* **1995**, *28*, 383–389. On biochemistry, biology, medicine & related subjects: e) P. E. Fleming, H. G. Floss, M. Haertel, A. R. Knaggs, A. Lansing, U. Mocek, K. D. Walker, *Pure Appl. Chem.* **1994**, *66*, 2045–2048; f) G. Appendino, *Nat. Prod. Rep.* **1995**, *12*, 349–360; g) G. M. Gragg, S. A. Schepartz, M. Suffness, M. R. Grever, *J. Nat. Prod.* **1993**, *56*, 1657–1668; h) F. Lavelle, *Exp. Opin. Invest. Drugs* **1995**, *4*, 771–775; i) K. J. Böhm, K. W. Wolf, *Biol. unserer Zt.* **1997**, *27*, 87–95.

[2] The influence of taxol on microtubuli is so extraordinary that only recently and after an immense screening effort other compounds of similar properties have been found. However, at the time of this writing it is still open, if they will have an impact on cancer therapy cf. [1h] and L. Wessjohann, *Angew. Chem.* **1997**, *109*, 739–742; *Angew. Chem. Int. Ed. Engl.* **1997**, *36*, 715–718; cf. also [3].

[3] U. Klar (Schering AG, Berlin, Germany), Symposium *"Aktuelle Entwicklungen in der Naturstofforschung"* – 9. *Irseer Naturstofftage der DECHEMA e. V.* (Irsee, Germany) 26.–28. Feb. 1997, Lecture 6: "Inhibitoren der Tubulindepolymerisation – von Taxol zu einer neuen, nicht-taxoiden Leitstruktur".

[4] Simple aromatic taxoids have been synthesized e. g. by K. C. Nicolaou, C. F. Claiborne, K. Paulvannan, M. H. D. Postema, R. K. Guy, *Chem. Eur. J.* **1997**, *3*, 399–409; ref. cited therein; and Wender et al. [8b]

[5] R. A. Holton, C. Somoza, H.-B. Kim, F. Liang, R. J. Biediger, P. D. Boatman, M. Shindo, C. C. Smith, S. Kim, H. Nadizadeh, Y. Suzuki, C. Tao, P. Vu, S. Tang, P. Zhang, K. K. Murthi, L. N. Gentile, J. H. Liu, *J. Am. Chem. Soc.* **1994**, *116*, 1597–1598; R. A. Holton, H.-B. Kim, C. Somoza, F. Liang, R. J. Biediger, P. D. Boatman, M. Shindo, C. C. Smith, S. Kim, H. Nadizadeh, Y. Suzuki, C. Tao, P. Vu, S. Tang, P. Zhang, K. K. Murthi, L. N. Gentile, J. H. Liu, *J. Am. Chem. Soc.* **1994**, *116*, 1599–1600.

[6] K. C. Nicolaou, Z. Yang, J. J. Liu, H. Ueno, P. G. Nantermet, R. K. Guy, C. F. Claiborne, J. Renaud, E. A. Couladouros, K. Paulvannan, E. J. Sorensen, *Nature (London)* **1994**, *367*, 630–634; K. C. Nicolaou, P. G. Nantermet, H. Ueno, R. K. Guy, E. A. Couladouros, K. Paulvannan, E. J. Sorensen, *J. Am. Chem. Soc.* **1995**, *117*, 624–633; K. C. Nicolaou, J.-J. Liu, Z. Yang, H. Ueno, E. J. Sorensen, C. F. Claiborne, R. K. Guy, C.-K. Hwang, M. Nakada, P. G. Nantermet, *J. Am. Chem. Soc.* **1995**, *117*, 634–644; K. C. Nicolaou, Z. Yang, J.-J. Liu, P. G. Nantermet, C. F. Claiborne, J. Renaud, R. K. Guy, K. Shibayama, *J. Am. Chem. Soc.* **1995**, *117*, 645–652.

[7] S. J. Danishefsky, J. J. Masters, W. B. Young, J. T. Link, L. B. Snyder, T. V. Magee, D. K. Jung, R. C. A. Isaacs, W. G. Bornmann, C. A. Alaimo, C. A. Coburn, M. J. Di Grandi, *J. Am. Chem. Soc.* **1996**, *118*, 2843–2859; and references cited.

[8] a) P. A. Wender, N. F. Badham, S. P. Conway, P. E. Floreancig, T. E. Glass, J. B. Houze, N. E. Krauss, D. Lee, D. G. Marquess, P. L. McGrane, W. Meng, M. G. Natchus, A. J. Shuker, J. C. Sutton, R. E. Taylor, *J. Am. Chem. Soc.* **1997**, *119*, 2757–2758; P. A. Wender, N. F. Badham, S. P. Conway, P. E. Floreancig, T. E. Glass, C. Gränicher, J. B. Houze, J. Jänichen, D. Lee, D. G. Marquess, P. L. McGrane, W. Meng, T. P. Mucciaro, M. Mühlebach, M. G. Natchus, H. Paulsen, D. B. Rawlins, J. Satkofsky, A. J. Shuker, J. C. Sutton, R. E. Taylor, K. Tomooka, *J. Am. Chem. Soc.* **1997**, *119*, 2755–2756; b) P. A. Wender, T. P. Mucciaro, *J. Am. Chem. Soc.* **1992**, *114*, 5878–5879; c) P. A. Wender, L. A. Wessjohann, B. Peschke, D. B. Rawlins, *Tetrahedron Lett.* **1995**, *36*, 7181–7184; d) P. A. Wender, N. F. Badham, S. P. Conway, P. E. Floreancig, T. E. Glass, J. B. Houze, N. E. Krauss, D. Lee, D. G. Marquess, P. L. McGrane, W. Meng, T. P. Mucciaro, M. Mühlebach, M. G. Natchus, T. Ohkuma, B. Peschke, D. B. Rawlins, A. J. Shuker, J. C. Sutton, R. E. Taylor, K. Tomooka, L. A. Wessjohann in *Taxane Anticancer Agents – Basic Science and Current Status*, Vol. 583 (Eds.: G. I. Georg, T. T. Chen, I. Ojima, D. M. Vyas), American Chemical Society, Washington, **1995**, pp. 326–339.

[9] The man years contributed by all the research groups presumably well exceed those required for the synthesis of, for example, vitamin B_{12} or any other single substance.

[10] The addition, alteration, or removal of protecting groups will not be discussed here: Ac = Acetyl, Ar = 2,4,6-triisopropylphenyl, Bn = benzyl, Bz = benzoyl, BOC = *tert*-butoxycarbonyl, BOM = benzyloxymethyl, TBPS = *tert*-butyldiphenylsilyl, TBS = *tert*-butyldimethylsilyl, TES = triethylsilyl, TIPS = triisopropylsilyl, TMS = trimethylsilyl. Actual yields of the first published procedures are given, conversions and loss to formation of isomers were included in the calculations where necessary. For precursors the yields were taken from the references given by the authors and used to

calculate back to commercially available starting materials. This leads to lower values than those presented in the references.

[11] This important partial reaction has been registered as a patent by Nicolaou et al., and has been published in detail: K. C. Nicolaou, P. G. Nantermet, H. Ueno, R. K. Guy, *J. Chem. Soc., Chem. Commun.* **1994**, 295–296; K. C. Nicolaou, E. A. Couladouros, P. G. Nantermet, J. Renaud, R. K. Guy, W. Wrasidlo, *Angew. Chem.* **1994**, *106,* 1669–1671; *Angew. Chem. Int. Ed. Engl.* **1994**, *33,* 1581–1583.

[12] *Note added in proof:*
A new total synthesis of Baccatin III and Taxol including a new method for the introduction of the phenylisoserine side chain was published after the typesetting for this book: I. Shiina, H. Iwadare, H. Sakoh, M. Hasegawa, Y.-I. Tani, T. Mukaiyama, *Chem Lett.* **1998**, 1–2; I. Shiina, K. Saitoh, I. Fréchard-Ortuno, T. Mukaiyama, *Chem. Lett.* **1998**, 3–4.

Erythromycin Synthesis – A Never-ending Story?

Johann Mulzer

The synthesis of the macrolide antibiotics, erythromycin A (**1**) and B (**2**), which are highly effective against gram-positive pathogens, has been one of the most extensive project in the history of synthetic organic chemistry. [1] This phenomenon is not rational as **1** and **2** are available in large quantities from fermentation of the microorganism *Streptomyces erythreus*. It is the complexity of the molecular structure, the plethora of stereocenters and functional groups, and the magic of the medium ring that has kept about 15 large research groups busy worldwide for more than a decade. All total syntheses follow the same pattern. They first aim at the aglycons (erythronolides A and B, **3** and **4** respectively) in protected form, which are synthesized from the corresponding seco acids by lactonization; for example, **3** is prepared from **5–7**, and **4** from **8–11**. The seco acids are assembled from smaller chiral fragments which are obtained from the chiral carbon pool, by optical resolution, or by enantioselective synthesis. The final glycosidation, and with it the synthesis of **1** and **2**, has been achieved three times only. [2, 16, 18]

Over the years the priorities have been shifted several times. Initially the attention focused on the seco acid fragments and therefore on the stereocontrolled construction of the often recurring β-hydroxycarbonyl and 1,3-diol units of **3/4**. In this connection the development of methods for enantio- and diastereomeric control (acyclic stereoselection [9]) proved to be unbelievably fruitful. The first syntheses of seco acids [2, 6] required cyclic intermediates and optical resolution, and often resorted to "relay compounds" recovered from the degradation of the natural

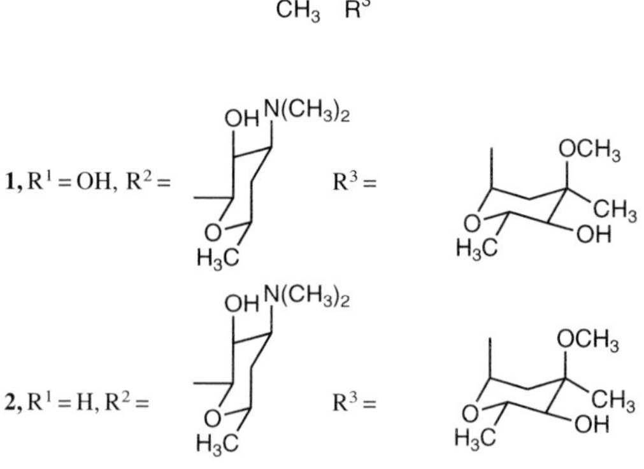

1, R¹ = OH, R² =

2, R¹ = H, R² =

3, R¹ = OH, R² = R³ = H

4, R¹ = R² = R³ = H

Scheme 1.

product. Nowadays derivatives such as **10** and **11** can be prepared stereoisomerically pure in ca. 15 steps and in gram amounts. [7, 8]

In all these synthesis it has been a tacit dogma, that the carbon framework of the seco acids be assembled in a convergent manner from two major fragments ("eastern" and "western" zones). R. W. Hoffmann et al. were among the first to show that even a highly functionalized seco acid such as **21** can be constructed in a linear fashion. [15a] Starting from the C_{13}–C_{11} fragment **20** the chain was elongated successively by aldehyde borocrotylation using the chiral reagents **23** and **24** respectively. This process was combined with other stereoselective reactions such

as Wittig olefination and Sharpless epoxidation to generate the protected seco acid **21**, and after lactonization and deprotection (9S)-dihydroerythronolide A (**22**) was formed. In a related fashion, Evans has made use of his iterative polypropionate methodology to prepare 6-deoxyerythronolide B. [15b]

In view of the growing number of easily accessible seco acids the attention was directed more and more toward the lactonization step, which is routinely carried out via an activated ester intermediate (**12**, Table 1). However, the formation of a 14-membered ring proved to be not straightforward; dimerization and polymerization are significant side reactions. Corey's thiopyridyl activation (via the

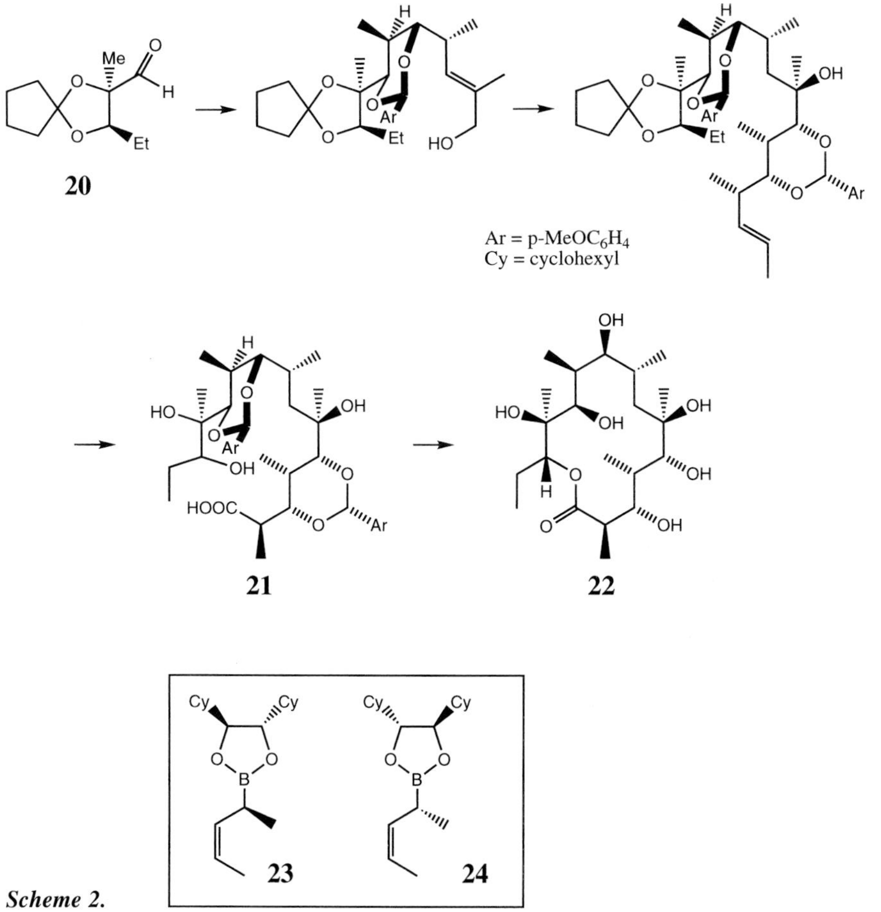

20

Ar = p-MeOC$_6$H$_4$
Cy = cyclohexyl

21 → **22**

23   **24**

Scheme 2.

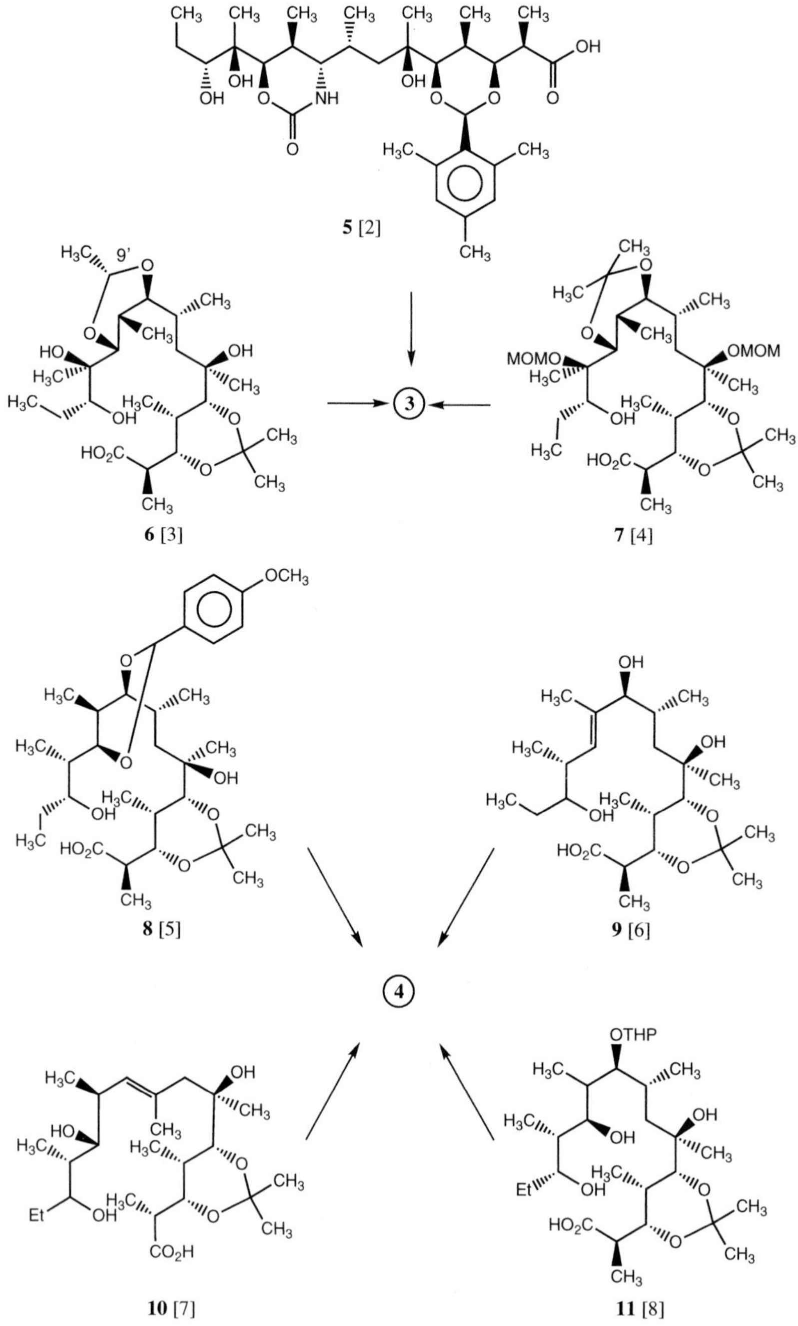

5 [2]

6 [3]

7 [4]

8 [5]

9 [6]

10 [7]

11 [8]

Scheme 3.

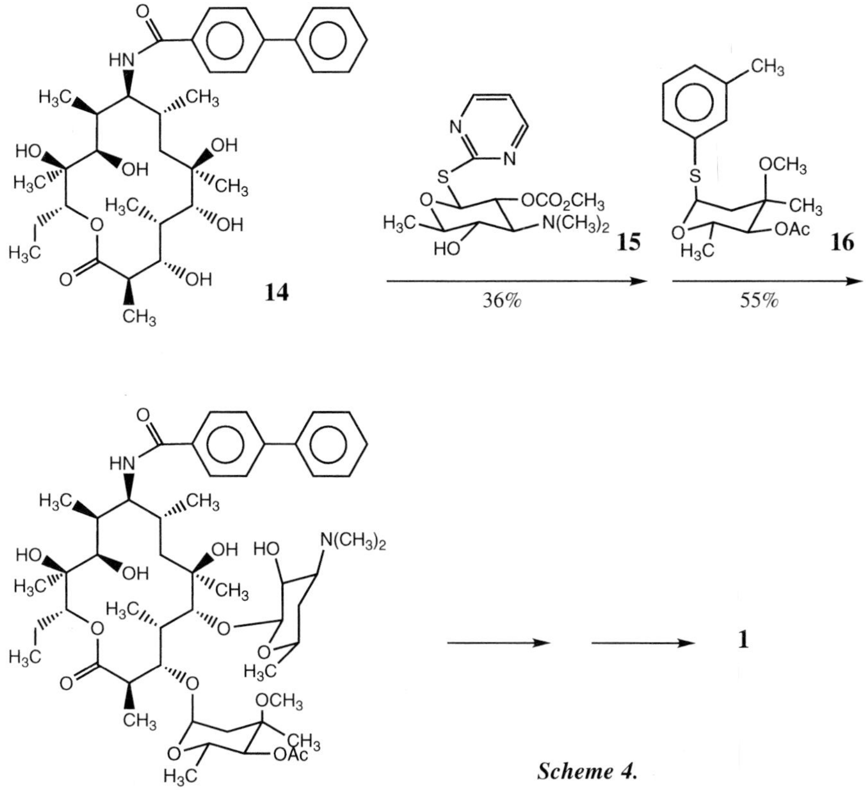

14   36% →   **15**   55% →   **16**

Scheme 4.

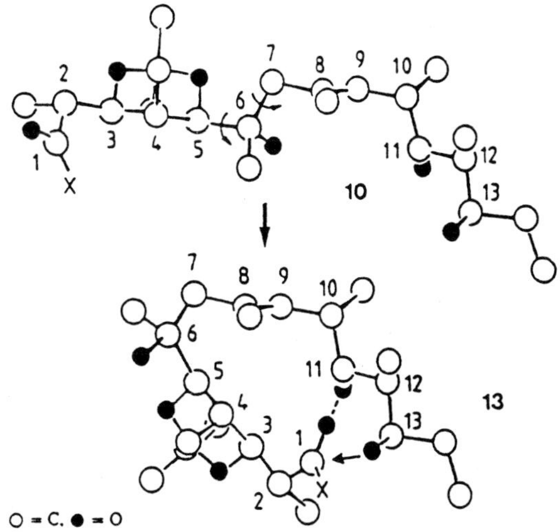

Figure 1. Conformational Change of the Activated Seco Acid **10** to an Optimum Conformation **13** for Macrolactonization.

intermediates **12a, b**) [10] was the standard macrolactonization procedure for ten years, but has now been replaced by the Yamaguchi lactonization which uses the mixed anhydride **12c**. Next to carboxyl group activation the hydroxyl protecting groups play a central role. O-protective groups not only suppress the formation of rings of unwanted size, they may also be used to stabilize conformations which are particularly favorable for cyclization. For instance, the 3,5-acetal or ketal unit in **5–11** freezes the C2–C6 fragment of the molecule in a rigid, extended conformation, due to the diequatorial arrangement around the 1,3- dioxan chair. This allows the 6-OH function to remain unprotected, because it could only be lactonized after flipping the acetonide to the highly unstable 3,5-diaxial conformation. The 9- and 11-hydroxyl groups

Table 1. Macrolactonization to form the 14-membered ring in the erythromycin synthesis.

12	Activating group X	Product yield in [%]
a [10]	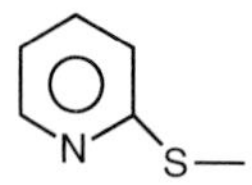	5 (70)[2]
b [10]	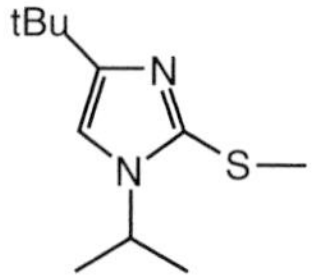	8 (65)[5], 9 (50)[6]
c [11]	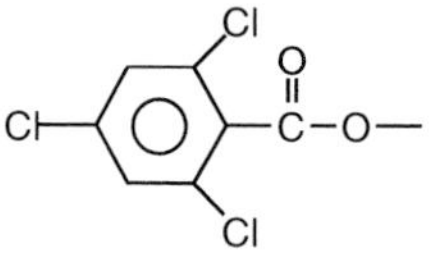	7 (27)[4], 10 (89)[7], 11 (>95)[8]
d [3]	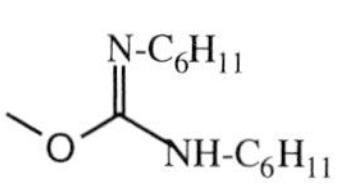	6 (64)[3]

are often protected by a cyclic acetal (such as in **5–8**), in which the 9′-substituent induces considerable transannular strain. In fact, 9′-disubstituted derivatives such as **7** are substantially more difficult to cyclize than the corresponding 9′-monosubstituted seco acids **6** and **8** (Table 1). [3, 4] Moreover, the configuration at C9 is important: (9S) seco acids cyclize much more readily than the (9R) isomers do. [2] Many of these hitches may be avoided by the introduction of a C,C-double bond in the critical region between C7 and C11 (see **9–11**). At least one OH function and its protecting group is then missing; in addition the two sp² centers on the ring periphery decrease the transannular ring strain. Trisubstituted olefins display the phenomenon of allylic 1,3 strain. [12] In seco acid **10** this leads to a sickle-like conformation of the C7–10-carbon chain. Simple rotations of the

rigid C1–C6 fragment around the C5/6 and C6/7 axes lead to an optimal conformation for cyclization (**13**). The 11-OH group is out of reach of the carboxyl function and can remain unprotected. [7]

In view of the high macrolactonization yields for **10/11** (Table 1) the ultimate problem of the erythromycin synthesis are the **3–1** and **4–2** glycosylations, which are indispensible for the physiological activity. As shown early by the Woodward group [2] the monosaccharide blocks desosamine (**15**) and cladinose (**16**), both in suitably protected and activated form, can be coupled with an almost "naked" aglycon (**14**). Only the 5- and the 3-OH groups are reactive towards the glycosyl donors, the 5-OH group showing a clear kinetic advantage over the 3-OH function (scheme 4).

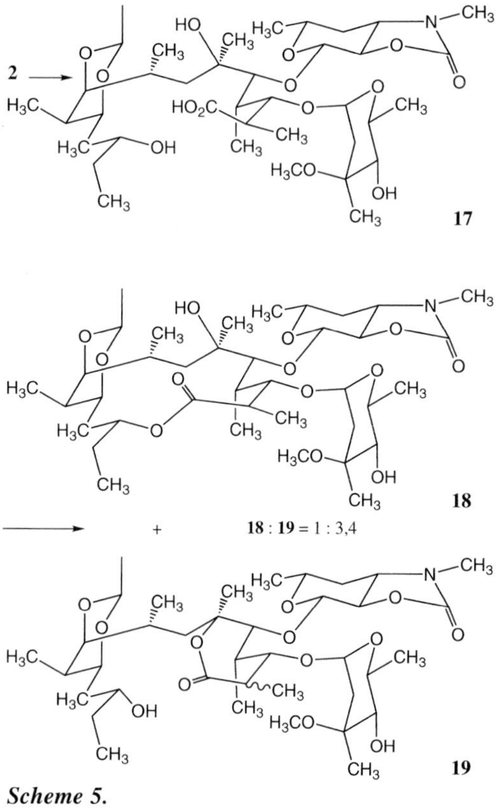

18 : 19 = 1 : 3,4

Scheme 5.

In a remarkable experiment (scheme 5) *S. F. Martin* and *M. Yamashita* [13] have parted with the accepted philosophy of erythromycin synthesis, namely to perform first the lactonization and then the glycosidation. They prepared diglycosidated seco acid **17** from **2** via partial synthesis in ten steps and subjected it to Yamaguchi's macrolactonization conditions. However, this protocol afforded only minor amounts of the desired macrolide **18**,

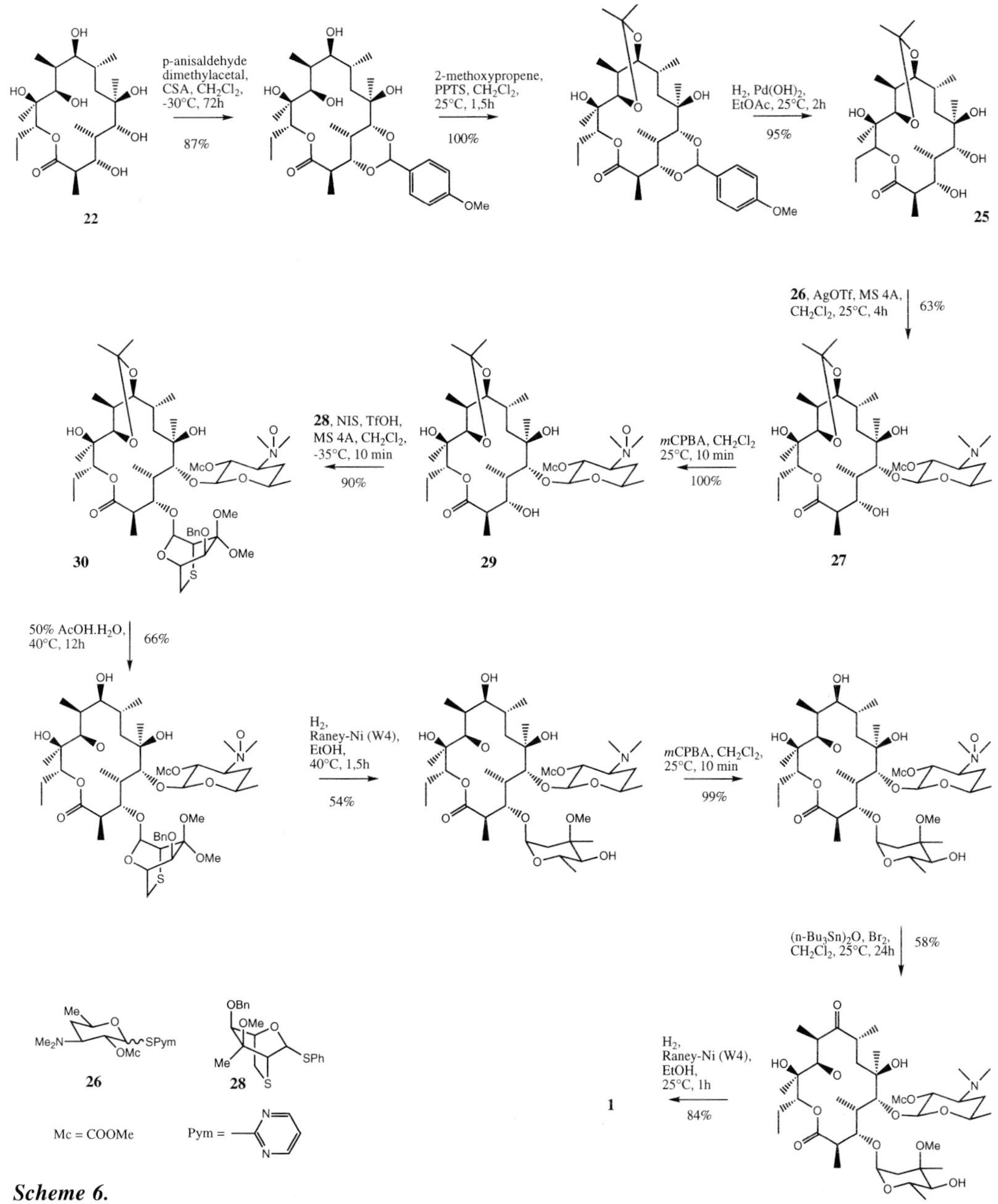

Scheme 6.

the major products being the two C2 epimers of the seven-membered lactone **19**. This is not surprising because the formation of seven-membered rings is a favorable process and can readily occur, since the 6-OH function is not protected and it is not conformationally deactivated by the presence of a 3,5-ketal. As predicted, the Yamaguchi lactonization of the 6-OMe derivative of **17** (14 steps from **2**) exclusively affords the 6-OMe derivative of **18**, but in only 53 % yield.

After a longer hiatus, the Martin group has, quite recently, reported the first synthesis of erythromycin B (**2**). Remarkably, in their communication [16] the authors did not comment on their previous strategy, but, quite matter-of-factly, they used the conventional procedure developed by the Woodward group for erythromycin A, instead. [2] Thus, O-protected 9(*S*)-dihydro-erythronolide B was prepared by total synthesis, deprotected at the 3- and 5-positions and successively submitted to 5- and 3-glycosylation to give after the usual manipulations, erythromycin B (**2**). Similarly, in the synthesis of **1** by Toshima et al. [17] the aglycon **22** was converted into the 9,11-acetonide **25**, which was than selectively glycosylated at the 5-position with the protected desosamine thioglycoside **26** to give the monoglycoside **27**. After N-oxidation the cladinose anhydrothioglycosyl donor **28** was used to convert **29** into **30**. Further manipulation eventually gave **1**. The crucial problem of stereoselective O-3-*a*-glycosylation was thus successfully solved.

In the light of these recent results it appears advisable to concentrate on further optimization of the glycosilation of **3/4** derivatives, perhaps by enzyme catalysis. After all, the biosynthesis follows this pathway: first aglycon **4** is generated, and from there **2** is formed. [14]

In this connection it is remarkable that the cladinose residue is not essential for the antibiotic activity of erythromycin A. Apparently the cladinose can be replaced by a 3-keto

function to give "ketolides" which are highly potent antibiotics. Additionally, two additional rings are fused to the "north western" hemisphere of the ketolide to enhance stability, bioavailability and antimicrobial activity ("tricyclic ketolides"). [18]

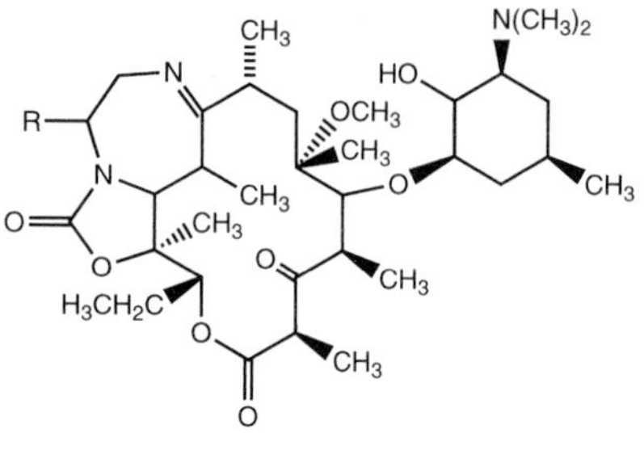

"Tricyclic Ketolide"

Scheme 7.

References

[1] Review: I. Paterson, M. M. Mansuri, *Tetrahedron* **1985**, *41*, 3569.

[2] R. B. Woodward et al. *J. Am. Chem. Soc.* **1981**, 103, 3210, 3213, 3215.

[3] G. Stork, S. D. Rychnovsky, *J. Am. Chem. Soc.* **1987**, *109*, 1564, 1565.

[4] H. Tone, T. Nishi, Y. Oikawa, M. Hikota, O. Yonemitsu, *Tetrahedron Lett.* **1987**, *28*, 4569; M. Hikota, H. Tone, K. Horita, O. Yonemitsu, *Tetrahedron* **1990**, *46*, 4613.

[5] N. K. Kochetkov, A. Sviridov, M. S. Ermolenko, D. V. Yashunsky, V. S. Borodkin, *Tetrahedron* **1989**, *45*, 5109.

[6] E. J. Corey et al., *J. Am. Chem. Soc.* **1978**, *100*, 4618.

[7] J. Mulzer, H. M. Kirstein, J. Buschmann, C. Lehmann, P. Luger, *J. Am. Chem. Soc.* **1991**, *113*, 910.

[8] J. Mulzer, P. A. Mareski, J. Buschmann, P. Luger, *Synthesis* **1992**, 215.

[9] P. A. Bartkett, *Tetrahedron* **1980**, *36*, 1.

[10] E. J. Corey, K. C. Nicolaou, *J. Am. Chem. Soc.* **1974**, *96*, 5614; E. J. Corey, D. J. Brunelle, *Tetrahedron Lett.* **1976**, 3409.

[11] J. Inanaga, K. Hirata, H. Saeki, T. Katsuki, M. Yamaguchi, *Bull. Chem. Soc. Jpn.* **1979**, *52*, 1989.

[12] R. W. Hoffmann, *Chem. Rev.* **1989**, *89*, 1841.

[13] S. F. Martin, M. Yamashita, *J. Am. Chem. Soc.* **1991**, *113*, 5478.

[14] J. Staunton, *Angew. Chem.* **1991**, *103*, 1331; *Angew. Chem. Int. Ed. Engl.* **1991**, *30*, 1302; R. Pieper, C. Kao, C. Khosla, G. Luo, D. E. Cane, *Chem. Soc. Rev.* **1996**, 297.

[15] a) R. Stürmer, K. Ritter, R. W. Hoffmann, *Angew. Chem.* **1993**, *105*, 112; *Angew. Chem. Int. Ed. Engl. 32*, 101; b) D. A. Evans, A. S. Kim, *Tetrahedron Lett.* **1997**, *38*, 53.

[16] S. F. Martin, T. Hida, P. R. Kym, M. Loft, A. Hodgson, *J. Am. Chem. Soc.* **1997**, *119*, 3193.

[17] K. Toshima, J. Nozaki, S. Mukaiyama, T. Tamai, M. Nakata, K. Tatsuta, M. Kiroshita, *J. Am. Chem. Soc.* **1995**, *117*, 3717.

[18] S. C. Stinson, *Chem. Eng. News,* **1996**, 75.

Great Expectations for a Total Synthesis of Vancomycin

Kevin Burgess, Dongyeol Lim, and Carlos I. Martinez

Vancomycin has a complicated molecular architecture and clinical significance as an antibiotic for treatment of infections caused by gram positive bacteria. These two attributes have inticed several groups of organic chemists into attempting to prepare this molecule, but no total synthesis has been reported to date.

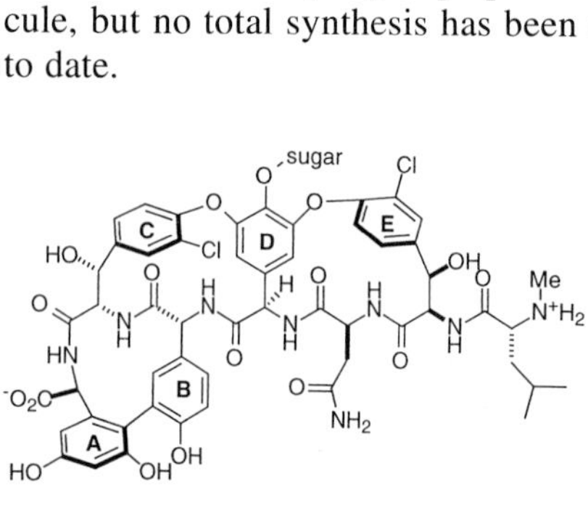

vancomycin

In 1993 there were two reasons to suppose that a total synthesis of vancomycin might be achieved soon. [1] First, the cyclic molecule **1** had been prepared. [2] This closely resembles the "southwestern fragment" of vancomycin which includes the **AB** biaryl linkage. Second, the bicyclic molecule **3** also had been made; [3] this is closely related to the central region of vancomycin which incorporates the **C**, **D**, and **E** rings.

Despite this optimism, it has not been convenient to exploit the syntheses of fragments **1** and **3** in a total synthesis, presumably for the following reasons. Vancomycin has only one chlorine atom on rings **C** and **E**, whereas both *ortho* positions on the corresponding aryl rings of synthon **3** are chlorinated. These extra chlorines were critical in the synthesis of that material. Synthon **3** was prepared by two thallium trinitrate-mediated couplings, which absolutely require both *ortho* positions relative to the phenolic-OH to be blocked. Furthermore, the formation of **1**, an oxidative coupling of the carbons marked **a** and **b** in structure **2**, required harsh conditions (VOF$_3$, BF$_3$·OEt$_2$, TFA, TFAA, CH$_2$Cl$_2$, 0 °C, then Zn) that may not be suitable in the absence of chlorine blocking groups at both *ortho* positions of ring **C**.

Reduction of a single C–Cl bond to a C–H on a dichlorinated aromatic ring is a challenging transformation. When two such monosubstitutions must be made in a molecule containing two dichlorinated aryl rings, it is more difficult; if the chlorines on each ring are inequivalent due to atropisomerism, [4] then the task is formidable. Little activity was then reported in this area, implying that a synthetic stalemate has been reached. Very recently, however, the oxidative route has been improved by using 2-bromo-6-chlorophenols allowing selective reduction of the C–Br bond after cyclization in a synthesis of the

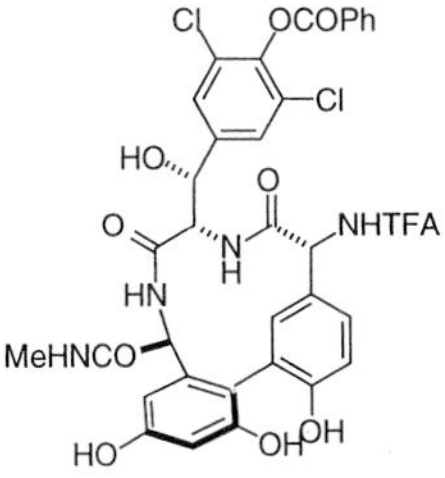

1

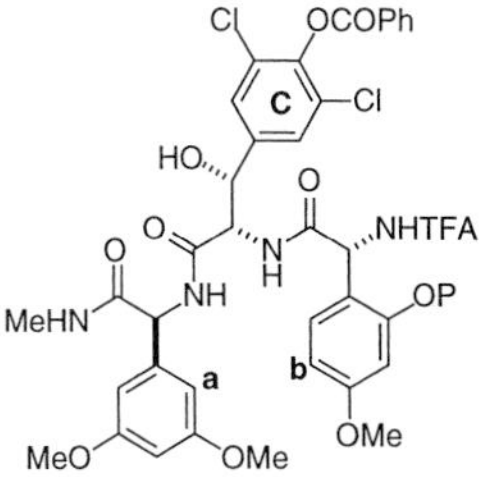

2
(P = 3,4-dichlorobenzyl)

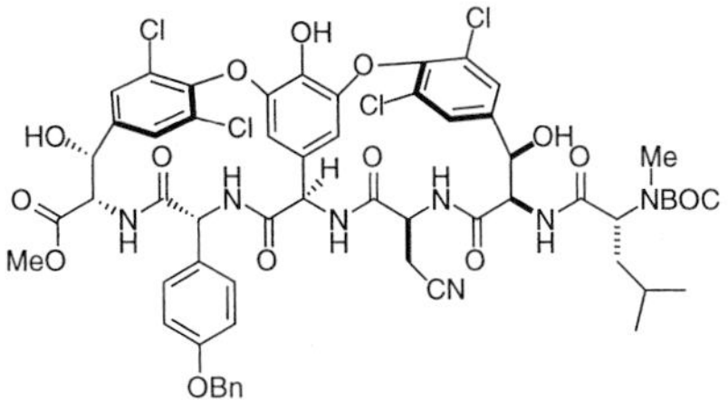

3

related reactions to prepare compound **5**, which has a 14-membered ring, then the 14,16 ring fused system **6** similar to that in teicoplanin (a natural product structurally related to vancomycin). [8–10] Model studies in the vancomycin series showed that 16-membered peptidic biaryl ethers also could be prepared, [4, 9] an observation which Rao and co-workers confirmed shortly after. [11] These important studies proved that the closure would work even for 16-membered rings of the type found in vancomycin, which had been shown to be difficult to form via other conditions (*vide supra*). Moreover, Zhu's research demonstrated the cyclization could be performed with minimal racemization at an arylglycine residue within the incipient ring. Indeed, the conditions are sufficiently

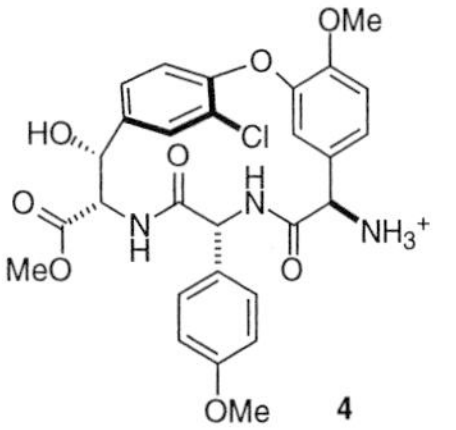

C–O–D ring derivative **4**. [5] Nevertheless, alternatives to these oxidative routes are highly desirable.

Since 1993, relatively new competitors in the vancomycin synthesis arena have provided an impetus which may have sufficient momentum to bring about a total synthesis of vancomycin. Notably, a series of papers from Zhu and co-workers have focused on intramolecular nucleophilic aromatic substitution (S$_N$Ar) reactions as a means to cyclize peptide-based systems containing biaryl ether linkages. This approach is a notable improvement on Ullmann coupling procedures using aromatic compounds which are not activated to nucleophilic attack. The latter approach had been explored by Boger and co-workers, [6] but the extreme temperatures and prolonged reaction times preclude application of such "classical" Ullmann couplings in vancomycin syntheses.

Zhu's first efforts in this area included a synthesis of the 17-membered ring compound K-13 via biaryl ether formation as the ring closure step. [7] Since then, his group has used

4

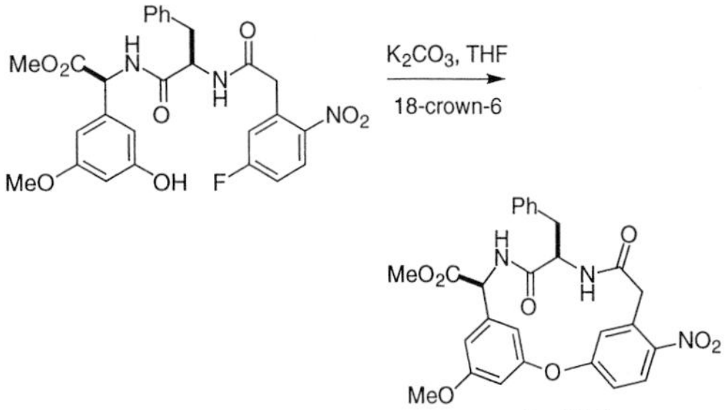

5   71 %

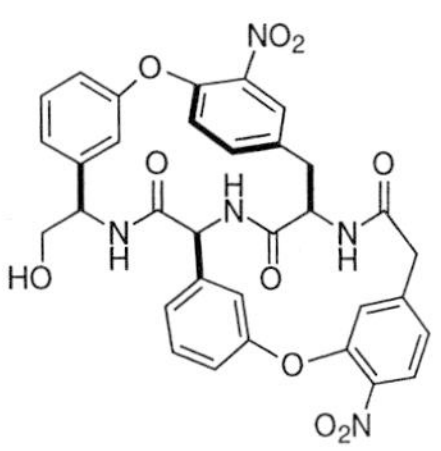

6

mild (e.g. K_2CO_3, DMF, 25 °C) that they should be suitable for more complicated substrates. A recent publication from Zhu's group proved that this methodology could be applied to prepare a close analog of the vancomycin **C–O–D** ring. [12] Shortly after these studies were published, Boger and co-workers reported the same type of cyclization method to prepare synthons for the **C–O–D** and **D–O–E** rings of vancomycin, compounds **7** and **8** respectively. [13] Boger's intermediate **7** is notable as it has functionality appropriate for attempted formation of the vancomycin **AB** biaryl fragment. Later, Evans and Watson also used the S_NAr to generate functionalized **C–O–D** and **D–O–E** ring analogs with protecting groups presumably chosen for elaboration into more complex intermediates in the vancomycin series. [14] The Rao Group have also expanded their studies; in recent work they prepared a teicoplanin fragment via a similar S_NAr route. [15]

The next landmark in the evolution of this story was to produce molecules with both the **C–O–D** and **D–O–E** rings, i.e. **C–O–D–O–E** vancomycin fragments. The Zhu group were the first to report this, and they demonstrated two logical approaches. [16, 17] In the first, they prepared an acyclic precursor and attempted to form the **C–O–D–O–E** ring system via two cyclizations in one reaction vessel. All four atropisomers formed when the reaction was run at room temperature, but compound **9** was isolated in 60% yield after a reaction at −5 °C. This is an encouraging result even though this atropisomer does not correspond to that in vancomycin. Zhu's stepwise couplings were less stereoselective. Cyclization of the appropriate aryl fluoride gave compounds **10** in a 2 : 3 ratio favoring the unnatural atropisomer series. These were separated, elaborated to other aryl fluorides, and cyclized to give **11a/11b** (3 : 1) and **11c/11d** (1 : 2). Thus, all four stereoisomers of the **C–O–D–O–E** ring system could be obtained.

10a X = NO_2, Y = H
10b X = H, Y = NO_2

Currently, it is the Evan's group, however, who can claim to have produced the most advanced compound in a vancomycin synthesis. They used an oxidative coupling to produce the **C–O–D** ring in compound **12**, [18] and a S_NAr cyclization to produce the complete **AB–C–O–D–O–E** system **13**. [19] The latter transformation favors the atropisomer corresponding to the vancomycin **D–O–E** atropisomer by a 7 : 1 factor, in a reaction that was surprisingly efficient (90% yield of combined cyclization products).

Based on the current literature, a total synthesis of vancomycin is imminent. The **AB–C–O–D–O–E** core has been produced, and coupling of the carbohydrate residues to one of the phenolic oxygen atoms of vancomycin is a known transformation. [20] Other methods for the production of biaryl ethers are emerging, [21, 22] but it seems that S_NAr couplings of nitroaryl fluorides will be featured in the first vancomycin synthesis.

11a X = Y' = NO₂, Y = X' = H
11b X = X' = NO₂, Y = Y' = H
11c X = X' = H, Y = Y' = NO₂
11d X = Y' = H , Y = X' = NO₂

12

13

References

[1] A. V. R. Rao, M. K. Gurjar, K. L. Reddy, A. S. Rao, *Chem. Rev.* **1995**, *95*, 2135–67.

[2] D. A. Evans, C. J. Dinsmore, D. A. Evrard, K. M. DeVries, *J. Am. Chem. Soc.* **1993**, *115*, 6426–7.

[3] D. A. Evans, J. A. Ellman, K. M. DeVries, *J. Am. Chem. Soc.* **1989**, *111*, 8912–4.

[4] R. Beugelmans, G. P. Singh, M. Bois-Choussy, J. Chastanet, J. Zhu, *J. Org. Chem.* **1994**, *59*, 5535–42.

[5] H. Konishi, T. Okuno, S. Nishiyama, S. Yamamura, K. Koyasu, Y. Terada, *Tetrahedron Lett.* **1996**, *37*, 8791–4.

[6] D. L. Boger, Y. Nomoto, B. R. Teegarden, *J. Org. Chem.* **1993**, *58*, 1425–33.

[7] R. Beugelmans, A. Bigot, J. Zhu, *Tetrahedron Lett.* **1994**, *35*, 7391–4.

[8] J. Zhu, R. Beugelmans, S. Bourdet, J. Chastanet, G. Roussi, *J. Org. Chem.* **1995**, *60*, 6389–96.

[9] R. Beugelmans, L. Neuville, M. Bois-Choussy, J. Zhu, *Tetrahedron Lett.* **1995**, *36*, 8787–90.

[10] R. Beugelmans, S. Bourdet, J. Zhu, *Tetrahedron Lett.* **1995**, *36*, 1279–82.

[11] A. V. R. Rao, K. L. Reddy, A. S. Rao, *Tetrahedron Lett.* **1994**, *35*, 8465–8.

[12] J. Zhu, J.-P. Bouillon, G. P. Singh, J. Chastanet, R. Beugelmans, *Tetrahedron Lett.* **1995**, *36*, 7081–4.

[13] D. L. Boger, R. M. Borzilleri, S. Nukui, *Biorg. Med. Chem. Lett.* **1995**, *5*, 3091–6.

[14] D. A. Evans, P. S. Watson, *Tetrahedron Lett.* **1996**, *19*, 3251–4.

[15] A. V. R. Rao, K. L. Reddy, A. S. Rao, T. V. S. K. Vittal, M. M. Reddy, P. L. Pathi, *Tetrahedron Lett.* **1996**, *37*, 3023–6.

[16] R. Beugelmans, M. Bois-Choussy, C. Vergne, J.-P. Bouillon, J. Zhu, *J. Chem. Soc., Chem. Commun.* **1996**, 1029–30.

[17] C. Vergne, M. Bois-Choussy, R. Beugelmans, J. Zhu, *Tetrahedron Lett.* **1997**, *38*, 1403–6.

[18] D. A. Evans, C. J. Dinsmore, A. M. Ratz, D. A. Evrard, J. C. Barrow, *J. Am. Chem. Soc.* **1997**, *119*, 3417–8.

[19] D. A. Evans, J. C. Barrow, P. S. Watson, A. M. Ratz, C. J. Dinsmore, D. A. Evrard, K. M. DeVries, J. A. Ellman, S. D. Rychnovsky, J. Lacour, *J. Am. Chem. Soc.* **1997**, *119*, 3419–20.

[20] R. G. Dushin, S. J. Danishefsky, *J. Am. Chem. Soc.* **1992**, *114*, 3471–5.

[21] A. J. Pearson, G. Bignan, P. Zhang, M. Chelliah, *J. Org. Chem.* **1996**, *61*, 3940–1.

[22] K. C. Nicolaou, C. N. C. Boddy, S. Natarajan, T.-Y. Yue, H. Li, S. Bršsse, J. M. Ramanjulu, *J. Am. Chem. Soc.* **1997**, *119*, 3421–2.

The Dimeric Steroid-Pyrazine Marine Alkaloids: Challenges for Isolation, Synthesis, and Biology

Arasu Ganesan

Isolation of Cephalostatins and Ritterazines

Since 1955, the American National Cancer Institute (NCI) has conducted a massive search for compounds with antitumor activity, with successes including the *Vinca* alkaloids from the Madagascar periwinkle and taxol from the Pacific yew tree. This highlight [1] covers a family of steroids also discovered by the NCI efforts.

In 1974, extracts of the tiny (~5 mm) marine tube worm *Cephalodiscus gilchristi* collected off South Africa were found to be active in the NCI's primary assay, the murine lymphocytic leukemia P388. Fifteen years of "relentless research" by Pettit's group culminated in the isolation of 139 mg of the major bioactive component, cephalostatin 1, from 166 kg wet weight of the tube worm, and its structural elucidation. [2] This phase was summarized [3] as follows: "Interest in such a powerfully antileukemic agent as cephalostatin 1 ... has prompted Americans to dive extensively at a depth of 20 meters to collect *C. gilchristi* in open seawaters controlled by the white shark."

Cephalostatin 1 is extremely potent against P388, with an ED_{50} of 10^{-7}–10^{-9} µg mL^{-1}. Over the years, the Pettit group has characterized [4] sixteen other cephalostatins. All con-tain the novel structural framework of two steroids linked by a pyrazine at the A ring (Fig. 1). The "right half" is identical in fifteen of the alkaloids, including the C_2-symmetric homodimer cephalostatin 12.

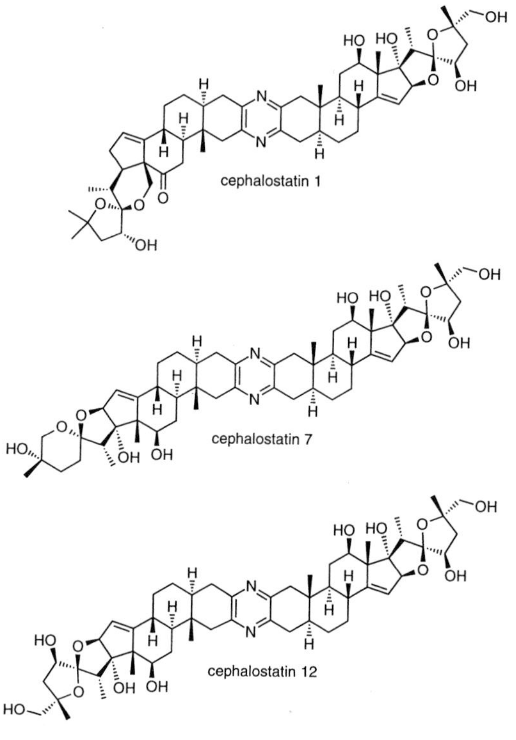

Figure 1. Three of the cephalostatins.

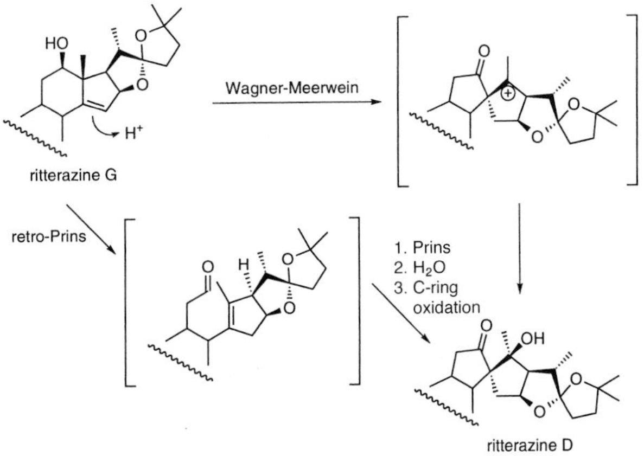

Scheme 1. Possible mechanism of the ritterazine CD-ring rearrangement.

In 1994, a steroid-pyrazine dimer was reported [5] from an unexpected source. Fusetani's group isolated ritterazine A, with an ED_{50} of 10^{-3} µg mL^{-1} against P388, from the tunicate *Ritterella tokioka* collected off Japan. Twelve other ritterazines have since been described [6] (Fig. 2), including one homodimer, ritterazine K. A possible mecha-

nism for the rearranged "right half" in some of the ritterazines may involve (Scheme 1) protonation of the D-ring alkene followed by either a 1,2-Wagner-Meerwein shift or retro-Prins and alternative Prins ring closure.

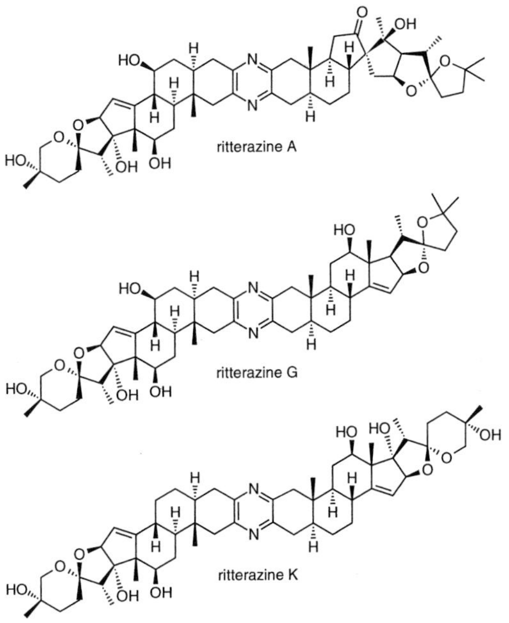

The compounds from *C. gilchristi* and *R. tokioka* are clearly related (ritterazine K, for example, contains the "left half" of cephalostatin 7), although the same alkaloid has yet to be identified from both. Meanwhile, their presence in two phyla suggests they may be produced by a symbiotic microorganism.

Figure 2. Three of the ritterazines.

Synthetic Studies

Initial synthetic attention focused on preparation [7] of symmetrical steroid-pyrazine dimers by the classical condensation of α-amino ketones (Scheme 2). Interestingly, dimer **1** is itself weakly cytotoxic, with an ED_{50} around 10 µg mL^{-1} against tumor cell lines.

Most of the natural products are unsymmetrical dimers. Novel methods have had to be developed for the synthesis of such compounds. At Berkeley, Smith and Heathcock [7b] used the combination of an α-amino oxime ether and an α-acetoxy ketone (Scheme 3). At 90 °C, **2** and **3** preferentially react with

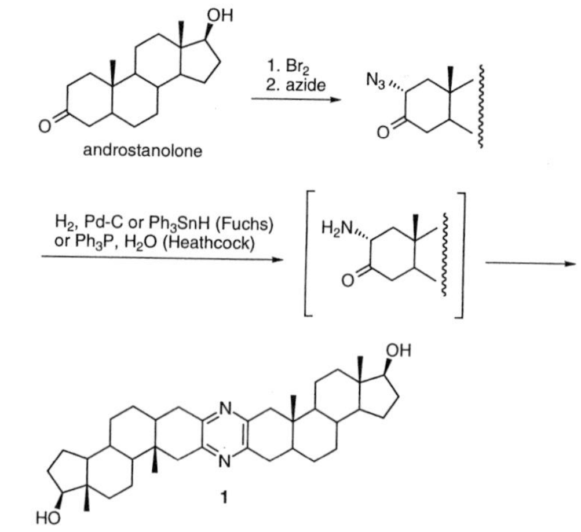

androstanolone

Scheme 2. Synthesis of symmetrical steroid-pyra-zine dimers.

each other to give intermediates which aro-matize to pyrazine **4** at higher temperature. The two-stage protocol is necessary as the α-amino oxime ether condenses with itself at 140 °C.

Recently, the Fuchs group at Purdue has introduced [8] a modification of the Heath-cock protocol, whereby an α-azido ketone is substituted for the α-acetoxy ketone. In this

90 °C, 24 h
140 °C, 24 h   43 %

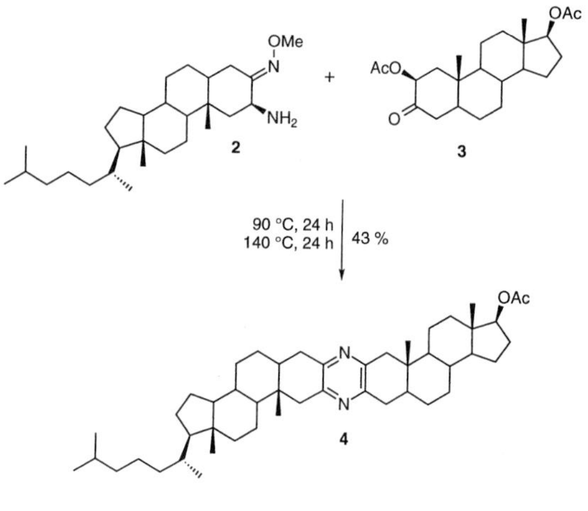

Scheme 3. Smith and Heathcock's unsymmetrical dimer synthesis.

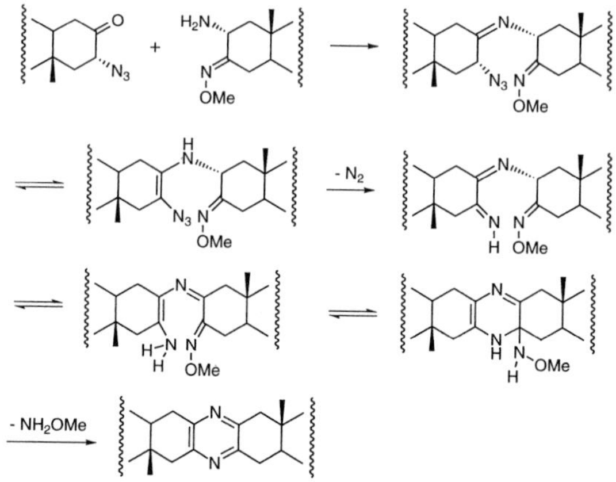

version, yields are significantly improved. Pyrazine **4**, for example, was formed in 78 % yield. The proposed mechanism (Scheme 4) involves imine formation followed by a series of prototropic shifts and the irreversible loss of nitrogen and methoxylamine.

Scheme 4. Proposed mechanism for the Fuchs dimerization.

Introduction of asymmetry into a homodi-mer was demonstrated [9] by Winterfeldt and coworkers at Hannover (Scheme 5). Commer-cially available hecogenin acetate was trans-formed [10] to **5** and dimerized to symmetrical diketone **6**. The two carbonyls could be sta-tistically differentiated by controlled reduction or trapping of the dienolate to yield hydroxy ketone **7**. Compounds **6** and **7** were evaluated at the NCI. The symmetrical **6** affects 32 of the 58 cell lines tested, while a saturated ver-sion lacking the D ring alkene shows weak activity. Meanwhile, unsymmetrical **7** is signi-ficantly cytotoxic against all 58 cell lines, being approximately 4,000 times weaker than cephalostatin 1.

The Winterfeldt group has also developed [11] an unsymmetrical dimerization procedure (Scheme 6). The two reacting partners are vinyl azide **8** and enamino ketone **9**. Thermo-lysis of **8** generates an azirine *in situ*, which

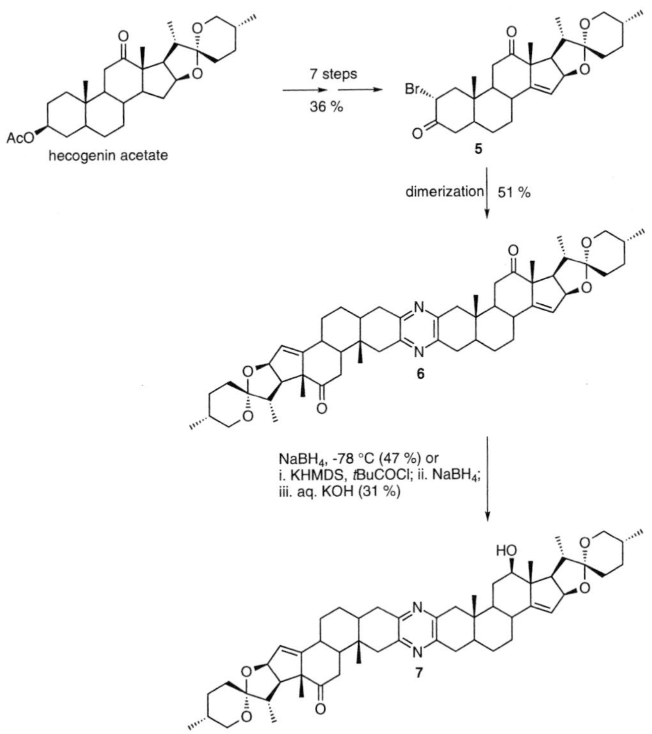
hecogenin acetate
AcO
7 steps
36 %
Br
5
dimerization
51 %
6
NaBH4, -78 °C (47 %) or
i. KHMDS, tBuCOCl; ii. NaBH4;
iii. aq. KOH (31 %)
HO
7

Scheme 5.
Desymmetrization of a homodimer by Winterfeldt et al. [9].

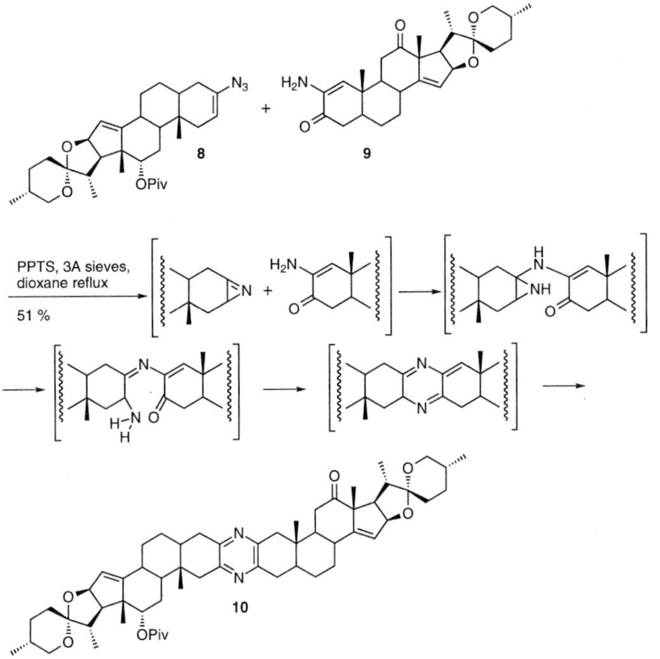
N3
8
OPiv
H2N
9
PPTS, 3A sieves,
dioxane reflux
51 %
10

Scheme 6.
Winterfeldt et al.'s unsymmetrical dimerization.

reacts by the likely mechanism shown to give dimer **10**. The enamino ketone is itself stable and not prone to self-dimerization.

So far, all the publications related to total synthesis have been from Fuchs' group, who first reported [12] a symmetrical dimer corresponding to the "right half" of cephalostatin 1 but lacking the D ring alkene and C17 alcohol. The dimer can be considered an analog of cephalostatin 12, whose existence had yet to be disclosed then.

The synthesis began with a derivative of hecogenin acetate, which already contains all the carbons needed. Classical Marker sapogenin spiroketal ring-opening gave **11** (Scheme 7) which was carried forward to **12**.

Acid-catalyzed cyclization of **12** yielded a mixture of 5/5 and 6/5 spiroketals epimeric to the natural product. In general, the ratio of spiroketal products with these intermediates is highly sensitive to the specific substitution pattern. The natural products also reflect this

fine balance, as there are four pairs of ritterazines differing only in configuration at the spiroketal carbon. In this case, the desired product **13** was obtained by the sequence of electrophilic bromoetherification and halogen removal. Dimerization afforded the cephalostatin 12 analog, whose biological activity was not revealed.

A series of communications from the Fuchs group climaxed [13] in the total synthesis of three of the alkaloids. The presence of the D ring alkene and C17 alcohol in the natural products necessitated a different route from the above. First, Marker sidechain degradation of hecogenin gave enone **14** (Scheme 8). The D ring functionality was introduced, and ring E constructed by intramolecular Wadsworth-Emmons reaction of **15**. The rest of the sidechain was reintroduced by methallyl stannane addition to **16**, yielding **17**.

The two epimers of **17** were separated, and individually processed to eventually yield

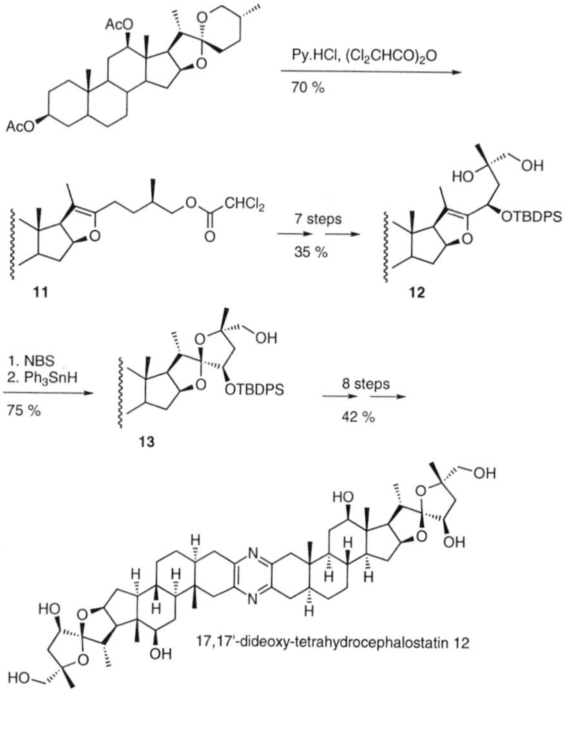

11

12

13

17,17'-dideoxy-tetrahydrocephalostatin 12

Scheme 7. Synthesis of a cephalostatin 12 analog by Fuchs et al. [12].

monomers **18** and **19**. In the grand finale, a 1 : 1 mixture of the *a*-azido ketones was reduced to give roughly the expected 1 : 2 : 1 mixture of three dimers, with the heterodimer predominating. Separation and deprotection afforded the natural products. Fuchs et al. suggest that similar statistical coupling occurs in nature, which implies that ritterazine K is also produced by *C. gilchristi*. Indeed, examination of residual material in the Pettit group identified a compound in microgram quantities with the same chromatographic profile.

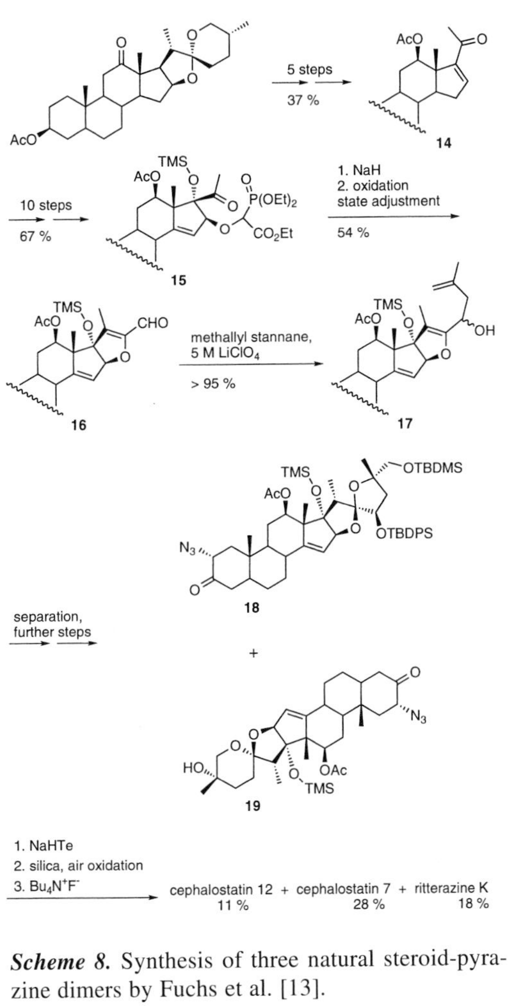

Scheme 8. Synthesis of three natural steroid-pyrazine dimers by Fuchs et al. [13].

This work represents a milestone in the synthesis of complex marine natural products. At the same time, it reflects on the state of organic synthesis, which some consider to be a fully mature science. While total synthesis is *effective*, in the sense that any stable compound can probably be made, it is seldom *efficient*. [14] The preparation of the two halves from hecogenin acetate required over 20 steps, the majority of which involve functional group interchange and protecting groups. This is typical of targets containing a multiplicity of functional groups at various oxidation levels. Clearly, there are worthy challenges left for the organic chemist of the next century!

The Fuchs group has also prepared [15] "dihydrocephalostatin 1". In this synthesis (Scheme 9), the angular methyl group of keto alcohol **20** (obtained from hecogenin acetate) was functionalized by Meystre's hypoiodite method and oxidized to yield **21**. Further reactions via intermediates **22** and **23** afforded **24**, which was coupled with *a*-amino oxime ether **25** (prepared from **18**) to give dihydrocephalostatin 1 after deprotection. This dihydro derivative had biological activity comparable to cephalostatin 1.

Biology

These alkaloids do not contain functional groups such as alkylation sites, Michael acceptors, intercalators, or redox-active quinones commonly associated with cytotoxicity, while their scarcity has hindered investigation of the mechanism of action. An early proposal, [16] taking into account the steroidal and dimeric nature of the alkaloids, was that these compounds span the lipid bilayer (cephalostatin 1 is 30 Å long) and perturb the eukaryotic cell membrane.

Fuchs has considered [7a] that the alkaloids are enzyme inhibitors forming hydrogen bonds with a specific target. More recently,

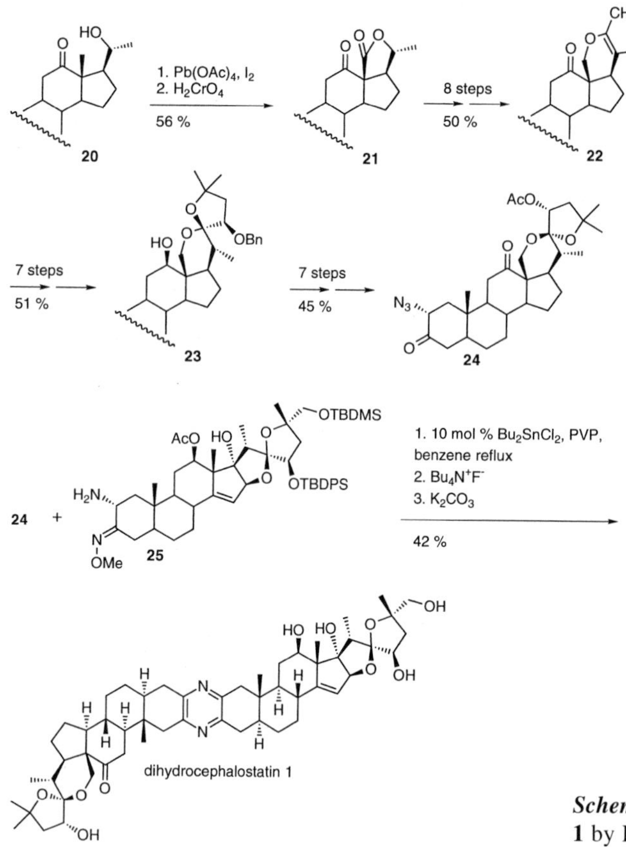

Scheme 9. Preparation of dihydrocephalostatin 1 by Fuchs et al. [8].

he has suggested [15a] that a process similar to Scheme 1 (vide infra) may occur *in vivo*, in which protonation or epoxidation of the D-ring alkene generates reactive electrophilic intermediates. This seems consistent with the fact that the natural products and biologically active synthetic analogs contain at least one alkene, while the tetrahydro-derivatives appear inactive. The advantage for the alkaloids to be dimeric is also unclear. An analogy may be drawn with the potent immunosuppressive agent FK506, where a synthetic dimer has unique properties. [17]

Steroids have important roles in providing membrane rigidity, as components of lipoproteins, and as ligands which dimerize the nuclear hormone receptor superfamily, result-

ing in gene transcription or repression. The effects of steroids can also be indirect, for example, in triggering apoptosis (programmed cell death), while the antiinflammatory glucocorticoids have recently been shown to interfere with NF-κB signaling besides binding their hormone response element. [18]

The determination of the precise biological target of these alkaloids requires larger amounts of material. Thus, it is particularly exciting that Winterfeldt's synthetic dimer **7**, prepared in 10 steps from hecogenin acetate, displays high activity. According to Pettit, [19] cephalostatins 1 and 7 are undergoing preclinical development.

Random screening continues to be an important source of leads for drug discovery.

The cephalostatins and ritterazines highlight the power of screening natural product extracts over currently fashionable synthetic combinatorial libraries in terms of structural novelty and complexity. Today, libraries are constructed by short sequences of synthetic transformations, whereas many natural products are the result of much longer and more creative pathways.

Finally, these alkaloids illustrate the ability of nature to assemble unusual arrays of functional groups that are beyond our imagination. Before the isolation of the natural products, methodology for the preparation of unsymmetrical pyrazines was unknown. The synthetic chemists have risen to the occasion, and three separate protocols are now available.

Note added in proof

The Fusetani group has reported [1] the structures of 13 new ritterazines. Some chemical transformations of ritterazine B and their effect on antitumor activity are also described. The Fuchs group has published [2] a full paper on the total synthesis of cephalostatin 1. They have also prepared two 'hybrid' dimers. The dimer containing the ritterazine G "right half" coupled to the cephalostatin 1 "right half" has similar activity to cephalostatin 1. A review has appeared [3] on steroid dimers and oligomers.

[1] S. Fukuzawa, S. Matsunaga, N. Fusetani, *J. Org. Chem.* **1997**, *62*, 4484–4491.
[2] T. G. LaCour, C. Guo, S. Bhandaru, M. R. Boyd, P. L. Fuchs, *J. Am. Chem. Soc.* **1998**, *120*, 692–707.
[3] Y. Li, J. R. Dias, *Chem. Rev.* **1997**, *97*, 283–304.

References

[1] For a more comprehensive review covering the work up to 1995, see: A Ganesan in *Studies in Natural Products Chemistry*, Vol. 18, *Stereoselective Synthesis* (Part K) (Ed.: Atta-ur-Rahman), Elsevier, Amsterdam, **1996**, p. 875–906.
[2] G. R. Pettit, M. Inoue, Y. Kamano, D. L. Herald, C. Arm, C. Dufresne, N. D. Christie, J. M. Schmidt, D. L. Doubek, T. S. Krupa, *J. Am. Chem. Soc.* **1988**, *110*, 2006–2007.
[3] F. Pietra, *A Secret World: Natural Products of Marine Life*, Birkhauser, Basel, **1990**, p. 149.
[4] For the most recent reference, see: G. R. Pettit, J.-P. Xu, J. M. Schmidt, *Bioorg. Med. Chem. Lett.* **1995**, *5*, 2027–2032.
[5] S. Fukuzawa, S. Matsunaga, N. Fusetani, *J. Org. Chem.* **1994**, *59*, 6164–6166.
[6] For the most recent reference, see: S. Fukuzawa, S. Matsunaga, N. Fusetani, *Tetrahedron* **1995**, *51*, 6707–6716.
[7] a) Y. Pan, R. L. Merriman, L. R. Tanzer, P. L. Fuchs, *Bioorg. Med. Chem. Lett.* **1992**, *2*, 967–972; b) C. H. Heathcock, S. C. Smith, *J. Org. Chem.* **1994**, *59*, 6828–6839.
[8] C. Guo, S. Bhandaru, P. L. Fuchs, M. R. Boyd, *J. Am. Chem. Soc.* **1996**, *118*, 10672–10673.
[9] A. Kramer, U. Ullmann, E. Winterfeldt, *J. Chem. Soc. Perkin Trans. I* **1993**, 2865–2867.
[10] R. Jautelat, E. Winterfeldt, A. Müllerfahrnow, *J. Praktische Chem.* **1996**, *338*, 695–701.
[11] M. Drögemüller, R. Jautelat, E. Winterfeldt, *Angew. Chem. Int. Ed. Engl.* **1996**, *35*, 1572–1574.
[12] J. U. Jeong, P. L. Fuchs, *J. Am. Chem. Soc.* **1994**, *116*, 773–774.
[13] J. U. Jeong, S. C. Sutton, S. Kim, P. L. Fuchs, *J. Am. Chem. Soc.* **1995**, *117*, 10157–10158.
[14] This distinction has been elegantly stated by C. H. Heathcock: *Angew. Chem. Int. Ed. Engl.* **1992**, *31*, 665–681; *Proc. Natl. Acad. Sci. USA* **1996**, *93*, 14323–14327.
[15] a) S. Bhandaru, P. L. Fuchs, *Tetrahedron Lett.* **1995**, *36*, 8347–8350; b) *ibid.* **1995**, *36*, 8351–8354; c) reference 8.
[16] A. Ganesan, C. H. Heathcock, *Chemtracts-Org. Chem.* **1988**, *1*, 311–312.

[17] D. J. Austin, G. R. Crabtree, S. L. Schreiber, *Chem. Biol.* **1994**, *1*, 131–136.

[18] B. van der Burg, J. Liden, S. Okret, F. Delaunay, S. Wissink, P. van der Saag, J.-A. Gustafsson, *Trends Endocrinol. Metabol.* **1997**, *8*, 152–157.

[19] G. R. Pettit, *J. Nat. Prod.* **1996**, *59*, 812–821.

New, efficient routes to cyclic enediynes

B. König

The special fascination of strained cyclic enediynes lies in their intrinsic tendency to undergo thermal cyclisation. This aspect has been stimulating the research activities on enediynes since their beginnings more than 20 years ago, [1] and it is this reactivity of the enediyne structure that brings about the pharmacological effectiveness of natural products, like Dynemicin A. [2]

The synthesis of an enediyne parent system calls for mild conditions with, at the same time, strong driving forces of the ring closure reaction. Various approaches have been taken to generate strained cyclic enediynes. Scheme 1 summarizes the major ones: Most often, an acetylide carbonyl addition is used as the ring closure reaction (Scheme 1, routes a and b). [3] Danishefsky, however, efficiently obtained highly substituted, 10-membered enediyne cycles by two-fold palladium-catalyzed coupling of iodoacetylene with *cis*-1,2-bis(trimethyltinethylene) (route c). [4] Substitution reactions can yield 10-membered enediyne cycles as well, though only after the ring strain has been reduced (route d). [5] In difference to a ring closure, Nicolaou employed a ring contraction of cyclic sulfones to synthesize strained cyclic enediynes (route e). [6]

Jones, Huber and Mathews were now able to demonstrate that strained cyclic enediynes are accessible via carbenoid intermediates,

too. [7] Their method is based on a sequence of tandem carbenoid coupling-elimination of two propynylic bromide units, a strategy pub-

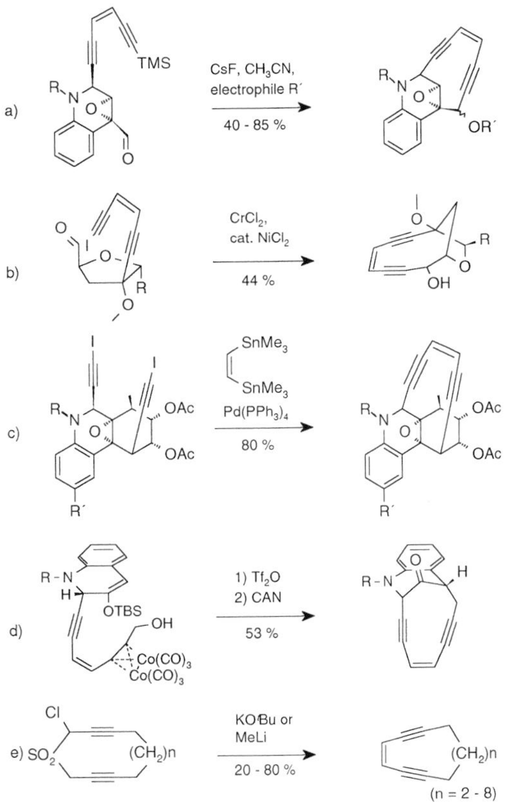

Scheme 1.

lished by the same group [8]: Here, the synthesis of the acyclic enediyne bis(trimethylsilyl)hexadiyne (**3**) was achieved by reaction of trimethylsilylpropynylic bromide (**1**) with lithium hexamethyldisilazide (LiHMDS) in THF/HMPT (HMPT = hexamethylphosphoric triamide) at −85 °C. A *cis/trans*-selectivity of about 2:1 was attained in good yields. Probably, a propynylic carbenoid is formed which can insert intermolecularly into the C–H bond of another propargyl bromide. The E2-elimination of HBr from the intermediate bromide (**2**) then leads to the product. As this method proved to be broadly applicable for substituted propynylic substrates, Jones et al. even had it patented. [9] Recently, still higher substituted acyclic enediynes like **4** and **5** could be obtained with this approach. [10]

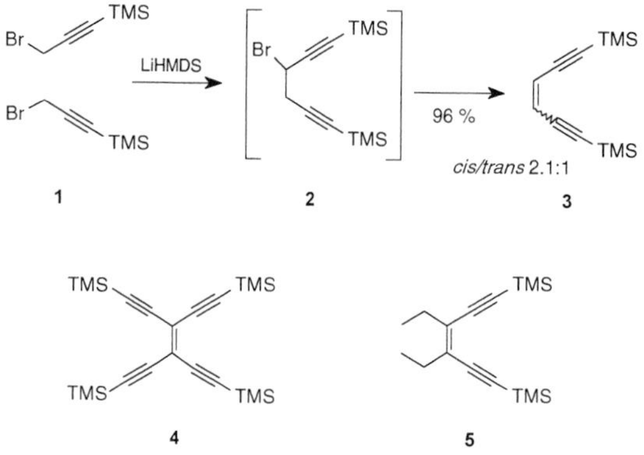

diynes linked to DNA intercalators, like **9**, have been synthesized already. [12]

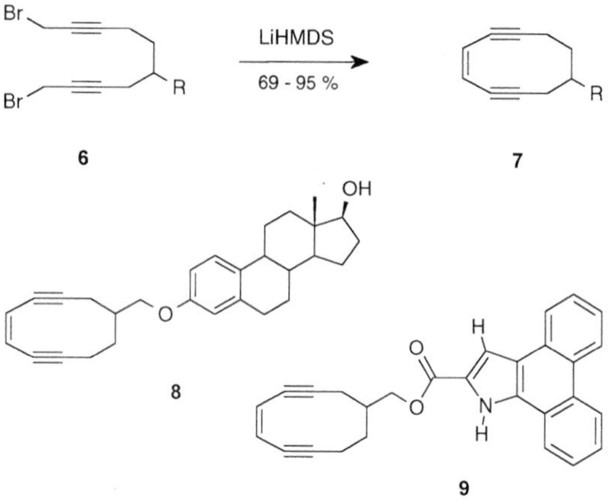

Scheme 3.

Scheme 2.

The intramolecular variant of this synthesis renders cyclic enediynes. Ten-membered functionalized enediyne cycles (**7**) were generated in yields of up to 95% using slightly altered reaction conditions. Since the starting materials are easily accessible, a synthesis of enediynes in multigramm scale can be achieved. Thus, it should be possible to use those functionalized enediyne components to build libraries of enediyne diversomers, by the approach of Nicolaou and others. [11] Enediyne-steroide conjugates, like **8**, and ene-

While most attention has been directed towards the preparation of carbocyclic enediynes, ten-membered heterocyclic enediynes

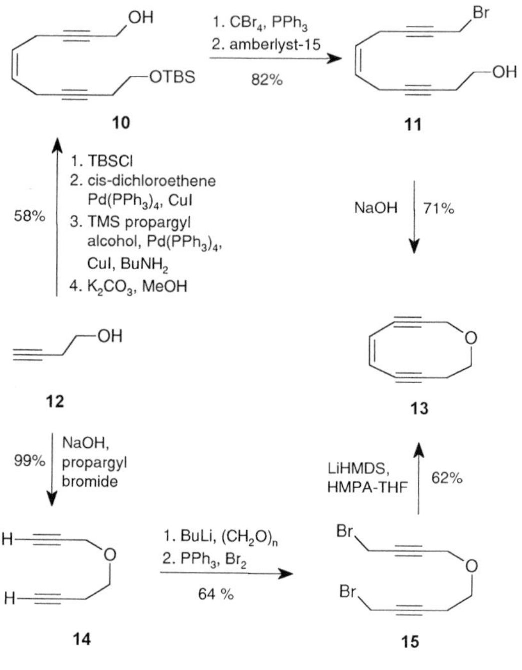

Scheme 4.

(**13**) have been prepared, too. [13] Two independent syntheses were investigated, either involving a Williamson type ring closure or the intramolecular carbenoid route. Overall, the carbenoid route proved far superior.

Starting from the propynylic bromide **16** Hopf, Werner, et al. successfully synthesized an only 8-membered cyclic aryl-1,2-diyne **17**. [14] The remarkably stable molecule was obtained by cyclisation with Bu$_3$SnSiMe$_3$/CsF [15] in 60 % yield.

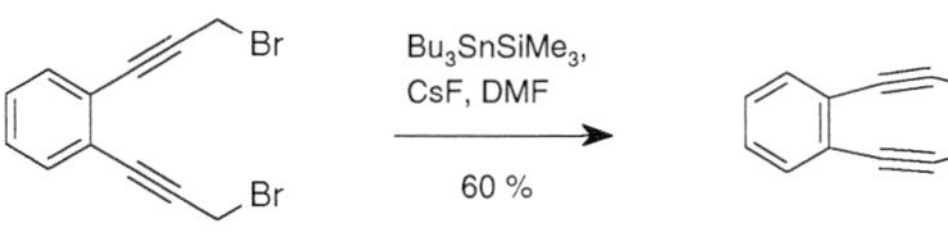

Scheme 5.

With the mechanistically interesting new approaches described here the spectrum of methods for the synthesis of strained cyclic enediynes could be enlarged markedly. It is to be expected that the search for simple pharmacologically active enediyne derivatives will benefit accordingly.

References

[1] a) R. R. Jones, R. G. Bergman, *J. Am. Chem. Soc.* **1972**, *94*, 660–661; b) R. G. Bergman, *Acc. Chem. Res.* **1973**, *6*, 25–31.

[2] Recent reviews on the synthesis and properties of enediyne-prodrugs: a) M. E. Maier, *Synlett*, **1995**, 13–26; b) K. C. Nicolaou, W.-M. Dai *Angew. Chem.* **1991**, *103*, 1453–1481; *Angew. Chem. Int. Ed. Engl.* **1991**, *30*, 1387–1415; c) enediynes as building blocks for new carbon allotropes: F. Diederich, *Nature* **1994**, *369*, 199–207; d) R. Gleiter, D. Katz, *Angew. Chem.* **1993**, *105*, 884–887; *Angew. Chem. Int. Ed. Engl.* **1993**, *32*, 842–845.

[3] a) P. A. Wender, S. Beckham, D. L. Mohler, *Tetrahedron Lett.* **1995**, *36*, 209–212; b) T. Brandstetter, M. E. Maier, *Tetrahedron* **1994**, *50*, 1435–1448; further examples using this ring-closure method: H. Mastalerz, T. W. Doyle, J. F. Kadow, D. M. Vyas, *Tetrahedron Lett.* **1996**, *37*, 8683–8686; H. Mastalerz, T. Doyle, J. Kadow, K. Leung, D. Vyas, *Tetrahedron Lett.* **1995**, *36*, 4927–4930; A. G. Myers, M. Hammond, Y. Wu, J.-N. Xiang, P. M. Harrington, E. Y. Kuo, *J. Am. Chem. Soc.* **1996**, *118*, 10006–10007; M. F. Brana, M. Morán, M. J. P. de Vega, I. Pita-Romero, *J. Org. Chem.* **1996**, *61*, 1369–1374; Y.-F. Lu, C. W. Harwig, A. G. Fallis, *Can. J. Chem.* **1995**, *73*, 2253–2262; L. Banfi, G. Guanti, *Angew. Chem.* **1995**, *107*, 2613–2615; *Angew. Chem. Int. Ed. Engl.* **1995**, *34*, 2393–2395.

[4] M. D. Shair, T. Yoon, S. J. Danishefsky, *J. Org. Chem.* **1994**, *59*, 3755–3757; synthesis of acyclic enediynes from alkenylbis(phenyliodonium)triflates: J. H. Ryan, P. J. Stang, *J. Org. Chem.* **1996**, *61*, 6162–6165.

[5] P. Magnus, *Tetrahedron*, **1994**, *50*, 1397–1418; T. Takahashi, H. Tanaka, A. Matsuda, H. Yamada, T. Matsumoto, Y. Sugiura, *Tetrahedron Lett.* **1996**, *37*, 2433–2436; synthesis of an eleven-membered cyclic enediyne *via* S$_N$-ring-closure: A. Basak, U. K. Khamrai, U. Mallik, *Chem. Commun.* **1996**, 749–750; synthesis of macrocyclic enediynes: B. König, W. Pitsch, I. Thondorf, *J. Org. Chem.* **1996**, *61*, 4258–4261; B. König, W. Pitsch, I. Dix, P. G. Jones, *Synthesis*, **1996**, 446–448.

[6] K. C. Nicolaou, G. Zuccarello, Y. Ogawa, E. J. Schweiger, T. Kumazawa, *J. Am. Chem. Soc.* **1988**, *110*, 4866–4868; synthesis of cyclic dihydroxy enediynes *via* Sm- or Ti-mediated pinacol coupling: K. C. Nicolaou, E. J. Sorensen, R. Discordia, C.-K. Hwang, R. E. Minto, K. N. Baharucha, R. G. Bergman, *Angew. Chem.* **1992**, *104*, 1094–1096; *Angew. Chem. Int. Ed. Engl.* **1992**, *31*, 1044–1046.

[7] G. B. Jones, R. S. Huber, J. E. Mathews, *J. Chem. Soc., Chem. Commun.* **1995**, 1791–1792.

[8] a) G. B. Jones, R. S. Huber, *Tetrahedron Lett.* **1994**, *35*, 2655–2658; b) the synthesis of **3** from **1** was reproduced in our laboratory without difficulties. However, the yield and stereoselectivity were slightly lower than reported.

[9] *A New Route to Enediynes* G. B. Jones, R. S. Huber, US-A 5436361, **1995**.

[10] G. B. Jones, Clemson University, Clemson, SC (USA), personal communication; for an alternative route to compound **4**, see: R. R. Tykwinski, F. Diederich, V. Gramlich, P. Seiler, *Helv. Chim. Acta* **1996**, *79*, 634–645.

[11] a) K. C. Nicolaou, A. L. Smith, E. W. Yue, *Proc. Natl. Acad. Sci. USA* **1993**, *90*, 5881–5892; b) K. Toshima, K. Otha, A. Ohahsi, T. Nakamura, M. Nakata, S. Matsumura, *J. Chem. Soc., Chem. Commun.* **1993**, 1525–1527; c) M. Tokuda, K. Fujiwara, T. Gomibuchi, M. Hirama, M. Uesugi, Y. Sugiura, *Tetrahedron Lett.* **1993**, *34*, 669–672; d) M. F. Semmelhack, J. J. Gallagher, W. Ding, G. Krishnamurthy, R. Babine, G. A. Ellestad, *J. Org. Chem.* **1994**, *59*, 4357–4359; e) D. L. Boger, J. Zhou, *J. Org. Chem.* **1993**, *58*, 3018–3024.

[12] G. B. Jones, R. S. Huber, J. E. Mathews, A. Li, *Tetrahedron Lett.* **1996**, *37*, 3643–3646; G. B. Jones, J. E. Mathews, *Tetrahedron* **1997**, *53*, 14599–14614.

[13] G. B. Jones, M. W. Kilgore, R. S. Pollenz, A. Li, J. E. Mathews, J. M. Wright, R. S. Huber, P. L. Tate, T. L. Price, R. P. Sticca, *Bio. Med. Chem. Lett.* **1996**, *6*, 1971–1976.

[14] a) H. Hopf, C. Werner, P. Bubenitschek, P. G. Jones, *Angew. Chem.* **1995**, *107*, 2592–2594; *Angew. Chem. Int. Ed. Engl.* **1995**, *34*, 2367–2368; for the first synthesis of a cyclooctaenediyne, stabilized as the alkynecobaltcarbonyl complex, see: b) G. G. Melikyan, M. A. Khan, K. M. Nicholas, *Organometallics* **1995**, *14*, 2170–2172.

[15] H. Sato, N. Isono, K. Okamura, T. Date, M. Mori, *Tetrahedron Lett.* **1994**, *35*, 2035–2038.

Conocurvone – Prototype of a New Class of Anti-HIV Active Compounds?

Hartmut Laatsch

Workers at the National Cancer Institute recently achieved a remarkable success in their search for active plant metabolites. During routine screening, extracts of a shrub indigenous to Australia were noted for their exceptionally high, and, more importantly, selective anti-HIV activity in a variety of cellular *in vitro* tests. The substance responsible for this effect was the naphthoquinone conocurvone **1a**, which was isolated after an elaborate purification procedure in a yield of 22 mg per kg plant material (*Conospermum sp., Proteaceae*). [1, 2]

As fluorescence tests with BCECF (2′,7′-bis-2(2-carboxyethyl)-4(5)-carboxyfluorescein), the XTT tetrazolium test, and a test with the intercalating dye DAPI (4′,6-diamidino-2-phenylindole dihydrochloride) showed, the presence of conocurvone **1a** at a concentration of $EC_{50} \leq 0.02$ μM completely averted the death of HIV l-infected human lymphoblastoid cells (CEM-SS). Measurements of viral reverse transcriptase, the viral P24 antigen, and the syncytium forming units (SFU) revealed that virus replication came to a standstill at the same time. Since

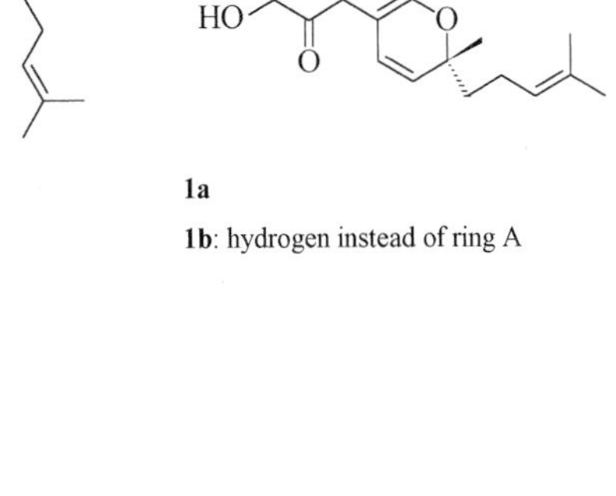

1a

1b: hydrogen instead of ring A

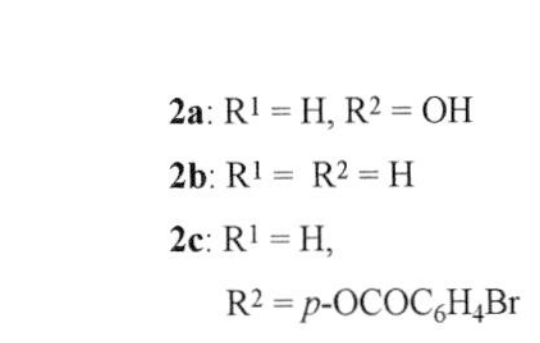

2a: $R^1 = H$, $R^2 = OH$

2b: $R^1 = R^2 = H$

2c: $R^1 = H$,

$R^2 = p\text{-OCOC}_6\text{H}_4\text{Br}$

2d: $R^1 = R^2 = SC_6H_5$

conocurvone **1a** is cytotoxic and inhibits growth only at or above a concentration of 50 μM, the therapeutic index has the unusually high value for a virostatic of 2500. Whether this value also holds for other viruses, or is limited to HIV-l viruses, was not reported. [3]

Conocurvone **1a** is a deoxy-trimer of teretifolion B (**2a**), a compound that has been known for longer and was first isolated from *Conospermum teretifolium*. The fast atom bombardment (FAB) mass spectrum of the trimeric quinone revealed a molecular ion corresponding to the formula $C_{60}H_{56}O_{11}$, but the structure of **1a** was in the end only fully elucidated by synthesis. The reason for this was that atropoisomeric equilibria were formed and led to more or less complex ^{1}H-NMR spectra that varied with solvent and with temperature. The compound therefore appeared to be a complex mixture. However, the synthetic product was identical to the natural material, even in its chiroptical parameters, thus confirming the structure and also the low rotation barrier around the quinone-quinone axis, a property which has also been found for other quinonoid–quinonoid-coupled oligomers.

Quinones constitute an extremely important group of natural products. They are one of the oldest known classes of compounds, have achieved importance as dyestuffs, and in addition possess many different biological activities: many quinones are active as anti-

biotics, are cytotoxic [like daunomycin (**3**)], or are (weakly) antiviral. Even simple compounds such as plumbagin (**4a**) prove to be highly effective and selective as enzyme inhibitors. For other quinones the activity derives from interaction with physiological redox systems, or – at least for unsubstituted quinonoid double bonds – from the reaction with nucleophiles. [4]

In the case of conocurvone (**1a**) a physiological effect has once again been found for an oligomeric quinone, which is completely absent for the monomer **2a** or related compounds. [5] This must therefore be a new property that arises from the *oligomerization*, and should thus stimulate renewed interest in oligonaphthoquinone synthesis. A brief outline of the current state of knowledge about this class of substances is hence justified.

In all but a few cases (one of which is **1a**), the over seventy naturally occurring dimeric, trimeric, *cyclo*-trimeric, and higher oligomers are all constructed of identical units derived from juglone (**4b**, 5-hydroxynaphthoquinone). These units are linked together in various ways, symmetrically or asymmetrically with respect to the C–C framework. With the exception of crisamicin A, all these compounds have an oxygen atom in the *ortho* position to the linkage site.

The reason for this variety lies in the mechanism of formation of dimeric quinones, which may be regarded as a phenol oxidation. The synthesis starts from substituted naphthols, such as **5b**, and leads, depending on the initial substrate and the enzymes of the organism concerned, not only to the monomeric quinone but via binaphthols [e. g. vioxanthin (**5c**)] also to dimeric [e. g. xanthomegnin (**5d**)] or oligomeric quinones. It also yields the polymeric, black allomelanin, which gives ebony, for instance, its characteristic color. This pathway is confirmed by the presence – often in the same organism – of dimers with lower oxidation states, which are formal intermediates in diquinone biosynthesis:

3

4a: R^1 = Me, R^2 = OH

4b: R^1 = H, R^2 = OH

4c: R^1 = OH, R^2 = H

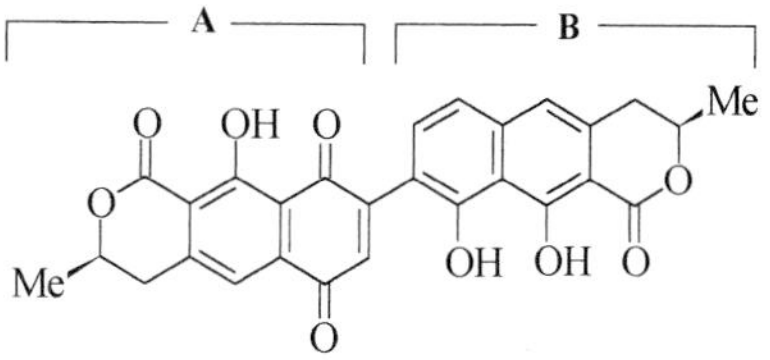

5a = A-B
5b: only component B
5c: B instead of A in 5a
5d: A instead of B in 5a

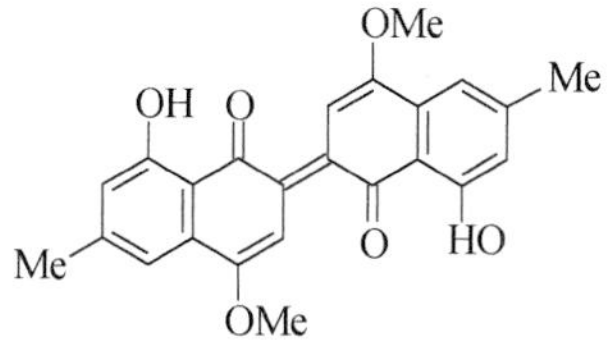

6

examples are the monoquinone viomellein **5a** and the binaphthylidenedione **6** (diosindigo A).

In nature, the oxidative dimerization of phenols is controlled by enzymes, as is demonstrated by the axial chirality of the 6,8'-coupled juglone derivative isodiospyrin. In synthesis, however, phenol oxidation only proceeds in high yields when the enzymatic reaction control is replaced by substituent control, that is, if all but one of the positions with high spin density in the radical (*ortho-* and *para* positions) are blocked.

The synthetic usefulness of this principle is well documented by numerous examples, even under biomimetic conditions. [6] In this way, using a similar synthetic sequence by co-oxidation of **7a** and **7b**, we have obtained not only the dimers, but also quinones **8a** and **8b**. [7] These are related to conocurvone **1a**, but their antiviral properties have not yet been studied.

Unlike in the case of phenols, direct oxidative dimerization of quinones only takes place under drastic conditions and requires a hydroxy or amino group on the quinonoid double bond, as in Lawson **4c**, the dye in Arabian henna. Since the work of Pummerer (1937), however, it has been known that monomeric 1,4-quinones can be oligomerized much more easily under acidic or basic conditions. [8] In pyridine/ethanol, or by warming in glacial acetic acid, we have converted naphthoquinone, 1,4-anthraquinone, and nu-

merous derivatives – though not benzoquinones – smoothly into dimers and *cyclo*-trimers. In this highly regioselective so-called phenol/quinone addition, juglone (**4b**) and several of its derivatives afford exclusively the symmetrical 3,3'-linked dimers, several of which were already known as natural products. [9] It was only recently discovered that *o*-quinones may also be dimerized by this principle. [10]

According to Brockmann, this process is an autocatalytic one, in which traces of hydroquinone, which are always present, undergo a Michael addition to the excess of quinone to form a biaryltetrol (**12**), [11] an intermediate which also occurs during phenol oxidation. Dehydrogenation of this intermediate by the monomeric quinone yields the biaryldiquinone and further hydroquinone, until all the monomeric quinone has been transformed (3 **9** + **10c** → **12c** + **10b**). When air is excluded, conversions of naphthoquinone reach

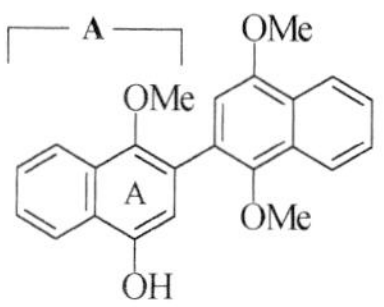

7a
7b: only component A

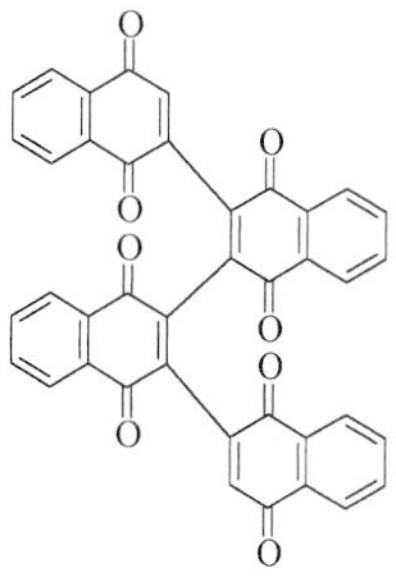

8a
8b: trimer instead of tetramer

90 %; when air is present, or when nitrobenzene is used as solvent, the yield of dimer can be increased even more through reoxidation of the hydroquinone.

Under suitable reaction conditions, the dimer that is initially formed can react further to yield trimers of type **8b**. It has not yet been possible to isolate these compounds, since they are always converted into the trimeric cyclic quinones **13a**. Indeed, **8b** is also converted into **13a** by phenol oxidation under weakly basic conditions. The stability of conocurvone **1a** is therefore due to the presence of hydroxy groups at position 2 of the monomer, which prevent cyclization. This cyclization via a symmetrical dimer of type **12b** explains why none of the known hydroxylated *cyclo*-trimers (e. g. *cyclo*-trijuglone **13b**) display C_3-symmetry with respect to the *peri*-hydroxy groups.

The synthesis of conocurvone (**1a**) also makes use of the well established principle of phenol/quinone addition, even though the term was inspired by the reaction of quinones with hydroxy-2*H*-1-benzopyran-2-ones, a reaction that was discovered later but proceeds in a similar manner. Brief warming of two equivalents of teretifolion B (**2a**) with naphthoquinone in glacial acetic acid afforded the predicted **1b** in 9 % yield. For the synthesis of conocurvone (**1a**), 2-deoxyteretifolione (**2b**) was warmed with two equivalents of **2a** in pyridine, analogously to the synthesis of 2,2′-binaphthyldiquinone. All the spectroscopic and pharmacological properties of the synthetic material agreed with those of the natural product.

The main difficulty in the synthesis of **1a** lay in the deoxygenation of **2a** to **2b**. This was eventually carried out by transforming the *p*-bromobenzoate **2c** to **2d** with thiophenol, with subsequent catalytic reduction with Raney nickel.

Nothing is yet known about the site of action or structure-activity relationship for derivatives of **1a** (as, for example, **1b**), and any discussion is therefore very speculative. However, it is conceivable that **1a** assumes a helical conformation which winds into the groove of the DNA strand. Similar conformations are also expected for **8a**, and in particular for **8b** and higher oligomers. As we have shown, partial reduction of these compounds leads to deep blue, intramolecular quinhydro-

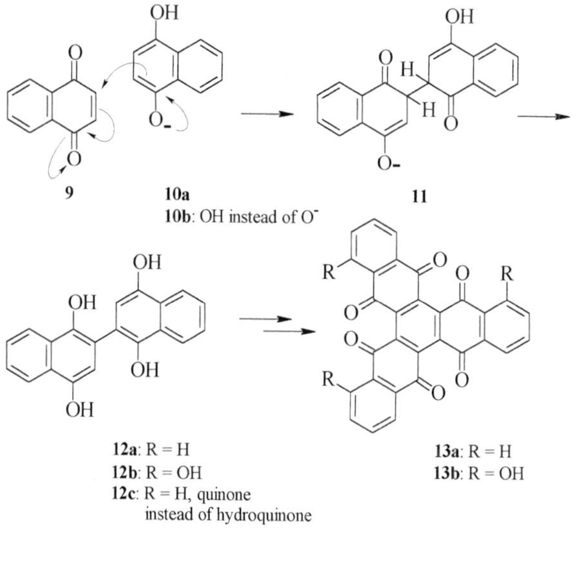

9   **10a**
     10b: OH instead of O⁻

11

12a: R = H
12b: R = OH
12c: R = H, quinone
        instead of hydroquinone

13a: R = H
13b: R = OH

nes, which are also stable in solution and can be shown by molecular modelling to be stabilized in a helical conformation. For **1a**, additional interactions may be expected between the quinonoid hydroxy groups and peptides, like those that play a role in coloring hair and skin with henna (**4c**). It will be fascinating to see whether these hypotheses are confirmed.

References

[1] L. A. Decosterd, I. C. Parsons, K. R. Gustafson, J. H. Cardellina II, J. B. McMahon, G. M. Cragg, Y. Murata, L. K. Pannell, J. R. Steiner, J. Clardy, M. R. Boyd, *J. Am. Chem. Soc.* **1993**, *115*, 6673–6679.

[2] Y. Kurimura, Bio Ind. **1996**, *13*, 71–7.

[3] M. R. Boyd, J. H. Cardelina II, K. R. Gustafson, L. A. Decosterd, I. Parsons, L. Pannel, J. B. Mcmahon, G. M. Cragg, PCT Int. Appl., WO 9417055 A1 940804 (1994).

[4] R. H. Thomson, *Naturally Occurring Quinones,* Vol. 2, Academic Press, London, **1971**; *Naturally Occurring Quinones,* Vol. 3, Chapman and Hall, London, **1987**.

[5] J.-R. Dai, L. A. Decosterd, K. R. Gustafson, J. H. Cardellina II, G. N. Gray, M. R. Boyd, *J. Nat. Prod.* **1994**, *57*, 1511–1516.

[6] H. Laatsch, *Liebigs Ann. Chem.* **1987**, 297–304, and preceeding publications on the same theme.

[7] H. Laatsch, *Liebigs Ann. Chem.* **1990**, 433–440.

[8] E. Rosenhauer, F. Braun, R. Pummerer, G. Riegelbauer, *Ber. Dtsch. Chem. Ges.* **1937**, *70*, 2281–2295.

[9] H. Brockmann, H. Laatsch, *Liebigs Ann. Chem.* **1983**, 433–447.

[10] K. Krohn, K. Khanbabaee, *Liebigs Ann. Chem.* **1993**, 905–909.

[11] H. Brockman, *Liebigs Ann. Chem.* **1988**, 1–7.

After proofreading of this contribution, J. Y. Yin and L. S. Liebeskind submitted a paper on a novel access to trisquinones (*Angew. Chem.* **1998**, in press).

Progress in Oligosaccharide Synthesis through a New Orthogonal Glycosylation Strategy

Hans Paulsen

Glycoconjugates have numerous important biological functions. [1, 2] For a better understanding of these functions, defined natural and modified glycoconjugates are needed. To this end, intensive efforts in the last few years have aimed at the synthesis of glycoconjugates, resulting in remarkable progress in oligosaccharide synthesis. New glycosylation methods have been developed with efficient leaving groups, which lead to good yields and high steroselectivity under mild conditions. The halide, thioglycoside, and trichloroacetimidate methods, [3–5] in modified forms, have proved to be especially effective. Chemical and enzymatic reactions can also be combined, as long as the glycosyl transferases and the activated nucleotide sugar are available. [6]

The most effective methods for the construction of larger oligosaccharides is block synthesis, in which blocks of di- or trisaccharides are coupled. The configuration of the anomeric centers is already fixed in the non-reducing block sequences. [7] One drawback here is that the anomeric center at the reducing unit of a block must be reactivated for subsequent coupling. Activation of the anomeric center should not require many steps or, better yet, should be direct. Thus 2-(trimethylsilyl)-ethylglycosides, for example, have been prepared from blocks, [8] which are readily

deprotected and converted into activated trichloroacetimidates. [5, 9] Glycal-type structures can also be activated by expoxidation, making renewed coupling possible. [10] However, use of stable thioglycosides at the anomeric center of the block is best, since a wide variety of structures and promoters is possible.

For the construction of an oligosaccharide building block a glycosylation reaction is also necessary. The synthesis of a thioglycoside-containing unit, for example, requires a different glycosylation procedure with a different leaving group for the donor, since the potentially activatable thioglycoside would be destroyed otherwise. Numerous strategies are available, which involve several or more additional reaction steps, depending on the blocking pattern. The coupling of donors activated by benzyl substituents with acceptors deactivated by acetyl groups has also been suggested. [11, 12] Ogawa et al. recently described an interesting and provocative strategy that starts with a set of two donors, which can be extended in an orthogonal glycosylation reaction. [13]

The approach was demonstrated with a model reaction sequence leadind to long-chain $\beta(1 \rightarrow 4)$-glycosidically linked 2-acetamido-2-deoxy-D-glycose units (Scheme 1). Thioglycoside **1** and glycosyl fluoride **4** serve

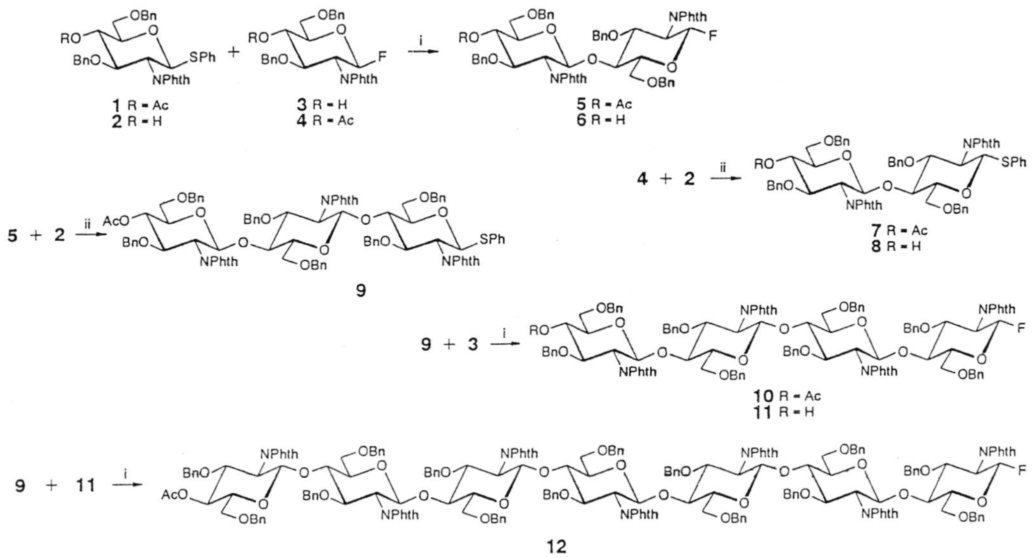

Scheme 1. i) NIS–TfOH (AgOTf); ii) Cp$_2$HfCl$_2$–AgClO$_4$.

as the two building blocks. Thioglycoside **1** can be coupled with the deactylated glycosyl fluoride **3** without affecting fluorine at the anomeric center, and the resulting disaccharide **5** with the activated anomeric center can be used directly for further glycoside synthesis. Inversely, the reaction of glycosyl fluoride **4** with the deacetylated thioglycoside **2** yields disaccharide **7** in which the thioglycoside is unchanged. This disaccharide can also be employed directly in further glycosidation without any additional steps. This flexible method thus enables glycosidation without destruction of the active anomeric center at the acceptor. This promising strategy can be applied to other systems.

The phenylthioglycosides employed are activated with *N*-iodosuccinimide trifluoromethanesulfonic acid (NIS-TfOH) (or NIS-AgOTf), and the phenythiogroup acts as a leaving group. In reactions with the glycosyl fluorides the promotor is the reagent Cp$_2$HfCl$_2$–AgClO$_4$. Thus larger oligosaccharide blocks can be constructed from the two monosaccharides **1** and **4** with these two rea-

gents. Compounds **1** and **4** contain an *N*-phthaloyl group at C2, which induces uniform *β*-glycosidic linkage.

The approach was demonstrated with the following synthesis. Coupling of **1** and **3** affords disaccharide **5** (85 %), whereas the reaction of donor **4** with **2** gives the disaccharide **7** (81 %). Disaccharide **5** is treated with **2** to furnish trisaccharide **9** (72 %). The subsequent coupling of **9** with **3** yields tetrasaccharide **10** (65 %). Deacetylation of **10** to **11** can be followed by block condensation of trisaccharide **9** with **11**, which yields heptasaccharide **12** (67 %). Phenylthioglycosides and glycosyl fluorides are used both as donors and as acceptors, and additional steps for activation of the anomeric center are not required in either case.

Kahne et al. [14] used a similar strategy in a very elegant synthesis of a trisaccharide in partically one step. The building blocks they used, the S-glycosides **13**, **14**, and **16**, have considerably different reactivities at the anomeric center (Scheme 2). Phenylthioglycoside **14** is definitely the least reactive,

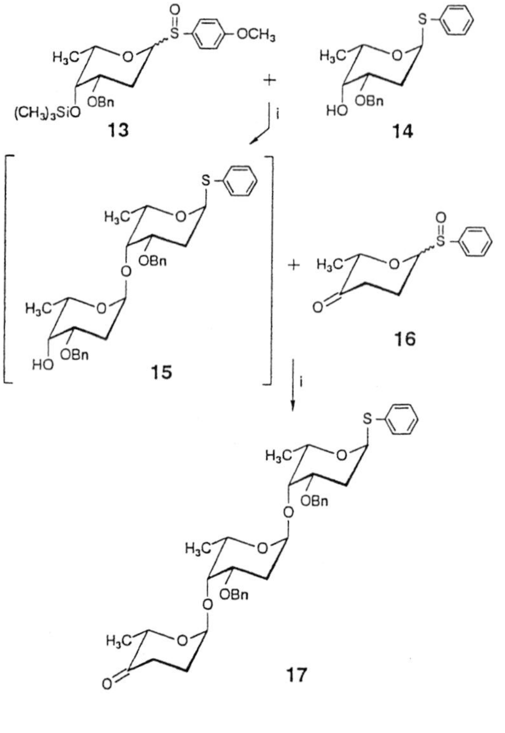

Scheme 2. i) TfOH, HC≡CCO$_2$CH$_3$, −78 °C.

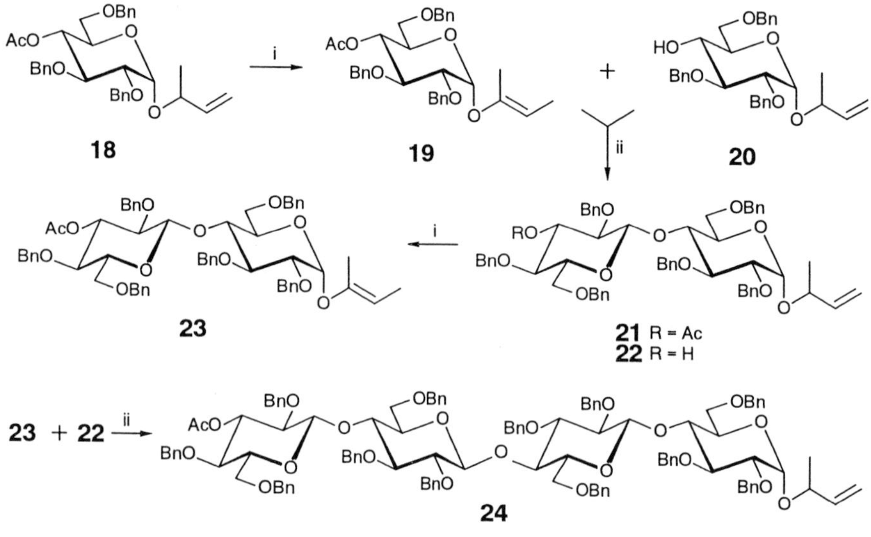

while phenyl sulfoxide **16** is considerably more reactive. The reactivity of **16** can be increased yet further by introduction of a methoxy substituent on the phenyl group, thus giving the highly reactive *p*-methoxy-phenylsulfoxide **13**.

A solution of building blocks **13**, **14**, and **16** in Et$_2$O/CH$_2$Cl$_2$ is simply mixed in the presence of TfOH and HC≡CCO$_2$CH$_3$ at −78 to −70 °C, and trisaccharide **17** is isolated in 25 % after 45 min. It is assumed that the modestly reactive **14** react first with the highly reactive **13** to give disaccharide **15**; the silyl ether is cleaved simultaneously, and **15** reacts with the moderately reactive **16** to provide **17**.

An interesting, new strategy which can also be used for an orthogonal oligosaccharide synthesis has been developed by Boons. [15, 16] It calls for the use of substituted type **18** allylglycosides with latent active group at the anomeric centre. Using (PhP$_3$)RhCl as the catalyst, compound **18** can be rearranged to the 2-isobutenyl glycoside **19**. Since the vinyl ether moiety is a good leaving group, compound **19** can serve as a glycosyl donor.

Scheme 3. i) (PhP$_3$)RhCl; ii) TMSOTf/CH$_3$CN.

The building block **19** can therefore be converted to the disaccharide **21** in the presence of TMSOTf by reaction with the acceptor **20**, which can also be obtained from compound **18** by deacetylation. This reaction sequence could be repeated analogously. Rearrangement of compound **21** in the presence of Wilkinson catalyst furnished the corresponding activated disaccharide donor **23**. This could then be coupled in like manner with acceptor **22**, which can be obtained by deacetylation of compound **21**, to give the tetrasaccharide **24**. The yields were over 80 % in all cases. In compound **24**, a rearrangement of the allylglycoside to the vinyl ether and hence to the glycosyl donor would be possible. However, compound **24** could also be converted at the nonreducing end into a glycosyl acceptor by cleavage of the acetyl group. Consequently, all the oligosaccharides could be prepared from one starting product **18**.

In principle, this strategy can also be extended to other glycosylation reactions. However, it must be assured that the anomeric centre can be readily activated, if possible in one step, to the glycosyl donor and that a

hydroxyl group can also be set free so that a glycosyl acceptor can be obtained from the same molecule as well. An example from the area of trichloracetimidate methodology is shown in Scheme 4. Compound **25** could be obtained by trichloracetimidate coupling according to Schmidt. [15, 17, 18] From compound **25**, the glycosyl donor **26** could be obtained after cleavage of the TBDMS group and treatment with trichloracetonitrile. Hydrolysis of the isopropylidene group of compound **25** gave the glycosyl acceptor **27**. Coupling of **26** with **27** furnished the hexasaccharide **28**, which could then be activated at the anomeric centre or just as easily deblocked selectively at the non-reducing end. Further coupling steps would then be possible. The compounds are intermediates of a Lewis[X] synthesis. [17]

It was also possible to reactivate the anomeric centre through a differentiation of activity at the anomeric centre during the synthesis of di- and oligosaccharides. A corresponding example is shown in the reaction sequence of Scheme 5. Pinto [19, 20] observed that selenoglycosides are more reactive and could there-

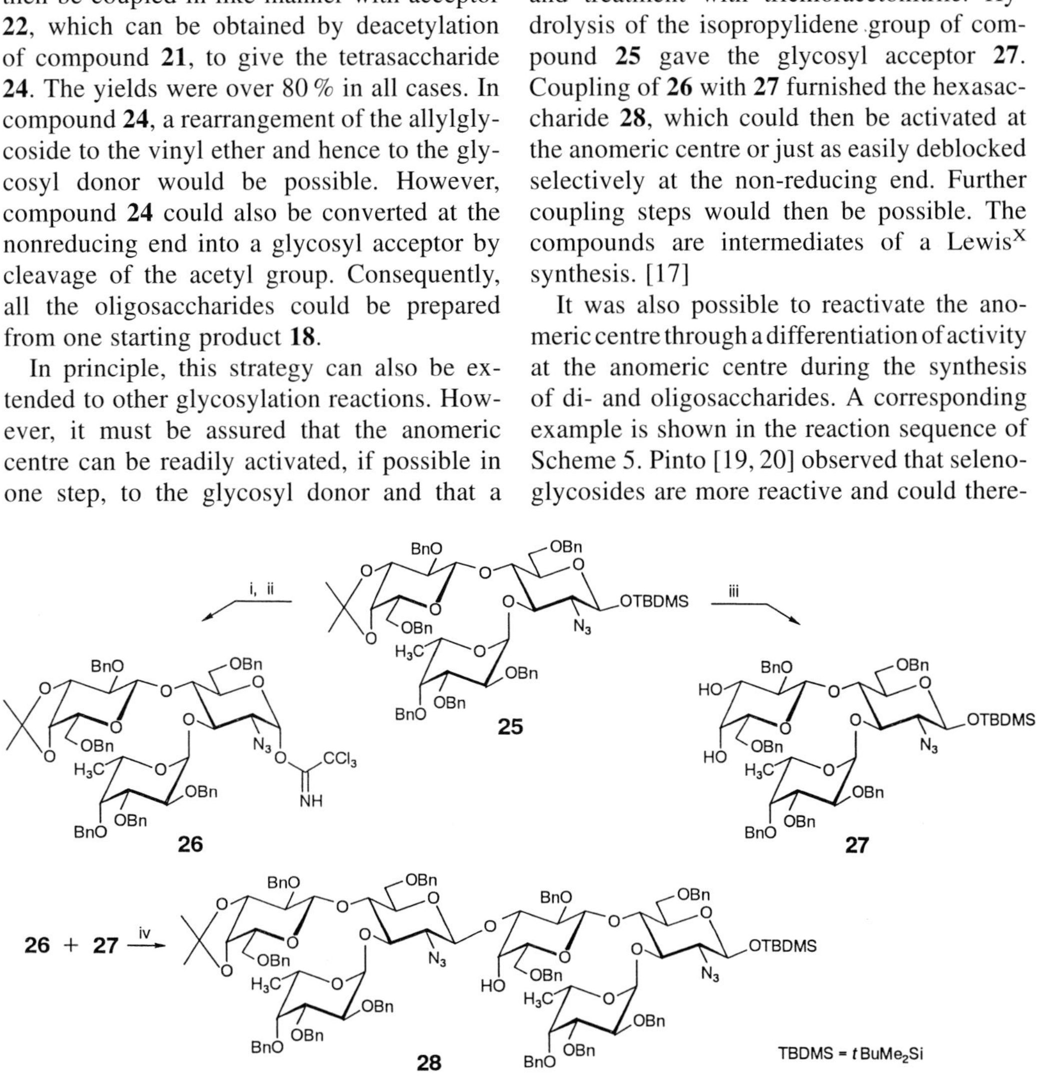

Scheme 4. i) Bu$_4$NF; ii) CCl$_3$CN/DBU; iii) TFA/CH$_2$Cl$_2$; iv) BF$_3 \cdot$ OEt$_2$.

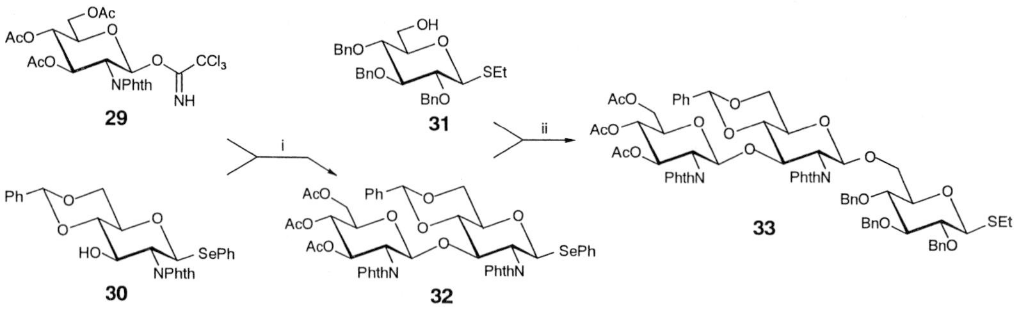

Scheme 5. i) Et$_3$SiOTf/–78 °C; ii) AgOTf/K$_2$CO$_3$.

fore be more readily activated than thioglycosides. In the presence of Et$_3$SiOTf at –78 °C, the imidate **29** could be coupled with the selenoglycoside **30** to give the disaccharide **32**. The selenodisaccharide **32** could then be coupled with the thioglycoside **31** in the presence of AgOTf to give the trisaccharide **33** without the thio group of compound **31** being attacked during the reaction. With compound **33**, another coupling at the anomeric centre with thiophilic reagents would be possible. It would also be possible to elongate both compounds **32** and **33** at the OH groups after cleavage of the benzylidene group.

The long known influence of the substituent pattern on the reactivity at the anomeric centre [21] can also be used to advantage for a differentiation during synthesis of an oligosaccharide. Ether substituents on the hydroxyl groups of the pyranose residue always lead to a higher reactivity at the anomeric centre, whereas acyl groups on the hydroxyl groups lower the reactivity. Fraser-Reid [11] has used the terms armed and disarmed to describe this effect. A combination of this effect with the seleno effect is shown in the reaction sequence in Scheme 6 performed by Ley. [22] The highly reactive (armed) seleno-

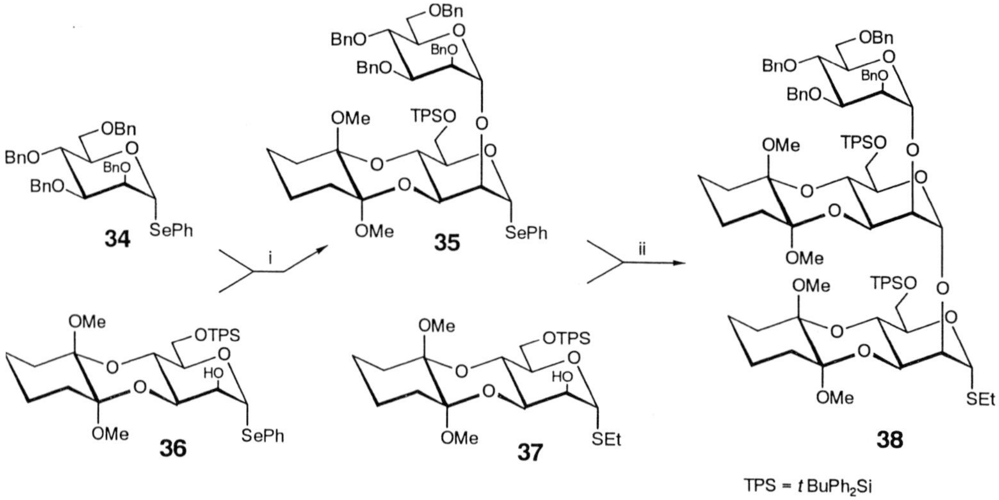

TPS = tBuPh$_2$Si

Scheme 6. i) *N*-Iodsuccinimide/TfOH (10 min); ii) *N*-Iodsuccinimide/TfOH.

glycoside **34** was treated for a very short reaction time with the acceptor **36** to give the disaccharide **35**. Coupling of the selenoglycoside **35** with the thioglycoside **37** resulted in the corresponding trisaccharide **38**. In **38**, an activation and coupling at the anomeric centre is possible with thiophilic reagents. Cleavage of the TPS groups would give free OH groups for an acceptor reaction.

In view of the available arsenal of glycosidation reactions, these examples should point the way for the development of similar strategies.

References

[1] A. Varki, *Glycobiology* **1993**, *3*, 97.

[2] R. A. Dwek, *Biochem. Soc. Trans.* **1995**, *23*, 1.

[3] O. Lockhoff, *Houben-Weyl: Methoden der Organischen Chemie*, Vol. E14a/3, Thieme, Stuttgart, **1992**, 621.

[4] K. Toshima, K. Tatsuda, *Chem. Rev.* **1993**, *93*, 1503.

[5] R. R. Schmidt, W. Kinzy, *Adv. Carbohydr. Chem. Biochem.* **1994**, *50*, 21.

[6] C.-H. Wong, R. L. Halcomb, Y. Ichikawa, T. Kajimoto, *Angew. Chem.* **1995**, *107*, 569; *Angew. Chem. Int. Ed. Engl.* **1995**, *34*, 521.

[7] H. Paulsen, *Angew. Chem.* **1990**, *102*, 851; *Angew. Chem. Int. Ed. Engl.* **1990**, *29*, 823.

[8] J. Jansson, G. Naori, G. Magnusson, *J. Org. Chem.* **1990**, *55*, 3181.

[9] H. Paulsen, M. Pries, J. P. Lorentzen, *Liebigs Ann. Chem.* **1994**, 389.

[10] S. J. Danishefsky, K. F. McClure, J. T. Randolph, R. B. Ruggeri, *Science* **1993**, *260*, 1307; J. T. Randolph, S. J. Danishefsky, *Angew. Chem.* **1994**, *106*, 1538; *Angew. Chem. Int. Ed. Engl.* **1994**, *33*, 1470.

[11] D. R. Mootoo, P. Konradsson, U. Udodong, B. Fraser-Reid, *J. Am. Chem. Soc.* **1988**, *110*, 5583.

[12] L. A. J. M. Sliedregt, K. Zegelaar-Jaarsveld, G. A. van der Marel, J. H. van Boom, *Synlett* **1993**, 335.

[13] O. Kanie, Y. Ito, T. Ogawa, *J. Am. Chem. Soc.* **1994**, *116*, 12073.

[14] S. Raghavan, D. Kahne, *J. Am. Chem. Soc.* **1993**, *115*, 1580.

[15] H.-J. Boons, *Tetrahedron* **1996**, *52*, 1095.

[16] G.-J. Boons, B. Heskamp, F. Hout, *Angew. Chem.* **1996**, *108*, 3053; *Angew. Chem. Int. Ed. Engl.* **1996**, *35*, 2845.

[17] A. Toepfer, W. Kinzy, R. R. Schmidt, *Liebigs Ann. Chem.* **1994**, 449.

[18] R. R. Schmidt in *Methods in Carbohydrate Chemistry* (Eds.: S. E. Khan, R. A. O'Neill), p. 20, Harwood Academic Publisher, Amsterdam, **1996**.

[19] S. Mehta, B. M. Pinto, *J. Org. Chem.* **1993**, *58*, 3269.

[20] S. Mehta, B. M. Pinto in *Methods in Carbohydrate Chemistry* (Eds.: S. E. Khan, R. A. O'Neill), p. 107, Harwood Academic Publisher, Amsterdam, **1996**.

[21] H. Paulsen, A. Richter, V. Sinnwell, W. Stenzel, *Carbohydr. Res.* **1978**, *64*, 339.

[22] P. Grice, S. V. Ley, J. Pietruszka, H. W. M. Priepke, *Angew. Chem.* **1996**, *108*, 206; *Angew. Int. Ed. Engl.* **1996**, *35*, 206.

Analogues of the Sialyl Lewis^X Group and of the N-Acetylneuraminic Acid in the Antiadhesion Therapy

Athanassios Giannis

In the past years our understanding of carbohydrate-protein interactions inreased dramatically [1]. Do to significant advances in molecular biology and carbohydrate chemistry the implication and importance of cell surface glycoconjugates in cell-cell interaction and in cell adhesion phenomena was elucidated. A great amount of literature on carbohydrate-binding proteins (lectins) and their ligands and carbohydrate-metabolizing enzymes was published. For these reasons carbohydrates are attractive starting points for drug development [2]. In order to demostrate this fact I will discuss below examples of carbohydrate-based antiinflammatory and antiviral compounds.

The Sialyl Lewis^X Group and its Analogues as Ligands for Selectins: Chemoenzymatic Synthesis and Biological Functions

White blood cells or leukocytes are components of blood that play a vital role in the immune response. Neutrophils, for example, are responsible for rapid and unspecific immune defense by digesting invading bacteria, while B and T lymphocytes are responsible for the specific immune response. As part of the immune response leukocytes leave the blood vessels and migrate to the zone of infection and the secondary lymphatic organs. The cells leave the blood vessels through the endothelium of the postcapillary venules in the area of the target tissue.

The complex multistage event leading to migration of leukocytes into the zone of infection occurs during normal chronic inflammation as well during acute inflammation and is mediated by a specific and regulated cell-cell recognition process between leukocytes and endothelial cells. This cell-cell recognition takes place by means of the interaction of a number of membrane-bound adhesion receptors with the corresponding membrane-bound ligands. The following receptors are known to be involved [3]: a) selectins, b) integrins, and c) members of the immunoglobulin superfamily.

The selectin family consists of three members, which are all glycoproteins and which mediate the initial stage of leukocyte adhesion (Fig. 1), namely, L-, E-, and P-selectin. L-selectin is constitutively expressed on the surface of leukocytes. It is involved in the recirculation of lymphocytes in peripheral lymph nodes as well as in the recruitment of leukocytes at the zone of inflammation. E-selectin

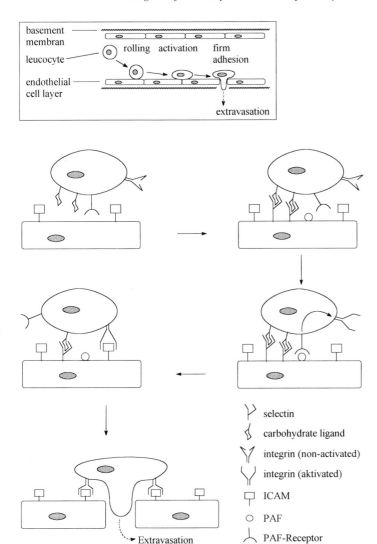

selectin

carbohydrate ligand

integrin (non-activated)

integrin (aktivated)

ICAM

PAF

PAF-Receptor

Extravasation

Figure 1. Interaction between leukocytes and endothelial cells of blood vessels (modified from [3e]). Top: overview; bottom: single steps with explanation of the symbols. Stimulation of endothelial cells by various mediators and factors leads to the expression of selectins and platelet-activating factor (PAF). The extra-vasation of leukocytes from the postcapillary venules starts with a weak, low-affinity, selectin-mediated adhesion of the leukocytes to endothelial cells, which leads to a slowing down of the leukocyte velocity (rolling). In the second phase a PAF-mediated activation of the leukocyte integrins occurs. In the third phase the activated integrins bind with high affinity to their ligands which are located on the endothelial cell surface. These ligands are called intercellular cell adhesion molecules (ICAM) and are members of the immunoglobulin superfamily. The firm adhesion of the leukocyte cells to the endothelium and finally their departure from the blood vessel (extravasation) results.

is expressed on the surface of endothelial cells a few hours after stimulation with interleukin 1 or tumor necrosis factor *a*. P-selectin is kept in intercellular stores in blood platelets (thrombocytes) and endothelial cells. Within a few minutes of exposure to thrombin, histamin, substance P, peroxide radicals, or complement factors, P-selectin is mobilized from these stores and transported to the cell surface. The exact structure of the membrane-associated carbohydrate ligand for L-selectin is not known. However, it is known that the L-selectin ligand is located at the postcapillary endothelium of the high endothelial venules (HEV) of the peripheral lymph nodes. According to the available data it is a sialic acid containing, sulfated, and fucosylated oligosaccharide [4].

The sialyl Lewis^X group (SLe^X) (**1** without R in Fig. 1) serves as a common ligand for all three selectins and is a component of cell surface glycosphingolipids and particularly glycoproteins. The physiological SLe^X-containing glycoproteins ligands for L-, P- and E-selectin are GlyCAM-1, PSGL-1 and ESL-1 respectively. In the case of PSGL-1, recent data indicate that high affinity binding is mediated at least in part by the comined recognition of sialylated oligosaccharides and tyrosine sulfate residues [5].

The importance of this oligosaccharide group for leukocyte adhesion was recently underlined by investigations into a hereditary disease, namely, leukocyte adhesion defiency type 2 [6]. Owing to a defect in the biosynthesis of the SLe^X group, the neutrophils are not able to adhere to stimulated endothelium and thus to migrate into the zone of inflammation. Patients with this condition suffer from recurrent severe bacterial infections. On the other hand, there is also a series of acute and chronic diseases whose course is negatively influenced when too many leukocytes migrate to the area of the infection or in the area of the pathological process itself. These include cardiogenic shock, stroke, thrombosis, rheumatism, psoriasis, dermatitis, adult respiratory

distress syndrome, bacterial meningitis, and encephalitis. In addition, metastasis in a number of tumors seems to be linked to expression of SLe^X and other closely related oligosaccharides on the membrane surface of the tumor cells and the interaction of these groups with endothelial selections. Efforts have thus been directed towards the development of adhesion blockers as new antiphlogistics, antithrombotics, immunosuppressants, and as drugs to prevent the formation of metastases.

A topic of high priority in the last years was the synthesis of the SLe^X group and analogues thereof, which could act as alternative selectin ligands and thus block the interaction between endothelium and leukocytes. Oligosaccharide synthesis must meet a number of criteria; it must be efficient, highly regio- and stereoselective, and suitable for large-scale production. In addition, the synthesis of a potential pharmaceutical must fulfill economic and ecological requirements (e. g. reagents containing heavy metals should be avoided). The groups of J. C. Paulson and C.-H. Wong have gone a long way towards solving these problems through the development of a clever (though not uncomplicated) enzyme-catalyzed synthesis of SLe^X (Fig. 2) [7]. They employ the glycosyl transferases involved in the biosynthesis of SLe^X. A number of difficulties had to be overcome before the synthesis became viable:

a) although galactosyl transferase was available, neither the Neu5Ac transferase (Neu5AcT) nor the Fuc transferase (FucT) were;

b) the pertinent glycosyl donors (sugar nucleotides) are too expensive to be used in stoichiometric amounts in large-scale synthesis;

c) the glycosyl transferases are inhibited by final products and/or intermediates.

The first problem was overcome by cloning and expressing the genes for Neu5Ac transferase and Fuc transferase [8] in Sf9 cells. In this way both transferases could be produced

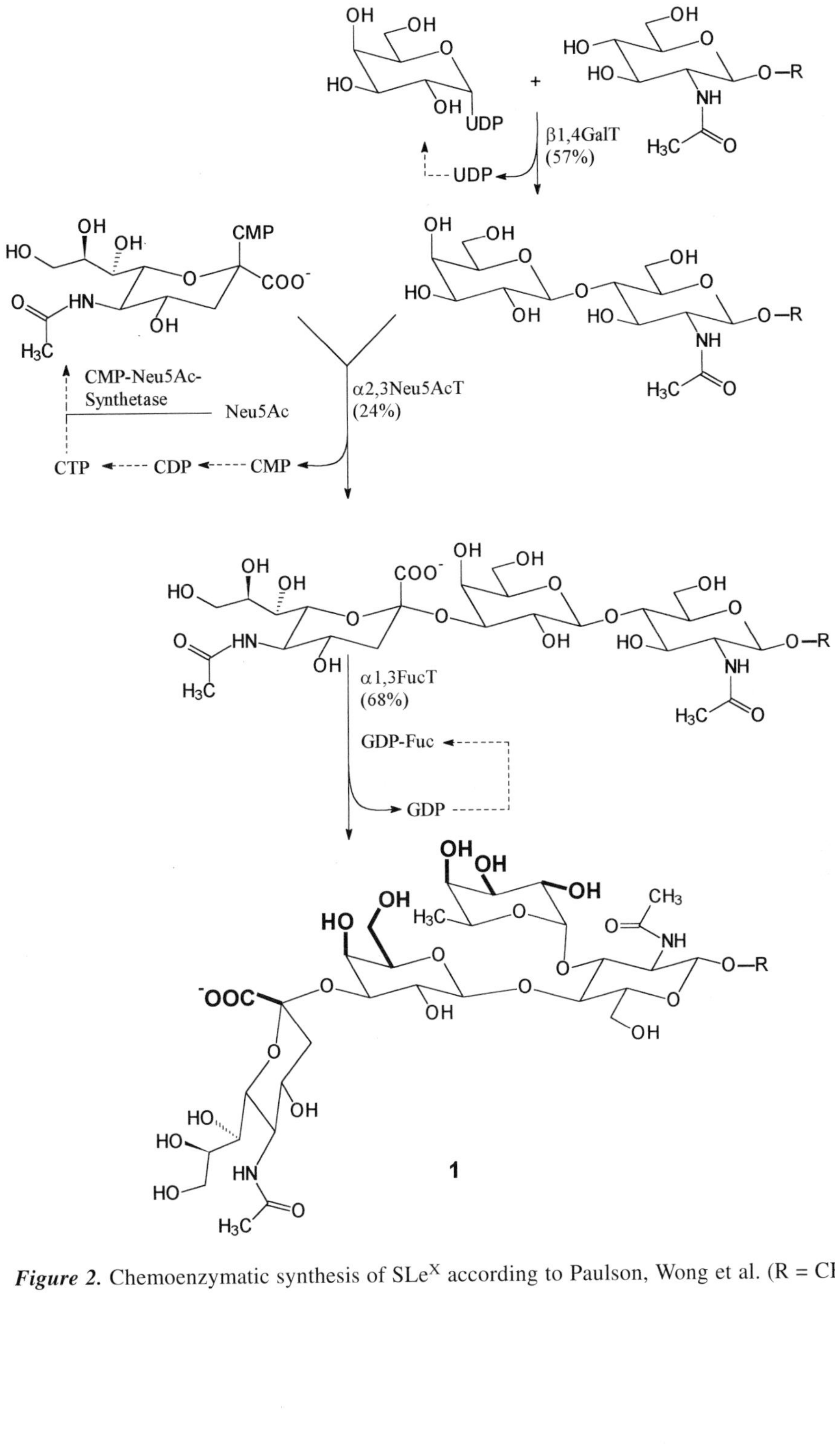

Figure 2. Chemoenzymatic synthesis of SLe^X according to Paulson, Wong et al. (R = CH₂–CH=CH₃).

in large amounts. A new method for cofactor regeneration *in situ* allowed the sugar nucleotides (UDP Gal, CMP Neu5Ac, GDP Fuc) to be easily generated and also largely suppressed the inhibition of the enzymes by products. To enable easy isolation, the SLe^X group was synthesized as the O-allylglycoside **1** (R=CH$_2$CHCH$_2$, Fig. 2). ^{13}C-labeled SLe^X derivatives were also synthesized for NMR spectroscopic studies. These studies showed that the Sle^X group has a folded, low-energy conformation, in which the L-fucose residue lies above the D-galactose unit. Therefore it appears that the tetrasaccharide forms a well-defined hydrophilic surface along the Neu5Ac-Gal-Fuc structural unit, while a hydrophobic region exists below the Neu5Ac-Gal-GlcNAc structural unit.

The fact that the transferases show some flexibility with regard to the glycosyl residues they will accept allowed the synthesis of several SLe^X analogues such as the glucal derivative **2** [9] (Scheme 1), which were used for investigations into structure-activity relationships [10]. The affinity of **2** for E-selectin was found to be similar to that of SLe^X. Based on these investigations and the fact that the sulfatides **3** and **4** (Scheme 1) isolated from biological material [11] demonstrated a higher affinity for E-selectin DeFrees, Gaeta et al. postulated a model of the E-selectin-

SLe^X interaction: the structural elements of the SLe^X unit essential for the recognition by E-selectin include the carboxyl function of *N*-acetylneuraminic acid, part of the galactose residue (possibly only the 4- and 6-hydroxyl functions), and the three hydroxyl groups of the L-fucose residue. The CH$_3$ group of this L-sugar seems to be not essential. These structure-activity relationships are largely in accordance with those found by Tyrrell et al. [12] The pharmacophore are highlighted in the chemical formula of SLe^X–R **1** (Fig. 2). It is important to note here that according to mutagenesis studies and crystallographic data the selectins have only one [13] binding site for the SLe^X group. However multivalency of both oligosaccharide and selectins on the surface of cells (cf. Fig. 1) generate high affinity binding.

Recently the conformations of the NeuAcα2(I)$\rightarrow$3Galβ1(II)$\rightarrow$4[Fucα1(III)$\rightarrow$3]Glc-NAc-O–CH$_3$ tetrasaccharide (SLe^X-OCH$_3$), in aqueous solution and bound to recombinant, soluble E-, P-, and L-selectin have been determined using high resolution NMR spectroscopy [14]. In the free ligand, the conformation of glycosidic linkage I is disordered with $\{\Phi_I, \Psi_I\}$ sampling values close to $\{-60, 0\}$, $\{-100, -50\}$, and $\{180, 0\}$. The trisaccharide portion is rigid and characterized by $\{\Phi_{II}, \Psi_{II}; \Phi_{III}, \Psi_{III}\} = \{46°, 18°; 48°, 24°\}$. The

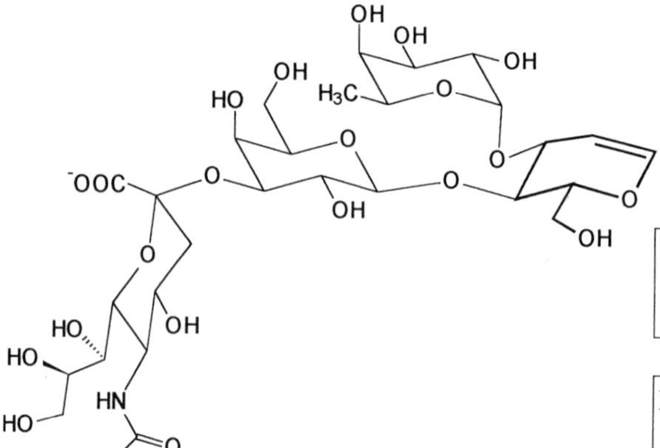

HSO$_3$-3Galβ1,4GlcNAc	β1,3Gal	
	α1,3	
	Fuc	**3**

HSO$_3$-3Galβ1,3GlcNAc	β1,3Gal	
	α1,4	
	Fuc	**4**

Scheme 1.

bound conformations (bioactive conformation) of the ligand were calculated from the full relaxation matrix analysis of transferred-NOE spectra for E- and P-selectin or by using a two-spin approximation for the L-selectin complex. Both E- and P-selectin recognize the {−60°, 0°} conformation of SLe^X while the {−100°, −50°} conformer is probably recognized by L-selectin. The conformation of the branched trisaccharide portion in the bound state remains close to the conformation of the free ligand. In the E-, P-, and L-selectin complexes the GalH4 proton is in the vicinity of protein aromatic protons, most likely Tyr94 and/or Tyr48. The measured equilibrium binding constants (K_D) of SLe^X, were 0.4 mM, 1.0 mM, and 0.6 mM for E-, P-, and L-selectin, respectively. On the other hand, the physiological ligands seem to have much higher affinities for the selectins. For example, a K_D value of 70 nM for monomeric, soluble P-selectin interacting with its native ligand on HL-60 cells was measured [15]. The final confirmation of the efficacy of **1** (R = H) as an antiinflammatory drug was provided by Mulligan et al. [16] with animal models. Rats were injected with a poison isolated from cobras (cobra venom factor, CVF), which led to a P-selectin induced adhesion of neutrophils to the endothelium of blood vessels of the lungs. This in turn resulted in a massive accumulation of neutrophils in the tissues of the lungs as neutrophils left the blood vessels and invaded the surrounding tissues. The accompanying increase in the permeability of the blood vessels led to lung edemas, bleeding in the alveoli, and severe damage to the lungs. An intravenous injection of SLe^X **1** (R = H) was given shortly before application of the snake venom, a 50 % reduction in the amount of bleeding in the lungs and accumulation of neutrophils in lung tissues was observed compared to that in untreated animals. In contrast, the same dose given after the application of snake venom was less effective in reducing the damage. Other, non-

fucosylated oligosaccharides such as Neu5-Acα2,3Galβ1,4GlcNAc proved to be inactive. The fact that such a low dose induces this protective effect is quite surprising: a single dose of 200 μg of **1** (R = H) corresponds to a concentration of less than 1 μmol L^{-1} in the blood! Based on these results, the hope that SLe^X-containing oligosaccharides and their analogues may prove to be effective antiinflammatory drugs was borne. Disappointingly, a phase II clinical trial on the treatment of reperfusion injury following myocardial infarction with Cylexin™ (a tetrasaccharide glycoside of SLe^X) produced by Cytel Corporation was recently terminated. A scheduled interim analysis of results showed that Cylexin™ was safe, but it had no benefit compared with a placebo. [17] The reason for this may be the low affinity of SLe^X for the selectins. A possible solution for this problem may be the administration of potent SLe^X ligands. For example compound Arg-Gly-Asp-Ala-NH-SLe^X was recently synthesized and proved to be a high affinity ligand of P-selectin [18]. In an assay employing P-selectin-IgG and HL-60 tumor cells this glycopeptide have an IC$_{50}$ value of 26 μM. A long-range goal continue to be the develpment of SLe^X group with mimetics that are potent selectin ligands and are effective when taken orally. The discovery that the sulfooligosaccharides [19] **4** and **5** can serve as selectin ligands also indicates that the SLe^X group may be replaced by simpler analogues that would be easier to synthesize. Crucial to the development of such carbohydrate mimetics may be the above mentioned determination of the SLe^X bioactive conformation. A suitable molecule could then be created by attaching the structural units essential for recognition to an appropriate scaffold. Several rationally designed SLe^X mimetics are shown in Scheme 2. As yet no realy potent selectin low molecular mass ligand with affinity in the low nanomolar range has been developed from SLe^X pharmacophore.

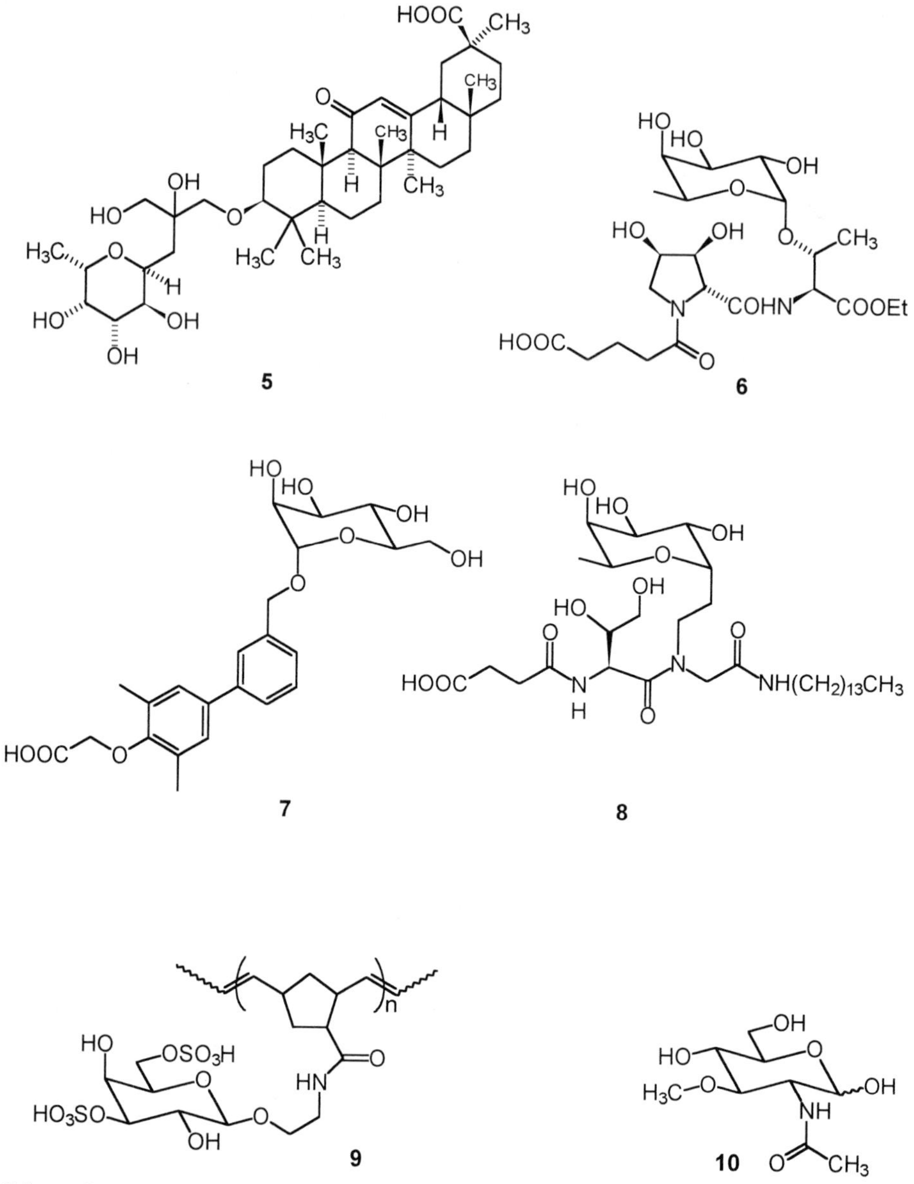

Scheme 2.

Compound **5** [20] was recently designed and synthesized and proved to be a selectin ligand with weak affinity (IC$_{50}$ ~ 0.5 mM for all three selectins). For its development the authors used conformational energy computations, high field NMR, and structure-function studies in order to define distance parameters of critical functional groups of SLe^x. This pharmacophore was used to search a three-dimensional data base of chemical structures. Compounds that had a similar spatial relationship of functional groups were tested as inhibitors of selectin binding. Glycyrrhizin, a natural occuring triterpene glycoside, was identified and found to block selectin binding to SLe^X *in vitro*. Subsequently they substituted different sugars for the glucuronic acids of glycyrrhizin and found the L-fucose derivative to be the most active in vitro and in vivo. A C-fucoside derivative, synthesized on a linker designed for stability and to more closely approximate the original pharmacophore, resulted in an easily synthesized, effective selectin blocker with anti-inflammatory activity in vitro. Other interesting and rationally designed selectin ligands with apparently little resemblance to the parent oligosaccharide SLe^X are the monosaccharide derivatives **6** [21], **7** [22] and **8** [23] (Scheme 2). These compounds display critical elements for recognition by the selectins and have higher affinity to E- and P-selectin than SLe^X. Another multivalent P-selectin ligand is the neoglycopolymer **9** [24] (IC$_{50}$ = 7 μM). This compound was developed based on the fact that sulfatides (3-sulfogalactosylceramide) is a P-selectin ligand [25]. Finally, since the SLe^X-induced leukocyte adhesion could also be blocked by inhibiting the enzymes responsible for the biosynthesis of this oligosaccharide group, intensive efforts are underway to develop such inhibitors [26]. In this context it is of interest that an inhibitor of Neu5Ac biosynthesis has recently become available: the simple D-glucosamine derivative **10** (Scheme 3) is a specific and potent inhibitor of both GlcNAc kinase and ManNAc kinase (K_i = 17 and 80 μM respectively) and thus inhibits Neu5Ac biosynthesis [27].

N-Acetylneuraminic Acid Analogues as Weapons against the Influenza Virus Infection

Influenza virus infection (flu) is the most serious respiratory tract disease and is poorly controlled by modern medicine [28]. Influenza viruses are divided into several types on the the basis of antigenic differences between their nucleoproteins. Haemagglutinin and neuraminidase are two major glycoproteins expressed in the surface of influenza A and B viruses. Haemagglutinin is a trimeric protein that binds to terminal N-acetylneuraminic acid residues of cell surface glycoconjugates. Whereas its role in the pathogenesis of the influenza virus infection has been elucidated [29] the importance of the viral enzyme neuraminidase is not clear. Influenza neuraminidase is a glycoprotein consisting of four identical subunits held together by disulfide bonds [30]. It is a glycosylhydrolase that cleaves terminal sialic acid from several glycoconjugates (glycoproteins and glycolipids). It has been postulated that this enzyme helps the release of newly formed viruses from infected cells, and it may also assist the movement of viruses through the sialic acid containing glycoconjugates of the mucus within the respiratory tract. The determination of the crystal structure of the influenza sialidase opened interesting opportunities for the design of new drugs for the therapy as well as for the prevention of the influenza virus infection. Early studies revealed 2,3-didehydro-2-deoxy-N-acetylneuraminic acid (Neu5Ac2en, **13**, Scheme 3) to be an weak and non-specific neuraminidase inhibitor [31] with a K_i value of 4 μM.

Compound **13** represent a transition state analogue of the *N*-acetylneuraminyl cation during the enzymatic reaction (Scheme 3). According to a postulate by L. Pauling compounds that closely resemble the transition state have higher binding affinity toward the

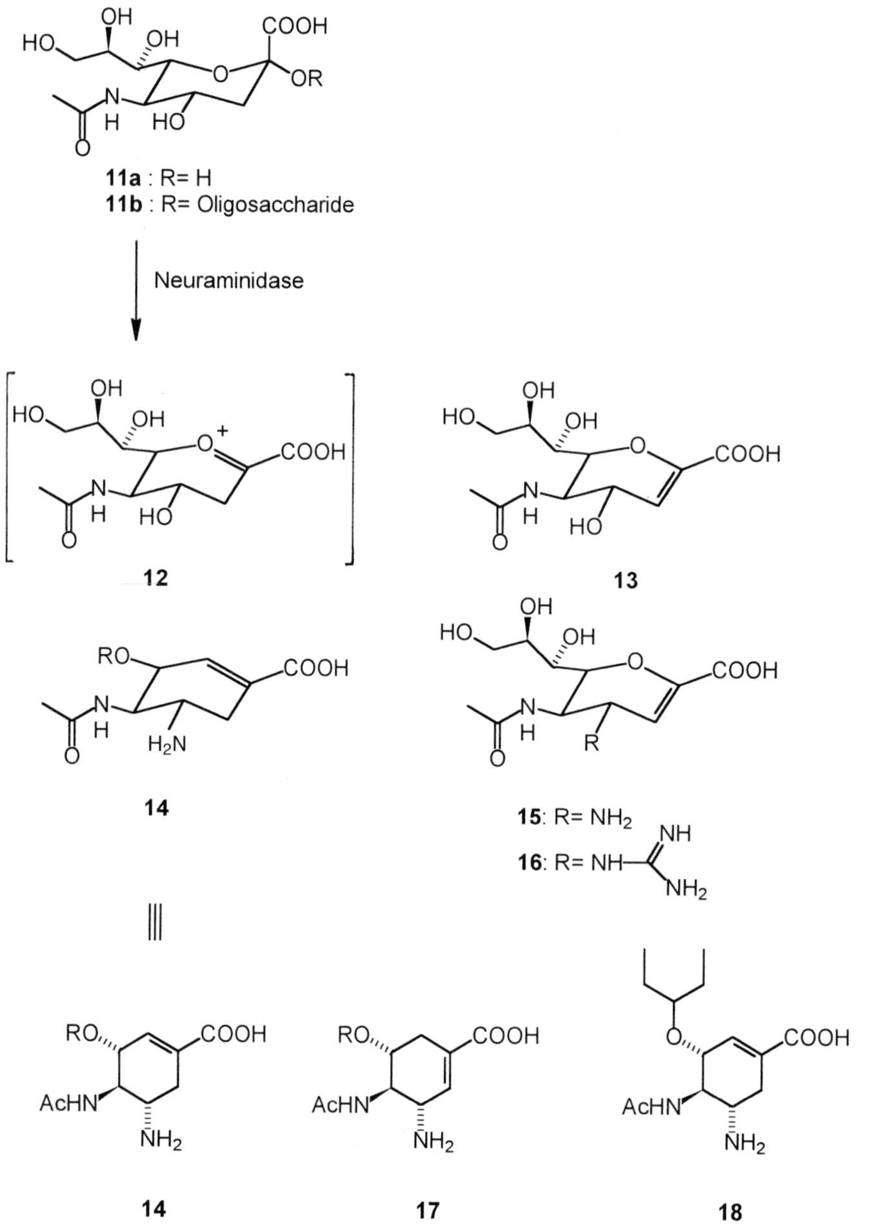

11a : R= H
11b : R= Oligosaccharide

Neuraminidase

12

13

14

15: R= NH$_2$

16: R= NH—

14

17

18

Scheme 3.

target enzyme in comparison to the enzyme substrate [32]. On the basis of structural information obtained from X-ray crystallographic investigation [33] of influenza neuraminidase complexed with **13** the derivatives 2,3-di- dehydro-2,4-dideoxy-4-amino-*N*-acetylneuraminic acid **15** and its guanidino analogue **16** were rationally designed and synthesized [34]. These compounds proved to be specific and potent highly potent inhibitors of the influenza virus neuraminidase: the corresponding K_i values are 10^{-8} and 10^{-10} M. The 4-amino group of compound **15** interacts with the carboxyl group of Glu 119 residue of the neuraminidase forming a salt bridge whereas the guanidine function of **16** interacts not only with Glu 119 but also with Glu 227 in a lateral bonding mode. Interestingly compound **16** (zanamivir) exhibited strong antiviral activity against several influenza A and B strains in the cell culture assay. Zanamivir is currently being in phase II clinical trials and has shown efficacy in in both prophylaxis and therapy of influenza virus infection. However do to its poor oral bioavailability this compound has to be administered directly to the upper respiratory tract by either intranasal spray or by inhalation. Because oral administration is a more convenient method the development of a new class of orally active neuraminidase inhibitors was recently initiated and succesfully completed [35]. Considering the flat oxonium ion in **12** as an double bond isostere, the cyclohexene scaffold was selected as a appropriate replacement of **12** which would keep the conformational changes to a minimum. Furthermore it was expected that the cyclohexenyl derivative would be chemically and metabolically more stable than the dihydropyran ring present in compounds **15** and **16** and easier for optimization of the pharmacological properties. The olefinic isomers **14** and **17** (R = H) were initially considered as two possible transition-state analogues and subsequently synthesized. The isomer **14** is structurally closer to transition-state intermediate **12** than the isomer **17** which is related to the known inhibitors **13**, **15** and **16**. Compound **14** (R = H) proved to be a moderate neuraminidase inhibitor whereas compound **17** (R = H) was inactive. On the basis of the above discussed X-ray crystallographic investigations of influenza neuraminidase complexed with N-acetylneuraminic acid analogues it was revealed that the glycerol side chain makes hydrophobic contacts with the hydrocarbon side chain of Arg-224 of the viral neuraminidase. This suggested that the C7 hydroxyl group of the known neuraminidase inhibitors could be replaced by lipophilic groups without loosing affinity to neuraminidase. Furthermore it was considered that in the transition-state intermediate, the oxonium double bond is electron deficient and highly polarized. For these reasons the glycerol side chain was replaced by an ether moiety at C3 position (cyclohexene numbering) to give compounds of the general formula **14** (R = Alkyl). Structure-activity studies identified the 3-pentyloxy moiety as an apparent optimal group at the C3 position: compound **18** proved to be a very potent inhibitor the influenza virus neuraminidase (IC_{50} = 1 nM). The X-ray crystallographic structure of **18** bound to viral neuraminidase clearly showed the presence of a large hydrophobic pocket in the region corresponding to the glycerol subsite of neuraminic acid. The high antiviral potency observed for **18** appears to be attributed to a highly favorable hydrophobic interaction in this pocket. The practical synthesis of **18** starting from (−)-quinic acid was also described. Finally the ethyl ester of derivative of this inhibitor (named GS4104) exhibited good oral bioavailability in several animals. More interesting GS4104 demonstrated oral efficacy in an animal influenza model. In a preclinical model, GS4104 was given orally to mice and demonstrated antiviral activity against multiple strains of influenza, increased survival, and decreased levels of the virus in lung tissue, without any reported toxicities. Do to these facts this compound has been selected as a clinical candidate for treatment and prophylaxis of influenza virus infection.

The examples discussed above show clearly that "knowledge of carbohydrate synthesis, structure and function can serve the medicinal chemist well in the design of new drug candidates". [36]

References

[1] M. Fukuda, O. Hindsgaul (Eds.) in *Molecular Glycobiology*, Oxford University Press, New York, **1994**.

[2] M. von Itzstein, P. Colman, *Curr. Opin. Struct. Biol.* **1996**, *6*, 703.

[3] a) L. A. Lasky, *Science* **1992**, *258*, 964; b) T. A. Springer, L. A. Lasky, *Nature* **1991**, *349*, 196; c) T. A. Springer, *Nature* **1990**, *346*, 425; d) E. C. Butcher, *Cell* **1991**, *67*, 1033; e) R. O. Hynes, A. D. Lander *Cell* **1992**, *68*, 303; f) L. Stoolman, in *Cell Surface Carbohydrates and Cell Development* (Ed.: M. Fukuda), CRC, Boca Raton-FL, USA, **1992**, p. 71; g) R. P. McEver, K. L. Moore, R. D. Cummings, *J. Biol. Chem.* **1995**, *270*, 11025.

[4] Y. Imai, L. A. Lasky, S. D. Rosen, *Nature* **1993**, *361*, 555.

[5] P. P. Wilkins, K. L. Moore, R. P. McEver, R. D. Cummings, *J. Biol. Chem.* **1995**, *270*, 22677.

[6] A. Etzioni, M. Frydman, S. Pollack, I. Avidor, M. L. Phillips, J. C. Paulson, R. Gershoni-Baruch, *N. Engl. J. Med.* **1992**, *327*, 1789.

[7] Y. Ichikawa, Y.-C. Lin, D. P. Dumas, G.-J. Shen, E. Garcia-Junceda, M. A. Williams, R. Bayer, C. Ketcham, L. E. Walker, J. C. Paulson, C.-H. Wong, *J. Am. Chem. Soc.* **1992**, *114*, 9283.

[8] B. W. Weston, R. P. Nair, R. D. Larsen, J. B. Lowe, *J. Biol. Chem.* **1992**, *267*, 4152.

[9] S. D. Danishefsky, K. Koseki, D. A. Griffith, J. Gervay, J. M. Peterson, F. McDonald, T. Oriyama, *J. Am. Chem. Soc.* **1992**, *114*, 8331.

[10] S. A. DeFrees, F. C. A. Gaeta, Y.-C. Lin, Y. Ichikawa, C.-H. Wong, *J. Am. Chem. Soc.* **1992**, *115*, 7549.

[11] a) C.-T. Yuen, A. M. Lawson, W. Chai, M. Larkin, M. S. Stoll, A. C. Stuart, F. X. Sullivan, T. J. Ahern, T. Feizi, *Biochemistry* **1992**, *31*, 9126.

[12] D. Tyrrell, P. James, N. Rao, C. Foxall, S. Abbas, F. Dasgupta, M. Nashed, A. Hasegawa, M. Kiso, D. Asa, J. Kidd, B. K. Brandley, *Proc. Natl. Acad. Sci. USA* **1991**, *88*, 10372.

[13] B. J. Graves, R. L. Crowther, C. Chandran, J. M. Rumberger, S. Li, K. S. Huang, D. H. Presky, P. C. Familletti, B. A. Wolitzky, D. K. Burns, *Nature* **1994**, *367*, 532.

[14] L. Poppe, G. S. Brown, J. S. Philo, P. V. Nikrad, B. H. Shah, *J. Am. Chem. Soc.* **1997**, *119*, 1727.

[15] S. Ushiyama, T. M. Laue, K. L. Moore, H. P. Erickson, R. P. McEver, *J. Biol. Chem.* **1993**, *268*, 15229.

[16] M. S. Mulligan, J. C. Paulson, S. DeFrees, Z.-L. Zheng, J. B. Lowe, P. A. Ward, *Nature* **1993**, *364*, 149.

[17] T. Feizi, D. Bundle, *Curr. Opin. Struct. Biol.* **1996**, *6*, 659.

[18] U. Sprengard, G. Kretzschmar, E. Bartnik, C. Hüls, H. Kunz, *Angew. Chem.* **1995**, *107*, 1104; *Angew. Chem. Int. Ed. Engl.* **1995**, *34*, 990.

[19] Sulfatides **4** und **5** are already synthesized: K. C. Nicolaou, N. J. Bockovich, D. R. Carcanague, *J. Am. Chem. Soc.* **1993**, *115*, 8843.

[20] B. N. Rao, M. B. Anderson, J. H. Musser; J. H. Gilbert, M. E. Schaefer, *J. Biol. Chem.* **1994**, *269*, 19663.

[21] C.-C. Lin, M. Shimazaki, M.-P. Heck, S. Aoki, R. Wang, T. Kimura, H. Ritzen, S. Takayama, S.-H.-Wu, G. Weitz-Schmidt, C.-H. Wong, *J. Am. Chem. Soc.* **1996**, *118*, 6826.

[22] B. Duppe, H. Bui, I. L. Scott, R. V. Market, K. M. Keller, P. J. Beck, T. P. Logan, *Bioorg. Med. Chem. Lett.* **1996**, *6*, 569.

[23] P. Sears, C.-H. Wong, *Proc. Natl. Acad. USA* **1996**, *93*, 12086.

[24] D. D. Manning, X. Hu, P. Beck, L. L. Kiessling, *J. Am. Chem. Soc.* **1997**, *119*, 3161.

[25] A. Aruffo, W. Kolanus, G. Walz, P. Fredman, B. Seed, *Cell* **1991**, *67*, 3161.

[26] a) M. M. Palcic, L. D. Heerze, O. P. Srivastava, O. Hindsgaul, *J. Biol. Chem.* **1989**, *264*, 17174; b) S. Cai, M. R. Stroud, S. Hakomori, T. Toyokumi, *J. Org. Chem.* **1992**, *57*, 6693 and literature cited therein; c) C.-H. Wong, D. P. Dumas, Y. Ichikawa, K. Koseki, S. J. Danishefski, B. W. Weston, J. B. Lowe, *J. Am. Chem. Soc.* **1992**, *114*, 7321.

[27] R. Zeitler, A. Giannis, D. Danneschewski, E. Henk, T. Henk, C. Bauer, W. Reutter,

K. Sandhoff, *Eur. J. Biochem.* **1992**, *204*, 1165.

[28] K. G. Nicholson, *Curr Opin. Infect. Dis.* **1994**, *7*, 168.

[29] J. M. White, *Science* **1992**, 258, 917.

[30] J. N. Varghese, W. G. Laver, P. M. Colman, *Nature* **1983**, *303*, 35.

[31] P. Meinal, G. Bodo, P. Palese, J. Schulman, H. Tuppy, *Virology* **1975**, *58*, 457.

[32] L. Pauling, *Chem Eng. News* **1946**, *24*, 1375.

[33] J. N. Varghese, J. L. McKimm-Breshkin, J. B. Caldwell, A. A. Kortt, P. M. Colman, *Proteins* **1992**, *14*, 327.

[34] a) M. von Itzstein, W-Y. Wu, G. B. Kok, M. S. Pegg, J. C. Dyason, B. Jinn, T. van Phan, M. L. Smythe, H. F. White, S. W. Oliver, P. M. Colman, J. N. Varghese, D. M. Ryan, J. N. Woods, R. C. Bethell, V. J. Hotham, J. M. Cameron, C. R. Penn, *Nature* **1993**, *363*, 418; b) M. von Itzstein, W. Y. Wu, B. Jin, *Carbohydr. Res.* **1994**, *259*, 301.

[35] C. U. Kim, W. Lew, M. A. Williams, H. Liu, L. Zhang, S. Swaminathan, N. Bischofberger, M. S. Chen, D. B. Mendel, C. Y. Tai, W. G. Laver, R. C. Stevens, *J. Am. Chem. Soc.* **1997**, *119*, 681.

[36] J. H. Musser, *Annu. Rep. Med. Chem.* **1992**, *27*, 301.

Peptidomimetics: Modern Approaches and Medical Perspectives

Athanassios Giannis

Introduction

A great variety of peptides acting as neuro-transmitters, hormones, autocrine and para-crine factors has been discovered and charac-terized during the last decades [1, 2]. After binding to their membrane-bound receptors, which belong mainly to the category of G-pro-tein coupled receptors [3, 4] they elicit chan-ges in cellular metabolism and control a series of vital functions such as blood pressure, diges-tion, immune defense, perception of pain, reproduction, behavior, tissue development, and cell proliferation. Furthermore, peptides as segments of proteins serve as recognition sites for enzymes, for the immune system, and are involved in cell-cell- and cell-extra-cellular matrix adhesion. Selective agonists and particularly antagonists of peptide recep-tors are indispensable for the investigation of peptidergic systems and are attractive starting points for drug development.

Already most bioactive peptides have been prepared in larger quantities and made avail-able for pharmacological and clinical investi-gations. Subsequently it became clear that the use of peptides as drugs is limited by the following factors a) their low metabolic stabil-ity towards proteolysis in the gastrointestinal tract and in serum; b) their poor transport from the gastrointestinal tract to the blood as well as their poor penetration into the central nervous system, in particular due to their rela-tively high molecular mass or the lack of spe-cific transport systems or both; c) their rapid excretion through liver and kidneys; and d) their side-effects caused by interaction of the conformationally flexible peptides with distinct receptors. In addition, a bioactive pep-tide can induce effects on several types of cells and organs, since peptide receptors and/or isoreceptors can be widely distributed in an organism. In recent years intensive efforts have been made to develop peptidomimetics [5–7] which display more favorable pharma-cological properties than their endogenous prototypes. A peptidomimetic is a compound that, as the ligand of the corresponding pep-tide receptor, can mimic or block the biolo-gical effects of a peptide. In this chapter I will discuss basic principles of peptidomi-metic design and discovery presenting select-ed examples of ligands developed for several G-protein coupled receptors as well as ligands for proteins involved in cell adhesion. Empha-sis will be given to small nonpeptide ligands. For the development of modified peptides as ligands for peptide receptors and for the design of inhibitors of peptidases the inter-ested reader is refered to several recent reviews.

Design of Peptidomimetics

As for any drug a peptidomimetic must fulfill the following requirements: a) metabolic stability, b) good bioavailability, c) high receptor affinity and receptor selectivity, and d) minimal side-effects. For the rational design of such compounds knowledge of the bioactive conformation of the endogenous peptide (i. e. the receptor-bound conformation) is of crucial importance. In aqueous solution and in the absence of the receptor, the biologically active conformation may be poorly populated and is frequently quite different from the conformation obtained by, for example, X-ray or NMR methods [8, 9]. However, due to the hydrophobic nature and/or size of most peptide receptors the detailed determination of the three-dimensional structures of receptor-ligand complexes has not yet been possible.

The rational design of small nonpeptide ligands i. e. the transformation of a peptide into a nonpeptide ligand is one of the most challenging and exciting fields in medicinal chemistry. Two long known examples clearly demonstrate that it is possible in principle to find small nonpeptide compounds which act as agonists or antagonists for peptide receptors: First, the alkaloid morphine is a ligand for the opioid receptor (a G-protein coupled receptor) and imitates the pharmacological effects of β-endorphin, an endogenous peptide composed of 31 amino acids, and of the tetrapeptide Met-enkephalin. On the other hand the structurally related morphine derivative naloxon represents an universal opioid receptor antagonist.

From the study of many peptide analogues it became apparent that a) for receptor recognition the side-chains of the amino acid residues are of crucial significance whereas for proteolytic enzymes the peptide backbone of the substrate participates considerably in binding affinity [10], b) frequently only a small number of three to eight amino acids in the peptide are responsible for the biological

activity ("message"), and c) conformation-activity studies of peptides suggested that their bioactive conformations are folded (turn-like) having molecular dimensions [11] in the range of 10×15 Å. In this context it should be pointet out that β-turns are the most frequently imitated secondary structures because they are a structural motif common to many proteins (including the complementarity determining region of antibodies) as well as cyclic peptides and has been postulated in many cases for the bioactive conformation of linear peptides [12]. Furthermore, β-turns are often localized at the surface of proteins where they serve as sites for molecular recognition. In linear biologically active peptides, turns arise mainly due to the tendency to form intramolecular hydrogen bonds. They may be also induced by interaction with biological membranes as well as by complexation with Ca^{2+} which is present in millimolar concentration in the serum.

A general rational approach to peptidomimetics is shown in Figure 1 (modified according to [13]). Using this approach it is possible to create a pharmacophore concept and to select an appropriate scaffold that carries the functional groups necessary for receptor binding and/or receptor activation in their correct spatial arrangement. However, despite of the emerging rational approaches it must be keept in mind that most of the peptidomimetics were discovered by screening of natural products and compound libraries. Recently very promising new techniques for the controlled synthesis of a great variety of compounds (combinatorial chemistry) have been established [14]. Importantly, the rapid examination of compound libraries is also possible by the so called high-throughput screening (HTS). HTS is the process by which large numbers of compounds can be tested, in an automated fashion, for agonistic or antagonistic activity at a biological target, for example a cell-membrane receptor or a enzyme [15]. By the aid of these new techniques as well as

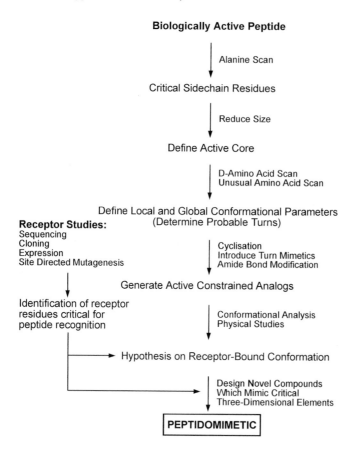

Figure 1. General approach to peptidomimetics.

impressive advances in cell culture and molecular biology (cloning and expression of the target receptors) several lead structures can be identified and subsequently optimized.

Biologically Active Mimetics of β-Turns

Compounds of the general formula **1** (Scheme 1) containing a hydrazine amino acid (Saragovi et al., 1991; Kahn, 1993) represent a landmark in the development of biologically active β-turn mimetics [16]. Compound **2** imitates a β-turn in the sequence Tyr–Ser–Gly–Ser–Thr **3**, a component of the hypervariable region of a monoclonal antibody against the reovirus type-3 receptor and is the first example of a low molecular mass immunoglobulin mimetic developed on the basis of an X-ray structure analysis of the antigen-antibody complex. Cyclopeptide **3** is resistant towards proteases and imitates the binding and functional properties of the native antibody!

The macrocycle **1** resemble compounds of type **4** and **5** (Scheme 2) originally developed by Olson et al. [17] and Kemp and Stites [18], respectively. A simple and efficient synthesis of a large library of chiral substituted hetero-

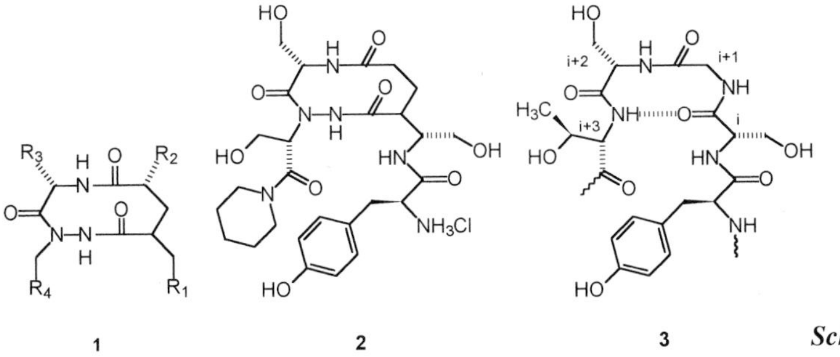

Scheme 1.

cycles of the type **6** as β-turn mimetics containing the side-chains R_{i+1} and R_{i+2} of the parent peptide was also reported [19]. For the same purposes macrocycles of the general formula **7** were designed [20]. A series of compounds containing the dipeptides Ala–Gly and cyclized with all stereoisomers of 6-amino-3,5-dimethylcaproic acid was prepared. A preliminary examination of these and other related compounds by NMR spectroscopy, circular dichroism and X-ray crystallography revealed that, depending on linker stereochemistry, different proportions of type II and type I exist in solution. Both type I and type II β-turns were observed also in the solid state. Although the use of synthetic linkers for the constraint of a dipeptide into various turns is not new, the ability of substituted linkers to affect the type of the turn is

novel and may be useful in fine-tuning of biologically active peptidomimetics.

Recently the design and synthesis of the 1,2,3-trisubstituted cyclohexane **9** (Scheme 3) as an analogue of the thyreotropin releasing hormone (TRH) with agonistic properties on the TRH receptor was reported [21]. On the basis of crystal and solution structures of TRH **8** (Scheme 3) they proposed a model for the pharmacophore which includes the lactam moiety of the pyroglutamyl group, the histidine imidazole ring, and the carboxamide function of the terminal prolineamide. For mimetic design they chose a starting conformation in which the peptide backbone approximates the Y-shaped X-ray structure of TRH. Subsequently the cyclohexane ring has been used as a scaffold for placing the pharmacophoric groups in the correct spatial arrangement. The most active compound was found to be the *N*-benzyl derivative **9**. The rational design of this derivative clearly demonstrates a) the value of X-ray and NMR spectroscopical studies for generating a fruitful hypothesis concerning the bioactive conformation of a ligand which is not always different from the solution or crystal structure, and b) the possibility for the use of scaffolds for replacing the peptide backbone.

One of the earliest examples of rationally designed low-molecular mass non-peptide compounds as turn mimetics is the bicyclic lactam derivative **11** [22]. This compound is a mimic of the immunosuppressing tripeptide Lys–Pro–Arg **10** which antagonizes the biolog-

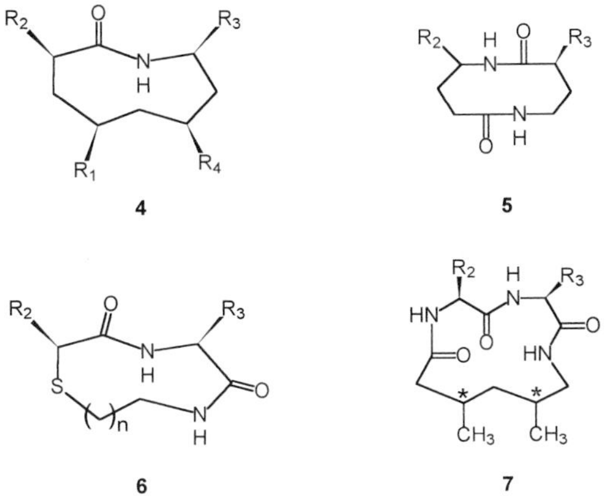

Scheme 2.

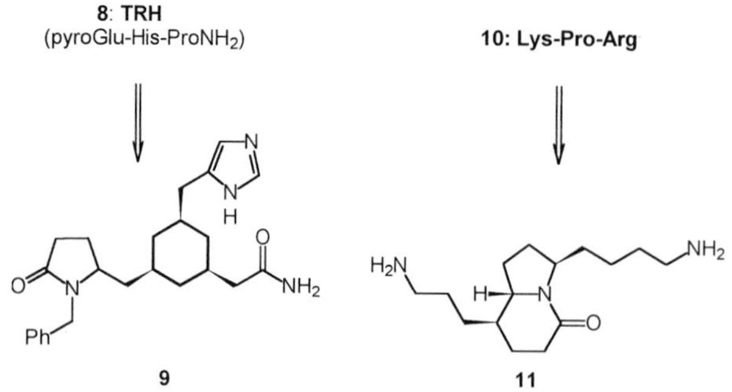

8: TRH
(pyroGlu-His-ProNH₂)

10: Lys-Pro-Arg

9

11

Scheme 3.

ical effects of the endogenous peptide tuftsin (Thr–Lys–Pro–Arg). This example clearly demonstrates that rational development of non-peptidal *antagonists* of peptide receptors is also possible using scaffolds carrying some critical side-chain groups of an endogenous peptide acting as *agonist*.

Peptidomimetics for G-Protein Coupled Receptors

As mentioned above, most of the receptors for neuropeptides and peptide hormones belong to the seven-transmembrane spanning G-protein coupled receptors. Binding of the endogenous peptide agonist results in a conformational change in the intracellular loops of the receptor leading to activation of a trimeric G-protein [3]. This in turn activates phospholipase C which cleaves phosphatidylinositol-4,5-bisphosphate (PIP₂) to inositol 1,4,5-trisphosphate (IP₃) and 1,2-diacylglycerol (DAG). IP₃ causes release of Ca^{2+} from endoplasmatic reticulum leading to a cellular response. Furthermore DAG activates protein kinase C which phosphorylates several intacellular receptors and enzymes thereby influencing their activity. Below development of peptidomimetics for two representative G-protein coupled receptors, namely the receptors for cholecystokinin and angiotensin will be discussed. For further examples the

interested reader is reffered to several detailed reviews [5, 6].

Peptidomimetics for the Cholecystokin Receptor

Cholecystokinin (CCK) exists in numerous biologically active forms (CCK-58, CCK-39, CCK-33, CCK-8, CCK-4,) having a common C-terminus which is essential for biological activity [23]. It exists in the nervous system both centrally and peripherally. CCK-8 (Asp–Tyr(SO₃H)–Met–Gly–Trp–Met–Asp–Phe–NH₂) is the most common neuropeptide in the brain. CCK was originally demonstrated to be one of the hormones responsible for regulating the function of the digestive tract. Peripherally cholecystokinin is released from nerve endings in many regions of the body. It is also synthesized in neuroendocrine cells in the upper gastrointestinal tract. In the latter it stimulates the contraction of the gall bladder by simultaneous relaxation of the sphincter oddi, increases the release of insulin, enhances the secretion of enzymes from the pancreas, and inhibits the secretion of gastrin and gastric emptying. In the central nervous system cholecystokinin acts as a neuromodulator/neurotransmitter, has an anxiogenic (anxiety generating) and appetite-suppressing effect, whilst at the spinal level it antagonizes the effect of opiates and thereby acts as an antianalgesic [24]. Two receptor cholecystoki-

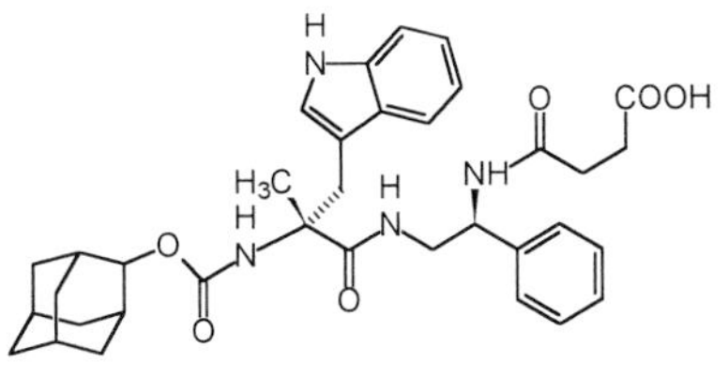

12

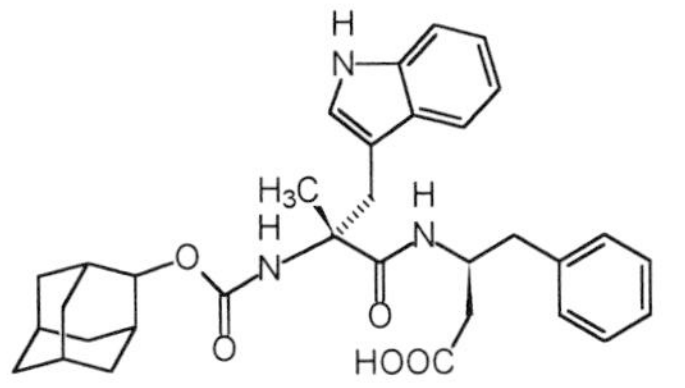

Scheme 4.   **13**

nin receptor subtypes have been identified: the predominantly peripheral CCK-A receptor and the largely centrally located CCK-B receptor. The C-terminal tetrapeptide sequence of CCK (Trp–Met–Asp–Phe–NH₂) and particularly the side chains of Trp, Asp and Phe are critical for biological activity. Early studies identified the dipeptide derivative Boc–Trp–Phe–NH₂ as a weak CCK-B agonist. The modification of this structure, supported by investigations of conformational energy of the tetrapeptide Trp–Met–Asp–Phe–NH₂, culminated in the synthesis of the orally active derivative CI-988 **12** (Scheme 4), which proved to be a selective CCK-B antagonist [25] (IC$_{50}$ = 1.7 nM). Compound CI-988 is the first example of a rationally developed nonpeptide ligand for a neuropeptide receptor and display interesting pharmacological prop-

erties in animal experiments. It has strong anxiolytic activity, but show no sedatory effects.

Another, structurally related CCK-B receptor antagonist is compound **13** (PD-135666). Interestingly, the corresponding enantiomeric derivative (*ent-***13**) is a potent and selective CCK-A receptor antagonist! [26] Using the substituted pyrrolidinone of the general formula **14** (Scheme 5) as scaffold and incorporating the side-chains of Trp, Asp, and Phe of the C-terminal region of the CCK peptides the derivative **15** which is a potent and orally active CCK-A antagonist (IC$_{50}$ = 16 nM) was developed [27]. For the antagonist activity the 3,4-*cis* stereochemical arrangement is necessary.

Angiotensin Receptor Antagonists

The octapeptide angiotensin II (A II: Asp–Arg–Val–**Tyr**–Ile–**His**–Pro–**Phe**) exerts its effects by interacting two receptors subtypes (AT1 and AT2). While AT2 receptors are important in embryonic development, AT1 receptors mediate the cardiovascular effects of angiotensin II and has been implicated in several cardiovascular diseases, including hypertension, cardiac hypertrophy, heart failure and myocardial infarction [28]. For these reasons A II antagonists are of enormous therapeutic potential.

On the basis of physicochemical and spectroscopical investigations of A II and the peptidic superagonist [Sar¹]A II Matsoukas et al. [29] suggested a conformational model for AII characterized by clustering of the three

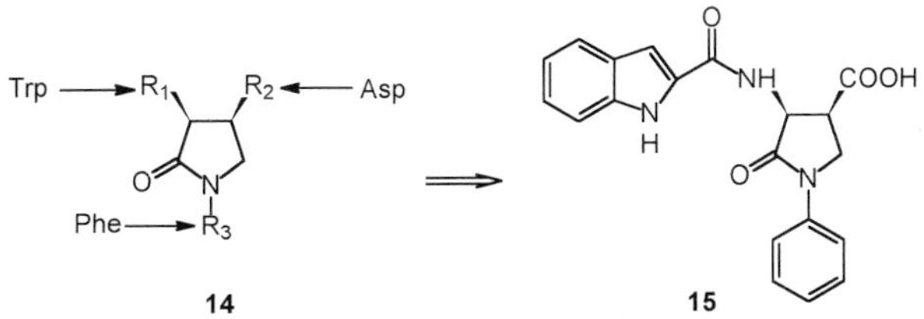

14   **15**   *Scheme 5.*

aromatic rings and a charge relay system involving the triad Tyr hydroxyl-His imidazole-Phe carboxylate. According to these studies, the N-terminal domain of A II appears to play a crucial role in generating the biologically active charge relay conformation of the peptide hormone. In addition these investigations confirmed that a Tyr–Ile–His bend is a predominant feature of the conformation of A II and [Sar^1]A II in the relatively non-polar "receptor simulating" environment provided by dimethyl sulfoxide.

The first potent, orally active nonpeptide antagonist of the AT1 receptor is losartan **16** (Lorzaar®, Scheme 6) with an IC$_{50}$ value of 19 nM [30]. Losartan was developed by overlaying the lead structure of **17** (S-8308), a weak but specific antagonist of the AT1 receptor discovered by screening, with the structure of A II, using molecular modeling techniques. It was postulated that the imidazole moiety of **17** serves as a template to present mimetics of the Tyr [4] and Ile [5] side-chains as well as the C-terminal carboxyl group of A II.

Losartan blocks the vascular constrictor effect of Ang II, the Ang II-induced aldosterone synthesis and/or release, and the AII-induced cardiovascular growth. In various models of experimental hypertension, losartan prevents or reverses the elevated blood pressure and the associated cardiovascular hypertrophy similar to angiotensin converting enzyme (ACE) inhibitors. Subsequently contolled clinical trials revealed that losartan is a new and valuable drug for treatment of hypertension. It has been well tolerated [31] and, in contrast with ACE inhibitors, does not cause cough and is a promising drug.

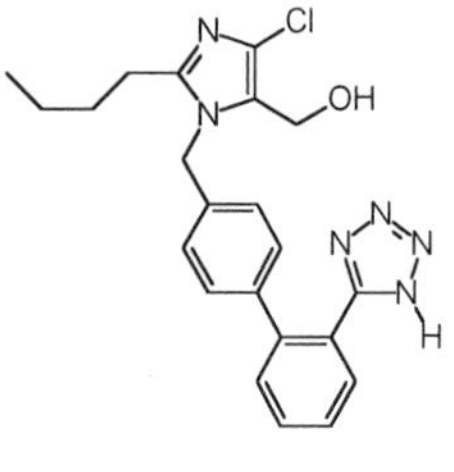

Integrin Ligands as Modulators of the Cell Adhesion

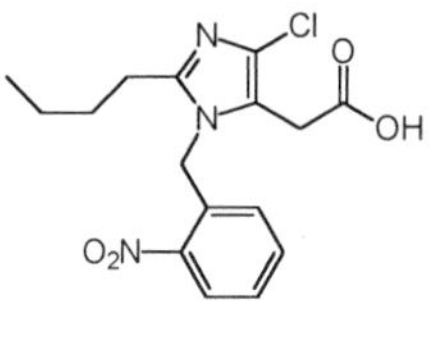

Integrins are noncovalently linked α/β heterodimeric proteins which play a critical role in cell-cell adhesion as well as in cell-extracellular matrix adhesion and interaction [32]. Such interactions determine important biological phenomena such as cell differentiation and cell viability, cellular traffic and organogenesis, angiogenesis and blood clotting. 15 α subunits and 8 β subunits have been identified to date. Different ab complexes are expressed on different cells [33]. For example the integrins $\alpha_L\beta_2$, $\alpha_M\beta_2$, and $\alpha_x\beta_2$ are expressed only on leucocytes whereas the integrins $\alpha_2\beta_1$, $\alpha_3\beta_1$, $\alpha_5\beta_1$, $\alpha_6\beta_1$, and $\alpha_v\beta_3$ are expressed on endothelial cells. In platelets, $\alpha_{IIb}\beta_3$ (known also as GPIIb/IIIa) is the major integrin. Furthermore tumor cells as for example human melanoma cells express $\alpha_v\beta_3$ (vitronectin receptor).

The Arg–Gly–Asp (RGD) sequence on several proteins such as fibrinogen, fibronectin, vitronectin, laminin, osteopontin, and von Willebrand factor serves often as an endogenous ligand for the integrins. The binding is Ca^{2+}-dependent. The specificity of the RGD-integrin interaction is generated by a combination of variations in the RGD conformation in different proteins and contributions of sequences near the RGD moiety [34]. Structure-activity investigations have revealed that in linear RGD peptides small structural modifications such as the exchange of alanine for glycine or glutamic acid for aspartic acid abolish the binding of the resulting peptides by integrins [35].

RGD analogues are potential candidates for example for treatment of a) thromboembolic

16   **17**

Scheme 6.

diseases (GpIIb-IIIa receptor antagonists) and b) as antiangiogenic agents (vitronectin receptor antagonists) i. e. agents that prevent the formation of new blood vessels.

a) The steroid derivative **18** was designed and synthesized as an analogue of a postulated [36] type I β-turn **19** of the glycoprotein GpIIb-IIIa-bound sequence Arg–Gly–Asp of fibrinogen.

Compound **18** binds to the GpIIb-IIIa receptor and shows an moderate IC_{50} value of 100 μM when fibrinogen is used as ligand. In comparison the peptide cyclo(Arg–Gly–Asp–Phe–D–Val), displays in IC_{50} value of 2 μM [37]. Apparently the glycoprotein GpIIb-IIIa bound RGD conformation is better imitated by the cyclopeptide. According to NMR spectroscopical investigations in dimethylsulfoxide solution, the Arg residue of this cyclopeptide lies in the *i*+2 position of an extended βII′-turn, whereas the Asp residue lies in the central position of a γ-turn, so that the Arg and Asp side chains are nearly parallelly oriented [38]. However it must be emphasized here that the receptor-bound conformation of RGD-containing peptides and RGD nonpeptide mimetics are still unknown. On the basis of an working hypothesis that the GpIIb-IIIa-bound conformation of the lead peptide Arg–Gly–Asp–Phe includes either a β- or a γ-turn a

γ-lactam was used as template [39]. The critical side chains of the peptide groups were attached to the γ-lactam via flexible linkers the length of which was optimized to afford a weakly active antagonist **20**. After further optimization the very rigid lactam derivative **21** (BIBU 52) with an EC_{50} value of 80 nM was obtained. This drug candidate possesses a negatively and a positively charged group in the appropriate distance and is a high-affinity ligand for GpIIb-IIIa and is also active *in vivo*. In an animal thrombosis model 1 mg/kg i. v. of BIBU 52 completely abolished thrombus formation for one hour.

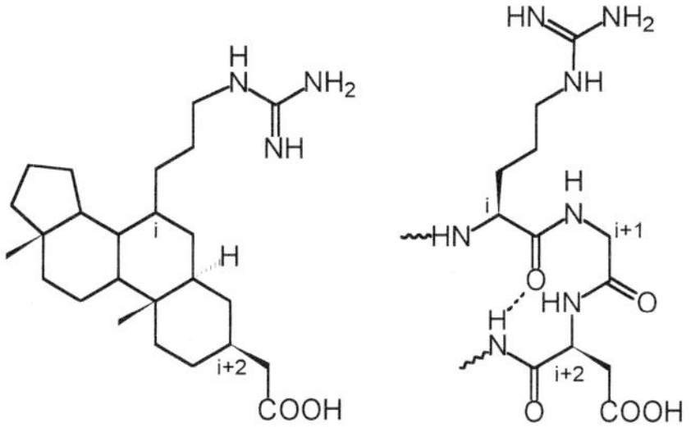

NMR studies of a rigid, potent GpIIb-IIIa antagonist **22** (G4120) and molecular dynamic simulations of flexible active analogues suggested a cupped [40] (U-shaped) bioactive conformation of the RGD moiety of G4120. The D-Tyr side chain of this cyclopeptide is spatially positioned nearly in the middle between the Arg and Asp side chain. Subsequently the pyrrolo[1,4]-benzodiazepine-2,5-dione [41] was selected as a template to fit the contour and volume of the peptide backbone of G4120. This template enabled the attachment of the critical Arg and Asp side-chains. The rational approach led to the benzodiazepine derivative **23** with an IC_{50} value of 9 nM.

The structural characteristics of G4120 suggested that a Arg–Tyr–Asp (RYD) group could be an alternative ligand for GpIIb/IIIa. Indeed the RYD mimetic lamifiban **24** (Ro 44–9883), which is currently in phase 3 clinical trials, is a potent inhibitor of the ADP-induced platelet aggregation determined in human platelet-rich plasma with an IC_{50} value of 30 nM [42, 43].

b) Malignant metastases are highly dependent on blood supply. For their survival neovascularization is essential. Neovasculariza-

18   **19**

Scheme 7.

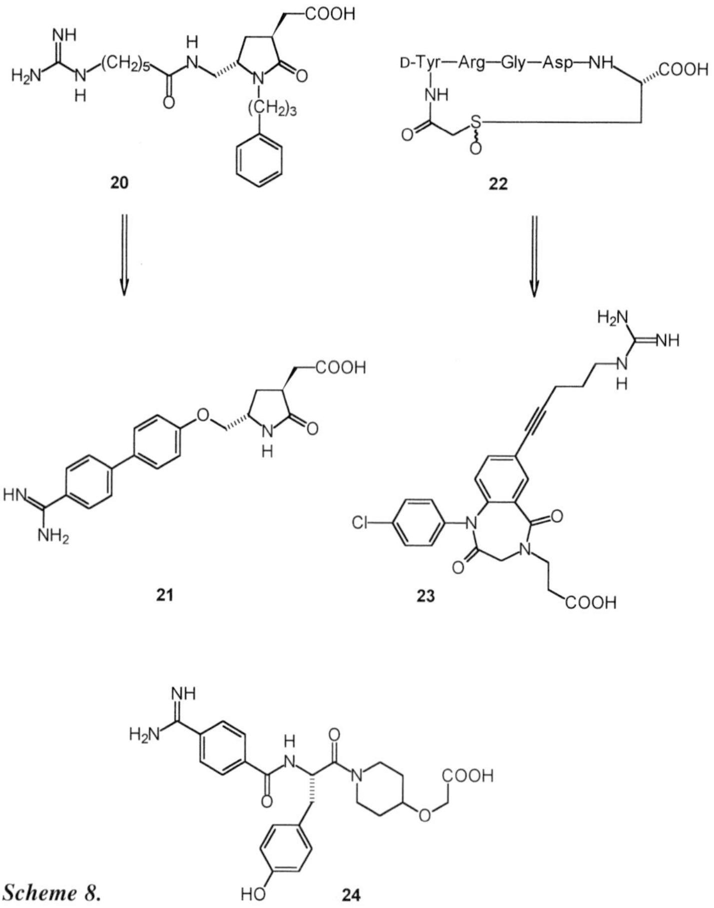

Scheme 8.

tion begins with vasodilatation of the parent vessel (i.e. the vessel from which a new capillary sprout originates) followed by a protease-mediated degradation of the basement membrane of the vessel. Thereafter endothelial cells migrate toward the angiogenic stimulus [44, 45] (Fig. 2).

Tumor cells promote vascular endothelial cells entry into the cell cycle and expression of integrin $a_v\beta_3$. After endothelial cells begin to move toward the angiogenic stimulus $a_v\beta_3$ ligation provides a survival signal which finally results in differentiation and formation of mature blood vessel. A disruption of the $a_v\beta_3$ liga-

tion may lead to apoptosis with subsequent tumor regression because of shortage in blood supply [46]. In order to test this hypothesis Brooks et al. [47] used the cyclopeptide RGDfV **25** which was previous identified as a potent inhibitor of $a_v\beta_3$-mediated cell adhesion [48].

The IC$_{50}$ value for the binding of **25** to both, soluble as well as immobilized $a_v\beta_3$-integrin is 50 nM. A single intravascular injection of this peptide disrupts ongoing angiogenesis on the chick chorioallantoic membrane. This leads to the rapid regression of histologically distinct human tumors transplanted onto the chick cho-

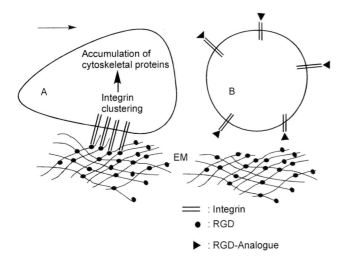

─── : Integrin

● : RGD

▶ : RGD-Analogue

Figure 2. $a_v\beta_3$-Integrins and angiogenesis. The action of different growth factors stimulates the expression of integrin $a_v\beta_3$ on endothelial cells. During the subsequent migration of endothelial cells in the direction of the angiogenic stimulator (arrow), the integrin $a_v\beta_3$ binds to RGD sequences present in multivalent form on the extracellular matrix (EM). As a consequence, integrin receptors aggregate on the cell membrane (A) and proteins of the cytoskeleton, like talin, paxilin, a-actinin, tensin, vinculin, and F-actin accumulate. This in turn results in maintenance of the migration process, serves as a signal for the survival of endothelial cells and finally leads to the formation of a new blood vessel. Prevention of integrin aggregation by soluble monovalent RGD analogues (B) leads to programmed cell death (apoptosis) of the migrating endothelial cells and therefore prevents vessel formation.

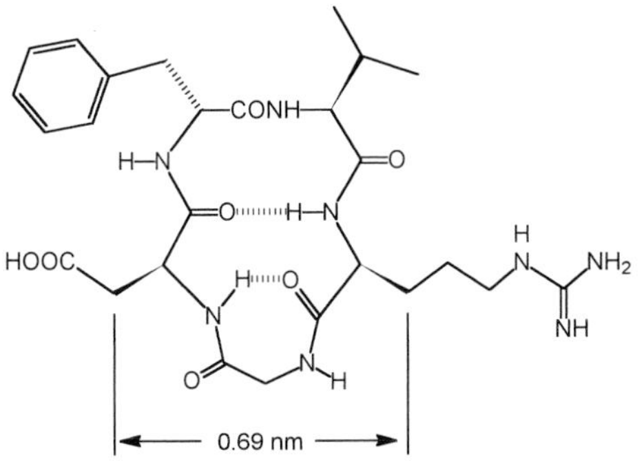

25

Scheme 9.

rioallantoic membrane (CAM assay). These results clearly demonstrate that $a_v\beta_3$ antagonists may provide a new alternative approach for the treatment of malignancies or other diseases characterized by angioge-

nesis such as rheumatoid arthritis, psoriasis, haemangiomas and corneal neovascularization. In the past it has been shown that antiangiogenic therapy with agents such as platelet factor 4, angiostatin and the fumagilin derivative AGM 1470 generally has low toxicity and drug resistance does not develop. In the future further $a_v\beta_3$ antagonists (particularly non-peptidal compounds) may be developed by optimization of known RGD analogues. In compound **25** a γ-turn with glycine in the central position is formed by the RGD sequence. According to NMR-spectroscopic investigations in solution, this leads to a parallel arrangement of the side chains of arginine and aspartate. As expected, the D-amino acid occupies the $i+1$ position of a βII'-turn. The distance between the β-carbon atoms of the Asp- and Arg-side chains within peptide **25** was determined by NMR spectroscopy to be 0.69 nm. Therefore, it is considerably shorter than the distance that was thought to be optimal for recognition by the fibrinogen receptor (0.75–0.85 nm). Recently, Kessler et al. synthesized all possible stereoisomers of peptide **25** and of its retro-sequence (32 peptides). An important result of the subsequent biochemical and NMR-spectroscopic investigations was that the retro-inverso compound c(-vFdGr-) shows a drastically reduced affinity towards the $a_v\beta_3$-integrin. This is explained by a conformation different from that of peptide **25**. More importantly, another peptide, c(-VfdGr-), shows nearly no affinity to the vitronectin receptor albeit its identical side chain orientation compared with **25**. This means that not only the side chains but also the peptide backbone contribute to receptor binding, at least by formation of one hydrogen bond. It is shown in the same study that a sterically demanding group in position 4 (D-Phe) is necessary for the biological activity of cyclopeptide **25**. In contrast to this, the L-valine residue can be replaced by any other amino acid. These structure-activity relationships constitute a valuable basis for the rational development of potent small nonpeptide $a_v\beta_3$ antagonists. In this regard one will certainly refer to the general principles in the design of peptidomimetics [5–7] and to the experience in the design of $a_{IIb}\beta_3$ antagonists .

In summary, after recognition of the importance of the RGD group for the integrin mediated cell adhesion several linear and cyclic RGD analogues were synthesized and their ability to act as alternative integrin ligands was investigated. On the basis of the subsequent pharmacological and physicochemical studies different hypotheses concerning the bioactive RGD conformation were generated leading to numerous interesting and exciting new agents for the treatment of thromboembolic diseases. The design and synthesis of RGD mimetics was initiated after early studies in patients with the rare, inherited, autosomal bleeding disorder Glanzmann's thrombasthenia revealed the role of GpIIb/IIIa in platelet aggregation. The GpIIb/IIIa antagonists will be probably the first clinically useful "anti-integrins" and should initiate the development of specific and potent low-molecular mimetics for other integrin for use in anti-adhesion therapy [49].

References

[1] D. T. Krieger, *Science* **1983**, *222*, 975.

[2] H. D. Jakubke in *Peptide*, Spektrum Akademischer Verlag, Heidelberg, Berlin, Oxford, **1996**.

[3] M. J. Berridge, *Nature* **1993**, *361*, 315.

[4] T. M. Savarese, C. M. Fraser, *Biochem. J.* **1992**, *283*, 1.

[5] A. Giannis, F. Rübsam, *Adv. Drug. Res.*, **1997**, *29*, 1.

[6] A. Giannis, T. Kolter, *Angew. Chem.* **1993**, *105*, 1303; *Angew. Chem. Int. Ed. Engl.* **1993**, *32*, 1244.

[7] J. Gante, *Angew. Chem.* **1994**, *106*, 1780; *Angew. Chem. Int. Ed. Engl.* **1994**, *33*, 1699.

[8] W. L. Jorgensen, *Science* **1991**, *254*, 954.

[9] a) K. Wüthrich, B. von Freiberg, C. Weber, G. Wider, R. Traber, H. Widmer, W. Braun, *Science* **1991**, *254*, 953; b) K. Wüthrich, *Acta Cryst.* **1995**, *D51*, 249.

[10] R. Hirschmann, *Angew. Chem.* **1991**, *103*, 1305; *Angew. Chem. Int. Ed. Engl.* **1991**, *30*, 1278.

[11] R. M. Freidinger, *Trends Pharmacol. Sci.* **1989**, *10*, 270.

[12] G. D. Rose, L. M. Gierasch, J. A. Smith, *Adv. Protein Chem.* **1985**, *37*, 1.

[13] G. R. Marshall, *Tetrahedron* **1993**, *49*, 3547.

[14] F. Balkenpohl, C. von dem Bussche-Hünnefeld, A. Lansky, C. Zechel, *Angew. Chem.* **1996**, *108*, 2436; *Angew. Chem. Int. Ed. Engl.* **1996**, *35*, 2288.

[15] J. R. Broach, J. Thorner, *Nature* **1996**, *384*, Suppl., 14.

[16] a) H. U. Saragovi, D. Fitzpatrick, A. Raktabutr, H. Nakanishi, M. Kahn, M. I. Greene, *Science* **1991**, *253*, 792: b) M. Kahn, *Synlett* **1993**, 821.

[17] G. L. Olson, M. E. Voss, D. E. Hill, M. Kahn, V. S. Madison, C. M. Cook, *J. Am. Chem. Soc.* **1990**, *112*, 323.

[18] D. S. Kemp, W. E. Stites, *Tetrahedron Lett.* **1988**, *29*, 5057.

[19] A. A. Virgilio, J. A. Ellman, *J. Am. Chem. Soc.* **1994**, *116*, 11580.

[20] O. Kitagawa, V. D. Velde, D. Dutta, M. Morton, F. Takusagawa, J. Aubé, *J. Am. Chem. Soc.* **1995**, *117*, 5169.

[21] a) G. L. Olson, D. R. Bolin, M. P. Bonner, M. Bös, C. M. Cook, D. C. Fry, B. J. Graves, M. Hatada, D. E. Hill, M. Kahn, V. S. Madison, V. K. Rusciecki, R. Sarabu, J. Sepinwall, G. P. Vincent, Voss, M. E., *J. Med. Chem.* **1993**, *36*, 3039; b) G. L. Olson, H. C. Cheung, E. Chiang; V. S. Madison, J. Sepinwall, G. P. Vincent, A. Winokur, K. A. Gary, *J. Med. Chem.* **1995**, *38*, 2866.

[22] M. Kahn, B. Chen, *Tetrahedron Lett.* **1987**, *28*, 1623.

[23] H. F. Bradford, in *Chemical Neurobiology*, p. 265–310, Freeman, New York, 1986.

[24] P. L. Faris, B. R. Komisaruk, L. R. Watkins, D. J. Mayer, *Science* **1983**, *219*, 310.

[25] D. C. Horwell, J. Hughes, J. C. Hunter, M. C. Pritchard, R. S. Richardson, E. Roberts, G. N. Woodruff, *J. Med. Chem.* **1991**, *34*, 404.

[26] M. Higginbottom, D. C. Horwell, E. Roberts, *Bioorg. Med. Chem. Lett.* **1993**, *3*, 881.

[27] D. L. Flynn, C. I. Villamil, D. P. Becker, G. W. Gullikson, C. Moummi, D.-C. Yang, *Bioorg. Med. Chem. Lett.* **1992**, *2*, 1251.

[28] K. K. Griendling, B. Lassegue, R. W. Alexander, *Annu. Rev. Pharmacol. Toxicol.* **1996**, *36*, 281.

[29] J. M. Matsoukas, J. Hondrelis, M. Keramida, T. Mavromoustakos, A. Makriyannis, R. Yamdagni, Q. Wu, G. J. Moore, *J. Biol. Chem.* **1994**, *269*, 5303.

[30] D. J. Carini, J. V. Duncia, P. E. Aldrich, A. T. Chiu, A. L. Johnson, M. E. Pierce, W. A. Price, J. B. Santella III, G. J. Wells, R. R. Wexler, P. C. Wong, S. Yoo, P. B. M. W. M. Timmermans, *J. Med. Chem.* **1991**, *34*, 2525.

[31] R. Davis, P. Benfield, *Disease Management & Health Outcomes* **1997**, *1*, 210.

[32] R. O. Hynes, *Cell* **1992**, *69*, 11.

[33] F. W. Lucinscas, J. Lawler, *FASEB J.* **1994**, *8*, 929.

[34] E. Ruoslati, M. D. Pierschbacher, *Science* **1987**, *238*, 491.

[35] E. Ruoslati, M. D. Pierschbacher, W. A. Border, in *"The Liver: Biology, and Pathobiology"*, 3rd ed. (Eds.: I. M. Arias, J. L. Boyer, N. Fausto, W. B. Jacoby, D. Schachter, D. A. Shafritz), pp. 889, Raven Press, New York, **1994**.

[36] R. Hirschmann, P. A. Sprengler, T. Kawasaki, J. W. Leahy, W. C. Shakespeare, A. B. Smith III, *J. Am. Chem. Soc.* **1992**, *114*, 9699–9701.

[37] M. Aumailley, M. Gurrath, G. Müller, J. Calvete, R. Timpl, H. Kessler, *FEBS Lett.* **1991**, *291*, 50.

[38] G. Müller, M. Gurrath, H. Kessler, R. Timpl, *Angew. Chem.* **1992**, *104*, 341; *Angew. Chem. Int. Ed. Engl.* **1992**, *31*, 326.

[39] V. Austel, F. Himmelsbach, T. Müller, *Drugs Fut.* **1994**, *19*, 757.

[40] R. S. McDowel, T. R. Gadek, *J. Am. Chem. Soc.* **1992**, *114*, 9243.

[41] R. S. McDowel, B. K. Blackburn, T. R. Gadek, L. R. McGee, T. Rawson, M. E. Reynolds, K. D. Robarge, T. C. Somers, E. D. Thorsett, M. Tischler, R. R. Webb, M. C. Venutti, *J. Am. Chem. Soc.* **1994**, *116*, 5077.

[42] L. Alig, A. Edenhofer, P. Hadváry, M. Hürzeler, D. Knopp, M. Müller, B. Steiner, A. Trzeciak, T. Weller, *J. Med. Chem.* **1992**, *35*, 4393.

[43] T. Weller, L. Alig, M. Hürzeler-Müller, W. C. Kounus, B. Steiner, *Drugs Fut.* **1994**, *19*, 461.

[44] a) J. Folkman, *Nature Med.* **1995**, *1*, 27; b) J. Folkman, *N. Engl. J. Med.* **1995**, *333*, 1757.

[45] A. Giannis, F. Rübsam, *Angew. Chem.* **1997**, *109*, 606; *Angew. Chem. Int. Ed. Engl.* **1997**, *36*, 588.

[46] S. Miyamoto, S. K. Akiyama, K. M. Yamada, *Science* **1995**, *267*, 883.

[47] P. C. Brooks, R. A. F. Clark, D. A. Cheresh, *Science* **1994**, *264*, 569.

[48] M. Pfaff, K. Tangemann, B. Müller, M. Gurrath, G. Müller, H. Kessler, R. Timpl, J. Engel, *J. Biol. Chem.* **1994**, *269*, 20233.

[49] J. Lefkovits, E. F. Plow, E. J. Topol, *N. Engl. J. Med.* **1995**, *332*, 1553.

A New Application of Modified Peptides and Peptidomimetics: Potential Anticancer Agents

R. M. J. Liskamp

Wide applications of (unmodified) peptides as drugs are limited by the disadvantages affiliated with their use in biological systems. An unmodified linear peptide is easily degraded by e. g. proteases, it is often too water-soluble to pass the cell membrane and therefore unable to pass the blood-brain barrier and rapidly excreted. In addition, a peptide is a flexible structure of which the bioactive conformation is usually hidden in a population of thousands of other conformers.

The desire to remedy these disadvantages and therefore to be able to obtain meaningful pharmaceutical compounds led to the development of *modified peptides* and *peptidomimetics*. [1, 2] A relatively new promising application of *modified peptides* and *peptidomimetics* is as potential anticancer agents for the treatment of tumors in which oncogenic Ras proteins contribute to transformation and abnormal growth.

Two aspects of this potential application are especially noteworthy: (1) the inhibition of malignant transformation is achieved on a level other than the DNA-level; and (2) the inhibition originates from inhibition of a post-translational modification process of proteins.

In many conventional chemotherapeutic regimens of cancer, the drug displays its activity on the DNA-level. The interference with or damage to the genetic material leads to the arrest of cellular growth and halting of uncontrolled cell division. However, these drugs are often deleterious for other rapidly dividing cells involved in the formation of e. g. bone marrow, gastrointestinal mucosa and hair follicles, causing the well-known side effects. In addition, interference with the genetic code means that these reagents are inherently carcinogenic and/or mutagenic. Therefore attempts directed towards the development of drugs acting on other cellular levels are especially interesting.

Post-translational modification processes of proteins include phosphorylation [3] glycosylation, methylation, acetylation, fatty acid acylation and prenylation. Among the prenylation processes *farnesylation* is especially noteworthy since it is implicated in a considerable number of human cancers. As a consequence interference with the post-translational modification process of *farnesylation* seems a promising and perhaps even viable approach for the development of anticancer drugs.

Eukaryotic polypeptides with a cysteine residue located at the fourth position counting from the C-terminus are in principle candidates for a post-translational modification involving the cysteine-SH. [4]

In a reaction catalyzed by farnesyl protein transferase (FPTase) this SH can be alkylated by farnesylpyrophosphate (FPP), an iso-prenoid compound also involved in the bio-synthesis of cholesterol. The hydrophobic iso-prenyl tail may be responsible for mem-brane association of proteins. Alternatively, the isoprenyl tail and methyl ester may promote binding of the protein to specific membrane associated receptors. [5]

The closely related forms of Ras proteins (H-Ras, N-Ras, K-Ras-A, K-Ras-B) are in-triguing examples of eukaryotic proteins, which are farnesylated (Scheme 1). Ras is a guanosine triphosphate (GTP) binding pro-tein. When GTP is bound to Ras, the cell di-vision is triggered. Normal Ras possesses GTPase activity, which hydrolyzes the bound GTP to GDP, so that the mitogenic signal is timely terminated. There are however, mutant forms i.e. oncogenic versions of Ras, found in about 50 % of human colon carcinomas and 90 % pancreatic carcinomas, which have an impaired GTPase activity. This impaired GTPase activity has as a consequence that the mutant Ras protein remains constitutively complexed to GTP, leading to unregulated cell proliferation and malignant transforma-tion. Farnesylation of both normal and onco-genic Ras proteins is necessary for attachment to the cell membrane and the resulting biolo-

gical effects. In case of the mutant Ras protein it appears to be required for efficient cell transforming activity. Therefore a straight-forward, albeit elegant, approach to inhibit transformation would be to interfere with Ras membrane localization by preventing farnesylation. [6]

Tetrapeptides with amino acid sequences identical to the COOH terminus of protein substrates for farnesylation by FPTase can compete with Ras for farnesylation. Selectiv-ity of inhibition of farnesylation as opposed to inhibition of the other isoprenylation pro-cess of attachment of a geranylgeranyl moiety can be obtained by using peptides of the consensus sequence CAAX in which C is cysteine, A is an aliphatic amino acid and X is serine, methionine or glutamine.

Modified peptides containing reduced amide bonds, i.e. compounds L-731,735 and L-731,734, were designed by the Merck groups [7] (Fig. 1), and have shown to be potent inhibitors of partially purified FPTase, the homoserine compound being the more active inhibitor *in vitro*. Subsequent inhibitors include L-739,749 and L-739,750 [8] and even truncated versions of the C-terminal tetrapeptide CAAX motif were prepared which do not have a C-terminal carboxyl

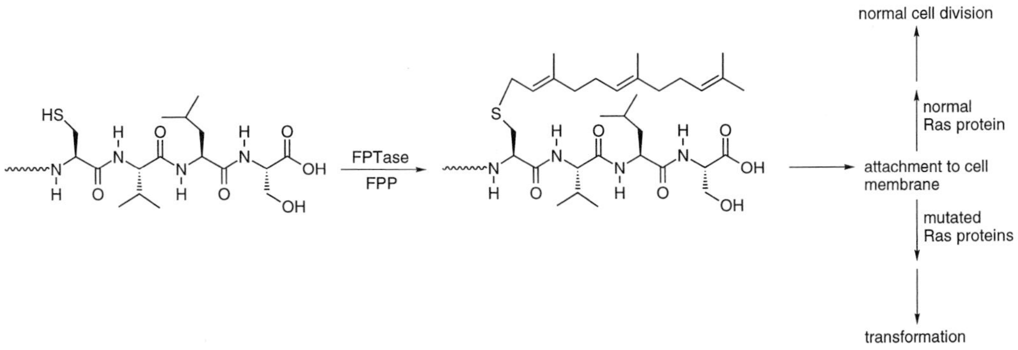

Scheme 1. Post-translational modification by farnesylation. The cysteine as part of the consensus sequence CAAX (C, cysteine; A, aliphatic amino acid residue; X, serine or methionine) near the C-terminus in the Ras protein is farnesylated by farnesylpyrophosphate (FPP) catalyzed by farnesyl protein transferase (FPTase). The farnesylated protein can then attach itself to the plasma membrane. If mutated Ras proteins are farnesylated and attached to the cell membrane this will lead to transformation.

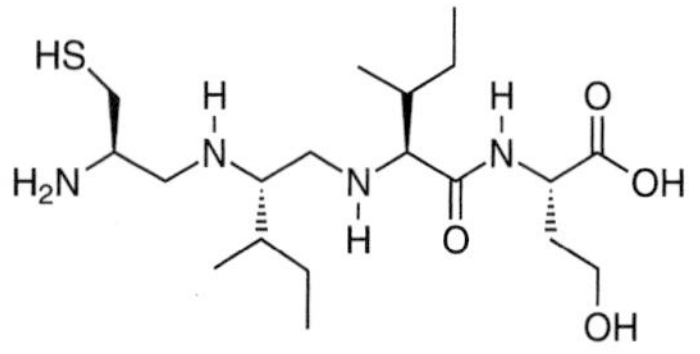

L-731,735

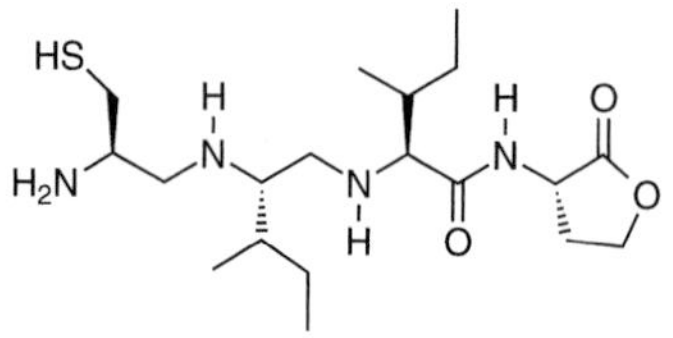

L-731,734

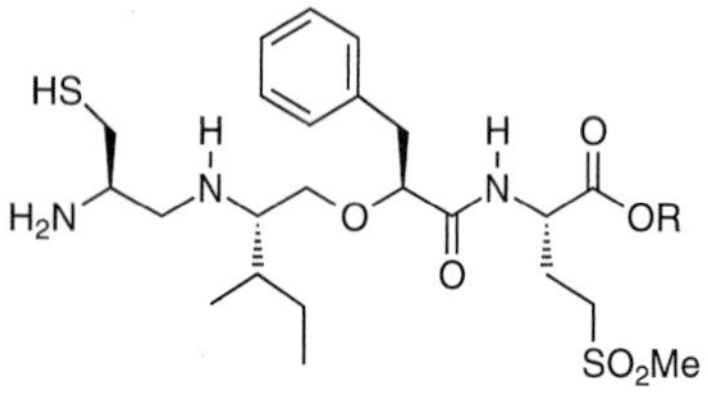

R = Me: L-739,749
R = H: L-739,750

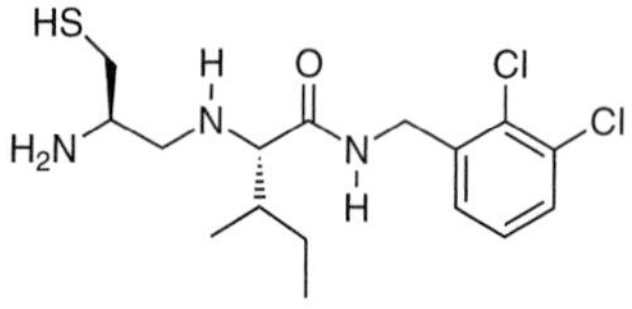

Figure 1. Reduced amide containing modified peptides L-731,735, L-731,734, L-739,749, L-739,750 and a truncated CAAX – analog.

moiety, yet they inhibit farnesyltransferase. [9]

The University of Texas and Genentech groups used the benzodiazepine skeleton to mimic a turn-like structure of the two middle amino acids in the consensus sequence, to which the N-terminal cysteine and a C-terminal methionine, serine or leucine were attached [10, 11] (Fig. 2). Although the free acid (BZA-2B) was best in the in vitro assay, the ester (BZA-5B) and to a somewhat lesser extent the amide were more active in a Chinese hamster ovary cell line, suggesting that the latter compounds may more easily penetrate into cells, because of their reduced polarity.

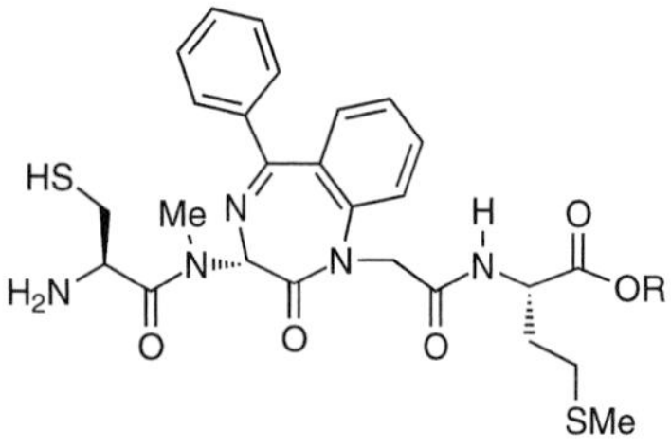

BZA-2B: R = H
BZA-5B: R = Me

Figure 2. BZA-2B and BZA-5B.

This set the stage for the development of other peptidomimetic farnesyltransferase inhibitors and shortly thereafter a peptidomimetic in which the two aliphatic amino acids were replaced by a relative rigid aromatic spacer was introduced. [12] Other aromatic spacers were introduced [13] and it was even possible to replace the C-terminal amino acid ("X") by an aromatic residue containing a carboxylic acid moiety and to obtain potent inhibitors. [14] The synthesis of a potent representative of this class of farnesyltransferase inhibitors is delineated in Scheme 2.

This approach of using an aromatic residue can be considered as a scaffold approach in which functional groups are attached to a relatively rigid core ('scaffold') thereby position-

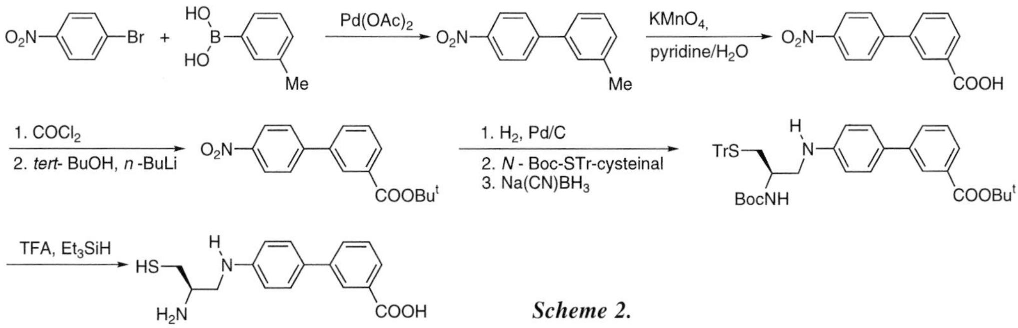

Scheme 2.

ing these groups with correct spatial orientations for the biomolecular interactions leading to inhibition. Another interesting scaffold in this respect turned out to be the piperazine ring leading to potent non carboxylic acid inhibitors of farnesyltransferase. [15] These inhibitors are accessible in a straightforward manner starting from an amino acid derivative (Scheme 3).

One of the most conservative modifications and as a consequence perhaps one of the 'safest' structural modifications with respect to the biological activity is the replacement of an amino acid in the CAAX motif sequence by conformationally restricted amino acids. [16] Despite the peptidic nature of these modified peptides, a number of them (Fig. 3) showed a considerable activity in vivo. [17, 18]

However, if one wishes to reduce proteolysis and increase lipophilicity, replacement of the amide bonds in the CAAX peptide by an isosteric replacement is an approach of considerable importance. The trans alkene moiety is a very suitable amide surrogate in terms of mimicking the rigidity, bond angles and bond length of the amide bond. [19] The alkene moiety was incorporated through elegant approaches into peptidomimetic compounds in which one [20] (Scheme 4) or even two amide bonds were replaced by double bonds as is the case in compound B956 [21] (Fig. 4). In this way powerful inhibitors were obtained.

Peptoids [22] form a particular promising class of oligomeric peptidomimetics, which consists of *N*-substituted-glycine derivatives. Thus, the concepts to translate a particular peptide sequence, in this case CAAX to a corresponding peptoid sequence is a perfectly suitable approach to obtain potential farnesyl inhibitors. This approach was adopted by

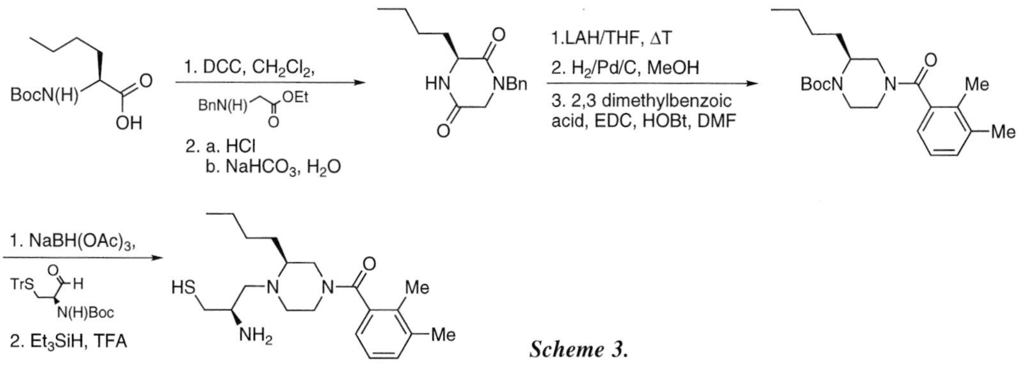

Scheme 3.

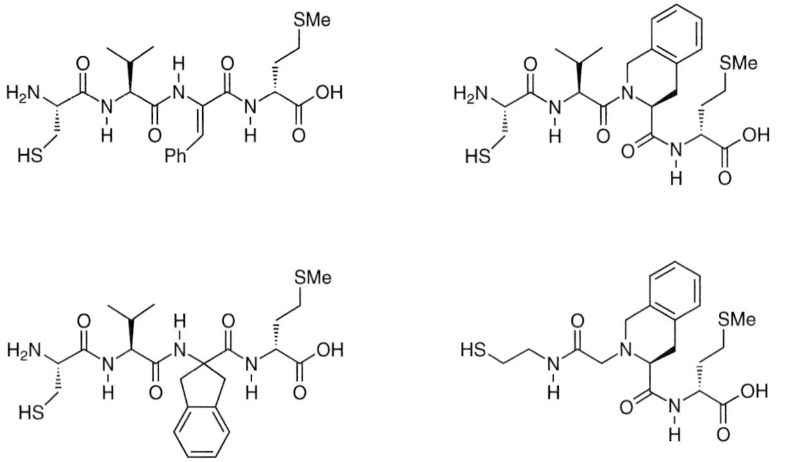

Figure 3. Farnesyltransferase inhibitors containing conformationally restricted amino acids.

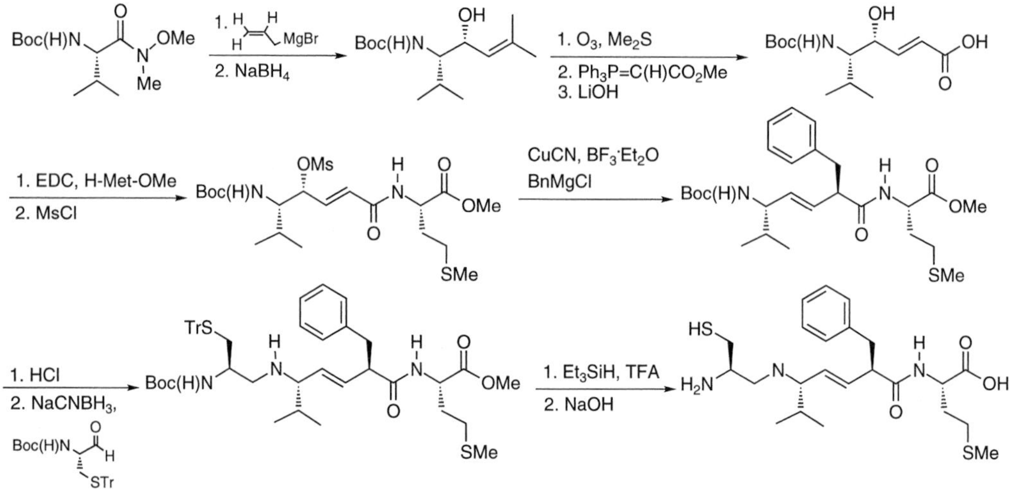

Scheme 4. Syntheses of an alkene moiety containing CAAX peptidomimetic.

Levitzki et al. who prepared a semipeptoid sequence by replacement of 'A' or 'AA' by a peptoid residue or dipeptoide sequence, respectively [23] (Scheme 5).

These examples show that state of the art organic chemistry is very effective in order to translate a peptide sequence into modified peptides or peptidomimetics having much

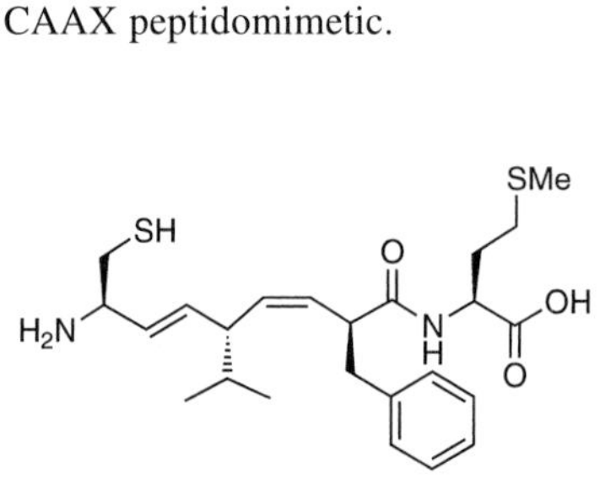

Figure 4. B956.

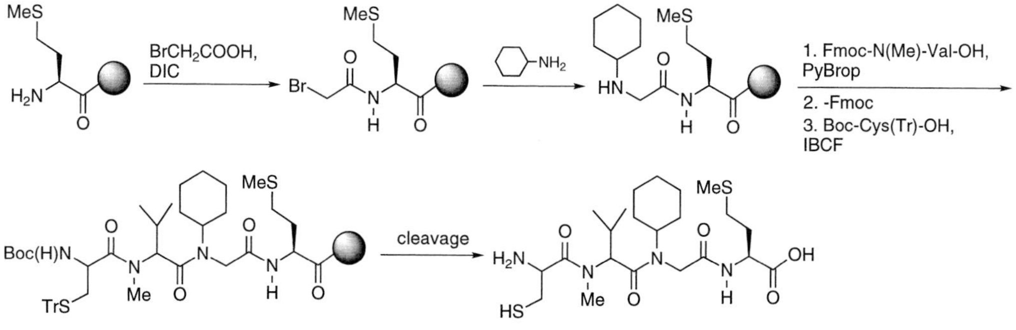

Scheme 5.

more favorable properties than the parent peptides. It is expected that the development of farnesyl inhibitors will continue to flourish. The preparation of tetrapeptides [24] and libraries of pentapeptides already led to non-thiol containing inhibitors of farnesyltransferase. [25] Furthermore, the isolation of naturally occurring farnesyltransferase inhibitors will undoubtedly lead to the synthesis of entirely new ones, [26] whereas the recent elucidation of the crystal structure of farnesyltransferase will surely lead to structure-based design of inhibitors. [27]

The use of modified peptides and peptidomimetics to selectively inhibit Ras farnesylation and thereby transformation and ultimately possibly the carcinogenic process represents an important and recent application of these compounds. [28, 29] It also shows that they can be employed for a finely tuned intervention in complex biochemical and biological processes. It is therefore expected that as the understanding of cellular processes (e. g. protein trafficking, intracellular signal transduction) progresses, it will be possible to intervene in, or selectively inhibit these processes using modified peptides and peptidomimetics, which are obtained by structure based design and/or combinatorial chemistry approaches.

References

[1] *Modified peptides* are being defined as peptides in which in essence the amino acid sequence is unchanged but the peptide contains e. g. some unnatural amino acids, a modification of a cysteine residue or a phospho amino acid. *Peptidomimetics* are compounds which imitate the structure and/or imitate or block the biological effect of a peptide at the receptor level [2a]. As a consequence peptidomimetics span the whole range of compounds varying from peptide isosteres to compounds without an identifiable amino acid or peptide moiety.,

[2] For reviews on peptidomimetics see e. g.: a) A. Giannis, T. Kolter, *Angew. Chem. Int. Ed. Engl.* **1993**, *32*, 1244–1267; b) R. M. J. Liskamp, *Recl. Trav. Chim. Pays-Bas*, **1994**, *113*, 1–19; c) A. E. P. Adang, P. H. H. Hermkens, J. T. M. Linders, H. C. J. Ottenheijm, C. J. van Staveren, *Recl. Trav. Chim. Pays-Bas*, **1994**, *113*, 63–78; d) J. Gante, *Angew. Chem. Int. Ed. Engl.* **1994**, *33*, 1699–1720; e) D. C. Rees, *Curr. Med. Chem.* **1994**, *1*, 145–158.

[3] Glycosylation see e. g.: M. A. Kukuruzinska, M. L. E. Bergh, B. J. Jackson, *Annu. Rev. Biochem.* **1987**, *56*, 915–944; acylation: D. A. Towler, J. I. Gordon, S. P. Adams, L. Glaser, *Annu. Rev. Biochem.* **1988**, *57*, 69–99; phosphorylation: E. G. Krebs, J. A. Beavo, *Annu. Rev. Biochem.* **1979**, *48*, 923–959; isoprenylation: [6]

[4] S. Clarke, *Annu. Rev. Biochem.* **1992**, *61*, 355–386.

[5] J. A. Glomset, M. H. Gelb, C. C. Farnsworth, *Trends Biochem. Sci.* **1990**, 139–142.

[6] J. B. Gibbs, *Cell*, **1991**, *65*, 1–4.

[7] N. E. Kohl, S. D. Mosser, S. J. DeSolms, E. A. Giuliani, D. L. Pompliano, S. L. Graham, R. L. Smith, E. M. Scolnick, A. Oliff, J. B. Gibbs, *Science* **1993**, *260*, 1934–1937; S. L. Graham, S. J. DeSolms, E. A. Giuliani, N. E. Kohl, S. D. Mosser, A. I. Oliff, D. L. Pompliano, E. Rands, M. J. Breslin, A. A. Deana, V. M. Garsky, T. H. Scholz, J. B. Gibbs, R. L. Smith, *J. Med. Chem.* **1994**, *37*, 725–732.

[8] N. E. Kohl, F. R. Wilson, S. D. Mosser, E. Giuliani, S. J. DeSolms, M. E. Conner, N. J. Anthony, W. J. Holtz, R. P. Gomez, Ta-Jyh. Lee, R. L. Smith, S. L. Graham, G. D. Hartman, J. B. Gibbs, A. Oliff, *Proc. Natl. Acad. Sci. USA* **1994**, *91*, 9141–9145.

[9] S. J. DeSolms, A. A. Deana, E. Giuliani, S. L. Graham, N. E. Kohl, S. D. Mosser, A. I. Oliff, D. L. Pompliano, E. Rands, T. H. Scholz, J. M. Wiggins, J. B. Gibbs, R. L. Smith, *J. Med. Chem.* **1995**, *38*, 3967–3971.

[10] G. L. James, J. L. Goldstein, M. S. Brown, T. E. Rawson, T. C. Somers, R. S. McDowell, G. W. Crowley, B. K. Lucas, A. D. Levinson, J. C. Marsters, *Science* **1993**, *260*, 1937–1942; J. C. Marsters Jr, R. S. McDowell, M. E. Reynolds, D. A. Oare, T. C. Somers, M. S. Stanley, T. E. Rawson, M. E. Struble, D. J. Burdick, K. S. Chan, C. M. Duarte, K. E. Paris, J. Y. F. Tom, D. T. Wan, Y. Xue, J. P. Burnier, *Bioorg. Med. Chem. Lett.* **1994**, *2*, 949–957; T. E. Rawson, T. C. Somers, J. C. Marsters Jr, D. T. Wan, M. E. Reynolds, D. J. Burdick, *Bioorg. Med. Chem. Lett.* **1995**, *5*, 1335–1338.

[11] For use of the benzodiazepine skeleton as *β*-turn mimetic see: W. C. Ripka, G. V. De Lucca, A. C. Bach II, R. S. Pottorf, J. M. Blaney, *Tetrahedron*, **1993**, *49*, 3593–3608.

[12] M. Nigam, C.-M. Seong, Y. Qian, A. D. Hamilton, S. M. Sebti, *J. Biol. Chem.*, **1993**, *268*, 20695–20698; Y. Qian, M. A. Blaskovich, C.-M. Seong, A. Vogt, A. D. Hamilton, S. M. Sebti, *Bioorg. Med. Chem. Lett.* **1994**, *4*, 2579–2584.

[13] Y. Qiam, M. A. Blaskovich, M. Saleem, C.-M. Seong, S. P. Wathen, A. D. Hamilton, S. M. Sebti, *J. Biol. Chem.* **1994**, *269*, 12410–12413; A. Vogt, Y. Qian, Blaskovich, R. D. Fossum, A. D. Hamilton, S. M. Sebti, *J. Biol. Chem.* **1995**, *270*, 660–664; E. Lerner, Y. Qian, A. D. Hamilton, S. M. Sebti, *J. Biol. Chem.* **1995**, *270*, 26770–26773.

[14] Y. Qiam, A. Vogt, S. M. Sebti, A. D. Hamilton, *J. Med. Chem.* **1996**, *39*, 217–223.

[15] T. M. Williams, T. M. Ciccarone, S. C. MacTough, R. L. Bock, M. W. Conner, J. P. Davide, K. Hamilton, K. S. Koblan, N. E. Kohl, A. M. Kral, S. D. Mosser, C. A. Omer, D. L. Pompliano, E. Rands, M. D. Schaber, D. Shah, F. R. Wilson, J. B. Gibbs, S. L. Graham, S. L. Hartman, A. I. Oliff, R. L. Smith, *J. Med. Chem.* **1996**, *39*, 1345–1348.

[16] See e. g. [2b].

[17] K. Leftheris, T. Kline, G. D. Vite, Y. H. Cho, R. S. Bhide, D. V. Patel, M. M. Patel, R. J. Schmidt, H. N. Weller, M. L. Andahazy, J. M. Carboni, J. L. Gullo-Brown, F. Y. F. Lee, C. Ricca, W. C. Rose, N. Yan, M. Barbacid, J. T. Hunt, C. A. Meyers, B. R. Seizinger, R. Zahler, V. Manne, *J. Med. Chem.* **1996**, *39*, 224–236.

[18] F.-F. Clerc, J.-D. Guitton, N. Fromage, Y. Lelièvre, M. Duchesne, B. Tocqué, E. James-Surcouf, A. Commerçon, J. Becquart, *Bioorg. Med. Chem. Lett.* **1995**, *5*, 1779–1784; G. Byk, M. Duchesne, F. Parker, Y. Lelievre, J. D. Guitton, F. F. Clerc, J. Becquart, J. Tocque, D. Scherman, *Bioorg. Med. Chem. Lett.* **1995**, *5*, 2677–2682; G. Byk, Y. Leleivre, F. F. Clerc, D. Scherman, J. D. Guitton, *Bioorg. Med. Chem.* **1997**, *5*, 115–124.

[19] For other recent examples of incorporation of the alkene moiety into biologically active peptides, see e. g.: J. S. Kaltenbronn, J. P. Hudspeth, E. A. Lunney, B. M. Michniewicz, E. D. Nicolaides, J. T. Repine, W. H. Roark, M. A. Stier, F. J. Tinney, P. K. W. Woo, A. D. Essenburg, *J. Med. Chem.* **1990**, *33*, 838–845; D. Tourwé, J. Couder, M. Ceusters, D. Meert, T. F. Burks, T. H. Kramer, P. Davis, R. Knapp, H. I. Yamamura, J. E. Leysen, G. van Binst, *Int. J. Pept. Prot. Res.* **1992**, *39*, 131–136; Y.-K. Shue, M. D. Tufano, G. M. Carrera Jr., H. Kopecka, S. L. Kuyper, M. W. Holladay,

C. W. Lin, D. G. Witte, T. R. Miller, M. Stashko, A. M. Nadzan, *Bioorg. Med. Chem.* **1993**, *1*, 161–171; M. Grommé, R. van der Valk, K. Sliedregt, R. Liskamp, G. Hämmerling, J. O. Koopmann, F. Momburg, J. Neefjes, *Eur. J. Immunol.* **1997**, *27*, 898–904.

[20] J. S. Wai, D. L. Bamberger, T. E. Fisher, S. L. Graham, R. L. Smith, J. B. Gibbs, S. D. Mosser, A. I. Oliff, D. L. Pompliano, E. Rands, N. E. Kohl, *Bioorg. Med. Chem.* **1994**, *2*, 939–947.

[21] M. D. Lewis, J. J. Kowalczyck, A. E. Christuk, R. Fan, E. M. Harrington, X. C. Sheng, Y. Hu, A. M. Carcia, I. Hishunuma, A. Et, Patent Application WO 95-US3387, *Chem. Abstr.* **124**, 146855; T. Nagusu, K. Yashimatsu, C. Rowell, M. D. Lewis, A. M. Garcia, *Cancer Res.* **1995**, *55*, 5310–5314.

[22] R. J. Simon, R. S. Kania, R. N. Zuckermann, V. D. Huebner, D. A. Jewell, S. Banville, S. Ng, L. Wang, S. Rosenberg, C. K. Marlowe, D. C. Spellmeyer, R. Tan, A. D. Frankel, D. V. Santi, F. E. Cohen, P. A. Bartlett, *Proc. Natl. Acad. Sci. USA*, **1992**, *89*, 9367–9371; H. Kessler, *Angew. Chem. Int. Ed. Engl.* **1993**, *32*, 543–544; S. M. Miller, R. J. Simon, S. Ng, R. N. Zuckermann, J. M. Kerr, W. H. Moos, *Bioorgan. Med. Chem. Lett.* **1994**, *22*, 2657–2662; R. N. Zuckermann, E. J. Martin, D. C. Spellmeyer, G. B. Stauber, K. R. Shoemaker, J. M. Kerr, G. M. Figliozzi, D. A. Goff, M. A. Siani, R. J. Simon, S. C. Banville, E. G. Brown, L. Wang, L. S. Richter, W. H. Moos, *J. Med. Chem.* **1994**, *37*, 2678–2685.

[23] H. Reuvini, A. Gitler, E. Poradosu, C. Gilon, A. Levitzki, *Bioorg. Med. Chem.* **1997**, *5*, 85–92.

[24] J. T. Hunt, V. G. Lee, K. Leftheris, B. Seizinger, J. Carboni, J. Mabus, C. Ricca, N. Yan, V. Manne, *J. Med. Chem.* **1996**, *39*, 353–358.

[25] D. M. Leonard, K. R. Shuler, C. J. Poulter, S. R. Eaton, T. K. Sawyer, J. C. Hodges, T.-Z. Su, J. D. Scholten, R. C. Gowan, J. S. Sebolt-Leopold, A. M. Doherty, *J. Med. Chem,.* **1997**, *40*, 192–200.

[26] R. Sekizawa, H. Iinuma, Y. Muraoka, H. Naganawa, N. Kinoshita, H. Nakamura, M. Hamada, T. Takeuchi, K. I. Umezawa, *J. Nat. Prod.* **1996**, *59*, 232–236.

[27] H. W. Park, S. R. Boduluri, J. F. Moomaw, P. J. Casey, L. S. Beese, *Science*, **1997**, *275*, 1800–1804.

[28] J. Travis, *Science* **1993**, *260*, 1877–1878.

[29] For recent reviews on inhibitors of Ras farnesyltransferases, including microbial products identified as such see: F. Tamanoi, *Trends Biochem. Sci.* **1993**, 349–353; G. L. Bolton, S. Sebolt-Leopold, Hodges, *Ann. Rep. Med. Chem.* **1994**, *29*, 165–174; J.C.; S. Ayral-Kaloustian, J. S. Skotnicki, *Ann. Rep. Med. Chem.* **1996**, *31*, 171–179; J. D. Scholten, K. Zimmerman, M. Oxender, J. Sebolt-Leopold, R. Gowan, D. Leonard, D. J. Hupe, *Bioorg. Med. Chem.* **1996**, *4*, 1537–1543; D. M. Leonard, *J. Med. Chem.* **1997**, *40*, 2971–2990; Y. Qian, S. M. Sebti, A. D. Hamilton, *Biopolym.* **1997**, *43*, 25–41.

Mechanically-Interlocked Molecular Systems Incorporating Cyclodextrins

Sergey A. Nepogodiev and J. Fraser Stoddart

If any theme is beginning to dominate the development of synthetic chemistry these days, it must surely be self-assembly. [1] The spread of this concept, though somewhat belatedly, into all branches of chemistry is mainly a result of the rapidly accelerating growth and acceptance of supramolecular chemistry. [2] Long before chemists were thinking and practising their own brands of chemistry beyond the molecule with wholly synthetic systems, nature had been extolling the virtues of molecular recognition in the production of sophisticated molecular assemblies and supramolecular arrays with linked forms and functions.

If one class of compounds has provided the bridge between the worlds of natural and unnatural hosts, it has been the cyclodextrins [3] – or CDs (Fig. 1a) as the aficionados call them. Yet, even as they enter their second century [3b] of cultivation by chemists, they continue to fascinate the chemical community. [3c] The reason is quite simple: they are aesthetically-appealing molecules, and they also lend themselves to novel experiments. The series of reports from different research laboratories [4–12] in recent years on the self-assembly of CDs into [2]rotaxanes [13] and polyrotaxanes (Fig. 1b,c) serves to remind us all that there is still a lot of innovation to be forged in and around CDs in the coming one-hundred years.

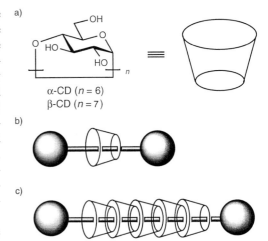

Figure 1. Schematic representations of (a) α-cyclodextrin (α-CD) and β-cyclodextrin (β-CD), (b) a [2]rotaxane, and (c) a polyrotaxane.

The self-assembly of CD rotaxanes is expected on account of the strongly pronounced ability of CDs to form inclusion complexes with organic molecules. In principle, any through-ring CD complex can be considered as a pseudorotaxane. Thus, polymethylene compounds, terminated with appropriate end groups such as pyridinium, [14] bipyridinium, [15] or carboxyl groups [16] could form rather stable complexes with α-CD where the hydrocarbon chain is located inside the CD cavity.

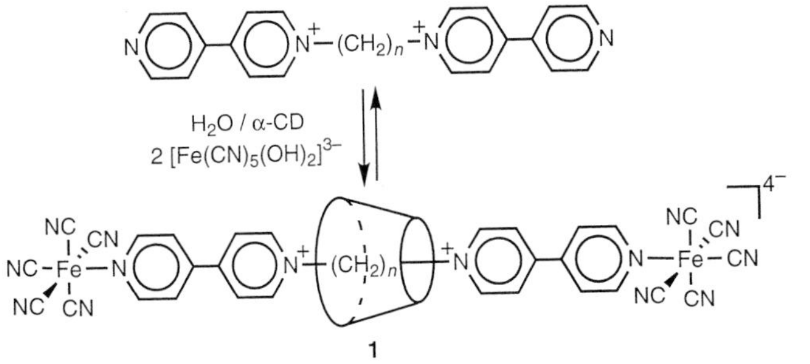

Scheme 1. The [2]rotaxane-metal complex **1** of Macartney et al. [5]

Extending the lifetime of such complexes may be achieved by using more bulky end groups like 4-*tert*-butylpyridinium [17] stoppers and the slippage mechanism for rotaxane formation. However, even in this case, the formation of the rotaxane is reversible. Therefore, the task of creating CD rotaxanes comes down to the problem of putting bulky stoppers on the ends of a molecule threaded through the CD ring. This task is not a trivial one because of the necessity of carrying out reactions in aqueous media where the incorporation of guest molecules by CDs normally takes place. One of the solutions to the problem of capturing CDs on included dumbbell-shaped molecules lies in the construction of coordination linkages – an approach first reported by Ogino in 1981. [4]

More recently, the construction of a series α-CD [2]rotaxanes, based on the attachment of stoppers to threads by means of metal complex formation, was described by Macartney et al. [5] They have shown that the self-assembly of rotaxanes of the type $[(NC)_5Fe\{R(CH_2)_nR' \cdot \alpha\text{-}CD\}Fe(CN)_5]^{4-}$ where R and R' are bipyridinium or pyrazinium (e. g., **1** in Scheme 1) happens irrespective of the order of the addition of the components to the aqueous solution. This observation infers a slow dissociation of one of the $[Fe(CN)_5]^{3-}$ ions, followed by α-CD inclusion and the recomplexation of $[Fe(CN)_5]^{3-}$ ion. Evidence for the formation of [2]rotaxanes is provided by the ^1H NMR spectra, in which the signals of the symme-

try-related protons of the $[R(CH_2)_nR']^{2+}$ unit are separated on account of the end-to-end asymmetry associated with the trapped α-CD ring. It is conceivable that this methodology could be applied to the preparation of a family of similar [2]rotaxanes with different threads and a broad range of stoppers, such as $[M(CN)_5]^{n-}$ and $[M(NH_3)_5]^{m+}$ ions with redox active d^6 metal (M) centers like Co, Ru, Os, and Fe.

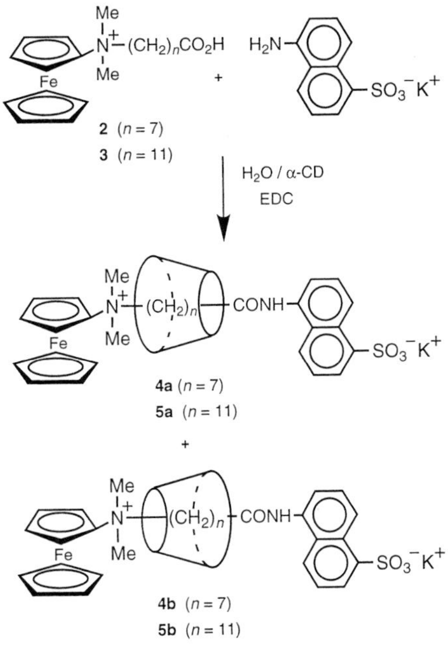

Scheme 2. The zwitterionic [2]rotaxanes **4** and **5** of Isnin and Kaifer [6]

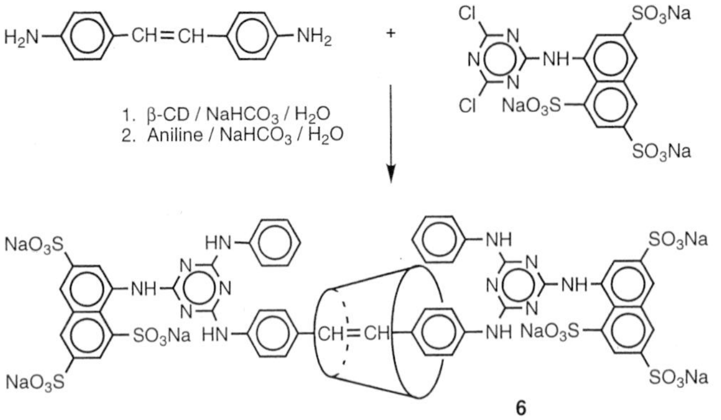

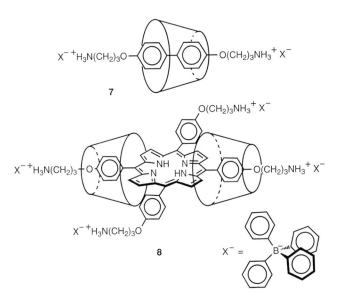

Scheme 3. The β-CD [2]rotaxane **6** with a covalently-stoppered diaminostilbene dumbbell described by Nakashima et al. [7]

In 1991, Isnin and Kaifer [6] announced the synthesis of the asymmetric zwitterionic [2]rotaxanes **4a, b** and **5a, b** incorporating α-CD rings *via* amide bond formation promoted by the water-soluble coupling reagent – 1-[3-dimethylamino)-propyl]-3-ethylcarbodiimide (EDC) (Scheme 2). In these cases, α-CD first of all binds oligomethylene chains of **2** (*n* = 7) or **3** (*n* = 11) bearing carboxyl groups which were subsequently capped with aminonaphthyl residues to afford isomeric [2]rotaxanes **4a, b** and **5a, b** in 15 % yields. The isomers were subsequently separated

[18] and the relative orientations of the α-CD rings on the asymmetric dumbbell-shaped components have been established: one isomer is stable but the other one unthreads slowly!

Another example of a CD [2]rotaxane in which the thread is capped with covalently-linked stoppers have been reported by Nakashima et al. [7] Using a host-guest interaction between β-CD and stilbene and the reactivity of the cyanuric chloride whereby chlorine substituents are displaced with amines in aqueous solution, they performed

Figure 2. The β-CD rotaxanes **7** and **8** with ammonium tetraphenylborate stoppers synthesized by Lawrence et al. [8]

the two-step synthesis (Scheme 3) of the [2]rotaxane **6**.

A very simple way of stoppering a bisammonium thread by using NaBPh₄ was proposed by Lawrence and Rao. [8a] They managed to obtain a "threaded molecular loop" – the per-2,6-dimethyl-β-CD [2]rotaxane **7** – in an excellent 71% yield by doing the self-assembly in aqueous solution (Fig. 2). Furthermore, the same methodology has been employed successfully [8b] for the construction of the [3]rotaxane **8** incorporating a tetraarylporphyrin unit.

The possibility of carrying out various chemical modifications on CDs has been used by Wenz et al. [9] for creating a series of lipophilic derivatives, which are not only readily soluble in most organic solvents, but also can bind cationic species. A new CD host molecule was used (Scheme 4) for self-assembling the CD [2]rotaxanes **10a** and **10b** incorporating the 4,4′-bipyridinium unit. The formation of these rotaxanes occurs as a result of the quaternization of the pyridine nitrogen of the monocations **9a, b** when they are complexed within the CD host. It has been proved that this process could take place only with a certain orientation of the CD ring with respect to the thread. Therefore, the selective formation of a single orientational isomer could be predicted in cases of [2]rotaxanes of the type **10** incorporating asymmetrical dumbbell units.

The isomer phenomena literally takes on a further dimension in polyrotaxanes, for

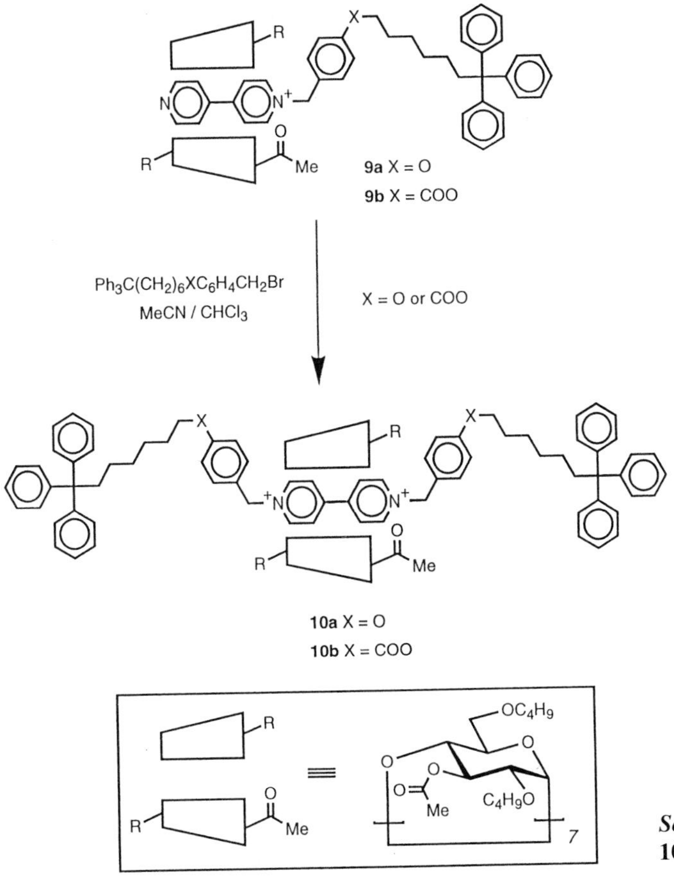

Scheme 4. The lipophilic [2]rotaxanes **10** constructed by Wenz et al. [9]

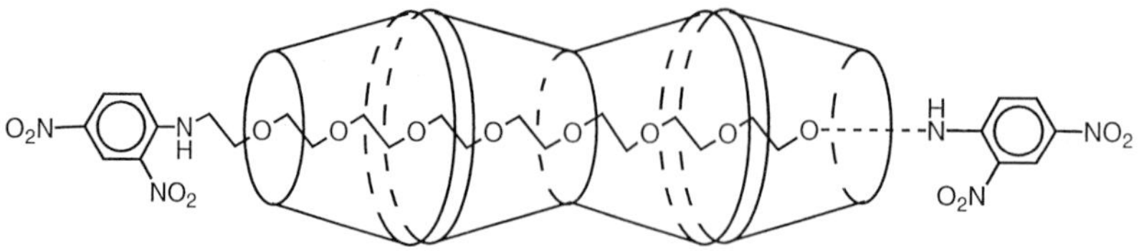

Figure 3. Schematic representation of polyrotaxane constructed by Harada et al. [10]

which the view has grown that the CDs might thread under conditions of equilibrium control with alternating orientations, wherein adjacent rings are matched head-to-head/tail-to-tail in order to optimize hydrogen bonding between the neighboring CD units.

In the knowledge that chains of poly(ethyleneglycol) (PEG) thread α-CD beads like a necklace, [19] Harada and co-workers [10] have succeeded in capping the chain ends of a poly(ethyleneglycol)diamine (PEG-DA, M_w 3450) with dinitrophenyl stoppers. The reac-

tion was carried out in a solution of PEG-DA in dimethylformamide, saturated with α-CD and laced with a gross excess (46 mol equiv) of 2,4-dinitrofluorobenzene. After an extensive purification procedure, full characterization of the product, which was obtained in high yield (60 %), indicated average molecular weights of 23 200 (by ^{1}H NMR) and 24 600 (by UV) that are commensurate with the respective threading of 20 and 23 α-CD rings per polyrotaxane molecule (Fig. 3). This type of molecular structure

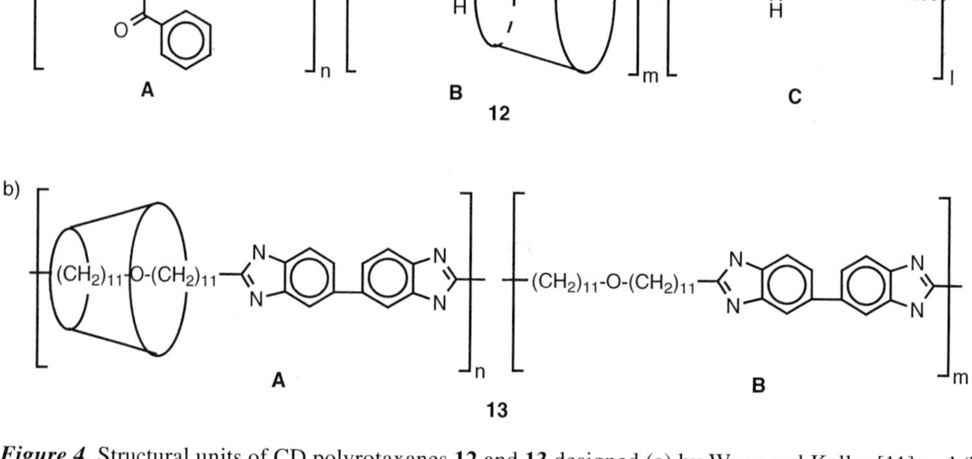

Figure 4. Structural units of CD polyrotaxanes **12** and **13** designed (a) by Wenz and Keller [11] and (b) by Osakada and Yamamoto [12]

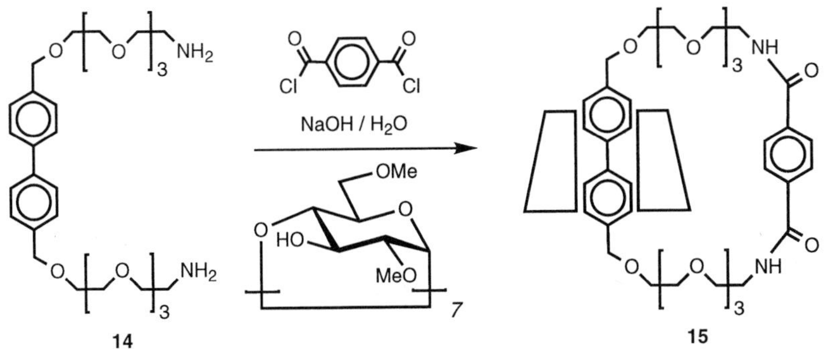

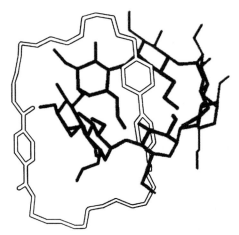

Scheme 5. Synthesis of [2]catenane **15** [22]

conjures up the image of a molecular abacus [20] in terms of both the structure and dynamics.

Quite independently, Wenz and Keller [11] have demonstrated that threading of CD molecules on to polymeric secondary amines affords water-soluble polymeric inclusion complexes. The complex of the poly(imine-trimethyleneimino-decamethylene) **11** with $P_n = 23 \pm 5$ is very stable: its equilibrium dialysis into separate components was far from complete after two weeks. When $11 \cdot (a\text{-CD})_n$ was treated with nicotinoyl chloride to introduce at least two hydrophilic nicotonoyl blocking groups at arbitrary positions along the poly(iminooligomethylene) chain, a polyrotaxane **12** (Fig. 4a) of M_w 55 000 $\pm$ 5000 (determined by laser light-scattering) was isolated with an average of 37 a-CD rings permanently threaded on to the chain, a number that corresponds to 67% of the monomer units with the polymer covered by CD. A similar type of polyrotaxane, with CD rings randomly clipped between blocking groups along the polymer chain, has been prepared by Osakado, Yamamoto and Yamaguchi. [12] Polycondensation of 3,3'-diaminobenzidine and 1,12-dodecanediol in the presence of a-CD (1 : 3 : 0.5 molar ratio), a reaction catalyzed by RuCl$_2$(PPh$_3$)$_3$, led to the irregular copolymer **13** containing structural units A and B in a 16 : 84 ratio (Fig. 4b).

Another possible way of forming interlocked molecular systems incorporating CDs involves in the macrocyclization of an acyclic guest threaded through a CD in a process which leads to catenated molecules. [21] In 1993, such a series of catenanes, based on dimethyl-β-CD (DM-β-CD), was realized for the first time in our research laboratories. [22] Simple Schotten-Baumann condensation of terephtaloyl chloride with the diamine **14** in a basic aqueous solution of DM-β-CD afforded (Scheme 5) the [2]catenane **15** (see Fig. 5 for an X-ray crystal structure) and

Figure 5. A framework representation of the crystal structure of the catenated CD **15**. The DM-β-CD ring component is shown black.

three other catenanes – a [2]catenane and two [3]catenanes involving the corresponding macrocyclic dimer.

These exciting developments surrounding CDs [23] have been going on against a background of activity in other laboratories on self-assembling wholly synthetic rotaxanes, polyrotaxanes [24] and catenanes. This area was reviewed recently. [1, 13, 21] Self-assembly is the unifying concept in the construction of all these mechanically-interlocked molecular systems. As we begin to comprehend the rules of the game for different systems in a range of media, our fundamental understanding of molecular recognition benefits by leaps and bounds. Applications will begin to surface on a somewhat longer time-scale. Suffice it to say at present that there are high hopes for the advent of new materials with both novel forms and functions.

References

[1] D. Philp, J. F. Stoddart, *Angew. Chem. Int. Ed. Engl.* **1996**, *35*, 1154.

[2] J.-M. Lehn, *Supramolecular Chemistry*, VCH, Weinheim, 1995; J.-M. Lehn, *Angew. Chem. Int. Ed. Engl.* **1988**, *27*, 89; *ibid.* **1990**, *29*, 1304.

[3] a) *Comprehensive Supramolecular Chemistry*; J. L. Atwood, J. E. D. Davies, D. D. MacNicol, F. Vögtle (Eds.), *Cyclodextrins*; J. Szejtli, T. Osa (Eds.), Elsevier Sci.: Oxford, 1996, Vol. 3; b) J. F. Stoddart, *Carbohydr. Res.* **1992**, *192*, xii; c) G. Wenz, *Angew. Chem. Int. Ed. Engl.* **1994**, *33*, 803.

[4] H, Ogino, *J. Am. Chem. Soc.* **1981**, *103*, 1303; H, Ogino, K. Ohata, *Inorg. Chem.* **1984**, *23*, 2312; H, Ogino, *New. J. Chem.* **1993**, *17*, 683.

[5] R. S. Wylie, D. H. Macartney, *J. Am. Chem. Soc.* **1992**, *114*, 3136; R. S. Wylie, D. H. Macartney, *Supramol. Chem.* **1993**, *3*, 29; D. H. Macartney, C. A. Wadding, *Inorg. Chem.* **1994**, *33*, 5912.

[6] R. Isnin, A. E. Kaifer, *J. Am. Chem. Soc.* **1991**, *113*, 8188.

[7] M. Kunitake, K. Kotoo, O. Manabe, T. Muramatsu, N. Nakashima, *Chem. Lett.* **1993**, 1033.

[8] a) T. V. S. Rao, D. S. Lawrence. *J. Am. Chem. Soc.* **1990**, *112*, 3614; b) J. S. Manka, D. S. Lawrence. *J. Am. Chem. Soc.* **1990**, *112*, 2440.

[9] G. Wenz, E. von der Bey, L. Schmidt, *Angew. Chem. Int. Ed. Engl.* **1992**, *31*, 783; G. Wenz, F. Wolf, M. Wagner, S. Kubik, *New J. Chem.* **1993**, *17*, 729.

[10] A. Harada, J. Li, M. Kamachi, *Nature*, **1992**, *356*, 325; A. Harada, T. Nakamitsu, J. Li, M. Kamachi, *J. Org. Chem.* **1993**, *58*, 7524; A. Harada, J. Li, M. Kamachi, *J. Am. Chem. Soc.* **1994**, *116*, 3192.

[11] G. Wenz, B. Keller, *Angew. Chem. Int. Ed. Engl.* **1992**, *31*, 197.

[12] I. Yamaguchi, K. Osakada, T. Yamamoto, *J. Am. Chem. Soc.* **1996**, *118*, 1811.

[13] For review on rotaxanes and other interlocked structures, see: D. B. Amabilino, J. F. Stoddart, *Chem. Rev.* **1995**, *95*, 2725.

[14] H. Saito, H. Yonemura, H. Nakamura, T. Matsuo, *Chem. Lett.* **1990**, 535.

[15] H. Yonemura, H. Saito, S. Matsushima, H. Nakamura, T. Matsuo, *Tetrahedron. Lett.* **1989**, *30*, 3143; H. Yonemura, M. Kasahara, H. Saito, H. Nakamura, T. Matsuo, *J. Phys. Chem.*, **1992**, *96*, 5765.

[16] M. Watanabe, H. Nakamura, T. Matsuo, *Bull. Chem. Soc. Jpn.*, **1992**, *65*, 164.

[17] D. H. Macartney, *J. Chem. Soc., Perkin Trans. 2* **1996**, 2775.

[18] See the comments in an article by R. Dagani, *Chem. Eng. News,* **1992**, *70*, (15) 39.

[19] Complexes of CDs with different polymers have been reported: *a*-CD forms crystalline complexes with PEG: A. Harada, M. Kamachi, *Macromolecules* **1990**, *23*, 2821; *β*-CD and *γ*-CD give crystalline and stoichiometric complexes with poly(propyleneglycol): A. Harada, M. Okada, J. Li, M. Kamachi, *Macromolecules* **1995**, *28*, 8406.

[20] M. V. Reddington, A. M. Z. Slawin, N. Spencer, J. F. Stoddart, C. Vicent, D. J. Williams, *J. Chem. Soc. Chem. Commun.* **1991**, 630.

[21] For recent review on catenanes see: M. Belohradsky, F. M. Raymo, J. F. Stoddart, *Collect. Czech. Chem. Commun.* **1997**, *62*, 527.

[22] a) D. Armspach, P. R. Ashton, C. P. Moore, N. Spencer, J. F. Stoddart, T. J. Wear, D. J.

Williams, *Angew. Chem. Int. Ed. Engl.* **1993**, *32*, 854; b) D. Armspach, P. R. Ashton, R. Ballardini, V. Balzani, A. Godi, C. P. Moore, L. Prodi, N. Spencer, J. F. Stoddart, M. S. Tolley, T. J. Wear, D. J. Williams, *Chem. Eur. J.* **1995**, *1*, 34.

[23] Apart from the parent CDs, fully synthetic CD analogs may have some advantages in the construction of mechanically-interlocked systems. Thus, D/L-alternating cyclic oligosaccharides designed and synthesized in our laboratories have S_n symmetry and so behave as molecular cylinders rather than as molecular lampshades.

a) P. R. Ashton, C. L. Brown, S. Menzer, S. A. Nepogodiev, J. F. Stoddart, D. J. Williams, *Chem. Eur. J.* **1996**, *2*, 580; b) P. R. Ashton, S. J. Cantrill, G. Gattuso, S. Menzer, S. A. Nepogodiev, A. N. Shipway, J. F. Stoddart, D. J. Williams, *Chem. Eur. J.* **1997**, *3*, 1299; c) G. Gattuso, S. Menzer, S. A. Nepogodiev, J. F. Stoddart, D. J. Williams, *Angew. Chem. Int. Ed. Engl.* **1997**, *36*, 1451.

[24] For examples of comb-shaped CD rotaxane polymers, see M. Born, H. Ritter, *Macromol. Chem. Rapid Commun.* **1991**, 471; M. Born, H. Ritter, *Adv. Mater.* **1996**, *8*, 149.

Bolaamphiphiles: Golf Balls to Fibers*

Gregory H. Escamilla and George R. Newkome

A bolaamphiphile is simply defined as a molecule in which two or more hydrophilic groups are connected by hydrophobic functionality (Scheme 1). Other terms such as bolaform amphiphile [1] or arborol [2] have been used to describe molecules possessing this general architecture. Bolaform electrolytes or bolytes, a term introduced [1] by Fuoss and Edelson in 1951, are structurally similar except the hydrophilic groups are ionic. [3] Since Fuhrhop and Mathieu [4] reported in 1984 the synthesis and self-assembly of several bolaamphiphiles, other researchers have explored the applications of this basic architecture to a variety of situations. A search of current literature demonstrates the potential of bolaphiles, short for bolaamphiphiles, for the preparation of ultrathin monolayer membranes and the disruption of biological membranes, which could lead to therapeutic agents, [5, 6] as well as their use as "antisense" agents; [7, 8] as catalysts in reactions; [9–11] and as models for the membranes found in thermophilic archaebacteria. [12]

In natural membranes bolaform amphiphiles have chiral head groups, which are different in size and thus react stereoselectively with guest molecules. [3] In non-natural membranes binding sites can also be present, [9, 13] or other functional groups can be incorporated such as redox- and/or photoactive moieties. [4] The repertoire of bolaphiles is not limited to assemblies such as membranes and vesicles; specific two-directional arborols can form gels, [2, 14–16] and a-L-lysine-ω-aminobolaphiles form rods and tubules. [17] When polymerizable functionalities are included either in the head groups or within the hydrophobic spacer, routes to extended covalently linked domains are created. [18, 19]

Nontraditional structures (e. g., aqueous gels) result when non-covalent forces are coupled with a proper molecular architecture. The series of two-directional arborols, synthesized by Newkome et al., demonstrates this variation on the theme of the self-assembly of ordered arrays. [13, 14] In the proposed model of aggregated structure, or automorphogenesis, [20] one arborol is stacked upon the next; the hydrophobic chains overlap and the spherical, connecting hydrophilic hydroxyl termini are oriented out toward the aqueous solution (Fig. 1). Hydrogen bonding between the polar end groups and the water results in the formation of a thermoreversible gel. This proposed model is supported by the long fiber-like structures seen in electron micro-

---

* Chemistry of Micelles, Part 76. Support for this work was provided by the National Science Foundation (DMR-92-17331; 92-08925, 96-22609).

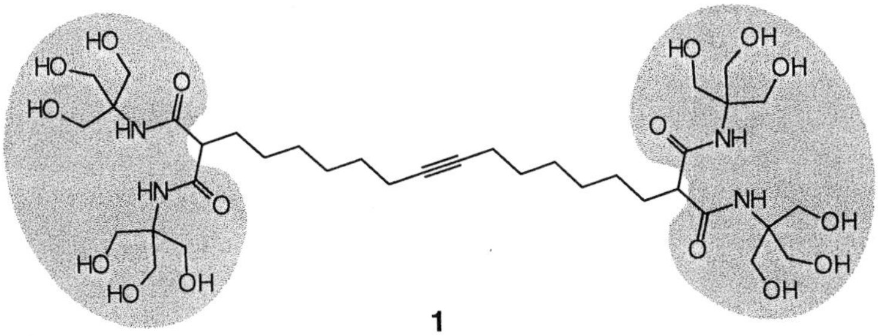

Scheme 1. Examples of bolaamphiphiles; the basic elements are hydyrophilic head groups (shaded) and the hydrophobic spacers (**1** from [14], **2** from [5], **3** from [3])

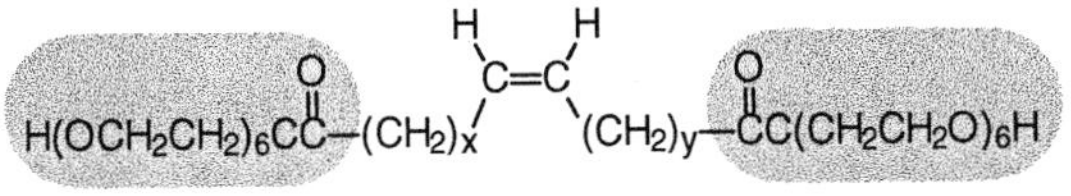

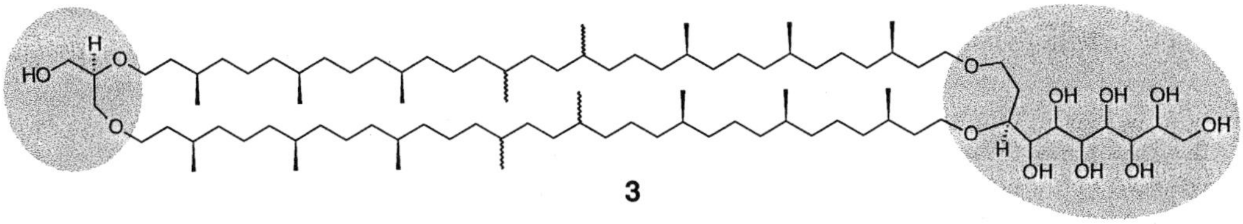

graphs of these molecular assemblages. [13, 21, 22] As with micelle formation, there is a minimum hydrophobic moiety needed for gelation to occur; for a saturated spacer this length is eight methylene units. [14] Aqueous *N,N',N'',N'''*-tetrakis[2-hydroxy-1,1-bis(hy-

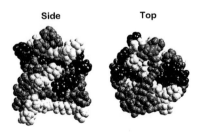

Side   **Top**

Figure 1. Spatial (side and top) representations of the stacking pattern calculated for **1**.

droxymethyl)ethyl]hexadec-8-yne-1,1,16,16-tetracarboxamide formed a novel helical thread and macroassemblies; whereas with added [18] crown-6-ether, particles are formed rather than fibers, as evidenced in electron micrographs of resulting gels. [13]

Related self-assembled, supramolecular assemblies have been reported by Jørgensen et al., [23] in which tetrathiafulvene (TTF) was synthetically incorporated into the inner hydrophobic region of the bolamphile. Aggregates, created from packed TTF-2-directional arborols, were shown to be thin string-like molecular stacks with lengths in the order of tens of microns and diameters of approaching 100 microns; it would appear that these superstructures are hydrogen bonded aggregates of the single strands.

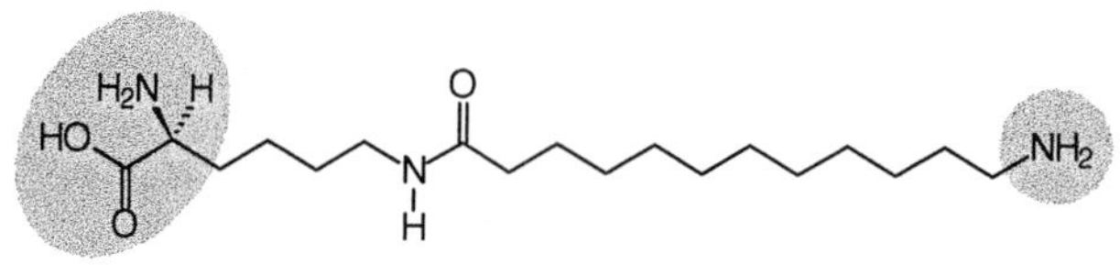

Scheme 2. The unsymmetrical bolaphiles synthesized by Fuhrhop et al. [17] for the generation of monolayer rods and tubules.

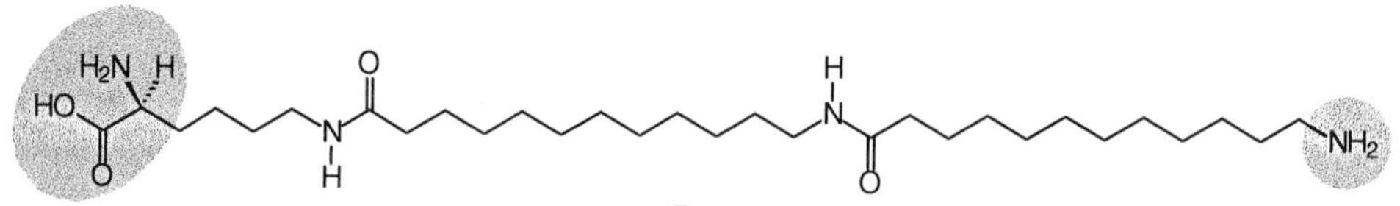

The rods and tubules, described by Fuhrhop et al. [17] follow an analogous pattern to that of these the 2-directional arborols. When the racemic form of the lysine bolaphile **4** was used (Scheme 2), no rod or tubule formation was noted; in contrast, the racemate of the extended bolaphile **5** produced supramolecular assemblies identical to those formed by the pure enantiomers. The effect of configuration in molecular monolayers was shown to depend on membrane curvature in the same manner as in bilayers. [17]

A related bipolar dodecane possessing *N*-benzyloxycarbonyl-L-phenylalanine terminal units has been shown [24] to form bundles of braided tapes in a hexane/EtOAc mixture. In order to probe the function of the urethane moieties, deprotection afforded the corresponding bis(*a*-aminoamide), which gave solids lacking any microstructural character. Bolaphiles composed of straight chain alkane termini have been reported [25] to afford self-assembled supramolecular structures in water.

Fuhrhop et al. demonstrated [26] that placing quinone moieties in either the head groups or within the hydrophobic spacer, formed redox-active membranes; thus, unsymmetrical bolaphile **6** (Scheme 3) is a model of electron acceptors in photosynthesis and forms vesi-cles as shown by electron micrographs. In the anticipated membrane organization, it was assumed that all of the quinone moieties must be located on the vesicle exterior. This assumption was confirmed by quantitative reduction of the quinone moieties with borohydride; since the reagent does not diffuse through lipid membranes, these "membrane quinones" must be part of the vesicle exterior.

The anthraquinone-based bolaphile **7** or **8** could not form homogeneous membranes due to steric interactions (Scheme 3); however, they could be integrated into host membranes formed by dihexadecylphosphate and dimethyldioctadecylammonium bromide. The bolaamphiphile fixed the quinone functionality in the center of the membrane and served as a model for pool quinones. [20] Electron transfer was further demonstrated by light-induced charge transfer between cationic porphyrins dissolved in water and membrane-bound anthraquinone bolaphiles. [27]

Bunton et al. proved that alkane a,ω-bis(trimethylammonium) bolaphiles catalyze the spontaneous hydrolysis of 2,4-dinitrophenyl-phosphates. [11] The rate of hydrolysis was enhanced by micellar bolaphiles; this rate enhancement followed the greater organiza-

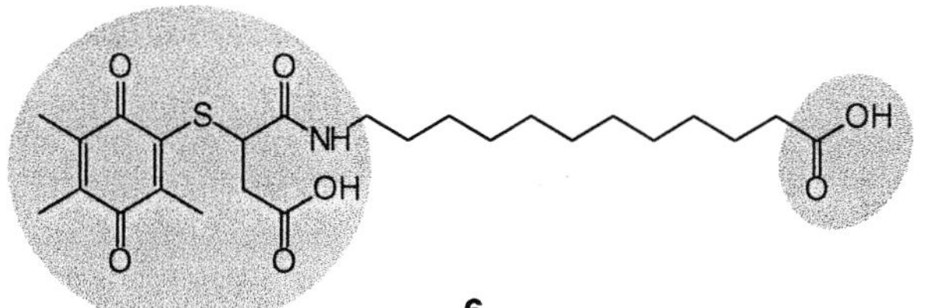

6

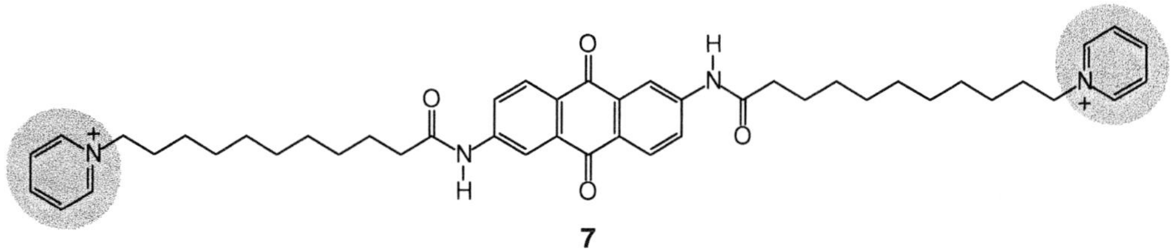

7

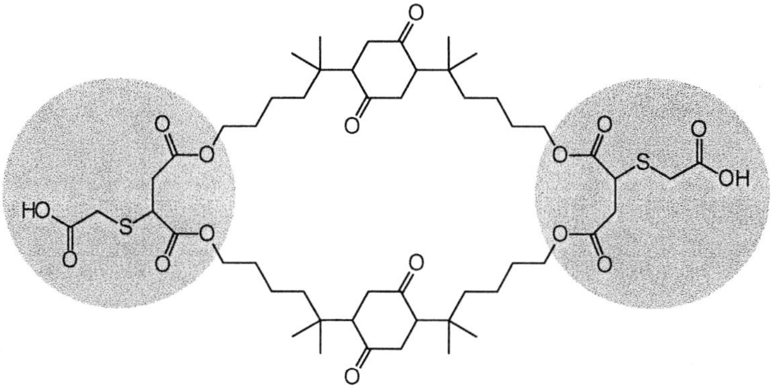

8

Scheme 3. The unsymmetrical bolaphile **6** was used to fix the quinone functionality at the surface of the micelle; whereas, **7** and **8** were used to position the quinone moiety within the membrane.

tion seen in vesicles formed by surfactants possessing longer spacers. Bolaphiles with hexadecane and dodecane spacers did not form micelles but rather small clusters instead, and the rate enhancement achieved was much lower than that with micelle-forming surfactants. [11] The aggregation and microstructure of other alkane α,ω-bis(ammonium) salts have been studied. [28]

Fornasier et al. used metallomicelles formed from bolaphiles and surfactants with only one head group as catalysts for the cleavage for *para*-nitrophenyl picolinate. [9]

Micelles generated from pyridine bolaphiles (Scheme 4) and Cu^{II} or Zn^{II} enhanced the rate of cleavage. As a comparison, the micelles generated from 6-{[(2-(*n*-hexadecyl)dimethylamino)ethyl]thio}methyl-2-(hydroxymethyl)pyridine bromide were not as effective catalysts. Increased electrostatic repulsion between the ammonium group and the metal ion site rationalized the difference in effectiveness of the bolaform versus classical micelle. The crux of the catalytic process was formation of a ternary complex consisting of substrate, metal, and bolaphile, as ligand. [9]

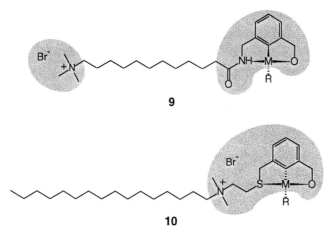

9

10

Scheme 4. The proximity of the charged head group to the binding site in **10** slows chelation of the metal relative to that of bolaphile **9**. M=Cu^{II}, Zn^{II}; R=*para*-nitrophenyl picolinate.

These micelles demonstrate substrate discrimination, in other words, the correct substrate geometry is required for catalysis. Accordingly, no catalytic enhancement was found for the hydrolysis of the isomeric nicotinate or isonicotinate esters.

Nolte and coworkers reported [29] the generation of a functional vesicle formed by an assembly of hosts (Scheme 5) capable of binding guests such as resorcinol and 4-(4'-nitrophenylazo)resorcinol. In order to maintain the hydrophobic binding site at the exterior of the vesicle, two quaternary nitrogen atoms each having a long alkyl chain were utilized. The rigidity of the binding site held the charged nitrogen atoms apart and fixed the binding site near the plane of the nitrogen atoms and therefore the vesicle surface. Typical for lipids, the hydrophobic alkyl chains served to form the membrane interior. Interactions of solvent, binding sites, charged nitrogen atoms, and hydrophobic alkyl chains together govern the vesicular morphology, which was likened to a golf ball because of its spherical shape and dimpled surface. [29] The golf ball analogy is completed by the concave shape of the binding site in the host. Examination of a cast film by X-ray diffraction yields a clear periodicity of 53 Å, which is approximately the length of two fully ex-

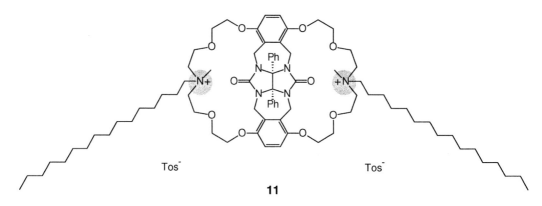

11

Scheme 5. Representation of bolamphile **11** used by Nolte et al. [29] to prepare a functionalized vesicle.

tended hexyldecylamine chains. Binding studies suggest that only half the total number of binding sites are able to interact with the guest. These data indicate a bilayer membrane with a thickness of approximately 53 Å.

Jayasuriya et al. [5] hypothesized that bolaform amphiphiles could be tailored to provide an optimal geometry for membrane disruption. Variation of either the head groups or spacer could provide a "tunability" to target certain microorganisms. The bolaphiles were proposed [5] to disrupt membranes by insertion of their hydrophobic spacers into the lipid layer, which would cause a mismatch of the preferred geometry of the bolaphile and lipid. When increased interruptions occur, the membrane would become more destabilized (Scheme 6). Spacers with central triple or double bonds were made with the expectation that these functionalities would increase the disruptive capability of the bolaform amphiphiles. Alkane spacers ranged from decane to eicosane, from which maximum activity was observed with pentadecane. In the olefinic analogues, the geometry and position of the double bond were less important in determining membrane disruption. Similarly, the position of unsaturation was not important for the acetylenic bolaphiles. [5] A general

trend was that maximum activity required an increase in spacer length when moving from saturated to olefinic to acetylenic spacers.

The use of polymeric strings of bolaphiles in membrane disruption was described by Jayasuriya et al. (Scheme 7). [6] The membrane disruptive activity of the polymeric bolaamphiphiles or "supramolecular surfactants" is up to 10^3 times greater than that of the monomeric analogues. The precise origin of this amplification is not yet understood; however, several factors were proposed. Since the polymeric bolaphiles have covalent linkages, the "bolaamphiphile defects" will be localized in the membrane and a high local concentration will be achieved. Domains of the supramolecular surfactant within the bilayer will be in equilibrium with nonaggregated membrane-bound polymers. Repeat unit defects in the polymer are intrinsically more disruptive than the free monomer. A polyester, very similar to these "supramolecular surfactants", exhibited substantial protection for human CD4+ lymphocytes against HIV-1 during *in vitro* studies. [6]

Thus a new route into nonionic membrane-disrupting agents is opened, in which activity and specificity could be tailored by molecular design.

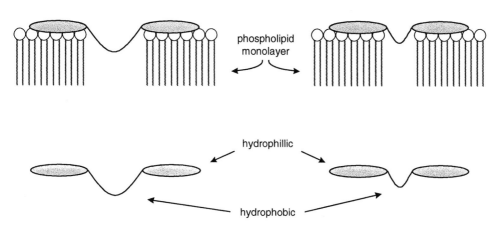

Scheme 6. Mechanism proposed by Jayasuriya et al. [6] for the destabilization of a membrane by the insertion of the hydrophobic spacer of a bolaphile into the phospholipid layer and the resulting local defect.

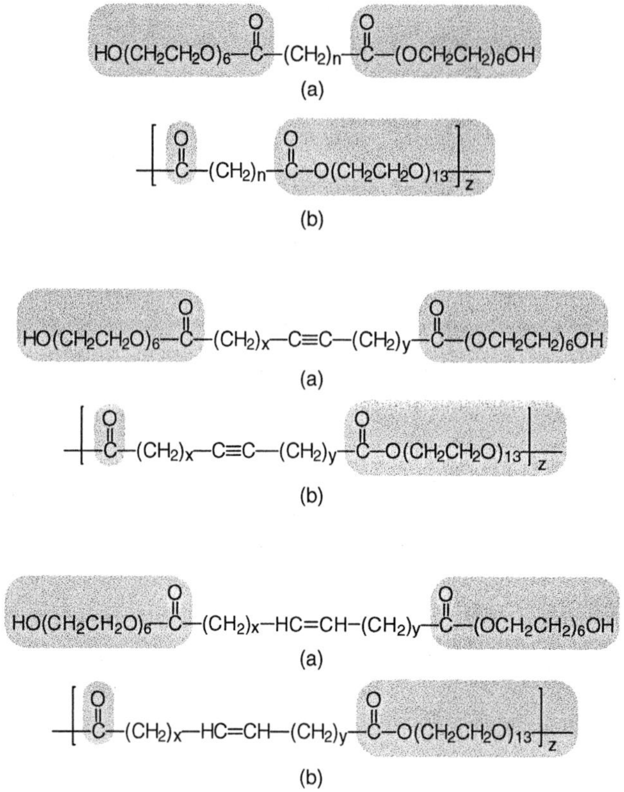

(1)

(a)

(b)

(2)

(a)

(b)

(3)

(a)

(b)

Scheme 7. Membrane-disruptive bolamphiles (a) and their polymeric analogues (b).

Hudson and Damha have succeeded in making branched RNA in a convergent-growth approach (Scheme 8). [8] The resulting compound composed of oligothymine units, modified adenosine spacers, and terminal nucleotides as head groups fits the description of a bolaamphiphile. These branched RNA molecules could be useful in capturing and binding matching nucleotide sequences.

This would allow access to potential antisense agents or ligands for affinity purification of the branch recognition factors, which catalyze the maturation or splicing of precursor messenger RNA. [8]

Beyond these initial examples, basic questions remain concerning the conformation of bolaamphiphile within micelles. [30] There is evidence that a bent conformation is adopted,

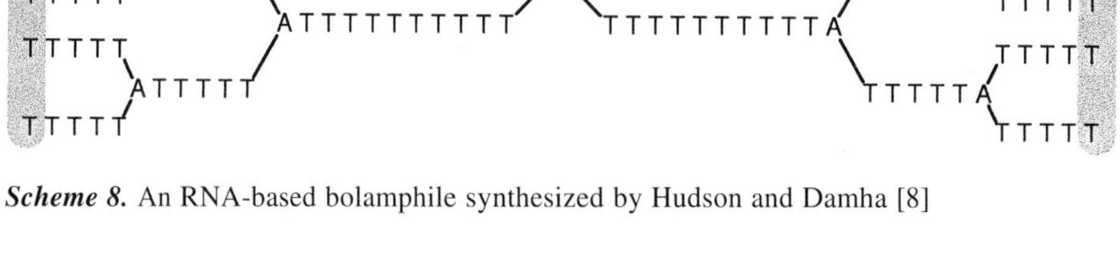

Scheme 8. An RNA-based bolamphile synthesized by Hudson and Damha [8]

[31] but micelles have been observed with very small hydrophobic cores and others are sometimes seen on electron micrographs with diameters that correspond to the length of the bolaphile. [30, 32] Among the many subjects for further exploration in the area of molecular self-assembly are the naturally occurring bolaamphiphiles found in thermophilic bacteria with their stereochemistry and stability in harsh environments. More robust membranes or chiral bolaphiles with their resulting chiral membranes could provide avenues to molecular recognition based on stereochemistry not only at one site, but over an entire membrane. [3] Rhizoferrin, N', N^4-bis(4-oxo-3-hydroxy-3,4-dicarboxybutyl)diaminobutane, isolated from an organism associated with mucormycosis observed in dialysis patients [33] and occurs in Zygomycetes strains of fungus [34] has been recently synthesized [35] and may possess bolaphilic properties.

Bolaamphiphiles, [36] like dendritic macromolecules, [37] have gained increasing attention recently, since they offer insight to micellar systems that can mimic enzymes, act as therapeutic agents, or as tools to investigate molecular recognition and assemblages. Nolte's golf ball [29] and the other specifically shaped examples described here illustrate the goal Lehn set forth for supramolecular chemistry – structures alone are not the goal but rather the expression of a desired chemical, biological, or physical property. [20]

References

[1] R. M. Fuoss, D. J. Edelson, *J. Am. Chem. Soc.* **1951**, *73*, 269.

[2] G. R. Newkome, G. R. Baker, S. Arai, M. J. Saunders, P. S. Russo, V. K. Gupta, Z.-Q. Yao, J. E. Miller, K. Bouillon, *J. Chem. Soc., Chem. Commun.* **1986**, 752.

[3] J.-H. Fuhrhop, D. Fritsch, *Acc. Chem. Res.* **1986**, *19*, 130.

[4] J.-H. Fuhrhop, J. Mathieu, *Angew. Chem.* **1984**, *96*, 124; *Angew. Chem., Int. Ed. Engl.* **1984**, *23*, 100.

[5] N. Jayasuriya, S. Bosak, S. L. Regen, *J. Am. Chem. Soc.* **1990**, *112*, 5844.

[6] N. Jayasuriya, S. Bosak, S. L. Regen, *J. Am. Chem. Soc.* **1990**, *112*, 5851.

[7] I. Amato, *Science* **1993**, *260*, 491.

[8] R. H. E. Hudson, M. J. Damha, *J. Am. Chem. Soc.* **1993**, *115*, 2119.

[9] R. Fornasier, P. Scrimin, P. Tecilla, U. Tonellato, *J. Am. Chem. Soc.* **1989**, *111*, 224.

[10] A. Cipiciani, M. C. Fracassini, R. Germani, G. Savelli, C. A. Bunton, *J. Chem. Soc., Perkin Trans.* **1987**, 547.

[11] C. A. Bunton, E. L. Dorwin, G. Savelli, V. C. Si, *Recl. Trav. Chim. Pays-Bas.* **1990**, *109*, 64.

[12] J.-M. Kim, D. H. Thompson, *Langmuir* **1992**, *8*, 637.

[13] G. R. Newkome, G. H. Escamilla, M. J. Saunders, **1997**, unpublished results.

[14] G. R. Newkome, G. R. Baker, S. Arai, M. J. Saunders, P. S. Russo, K. J. Theriot, C. N. Moorefield, J. E. Miller, K. Bouillon, *J. Am. Chem. Soc.* **1990**, *112*, 8458.

[15] G. R. Newkome, X. Lin, C. Yaxiong, G. H. Escamilla, *J. Org. Chem.* **1993**, *58*, 3123.

[16] G. R. Newkome, C. N. Moorefield, G. R. Baker, R. K. Behera, G. H. Escamilla, M. J. Saunders, *Angew. Chem.* **1992**, *104*, 901; *Angew. Chem., Int. Ed. Engl.* **1992**, *31*, 917.

[17] J.-H. Fuhrhop, D. Spiroski, C. Boettcher, *J. Am. Chem. Soc.* **1993**, *115*, 1600.

[18] L. Gros, H. Ringsdorf, H. Schupp, *Angew. Chem.* **1981**, *93*, 311; *Angew. Chem., Int. Ed. Engl.* **1981**, *20*, 305.

[19] J.-H. Fendler, P. Tundo, *Acc. Chem. Res.* **1984**, *17*, 3.

[20] J.-M. Lehn, *Angew. Chem.* **1990**, *102*, 1347; *Angew. Chem., Int. Ed. Engl.* **1990**, *29*, 1304.

[21] T.-P. Engelhardt, L. Belkoura, D. Woermann, *Ber. Bunsenges. Phys. Chem.* **1996**, *100*, 1064.

[22] K. H. Yu, P. S. Russo, L. Younger, W. G. Henk, D.-W. Hua, G. R. Newkome, G. R. Baker, *J. Polym. Sci.-Polym. Phys.* **1997**, *35*, 2787.

[23] M. Jørgensen, K. Bechgaard, R. Bjørnholm, P. Sommer-Larsen, L. G. Hansen, K. Schaumburg, *J. Org. Chem.* **1994**, *59*, 5877.

[24] S. Bhattacharya, S. N. Ghanashyam, A. R. Raju, *J. Chem. Soc., Chem. Commun.* **1996**, 2101.

[25] T. Shimizu, M. Masuda, *J. Am. Chem. Soc.* **1997**, *119*, 2812; M. Masuda, T. Shimizu, *J. Chem. Soc., Chem. Commun.* **1996**, 1057. Also see: F. Brisset, R. Garelli-Calvet, J. Azema, C. Chebli, I. Rico-Lattes, A. Lattes, A. Moisand, *New. J. Chem.* **1996**, *20*, 595; D. Lafont, P. Boullanger, Y. Chevalier, *J. Carbohydr. Chem.* **1995**, *14*, 533; P. Goueth, A. Ramez, G. Ronco, G. Mackenzie, P. Villa, *Carbohyydr. Res.* **1995**, *266*, 171; A. Mueller-Fahrow, W. Saenger, D. Fritsch, P. Schneider, J. H. Fuhrhop, *Carbohydr. Res.* **1993**, *242*, 11; R. Garelli-Calvet, F. Brisset, I. Rico, H. Lattes, *Synth. Commun.* **1993**, *23*, 35.

[26] J.-H. Fuhrhop, H. Hungerbuhler, U. Siggel, *Langmuir* **1996**, *6*, 1295.

[27] U. Siggel, H. Hungerbuhler, J.-H. Fuhrhop, *J. Chim. Phys.* **1987**, *84*, 1055.

[28] D. Danino, Y. Talmon, R. Zana, *Langmuir* **1995**, *11*, 148; M. Frindi, B. Michels, H. Levy, R. Zana, *Langmuir* **1994**, *10*, 1140; E. Alami, G. Beinert, P. Marie, R. Zana, *Langmuir* **1993**, *9*, 1465; E. Alami, H. Levi, R. Zana, A. Skoulios, *Langmuir* **1993**, *9*, 940.

[29] A. P. H. J. Schenning, B. de Bruin, M. C. Feiters, R. J. M. Nolte, *Angew. Chem.* **1994**, *106*, 1741; *Angew. Chem., Int. Ed. Engl.* **1994**, *33*, 1662.

[30] J.-H. Fuhrhop, R. Bach in *Advances in Supramolecular Chemistry*, Vol. 2 (Ed.: G. W. Gokel) JAI Press, Greenwich, **1992**, p. 25.

[31] R. Zana, S. Yiv, K. M. Kale, *J. Colloid Interface Sci.* **1980**, *77*, 456.

[32] S. Yiv, R. Zana, *J. Colloid Interface Sci.* **1980**, *77*, 449.

[33] H. Drechsel, J. Metzger, S. Freund, G. Jung, J. R. Boelaert, G. Winklmann, *Biol. Mat.* **1991**, *4*, 238.

[34] A. Thieken, G. Wienkelmann, *FEMS Microbiol. Lett.* **1992**, *94*, 37.

[35] R. J. Bergeron, M.-G. Xin, R. E. Smith, M. Wollenweber, J. S. Mmanis, C. Ludin, K. A. Abboud, *Tetrahedron* **1997**, *53*, 427.

[36] J.-H. Fuhrhop, W. Helfrich, *Chem. Rev.* **1993**, *93*, 1565; G. H. Escamilla, "Dendritic Bolaamphiphiles and Related Molecules", in *Advances in Dendritic Macromolecules*, Vol. 2, (Ed.: G. R. Newkome) JAI Press, Greenwich, **1995**, 157.

[37] G. R. Newkome, C. N. Moorefield, F. Vögtle, *Dendritic Molecules: Concepts, Syntheses, Perspectives*, VCH, Weinheim, **1996**.

Dendrimers, Arborols, and Cascade Molecules: Breakthrough into Generations of New Materials

Andreas Archut, Jörg Issberner and Fritz Vögtle

When a new idea in science is particularly appealing or "contagious", activity in that area may rapidly reach epidemic proportions. A recent example is the development of dendrimer research in the past few years. [1, 2] CAS online searches [3] shows that the number of publications in this field has increased nearly exponentially over the past decade, and the limit has not yet been reached (Fig. 1).

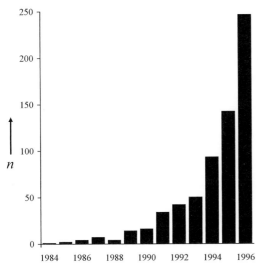

Figure 1. The number of publications on dendrimers has increased almost exponentially within the last decade.

The initial impetus can be traced to our cascadelike synthesis of noncyclic, branched polyamines in 1978. [4] Thereafter Denkewalter, Newkome, Tomalia, and others expanded this theme considerably. [1, 5] The extremely branched molecules are mainly synthesized from identical building blocks that contain branching sites, and often have a variety of functional groups on the periphery. These dendrimers are constructed in stages in repeatable synthetic steps (repetitive strategy). Each reaction cycle creates a new "generation" of branches. Unlike the divergent method, i.e. the dendrimer is built up shellwise from the core, the convergent approach is to create larger fragments first and then to couple these in a final step to the core building block.

Polypropylene amines, poly(amidoamines) (PAMAM) and silicon dendrimers set "records" for the number of generations, whereas the polyacetylene dendrimer prepared by Moore et al. [6] with the formula $C_{1398}H_{1278}$ and a mass of 18079.53 holds the "heavyweight championship title" of pure hydrocarbons. Müllen et al. have recently published a series of aromatic hydrocarbon dendrimers consisting of phenylene-bridged benzene rings. [7] Interestingly, these species form two-dimensional planes rather than three-dimensionally extended structures.

The nomenclature of cascade molecules according to the common rules has its difficulties. The names became extremely long and the fundamental structure of the molecule cannot be quickly derived from them. Newkome et al. therefore proposed a new nomenclature that reflects the structure in the sense that it names the molecule from the core towards the periphery. [8] Furthermore, the class of compound becomes clear, since the names begin with "Z-Cascade", where Z is the number of functional groups on the periphery.

Today research does no longer focus on the dendrimer itself but on the multiplication of functional components attached to a dendritic skeleton and new materials with specific properties (redox, ligand, and liquid crystalline properties, biochemical activity ...) are anticipated. [9] Industry has also shown increasing interest in functional cascade molecules for applications in diverse areas such as medical engineering, agrochemistry, and the development of photocopier toner additives. Concrete applications include *nanoscale* catalysts [2a],

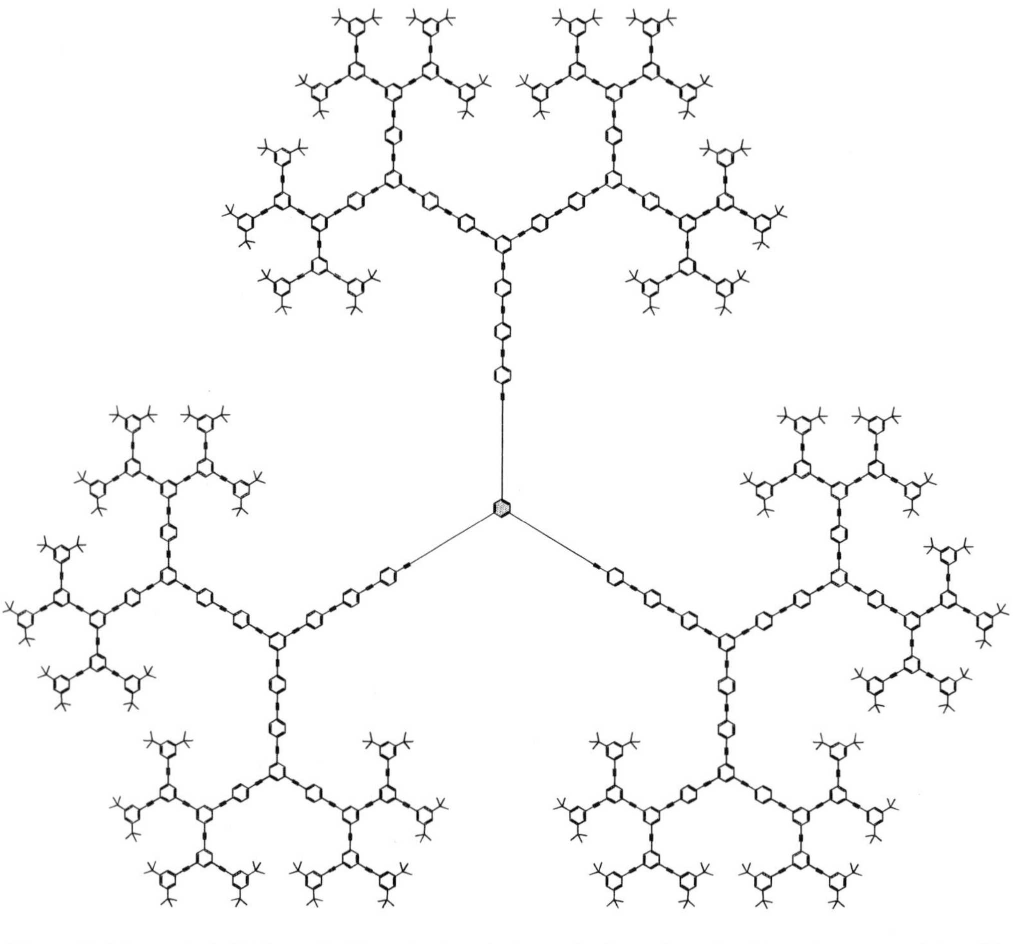

Figure 2. Moore et al. [6] have built up the heaviest pure hydrocarbon dendrimer known to date with a molecular weight larger than 18 000 u.

reaction vessels for chemical reactions [10], biomimetics of cells [11], diagnostic imaging [12], radio therapy [13], immunoassays [14], antiviral drugs [15], drug targeting [16, 17], sensors [18], conducting polymers [19], co-olant additives [20], column materials [2a], metal absorbers [21], calibration standards of submicron apertures [22] and molecular antennas [2a,f] as documented by many patents.

The birth of cascade construction and cascade molecules was in 1978 when we reported the first preparation, separation, and mass spectrometric characterization of dendritic structures by repetition of certain synthetic steps. [4] Treating primary amines with acrylonitril and using a Michael-type addition afforded dendritic polynitriles with twice as many functionalities as the starting amines. After reduction and purification the product of the first reaction cycle was subject to the same sequence to generate the next-generation cascade molecule again with the doubled number of functional groups. Meanwhile the reduction conditions that are reliable if properly applied have been modernized by using

diisobutylaluminum hydride (DIBAH) supporting the original structural conclusions. [23, 24] In a further attempt we later prepared readily soluble dendrimers with large monomeric units up to the third generation. This general synthesis can be extended to other core units and to higher generations. [25] In addition it is possible to easily functionalize these dendrimers.

Tomalia, Denkewalter and Newkome followed the cascade synthesis for the preparation of "Starburst" polyamidoamines, polylysines, and "arborols" during the 80's and developed the dendrimer chemistry to a high skill. [1]

Fréchet et al. employed the convergent method for the first time to synthesize high-generation dendritic polyethers. [26] The attraction of the convergent method is the small number of molecules that are involved in the reaction steps to give each successive generation. Large excess of reagents can be avoided resulting in good yields. However, for higher generations the yields decrease because of increasing steric hindrance at the reacting functional groups. Also the multiple

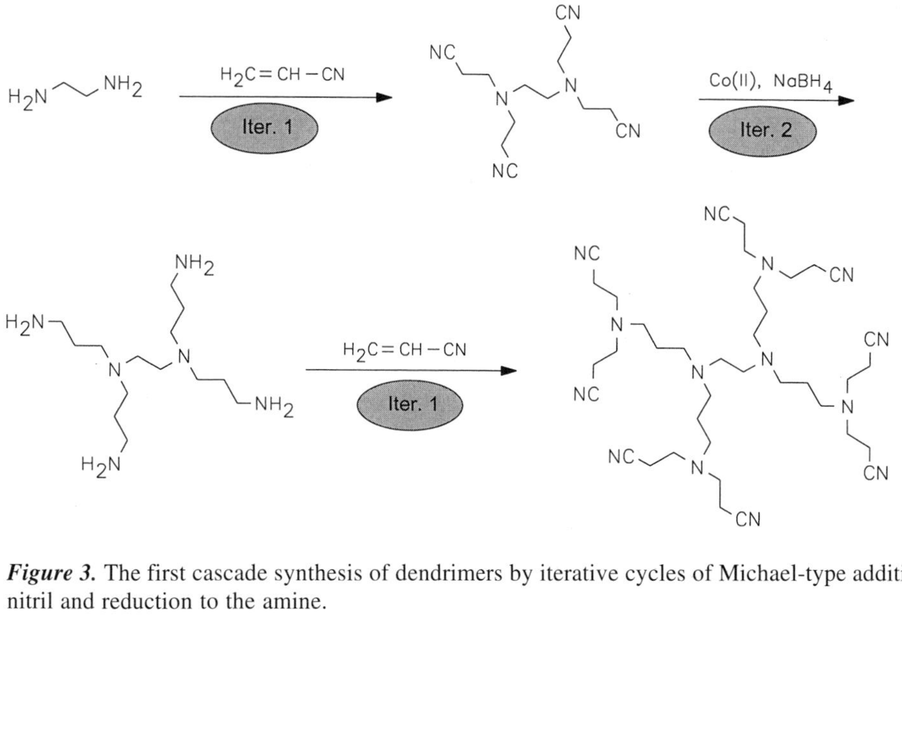

Figure 3. The first cascade synthesis of dendrimers by iterative cycles of Michael-type addition of acrylonitril and reduction to the amine.

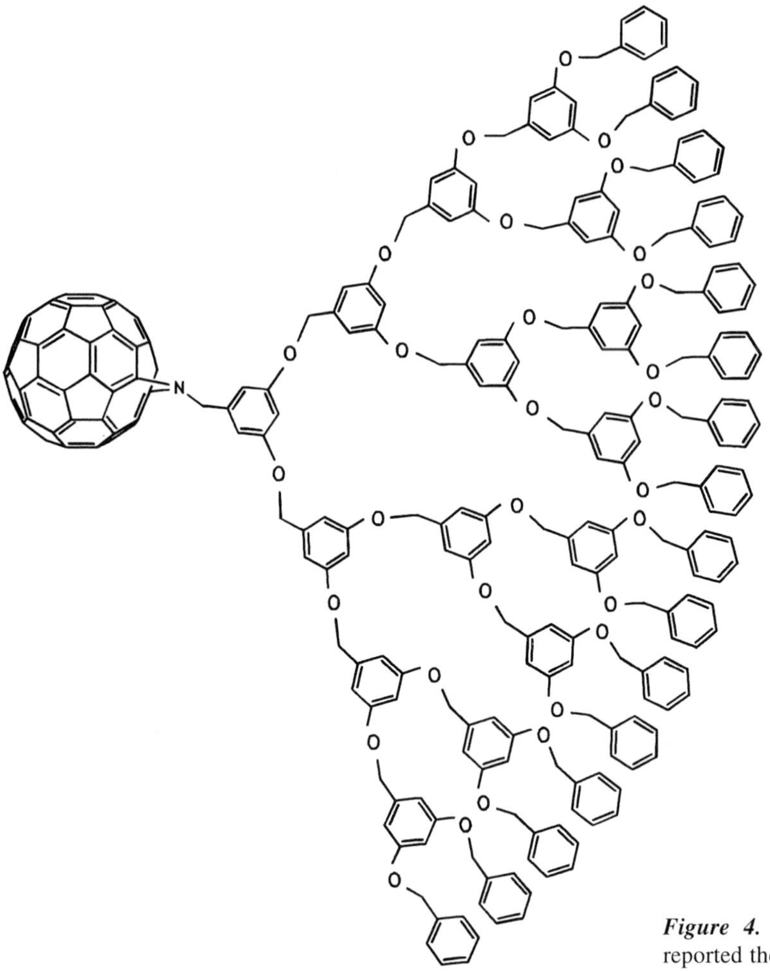

Figure 4. Fréchet et al. [26] have reported the first fullerene dendrimer.

benzylic ether bonds require non-acidic conditions which narrows chemical handling of this type of dendrimers. [27b] The preparation of the first fullerene dendrimer by Fréchet et al. combined two topics of supramolecular chemistry when they linked C_{60} and a dendritic benzylic bromide by Williamson ether synthesis. [28] Just recently, Hirsch et al. managed to use C_{60} as the core building block of a dendrimer with as many as 12 dendritic branches attached to the fullerene center. [29]

Shinkai et al. have reported the synthesis of potential complexing arborols bearing crown

ether units in the core and on the periphery. [30] The synthesis of a dendrimer with a complexing inner core (hexacyclene = hexaaza[18]crown-6) succeeded in Bonn; hexacyclene was linked with a branching unit which was obtained in a convergent manner. [31]

In most cases the functional units are located on the dendrimer surface, but dendrimers are now accessible containing a nucleus capable of performing special functions (luminescence, complexation) which are influenced to some extent (sterically or electronically) by the periphery. Inoue et al. were the first to describe a dendrimer with a metal porphyrin at its center. The convergent synthetic method of Fréchet was used to prepare a dendrimer in which the photoactive metal porphyrin center is sterically shielded. [32] Diederich et al. investigated the influence of the surface groups on the electrochemical behavior of metal porphyrins. Following a divergent synthesis devised by Newkome et al. they prepared the third generation porphyrin dendrimer shown in Figure 5, which has a molecular weight of over 19 000. The redox chemistry of this dendritic porphyrin strongly contrasts that of non-dendritic zinc porphyrins. [33]

The porphyrin nucleus is shielded by a core of electronegative oxygen atoms such that reduction by electrons from the outside is hindered. [34] We ourselves have explored several strategies to prepare Ru(II) complexes of

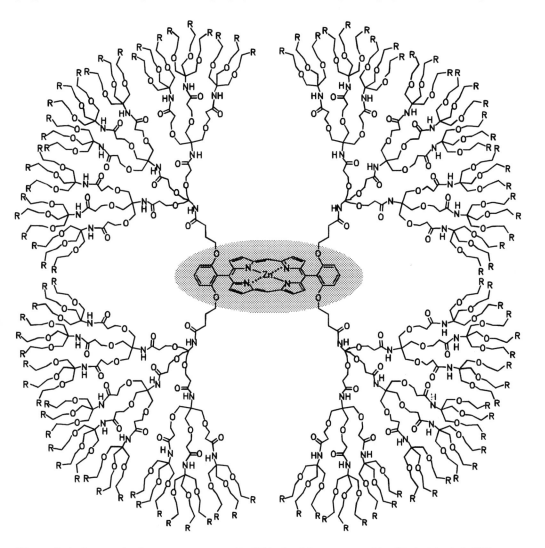

Figure 5. A zinc porphyrin serves as core unit of this dendrimer prepared by Diederich et al. [33]

dendritic bipyridine ligands by an analogous way. [27] Such complexes exhibit the well known absorption and emission properties of Ru(II)-polypyridine complexes. However, their excited state lifetime in air-equilibrated solutions is longer than expected. Indeed the dendrimer branches protect the Ru-bipy based core from oxygen quenching. A long lifetime of the luminescent excited state is important e.g. for immunoassay applications since the signal of the label can be read after the decay of the background fluorescence of the sample, whose lifetime usually is on the nanosecond time scale.

The first studies on dendritic metal and non-metal complexes demonstrate that dendrimers/arborols can be attached to metal centers for the preparation of "supramolecular aggregates". The controlled complexation of metal ions at specific binding sites in dendrimer cavities was achieved by Newkome et al., who prepared a dendrimer framework containing triple bonds, which could be complexed by dicobalthexacarbonyl units. [35]

Puddephatt et al. described the synthesis of dendrimers containing metal ions by coordinating bipyridine derivatives to transition metals, in analogy to the synthetic strategy developed by Balzani et al. In a convergent synthesis, in other words by repetition of alternating steps, (oxidative addition of benzylic bromides to platinum with bipyridine) Puddephatt et al. obtained a cluster with 28 Pt centers (Fig. 6), which is sterically that crowded that further conversion to yield a Pt_{30} cluster has failed. [36]

The attachment of natural products or drugs to a dendritic skeleton is a promising concept as well. Multiplication of specific sugar epitopes in one molecule results in highly increased avidities [37] in adhesion processes and is called the "cluster effect". [38] This is of importance where carbohydrate-protein interactions are under investigation. Sugar units were multiply built into polymers via copolymerization or telomerization to give glucomimetica. Den-

drimers are currently more and more used as core building blocks of taylor-made cluster glucosides. Roy et al. [39], Lindhorst et al. [40], and Stoddart et al. [41] have published new gluco dendrimers. In another study Roy et al. prepared a second generation poly(lysine) dendrimer functionalized with disaccharides. [39]

Preliminary tests with the influenza A virus indicate that the dendritic cluster depicted in Figure 7 is a strong inhibitor of erythrocyte hemagglutination. Similar systems among the star polymers were synthesized with peptide residues. Even nucleic acids can be constructed in a dendritic way, as Hudson and Damha have shown. [42] In an automated procedure the nucleic acid chains were first prepared and then divergently connected to give a cascade molecule containing 87 nucleic acid residues and having an approximate molecular weight of 25 000. It was necessary to use longer branches for the inner core of the dendrimer than for the periphery.

Rao and Tam proved that even peptides can be connected in a dendritic manner. [43] An octameric peptide dendrimer, which could be useful as a synthetic protein, was prepared and characterized by laser-desorption mass spectrometry.

First investigations with cell cultures showed that certain dendrimers support the transfection of mammalian cells by plasmids. The controlled synthesis, low toxicity, and pH buffering effect of dendrimers are the main criteria for dendrimers suitable for gene-transfer experiments. [44]

Seebach et al. met the challenge of synthesizing chiral dendrimers to investigate the influence of chiral building blocks on the chirality of the whole molecule and to determine whether enantioselective complexation was possible. [45] They achieved dendrimers with a chiral nucleus as well as dendrimers with additional chiral branches. The optical activity of dendrimers with only a chiral nucleus decreases with increasing size of the dendrimer whereas the optical activity of dendrimers that are "fully chiral" correspond to the optical activity of the nucleus. These dendrimers also form quite stable clathrates.

Meijer et al. recently reported the synthesis of dendrimers thereof with chiral terminal units. [46] When only the chiral surface unit contains a stereogenic carbon atom the dendrimers exhibit low or vanishing optical activity with increasing generation number. In contrast, dendrimers with rigid chiral units show increasing optical activity with increasing generation number.

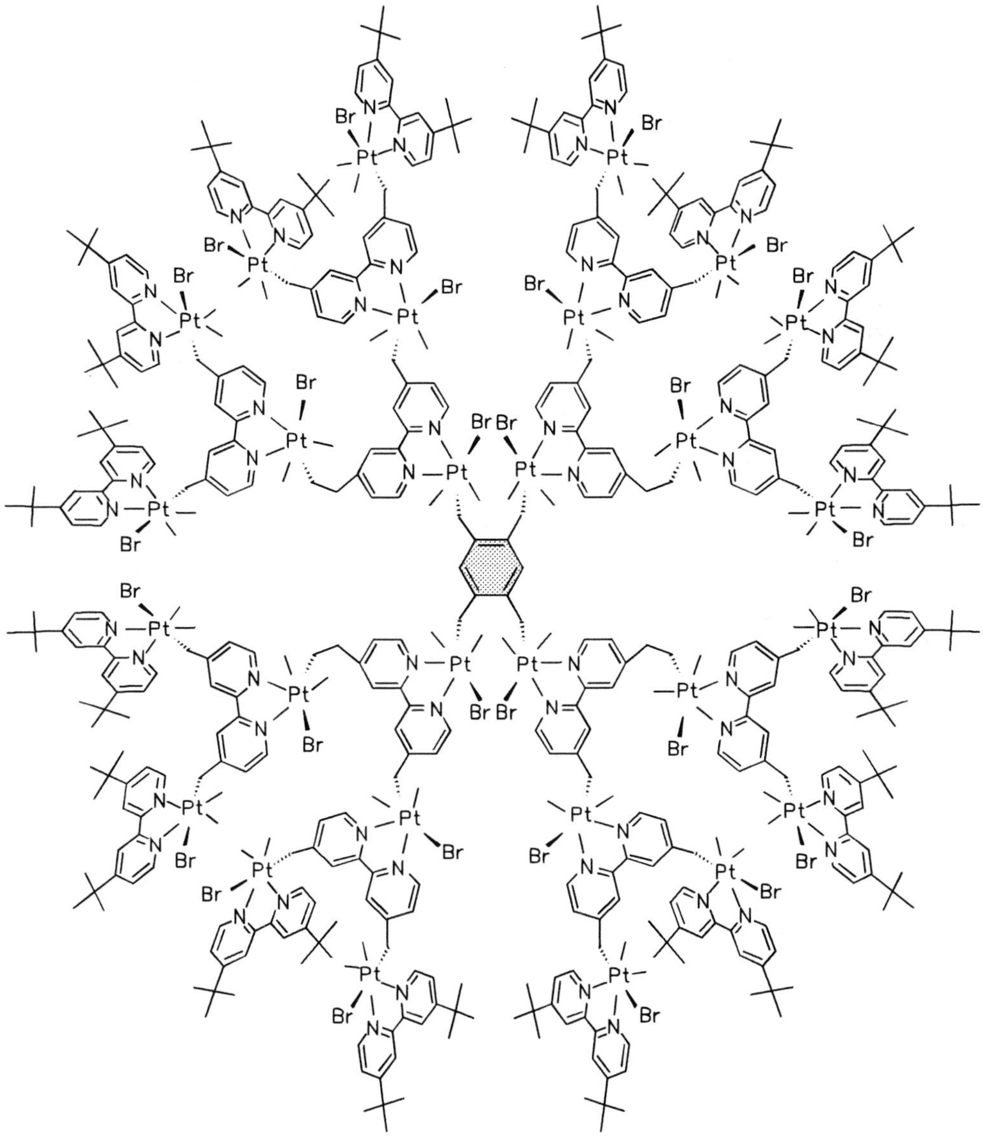

Figure 6. 28 platinum centers are clustered in this metal-complex dendrimer (Puddephatt et al.) [36]

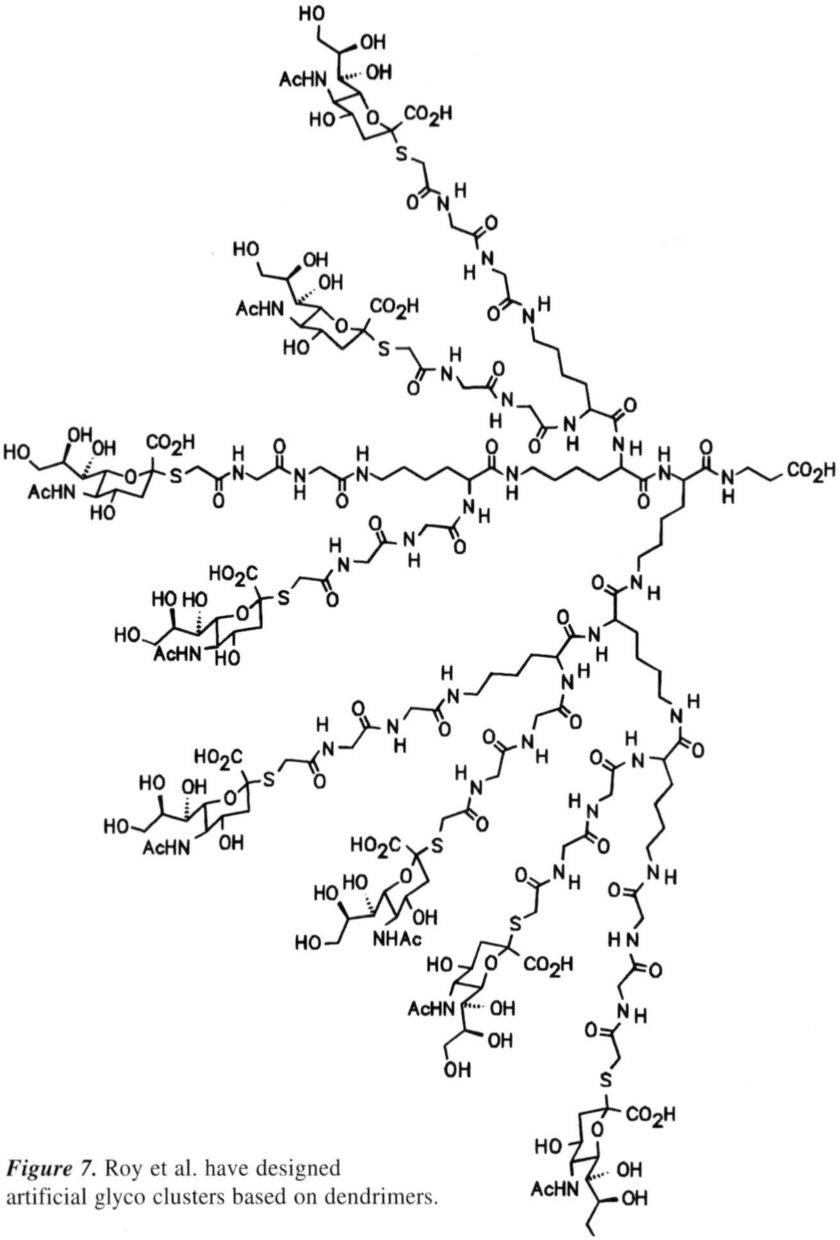

Figure 7. Roy et al. have designed artificial glyco clusters based on dendrimers.

We were able to synthesize chiral dendrimers with stable planar-chiral building blocks to avoid racemisation. [47] In contrast to the results of Seebach et al. these dendrimers show increasing chirality with inreasing gene-rations. In addition, the circular dichrograms clearly indicate that chiral dendrimers based on derivatives of [2.2]paracyclophanes can be employed for complexation of certain metal cations.

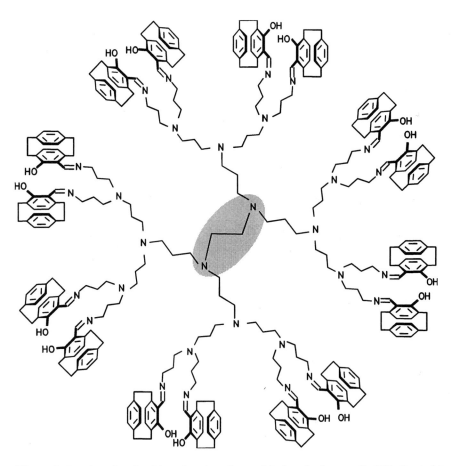

Figure 8. A polyamine dendrimer bearing planar-chiral cyclophane units (Vögtle et al.)

These dendrimers are precisely built nanoscopic molecules with physical characteristics such as size, solubility and dispersity of complexation sites. The chiral information should be useful for further reactions, e. g. in the field of homogeneous catalysis. There is a strong interest in the developing of new materials that combine the advantages and/or minimize disadvantages associated with individual homogeneous and heterogeneous catalysts. [48] Dendrimer construction might offer a

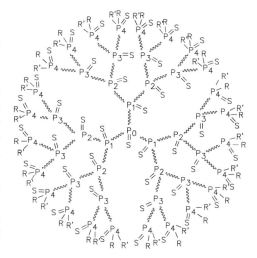

Figure 9. Majoral et al. [50] have devised the synthesis of a new class of phosphorus-containing dendrimers.

better means of controlling the disposition of pendant metal-containing catalytic sites in soluble, polymer-based catalysts. [49] Catalytic dendrimers with nanoscale dimensions may be recycled by using simple filtration methods to remove the catalyst from the reaction mixture.

Majoral et al. have prepared the first neutral phosphorus-containing dendrimers. The forth-generation dendrimer (Fig. 9) was obtained by reacting $PSCl_3$ and the sodium salt of para-hydroxybenzaldehyde and H_2N-$N(Me)P(S)CCl_2$. [50] Analogous reactions with $POCl_3$ instead of $PSCl_3$ are possible as

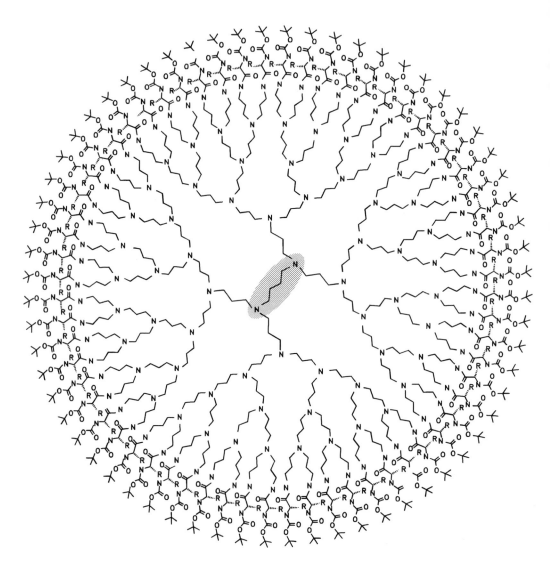

Figure 10. Meijer et al. [53] have discovered the "dendritic box", a fifth-generation polyamine dendrimer with bulky substituents, that can be used to include various types of guest molecules (amide hydrogen atoms have been omitted in the picture).

well. The only byproducts are sodium chloride and water, and the dendrimer is formed almost quantitatively. The monodispersity of the products could be easily monitored by phosphorus NMR.

In addition to forming clathrates, dendrimers have been found to encapsulate guest molecules inside their cavities ("dendritic box", "container dendrimers"). Suitable methods for inclusion of even large molecules (dyes, spin markers, fluorescence markers) have been developed. [51] After a guest has been bound in a certain generation of the dendrimer, the next generation is covalently linked closing the remaining gaps in the surface and thus confining the guest molecule in the dendrimer's inner cavities. Such dendrimers are clearly supramolecular hosts.

If the outer shell that closes the cavities containing guests is linked to the dendrimer supramolecularly (e.g. via hydrogen bonding), this linkage could later be broken more easily. [52, 53] Thus even without guests inside dendrimers can be supramolecules. Dendrimers that release guests upon changing pH hold promise for the directed application of pharmaceuticals (drug release, drug targeting).

Newkome et al. successfully prepared dendrimers with terpyridine units as linkers that are able to form supramolecular network assemblies. [54] The same authors have suggested recently to incorporate H-bonding moieties within a dendrimer to allow supramolecular network formation. [55]

Zimmerman et al. have disclosed the convergent preparation of dendritic wedges possessing tetraacid moieties that self-assemble into the hexameric, disk-like network. [56] The tetraacid unit is known to form cyclic as well as linear structures in solution via carboxylic-acid dimerization. It is postulated that the cyclic structure forms predominantly since it is sterically less demanding than the linear aggregates would be. This suggestion has been supported by size-exclusion chromatography (SEC).

Above all, the synthesis of large, branched molecules bordering on polymers challenges synthetic and analytical chemists alike. A large number of reaction centers or considerab-

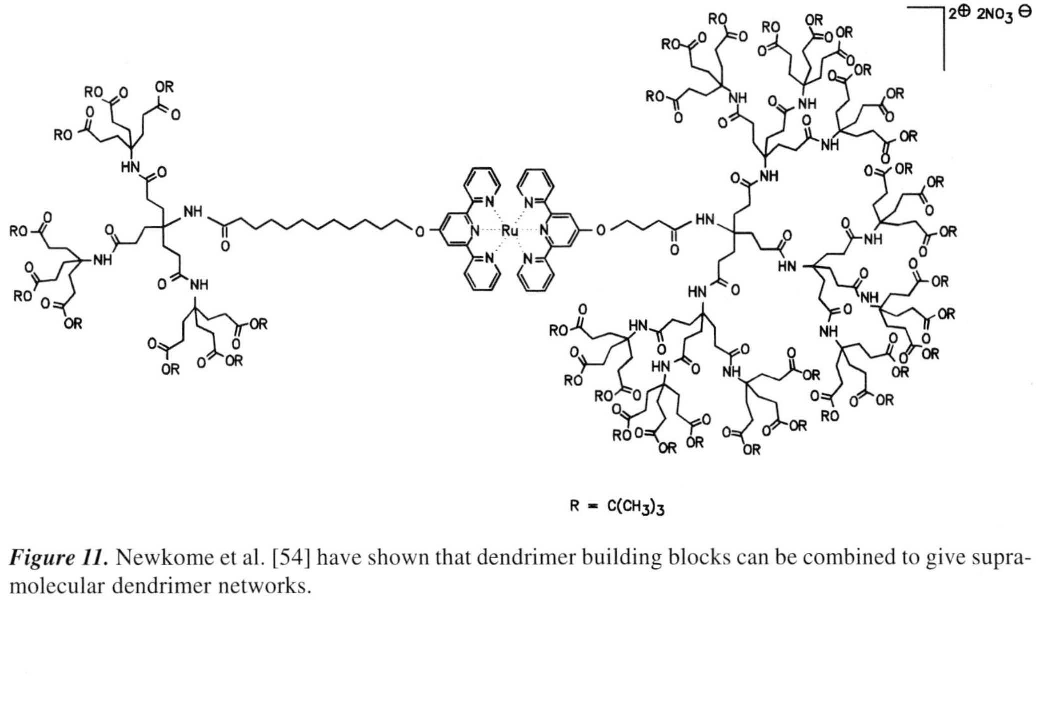

R = C(CH₃)₃

Figure 11. Newkome et al. [54] have shown that dendrimer building blocks can be combined to give supramolecular dendrimer networks.

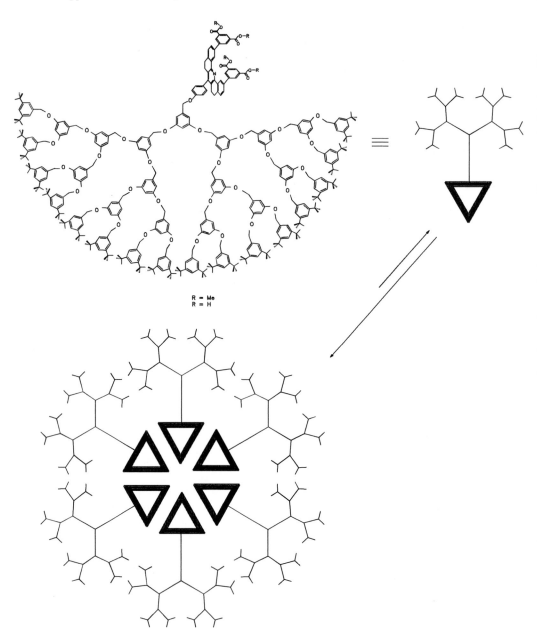

R = Me
R = H

Figure 12. A supramolecular cluster consisting of six dendritic wedges has been reported by Zimmerman et al. [56]

le steric congestion hinders the synthesis of monodisperse molecules, i.e. having all exactly the same structure. The classical methods of analysis in organic chemistry are sometimes pushed to their limits. NMR spectra and elemental analysis are becoming increasingly less informative, and mass spectrometry of very heavy molecules creates problems as well.

The variety of the examples listed above documents that dendrimer chemistry has attained increasing interest. In this compilation of recent results the trend towards functional and application-oriented molecules including biochemically active, photoswitchable, and polymerlike dendrimers is particularly apparent. Dendrimers cross the boundaries of classical organic chemistry and as new materials will penetrate deeper into the topical fields of "nanostructures", supramolecules and polymers in the future. Increasing industrial research on dendrimers and the commercial availability of PAMAM and polyamine dendrimers should stimulate further investigations in this field.

References

[1] G. R. Newkome, C. N. Moorefield, F. Vögtle in *Dendritic Molecules*, VCH, Weinheim, **1996**.

[2] a) R. Dagani, *Chem. Eng. News* **1996**, *June 3*, 30–38; b) R. F. Service, *Science* **1995**, *267*, 458–459; c) T. W. Bell, *Science* **1996**, *271*, 1077–1078; d) M. Groß, *Spektrum der Wissensch.* **1995**, *6*, 30–32; e) J. Breitenbach, *Spektrum der Wissensch.* **1993**, 96–97; R. Dagani, *Chem. Eng. News* **1993**, *February 1*, 28–32; V. Balzani, S. Campagna, G. Denti, A. Juris, S. Serroni, M. Venturi, *Acc. Chem. Res.* **1998**, *31*, 26–34..

[3] The CAS online search was done with the terms "dendrimer" or "cascade molecule" or "cascade polymer" or "starburst dendrimer" or "arborol" or "dendritic molecule" or repetitive synthes? or "pamam" (1986–1993) and "dendrimer" (1994–1996).

[4] E. Buhleier, W. Wehner, F. Vögtle, *Synthesis* **1978**, 155–158.

[5] a) D. A. Tomalia, A. M. Naylor, W. A. Goddard III, *Angew Chem.* **1990**, *102*, 119–157; *Angew. Chem. Int. Ed. Engl.* **1990**, *29*, 113–151; b) H. B. Mekelburger, W. Jaworek, F. Vögtle, *Angew. Chem.* **1992**, *104*, 1609–1614; *Angew. Chem. Int. Ed. Engl.* **1992**, *31*, 1571–1576; c) D. A. Tomalia, H. D. Durst, *Top. Curr. Chem.* **1993**, *165*, 193–313; d) H. Frey, K. Lorenz, Ch. Lach, *Chem. in unserer Zeit* **1996**, *30*, 75–85; e) N. Ardoin, D. Astruc, *Bull. Soc. Chim. Fr.* **1995**, *132*, 875–909; f) D. A. Tomalia, *Aldrichim. Acta* **1993**, *26*, 91–101; g) J. M. J. Fréchet, *Science* **1994**, *263*, 1710–1715.

[6] a) Z. Xu, J. S. Moore, *Angew. Chem.* **1993**, *105*, 261–264; *Angew. Chem. Int. Ed. Engl.* **1993**, *32*, 246–248; b) Z. Xu, J. S. Moore, *Angew. Chem.* **1993**, *105*, 1394–1396; *Angew. Chem. Int. Ed. Engl.* **1993**, *32*, 1354–1356.

[7] F. Morgenroth, E. Reuther, K. Müllen, *Angew. Chem.* **1997**, *109*, 647–649; *Angew. Chem. Int. Ed. Engl.* **1997**, *36*, 661–634.

[8] G. R. Newkome, G. R. Baker, J. K. Young, J. G. Trayham, *J. Polym. Sci. Part A: Polymer Chemistry* **1993**, *31*, 641–651.

[9] J. Issberner, R. Moors, F. Vögtle, *Angew. Chem.* **1994**, *106*, 2507–2514; *Angew. Chem. Int. Ed. Engl.* **1994**, *33*, 2413–2420.

[10] F. Svec, J. M. J. Fréchet, *Science* **1996**, *273*, 205–211.

[11] D. A. Tomalia, D. M. Hedstrand (The Dow Chemical Company), WO 9417130 A1 940804, **1994**.

[12] a) E. C. Wiener, M. W. Brechbiel, H. Brothers, R. L. Magin, O. A. Gansow, D. A. Tomalia, P. C. Lauterbur, *Mag. Res. Med.* **1994**, *31*, 1–8; b) E. C. Wiener, F. P. Auteri, J. W. Chen, M. W. Brechbiel, O. A. Gansow, D. S. Schneider, R. L. Belford, R. B. Clarkson, P. C. Lauterbur, *J. Am. Chem. Soc.* **1996**, *118*, 7774–7782.

[13] B. Qualmann, M. M. Kessels, H.-J. Musiol, W. D. Sierralata, P. W. Jungblut, L. Moroder, *Angew. Chem.* **1996**, *108*, 970–973; *Angew. Chem. Int. Ed. Engl.* **1996**, *35*, 909–911.

[14] a) F. Moll, C. Ferzil, S. Lin, P. Singh (Dade International Inc.), WO 9527902, **1995**; b)

F. Moll, C. Ferzil, S. Lin, P. Singh, K. Koshi, P. Cronin (Dade International Inc.), WO 9528641, **1995**.

[15] a) B. R. Matthews, G. Holan (Biomolecular Research Institute LTD.), WO 9534595, **1995**; b) J. R. M. Cockbain, L. Margerum, J. Carvalho, M. Garrity, J. D. Fellmann (Nycomed Salutar Inc.), WO 9524225, **1995**; c) J. P. Tam, F. P. Zavala (The Rockefeller University, New York University), WO 9011778, **1990**.

[16] a) D. Günther, *Pharm. in unserer Zeit* **1996**, 130–134; b) U. Bickel, *Chimica* **1995**, *49*, 386–395; c) F. Kratz, *Pharm. in unserer Zeit* **1995**, *24*, 14–26; d) R. Gref, Y. Minamitake, M. T. Peracchia, V. Trubetskoy, V. Torchilin, R. Langer, S*cience* **1994**, *263*, 1600–1603.

[17] R. H. Guy, Y. N. Kali, C. S. Lim, L. B. Nonato, N. G. Turner, *Chem. in Britain*, **1996**, 42–45.

[18] P. Jutzi, C. Batz, B. Neumann, H.-G. Stammler, *Angew. Chem.* **1996**, *108*, 2272–2274; *Angew. Chem. Int. Ed. Engl.* **1996**, *35*, 2118–2121.

[19] M. L. Daroux, D. G. Pucci, D. W. Kurz, M. Litt, A. Melissaris (Gould Electronics Inc.), EP 0682059 A1 951115, **1995**.

[20] R. D. Tack (Exxon Chemical Patents Inc.), WO 9612755, **1996**.

[21] D. M. Hedstrand, D. A. Tomalia, B. J. Helmer (The Dow Chemical Company), WO 9417125, **1994**.

[22] D. A. Tomalia, L. R. Wilson (The Dow Chemical Company), EP 0566165 A1 931020, **1993**.

[23] a) R. Moors, *Dissertation*, Universität Bonn, **1994**; b) R. Moors, F. Vögtle, *Adv. in Dendritic Macromolecules* **1995**, *2*, 41–71; c) R. Moors, F. Vögtle, *Chem. Ber.* **1993**, *126*, 2133–2135.

[24] a) C. Wörner, R. Mülhaupt, *Angew. Chem.* **1993**, *105*, 1367–1370; *Angew. Chem. Int. Ed. Engl.* **1993**, *32*, 1306–1308; b) E. M. M. de Brabander-van den Berg, E. W. Meijer; *Angew. Chem.* **1993**, *105*, 1370–1372; *Angew. Chem. Int. Ed. Engl.* **1993**, *31*, 1308–1311; c) E. M. M. De Brabander-van den Berg, E. W. Meijer, F. H. A. M. J. Vanderbooren, H. J. M. Bosman (DSM N. V.), WO 9314147, **1993**.

[25] H. B. Mekelburger, K. Rissanen, F. Vögtle, *Chem. Ber.* **1993**, *126*, 1161–1169.

[26] a) C. J. Hawker, J. M. J. Fréchet, *J. Chem. Soc., Chem. Commun.* **1990**, 1010–1013; b) C. J. Hawker, J. M. J. Fréchet, *J. Am. Chem. Soc.* **1990**, *112*, 7638–7647; c) K. L. Wooley, C. J. Hawker, J. M. J. Fréchet, *J. Am. Chem. Soc.* **1991**, *113*, 4252–4261; d) I. Gitsov, K. L. Wooley, J. M. J. Fréchet, *Angew. Chem.* **1992**, *104*, 1282–1285; *Angew. Chem. Int. Ed. Engl.* **1992**, *31*, 1203; e) J. W. Leon, M. Kawa, J. M. J. Fréchet, *J. Am. Chem. Soc.* **1996**, *118*, 8847–8859.

[27] a) J. Issberner, F. Vögtle, L. De Cola, V. Balzani, *Chem. Eur. J.*, **1997**, *3*, 706–712; b) M. Plevoets, F. Vögtle, unpublished results.

[28] K. L. Wooley, C. J. Hawker, J. M. J. Fréchet, F. Wudl, G. Srdanov, S. Shi, C. Li, M. Kao, *J. Am. Chem. Soc.* **1993**, *115*, 9836–9837.

[29] X. Camps, H. Schönberger, A. Hirsch, *Chem. Eur. J.* **1997**, *3*, 561–567.

[30] a) T. Nagasaki, M. Ukon, S. Arimori, S. Shinkai, *J. Chem. Soc., Chem. Commun.* **1992**, 608–610; b) T. Nagasaki, O. Kimura, M. Ukon, S. Arimori, I. Hamachi, S. Shinkai, *J. Chem. Soc., Perkin Trans. I* **1994**, 75–81.

[31] K. Kadei, R. Moors, F. Vögtle, *Chem. Ber.* **1994**, *127*, 897–903.

[32] R.-H. Jin, T. Aida, S. Inoue, *J. Chem. Soc., Chem. Commun.* **1993**, 1260 -1262.

[33] P. Wallimann, P. Seiler, F. Diederich, *Helv. Chim. Acta* **1996**, *79*, 779–788.

[34] P. J. Dandlinker, F. Diederich, J.-P. Gisselbrecht, A. Louati, M. Gross, *Angew. Chem.* **1995**, *107*, 2906–2909; *Angew. Chem. Int. Ed. Engl.* **1995**, *34*, 2906–2909.

[35] G. R. Newkome, C. N. Moorefield, *Polym. Reprints Am. Chem. Soc. Div. Poly. Chem.* **1993**, *34*, 75–76.

[36] a) S. Achar, R. J. Puddephatt, *Angew. Chem.* **1994**, *106*, 895–897; *Angew. Chem. Int. Ed. Engl.*, **1994**, *33*, 847–849; b) S. Achar, R. J. Puddephatt, *J. Chem. Soc., Chem. Commun.* **1994**, 1895–1896; c) S. Achar, J. J. Vittal, R. J. Puddephatt, *Organometallics*, **1996**, *15*, 43–50.

[37] Whereas the word affinity describes the strength of a reaction of a monovalent antigen (haptene) with a monovalent antibody (antigen docking unit), the word avidity is used to describe the total tendency of an antibody to bind an antigen, particularly that of antibodies

with several docking units and antigens with several epitopes: J. Klein, *Immunologie*, VCH, Weinheim, **1991**.

[38] E. A. L. Biessen, D. M. Beuting, H. C. P. F. Roelen, G. A. van de Marel, J. H. van Boom, T. J. C. van Berkel, *J. Med. Chem.* **1995**, *38*, 1538–1546.

[39] a) D. Zanini, W. K. C. Park, R. Roy, *Tetrahedron Lett.* **1995**, *36*, 7383–7386; b) R. Roy, W. K. C. Park, Q. Wu, S.-N. Wang, *Tetrahedron Lett.* **1995**, *36*, 4377–4380; c) D. Pagé, S. Aravind, R. Roy, *J. Chem. Soc., Chem. Commun.* **1996**, 1913–1914.

[40] a) T. K. Lindhorst, C. Kieburg, *Angew. Chem.* **1996**, *108*, 2083–2086; *Angew. Chem. Int. Ed. Engl.* **1996**, *35*, 1953–1956; b) Recent review: T. K. Lindhorst, *Nachr. Chem. Tech. Lab.* **1996**, *44*, 1073–1079.

[41] a) P. R. Ashton, S. E. Boyd, C. L. Brown, N. Jayaraman, S. A. Nepogodiev, J. F. Stoddart, *Chem. Eur. J.* **1996**, *2*, 1115–1128; b) P. R. Ashton, S. E. Boyd, C. L. Brown, N. Jayaraman, J. F. Stoddart, *Angew. Chem.* **1997**, *109*, 756–759; *Angew. Chem. Int. Ed. Engl.* **1997**, *36*, 756–759.

[42] R. H. E. Hudson, M. J. Damha, *J. Am. Chem. Soc.* **1993**, *115*, 2119–2124.

[43] C. Rao, J. P. Tam, *J. Am. Chem. Soc.* **1994**, *116*, 6975–6976.

[44] J. F. Kukowska-Latallo, A. U. Bielinska, J. Johnson, R. Spindler, D. A. Tomalia, J. R. Baker Jr., *Proc. Natl. Acad. Sci. USA* **1996**, *93*, 4897–4902.

[45] a) D. Seebach, J.-M. Lapierre, K. Skobridis, G. Greiveldinger, *Angew. Chem.* **1994**, *106*, 457–458; *Angew. Chem. Int. Ed. Engl.* **1994**, *33*, 440–441; b) H.-F. Chow, L. F. Fok, C. C. Mak, *Tetrahedron Lett.* **1994**, *35*, 3547–3550; c) L. J. Twyman, A. E. Beezer, J. C. Mitchell, *Tetrahedron Lett.* **1994**, *35*, 4423–4424.

[46] J. F. G. A. Jansen, H. W. I. Peerlings, E. M. M. de Brabander-van den Berg, E. W. Meijer, *Angew. Chem.* **1995**, *107*, 1321–1324; *Angew. Chem. Int. Ed. Engl.* **1995**, *34*, 1206–1209.

[47] J. Issberner, M. Böhme, S. Grimme, M. Nieger, W. Paulus, F. Vögtle, *Tetrahedron Asymmetry* **1996**, *7*, 2223–2232.

[48] J. W. J. Knapen, A. W. van der Made, J. C. de Wilde, P. W. N. M. van Leeuwen, P. Wijkens, D. M. Grove, G. van Koten, *Nature* **1994**, *372*, 659–663.

[49] a) B. B. De, B. B. Lohray, S. Sivaram, P. K. Dhal, *Tetrahedron Asymmetry* **1995**, 6, 2105–2108 and references therein; b) R. S. Drago, J. Gaul, A. Zombeck, D. K. Straub, *J. Am. Chem. Soc.* **1980**, *102*, 1033–1038; c) R. S. Drago, J. P. Cannady, K. A. Leslie, *J. Am. Chem. Soc.* **1980**, *102*, 6014–6019; d) G. Henrici-Olivé, S. Olivé, *Angew. Chem.* **1974**, *86*, 1–12; *Angew. Chem. Int. Ed. Engl.* **1974**, *13*, 49–60.

[50] a) N. Launay, A.-M. Caminade, R. Lahana, J. P. Majoral, *Angew. Chem.* **1994**, *106*, 1682–1684; *Angew. Chem. Int. Ed. Engl.* **1994**, *33*, 1589–1592; b) M.-L. Lartigue, M. Slany, A.-M. Caminade, J.-P. Majoral, *Eur. Chem. J.* **1996**, *2*, 1417–1426.

[51] a) E. W. Meijer, Talk COST-Workshop, Stockholm 28. 05. 1994. b) J. F. G. A. Jansen, E. M. M. de Brabander-van den Berg, E. W. Meijer, *Science* **1994**, *266*, 1226–1229; c) J. F. G. A. Jansen, R. A. J. Janssen, E. M. M. de Brabander-van den Berg, E. W. Meijer, *Adv. Mater.* **1995**, *7*, 561–564.

[52] J. F. G. A. Jansen, E. W. Meijer, *J. Am. Chem. Soc.* **1995**, *117*, 4417–4418.

[53] S. Stevelmans, J. C. M. van Hest, J. F. G. A. Jansen, D. A. F. J. van Boxtel, E. M. M. de Brabander-van den Berg, E. W. Meijer, *J. Am. Chem. Soc.* **1996**, *118*, 7398–7399.

[54] G. R. Newkome, R. Güther, C. N. Moorefield, F. Cardullo, L. Echegoyen, E. Pérez-Cordero, H. Luftmann, *Angew. Chem.* **1995**, *107*, 2159–2162; *Angew. Chem. Int. Ed. Engl.* **1995**, *34*, 2023–2026.

[55] G. R. Newkome, C. N. Moorefield, R. Güther, G. R. Baker, *Polym. Reprints* **1995**, *36*, 609.

[56] S. C. Zimmerman, F. Zeng, D. E. C. Reichert, S. V. Kolotuchin, *Science* **1996**, *271*, 1095–1098.

Carboranes, Anti-Crowns, Big Wheels, and Supersandwiches

Russell N. Grimes

Introduction

Carboranes – polyhedral boranes containing carbon in the framework – have been known for over 35 years, and their intrinsic stability, versatility, structural variety, and electronic properties have been put to use in a number of diverse areas, [1] for example in the synthesis of extraordinarily heat-stable polymers, in BNCT (boron neutron capture therapy), as ligands in metallacarborane catalysts, as complexing agents for extraction of metal ions, as precursors to ceramics, conducting polymers, and nonlinear optical materials, as anticancer agents, and as carriers for radioactive metals in radioimmunodetection and radioimmuno-therapy.

Although polyhedral carboranes having as few as five and as many as twelve vertices (fourteen if metal atoms are included) are known, [2] most research and applications have centered on the exceptionally stable 7- and 12-vertex cage systems. Largely for reasons of accessibility, the 12-vertex $C_2B_{10}H_{12}$ icosahedral carboranes have been most widely studied. The three known isomers, in which the carbon atoms occupy *ortho* (1,2), *meta* (1,7), or *para* (1,12) vertices (Fig. 1a–c), are

●CH ○BH

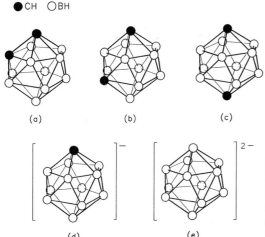

Figure 1. Icosahedral boron clusters:
(a) $1,2\text{-}C_2B_{10}H_{12}$; (b) $1,7\text{-}C_2B_{10}H_{12}$; (c) $1,12\text{-}C_2B_{10}H_{12}$; (d) $CB_{11}H_{12}^-$; (e) $B_{12}H_{12}^{2-}$. In these nonclassical electron-delocalized molecules, the connecting lines show bonding interactions, but do not necessarily represent electron pairs.

white solids that are among the most stable molecular compounds known.

Together with the isoelectronic anions $CB_{11}H_{12}^-$ and $B_{12}H_{12}^{2-}$ (Fig. 1d and e), these clusters are three-dimensional "super-aromatic" systems in which 26 electrons fill the 13 bonding molecular orbitals on the polyhedral framework; moreover, the cage volume approximates that displaced by a benzene molecule rotating on one of its twofold axes. The very high intrinsic stability of the $C_2B_{10}H_{12}$ isomers, together with the acid character of the hydrogens bound to the cage carbon atoms (which allows facile introduction of functional groups at carbon), is the basis of extensive development of icosahedral carborane chemistry over three decades and its application to practical problems. [1, 2]

Metallacarborands

The remarkable versatility of carboranes presents an almost limitless range of possible roles in designed synthesis, a fact that is drawing increasing attention in organic and inorganic chemistry, materials science, engineering, and biologically related areas. Particularly elegant examples in recent years have been afforded by the construction of novel carborane-based macrocycles, in which both the electron-withdrawing character of the car-

borane cage and the close geometric relationship of icosahedral $C_2B_{10}H_{12}$ to planar C_6H_6 are exploited. In recent work, Hawthorne and his associates have synthesized a series of novel complexes that feature host "mercuracarborands", which are metallacycles incorporating three or four $C_2B_{10}H_{10}$ cages linked by an equal number of mercury atoms. [3] As shown in Figure 2, the reaction of 1,2-dilithio-*o*-carborane ($Li_2C_2B_{10}H_{10}$) with mercury(II) halides generated Cl⁻, Br⁻, or I⁻ complexes of the $(C_2B_{10}H_{10})_4Hg_4$ host tetramer whose structure is illustrated in Figure 3a. This species binds to Cl⁻ in nearly square-planar geometry (Fig. 3b), an unprecedented coordination mode for halide ions. [3a] The mercury-chloride binding is proposed to involve a pair of orthogonal 3-center, 2-electron Hg–Cl–Hg bonds, [3a] involving overlap of two filled p orbitals on Cl⁻ with four empty Hg orbitals, as pictured in Figure 3c.

All of the mercuracarborand halide complexes react with silver acetate to form the silver halide and liberate the free macrocycle. X-ray crystallography on the THF complex of the host tetramercuracarborand [3d] and on an octa-B-ethyl derivative (prepared to facilitate solubility in hydrocarbon solvents) has revealed that both species have saddle-shaped S_4 symmetry with the centers of the icosahedra well outside the plane of the four mercury atoms. [3a, d] The square planar conformation that is found in the chloride com-

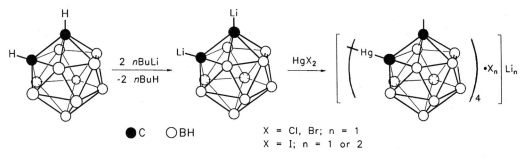

$\bullet$ C $\bigcirc$ BH

X = Cl, Br; n = 1
X = I; n = 1 or 2

Figure 2. Synthesis of mercuracarborand halide complexes from 1,2-$Li_2C_2B_{10}H_{10}$ and mercury(II) halides. [3a]

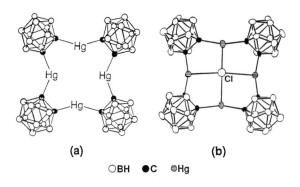

(a)   **(b)**

○BH   ●C   ⊖Hg

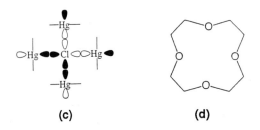

(c)   **(d)**

Figure 3. (a) Structure of the host tetramer $(1,2\text{-}C_2B_{10}H_{10})_4Hg_4$ in its planar conformation. [3a] (b) Structure of the tetramer with a bound chloride ion. [3a] (c) 3-Center, 2-electron Hg-Cl-Hg bonds in the mercuracarborand-4 chloride ion complex. [3a] (d) Structure of [12]-crown-4.

plex suggests that the halide anion functions as a template, enforcing the planar geometry. The carborane macrocycle is a Lewis acid and coordinates nucleophiles; it is therefore an "anti-crown", [4] i.e., a charge-reversed analogue of the well-known family of nucleophilic hosts such as [12]-crown-4 (Fig. 3d) that have figured prominently in molecular recognition studies. [5]

Bromide ion forms a complex analogous to that of the chloride, but treatment of the 1,2-dilithiocarborane with HgI_2 generated the dianionic species $(C_2B_{10}H_{10})_4Hg_4(I)_2{}^{2-}$ in which the two iodide ions are bound on each side of the Hg_4 plane. [3c] In contrast, the corresponding reaction of the 3-phenyl-1,2-dilithiocarborane gave a cyclic tetramer with only one I^- bound in the sterically encumbered cavity (Fig. 4); [3b] the only steroisomer formed was that having two phenyls directed "up" and two "down" with respect to the cavity. In this case the bulky phenyl groups prevent the macrocyclic host from accommodating more than one iodide. Similar steric effects have

been found in chloride complexes of this mercuracarborand. [3b]

Other electron-rich substrates also coordinate to mercuracarborands, as illustrated by the complexation of a polyhedral $B_{10}H_{10}{}^{2-}$ guest dianion that is bound to the four Hg atoms in $(9,12\text{-}Et_2\text{-}1,2\text{-}C_2B_{10}H_8)_4Hg_4$ via 3-center, 2-electron B–H–Hg bonds. [3a] Figure

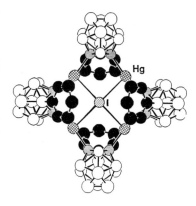

Figure 4. Structure of the $(3\text{-}Ph\text{-}1,2\text{-}C_2B_{10}H_9)_4\text{-}Hg_4 \cdot I^-$ anion (H atoms omitted). [3b]

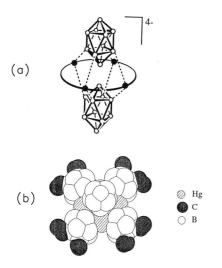

(a)

(b)

⊘	Hg
●	C
○	B

Figure 5. (a) Interaction of two $B_{10}H_{10}^{2-}$ dianions with the four mercury atoms in the (1,2-$C_2B_{10}H_8Et_2)_4Hg_4$ host. [3a] (b) Space-filling view of the complex from above the Hg_4 plane.

5a presents a schematic view of this interaction, with a space-filled drawing of the structure as seen from above one of the $B_{10}H_{10}^{2-}$ dianions shown in Figure 5b.

In contrast to its reactions with mercuric halides, the treatment of dilithiocarborane with mercuric acetate [3e] yielded a trimer (Fig. 6) which coordinates acetonitrile in a

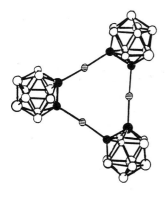

Figure 6. Structure of (1,2-$C_2B_{10}H_{10})_3Hg_3$ (bound CH_3CN molecules not shown). [3e]

most unusual manner: the solid-state structure features two cocrystallized adducts having three and five CH_3CN-bound molecules, respectively.

This anti-crown is analogous to trimeric *o*-phenylene mercury, but its mean Hg–Hg distance of more than 3.7 Å implies a larger central cavity than that of the latter compound. [3e] The reaction of $(C_2B_{10}H_{10})_3Hg_3$ with LiCl produced an anionic chloride complex whose Cl^- ion is proposed to reside at the center of the Hg_3 triangle. The binding ability of these mercuracarborands toward Lewis bases implies that nitrogen-containing bases of biological relevance – adenine, guanine, and the like – may be similarly bound. [3e]

Carboracycles

The geometries of the outward-pointing (exo-polyhedral) carbon orbitals in the *o*- and *m*-carborane cages can similarly be exploited to construct nonmetallic macrocycles. In a synthetic tour de force that combined chemistry with art, Wade and co-workers [6] prepared the aesthetically appealing molecule shown in Figure 7. In this case the starting reagent was the dicopper *m*-carborane, 1,7-$Cu_2C_2B_{10}H_{10}$, which reacted with *m*-diiodobenzene to give the desired trimeric product in low yield. As revealed by X-ray crystallography, the carborane cages are tilted away by 17 ° from the plane defined by their carbon atoms, while the benzene rings are tilted in the opposite direction by the same amount; consequently, the molecule has a dish-shaped structure whose central cavity is defined by three inward-directed carborane hydrogen atoms (mean separation 3.16 Å) and three phenylene hydrogens that lie almost in the same plane; three other carborane hydrogens are much further apart (ca. 4.48 Å). It may be possible to remove the boron atoms on the

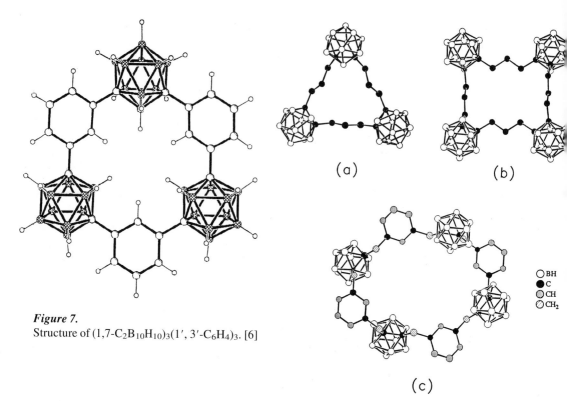

Figure 7.
Structure of $(1,7\text{-}C_2B_{10}H_{10})_3(1', 3'\text{-}C_6H_4)_3$. [6]

Figure 8. (a) Structure of $(1,2\text{-}C_2B_{10}H_{10})_3(C_3H_6)_3$. [7c] (b) Structure of $(1,2\text{-}C_2B_{10}H_{10})_4(C_3H_6)_4$. [7c] (c) Structure of $(1,2\text{-}C_2B_{10}H_{10})_4$ $(1', 3'\text{-}CH_2C_6H_4CH_2)_4$. [7a]

○ BH
● C
⊗ CH
○ CH₂

inside of the macrocycle and replace them with metals, [6] opening intriguing possibilities for catalysis and other applications wherein the three metal centers act in concert.

Macrocycles incorporating the $1,2\text{-}C_2B_{10}H_{10}$ cage and linked by trimethylene or 1,3-xylyl groups have been prepared in Hawthorne's laboratory, and include both trimers and tetramers (Fig. 8). [7] The crystallographically determined structure of a xylyl-linked tetramer that features an unusual 28-membered ring is shown in Figure 8c.

Carborods

While the geometries of the 1,2- and 1,7-C_2B_{10} cages are made to order for the synthesis of macrocycles via substitution at carbon, the 1,12 isomer (*p*-carborane) has car-

bons at opposite ends of the polyhedron and is ideally suited for the assembly of rigid linear "carborods" via direct C–C connections, as has been demonstrated in two research groups. [8] The synthesis of carborods via C-lithio or C-silyl *p*-carborane intermediates produced isolable products having up to four carborane units, such as that shown in Figure 9, but longer rods proved insoluble and could not be made in this way. This problem has been alleviated by the use of B-alkylated carboranes which afford greater solubility in hydrocarbon solvents and allow the synthesis of higher molecular weight carborod products. [8c]

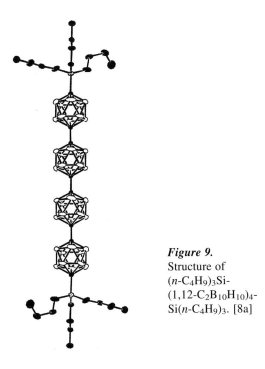

Figure 9.
Structure of
$(n\text{-}C_4H_9)_3Si\text{-}$
$(1,12\text{-}C_2B_{10}H_{10})_4\text{-}$
$Si(n\text{-}C_4H_9)_3$. [8a]

Multidecker Molecular Sandwiches

As the foregoing examples illustrate, the special stereochemistry of the icosahedral carboranes can be used to advantage in the synthesis of structurally novel molecules. Molecular engineering employing non-icosahedral carborane or organoborane units is also under active investigation, as in stacked assemblies that incorporate 7-vertex MC_2B_3M' or MC_3B_2M' pentagonal bipyramidal cages where M and M' are transition metals. [9, 10] Planar, aromatic C_2B_3 and C_3B_2 (diborolyl) rings have a remarkable ability to bind in η^5 fashion to metal atoms *on both sides of the ring plane*, allowing the construction of stable, neutral multidecker sandwich complexes such as those dipicted in Figure 10. [9] This type of double-metal coordination is rare for hydrocarbon ligands; there are only a few reported isolable triple-

decker complexes (and no species with more than three decks) bridged by C_5H_5, C_6H_6, or other monocyclic hydrocarbons. [11]

The formal $R_2C_2B_3H_3^{4-}$ and $R_3C_3B_2R'_2^{3-}$ ring ligands (R, R' = alkyl) and their B-substituted derivatives are isoelectronic analogues of $C_5H_5^-$ and form strong covalent bonds with metal ions, creating species that can be described equally well as metal sandwich complexes or as electron-delocalized polyhedral metallacarboranes linked via metal vertices. [9] Like their larger (e. g., icosahedral) metallacarborane homologues, [12] these carborane-metal sandwiches are considerably more robust than most metallocenes, allowing much greater versatility in their synthesis and modification. This has led to the preparation of a wide range of "molecular skyscrapers" that typically are air- and water-stable crystalline solids that dissolve in organic media, can accommodate first-, second, or third-row transition metals, and are readily derivatized via introduction of substituents. [9] Significantly, many of these complexes are paramagnetic mixed-valence species, whose electron-delocalization over several metal centers suggests potential utility as building-block units for conducting polymers and other applications. [9, 13]

Synthetic routes to the C_2B_3-bridged compounds are based on pentagonal-pyramidal *nido*-2,3-$R_2C_2B_4H_6$ carboranes, which are deprotonated with lithium alkyls in THF and metal ions inserted to generate 7-vertex MC_2B_4 clusters. [9, 14] Removal of the apex BH unit (decapitation) affords open-cage (nido) MC_2B_3 complexes which serve as building-blocks for the synthesis of higher-decker sandwiches via coordination to metal ions or metal-hydrocarbon units such as $CpCo^{2-}$. Diborolyl-bridged sandwich complexes that incorporate MC_3B_2M' cluster units are prepared directly from the neutral 1,3-diborolyl ring ligands. [10]

Work in the author's laboratory and in that of W. Siebert, both separately [9, 10] and in

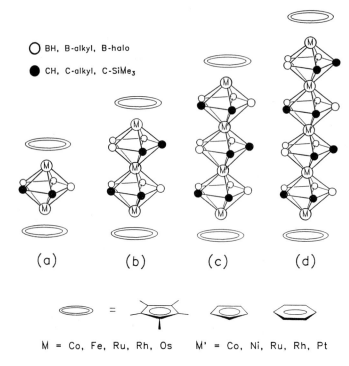

○ BH, B-alkyl, B-halo

● CH, C-alkyl, C-SiMe₃

(a)   (b)   (c)   (d)

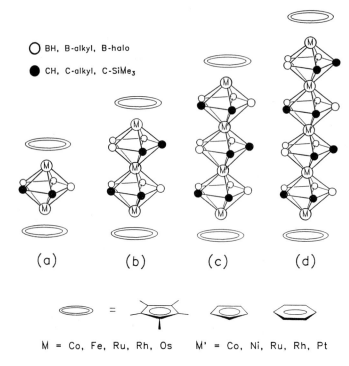

⬭ = ⬡   ⬭   ⬡

M = Co, Fe, Ru, Rh, Os   M' = Co, Ni, Ru, Rh, Pt

Figure 10. Multidecker metall-acarborane sandwich complexes. [9] The rotational orientations of the C_2B_3 rings vary in different complexes, and are arbitrarily depicted here.

collaboration [15] over the past decade, has produced numerous triple- to hexadecker sandwiches, many of which have been examined in detail via electrochemistry, EPR, correlated paramagnetic NMR, [15] and other techniques designed to probe their electronic character. Spectroelectrochemical studies on these compounds in cooperation with W. E. Geiger and his students have been particularly illuminating. [16] Larger systems have also been investigated, such as the linked-tetradecker oligomer [13] shown in Figure 11a and the nickel-diborolyl polymer [17] in Figure 11b. The electronic properties of the multidecker sandwich compounds vary considerably depending on structure, metals, external substituents, and other factors, suggesting that they can be tailored to suit particular needs. For example, in the fulvalene-bridged complex in Figure 11a, each tetradecker unit has one formal Co(IV) paramagnetic metal center, and exhibits electrochemical behavior that indi-cates electron-delocalization throughout the chain, [18] However, if the fulvalene units are replaced by 1,4-$(Me_4C_5)_2C_6H_4$ linking groups, electrochemical data indicate that delocalization of the unpaired electrons ocurs *within*, but not *between* the individual tetradecker units because the connecting phenylene rings are rotated out of coplanarity with the cyclopentadienyls, preventing effective π-conjugation. [19] Still different is the polydecker sandwich in Figure 11b, which is an insoluble semiconducting solid; the corresponding rhodium polymer is an insulator. [17]

A basic theme of the chemistry highlighted in this article is the exploitation of the distinctive features of icosahedral and planar carborane units in order to generate new types of stable molecular structures that are generally inaccessible via conventional organic or organometallic approaches. Many other examples of creative and imaginative synthesis in

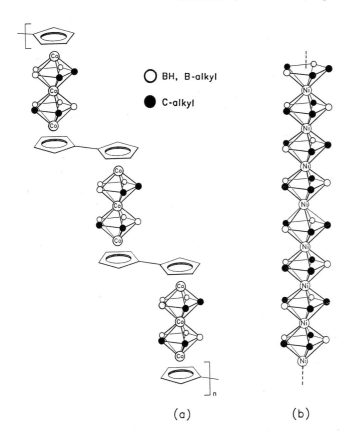

○ BH, B-alkyl

● C-alkyl

(a)

(b)

Figure 11. (a) Structure of a ful-valene-bridged linked-tetradecker oligomer. [12] (b) Structure of a diborolyl-nickel polydecker sand-wich. [17]

borane and carborane chemistry can be cited, and the reader is referred to a series of comprehensive recent reviews in this area. [20] A notable aspect of much of the work described here is its essential simplicity – for example, the generation of pre-organized cyclic host molecules in one or two steps from readily available reagents – an advantage not afforded by crown ethers and cryptands. These synthetic advances illustrate the rapidly evolving art of "designer chemistry" in which inorganic and organometallic assemblies of specified architectures are obtained in directed reactions from available building-block molecules. [21] Because of their unique steric and electronic properties and synthetic versatility, carboranes and other boron clusters seem destined to play a significant role in this field. Philosophically

and historically, designed synthesis derives from organic chemistry; but for inorganic chemists, given the entire periodic table of elements to work with, the scope of possibility seems far larger and the ultimate achievements hardly imaginable at present.

References

[1] J. Plesek, *Chem. Rev.* **1992**, *92*, 269.
[2] (a) V. I. Bregadze, *Chem. Rev.* **1992**, *92*, 209. (b) B. Stibr, *ibid.* **1992**, *92*, 225. (c) R. N. Grimes, *Carboranes*, Academic Press, New York, **1970**.
[3] Leading references: (a) X. Yang, C. B. Knobler, Z. Zheng, M. F. Hawthorne, *J. Am. Chem. Soc.* **1994**, *116*, 7142. (b) Z. Zheng,

C. B. Knobler, M. F. Hawthorne, *J. Am. Chem. Soc.* **1995**, *117*, 5105. (c) Z. Zheng, C. B. Knobler, M. D. Mortimer, G. Kong, Hawthorne, M. F., *Inorg. Chem.* **1996**, *35*, 1235. (d) X. Yang, S. E. Johnson, S. I. Khan, M. F. Hawthorne, *Angew. Chem. Int. Engl.* **1992**, *31*, 893. (e) X. Yang, Z. Zheng, C. B. Knobler, M. F. Hawthorne, *J. Am. Chem. Soc.* **1993**, *115*, 193.

[4] For a summary of published reports on anion complexation by multidentate Lewis acid hosts, see [3a].

[5] (a) D. J. Cram, *Science* **1983**, *219*, 1177. (b) F. Vögtle, E. Weber, *Host-Guest Complex Chemistry/Macrocycles*, (Eds.: I. Vögtle, E. Weber), Springer, Berlin, **1985**. (c) L. F. Lindoy, *The Chemistry of Macrocyclic Ligands*, Cambridge University Press, Cambridge, **1989**.

[6] W. Clegg, W. R. Gill, J. A. H. MacBride, K. Wade, *Angew. Chem. Int. Engl.* **1993**, *32*, 1328.

[7] (a) I. T. Chizhevsky, S. E. Johnson, C. B. Knobler, F. A. Gomez, M. F. Hawthorne, *J. Am. Chem. Soc.* **1993**, *115*, 6981. (b) W. Jiang, I. T. Chizhevsky, M. D. Mortimer, W. Chen, C. B. Knobler, S. E. Johnson, F. A. Gomez, M. F. Hawthorne, *Inorg. Chem.* **1996**, *35*, 5417.

[8] (a) X. Yang, W. Jiang, C. B. Knobler, M. F. Hawthorne, *J. Am. Chem. Soc.* **1992**, *114*, 9719. (b) J. Müller, K. Base, T. F. Magnera, J. Michl, *Ibid.* **1992**, *114*, 9721. (c) W. Jiang, C. B. Knobler, C. E. Curtis, M. D. Mortimer, M. F. Hawthorne, *Inorg. Chem.* **1995**, *34*, 3491. (d) W. Jiang, D. E. Harwell, M. D. Mortimer, C. B. Knobler, M. F. Hawthorne, *Inorg. Chem.* **1996**, *35*, 4355.

[9] (a) R. N. Grimes, *Applied Organometallic Chemistry* **1996**, *10*, 209, and references therein. (b) R. N. Grimes, *Chem. Rev.* **1992**, *92*, 251.

[10] (a) W. Siebert, *Adv. Organometal. Chem.* **1993**, *35*, 187. (b) W. Siebert, in *Current Topics in the Chemistry of Boron*, Kabalka, G. W., Ed., Royal Society of Chemistry, **1994**, 275, and references therein.

[11] Notable examples: (a) H. Werner, A. Salzer, *Synth. React. Inorg. Met. Org. Chem.* **1972**, *2*, 239. (b) A. W. Duff, K. Jonas, R. Goddard,

H.-J. Kraus, C. Krueger, *J. Am. Chem. Soc.* **1983**, *105*, 479. (c) W. M. Lamanna, *J. Am. Chem. Soc.* **1986**, *108*, 2096. (d) A. R. Kudinov, M. I. Rybinskaya, Yu. T. Struchkov, A. I. Yanovskii, P. V. Petrovskii, *J. Organometal. Chem.* **1987**, *336*, 187. (e) J. J. Schneider, R. Goddard, S. Werner, C. Krüger, *Angew. Chem. Int. Engl.* **1991**, *30*, 1124. (f) G. E. Herberich, U. Englert, F. Marken, P. Hofmann, *Organometallics* **1993**, *12*, 4039.

[12] M. F. Hawthorne, *Accounts Chem. Res.* **1968**, *1*, 281.

[13] X. Meng, M. Sabat, R. N. Grimes, *J. Am. Chem. Soc.* **1993**, *115*, 6143.

[14] N. S. Hosmane, N. S., J. A. Maguire, J. *Cluster Science* **1993**, *4*, 297.

[15] (a) M. Stephan, P. Müller, U. Zenneck, H. Pritzkow, W. Siebert, R. N. Grimes, *Inorg. Chem.* **1995**, *34*, 2058. (b) M. Stephan, J. Hauss, U. Zenneck, W. Siebert, R. N. Grimes, *Inorg. Chem.* **1994**, *33*, 4211, and references therein.

[16] (a) J. Merkert, J. H. Davis, Jr., W. E. Geiger, R. N. Grimes, *J. Am. Chem. Soc.* **1992**, *114*, 9846. (b) T. T. Chin, S. R. Lovelace, W. E. Geiger, C. M. Davis, R. N. Grimes, *J. Am. Chem. Soc.* **1994**, *116*, 9359.

[17] W. Siebert, *Pure Appl. Chem.* **1988**, *60*, 1345.

[18] W. E. Geiger, private communication.

[19] J. R. Pipal, R. N. Grimes, *Organometallics* **1993**, *12*, 4459.

[20] (a) G. E. Herberich, in E. Abel, F. G. A. Stone, G. Wilkinson (Eds.), *Comprehensive Organometallic Chemistry II*, Pergamon Press: Oxford, England, 1995; Volume 1, Chapter 5, pp. 197–216. (b) T. Onak, *ibid.*, Chapter 6, pp. 217–255. (c) L. J. Todd, *ibid.*, Chapter 7, pp. 257–273. (d) L. Barton, D. K. Srivastana, *ibid.*, Chapter 8, pp. 275–372. (e) R. N. Grimes, *ibid.*, Chapter 9, pp. 373–430. (f) A. K. Saxena, N. S. Hosmane, *Chem. Rev.* **1993**, *93*, 1081.

[21] For a beautifully illustrated brief account of this subject, see: M. F. Hawthorne and M. D. Mortimer, *Chem. in Britain*, **1996**, *32*, 32.

Framework Modifications of [60]Fullerene: Cluster Opening Reactions and Synthesis of Heterofullerenes

Andreas Hirsch

The accessibility of the fullerenes [1] in macroscopic quantities [2] opened up the unprecedented opportunity to develop a rich "three-dimensional" chemistry of spherical and polyfunctional all carbon molecules. [3–8] A large multitude of fullerene derivatives like exohedral covalent addition products, salts, cluster opened and defined degradation products, heterofullerenes and endohedral derivatives can be imagined and numerous examples, especially of covalent adducts have been synthesized and characterized. [3–8] Within a few years the fullerenes became essential building blocks in organic chemistry. Most of the chemistry of fullerenes has so far been carried out with C_{60} (**1**) with little work on C_{70} and few experiments with C_{76} and C_{84}. This is simply due to the fact that C_{60} (**1**) is the most abundant as well as the most symmetrical fullerene. The first period of preparative fullerene chemistry was dominated by the development of defined addition reactions to the unsaturated π-system of the fullerene cage, [3–4] like nucleophilic and radical additions, cycloadditions, hydrogenations, transition metal complex formations, halogenations, oxygenations and others, which together with the careful examination of the structural and electronic properties of C_{60} (**1**) allowed to deduce principles of fullerene chemistry. [3–4] The systematic development of

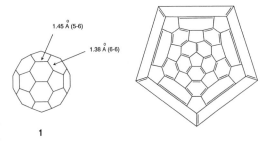

1.45 Å (5-6)

1.38 Å (6-6)

1

covalent fullerene chemistry provides a rich diversity of tailor-made three-dimensional building blocks for technological interesting materials. The various aspects that are associated with covalent fullerene chemistry, including the regioselective synthesis of multiple adducts, in which the fullerene serves as a versatile structure determining tecton, as well as the synthesis of fullerene derivatives with potential biological or materials applications have been summarized in a variety of monographs and reviews. [3–8]

Here, a first review is provided which summarizes the more recent development *framework modified fullerenes* like cluster opened structures and heterofullerenes. The key steps for such framework modifications are always defined activations of the fullerene cluster due to specific covalent addition reactions. Therefore, the principles of covalent fullerene chemistry [3–8] will be considered first:

Each fullerene contains $2(10 + M)$ carbon atoms corresponding to exactly 12 pentagons and M hexagons. This building principle is a simple consequence of the Euler's theorem. Starting at C_{20} any even-membered carbon cluster, except C_{22}, can form at least one fullerene structure. With increasing M the number of possible fullerene isomers rises dramatically. The soccer-ball shaped C_{60} isomer with icosahedral symmetry (I_h) is the smallest stable fullerene, because it is the first to obey the isolated pentagon rule (IPR) [3].

Three properties, which are due to the structure of C_{60} largely govern its chemical behavior:

1. *The bonds at the junctions of two hexagons ([6–6] bonds) are shorter than the bonds at the junctions of a hexagon and a pentagon ([5–6] bonds).* As a consequence, among the 12.500 possible, the lowest energy Kekulé structure of C_{60} is that with all the double bonds located at the junctions of two hexagons ([6–6] double bonds) and the single bonds at the junctions of a hexagon and a pentagon ([5–6] single bonds). Therefore, C_{60} can be considered a sphere built up of fused [5]radialene and cyclohexatriene units and a complete delocalization of the conjugated π-system leading to a reactivity closely related to aromatics can be ruled out.

2. *The highly pyramidalized sp^2 C-atoms in spherical C_{60} cause a large amount of strain energy within the molecule.* Mainly due to this strain energy, which is about 80 % of its heat of formation (ΔH_f), C_{60} ($\Delta H_f = 10.16$ kcal/mol per C-atom) is thermodynamically less stable than graphite ($\Delta H_f = 0$ kcal/mol).

3. *C_{60} is an electronegative molecule, which can be easily reduced but hardly oxidized.* This is reflected theoretically by the MO diagram of C_{60} showing low lying triply degenerate LUMOs and five-fold degenerate HOMOs as well as experimentally by the ease of reversible one-electron reductions up to the hexaanion.

The following rules of reactivity, which are based on these properties, can be deduced from the multitude of chemical transformations that have been carried out with C_{60}:

1. *The reactivity of C_{60} is that of a fairly localized electron deficient polyolefin.* The main type of chemical transformations are therefore additions to the [6–6] double bonds, especially, nucleophilic-, radical- and cycloadditions and the formation of η^2-transition metal complexes; but also, for example, hydroborations and hydrometalations, hydro-genations, halogenations and Lewis acid complex formations are possible.

2. *The driving force for exohedral addition reactions is the relief of strain in the fullerene cage.* Reactions leading to saturated tetrahedrally hybridized sp^3 C-atoms are strongly assisted by the strain of pyramidalization. In most cases, addition reactions to C_{60} are exothermic. The exothermicity of subsequent additions depends on the size and the number of addends already bound and decreases at a certain stage. Therefore, adducts with a high degree of addition become eventually unstable (elimination) or do not form at all, since new types of strain, for example, steric repulsion of addends or introduction of planar cyclohexane rings are increasingly built up. The interplay of these strain arguments largely determines the number of energetically favourable additions to the fullerene core. Also the reduction of C_{60} can be regarded as a strain-relief process, because many carbanions are known to prefer pyramidal geometries.

3. *The regiochemistry of exohedral addition reactions is governed by the minimization of [5–6] double bond within the fullerene framework.* The exclusive mode for typical

cycloadditions and the preferred mode for additions of sterically non demanding segregated addends is 1,2 (addition to a [6–6] double bond), since in this case no unfavourable double bonds within five membered rings have to be formed. The introduction of each [5–6] double bond costs about 8.5 kcal/mol. In 1,2-additions, however, eclipsing interactions between the addends are introduced. Thus, if bulkier segregated addends are allowed to react with C_{60}, a 1,4-mode avoiding eclipsing interactions may occur simultaneously or exclusively. In 1,4 additions eclipsing 1,2 interactions are avoided but the introduction of [5–6] double bonds is required.

Among the cycloadducts of C_{60} [3–8] the methano- [8] and imino derivatives [9–22] exhibit a special case, since the bridging of the addend can take place not only on the junctions between two six-membered rings but also on the junctions between a five-membered and a six-membered ring. The resulting constitutional isomers are denoted as [6,6]- and [5,6]-adducts. Without considering the specific properties of the fullerene moiety one could propose that in analogy to methano- and iminoannulenes [23] the corresponding transannular bonds can, for example, depending on the addend, either be open or closed. As a consequence, four different types of isomers (2–5) of monoaddition products, the open and closed [6,6]- as well as the open

and closed [5,6]-adducts can be imagined. So far, only the closed [6,6]-bridged adducts **2** and the open [5,6]-bridged adducts **4** have been found. Computational investigations [8, 22] confirm the experimental finding, that **2** and **4** are the most stable isomers. The simple rationale for this behaviour is that only in the structures **2** and **4** no unfavorable double bonds within five-membered rings have to be included. The [5,6]-bridged methano- and iminofullerenes were the first examples of cluster opened derivatives. The entire $60\text{-}\pi$-electron chromophore remains intact and the fullerene cage contains a bridged nine-membered ring, whereas in closed [6,6]-bridged adducts one double bond of the cages is removed. As a consequence the electronic properties of [5,6]-bridged methano- and imino[60]fullerenes are very similar to those of C_{60} itself.

The first syntheses of cluster open [5,6]-bridged methano- and iminofullerenes were elaborated by Wudl and co-workers [7, 9] by allowing diazo compounds or alkyl azides thermally to react with C_{60}. The scope of the thermal addition reaction of diazo compounds is quite broad. In addition to the parent and substituted diazomethanes, diazoacetates, diazomalonates and diazoamides have been employed. [3–8] In general next to the [5,6]-bridged isomers the corresponding closed [6,6]-bridged systems are formed as side products. Whereas the open iminofullerenes are stable, many of the [5,6]-bridged

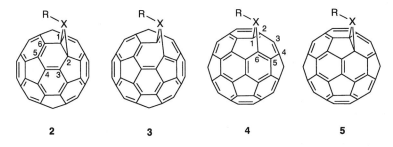

2   **3**   **4**   **5**

X = N, CH, CR'

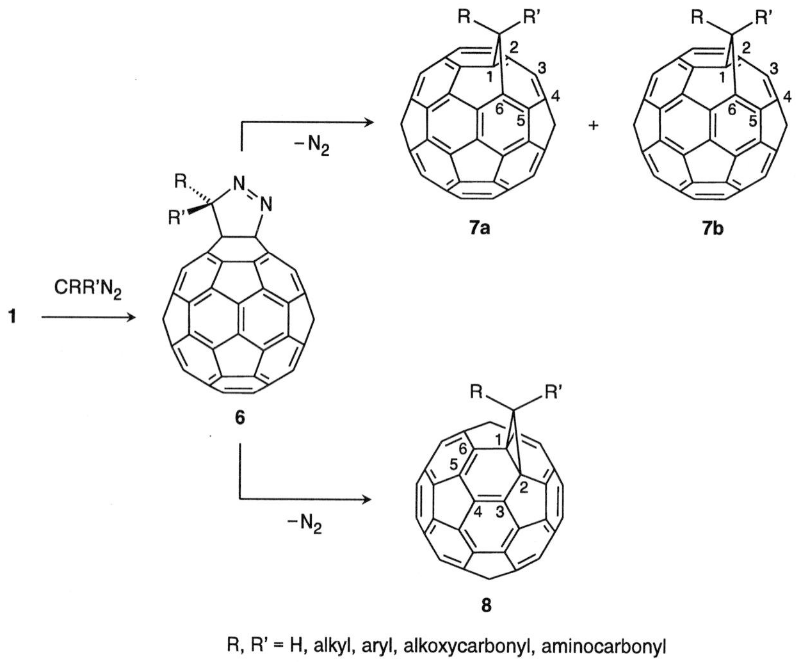

R, R' = H, alkyl, aryl, alkoxycarbonyl, aminocarbonyl

methano fullerenes undergo a facile isomerization to the closed [6,6]-adducts upon thermal or photochemical activation. [3–8] The first step of the addition of diazo compounds is the formation of pyrazolines 6. In the case of the parent diazomethane the corresponding pyrazoline intermediate could be isolated and characterized. The thermal extrusion of N_2 out of pyrazolines 6 in general leads to a mixture of [5,6]- and [6,6]-bridged structures 7 and 8 in different ratios. In the case of the reaction of the parent diazomethane with C_{60} the [6,6]-adduct 8 was formed only in trace. Here no thermal conversion of 7 to 8 was observed.

The addition of alkyl azides to C_{60} proceeds similarly via intermediate triazolines 9, which after extrusion of N_2 rearrange to [5,6]-adducts 10 as major and the [6,6]-adducts 11 as the minor monoaddition products. [3, 4, 9–22] These reactions were carried out for example in refluxing chlorobenzene. Under such conditions azidoformiates react prefer-

ably to [6,6]-adducts 11 since in this case stabilized nitrenes can be formed, which undergo in a competing pathway [2+1]-cycloadditions with the [6,6]-double bonds to form directly the corresponding aziridines 11. [18] If the reactions are carried out in highly concentrated solutions at temperatures not exceeding 60°C the triazolines can be isolated and characterized regardless of the nature of R. [18, 19, 24]

The introduction of two [5,6]-aza bridges shows a remarkable regioselectivity even if segregated alkyazides are used, [19] since bisadducts 13 are by far the major products and only traces of one other bisadduct 14 with unidentified structure are found, if, for example a twofold excess of azide is allowed to react with C_{60} at elevated temperatures. To obtain further information on this most regioselective bisadduct formation process in fullerene chemistry a concentrated solution of 10 was treated with methyl azidoacetate at room temperature. [19] Under these conditions, in

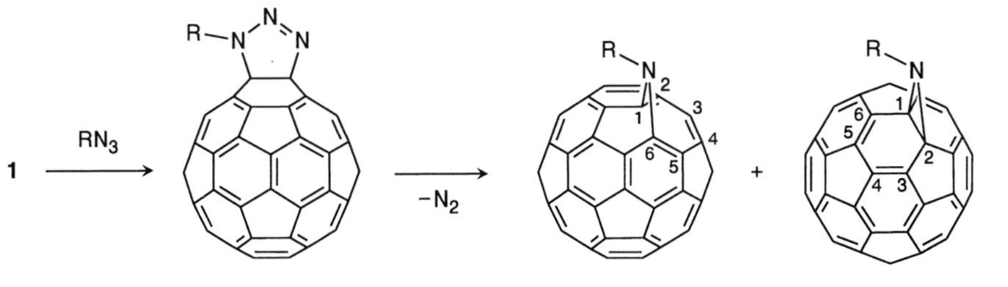

1 $\xrightarrow{RN_3}$ **9** $\xrightarrow{-N_2}$ **10** $+$ **11**

addition to a small amount of bisadducts **13** and **14**, only one mixed [6,6]-triazoline/[5,6]-iminofullerene isomer **12** was formed. The exclusive formation of **12** is explained by the fact that **10** behaves as a strained electron-poor enamine what the reactivity of the bonds between C1 and C2 as well as between C5 and C6 is concerned. The significantly highest Mulliken charge of 0.06 (AM1) is located at C1 and C6, and the lowest of -0.07 at C2 and C5. The most negatively polarized N-atom of the azide (AM1) is that bearing R. A kinetically controlled attack of the azide, therefore, leads predominantely to **12**. The further ring expanded doubly bridged bisiminofullerenes **13** exhibit three seven membered rings and one eleven membered ring within the fullerene cage. One C–atom is already halfway decoupled from the spherical carbon core. The pronounced relative reactivity of the vinylamine type double bonds within [5,6]-bridged iminofullerenes was also demonstrated by the facial addition

of mild nucleophile such as water and amines. For example, the first stable fullerenol was synthesized by treatment of a toluene solution of **10** in the presence of water and neutral alumina and a almost quantitative reaction. [19b] This finding that the reactivity of the vinylamine type [6,6]-double bonds in **10** is dramatically enhanced over the remaining [6,6]-bonds turned out to be a key for further cluster opening reactions and the formation of nitrogen hetereofullerenes as will be shown below. Subsequent attacks to [6,6]-bonds within [6,6]-adducts and also to [6,6]-bonds within the [5,6]-bridged methano-fullerenes **7** are by far less regioselective. [3, 4]

Bis-[5,6]-bridged iminofullerenes with another addition pattern can only be obtained in good yields if a tether directed synthesis is applied. If the tether between two azide groups is rigid enough than the second addition is forced to occur at specific regions of the fullerene cage. Using this approach Luh et al. [20] synthesized the doubly cluster

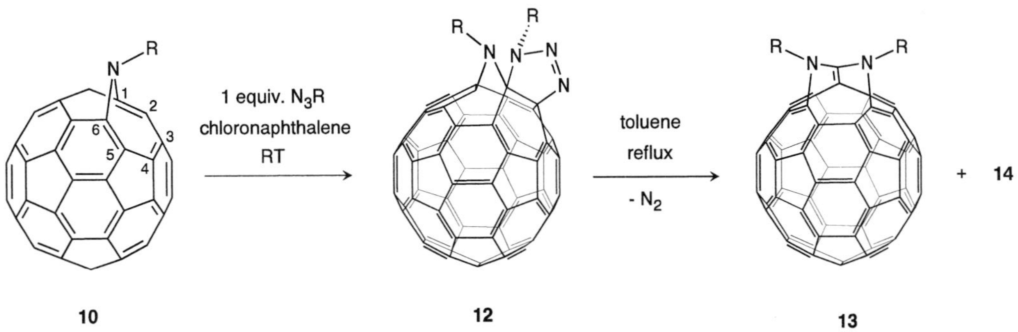

10 $\xrightarrow[\substack{\text{chloronaphthalene} \\ \text{RT}}]{\text{1 equiv. } N_3R}$ **12** $\xrightarrow[\substack{\text{reflux} \\ -N_2}]{\text{toluene}}$ **13** $+$ **14**

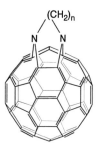

15

openend adducts **15** by allowing an excess of $N_3(CH_2)_nN_3$ (n = 2,3) to react with C_{60} in refluxing chlorobenzene. The addition occurs at two non adjacent [5,6]-bonds of the same five membered ring.

A new type of cluster opened iminofullerenes was recently described by Hirsch et al. [22] with the synthesis of the bisiminofullerenes **18**. The comparatively regioselective formation of **18** can be achieved starting from the closed [6,6]-bridged iminofullerenes **11**,

which was synthesized by allowing the corresponding azidoformiate to react with C_{60} in 1,1,2,2-tetrachloroethane or 1-chloronaphthalene at high temperatures (conditions for generation and [2+1]-additions of nitrenes). If **11** is allowed to react with another equivalent of the azidoformiate but under conditions, which would be typical for the formation of [5,6]-bridged adducts *via* intermediate triazolines (highly concentrated solutions, moderate temperatures), of the large number of regioisomeric bisadducts (doubly [6,6]-bridged as well as mixed [6,6]/[5,6]-bridged isomers), the most polar bisadducts **18** are formed as major products (> 50 % yield). In addition to a directing effect into *cis-1* positions caused by the attachment of the first addend [25] this result can be understood with a mechanism based on the AM1-calculated bond polarizations of the starting materials. [22] The reaction intermediate leading to **18** is the mixed triazoline/iminofullerene **16**, formed in a [3+2]-cycloaddition, where the negatively

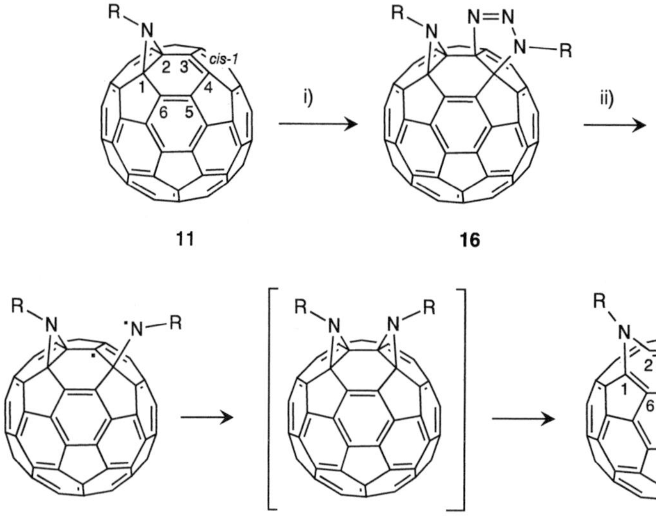

11   **16**

17   **18**

R = COOEt, COOtBu

i) 2 equiv. N$_3$R, 1-chloronaphthalene, 60°C, 4d
ii) toluene, reflux, 30 min.

polarized nitrogen (R–N–N₂) of the azide group forms a bond with the positively polarized C-atom (e. g. C4) of a *cis-1* bond. The thermal extrusion of N₂ leads to the diradical intermediate **17**.

Due to the location of the first addend in the same six-membered ring the usual delocalization of spin density within the six-membered ring, [3, 4] like transfer of spin density to C5 is impossible and radical recombination can only take place at C3. Hence, the formation of a heterotropilidene like structure (1,2,4,5-bisimino[60]fullerene) is suppressed. The ring opening from the non isolable closed form to **18** occurs *via* an inter ring retro-Diels-Alder reaction. *These compounds represent the first examples of [60]fullerene derivatives with open transannular [6,6]-bonds.* Characteristic features within in the fullerene framework of these valence isomer are the presence of i) a doubly bridged 14-membered ring with a phenanthrene perimeter and ii) a eight membered 1,4-diazocine heterocycle. Upon changing the addition pattern, as demonstrated by the investigation of the other possible regioisomers of C₆₀(NCOOR)₂ "regular" behaviour with closed transannular [6,6]-bonds is observed. [22] The fullerene cage can be re-closed again *via* an intra ring Diels-Alder reaction [22] by transferring *cis-1*-C₆₀(NCOO*t*-Bu)₂ **18a** into *cis-1*-C₆₀(NH)₂ **19** as cluster closed valence isomer.

The latter phenomenon clearly reveals the role of the addend. An extensive AM1- and DFT study revealed that only in *cis-1* adducts that prefer planar imino bridges (e. g. carba-mates or amides) are the open forms more stable than the closed forms. [22] These are the first chemical modifications of the fullerene core, which allow the synthesis of open as well as of closed valence isomers with the same addition pattern. The following conclusions can be deduced:

1. With the exception of a *cis-1* adduct, where upon the two-fold ring opening due to the location of the imino bridges in the same six-membered ring only three [5,6]-double bond have to be introduced, six of these energetically unfavourable bonds would be required for the hypothetical open structures of the other seven regioisomers (*trans-1* to *cis-2*). In the latter regioisomers the transannular [6,6]-bonds are always closed.

2. In a *cis-1* adduct a closed valence isomer bears a strained planar cyclohexene ring but the introduction of a unfavourable [5,6]-double bonds is avoided, whereas in an open valence isomer no strained planar cyclohexene but three [5,6]-double bonds are present.

3. For the *cis-1* adducts open valence isomers are favoured for imino addends with planar imino bridges like carbamates and the closed isomers are favoured for imino addends with pyramidalized imino bridges like alkylimines or HN.

4. Carbamates or amides prefer planar arrangements of the nitrogen due to resulting favourable conjugation of the free electron pair with the carbonyl group. This has consequences for imimo-[60]fullerenes, since

18a   **19**

the planar arrangement of the carbonate N-atoms and the required enlargement of the bond angles between C1, N, C2 or C3, N, C4 are most favourably realised if the transannular [6,6]-bonds are open.

A major breacktrough in the field of cluster opening reactions was achieved by Hummelen et al. who allowed **10** to react with singlet oxygen, which afforded the ring-opened keto-lactam **21**. [26] This synthesis takes advantage from 1) the pronounced susceptibility of the vinylamine type double bonds of **10** to addition reactions and 2) the ability of C_{60} and C_{60}-adducts to efficiently sensitize the photo-chemcially induced formation of singlet oxygen. [27] The reaction, using 1,2-dichloroben-zene (ODCB) as the solvent and a Kapton 500HN (DuPont) filtered flood lamp as the light source, is virtually complete within 3 h at 25–30 °C. The Kapton filter is a convenient alternative for the commonly used aqueous dichromate solution filter. It is assumed that the reaction proceeds via a 1,2-dioxetan inter-mediate **20**. This cluster opening reaction seems to be general, as analogous iminoful-lerenes give the corresponding products. The ketolactams are stable under the reaction con-ditions. No new products were detected after 4 h of photooxygenation in ODCB at 25–30 °C. Molecular modelling investigations suggest that at its minimum energy conformation, ketolactam **21** does not have an opening large enough to allow any guest molecule to enter the cage. Some flexibility in the keto and lactam moieties at elevated temperatures is expected and should increase the size of the opening, also, removal of the MEM protecting group should further increase access to the cavity.

Another oxidative opening of the a ful-lerene derivatives was achieved by Taylor et al., [28] who observed that the [70]fullerene derivative $C_{70}Ph_8$ sponataneously oxidizes in air to form a bis-lactone $C_{70}Ph_8O_4$, having an eleven membered hole in the fullerene cage. The reaction is believed to proceed *via* insertion of oxygen into [5,6]-double bonds followed by oxidative cleavage of the adjacent double bond. This process has precedent in the spontaneous oxidations of vinyl ethers to esters.

Triple scission of a six-membered ring on the surface of C_{60} *via* consecutive pericyclic reactions and oxidative cobalt insertion was achieved by Rubin et. al. [29] This work origi-nated from the discovery that diene **22** under-goes a very facile photochemically promoted rearrangement to the stable doubly [5,6]-bridged bismethano fullerene **23**, which is the first example for this adduct type. In **23** a twelve membered ring within the fullerene cage is open. The formation of **23** out of **22** occurs *via* an initial [4+4] photoadduct (not observed), which undergoes a thermally allowed [2+2+2] cycloreversion to afford the bis-methano[12]annulene structure within **23**. The treatment of **23** with $CpCo(CO)_2$ afforded

| 10 | 20 | 21 |

R = $CH_2OCH_2CH_2OCH_3$ (MEM)

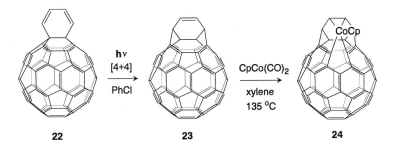

22   **23**   **24**

the stable compound **24**, whose structure was determined by X-ray crystallography. The bond within the five-membered ring adjacent to the bismethano[12]annulene ring of **24** has been broken by oxidative insertion of the cobalt. The "trismethano[15]annulene" opening within the C_{60} framework is the largest annulene ring created thus far in a fullerene.

At about the same time the groups of J. Mattay, [30] A. Hirsch [31] and F. Wudl [32] indenpently provided the first convincing mass spectrometric evidence for the formation of the heterofullerene ions $C_{59}NH^+$, $C_{59}NH_2^+$, $C_{59}N^+$ and $C_{69}N^+$ by fragmentation reactions of the [6,6]-bridged $C_{60}NH$ (**25**), [30] **18a**, [31] the *n*-butylamine adducts of **13a** [31] and **30** [31] as well as of **21**. [32] The first synthesis of the heterofullerene '$C_{59}N$' as its dimer $(C_{59}N)_2$ in macroscopic quantities was

achieved by J. C. Hummelen and F. Wudl. [32] When **21** was treated with a large excess (15 to 20 equivalents) of *p*-toluenesulfonic acid monohydrate in ODCB at reflux temperature under nitrogen a very fast reaction occurs with the formation of the very apolar **29** as major product accounting for 85 % of starting material. The absence of an ESR signal as well as the other spectroscopic results excluded the possibility that the product is a stable free radical. During the formation of **29** the acid protonates the MEM moiety, inducing the loss of 2-methoxyethanol. The presence of the reaction intermediate **26** was proven by trapping experiments with nucleophiles and characterization of the corresponding products. [33] **26** rearranges to the four-membered 1,3-oxazetidinium ring compound **27**, which in turn loses formaldehyde and carbon

21   **26**

27   **28**   **29**

monoxide to yield the azafulleronium ion **28**. The azafulleronium ion, expected to be a very strong oxidant, can apparently be reduced by either (2-methoxyethanol or water) to the azafullerenyl radical, which dimerizes to yield bisazafullerenyl **29**. The parent hydroazafullerene $C_{59}HN$ was synthesized either by the thermal treatment of **21** with *p*-toluenesulfonic acid monohydrate in presence of hydroquinone as reducing agent or by irradiation of $(C_{59}N)_2$ (**29**) through a Kapton filter in ODCB in the presence of tributyltin hydride. [33]

A new rout to nitrogen heterofullerens and the first synthesis of $(C_{69}N)_2$ was achieved by Hirsch et al. [34] starting from **13a** and **30** *via* their *n*-butylamin monoadducts. Forma-

tion of **29** and **31** was found to occur only if the substituents of the imino bridges can be easily removed by acids. For example, heterofullerenes were not obtained using the CH_2CO_2Me analogue of **13a** as starting material. This implies that during heterofullerene formation, which is accompanied by the elimination of one C-atom of the fullerene cage as an isonitrile or a carbodiimide species unprotected imino bridges are required. Next to the dimers **29** and **31** derivatives **32** and **33** are formed as by-products. Presumably, derivatives **32** and **33** are formed by trapping of the expected but unstable $C_{59}N^+$ or $C_{69}N^+$-intermediates with the acetal cleavage product $HO(CH_2)_2OMe$.

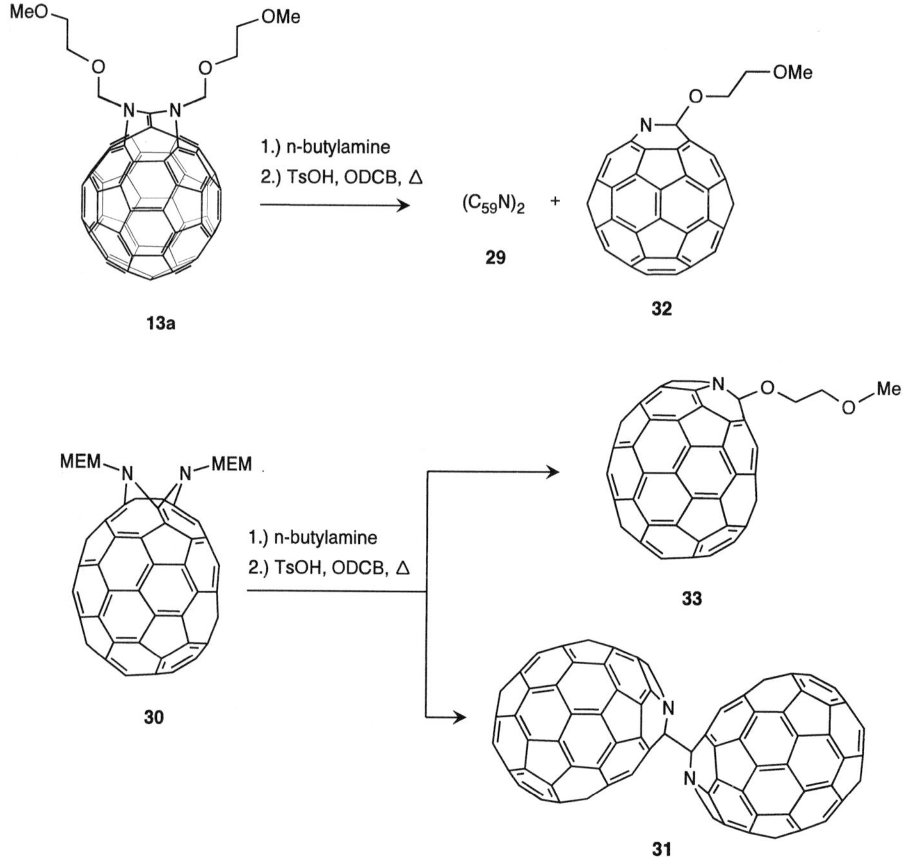

13a

1.) n-butylamine
2.) TsOH, ODCB, Δ

$(C_{59}N)_2$ +

29

32

30

1.) n-butylamine
2.) TsOH, ODCB, Δ

33

31

References

[1] H. W. Kroto, J. R. Heath, S. C. O'Brien, R. F. Curl, R. E. Smalley *Nature* **1985**, *318*, 162.

[2] W. Krätschmer, L. D. Lamb, K. Fostiropoulos, D. R. Huffman *Nature* **1990**, *347*, 354.

[3] A. Hirsch, *The Chemistry of the Fullerenes*, Thieme, Stuttgart, **1994.**

[4] A. Hirsch, *Synthesis* **1995**, 895.

[5] F. Diederich and C. Thilgen, *Science* **1996**, 271.

[6] *The Chemistry of Fullerenes*, R. Taylor (Ed.), World Scientific, Singapore, **1995**.

[7] F. Wudl, *Acc. Chem. Res.* **1992**, *25*, 157.

[8] F. Diederich, L. Isaacs, D. Philip, *Chem. Soc. Rev.* **1994**, 243.

[9] M. Prato, Q. Li, F. Wudl, V. Lucchini, *J. Am. Chem. Soc.* **1993**, *115*, 1148.

[10] M. R. Banks, J. I. G. Cadogan, I. Gosney, P. K. G. Hodgson, P. R. R. Langridge-Smith, D. W. H. Rankin, *J. Chem. Soc.; Chem. Commun.* **1994**, 1365.

[11] M. R. Banks, J. I. G. Cadogan, I. Gosney, P. K. G. Hodgson, P. R. R. Langridge-Smith, J. R. A. Millar, A. T. Taylor, *Tetrahedron Lett.* **1994**, *35*, 9067.

[12] T. Ishida, K. Tanaka, T. Nogami, *Chem Lett.* **1994**, 561.

[13] C. J. Hawker, K. L. Wooley, J. M. J. Frechet, *J. Chem. Soc.; Chem. Commun.*, **1994**, 925.

[14] M. Yan, S. X. Cai, J. F. W. Keana, *J. Org. Chem.* **1994**, *59*, 5951.

[15] J. Averdung, H. Luftmann, J. Mattay, K.-U. Claus, W. Abraham, *Tetrahedron Lett.* **1995**, *36*, 2957.

[16] M. R. Banks, J. I. G. Cadogan, I. Gosney, P. K. G. Hodgson, P. R. R. Langridge-Smith, J. R. A. Millar, J. A. S. Parkinson, D. W. H. Rankin, A. T. Taylor, *J. Chem. Soc.; Chem. Commun.* **1995**, 887.

[17] M. R. Banks, J. I. G. Cadogan, I. Gosney, P. K. G. Hodgson, P. R. R. Langridge-Smith, J. R. A. Millar, A. T. Taylor, *J. Chem. Soc.; Chem. Commun.* **1995**, 88.

[18] G. Schick, T. Grösser, A. Hirsch, *J. Chem. Soc.; Chem. Commun.* **1995**, 2289.

[19] a) T. Grösser, M. Prato, V. Lucchini, A. Hirsch, F. Wudl, *Angew. Chem.* 1995, **107**, 1462; *Angew. Chem. Int. Ed. Engl.* 1995, **34**, 1343, b) T. Grösser, A. Hirsch, unpublished results.

[20] L.-L. Shiu, K.-M. Chien, T.-Y.Liu, T.-I. Lin, G.-R. Her, T.-Y. Luh, *J. Chem. Soc.; Chem. Commun.* **1995**, 1159.

[21] G.-X. Dong, J.-S. Li, T.-H. Chang, *J. Chem. Soc.; Chem. Commun.* **1995**, 1725.

[22] G. Schick, A. Hirsch, H. Mauser, T. Clark, *Chem. Eur. J.* **1996**, *2*, 935.

[23] a) E. Vogel, *Aromaticity*, Spec. Publ. No. 21, The Chemical Society, London, **1967**, 113; b) E. Vogel, *Pure Appl. Chem.* **1969**, *20*, 237; c) E. Vogel, *Isr. J. Chem.* **1980**, *20*, 215; d) E. Vogel, *Pure Appl. Chem.* **1993**, *65*, 143.

[24] B. Nuber, F. Hampel, A. Hirsch, *J. Chem. Soc.; Chem. Commun.*, **1996**, 1799.

[25] F. Djojo, A. Herzog, I. Lamparth, F. Hampel, A. Hirsch, *Chem. Eur. J.* **1996**, *2*, 1537.

[26] J. C. Hummelen, M. Prato, F. Wudl, *J. Am. Chem. Soc.*, **1995**, *117*, 7003.

[27] X. Zhang, A. Romero, C. S. Foote, *J. Am. Chem. Soc.*, **1993**, *115*, 11024.

[28] P. R. Birkett, A. G. Avent, A. D. Darwish, H. W. Kroto, R. Taylor, D. R. M. Walton, *J. Chem. Soc.; Chem. Commun.* **1995**, 1869.

[29] M.-J. Arce, A. L. Viado, Y.-Z. An, S. I. Khan, Y. Rubin, *J. Am. Chem. Soc.* **1996**, *118*, 3775.

[30] I. Lamparth, B. Nuber, G. Schick, A. Skiebe, T. Grösser, A. Hirsch, *Angew. Chem.* **1995**, *107*, 2473; *Angew. Chem. Int. Ed. Engl.* **1995**, *34*, 2257.

[31] J. Averdung, H. Luftmann, I. Schlachter, J. Mattay, *Tetrahedron*, 1995, **51**, 6977.

[32] J. C. Hummelen, B. Knight, J. Pavlovich, R. Gonzalez, F. Wudl, *Science* **1995**, *269*, 1554.

[33] M. Keshavaraz, R. Gonzalez, R. G. Hicks, G. Srdanov, V. I. Srdanov, T. G. Collins, J. C. Hummelen, C. Bellavia-Lund, J. Pavlovich, F. Wudl, K. Holczer. *Nature*, **1996**, *383*, 147.

[34] B. Nuber, A. Hirsch, *J. Chem. Soc.; Chem. Commun.* **1996,** 1421.

Index

B

$B_{12}H_{12}^{2-}$ ion   406 f.
baccatin III   295 ff.
basic pancreatic trypsin inhibitor (BPTI)
  170
benzenediazonium
  – azide   265
  – diazoazide   266
  – ion   266
benzodiazepines   246 f., 368
benzoin condensation   205
benzophenone   212
benzyl
  – bromide   213
  – chloride   213
  – ketones   212
benzylation   4
biaryl ether   315
bicyclo[3.1.0]hexanones   37
bicyclo[4.2.0] systems   301
  – fragmentation   301
bimolecular process   259
BINAP   9, 53 ff., 121, 129
binding site   194, 283
bio-Diesel   217
bioactive conformation   347, 355
biomesogen systems   201
biopolymer mimetics   246, 248
biosynthesis   156 ff.
  – aromatic compounds   156
  – aspidosperma alkaloids   159
  – endiandric acids   158
  – fatty acids   164
  – iboga alkaloids   159
  – polyketides   164
  – steroids   162
biotin   75
bipentazole   269
bipyridine   193, 396
  – dendritic ligands   396
bipyridyl complex   212
bis(a-aminoamide)   384
bisazafullerenyl   424
2,2'-bis(diphenylphosphino)-1,1'-binaphthyl
  121

1,1'-bis(diphenylphosphino)-ferrocene   129
bis-homodiene   96
bisoxazolines (BOX)   5, 10
N,O-(bistrimethylsilyl)-acetamide)   11
bis(tri-*o*-tolylphosphine)-palladium(0)   127
bitropenyl   213
blocked quadrants   55
bolaamphiphiles (bolaphiles)   382 ff.
  – acetylenic   387
  – alkane-a,ω-bis(trimethylammonium)
    384 f.
  – anthraquinone-based   384
  – a-l-lysine-ω-amino-   382
  – lysine   384
  – RNA-based   388
  – unsymmetrical   385
bolytes   382
bond fixation   258
bond-alternation   259
boranes
  – polyhedral   406
boron clusters
  – icosahedral   406 ff.
BPE   51 ff.
branched molecules   391
Bredt's rule   296
bromination   94, 213
bromine atoms   213
bromoetherification   322
bromonium ion   256
1-bromo-1-phenylethane   213
N-bromosuccinimide   213
4-bromotoluene   213
BSA *see N,O*-(bistrimethyl-silyl)acetamide
building blocks   222
  – chiral   222
  – planar-chiral   398
  – vicinal diols   222
Burgess' reagent   298
Bürgi-Dunitz trajectory   302
butadiene
  – highly hindered   256
butenolide synthesis   98
n-Bu$_4$NBH$_4$ 254
butyrolactones   13, 216
butyrylcholinesterase   158